Lecture Notes in Computer Science 7338

Commenced Publication in 1973
Founding and Former Series Editors:
Gerhard Goos, Juris Hartmanis, and Jan van Leeuwen

Anastasia Ailamaki   Shawn Bowers (Eds.)

Scientific and Statistical Database Management

24th International Conference, SSDBM 2012
Chania, Crete, Greece, June 25-27, 2012
Proceedings

 Springer

Volume Editors

Anastasia Ailamaki
Ecole Polytechnique Federale de Lausanne
Computer Science, EPFL IC SIN-GE
Batiment BC 226, Station 14, 1015 Lausanne, Switzerland
E-mail: anastasia.ailamaki@epfl.ch

Shawn Bowers
Gonzaga University, Department of Computer Science
502 E. Boone Avenue, Spokane, WA 99258-0026, USA
E-mail: bowers@gonzaga.edu

ISSN 0302-9743   e-ISSN 1611-3349
ISBN 978-3-642-31234-2   e-ISBN 978-3-642-31235-9
DOI 10.1007/978-3-642-31235-9
Springer Heidelberg Dordrecht London New York

Library of Congress Control Number: 2012939731

CR Subject Classification (1998): H.2.7-8, H.2.4-5, H.2, H.3.3, H.3, H.4, E.1, G.2.2, C.2, F.2

LNCS Sublibrary: SL 3 – Information Systems and Application, incl. Internet/Web and HCI

Typesetting: Camera-ready by author, data conversion by Scientific Publishing Services, Chennai, India

Printed on acid-free paper

Springer is part of Springer Science+Business Media (www.springer.com)

Welcome from the General Chair

The 24th SSDBM conference has a special meaning for me. It was an honor to preside this conference for the second time. Having the conference in Greece in such a short time (the 16th SSDBM was on the island of Santorini in 2004) is an appreciation of the contribution of Greek researchers in the area of databases. I am also proud of organizing SSDBM 2012 in the historical and beautiful city of Chania, the "Venice of the East" as it is known.

The 2012 24th International Conference on Scientific and Statistical Database Management was held June 25–27 in Chania, Greece. The conference brought together researchers, practitioners, and developers for the presentation and exchange of current research on concepts, tools, and techniques for scientific and statistical database management.

I would like to thank first of all the authors who submitted, whether accepted or rejected, papers or proposed demos and panels. Without them it would be impossible to assemble such a technical program. I would also like to thank the 54 Program Committee members that had to work hard to meet short deadlines, and participated in electronic discussions to resolve conflicts, as well as the external referees. I also would like to thank the sponsors of SSDBM 2012, which included Piraeus Bank and the University of Athens. I would also like to express my appreciation to Springer for publishing the SSDBM 2012 proceedings.

Finally, I would like to express special thanks to the Program Chair, Anastasia Ailamaki. She organized the Program Committee and the entire review procedure. Besides this, she assisted in the organization and was always there to resolve any problems.

I believe all the participants had a wonderful and productive time. The first was guaranteed by the beauty of Chania and the second by the quality of the invited talks and the technical program.

M. Chatzopoulos

Message from the Program Chair

It is a real pleasure to welcome you to the proceedings of the 24th edition of SSDBM, which include work presented at the conference in beautiful Chania, Crete, Greece. As scientific datasets explode in size, turning data into information permanently influences the scientific method. SSDBM 2012 presented pioneering research on scientific and statistical data management, thereby bridging computer science with other domains. It comes as no surprise, therefore, that the favorite topics this year centered around innovative techniques for scientific data mining and scientific query evaluation, as well as targeted challenging scientific applications.

We were fortunate to attract three internationally known researchers to open each of the three days of the conference. David Maier from Portland State University gave a keynote on lessons and experiences from managing data coming from various data sources in scientific observatories. Ricardo Baeza-Yates from Yahoo! Research gave a keynote analyzing Web Search and discussed techniques to meet the ever-demanding user requirements. Finally, Yannis Ioannidis from the University of Athens called a controversial panel with different views on global data infrastructures. There is an invited paper describing each keynote as well as the panel in the proceedings.

SSDBM 2012 offered 25 full and ten short paper contributions, meticulously synthesized into a proceedings volume by Shawn Bowers and Springer. This rich collection of scientific work is selected from 68 submissions,with the invaluable organizational help by Dimitra Tsaoussi-Melissargos, a reviewing marathon by 53 reviewers and several additional external referees. Each paper had at least three reviews; most had four, and in some cases five. Most papers were exhaustively discussed, and as a result nine generous reviewers (Val Tannen, Apostolos Papadopoulos, Mario Nascimento, Tore Risch, Amelie Marian, Thomas Heinis, Zografoula Vagena, Gultekin Ozsoyoglu, and Mohamed Mokbel) further served as shepherds for 11 papers, to ensure a high-quality program.

In addition to the main track papers, we present nine posters and five system demonstrations (selected from 20 and nine submissions, respectively).The selection was made by to Ioana Manolescu, Bill Howe, Miguel Branco, and Thomas Heinis. All papers are featured on the SSDBM website along with extensive information about the conference thanks to the webmaster Sadegh Nobari.

I would like to cordially thank the excellent team of SSDBM 2012 for their dedication and hard work. Special thanks go to the General Chair, Mike Chatzopoulos, for his immediate responsiveness to all questions and emergencies during the process. Most importantly, all of us in the Program Committee express our sincere gratitude to all the authors who submitted their work in the form of a paper, poster, or demo to the conference. They are the reason for SSDBM's continued success!

June 2012   Anastasia Ailamaki

SSDBM 2012 Conference Organization

General Chair

Mike Chatzopoulos   University of Athens, Greece

Program Committee Chair

Anastasia Ailamaki   EPFL, Switzerland

Proceedings Editor

Shawn Bowers   Gonzaga University, USA

Webmaster

Sadegh Nobari   NUS, Singapore

Program Committee

Foto Afrati	National Technical University of Athens, Greece
Gagan Agrawal	Ohio State University, USA
Walid G. Aref	Purdue University, USA
Magdalena Balazinska	University of Washington, USA
Roger Barga	Microsoft Research, USA
Carlo Batini	University of Milano Bicocca, Italy
Elisa Bertino	Purdue University, USA
Paul Brown	SciDB, USA
Peter Buneman	University of Edinburgh, UK
Randal Burns	John Hopkins University, USA
Stefano Ceri	Politechnico di Milano, Italy
Judith Cushing	The Evergreen State College, USA
Alfredo Cuzzocrea	ICAR-CNR and University of Calabria, Italy
Lois Delcambre	Portland State University, USA
Alex Delis	University of Athens, Greece
Jim Dowling	Swedish Institute of Computer Science, Sweden
Johann Gamper	Free University of Bozen-Bolzano, Italy
Wolfgang Gatterbauer	Carnegie Mellon University, USA
Michael Gertz	Heidelberg University, Germany
Theo Haerder	TU Kaiserslautern, Germany
Thomas Heinis	EPFL, Switzerland

Panos Kalnis	KAUST, Saudi Arabia
Vana Kalogeraki	Athens University of Economics and Business, Greece
Verena Kantere	Cyprus University of Technology, Cyprus
Martin Kersten	CWI Amsterdam, The Netherlands
George Kollios	Boston University, USA
Hans-Peter Kriegel	Ludwig-Maximilians-Universität München, Germany
Alex Labrinidis	University of Pittsburgh, USA
Wolfgang Lehner	Technische Universität Dresden, Germany
Ulf Leser	Humboldt University of Berlin, Germany
Julio Lopez	Carnegie Mellon University, USA
Paolo Manghi	Istituto di Scienza e Tecnologie dell'Informazione, Italy
Amelie Marian	Rutgers University, USA
Mohamed Mokbel	University of Minnesota, USA
Mario Nascimento	University of Alberta, Canada
Silvia Nittel	University of Maine, USA
Gultekin Ozsoyoglu	Case Western Reserve University, USA
Apostolos Papadopoulos	Aristotle University of Thessaloniki, Greece
Olga Papaemmanouil	Brandeis University, USA
Thanasis Papaioannou	EPFL, Switzerland
Tore Risch	Uppsala University, Sweden
Domenico Sacca	University of Calabria, Italy
Heiko Schuldt	University of Basel, Switzerland
Thomas Seidl	RWTH Aachen University, Germany
Timos Sellis	Research Center "Athena" and National Technical University of Athens, Greece
Myra Spiliopoulou	Otto von Guericke University Magdeburg, Germany
Julia Stoyanovich	University of Pennsylvania, USA
Val Tannen	University of Pennsylvania, USA
Nesime Tatbul	ETH Zürich, Switzerland
Martin Theobald	Max Planck Institute for Informatics, Germany
Agma Traina	University of Sao Paulo, Brazil
Peter Triantafillou	University of Patras, Greece
Zografoula Vagena	Rice University, USA
Jeffrey Yu	Chinese University of Hong Kong, China

Posters and Demonstrations Committee

Ioana Manolescu	INRIA, France
Miguel Branco	EPFL, Switzerland
Bill Howe	University of Washington, USA

Additional Reviewers

Lory Al Moakar	Periklis Andritsos
Brigitte Boden	Leonardo Candela
Gianpaolo Coro	Kyriaki Dimitriadou
Christopher Dorr	Tobias Emrich
Sergej Fries	Filippo Furfaro
Gayatree Ganu	Antonella Guzzo
Thomas Jörg	Philipp Kranen
Hardy Kremer	Peer Kroeger
Elio Masciari	Massimiliano Mazzeo
Mohamed Nabeel	Panayiotis Neophytou
Nikos Ntarmos	Irene Ntoutsi
Pasquale Pagano	Thao N. Pham
Andrea Pugliese	Matthias Renz
Matthias Schubert	Erich Schubert
Silvia Stefanova	Salmin Sultana
Nikolaos Triandopoulos	Minji Wu
Arthur Zimek	Andreas Zuefle

SSDBM Steering Committee

Michael Gertz	University of Heidelberg, Germany
Judith Cushing	The Evergreen State College, USA
James French	CNRI and University of Virginia, USA
Arie Shoshani	Lawrence Berkeley National Laboratory, USA (Chair)
Marianne Winslett	University of Illinois, USA

SSDBM 2012 Conference Sponsors

Bank of Piraeus
University of Athens

Table of Contents

Graph Processing

Panel

Mining Multidimensional Data

Provenance and Workflows

Processing Scientific Queries

Keynote II

Support for Demanding Applications

Demonstration and Poster Papers

Navigating Oceans of Data

David Maier[1], V.M. Megler[1], António M. Baptista[2], Alex Jaramillo[2],
Charles Seaton[2], and Paul J. Turner[2]

[1] Computer Science Department, Portland State University
[2] Center for Coastal Margin Observation & Predication, Oregon Health & Science University
{maier,vmegler}@cs.pdx.edu,
{baptista,jaramilloa,cseaton,pturner}@stccmop.org

Abstract. Some science domains have the advantage that the bulk of the data comes from a single source instrument, such as a telescope or particle collider. More commonly, big data implies a big variety of data sources. For example, the Center for Coastal Margin Observation and Prediction (CMOP) has multiple kinds of sensors (salinity, temperature, pH, dissolved oxygen, chlorophyll A & B) on diverse platforms (fixed station, buoy, ship, underwater robot) coming in at different rates over various spatial scales and provided at several quality levels (raw, preliminary, curated). In addition, there are physical samples analyzed in the lab for biochemical and genetic properties, and simulation models for estuaries and near-ocean fluid dynamics and biogeochemical processes. Few people know the entire range of data holdings, much less their structures and how to access them. We present a variety of approaches CMOP has followed to help operational, science and resource managers locate, view and analyze data, including the Data Explorer, Data Near Here, and topical "watch pages." From these examples, and user experiences with them, we draw lessons about supporting users of collaborative "science observatories" and remaining challenges.

Keywords: environmental data, spatial-temporal data management, ocean observatories.

1 Introduction

The growth in the variety and numbers of sensors and instrument platforms for environmental observation shows no signs of abating. In the past, measuring an environmental variable (such as the chlorophyll level in water) might have required collection of a physical sample, followed by laboratory analysis (say on a monthly basis). Now an in-situ sensor can monitor the variable continuously. Laboratory analysis of samples still occurs, but some tests now generate gigabytes of data, such as high-throughput DNA sequencing. Observational and analytic data is itself dwarfed by outputs of simulation models. Oceans of data are upon us.

An equally important trend is a change in how science is performed. Traditionally, for much of ocean science, data collection was on a per-investigation basis, with the same researcher or group analyzing the data as gathered it. That data might be shared

A. Ailamaki and S. Bowers (Eds.): SSDBM 2012, LNCS 7338, pp. 1–19, 2012.

with other scientists, but typically months or years after initial collection. Answering questions about human effects on the environment, or influences of climate change, require data collection on spatial and temporal scales beyond the abilities of any individual or small group. Thus we are seeing shared environmental observatories (much like in astronomy [15]), such as that operated by the NSF-sponsored Center for Coastal Observation and Prediction (CMOP, www.stccmop.org). In such observatories, those planning and carrying out the data collection are often different from those using the data, and the common pool of data supports a "collaboratory" in which scientists from disparate disciplines work together on complex environmental questions. This shift in the nature of the scientific enterprise presents challenges for data dissemination and analysis. It is no longer reasonable to expect an individual scientist to have comprehensive knowledge of the complete type and extent of data holdings in an observatory such as CMOP's. Moreover, with an expanded base of users, providing display and analysis tools for each type of user separately would be challenging. Thus it is important to have a common base of capabilities that can help investigators locate and judge datasets relevant to their work, as well as carry out initial graphing and analysis tasks on line, without having to download and work locally with that data (though that mode of interaction must also be supported).

The cyber-infrastructure team of CMOP is charged with managing the storage and dissemination of data assets associated with observation and modeling activities, as well as producing web-based interfaces for navigating, accessing and analyzing those assets. We begin by surveying the main user groups that CMOP supports (Section 2), then briefly describe the data-collection and management process (Section 3), along with several of the tools that support these groups (Section 4). We touch on some of the techniques we are investigating to meet the performance demands of one of these tools, Data Near Here (Section 5) and recount some of our lessons learned (Section 6). Section 7 concludes by laying out current issues and challenges that could be the basis for future research on scientific data management.

2   The User Base

There are broadly three classes of users of CMOP systems.

Operations: This class consists of internal users with responsibility for the day-to-day operation of the CMOP observatory, from sensors and telemetry to data ingest to quality assurance to data download and display services. The needs of this class concern detecting problems in the data chain, such as fouled or failed sensors, records corrupted in transmission, failed loads and inoperative interfaces. In many cases, such problems are currently exposed via the tools used by "regular" users, such as the pages that display recent observations from sensor stations. However, this time-intensive approach can lead to delays in problem detection and in data quality assurance. Sometimes specialized interfaces are needed for status reporting, station viewing and quality-assessment tasks. In addition, recording information on collection of water and other samples, and the results of laboratory analyses, is also needed to support observatory operations.

Science: This class consists of researchers internal and external to CMOP. The data holdings of CMOP are becoming extensive enough that few scientists are aware of their totality, in terms of time, location and type of observation. Even when someone might know where a sensor station is located, and when it first became operational, he or she might be unaware for exactly what times data is available – some instruments are deployed only seasonally, some may be removed temporarily for repairs, some segments of data might be dropped during quality assurance. Thus, there need to be tools to help a scientist find data that is potentially relevant to his or her research question, and also to get a quick view of temporal coverage of a specific observation station. Once a dataset of possible interest has been identified, a scientist often wants a simple plot of it, to assess its suitability. She might be checking if there are dropouts during the period of particular interest, or if it contains some event she is seeking, say, unusually low dissolved-oxygen levels. Once a dataset is deemed useful, she might want to download all or part of it in a form suitable for use in a desktop tool, such as Excel or Matlab. However, there should be some capability to analyze the data online, such as charting several variables on the same graph, or plotting one variable against another. Finally, scientists want to comment on or annotate data or products of analysis, to point out suspected problems or to highlight interesting subsets.

While this paper focuses on observed data, there are places where observed data and simulated data intersect. One is in comparing observed and modeled behavior of an environmental variable, such as salinity. To judge model *skill* (a model's ability to reproduce real-world physical phenomena), it is useful to plot observed and model data together. Since the model data has much denser coverage than observed data, it must be sub-sampled to a dataset that matches the location and times of a corresponding sensor dataset. Such sub-sampling is essentially a "virtual sensor" operating in the simulated environment at a place and time that matches the corresponding physical sensor in the real environment. A second interaction between observed and modeled data is when the latter provides a context for sensed and sampled observations. To this end, *climatologies* are useful for comparing current conditions to historical trends. A climatology is an aggregation of a particular variable, generally over both time and space. Examples are the monthly average of maximum daily *plume* volume (the portion of the ocean at a river mouth with reduced salinity), and the average weekly temperature of the estuary. A scientist can then see, for example, if water samples were taken when temperatures were relatively high for the time of year.

Education: An important subset of science use is educational use. For students pursuing undergraduate or graduate research, the needs for data access and analysis largely match those of scientific staff. For classrooms and science camps, the user base is quite different in motivation and sophistication. Currently, we do not have interactive tools specifically for K-12 use. However, this class of users is considered in the design of interactive tools, particularly in choosing default settings that are likely to yield viewable results on initial encounter.

Resource Management: There is a growing class of users who use observatory data in reaching decisions, both in day-to-day resource management as well as for longer-range policy making. For example, the Quinault Indian Nation is highly interested in the timing and spatial extent of hypoxic (low-oxygen) conditions near their tribal lands, to understand the possible effects on the shellfish harvest. A second example is a manager at a fish hatchery deciding when to release juvenile fish to the estuary. Research points to a correlation between estuary conditions (properties of the fresh-water plume extending into the ocean from the river's mouth) with survivability of hatchlings [3]. Comparing predicted conditions for the coming week against the typi-cal range of conditions at the same time in past years might help optimize the release time. In general, for such users, it helps to organize data thematically, bringing to-gether data from a range of sources related to a theme (such as hypoxia) on a single web page, preferably with accompanying commentary that highlights important cur-rent trends or conditions. Such thematic pages are also useful to scientists studying a particular phenomenon or condition. For example, the Columbia River often exhibits red-water blooms in the late summer. It is useful to collect information that indicates the onset of such events so that, for example, additional sampling can take place.

Our Goal: The data and environment we support are complex (and our resources are bounded); the users have a wide spectrum of skill sets (K-12 students, resource man-agers, ocean scientists); and we have a huge range of scales of analysis and processing that we must deal with (models of the entire coastal shelf versus RNA in one water sample; decades of data versus phenomena that manifest in a few seconds). Further, a given line of research can require different levels of detail at different stages. We do not want to require people to learn (nor expend the resources to build) different, spe-cialized tools for each of these combinations (which has often been the norm in the past). We also do not want to enforce simplicity by dictating a single workflow or by limiting the user to only one set of data or analysis. Thus our goal is to find simple, consistent abstractions that expose the complexity in the data (which is relevant to scientists) while hiding the complexity in the infrastructure (generally not of concern).

3   The CMOP Observatory

CMOP is funded by the National Science Foundation's Science and Technology Cen-ters program, along with matching contributions from center participants. It studies conditions and processes in the estuary, plume (the jet of fresher water that protrudes from a river's mouth into the ocean) and near-ocean systems, trying, in particular, to anticipate and detect the influences of human activity and climate change. A major component of CMOP's common infrastructure is an environmental observatory fo-cused on the Columbia River Estuary, but also extending up river as well as to the near ocean off the Oregon and Washington coasts. (See Figure 1.) The observatory collects measurements of environmental variables (henceforth just *variables*) via sen-sors for physical (temperature, salinity), geochemical (turbidity, nitrate) and biologi-cal (chlorophyll, phycoerythrin) quantities [12]. These sensors are mounted on fixed (pier, buoy), profiling (moving through the water column) and mobile platforms. The mobile platforms include staffed research vessels as well as autonomous vehicles. The readings from many of the sensors are immediately relayed back, via wired and wire-less links, to CMOP servers. However, some information, particularly from mobile platforms, is downloaded in bulk, for example, at the end of a cruise or mission. Sensor readings are supplemented with laboratory tests of water and other samples for chemical and biological properties, including RNA and DNA assays. (However, technology is developing to allow in-situ performance of some of these tests [6].) While most sensors deliver a few floating-point numbers per reading, others can produce vectors of values (e.g., density profiles) or 2-D images (for example, of surface waves or micro-organisms [8]). Observation frequencies can be as often as every few microseconds, or as few as tens per year for DNA assays.

In terms of volume and growth, CMOP collected about 75K observations of physical variables in 1996 from fixed stations. A decade later, the rate was about 10M observations per year, and rising to 42M observations in 2011. Collection of biogeochemical variables began in 2008, with 38M observations collected in 2011. In 2002, total observations from mobile platforms was just 5K. Since then, it has been as high as 17M observations (2008), though it dropped off last year because of fewer cruises.

Fig. 1. An overview of the CMOP observation network, including both current and past positions of sensor stations. This map also serves as an interface for navigating to the information pages for specific stations.

While observational data is our main focus here, another major component of the shared CMOP infrastructure is a modeling capability for the Columbia estuary and other coastal systems. These simulation codes are used both to prepare near-term (days) forecasts of future conditions as well as long-term (decades) retrospective runs, called *hindcasts* [2]. Historically, these models have addressed the 4-D hydrodynamics of the river-ocean system, including velocity, temperature, salinity and elevation over time. More recently, the models are being extended to include geochemical and biological aspects. Currently, each forecast run produces almost 20GB of data, mostly as time series of values on a 3-D irregular mesh, but also including pre-generated images and animations. The hindcast databases are approaching 20TB of data.

Fixed stations daily report [1/19/2011]

Prev report	Next report

| << < | February 2012 | > >> |

M	T	W	T	F	S	S
30	31	1	2	3	4	5
6	7	8	9	10	11	12
13	14	15	16	17	18	19
20	21	22	23	24	25	26
27	28	29				

Choose date

Legend

	Active
	Problem-causing issues
	Not working
	Not currently installed
	Not applicable to station
	Not assessed today

	SATURN-01	SATURN-02	SATURN-03	SATURN-04	SATURN-05	SATURN-06
APNA						
CDOM Fluorometer						
CT						
CyclePO4						
FLNTU						
Fluorometer						
LOBO						
Oxygen						
Phycoerythrin						
Phytoflash						
Pressure						
Pump						
SAMI CO2						
SUNA						
Thermistor						
Turbidity						

	am169	cbnc3	dsdma	eliot	red26	grays	hmndb	jetta	sandi	marsh	ogi01	sveni	tansy	tnslh	coaof	woody
CT																
CTD																
Thermistor																
Tide Gauge																

Comments:

Cathlamet Bay North Channel (USCG day mark green 3): [10/12/13] Due for inst. switch

Desdemona Sands Light: [10/12/17] Last report 12/14 1921, will investigate at first opportunity

Grays Point (USCG day mark green 13): [10/12/20] reporting normal ct values

Jetty A: [10/11/12] station to be visited at first opportunity, sensor believed to be recording internally

Lower Sand Island light (USCG day mark green 5): [11/01/05] Possible low battery, unknown sensor status

SATURN-01: [11/01/19] winch back in place

SATURN-03: [11/01/19] midwater cable is being repaired, to be replaced 1/20/11

Fig. 2. The daily status page for fixed observatory stations. It indicates deployment and operational status for various instrument types at various CMOP stations.

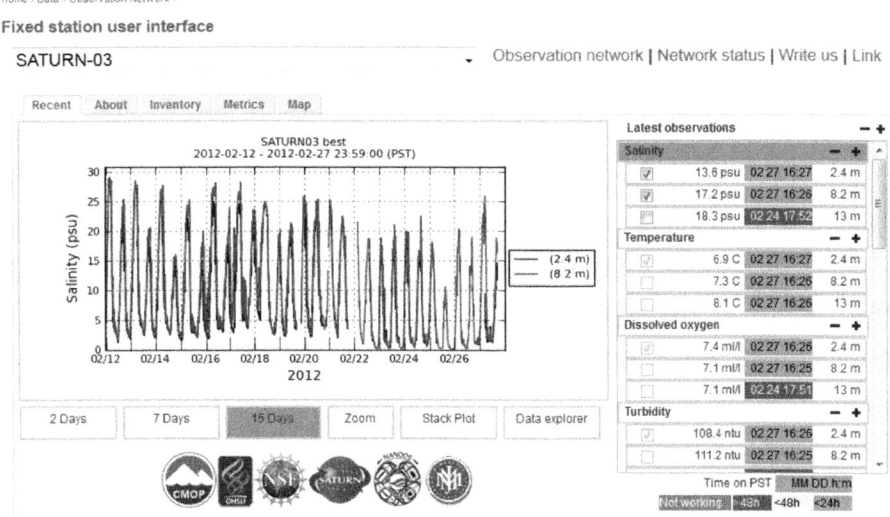

Fig. 3. The station page for SATURN03, plotting salinity offerings for two different depths

CMOP makes as much data as it can available online as soon as possible. The preponderance of the sensor data is routed into a relational DBMS. Ingest processes work directly with network feeds or through frequently polled remote files to get sensor records, which are parsed and inserted into database tables. The unit of designation for time-series observational data is the *offering*, which generally refers to the data for a particular environmental variable coming from a specific instrument at a particular position (often given as a station name, e.g., SATURN04, and a depth, e.g., 8.2 meters). There are also offerings from mobile platforms, where position is itself captured as a time series. A physical dataset can give rise to multiple offerings: a raw stream, as well as one or more corresponding streams that are the output of quality assurance and calibration procedures. High-frequency (multiple readings per second) data can be "binned" down to a coarser time step (1 minute, 5 minutes), and registered to a common time scale. Offerings also exist for derived variables (for example, conductivity and temperature used to compute salinity) and "virtual" observations from the simulation models. A few offerings provide monitoring information about the observatory infrastructure, such as the status of pumps, for use by operations staff.

4   CMOP Interfaces and Tools

While observed data are often available on CMOP database servers within minutes (if not seconds), they have little value if they are not easily accessible to CMOP scientists and other users. CMOP endorses the vision of a "collaboratory" where there is open sharing of data, and scientists of multiple disciplines can easily interact with each other and CMOP information resources. Often the easiest way for a scientist to get an initial impression of data is through a plot or graph. Thus, a key strategy is making plot production a basic service in the cyber-infrastructure. The CMOP *offe-ringPlot* service is available via a RESTful API, where a URL details both the offerings of interest and the plot parameters (kind of plot, extent of axes, aspect ratio, etc.). For example, the URL

```
http://amb6400a.stccmop.org/ws/product/offeringplot.py
?handlegaps=true&series=time,saturn03.240.A.CT.salt.PD0
&series=time,saturn03.820.A.CT.salt.PD0&series=time,
saturn03.1300.R.CT.salt.PD0&width=8.54&height=2.92
&days_back=2&endtime=2012-03-09
```

produces a scatter plot of two salinity offerings at the SATURN03 station versus time. The plot will be generated at a particular width and height, and will cover data going back two days from 9 March 2012.

Fig. 4. The first configuration screen for Data Explorer, where the offerings to be plotted are selected

Plotting as a service is used heavily by CMOP interfaces, but supporting interactive response times was a bit of a challenge. In theory, the plotting service could get its data directly from the RDBMS. However, our experience was that direct access often resulted in significant latency, likely influenced by the fairly constant load of ingest tasks. Also, users are often interested in the most recent data from a station, so as use increases, redundant access to the same data is likely. Thus, we moved to an information architecture where we maintain a cache of extracts from the database, and, in fact, pre-populate the cache. The cache consists of about 36GB of files in netCDF format [13], arranged in a directory structure with a file for each offering for each month. (We may switch to individual days in the future, to avoid regenerating files for the current month and to support extra detail in some of our tools.) While an interface can access the database directly—and some do—the netCDF caches satisfies much of the read load. As a side benefit, the cache also supports programmatic data download, via a THREDDS [5] server using the OpenDAP protocol [4].

We now turn to some of our existing interfaces, plus one under development. Figure 2 shows the observatory *status page*, as used for fixed and profiling stations. It reflects daily reports by field staff on the dispositions of various instruments installed at CMOP observation stations. Operations users record and report overall status, which CMOP management monitors via this page.

Operations staff and resource managers coordinate many of their day-to-day activities using the *station pages*, which are also a starting point for researchers with specific instruments supporting their studies. These pages provide immediate display of data transmitted from instruments. Figure 3 shows an example of a station page, for the station named SATURN03. Shown is a 15-day plot of the salinity offerings at the station for 2.4 meters and 8.2 meters. Such pages are designed to be easy to interact with, so present a limited set of choices for configuration. Along the right side are different offerings associated with the station, such as temperature at 2.4 meters and turbidity (cloudiness) at 8.2 meters, grouped by variable type; a user can quickly display other instruments' data using the checkboxes. For each offering, the time and value of the latest available measurement is shown. The colors over the time indicate if new data has appeared in the past day (green), last two days (yellow) or longer ago (red). Along the bottom are time periods. It is also possible to obtain all offerings in a single page of plots (a *stack plot*). The station pages are intentionally limited in their capabilities, to keep them simple to work with and give fast response times.

The simple plots of the station pages are limited in many ways: no arbitrary time periods, no combination of offerings for different variables or different stations in one plot, only plots of variables against time (as opposed to plotting one against another). The Data Explorer, accessible directly from this page, is a more sophisticated tool, that allows control of these aspects (and more), but with a more complicated interface. It supports both variable-time and variable-variable plots, optionally coloring the plot by an additional variable. The configuration process in Data Explorer involves sequencing through several set-up screens. The first screen (shown in Figure 4) is used to select the offerings to include in the plot. Here Chlorophyll and Salinity from station Saturn03 at 8.2m are selected for a scatter plot, to be colored by Turbidity at the same depth, perhaps to contrast the influences of the river and the ocean on the estuary.

Additional screens allow selection of a time period, axes limits, and aspect ratio of the plot. Figure 5 shows the requested plot. The Data Explorer supports saving and annotating plots, as well as downloading the underlying data. This powerful tool is used by researchers for everything from exploratory research to producing diagrams for publication. Operations staff use it as well, to identify the onset of instrument malfunctions, for further annotation and analysis during quality-control processing.

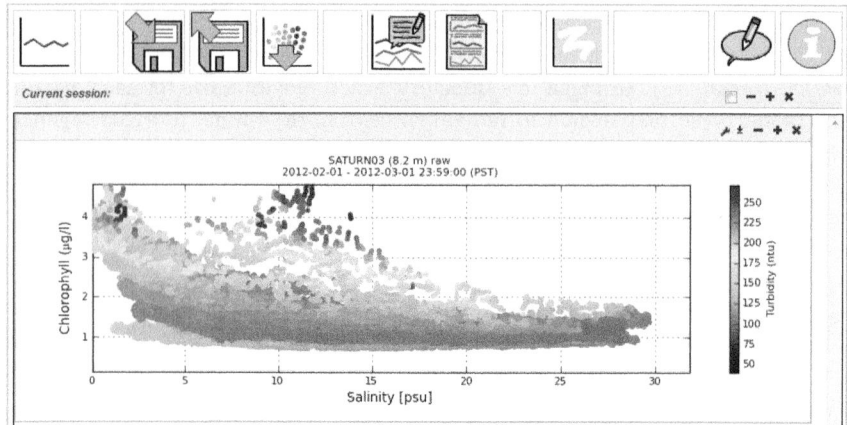

Fig. 5. The resulting plot from Data Explorer

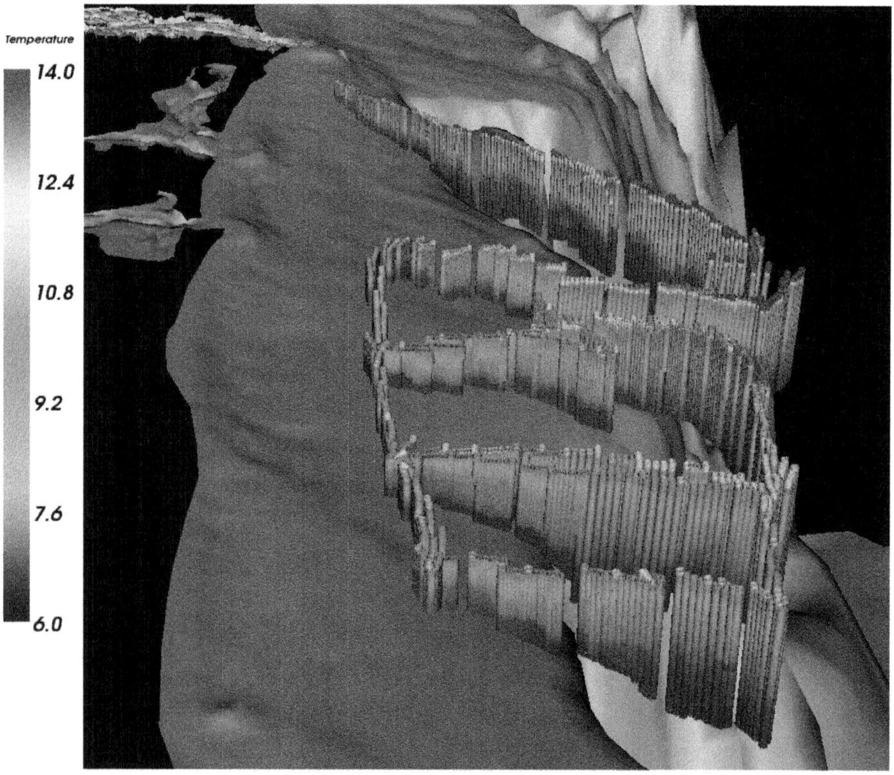

Fig. 6. Specialized plots for a glider mission, showing the trajectory for the glider, colored by temperature in this case, superimposed on the sea-bottom topography

The design of the plot-specification interface for Data Explorer has been challenging. One issue is avoiding creating plots where there is no data, usually due to selecting a time period before a sensor was deployed or during an outage, or choosing a data-quality level that has not yet been produced. Once an offering is selected, one can see an inventory of data for it (the "Availability" button). However, it might be helpful to default the time selection on the next screen to the most recent period with data.

A related issue is the order in which plot aspects are specified. Currently, the configuration interface aims at a work pattern where a user is first interested in one or more stations, then selects offerings from those stations, followed by choosing a time period. But there are certainly other patterns of work. A scientist might be interested in a particular variable, say dissolved oxygen, at a particular time (say corresponding to field work), and want any stations that have an offering for that variable at the time. We have not yet devised a means to simultaneously support a variety of work patterns with the Data Explorer interface.

Some of the data-collection platforms have specialized displays related to particular properties of the platform. For example, CMOP's underwater glider, called Phoebe, runs multi-day missions over a pre-programmed trajectory. The gathered data are time series, hence can be used with time-series oriented interfaces such as the Data Explorer However, because of the nature of the glider path (repeated dives from the sea surface to near the bottom), it often makes more sense to plot depth versus time, with the plot colored by the variable of interest, such as salinity. Thus we provide a special interface for specifying such plots (which are generated by the plot service). As shown in Figure 6, there are also renderings that depict the 3-dimensional trajectory of the glider. These plots are pre-computed for each glider mission and variable.

For resource managers (and scientists), CMOP provides *watch* pages for particular interest areas. A watch page has a selection of plots connected to the interest area, along with commentary. Figure 7 shows the Oxygen Watch page, which targets hypoxic (low-oxygen) conditions [14]. It contains a plot of dissolved oxygen from multiple observation stations, along with reference lines that reflect different definitions of hypoxic conditions from the literature. Additional plots show environment conditions (river discharge, north-south wind speed) that are often correlated with oxygen levels in the estuary. Commentary in the "Blog" tab interprets current conditions.

Other watch pages under development include one for Myrionecta rubra (a microorganism) causing red-water bloom in the river [9], and one directed at steelhead survivability. The latter features displays that compare predicted plume area, volume and distance off shore to historical conditions for the same day of year, which could provide hatchery managers with guidance on the best time to release young steelhead (a fish related to salmon) [3].

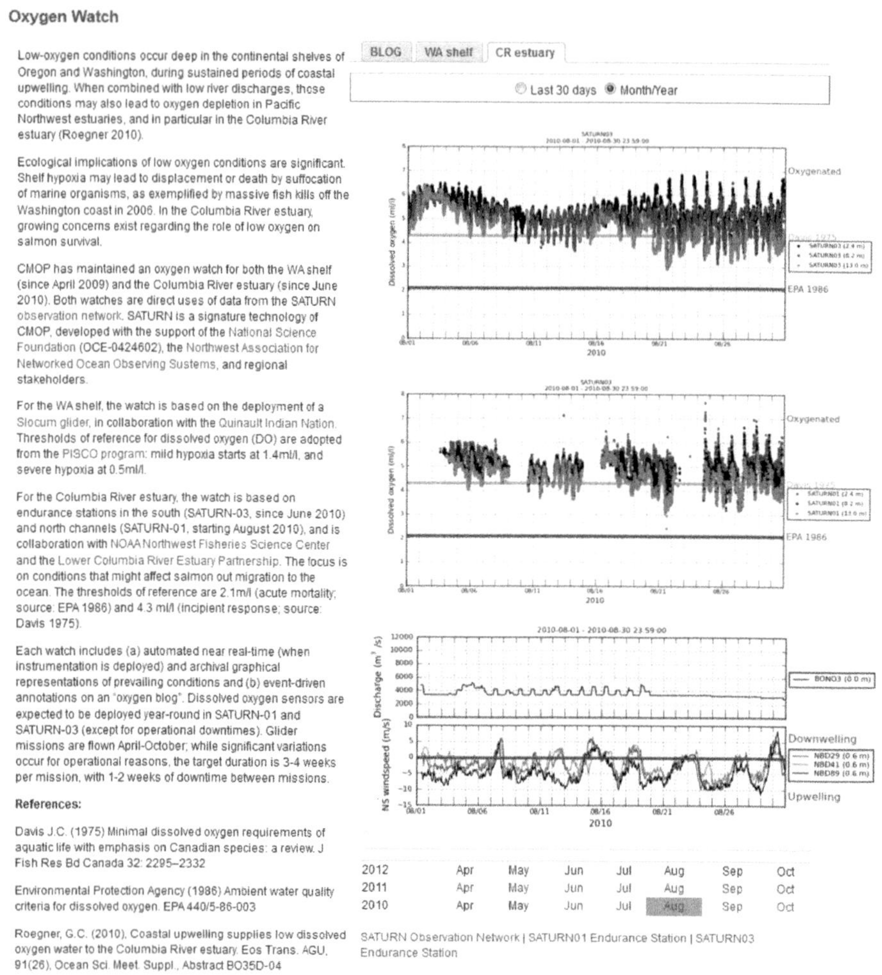

Oxygen Watch

Low-oxygen conditions occur deep in the continental shelves of Oregon and Washington, during sustained periods of coastal upwelling. When combined with low river discharges, those conditions may also lead to oxygen depletion in Pacific Northwest estuaries, and in particular in the Columbia River estuary (Roegner 2010).

Ecological implications of low oxygen conditions are significant. Shelf hypoxia may lead to displacement or death by suffocation of marine organisms, as exemplified by massive fish kills off the Washington coast in 2006. In the Columbia River estuary, growing concerns exist regarding the role of low oxygen on salmon survival.

CMOP has maintained an oxygen watch for both the WA shelf (since April 2009) and the Columbia River estuary (since June 2010). Both watches are direct uses of data from the SATURN observation network. SATURN is a signature technology of CMOP, developed with the support of the National Science Foundation (OCE-0424602), the Northwest Association for Networked Ocean Observing Systems, and regional stakeholders.

For the WA shelf, the watch is based on the deployment of a Slocum glider, in collaboration with the Quinault Indian Nation. Thresholds of reference for dissolved oxygen (DO) are adopted from the PISCO program: mild hypoxia starts at 1.4ml/l, and severe hypoxia at 0.5ml/l.

For the Columbia River estuary, the watch is based on endurance stations in the south (SATURN-03, since June 2010) and north channels (SATURN-01, starting August 2010), in collaboration with NOAA Northwest Fisheries Science Center and the Lower Columbia River Estuary Partnership. The focus is on conditions that might affect salmon out migration to the ocean. The thresholds of reference are 2.1ml/l (acute mortality; source: EPA 1986) and 4.3 ml/l (incipient response; source: Davis 1975).

Each watch includes (a) automated near real-time (when instrumentation is deployed) and archival graphical representations of prevailing conditions and (b) event-driven annotations on an "oxygen blog". Dissolved oxygen sensors are expected to be deployed year-round in SATURN-01 and SATURN-03 (except for operational downtimes). Glider missions are flown April-October; while significant variations occur for operational reasons, the target duration is 3-4 weeks per mission, with 1-2 weeks of downtime between missions.

References:

Davis J.C. (1975) Minimal dissolved oxygen requirements of aquatic life with emphasis on Canadian species: a review. J Fish Res Bd Canada 32: 2295–2332

Environmental Protection Agency (1986) Ambient water quality criteria for dissolved oxygen. EPA 440/5-86-003

Roegner, G.C. (2010). Coastal upwelling supplies low dissolved oxygen water to the Columbia River estuary. Eos Trans. AGU, 91(26), Ocean Sci. Meet. Suppl., Abstract BO35D-04

BLOG   WA shelf   CR estuary

Last 30 days   Month/Year

SATURN03 (2.4 m)
SATURN03 (8.2 m)
SATURN03 (11.0 m)
EPA 1986

SATURN01 (2.4 m)
SATURN01 (8.2 m)
SATURN01 (11.0 m)
EPA 1986

BONO3 (0.0 m)

Downwelling

NBD29 (0.6 m)
NBD41 (0.6 m)
NBD89 (0.6 m)

Upwelling

2012		Apr	May	Jun	Jul	Aug	Sep	Oct
2011		Apr	May	Jun	Jul	Aug	Sep	Oct
2010		Apr	May	Jun	Jul	Aug	Sep	Oct

SATURN Observation Network | SATURN01 Endurance Station | SATURN03 Endurance Station

Fig. 7. The Oxygen Watch page, showing conditions during August 2010

5   Supporting Ranked Search for Datasets

One challenge for CMOP scientists is knowing what datasets might be relevant to their current work. Database and basic spatial search techniques (contains, overlaps) often prove unsatisfactory, in that it is easy to get answers that return no datasets or thousands of them, requiring iterative tweaking of search conditions to get a candidate set of answers. As an alternative, we are developing an interface that applies Information Retrieval approaches to give ranked search of datasets. Data Near Here (inspired by the "search nearby" in map services) makes use of similarity search over spatial-temporal "footprints" that are computed from the datasets. Our initial work [10] focused on identifying a similarity measure that would balance geospatial and temporal search conditions in a way that resonated with our user community. At scientists' request, we have since added "dimensions" of depth, variable existence and variable values to our search capabilities. Figure 8 shows the results of a Data Near Here query, with the top few matching datasets shown.

Most data in the archive treats depth as a separate field; also, the currently used version of spatial tools (PostGIS 1.5) does not fully support three-dimensional spatial functions. As a result, depth is currently treated as a separate search condition, and the search condition is given the same weight as geospatial location. An alternate approach is to treat the geospatial locations, including depth, as true three-dimensional locations. The current spatial distance metric does not change if given fully three-dimensional data, although some implementation details will need to change.

Scientists may also wish to search for data based on variable values; for example, all places and times where low oxygen conditions occurred. A scientist may even be searching for places and times where a variable was collected, irrespective of the variable's values. We added the capability to search over variables and their values into the same metadata extraction and search framework. The metadata extraction tools were extended to identify and store the variable names for each dataset. The variables are generally represented by column names, and so we assume that each column represents a variable. For netCDF files, this information is available in the header; for comma-separated value files it is often in the first row, and for data served from CMOP's relational database, it is in the database catalog. If available, we also capture the data type and units for each variable. If the units for a variable cannot be inferred, they are shown in the catalog as "unknown". Data types are treated the same way; alternately, techniques exist (such as those used in Google Fusion Tables [7]) to infer likely data types from the data itself. We also read the data and store the maximum and minimum values found for each variable, handling character and numeric data similarly. We intend to provide search capabilities over the modeled data.

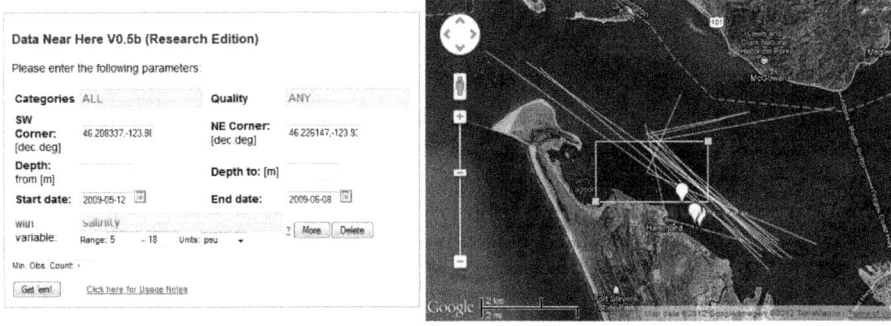

Fig. 8. The Data Near Here prototype, showing a search based on a particular X-Y region, with no constraint on depth, seeking datasets that contain salinity in a certain range. Results are ranked on a weighted combination of similarity to the search conditions, rather than on exact match. Data can be directly downloaded, or plotted in the Data Explorer.

Once we have extracted metadata for each dataset to identify contained variables and their values, we are able to search over it using extensions of the techniques and formulae we use for geospatial-temporal search. We provide two types of search conditions for variables. The first specifies a variable name and a desired range of values, in some specified units. For each dataset that contains the desired variable in the specified units, the range of values is compared to the desired range and a similarity score computed; the computed score contributes to the overall dataset similarity score. A dataset that does not contain the desired variable can still be returned in the query results if it has high scores on the other query conditions. However, a dataset that contains the desired variable with values similar to the desired data range is likely to receive a higher overall score, even if its scores on the geospatial and temporal query conditions are lower. Unit translation is possible in many cases, and we are experimenting with approaches to this problem. If the units for a dataset are unknown, we assume the values are in the desired units but substantially discount the score.

The second type looks for datasets that contain a certain variable but does not specify a range. In concept, this condition specifies a variable with an infinite range of values; thus, any dataset that contains a column of that name, with any values at all, is considered "closer" to that query condition than a dataset that lacks that variable. In effect, the resulting score is binary: a dataset is a perfect match to the query condition if the desired variable is found in that dataset, or a complete non-match if it does not.

At present, we only match on exact variable names; a search for "temperature" will not match "air temperature" or "airtemp". In a large archive built over more than a decade, inconsistencies and changes in variable names are common. We are considering methods to match on "close" variable names, as these inconsistencies frustrate our scientists. One possibility is to extend our approach to variable existence, so that the existence of a variable with a similar name is given a score reflecting the higher similarity, converting variable existence from a binary to a continuous similarity score.

We are finding that as Data Near Here queries become more sophisticated, it becomes expensive to apply the similarity function to the footprints of all the data sets. Figure 9 illustrates the problem for queries with increasing numbers of search conditions on variables. The "cast variables" are those typically measured by lowering an instrument package from a cruise vessel, whereas "station variables" are those typically seen at fixed stations. The alternating line is for queries that include queries asking for datasets containing both kinds of variables, which none of the existing datasets will match very closely. As can be seen, response times start to grow out of the interactive range rather quickly for this last category of query.

One technique we are investigating starts by selecting a cut-off on minimal similarity score, and incorporating a pre-filter into the query that can quickly rule out certain datasets being over that score without applying the full (and more expensive) similarity calculation. As can be seen in Figure 9, incorporating the cut-off does improve response times on the more expensive searches. An area for further work is determining how to initially set the cut-off threshold for a given query and limiting the number of expensive geospatial comparisons by using cheaper pre-filters.

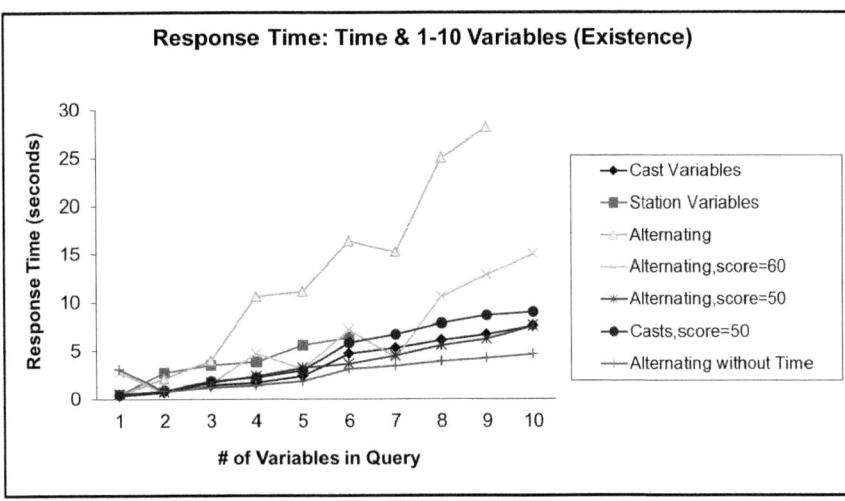

Fig. 9. The effect of incorporating an initial cut-off threshold on similarity score in Data Near Here Queries

Data Near Here currently provides access to more than 750B observations from fixed stations, glider, cruises, casts, and water samples, at three quality levels (raw, preliminary and verified). The largest dataset indexed has 11.5M values, the smallest, one. The mean dataset size is about 33K entries.

6   General Lessons

While the development of CMOP information access and analysis capabilities is an on-going process, we can identify some important guidance for similar endeavors.

1. *Don't make users repeat work.* For example, if a user has gone through a data-selection process in order to specify a plot or chart, do not make him or her repeat the specification to download the underlying data. Similarly, if a user has invested time in configuring a graph of an interesting segment of data using an on-line tool, there should be a way to share the result. At a minimum, the resulting image should be savable, but much better is providing a URL that can put the tool back into the same state. We are not perfect on avoiding repeated work–in some cases a user must re-specify some aspect of a plot to change it–but we are working to reduce such cases.

2. *Default to where the data is.* Upon initially coming to an interface it is useful to have default settings selected. It is tempting to choose these settings in a uniform way. For example, every station page could be set to display salinity at that station for the past two days. However, such settings can result in no data being displayed, because there is a problem with the salinity sensor or transmission of its data. In such cases we find it preferable to adjust the settings so data is available for display–for example, expanding the time period or selecting a different

variable. More generally, we try to not offer the user selections in the interface where no data is available. For instance, in Data Explorer, if the user has selected an observation station in Figure 4, only variables for that station are then listed to select from. (We could go further in this direction. For example, if a station and an offering are selected, then offer only choices of date range where data is present.)

3. *Give access to underlying data.* Any display of data should provide a ready means to download that data. While we hope in most cases a user can meet his or her data-location and analysis needs through our interfaces, in many cases a user will want to view the data using a plot style our tools do not provide, or carry out more advanced computation, say in Matlab. Thus, whenever a tool displays a data set it should be possible to download the underlying data, preferably in a choice of formats. Currently, data can be downloaded from any station page, such as shown in Figure 3 (via the "Inventory" tab), or from the other tools we discuss.

4. *Integrate the tools.* Each of the tools provided has a place in the scientists' workflows. A scientist can quickly search for or browse to a likely source of data using Data Near Here, use Data Explorer to plot some variables to confirm its relevance, and download data in a variety of formats directly from these tools. Such workflows are inherently iterative. By allowing multiple tools to operate over the same data and, where possible, pass settings and selections from tool to tool, we allow the scientists to focus on the research and not on the complexities of the tooling and infrastructure.

5. *Balance pre-computation with production on demand.* Ideally, we could provide any possible data display with zero delay. The realities are that there is a bounded amount of processing that has to support data ingest, quality assurance, model evaluation and servicing of analysis and retrieval requests. If the last grows to consume too great a share of resources, the observation system cannot keep up with the other functions. Even if we could upgrade to meet all these demands today, the continuing increasing volume and density of the data being collected would make this goal unattainable tomorrow. Obviously we can control the cost of analysis and display requests by how complex of processing we support in interactive mode. To do more resource-intensive operations, the user needs to download data and compute locally. We also pre-compute and cache display plots that are likely to be requested by multiple users, such as the plots displayed upon entering the station pages, as that shown in Figure 3. We also pre-compute and cache plots that are hard to produce at interactive speeds, for example the track plots for glider missions shown in Figure 6. The output of the simulation model is handled similarly. For forecast simulations we pre-compute various data products at the time of model generation. Many of these products are animations of a particular variable along a 2-D horizontal or vertical slice. (In fact, these animations can be computed incrementally as the model runs, and provide a means for diagnosing computations going awry.) However, it is also possible to produce map layers from model data (via a WMS [11]) on demand.

7   Issues and Challenges

While the various CMOP information interfaces described here have gone a long way towards meeting the needs of the various user groups, there are still areas that could be expanded and enhanced. Here are list of areas of work, ranging from ones where we are fairly certain how to handle to ones that will require extensive research.

1. With the wide range of interfaces, there can of course be inconsistencies. We have discussed how we try to use common components, such as the plot service, across interfaces for uniformity. We also try to drive menus and choices (such as available offerings from a station) out of a common database of metadata. However, there can still be variations in grouping or ordering of options, which could possibly become more table-driven.

2. While we have various means of showing the inventory (for different time periods) of holdings for a given offering, we lack means to depict "joint availability". For example, a scientist might want to know for what time periods is temperature available at both SATURN03 and SATURN04, in order to cross-compare them.

3. Our current plotting facility can deal with datasets spanning many months. However, we are only beginning to develop representations for multiple years of data that allow short-term trends and events to be discerned. Simple plots and aggregates can lose the fine detail.

4. As mentioned in Section 2, fault diagnosis and quality assurance are often handled with general purpose interfaces, requiring a fair amount of manual effort. We need more automated methods to allow limited staff to support continued growth in the sensor arrays. We have had some success in the past applying machine-learning techniques to detecting biofouling of sensors [1], but there remains a wide range of approaches to explore in specifying or learning normal reporting patterns and detecting divergence from them.

5. Another open area is the display and indication of uncertainty. While we are currently expanding our capabilities for flagging and suppressing problem data, we do not know of good methods to portray the inherent systemic uncertainty of our various datasets, nor can we propagate such knowledge through analysis and charting tools. We welcome the suggestions of other researchers here.

6. We have over a decade of historical simulated data, and one chief use for these "hindcasts" is *climatology queries*. Such a query aggregates possibly the whole hindcast database over time and space, for example, daily maximum temperatures over the estuary averaged by month, or fresh-water plume volume on a daily basis. A variety of these queries are pre computed and constitute the CMOP Climatological Atlas [16], but given the size of the hindcast database (tens of terabytes), we do not support climatological queries on demand. The size of the hindcasts similarly makes download of the database for local use generally infeasible. This problem will become more challenging as we include chemical and biological quantities in our models. We also contemplate producing hindcast databases for "what-if" scenarios, such as different river-discharge levels and

changes in bathymetry (bottom topography) of the estuary. While reduced-resolution databases might address on-demand climatologies for quick comparisons, detailed analysis of differences will require computation at full resolution. Putting the hindcast databases in the cloud, and having users pay for their processing is an intriguing possibility; especially as most climatology queries are easily parallelizable. However, current cost schedules for cloud storage are prohibitive for the amount of data contemplated. One issue is that even the "cheap" option at such services has availability guarantees (99.9%) beyond what we really require. (Even 90% availability would probably satisfy most of our demands.)

Going forward, the ocean of data will continue to swell and present greater challenges for navigation. On one hand, we want to minimize both the complexity of interfaces and their need for manual support. On the other, the questions scientists are trying to answer, and their processes for investigating them, are becoming more complicated. It will be a balancing act not to constrain them by making interfaces too limited to handle their needs or too difficult to work with efficiently.

Acknowledgments. This work is supported by NSF award OCE-0424602. We would like to thank the staff of CMOP for their support.

References

1. Archer, C., et al.: Fault detection for salinity sensors in the Columbia estuary. Water Resources Research 39(3), 1060 (2003)
2. Burla, M., et al.: Seasonal and Interannual Variability of the Columbia River Plume: A Perspective Enabled by Multiyear Simulation Databases. Journal of Geophysical Research 115(C2), C00B16 (2010)
3. Burla, M.: The Columbia River Estuary and Plume: Natural Variability, Anthropogenic Change and Physical Habitat for Salmon. Ph.D. Dissertation. Beaverton, OR: Division of Environmental and Biomolecular Systems, Oregon Health & Science University (2009)
4. Cornillon, P., et al.: OPeNDAP: Accessing Data in a Distributed, Heterogeneous Environment. Data Science Journal 2, 164–174 (2003)
5. Domenico, B., et al.: Thematic Real-time Environmental Distributed Data Services (THREDDS): Incorporating Interactive Analysis Tools into NSDL. Journal of Digital Information 2(4) (2006)
6. Ghindilis, A.L., et al.: Real-Time Biosensor Platform: Fully Integrated Device for Impedimetric Assays. ECS Transactions 33(8), 59–68 (2010)
7. Gonzalez, H., et al.: Google Fusion Tables: Data Management, Integration and Collaboration in the Cloud. In: Proceedings of the 1st ACM Symposium on Cloud Computing, pp. 175–180. ACM, New York (2010)
8. Haddock, T.: Submersible Microflow Cytometer for Quantitative Detection of Phytoplankton (2009), https://ehb8.gsfc.nasa.gov/sbir/docs/public/recent_selections/SBIR_09_P2/SBIR_09_P2_094226/briefchart.pdf
9. Herfort, L., et al.: Myrionecta rubra (Mesodinium rubrum) bloom initiation in the Columbia River Estuary. Estuarine, Coastal and Shelf Science (2011)

10. Megler, V.M., Maier, D.: Finding Haystacks with Needles: Ranked Search for Data Using Geospatial and Temporal Characteristics. In: Bayard Cushing, J., French, J., Bowers, S. (eds.) SSDBM 2011. LNCS, vol. 6809, pp. 55–72. Springer, Heidelberg (2011)
11. Open Geospatial Consortium, Inc.: OpenGIS® Web Map Server Implementation Specification Version: 1.3.0 (2006)
12. Plant, J., et al.: NH 4-Digiscan: an in situ and laboratory ammonium analyzer for estuarine, coastal and shelf waters. Limnology and Oceanography: Methods 7, 144–156 (2009)
13. Rew, R., Davis, G.: NetCDF: an interface for scientific data access. IEEE Computer Graphics and Applications 10(4), 76–82 (1990)
14. Roegner, G.C., et al.: Coastal Upwelling Supplies Oxygen-Depleted Water to the Columbia River Estuary. PLoS One 6(4), e18672 (2011)
15. Szalay, A.S., et al.: Designing and mining multi-terabyte astronomy archives: the Sloan Digital Sky Survey. In: Proceedings of the 2000 ACM SIGMOD International Conference on Management of Data, vol. 29(2), pp. 451–462 (2000)
16. Climatological Atlas, Center for Coastal Margin Observation & Prediction, http://www.stccmop.org/datamart/virtualcolumbiariver/simulat iondatabases/climatologicalatlas

Probabilistic Range Monitoring of Streaming Uncertain Positions in GeoSocial Networks

Kostas Patroumpas[1], Marios Papamichalis[1], and Timos Sellis[1,2]

[1] School of Electrical and Computer Engineering
National Technical University of Athens, Hellas
[2] Institute for the Management of Information Systems, R.C. "Athena", Hellas
{kpatro,timos}@dbnet.ece.ntua.gr, papamixmarios@gmail.com

Abstract. We consider a social networking service where numerous subscribers consent to disclose their current geographic location to a central server, but with a varying degree of uncertainty in order to protect their privacy. We aim to effectively provide instant response to multiple user requests, each focusing at continuously monitoring possible presence of their friends or followers in a time-varying region of interest. Every continuous range query must also specify a cutoff threshold for filtering out results with small appearance likelihood; for instance, a user may wish to identify her friends currently located somewhere in the city center with a probability no less than 75%. Assuming a continuous uncertainty model for streaming positional updates, we develop novel pruning heuristics based on spatial and probabilistic properties of the data so as to avoid examination of non-qualifying candidates. Approximate answers are reported with confidence margins, as a means of providing quality guarantees and suppressing useless messages. We complement our analysis with a comprehensive experimental study, which indicates that the proposed technique offers almost real-time notification with tolerable error for diverse query workloads under fluctuating uncertainty conditions.

1 Introduction

Over this decade, we have been witnessing the rising popularity of social networks. Connecting people who share interests or activities has an all-increasing impact on communication, education, and business; their role even in politics and social movements is indisputable. One of the latest trends heads for *GeoSocial Networking Services* [13,15], allowing location-aware mobile users to interact relative to their current positions. Platforms like Facebook Places, Google Latitude, or FireEagle[1], enable users to pinpoint friends on a map and share their whereabouts and preferences with the followers they choose. Despite their attraction, such features may put people's privacy at risk, revealing sensitive information about everyday habits, political affiliations, cultural interests etc. Hence, there has been strong legal and research interest on controlling the level of location precision, so as to prevent privacy threats and protect user anonymity.

---

[1] http://facebook.com/about/location; http://google.com/latitude;
http://fireeagle.yahoo.net

A. Ailamaki and S. Bowers (Eds.): SSDBM 2012, LNCS 7338, pp. 20–37, 2012.

Respecting privacy constraints, we turn our focus to real-time processing of *continuous range queries* against such imprecise user locations. In our proposed framework, a subscriber may receive instant notifications when a friend appears *with sufficient probability within her area of interest*. Mobile users are aware of their own exact location thanks to geopositional technologies (e.g., GPS, WiFi, Bluetooth), but they do not wish to disclose it to third parties. Instead, they consent to relay just a cloaked indication of their whereabouts [6] abstracted as an *uncertainty region* with Gaussian characteristics, enclosing (but apparently not centered at) their current position. Hence, the service provider accepts a geospatial stream of obfuscated, time-varying regions sent from numerous users at irregular intervals. Based on such massive, transient, imprecise data, the server attempts to give response to multiple search requests, which may also dynamically modify their spatial ranges and probability thresholds.

This query class may prove valuable to GeoSocial Networking and Location-based services (LBS). A typical request is "Notify me whenever it is highly likely (more than 75%) that any friends of mine are located somewhere in my neighborhood" just in case one wants to arrange a meeting. A micro-blogging enthusiast could be traveling or walking, and while on the move, may wish to post messages to followers nearby. Even virtual interactive games on smartphones could take advantage of such a service, e.g., assessing the risk of approaching "unfriendly territory" with several adversaries expectedly present in close proximity.

In a geostreaming context, identifying mobile users with varying degrees of uncertainty inside changing areas of interest poses particular challenges. Faced with strict privacy preferences and intrinsic positional inaccuracy, while also pursuing adaptivity to diverse query workloads for prompt reporting of results, we opt for an approximate evaluation scheme. We introduce optimizations based on inherent probabilistic and spatial properties of the uncertain streaming data. Thus, we can quickly determine whether an item possibly qualifies or safely skip examination of non-qualifying cases altogether. Inevitably, this probabilistic treatment returns approximate answers, along with confidence margins as a measure of their quality. Our contribution can be summarized as follows:

- We model uncertainty of incoming locations as a stream of moving regions with fluctuating extents under a Bivariate Gaussian distribution.
- We develop an online mechanism for evaluating range requests, employing lightweight, discretized verifiers amenable to Gaussian uncertainty.
- We introduce pruning criteria in order to avoid examination of objects most unlikely to fall inside query boundaries, with minimal false negatives.
- We empirically demonstrate that this methodology can provide approximate, yet timely response to continuous range queries with tolerable error margins.

The remainder of this paper proceeds as follows: Section 2 briefly reviews related work. Section 3 covers fundamentals of positional uncertainty and outlines application specifics. In Section 4, we develop an (ϵ, δ)-approximation algorithm for continuous range search against streaming Gaussian regions. In Section 5, we introduce heuristics for optimized range monitoring. Experimental results are reported in Section 6. Finally, Section 7 concludes the paper.

2   Related Work

Management of uncertain data has gained particular attention in applications like sensor networks, market surveillance, biological or moving objects databases etc. In terms of processing [14], besides range search, a variety of *probabilistic queries* have been studied: nearest-neighbors [10], reverse nearest neighbors [1], *k*-ranked [11], continuous inverse ranking [2], similarity joins [9,12], etc.

In contrast to traditional range search, a probabilistic one requires its answer to be assessed for quality. Among related techniques in uncertain databases, the notion of x-bounds in [5] clusters together one-dimensional features with similar degrees of uncertainty in an R-tree-like index. U-tree [16] is its generalization for multiple dimensions and arbitrary probability distributions. U-tree employs probabilistically constrained regions to prune or validate an object, avoiding computation of appearance probabilities. U-tree can be further useful for "fuzzy" search, when the query range itself becomes uncertain [17]. We utilize a pruning heuristic with a similar flavor, but our proposed minimal areas correspond to distinct threshold values and are independent of uncertainty specifications for any object. For predicting the location distribution of moving objects, an adapted B^x-tree [18] has been used to answer range and nearest neighbor queries. Especially for inexact Gaussian data, the Gauss-tree [3] (also belonging to the R-tree family) models the means and variances of such feature vectors instead of spatial coordinates. Such policies may be fine for databases with limited transactions, but are not equally fit for geostreaming uncertain data; the sheer massiveness and high frequency of updates could overwhelm any disk-based index, due to excessive overhead for node splits and tree rebalancing.

Note that spatial ranges may be uncertain as well, e.g., modeled as Gaussians [8], or due to query issuer's imprecise location when checking for objects within some distance [4]. In our case, spatial ranges are considered typical rectangles, yet subject to potential changes on their placement, shape and extent.

Privacy-aware query processing in LBS and GeoSocial networks has also attracted particular research interest. For exact and approximate search for nearest neighbors, the framework in [7] uses Private Information Retrieval protocols, thus eliminating the need for any trusted anonymizer. Shared processing for multiple concurrent continuous queries in [6] handles cloaked user areas independent of location anonymizers, offering tunable scalability versus answer optimality. A privacy-aware proximity detection service is proposed in [15], so that two users get notified whenever the vicinity region of each user includes the location of the other. Encryption and spatial cloaking methods enable the server to perform a blind evaluation with no positional knowledge. More sophisticated protocols [13] offer controllable levels of location privacy against the service provider and third parties, essentially trading off quality of service against communication cost. Nonetheless, such techniques principally address privacy concerns and assume uniform uncertainty distribution. Thus, they lack any probabilistic treatment of spatial queries and user whereabouts, as we attempt in this work. We stress that our approach is orthogonal to privacy preservation policies, focusing entirely on swift processing of continuous range requests at the service provider.

3   Managing Uncertain Moving Objects

3.1   Capturing Positional Uncertainty

Typical causes of data uncertainty [14,16] include communication delays, data randomness or incompleteness, limitations of measuring instruments etc. Apart from inherent imprecision of location representations, in this work we assume that mobile users have purposely sent an "inflated" positional update so as to conceal their precise coordinates from the server. Any anonymization technique that cloaks users' locations into *uncertainty regions* can be employed (e.g.,[6]). Typically, the larger the size of the region, the more the privacy achieved.

Positional uncertainty can be captured in a discrete or continuous fashion. A *discrete model* uses a probability mass function (*pmf*) to describe the location of an uncertain object. In essence, a finite number of alternative instances is obtained, each with an associated probability [10,14]. In contrast, a *continuous model* uses a probability density function (*pdf*), like Gaussian, uniform, Zipfian etc., to represent object locations over the space. Then, in order to estimate the appearance probability of an uncertain object in a bounded region, we have to integrate its pdf over this region [16]. In a geostreaming scenario, a discrete model should be considered rather inappropriate, as the cost of frequently transmitting even a small set of samples per object could not be affordable in the long run. Hence, we adopt a continuous model, which may be beneficial in terms of communication savings, but it poses strong challenges in terms of evaluation. Table 1 summarizes the notation used throughout the paper.

Table 1. Primary symbols and functions

Symbol	Description
ϵ	Error margin for appearance probability of qualifying objects
δ	Tolerance for reporting invalid answers
N	Total count of moving objects (i.e., users being monitored)
M	Total count of registered continuous range queries
μ_x, μ_y	Mean values of uncertainty pdf per object along axes x, y
σ_x, σ_y	Standard deviations of uncertainty pdf per object along axes x, y
Σ	Set of discrete uncertainty levels $\{\sigma_1, \sigma_2, \ldots, \sigma_k\}$ for regulating location privacy
$\mathcal{N}(\mathbf{0}, \mathbf{1})$	Bivariate Gaussian distribution with mean $(0,0)$ and standard deviation $\sigma_x = \sigma_y = 1$
r_q	Time-varying 2-d rectangular range specified by query q
r_o	Time-varying uncertainty area of moving object o
$MBB(r_o)$	Minimum Bounding Box of uncertainty area r_o with center at (μ_x, μ_y) and side 6σ
$V(r_o)$	Verifier for uncertainty area r_o, comprised of elementary boxes with known weights
L_q	Set of objects monitored by query q (i.e., *Contact list* of q)
C_q	Set of candidate objects that might qualify for query q
Q_q	Final set of objects estimated to qualify for query q
p_T, p_F, p_U	Total cumulative probability of elementary boxes marked with T, F, U respectively
λ	Granularity of subdivision for every discretized verifier along either axis x, y
$\beta(i, j)$	Weight (i.e., estimated cumulative probability) of elementary box $V(i, j)$
$P_{in}(o, q)$	Estimated probability that object o appears within range of query q
θ	Cutoff threshold for rejecting objects with insufficient appearance probability
Θ	A set $\{\theta_1, \theta_2, \ldots, \theta_m\}$ of m typical threshold values
$\bar{\alpha}_\theta$	Minimal uncertainty area representing cumulative probability barely less than θ
$\mathcal{A}$	Set of minimal areas $\{\bar{\alpha}_1, \bar{\alpha}_2, \ldots, \bar{\alpha}_m\}$ corresponding to indicative thresholds $\theta_i \in \Theta$

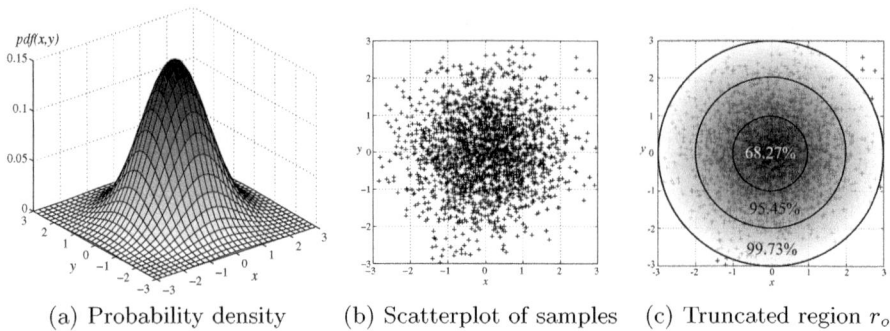

(a) Probability density   (b) Scatterplot of samples   (c) Truncated region r_o

Fig. 1. Standard Bivariate Gaussian distribution $\mathcal{N}(\mathbf{0}, \mathbf{1})$

3.2 Object Locations as Bivariate Gaussian Features

Locations of mobile users are modeled with Bivariate Gaussian random variables X, Y over the two dimensions of the Euclidean plane. Intuitively, the resulting uncertainty region implies higher probabilities closer to the mean values (i.e., the origin of the distribution), as illustrated with the familiar "bell-shaped" surface in Fig. 1a. For privacy preservation, the origin of the distribution should not coincide with the precise coordinates, known only by the user itself.

More specifically, let a Bivariate Gaussian (a.k.a. Normal) distribution with

$$\text{mean} \begin{bmatrix} \mu_x \\ \mu_y \end{bmatrix} \quad \text{and} \quad \text{covariance matrix} \begin{bmatrix} \sigma_x^2 & \rho\sigma_x\sigma_y \\ \rho\sigma_x\sigma_y & \sigma_y^2 \end{bmatrix},$$

where μ_x, μ_y are the mean values and σ_x, σ_y the standard deviations along axes x, y respectively, whereas ρ is the correlation of random variables X and Y. Assuming that objects are moving freely, X and Y are independent, hence $\rho = 0$. Of course, location coordinates may spread similarly along each axis, so $\sigma_x = \sigma_y = \sigma$. Thus, the joint probability density function (pdf) is simplified to:

$$pdf(x, y) = \frac{1}{2\pi\sigma^2} \cdot e^{-\frac{(x-\mu_x)^2+(y-\mu_y)^2}{2\sigma^2}} \tag{1}$$

As in the univariate case, we can define random variables $X' = \frac{X-\mu_x}{\sigma}$ and $Y' = \frac{Y-\mu_y}{\sigma}$ and derive the *standard bivariate Gaussian distribution* $\mathcal{N}(\mathbf{0}, \mathbf{1})$ with

$$pdf(x', y') = \frac{1}{2\pi} e^{-\frac{r^2}{2}} \tag{2}$$

where $r = \sqrt{x'^2 + y'^2}$ denotes the distance from the origin of the derived distribution at $(0, 0)$ with standard deviations $(1, 1)$. Figure 1b depicts a scatterplot of random samples under this distribution. As illustrated in Fig. 1c, there is 99.73% probability that the location is found within a radius 3σ from the origin. Depending on the variance, the density of a Gaussian random variable is rapidly diminishing with increasing distances from the mean. Thanks to its inherent simplicity, the uncertainty region can be truncated in a natural way *on the server side*, so the user itself does not need to specify a bounded area explicitly.

3.3 System Model

We consider a social networking service with a large number N of location-aware subscribers, each moving over the Euclidean plane and communicating with the provider. Messages transmitted from mobile users concern either cloaked positions or spatial requests. By convention, the former (termed *objects*) are being searched by the latter (*queries*). So, objects and queries alike represent mobile users of the service, but with distinct roles (passive or active) in terms of monitoring. All messages are timestamped according to a global clock at distinct instants τ (e.g., every minute).

Every object o relays to the centralized server its *uncertainty region* r_o, i.e., an imprecise indication of its current location. The server is not involved in the cloaking process, but passively receives vague positional information according to a privacy preserving protocol. Updates may be sent over at irregular intervals, e.g., when an object has gone far away from its previously known position or upon significant change at its speed. Although the server knows nothing about the exact (x, y) coordinates of a given object o, it can be sure that o is definitely found somewhere within its uncertainty region until further notice.

Each uncertainty region r_o follows a Bivariate Gaussian distribution, so an object o must sent the origin (μ_x, μ_y) of its own pdf and the standard deviation σ (common along both dimensions). Upon arrival to the server from numerous objects, these items constitute a unified *stream* of tuples $\langle o, \mu_x, \mu_y, \sigma, \tau \rangle$, ordered by their timestamps τ. Note that μ, σ are expressed in distance units of the coordinate system (e.g., meters). Larger σ values indicate that an object's location can be hidden in a greater area around its indicated mean (μ_x, μ_y). As object o is moving, it relays (μ_x, μ_y) updates. We prescribe k *uncertainty levels* $\Sigma = \{\sigma_1, \sigma_2, \dots, \sigma_k\}$, so any object can adjust its degree of privacy dynamically.

A set of M continuous queries are actually registered at the server, each specifying a *rectangular extent* r_q and a cutoff *threshold* $\theta \in (0, 1)$. During their lifetime, ranges r_q may be moving and also vary in size, whereas a query may arbitrarily change its own θ. Therefore, the server accepts query updates specifying $\langle q, r_q, \theta, \tau \rangle$, replacing any previous request with identifier q. As is typical in social networking [13], each query issuer states its *contact list* L_q declaring its friends, fans, or followers. Hence, the server retains a table with entries $\langle q, o \rangle$, which specifies that query q has an interest on monitoring object o, provided that the latter is consenting. Evaluation takes place periodically with execution cycles at each successive τ, upon reception of the corresponding updates. Query q identifies any object o from its contact list L_q currently within specified range r_q with *appearance probability* $P_{in}(o, q)$ at least θ. Analogously to [16,18]:

Definition 1. *A probabilistic range query q, at any timestamp τ reports objects* $\{o \in L_q \mid P_{in}(o, q) \geq \theta\}$ *with:*

$$P_{in}(o, q) = \int_{r_q \cap r_o | y} \int_{r_q \cap r_o | x} pdf(x, y) dx dy \tag{3}$$

where $r_q \cap r_o | x$ denotes the interval along x-axis where areas r_q and r_o spatially overlap (notation similar for the y-axis). For the example setting in Fig. 2a,

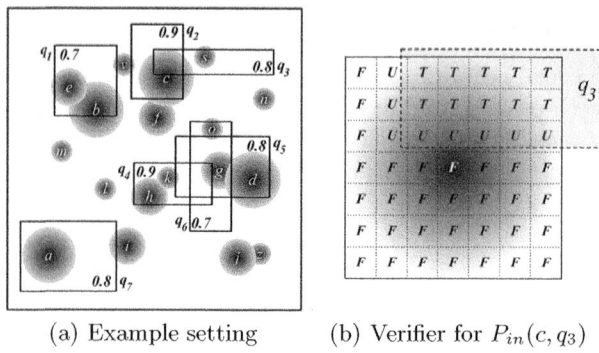

(a) Example setting   (b) Verifier for $P_{in}(c, q_3)$

Fig. 2. Probabilistic range search over uncertain objects

object c qualifies for query q_2, but not for q_3 since $P_{in}(c, q_3) < 0.8$, assuming that c belongs to the contact list of both queries.

The problem is that Gaussian distributions cannot be integrated analytically, so we need to resort to numerical methods like Monte-Carlo to get a fair estimation for Eq. (3). Yet, Monte-Carlo simulation incurs excessive CPU cost as it requires a sufficiently large number of samples (at the order of 10^6 [16]). Given the mobility and mutability of objects and queries, such a solution is clearly prohibitive for processing range requests in online fashion.

4   Approximation with Discretized Uncertainty Regions

4.1   Probing Objects through Probabilistic Verifiers

An object never specifies a bounded uncertainty region; still, the server may safely conjecture that its location is within a truncated density area of radius 3σ around its mean (μ_x, μ_y), as exemplified in Fig. 1c. To simplify computations, instead of such a circle, its *rectilinear circumscribed square* of side 6σ can stand for uncertainty region just as well. In fact, the cumulative probability of this Minimum Bounding Box (MBB) is greater than 99.73% and tends asymptotically to 1, although its area is $\pi/4$ times larger than the circle of radius 3σ.

Now suppose that for a known σ, we subdivide this MBB uniformly into $\lambda \times \lambda$ *elementary boxes*, $\lambda \in \mathbb{N}^*$. Boxes may have the same area, but represent diverse cumulative probabilities, as shown in Fig. 3. Once precomputed (e.g., by Monte-Carlo), these probabilities can be retained in a lookup table V. If λ is odd, the central box $V(\lceil \frac{\lambda}{2} \rceil, \lceil \frac{\lambda}{2} \rceil)$ is the one with the highest density. Anyway:

Lemma 1. *The cumulative probability in each of the $\lambda \times \lambda$ elementary boxes is independent of the parameters of the applied Bivariate Gaussian distribution.*

In other words, for a fixed λ, the contribution of each particular box in Fig. 3 remains intact for any σ value. The spatial area of a box is $(6\sigma/\lambda)^2$, so it expands quadratically with increasing σ. Yet, as a measure of its probability density, each

0.00003	0.00105	0.00328	0.00105	0.00003
0.00105	0.03707	0.11640	0.03707	0.00105
0.00328	0.11640	0.36448	0.11640	0.00328
0.00105	0.03707	0.11640	0.03707	0.00105
0.00003	0.00105	0.00328	0.00105	0.00003

(a) Box weights for $\lambda = 5$

0.00001	0.00018	0.00098	0.00098	0.00018	0.00001
0.00018	0.00582	0.03215	0.03215	0.00582	0.00018
0.00098	0.03215	0.17755	0.17755	0.03215	0.00098
0.00098	0.03215	0.17755	0.17755	0.03215	0.00098
0.00018	0.00582	0.03215	0.03215	0.00582	0.00018
0.00001	0.00018	0.00098	0.00098	0.00018	0.00001

(b) Box weights for $\lambda = 6$

0	0.00004	0.00029	0.00056	0.00029	0.00004	0
0.00004	0.00111	0.0079	0.0151	0.0079	0.00111	0.00004
0.00029	0.0079	0.0565	0.1083	0.0565	0.0079	0.00029
0.00056	0.0151	0.1083	0.2078	0.1083	0.0151	0.00056
0.00029	0.0079	0.0565	0.1083	0.0565	0.0079	0.00029
0.00004	0.00111	0.0079	0.0151	0.0079	0.00111	0.00004
0	0.00004	0.00029	0.00056	0.00029	0.00004	0

(c) Box weights for $\lambda = 7$

Fig. 3. Diverse subdivisions of the same uncertainty region into $\lambda \times \lambda$ elementary boxes

box $V(i,j)$ maintains its own characteristic *weight* $\beta(i,j)$, which depends entirely on λ and fluctuates with the placement of $V(i,j)$ in the MBB.

The rationale behind this subdivision is that it may be used as a *discretized verifier* V when probing uncertain Gaussians. Consider the case of query q_3 against object c, shown in detail in Fig. 2b. Depending on its topological relation with the given query, each elementary box of V can be easily characterized by one of three possible *states*: (i) T is assigned to elementary boxes totally within query range; (ii) F signifies disjoint boxes, i.e., those entirely outside the range; and (iii) U marks boxes partially overlapping with the specified range.

Then, summing up the respective cumulative probabilities for each subset of boxes returns three *indicators* p_T, p_F, p_U suitable for object validation:

(i) In case that $p_T \geq \theta$, there is no doubt that the object qualifies.
(ii) If $p_F \geq 1 - \theta$, then the object may be safely rejected, as by no means can its appearance probability exceed the query threshold. This is the case for object c in Fig. 2, since its $p_N = 0.72815 \geq 1 - 0.8$.
(iii) Otherwise, when $p_T + p_U \geq \theta$, eligibility is ambiguous. To avoid costly Monte-Carlo simulations, the object could be regarded as *reservedly qualifying*, but along with a confidence margin $[p_T, 1 - p_F)$ as a degree of its reliability.

Because $p_T + p_F + p_U \simeq 1$, only indicators p_T, p_F need be calculated. Still, in case (iii) the magnitude of the confidence margin equals the overall cumulative probability of the U-boxes, which depends entirely on granularity λ. The finer the subdivision into elementary boxes, the less the uncertainty in the emitted results. In contrast, a small λ can provide answers quickly, which is critical when coping with numerous objects. As a trade-off between timeliness and answer quality, next we turn this range search problem into an (ϵ, δ)-approximation one.

4.2 Towards Approximate Answering with Error Guarantees

Let $\bar{p}$ the exact[2] $P_{in}(o, q)$ appearance probability that object o of uncertainty region r_o lies within range r_q of query q. Also, let $\hat{p}$ the respective approximate

---

[2] $\bar{p}$ cannot be computed analytically, but can be estimated with numerical methods.

probability derived after probing the elementary boxes of verifier $V(r_o)$. Given parameters ϵ and δ, we say that object o *qualifies* for q, if approximate estimation $\hat{p}$ deviates less than ϵ from exact $\bar{p}$ with probability at least $1 - \delta$. Formally:

$$P(|\bar{p} - \hat{p}| \leq \epsilon) \geq 1 - \delta. \qquad (4)$$

Intuitively, $\epsilon \in (0, \theta)$ is the *error margin* of the allowed overestimation in $P_{in}(o, q)$ when reporting a qualifying object. In fact, ϵ relates to the size of the elementary boxes and controls the granularity of $V(r_o)$. On the other hand, $\delta \in (0,1)$ specifies the *tolerance* that an invalid answer may be given (i.e., a false positive). But in practice, given the arbitrary positions and extents of objects and queries, as well as the variability of threshold θ which determines qualifying results, it is hard to verify whether (4) actually holds for specific ϵ, δ values.

As it is difficult to tackle this problem, we opt for a relaxed approach with heuristics. Without loss of generality, we assume that extent r_q is never fully contained within uncertainty region r_o of any object o. According to the abstraction of uncertainty with MBB's, it suffices that any side of rectangle r_q is never less than 6σ, which is quite realistic. Thus, a query range either contains or intersects or is disjoint with an uncertainty region. In cases of full containment or clear separation, there is no ambiguity; the object is qualified or rejected with 100% confidence, respectively. As for intersections, among all cases discussed in Section 4.1, the trouble comes from partial overlaps of type (iii) that may lead to considerable overestimation. Indeed, rectangle r_q may only cover a tiny slice of elementary boxes marked as U, as it occurs with the three vertical U-boxes in Fig. 2b. The redundancy in the estimated cumulative probability owed to the non-covered area of U-boxes is evident.

Let us take a closer look at partial overlaps of type (iii) between a query rectangle r_q and the elementary boxes of an uncertainty region r_o, assuming a fixed λ. As illustrated in Fig. 4, there are three possible cases that r_q may intersect $V(r_o)$ and leave uncovered a particular stripe of the verifier. Of particular concern are *horizontal, vertical* or *L-shaped stripes*, comprised of consecutive slices of U-boxes (the red hatched bars in Fig. 4), which amplify the confidence margin. There are eight combinations in total, classified into two groups: (a) four cases concern a straight (horizontal or vertical) stripe, depending on which side of the verifier remains uncovered, and (b) other four create an L-shaped stripe touching the enclosed query corner ($\lrcorner, \llcorner, \urcorner, \ulcorner$). Due to the square shape of verifiers and the underlying symmetry of Gaussian features, the horizontal and vertical cases are equivalent; it also does not matter which corner of $MBB(r_o)$ is enclosed in range r_q. Hence, it suffices to examine an indicative combination from either group.

The worst case happens when r_q has just a tiny overlap with each U-box, hence overestimation $|\bar{p} - \hat{p}|$ becomes almost p_U. In contrast, when just a small area of all U-boxes is left uncovered, overestimation is minimized and upper bound $p_T + p_U$ of the margin is fairly reliable. In between, since objects and queries are not expected to follow any specific mobility pattern, there are infinitely many chances for such partial overlaps, leading to a variety of stripes with diverse cumulative probability. Each case has equal likelihood to occur, but incurs varying errors

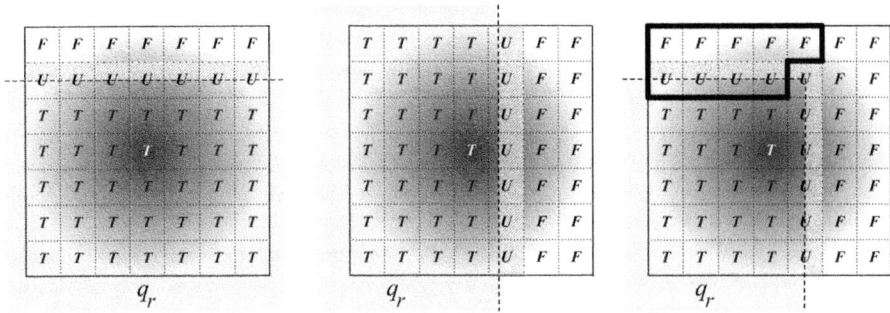

Fig. 4. Horizontal, Vertical and L-shaped stripes of U-box slices beside query boundary

in probability estimation. Nevertheless, for increasing λ values, each elementary box of the verifier steadily gets less and less weight, so the overestimation effect weakens drastically. In the average case, and for sufficiently large λ, we may approximately consider that each U-box contributes half of its density to the confidence margin. In other words, we assume that the query boundary crosses each U-box in the middle (especially for a corner box, it encloses a quarter of its area), as exemplified in Fig. 4c.

Under this discretized relaxation of the problem, we could evaluate the expected superfluous density for all possible arrangements of straight or L-shaped overlaps and estimate the chances that Formula (4) gets fulfilled. For a fixed subdivision of MBB's, there are 4λ possible instantiations for a straight stripe, considering that each side of the query rectangle r_q may be crossing a horizontal or vertical series of U-boxes (Fig. 4a, 4b). Similarly, any corner of r_q may be centered in any elementary box, giving $4\lambda^2$ potential instantiations of L-shaped stripes (Fig. 4a). In total, we consider $4\lambda + 4\lambda^2$ equiprobable instantiations, yet each one causes a varying overestimation. Suppose that for a given λ, it turns out that ν out of those $4\lambda(\lambda+1)$ cases incur an error less than ϵ. Then, if

$$P(|\bar{p} - \hat{p}| \leq \epsilon) = \frac{\nu}{4\lambda(\lambda+1)} \geq 1 - \delta, \tag{5}$$

we may *accept* that the object approximately qualifies under the aforementioned assumptions.

Since the quality of the approximate answer strongly depends on granularity λ of verifier $V(r_o)$, we wish to select the *minimal* λ^* value so that the resulting probabilities could fulfill inequality (5). In a brute-force preprocessing step based on Monte-Carlo simulation, we can estimate the cumulative probabilities of problematic stripes, starting from a small λ and steadily incrementing it until (5) eventually holds. Then, for the given ϵ, δ values, these fine-tuned $\lambda^* \times \lambda^*$ elementary boxes are expected to provide reliable results that only rarely digress from the given confidence margins, as we experimentally verify in Section 6.

Algorithm 1.   Probabilistic Range Monitoring

1: **Procedure** *RangeMonitor* (timestamp τ)
2: **Input:** Stream items $\langle o^j, \mu_x^j, \mu_y^j, \sigma^j, \tau \rangle$ from $j = 1..N$ Bivariate Gaussian objects.
3: **Input:** Specification updates $\langle q^i, r_q^i, \theta^i, \tau \rangle$ from $i = 1..M$ continuous range queries.
4: **Output:** Qualifying results $Q = \{\langle q^i, o^j, \theta^{min}, \theta^{max}, \tau \rangle : \bigcap (r_o^j, r_q^i) \neq \emptyset$ with confidence $(\theta^i \leq \theta^{min} < \theta^{max}) \vee (\theta^{min} < \theta^i \leq \theta^{max})\}$.
5:    $Q \leftarrow \{\}$;   *//Initial result set for all queries at execution cycle τ*
6: **for each** q^i **do**
7:      $\tilde{\alpha}^* \leftarrow$ minimal area looked up from $\mathcal{A}$, corresponding to maximal $\theta^* \in \Theta, \theta^* \leq \theta^i$
8:      $C_q^i \leftarrow \{o^j \in L_q^i \mid MBB(r_o^j) \cap r_q^i \neq \emptyset\}$; *//Candidates only from contact list of q^i*
9:      **for each** $o^j \in C_q^i$ **do**
10:        **if** (o^j is unchanged $\wedge$ q^i is unchanged) **then**
11:          continue;   *//Skip evaluation for unmodified entities*
12:        **else if** $MBB(r_o^j) \subset r_q^i$ **then**
13:          $Q \leftarrow Q \cup \langle q^i, o^j, 1, 1, \tau \rangle$;   *//Certain object due to full containment*
14:        **else if** $\|MBB(r_o^j) \cap r_q^i\| < \sigma^j \cdot \sigma^j \cdot \tilde{\alpha}^*$ **then**
15:          continue;   *//Pruning with respective minimal area of overlap*
16:        **else**
17:          $\langle \theta^{min}, \theta^{max} \rangle \leftarrow ProbeVerifier (r_q^i, MBB(r_o^j), \theta^i)$; *//Approximate indicators*
18:          **if** $\theta^i \leq \theta^{min}$ **then**
19:            $Q \leftarrow Q \cup \langle q^i, o^j, \theta^{min}, \theta^{max}, \tau \rangle$;   *//Object qualifies, margin $[\theta^{min}, \theta^{max})$*
20:          **else if** $\theta^{min} < \theta^i \leq \theta^{max}$ **then**
21:            $Q \leftarrow Q \cup \langle q^i, o^j, \theta^{min}, \theta^{max}, \tau \rangle$;   *//Reservedly qualifying object*
22:          **end if**
23:        **end if**
24:      **end for**
25:   **end for**
26: Report Q;   *//Disseminate results to each query for execution cycle τ*
27: **End Procedure**

28: **Function** *ProbeVerifier* (query range r_q^i, object region $MBB(r_o^j)$, threshold θ^i)
29: $V(r_o^j) \leftarrow$ verifier with symbols $\{T, F, U\}$ stating any overlap of r_q^i over $MBB(r_o^j)$;
30: $p_T \leftarrow 0$; $p_F \leftarrow 0$;   *//Initialize indicators for appearance probability $P_{in}(o^j, q^i)$*
31: **for each** box $b_k \in V(r_o^j)$ by spiroid (or ripplewise) visiting order **do**
32:      **if** $b_k = 'T'$ **then**
33:        $p_T \leftarrow p_T + \beta(k)$; *//$k^{th}$ elementary box of verifier V is completely inside r_q^i*
34:      **else if** $b_k = 'F'$ **then**
35:        $p_F \leftarrow p_F + \beta(k)$; *//$k^{th}$ elementary box of verifier V is completely outside r_q^i*
36:      **end if**
37:      **if** $p_F \geq 1 - \theta_i$ **then**
38:        **return** $\langle 0, 0 \rangle$;   *//Eager rejection for non-qualifying objects*
39:      **end if**
40: **end for**
41: **return** $\langle p_T, 1 - p_F \rangle$; *//Bounds for appearance probability $P_{in}(o^j, q^i)$*
42: **End Function**

5   Online Range Monitoring over Streaming Gaussians

5.1   Evaluation Strategy

The pseudocode for the core range monitoring process is given in Algorithm 1. Implicitly, query q does not wish to find every object in range r_q; searching concerns only those enrolled in its contact list L_q. During query evaluation at timestamp τ, the spatial predicate is examined first against items from contact list L_q, offering a set of *candidate objects* $C_q(\tau) = \{o \in L_q \mid MBB(r_o) \cap r_q \neq \emptyset\}$ with uncertainty regions currently overlapping with r_q (Line 8). At a second stage described next, candidates with a likelihood above θ to lie within range should be returned as *qualifying objects* $Q_q(\tau) = \{o \in C_q(\tau) \mid P_{in}(o, q) \geq \theta\}$.

In case that $MBB(r_o)$ is fully contained in rectangle r_q, object o clearly qualifies with confidence 100%, irrespective of any threshold θ the query may stipulate (Lines 12-13). Similarly, if $MBB(r_o)$ and r_q are spatially disjoint, then object o is rejected also with confidence 100%. Both cases involve no probabilistic reasoning, as simple geometric checks can safely determine eligible objects.

But as already pointed out, evaluation is mainly complicated because of partial overlaps between $MBB(r_o)$ and r_q. Since this is expectedly a very frequent case, employing indicators over discretized verifiers with precomputed cumulative probabilities can provide a tolerable approximation, instead of unaffordable Monte-Carlo simulations as analyzed in Section 4. Even so, probing a few hundred or maybe thousand elementary boxes per candidate object may still incur excessive CPU time. Moreover, such a task must be repeatedly applied at each execution cycle τ against changing query specifications and mutable uncertainty regions. Next, we propose heuristics that may substantially reduce processing cost, effectively filtering out improbable candidate objects (i.e., true negatives) and avoiding exhaustive investigation of discretized verifiers.

5.2   Pruning Candidates Using Indicative Minimal Areas

Suppose that we could identify the smallest possible area $\tilde{\alpha}_\theta$ inside an uncertainty region, such that $\tilde{\alpha}_\theta$ represents a cumulative probability barely less than threshold θ of a given query q. If $\|r_q \cap MBB(r_o)\| < \tilde{\alpha}_\theta$, object o cannot qualify for query q, as its appearance probability $P_{in}(o, q)$ is definitely below θ. Then, estimating $P_{in}(o, q)$ is not necessary at all, because $\tilde{\alpha}_\theta$ indicates the *minimal area of overlap* between object o and query q in order for o to qualify.

Ideally, this observation could eliminate candidate objects substantially, as no further examination is required for those of overlapping areas less than $\tilde{\alpha}_\theta$ with the given query q. However, applying such a pruning criterion, necessitates precomputation of the respective $\tilde{\alpha}_\theta$ values for every possible threshold $\theta \in (0,1)$ a query could specify. A second issue relates to the density of uncertainty regions. Let query q equally overlap two objects o_1, o_2 with uncertainty regions r_{o_1}, r_{o_2} of diverse standard deviations $\sigma_1 \neq \sigma_2$. Notwithstanding that $\|r_q \cap MBB(r_{o_1})\| = \|r_q \cap MBB(r_{o_2})\|$, it does not necessarily hold that these overlaps represent equal appearance likelihood, since $P_{in}(o_1, q)$ and $P_{in}(o_2, q)$ are derived from Eq. (3)

according to different pdf parametrization. Thus, even for a fixed θ, a single minimal value $\tilde{\alpha}_\theta$ cannot be used against every uncertainty region.

To address issues concerning computation of such minimal areas $\tilde{\alpha}_\theta$, let us start with a specific threshold θ, stipulating a Standard Bivariate Gaussian distribution $\mathcal{N}(\mathbf{0}, \mathbf{1})$ for the uncertainty region. Because the density of Gaussians is maximized around the mean and then decreases rapidly for increasing distances across all directions (Fig. 1a), the sought minimal area is always a circle centered at the origin (μ_x, μ_y) of the pdf with a radius $R \in (0, 3)$ that depends on the given θ. To discover that R value and hence compute $\tilde{\alpha}_\theta = \pi R^2$, we can perform successive Monte-Carlo simulations, increasing R by a small step until the cumulative probability inside the circle becomes only just below θ. For other $\sigma \neq 1$, it turns out that the respective area is $\tilde{\alpha}_\theta = \pi(\sigma R)^2$, as standard deviation σ actually dictates the spread of values, and hence the magnitude of the circle.

Due to the variety of possible thresholds specified by user requests, it makes sense to discover minimal areas only for a small set $\Theta = \{\theta_1, \theta_2, \ldots, \theta_m\}$ of m typical values, e.g., $\Theta = \{10\%, 20\%, \ldots, 100\%\}$. Catalogue $\mathcal{A} = \{\tilde{\alpha}_1, \tilde{\alpha}_2, \ldots, \tilde{\alpha}_m\}$ of respective area magnitudes can be computed offline by the aforesaid Monte-Carlo process assuming a distribution $\mathcal{N}(\mathbf{0}, \mathbf{1})$. Having Θ readily available during online evaluation, when a query specifies an arbitrary threshold $\theta \notin \Theta$, we can easily identify the maximal $\theta_i^* \in \Theta, \theta_i^* \leq \theta$ and safely choose its corresponding minimal area $\tilde{\alpha}_i^*$ from the precomputed set $\mathcal{A}$ (Line 7). For the pruning condition, it suffices to compare whether $\|MBB(r_o) \cap r_q\| < \sigma^2 \cdot \tilde{\alpha}_i^*$, so as to account for the magnitude of an uncertainty region r_o with any particular σ (Lines 14-15).

5.3   Optimized Examination of Elementary Boxes

As pointed out in Section 4.1, a discretized verifier can provide a fairly reliable approximate answer in case of partial overlaps between query rectangles and circumscribed uncertainty regions. Essentially, after iterating through each elementary box b_i and having updated indicators p_T, p_F, we can safely determine whether an object qualifies (if $p_T \geq \theta$) or must be rejected (when $p_F \geq 1 - \theta$). In addition, reservedly qualifying objects could be reported with a confidence margin $[p_T, 1 - p_F)$, in case that $1 - p_F \geq \theta$ (Lines 16-21).

However, for finer subdivisions of verifiers, probing at each execution cycle τ an increasing number $\lambda^* \times \lambda^*$ of elementary boxes could incur considerable cost, especially for objects having little chance to qualify. Note that elementary boxes towards the center have much more weight (i.e., greater cumulative probabilities) than peripheral ones. Therefore, we had better start visiting boxes from the center and progressively inspect others of less and less importance. Updating p_T and p_F accordingly, we can resolve object qualification much faster, since peripheral boxes have practically negligible weight (Fig. 3). Such *eager rejections* can be decided as soon as a (yet incomplete) p_F exceeds $1 - \theta$, thus avoiding an exhaustive investigation of the entire verifier, particularly when $\theta > 0.5$. Since p_F for a given object could never decrease with further box examinations at current τ, continuing calculation of indicators is pointless as the result cannot be altered. Considering that a query range might overlap many uncertainty areas only by a

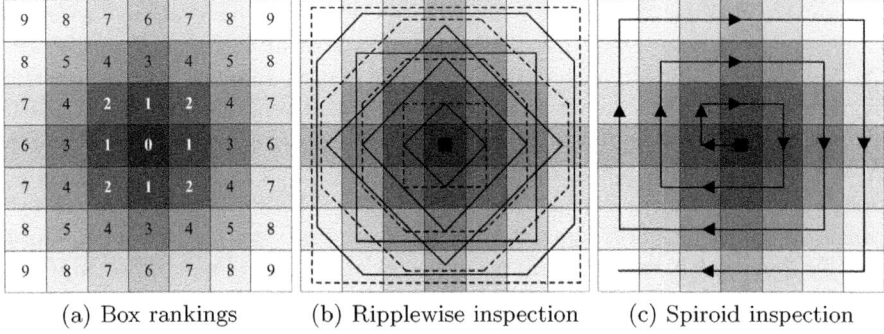

(a) Box rankings   (b) Ripplewise inspection   (c) Spiroid inspection

Fig. 5. Visiting order of elementary boxes for $\lambda = 7$

small fringe, the savings can be enormous, as rejections could be resolved soon after inspecting just a few boxes around the center. A similar argument holds when issuing qualifying objects, especially for relatively lower thresholds.

Based on this important observation, we could take advantage of the inherent ranking of elementary boxes for visiting them by decreasing weight, so as to progressively update indicators p_T and p_F. In Fig. 5, graduated gray color reflects a ranking (Fig. 5a) of elementary boxes classified by their inferred cumulative probability for the verifier depicted in Fig. 3c. Intuitively, we may opt for a *ripplewise order* regarding box inspections, distantly reminiscent of raindrops rippling on the water surface. But instead of forming circles, groups of perhaps nonadjacent, yet equi-ranked boxes are arranged as vertices of squares, diamonds, octagons etc. Inspection starts from the central box and continues in rippling waves (depicted with alternating solid and dashed lines in Fig. 5b) rushing outwards across the underlying uncertainty region. An alternative choice is the simplified visiting order illustrated in Fig. 5c, which takes a *squarish spiroid* pattern. Occasionally violating the strict succession of rankings, it follows a continuous *meander* line that again starts from the central box (or next to the center, in case of even subdivisions) and traverses the rest in rings of increasing radius and gradually diminishing weight.

Both orderings aim to give precedence to boxes with potentially significant contribution to appearance probabilities. Assuming a $\theta = 0.6$ and following a spiroid visiting order for the example shown in Fig. 4b, we can easily conclude that the object qualifies for query q after examining the nine central boxes only, which account for a $p_T \simeq 0.6457 \geq \theta$ based on cumulative probabilities (Fig. 3c). Function *ProbeVerifier* in Algorithm 1 outlines this optimized verification step.

6 Experimental Evaluation

6.1 Experimental Setup

Next, we report results from an empirical validation of our framework for probabilistic range monitoring against streams of Gaussian positional uncertainty. We generated synthetic datasets for objects and queries moving at diverse speeds along the road network of greater Athens in a spatial Universe $\mathcal{U}$ of 625 km^2. By calculating shortest paths between nodes chosen randomly across the network, we were able to create samples of 200 concurrent timestamps from each such route. In total, we obtained a point set representing mean locations for $N = 100\,000$ objects, and similarly, the centroids of $M = 10\,000$ query ranges. Spatial range of queries is expressed as percentage (%) of the entire $\mathcal{U}$. However, ranges are not necessarily squares with the given centroid, because we randomly modify their width and height in order to get arbitrarily elongated rectangles of equal area. Each object updates its uncertainty area regularly at every timestamp. Concerning query ranges, their *agility* of movement is set to 0.1, so a random 10% of them modify their specification at each timestamp.

However, contact list L_q per user must not be specified at random; otherwise, probabilistic search would hardly return any meaningful results with synthetic datasets. Thus, for each query we computed a preliminary list of all objects in its vicinity for the entire duration (200 timestamps). Only for data generation, we considered exact locations of N objects within circular areas of 1% of $\mathcal{U}$ centered at each query centroid. After calculating object frequencies for each query, we created two sets of contact lists nicknamed *MOD* and *POP*, respectively retaining the top 50% and top 75% of most recurrent objects per query. Indicatively, a query in *MOD* on average has a modest number of 87 subscribers (i.e., monitored objects) with a maximum of 713, whereas a *POP* list typically has 693 and at most 5911 members, i.e., is almost an order of magnitude more popular.

Evaluation algorithms were implemented in C++ and executed on an Intel Core 2 Duo 2.40GHz CPU running GNU/Linux with 3GB of main memory. Typically for data stream processing, we adhere to online in-memory computation, excluding any disk-bound techniques. We ran simulations using different parameter settings for each experiment. Due to space limitations, we show results just from some representative ones. All results are averages of the measured quantities for 200 time units. Table 2 summarizes experimentation parameters and their respective ranges; the default value (in bold) is used for most diagrams.

6.2   Experimental Results

Verifiers for uncertainty areas should strike a balance between approximation quality and timely resolution of appearance probabilities. So, we first attempt to

Table 2. Experiment parameters

Parameter	Values
Number N of objects	**100 000**
Number M of range queries	**10 000**
Range area (% of universe $\mathcal{U}$)	0.01, 0.1, **1**, 2, 5, 10
Standard deviation σ (meters)	50, 100, **200**, 300, 500
Cutoff threshold θ	0.5, 0.6, 0.7, **0.75**, 0.8, 0.9, 0.99
Error margin ϵ	0.02, 0.03, **0.05**, 0.1
Tolerance δ	0.01, 0.02, **0.03**, 0.05, 0.1

Table 3. Fine-tuning λ^*

ϵ	δ	λ^*	ϵ	δ	λ^*
0.02	0.01	103	0.05	0.02	38
0.02	0.02	97	**0.05**	**0.03**	**37**
0.03	0.01	67	0.05	0.05	35
0.03	0.02	**65**	0.1	0.02	19
0.03	0.03	63	0.1	0.05	**18**
0.05	0.01	41	0.1	0.1	17

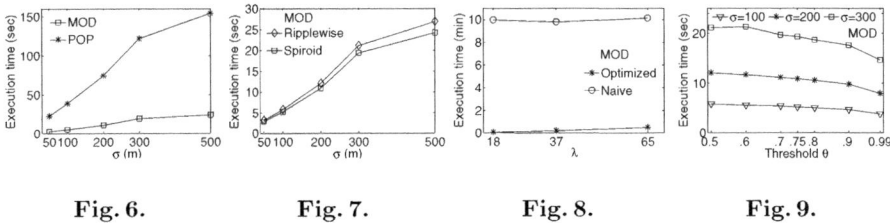

determine a fine-tuned subdivision according to the desired accuracy of answers. Table 3 lists the minimal granularity λ^* of verifiers so as to meet the bounds for tolerance δ and error ϵ, using a brute-force preprocessing step (Section 4.2). But a large λ around 100, would create verifiers with 10 000 tiny elementary boxes of questionable practical use, considering the numerous spatial arrangements of queries and objects. For our experiments, we have chosen three moderate values (in bold in Table 3) that represent distinct levels of indicative accuracy. Unless otherwise specified, we mostly set $\lambda = 37$, which dictates that qualifying objects must not deviate above $\epsilon = 5\%$ from their actual appearance likelihood, and these results can be trusted with $1 - \delta = 97\%$ probability at least.

Next, we examine the total query evaluation cost per timestamp for each of the two query workloads *MOD* and *POP*, assuming diverse sizes of Gaussian uncertainty regions. In fact, standard deviation σ controls the density as well as the extent of the region; e.g., a $\sigma = 200$ meters prescribes a square area of side $6\sigma = 1200$ meters, which is large enough for urban settings. Quite predictably, execution cost deteriorates with σ as plotted in Fig. 6, because larger uncertainty regions intersect more frequently with multiple query ranges. Despite the increasing number of such overlapping cases, the pruning heuristics can quickly discard improbable candidates, hence the total cost for all queries mostly remains at reasonable levels, particularly for the *MOD* workload. It only exacerbates for larger uncertainty regions with the *POP* dataset, but mainly due to the disproportionate size of its contact lists.

The choice of inspection order for elementary boxes is not critical, provided that they are visited by descending weight. Thanks to its simplicity, a spiroid ordering gives response slightly faster than its ripplewise counterpart (Fig. 7). Still, Fig. 8 demonstrates that such optimizations economize enormously by first examining important boxes as opposed to a naïve strategy. With more restrictions on accuracy (i.e., larger λ), execution time escalates linearly, but always remains under 30 sec per cycle for answering all queries. In contrast, blindly examining all boxes and employing expensive Monte-Carlo simulations for ambiguous cases incurs execution times utterly incompatible with online monitoring.

With respect to threshold values, Fig. 9 shows the effectiveness of pruning for diverse uncertainty levels. Clearly, the higher the threshold, the more frequent the cases of eager rejections, as examination of objects terminates very early. This trend gets even more pronounced with greater uncertainty ($\sigma = 300$m).

When specifying diverse areas of query range, execution cost fluctuates, as illustrated in Fig. 10. However, this phenomenon depends on the extent and spread of the uncertainty regions that may cause a mounting number of partial

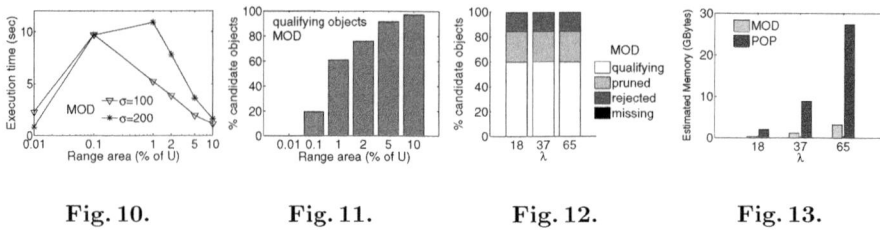

Fig. 10.   Fig. 11.   Fig. 12.   Fig. 13.

overlaps with the query rectangles, which require verification. For smaller query areas, such intersections are rare, so they incur negligible cost. Similarly, with ranges equal to 10% of the entire universe $\mathcal{U}$, many more objects fall completely within range and get directly qualified with less cost. This is also confirmed with statistical results in Fig. 11 regarding the fraction of candidate objects that finally get qualified for $\sigma = 200$m. For ranges with extent 1% of $\mathcal{U}$, about 60% of candidates are reported, so a lot many of the rest 40% have been disqualified after verification, which explains the respective peak in Fig. 10.

Concerning the quality of the reported results, Fig. 12 plots a breakdown of the candidates for varying accuracy levels. Compared with an exhaustive Monte-Carlo evaluation, about 15% of candidates are eagerly rejected, while another 25% is pruned. Most importantly, false negatives are less than 0.1% at all cases, which demonstrates the efficiency of our approach. Although qualitative results are similar for varying λ, they still incur differing execution costs (Fig. 8).

The astute reader may have observed that our approach is not incremental; at every cycle, each candidate must be examined from scratch for any query, no matter its previous state regarding the given query. This is a deliberate choice, if one considers the extreme mutability of both objects and queries. Apart from their continuous free movement and features that change in probabilistic fashion, there are also practical implications. Figure 13 plots the estimated memory consumption for maintaining states of every verifier for all combinations of queries and members of their contact lists. This cost may seem reasonable for fair accuracy constraints ($\lambda = 18$), but becomes unsustainable with stricter quality requirements, especially for query workloads with excessively large membership. Considering its maintenance overhead, a stateful approach would clearly become more a burden rather than an assistance in terms of probabilistic evaluation.

7 Conclusion

In this work, we proposed a probabilistic methodology for providing online response to multiple range requests over streams of Gaussian positional uncertainty in GeoSocial networks. Abiding to privacy preserving protocols, we introduced an (ϵ, δ)-approximation framework, as a trade-off between quality guarantees and timeliness of results. We also developed optimizations for effective pruning and eager rejection of improbable answers. Our evaluation strategy drastically reduces execution cost and offers answers of tolerable error, confirmed by an extensive experimental study over massive synthetic datasets.

References

1. Bernecker, T., Emrich, T., Kriegel, H.-P., Renz, M., Zankl, S., Züfle, A.: Efficient Probabilistic Reverse Nearest Neighbor Query Processing on Uncertain Data. PVLDB 4(10), 669–680 (2011)
2. Bernecker, T., Kriegel, H.-P., Mamoulis, N., Renz, M., Zuefle, A.: Continuous Inverse Ranking Queries in Uncertain Streams. In: Bayard Cushing, J., French, J., Bowers, S. (eds.) SSDBM 2011. LNCS, vol. 6809, pp. 37–54. Springer, Heidelberg (2011)
3. Böhm, C., Pryakhin, A., Schubert, M.: Probabilistic Ranking Queries on Gaussians. In: SSDBM, pp. 169–178 (2006)
4. Chen, J., Cheng, R.: Efficient Evaluation of Imprecise Location-Dependent Queries. In: ICDE, pp. 586–595 (2007)
5. Cheng, R., Xia, Y., Prabhakar, S., Shah, R., Vitter, J.S.: Efficient Indexing Methods for Probabilistic Threshold Queries over Uncertain Data. In: VLDB, pp. 876–887 (2004)
6. Chow, C.-Y., Mokbel, M.F., Aref, W.G.: Casper*: Query Processing for Location Services without Compromising Privacy. ACM TODS 34(4), 24 (2009)
7. Ghinita, G., Kalnis, P., Khoshgozaran, A., Shahabi, C., Tan, K.-L.: Private Queries in Location Based Services: Anonymizers are not Necessary. In: SIGMOD, pp. 121–132 (2008)
8. Ishikawa, Y., Iijima, Y., Xu Yu, J.: Spatial Range Querying for Gaussian-Based Imprecise Query Objects. In: ICDE, pp. 676–687 (2009)
9. Kriegel, H.-P., Kunath, P., Pfeifle, M., Renz, M.: Probabilistic Similarity Join on Uncertain Data. In: Li Lee, M., Tan, K.-L., Wuwongse, V. (eds.) DASFAA 2006. LNCS, vol. 3882, pp. 295–309. Springer, Heidelberg (2006)
10. Kriegel, H.-P., Kunath, P., Renz, M.: Probabilistic Nearest-Neighbor Query on Uncertain Objects. In: Kotagiri, R., Radha Krishna, P., Mohania, M., Nantajeewarawat, E. (eds.) DASFAA 2007. LNCS, vol. 4443, pp. 337–348. Springer, Heidelberg (2007)
11. Lian, X., Chen, L.: Ranked Query Processing in Uncertain Databases. IEEE TKDE 22(3), 420–436 (2010)
12. Lian, X., Chen, L.: Similarity Join Processing on Uncertain Data Streams. IEEE TKDE 23(11), 1718–1734 (2011)
13. Mascetti, S., Freni, D., Bettini, C., Wang, X.S., Jajodia, S.: Privacy in Geo-social Networks: Proximity Notification with Untrusted Service Providers and Curious Buddies. VLDB Journal 20(4), 541–566 (2011)
14. Pei, J., Hua, M., Tao, Y., Lin, X.: Query Answering Techniques on Uncertain and Probabilistic Data: Tutorial Summary. In: SIGMOD, pp. 1357–1364 (2008)
15. Šikšnys, L., Thomsen, J.R., Šaltenis, S., Yiu, M.L.: Private and Flexible Proximity Detection in Mobile Social Networks. In: MDM, pp. 75–84 (2010)
16. Tao, Y., Cheng, R., Xiao, X., Ngai, W., Kao, B., Prabhakar, S.: Indexing Multi-Dimensional Uncertain Data with Arbitrary Probability Density Functions. In: VLDB, pp. 922–933 (2005)
17. Tao, Y., Xiao, X., Cheng, R.: Range Search on Multidimensional Uncertain Data. ACM TODS 32(3), 15 (2007)
18. Zhang, M., Chen, S., Jensen, C.S., Ooi, B.C., Zhang, Z.: Effectively Indexing Uncertain Moving Objects for Predictive Queries. PVLDB 2(1), 1198–1209 (2009)

Probabilistic Frequent Pattern Growth for Itemset Mining in Uncertain Databases

Thomas Bernecker, Hans-Peter Kriegel, Matthias Renz,
Florian Verhein, and Andreas Züfle

Institute for Informatics, Ludwig-Maximilians-Universität München, Germany
{bernecker,kriegel,renz,verhein,zuefle}@dbs.ifi.lmu.de

Abstract. Frequent itemset mining in uncertain transaction databases semantically and computationally differs from traditional techniques applied on standard (certain) transaction databases. Uncertain transaction databases consist of sets of existentially uncertain items. The uncertainty of items in transactions makes traditional techniques inapplicable. In this paper, we tackle the problem of finding probabilistic frequent itemsets based on possible world semantics. In this context, an itemset X is called frequent if the *probability* that X occurs in at least *minSup* transactions is above a given threshold τ. We make the following contributions: We propose the first probabilistic FP-Growth algorithm (ProFP-Growth) and associated probabilistic FP-tree (ProFP-tree), which we use to mine all probabilistic frequent itemsets in uncertain transaction databases without candidate generation. In addition, we propose an efficient technique to compute the support probability distribution of an itemset in linear time using the concept of generating functions. An extensive experimental section evaluates our proposed techniques and shows that our approach is significantly faster than the current state-of-the-art algorithm.

1 Introduction

Association rule analysis is one of the most important fields in data mining. It is commonly applied to market-basket databases for analysis of consumer purchasing behavior. Such databases consist of a set of transactions, each containing the items a customer purchased. The most important and computationally intensive step in the mining process is the extraction of *frequent itemsets* – sets of items that occur in at least *minSup* transactions. It is generally assumed that the items occurring in a transaction are known for certain. However, this is not always the case. For instance;

- In many applications the data is inherently noisy, such as data collected by sensors or in satellite images.
- In privacy protection applications, artificial noise can be added deliberately [20]. Finding patterns despite this noise is a challenging problem.
- By aggregating transactions by customer, we can mine patterns across customers instead of transactions. This produces estimated purchase probabilities per item per customer rather than certain items per transaction.

A. Ailamaki and S. Bowers (Eds.): SSDBM 2012, LNCS 7338, pp. 38–55, 2012.

TID	Transaction
1	(A, 1.0), (B, 0.2), (C, 0.5)
2	(A, 0.1), (D, 1.0))
3	(A, 1.0), (B, 1.0), (C, 1.0), (D, 0.4)
4	(A, 1.0), (B, 1.0), (D, 0.5)
5	(B, 0.1), (C, 1.0)
6	(C, 0.1), (D, 0.5)
7	(A, 1.0), (B, 1.0), (C, 1.0)
8	(A, 0.5), (B, 1.0)

Fig. 1. Uncertain Transaction Database (running example)

In such applications, the information captured in transactions is *uncertain* since the existence of an item is associated with a likelihood measure or existential probability. Given an uncertain transaction database, it is not obvious how to identify whether an item or itemset is frequent because we generally cannot say for certain whether an itemset appears in a transaction. In a traditional (certain) transaction database on the other hand, we simply perform a database scan and count the transactions that include the itemset. This does not work in an uncertain transaction database. An example of a small uncertain transaction database is given in Figure 1, where for each transaction t_i, each item x is listed with its probability of existing in t_i. Items with an existential probability of zero can be omitted. We will use this dataset as a running example.

Prior to [6], expected support was used to deal with uncertain databases [8,9]. It was shown in [6] that the use of expected support had significant drawbacks which led to misleading results. The proposed alternative was based on computing the entire probability distribution of itemsets' support, and achieved this in the same runtime as the expected support approach by employing the Poisson binomial recurrence relation. [6] adopts an Apriori-like approach, which is based on an *anti-monotone* Apriori property [3] (if an itemset X is not frequent, then any itemset $X \cup Y$ is not frequent) and candidate generation. However, it is well known that Apriori-like algorithms suffer a number of disadvantages. First, all candidates generated must fit into main memory and the number of candidates can become prohibitively large. Secondly, checking whether a candidate is a subset of a transaction is non-trivial. Finally, the entire database needs to be scanned multiple times. In uncertain databases, the effective transaction width is typically larger than in a certain transaction database which in turn can increase the number of candidates generated and the resulting space and time costs.

In certain transaction databases, the FP-Growth Algorithm [12] has become the established alternative. By building an FP-tree – effectively a compressed and highly indexed structure storing the information in the database – candidate generation and multiple database scans can be avoided. However, extending this idea to mining probabilistic frequent patterns in uncertain transaction databases is non-trivial. It should be noted that previous extensions of FP-Growth to uncertain databases used the expected support approach [1,13]. This is much easier since these approaches ignore the probability distribution of support.

In this paper, we propose a compact data structure called the probabilistic frequent pattern tree (*ProFP-tree*) which compresses probabilistic databases and allows the efficient extraction of the existence probabilities required to compute the support probability distribution and frequentness probability. Additionally, we propose the novel *ProFPGrowth* algorithm for mining all probabilistic frequent itemsets without candidate generation.

1.1   Uncertain Data Model

The uncertain data model applied in this paper is based on the possible worlds semantic with existential *uncertain items*.

Definition 1. *An* uncertain item *is an item* $x \in I$ *whose presence in a transaction* $t \in T$ *is defined by an* existential probability $P(x \in t) \in (0, 1)$. *A* certain item *is an item where* $P(x \in t) \in \{0, 1\}$. *I is the set of all possible items.*

Definition 2. *An* uncertain transaction t *is a transaction that contains uncertain items. A transaction database T containing uncertain transactions is called an* uncertain transaction database.

An uncertain transaction t is represented in an uncertain transaction database by the items $x \in I$ associated with an existential probability value [1] $P(x \in t) \in (0, 1]$. An example of an uncertain transaction databases is depicted in Figure 1. To interpret an uncertain transaction database we apply the *possible world* model. An uncertain transaction database generates *possible worlds*, where each world is defined by a fixed set of (certain) transactions. A possible world is instantiated by generating each transaction $t_i \in T$ according to the occurrence probabilities $P(x \in t_i)$. Consequently, each probability $0 < P(x \in t_i) < 1$ derives two possible worlds *per transaction*: One possible world in which x exists in t_i, and one possible world where x does not exist in t_i. Thus, the number of possible worlds of a database increases exponentially in both the number of transactions and the number of uncertain items contained in it. Each possible world w is associated with a probability that that world exists, $P(w)$.

Independence between items both within the same transaction, as well as in different transaction is often assumed in the literature [9,8,1]. This can often be justified by the assumption that the items are observed independently.

In this case, the probability of a world w is given by:

$$P(w) = \prod_{t \in I}(\prod_{x \in t} P(x \in t) * \prod_{x \notin t}(1 - P(x \in t)))$$

Note that this assumption does not imply that the underlying instantiations of an uncertain transaction databases will result in uncorrelated items, since the set of items having non-zero probability in a transaction may be correlated.

---

[1] If an item x has an existential probability of 0, it does not appear in the transaction.

Example 1. In the database of Figure 1, the probability of the world existing in which t_1 contains only items A and C and t_2 contains only item D is $P(A \in t_1)*(1-P(B \in t_1))*P(C \in t_1)*(1-P(A \in t_2)*P(D \in t_2) = 1.0 \cdot 0.8 \cdot 0.5 \cdot 0.9 \cdot 1.0 = 0.36$. For simplicity we omit the consideration of other customers in this example.

1.2 Problem Definition

An itemset is a *frequent itemset* if it occurs in at least *minSup* transactions, where *minSup* is a user specified parameter. In uncertain transaction databases however, the support of an itemset is uncertain; it is defined by a discrete probability distribution function (p.d.f). Therefore, each itemset has a *frequentness probability*[2] – the probability that it is frequent. In this paper, we focus on the two distinct problems of efficiently calculating this p.d.f. and efficiently extracting all *probabilistic frequent itemsets*;

Definition 3. *A Probabilistic Frequent Itemset (PFI) is an itemset with a frequentness probability of at least τ.*

The parameter τ is the user specified minimum confidence in the frequentness of an itemset.

We are now able to specify the *Probabilistic Frequent Itemset Mining (PFIM) problem* as follows; Given an uncertain transaction database T, a minimum support scalar *minSup* and a frequentness probability threshold τ, find all probabilistic frequent itemsets.

1.3 Contributions

We make the following contributions:

– We introduce the probabilistic Frequent Pattern Tree, or ProFP-tree, which is the first FP-tree type approach for handling uncertain or probabilistic data. This tree efficiently stores a probabilistic database and enables efficient extraction of itemset occurrence probabilities and database projections.
– We propose ProFPGrowth, an algorithm based on the ProFPTree which mines all itemsets that are frequent with a probability of at least τ without using expensive candidate generation.
– We present an intuitive and efficient method based on generating functions for computing the probability that an itemset is frequent, as well as the entire probability distribution function of the support of an itemset, in $O(|T|)$ time[3]. Using our approach, our algorithm has the same time complexity as the approach based on the Poisson Binomial Recurrence (denoted as *dynamic programming technique*) in [6], but it is much more intuitive and thus offers various advantages, as we will show.

---

[2] Frequentness is the rarely used word describing the property of being frequent.
[3] Assuming *minSup* is a constant.

The remainder of this paper is organized as follows; Section 2 surveys related work. In Section 3 we present the ProFP-tree, explain how it is constructed and briefly introduce the concept of conditional ProFPTrees. Section 4 describes how probability information is extracted from a (conditional) ProFP-tree. Section 5 introduces our generating function approach for computing the frequentness probability and the support probability distribution in linear time. Section 6 describes how conditional ProFPT-rees are built. Finally, Section 7 describes the ProFP-Growth algorithm by drawing together the previous sections. We present our experiments in Section 8 and conclude in Section 9.

2   Related Work

There is a large body of research on Frequent Itemset Mining (FIM) but very little work addresses FIM in uncertain databases [8,9,15]. The approach proposed by Chui et. al [9] computes the expected support of itemsets by summing all itemset probabilities in their U-Apriori algorithm. Later, in [8], they additionally proposed a probabilistic filter in order to prune candidates early. In [15], the UF-growth algorithm is proposed. Like U-Apriori, UF-growth computes frequent itemsets by means of the expected support, but it uses the FP-tree [12] approach in order to avoid expensive candidate generation. In contrast to our probabilistic approach, itemsets are considered frequent if the expected support exceeds *minSup*. The main drawback of this estimator is that information about the uncertainty of the expected support is lost; [8,9,15] ignore the number of possible worlds in which an itemset is frequent. [22] proposes exact and sampling-based algorithms to find likely frequent items in streaming probabilistic data. However, they do not consider itemsets with more than one item. The current state-of-the-art (and only) approach for probabilistic frequent itemset mining (PFIM) in uncertain databases was proposed in [6]. Their approach uses an Apriori-like algorithm to mine all probabilistic frequent itemsets and the poisson binomial recurrence to compute the support probability distribution function (SPDF). We provide a faster solution by proposing the first probabilistic frequent pattern growth approach (ProFP-Growth), thus avoiding expensive candidate generation and allowing us to perform PFIM in large databases. Furthermore, we use a more intuitive generating function method to compute the SPDF.

Existing approaches in the field of uncertain data management and mining can be categorized into a number of research directions. Most related to our work are the two categories "*probabilistic databases*" [5,17,18,4] and "*probabilistic query processing*" [10,14,21,19].

The uncertainty model used in our approach is very close to the model used for probabilistic databases. A probabilistic database denotes a database composed of relations with uncertain tuples [10], where each tuple is associated with a probability denoting the likelihood that it exists in the relation. This model, called "*tuple uncertainty*", adopts the possible worlds semantics [4]. A probabilistic database represents a set of possible "certain" database instances (worlds), where a database instance corresponds to a subset of uncertain tuples. Each instance (world) is associated with the probability that the world is "true". The probabilities reflect the probability distribution of all possible database instances. In the general model description [18], the possible worlds are constrained by rules that are defined on the tuples in order to incorporate object (tuple) correlations. The ULDB model proposed in [5], which is used in *Trio*[2], supports uncertain tuples with alternative instances which are called x-tuples. Relations in ULDB are called x-relations containing a set of x-tuples. Each x-tuple corresponds to a set of tuple instances which are assumed to be mutually exclusive, i.e. no more than one instance of an x-tuple can appear in a possible world instance at the same time. Probabilistic top-k query approaches [19,21,17] are usually associated with uncertain databases using the tuple uncertainty model. The approach proposed in [21] was the first approach able to solve probabilistic queries efficiently under tuple independency by means of dynamic programming techniques. Recently, a novel approach was proposed in [16] to solve a wide class of queries in the same time complexity, but in a more elegant and also more powerful way using generating functions. In our paper, we adopt the generating function method for the efficient computation of frequent itemsets in a probabilistic way.

3   Probabilistic Frequent-Pattern Tree (ProFP-tree)

In this section we introduce a novel prefix-tree structure that enables fast detection of probabilistic frequent itemsets without the costly candidate generation or multiple database scans that plague Apriori style algorithms. The proposed structure is based on the frequent-pattern tree (FP-tree [12]). In contrast to the FP-tree, the ProFP-tree has the ability to compress uncertain transactions. If a dataset contains no uncertainty it reduces to the (certain) FP-tree.

Definition 4 (ProFP-tree). *A probabilistic frequent pattern tree is composed of the following three components:*

1. **Uncertain item prefix tree**: *A root labeled "null" pointing to a set of prefix trees each associated with uncertain item sequences. Each node n in a prefix tree is associated with an (uncertain) item a_i and consists of five fields:*
   - *n.item denotes the item label of the node. Let path(n) be the set of items on the path from root to n.*
   - *n.count is the number of* certain *occurrences of path(n) in the database.*
   - *n.uft, denoting "uncertain-from-this", is the set of transaction ids (tids). A transaction t is contained in uft if and only if n.item is uncertain in t (i.e. $0 < P(\text{n.item} \in t) < 1$) and $P(path(n) \subseteq t) > 0$.*
   - *n.ufp, denoting "uncertain-from-prefix", is a set of transaction ids. A transaction t is contained in ufp if and only if n.item is certain in t ($P(n.item \in t) = 1$) and $0 < P(path(n) \subseteq t) < 1$.*
   - *n.node − link links to the next node in the tree with the same item label if there exists one.*

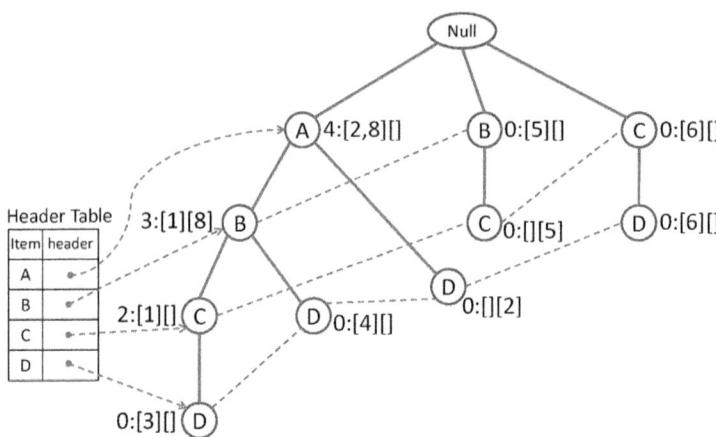

(a) *Uncertain item prefix tree* with *item header table.*

$(1, B) \to 0.2$	$(1, C) \to 0.5$	$(2, A) \to 0.1$
$(3, D) \to 0.4$	$(4, D) \to 0.5$	$(5, B) \to 0.1$
$(6, C) \to 0.1$	$(6, D) \to 0.5$	$(8, A) \to 0.5$

(b) Uncertain-item lookup table.

Fig. 2. *ProFPTree* generated from the uncertain transaction database given in Figure 1

2. **Item header table**: *This table maps all items to the first node in the Uncertain item prefix tree*
3. **Uncertain-item lookup table**: *This table maps item, tid pairs to the probability that item appears in t_{tid} for each transaction t_{tid} contained in a uft of a node n with n.item = item.*

The two sets, *uft* and *ufp*, are specialized fields required in order to handle the existential uncertainty of itemsets in transactions associated with *path(n)*. We need two sets in order to distinguish where the uncertainty of an itemset (path) comes from. Generally speaking, the entries in *n.uft* are used to keep track of existential uncertainties where the uncertainty is caused by *n.item*, while the entries in *ufp* keep track of uncertainties of itemsets caused by items in *path(n) − n.item* but where *n.item* is certain.

Figure 2 illustrates the ProFP-tree of our example database of Figure 1. Each node of the *uncertain item prefix tree* is labeled by the field *item*. The labels next to the nodes refer to the node fields *count: uft ufp*. The dotted lines denote the *node-links*.

The ProFP-tree has the same advantages as a FP-tree, in particular: It avoids repeatedly scanning the database since the uncertain item information is efficiently stored in a compact structure. Secondly, multiple transactions sharing identical prefixes can be merged into one with the number of certain occurrences registered by *count* and the uncertain occurrences reflected in the transaction sets *uft* and *ufp*.

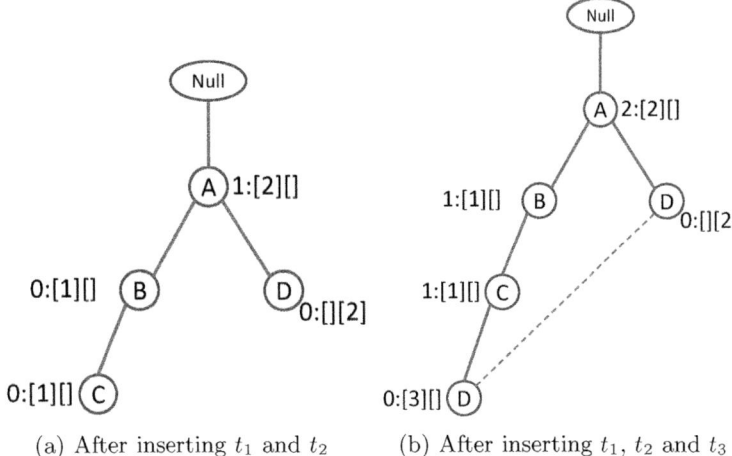

(a) After inserting t_1 and t_2 (b) After inserting t_1, t_2 and t_3

Fig. 3. *Uncertain item prefix tree* after insertion of the first transactions

3.1 ProFP-Tree Construction

For further illustration, we refer to our example database of Figure 1 and the corresponding ProFP-tree in Figure 2. We assume that the (uncertain) items in the transactions are lexicographically ordered, which is required for prefix tree construction.

We first create the root of the uncertain item prefix tree labeled "*null*". Then we read the uncertain transactions one at a time. While scanning the first transaction t_1, the first branch of the tree can be generated leading to the first path composing entries of the form (*item,count,uft,ufp,node-link*). In our example, the first branch of the tree is built by the following path:
    $<root,(A,1,[],[],null),(B,0,[1],[],null),(C,0,[1],[],null)>$.
Note that the entry "1" in the field *uft* of the nodes associated with B and C indicate that item B and C are uncertain in t_1.

Next, we scan the second transaction t_2 and update the tree structure accordingly. The itemset of transaction t_2 shares its prefix with the previous one, therefore we follow the existing path in the tree starting at the root. Since the first item in t_2 is existentially uncertain, i.e. it exists in t_2 with a probability of 0.1, *count* of the first node in the path is not incremented. Instead, the current transaction t_2 is added to *uft* of this node. The next item in t_2 does not match with the next node on the path and, thus, we have to build a new branch leading to the leaf node N with the entry (D,0,[],[2],*null*). Although item D is existentially certain in t_2 *count* of N is initialized with zero, because the itemset A,D associated with the path from the root to node N is existentially uncertain in t_2 due to the existential uncertainty of item A. Hence, we add transaction t_2 to the *uncertain-from-prefix (ufp)* field of n. The resulting tree is illustrated in Figure 3(a).

The next transaction to be scanned is transaction t_3. Again, due to matching prefixes we follow the already existing path $<$A,B,C$>$[4] while scanning the (uncertain) items in t_3. The resulting tree is illustrated in Figure 3(b). Since the first item A is existentially certain, *count* of the first node in the prefix path is incremented by one. The next items, item B and C, are registered in the tree in the same way by incrementing the *count* fields. The rational for these *count* increments is that the corresponding itemsets are existentially certain in t_3. The final item D is processed by adding a new branch below the node C leading to a new leaf node with the fields: (D,0,[3],[],*ptr*), where the link *ptr* points to the next node in the tree labeled with item label D. Since item D is existentially uncertain in t_3 the *count* field is initialized with 0 and t_3 is registered in the *uft* set. The *uncertain item prefix tree* is completed by scanning all remaining transactions in a similar fashion.

For details of the ProFP-tree construction algorithm, please refer to Algorithm 1 in the extended version of this paper [7].

3.2   Construction Analysis

The construction of the ProFP-tree requires a single scan of the uncertain transaction database $\mathcal{T}$. For each processed transaction we must follow and update or construct a single path of the tree, of length equal to the number of items in the corresponding transaction. Therefore the ProFP-tree is constructed in linear time w.r.t. to size of the database.

Since the ProFP-tree is based on the original FP-tree, it inherits its compactness properties. In particular, the size of a ProFP-tree is bounded by the overall occurrences of the (un)certain items in the database and its height is bounded by the maximal number of (un)certain items in a transaction. For any transaction t_i in $\mathcal{T}$, there exists exactly one path in the *uncertain item prefix tree* starting below the *root* node. Each item in the transaction database can create no more than one node in the tree and the height of the tree is bounded by the number of items in a transaction (path). Note that as with the FP-tree, the compression is obtained by sharing common prefixes.

We now show that the values stored at the nodes do not affect the bound on the size of the tree. In particular, in the following Lemma we bound the *uncertain-from-this (uft)* and *uncertain-from-prefix (ufp)* sets.

Lemma 5. *Let T be the* uncertain item prefix tree *generated from an uncertain transaction database $\mathcal{T}$. The total space required by all the transaction-id sets (*uft *and* ufp*) in all nodes in T is bounded by the the total number of uncertain occurrences[5] in $\mathcal{T}$.*

The rational for the above lemma is that each occurrence of an uncertain item (with existence probability in $(0,1)$) in the database yields at most one transaction-id entry in one of the transaction-id sets assigned to a node in the

---

[4] For illustration purposes, we use the *item* fields to address the nodes in a path.
[5] Entries in transactions with an existential probability in $(0, 1)$.

tree. In general there are three update possibilities for a node N: If the current item and all prefix items in the current transaction t_i are certain, there is no new entry in uft or ufp as $count$ is incremented. t_i is registered in $N.uft$ if and only if $N.item$ is existentially uncertain in t_i while t_i is registered in $N.ufp$ if and only if $N.item$ is existentially certain in in t_i but at least one of the prefix items in t_i is existentially uncertain. Therefore each occurrence of an item in $\mathcal{T}$ leads to either a count increment or a new entry in uft or ufp.

Finally, it should be clear that the size of the uncertain item lookup table is bounded by the number of uncertain (> 0 and < 1) entries in the database.

In this section we showed that the ProFP-tree inherits the compactness of the original FP-tree. In the following Section we show that the information stored in the ProFP-tree suffices to retrieve all probabilistic information required for PFIM, thus proving completeness.

4   Extracting Certain and Uncertain Support Probabilities

Unlike the (certain) FP-Growth approach, where extracting the support of an itemset X is easily achieved by summing the support counts along the node-links for X in a suitable conditional ProFPTree, we are interested in the support distribution of X in the probabilistic case. For that however, we first require both the number of certain occurrences as well as the probabilities $0 < P(X \in t_i) < 1$. Both can be efficiently obtained using the ProFP-tree. To obtain the certain support of an item x, follow the node-links from the header table and accumulate both the counts and the number of transactions in which x is uncertain-from-prefix. The latter is counted since we are interested in the support of x and by construction, transactions in ufp are known to be certain for x. To find the set of transaction ids in which x is uncertain, follow the node-links and accumulate all transactions that are in the uncertain-from-this (uft) list.

Example 2. By traversing the node-list, we can calculate the certain support for item C in the $ProFP$-tree in Figure 2 by $2 + |\emptyset| + |\{t_5\}| + |\emptyset| = 3$. Note there is one transaction in which C is uncertain-from-prefix (t_5). Similarly, we find that the only transactions in which C is uncertain are t_1 and t_6. The exact appearance probabilities in these transactions can be obtained from the uncertain-item lookup table. By comparing this to Figure 1 we see that the tree allows us to obtain the correct certain support and the transaction ids where C is uncertain.

To compute the support of an itemset $X = \{a, ..., k\}$, we use the conditional tree for items $b, ..., k$ and extract the certain support and uncertain transaction ids for a. Since it is somewhat involved, we defer the construction of conditional ProFP-trees to Section 6. By using the conditional tree, the above method provides the certain support of X and the exact set of transaction ids in which X is uncertain ($utids$). To compute the probabilities $P(X \in t_i) : t_i \in utids$ we use the independence assumption and multiply, for each $x \in X$ the probability that x appears in t_i. Recall that the probability that X appears in t_i is an $O(1)$

lookup in the uncertain-item lookup table. Recall that if additional information is given on the dependencies between items, this can be incorporated here.

We have now described how the certain support and all probabilities $P(X \in t) : X \text{ uncertain in } t$ can be efficiently computed from the ProFPTree (Algorithm 2 in the extended version of this paper [7]). Section 5 shows how we use this information to calculate the support distribution of X.

5 Efficient Computation of Probabilistic Frequent Itemsets

This section presents our linear-time technique for computing the probabilistic support of an itemset using generating functions. The problem is as follows:

Definition 6. *Given a set of N mutually independent but not necessarily identical Bernoulli (0/1) random variables $P(X \in t_i)$, $1 \leq i \leq N$, compute the probability distribution of the random variable $Sup = \sum_N^{i=1} X_i$*

A naive solution would be to count for each $0 \leq k \leq N$ all possible worlds in which exactly k items contain X and accumulate the respective probabilities. This approach however, shows a complexity of $O(2^N)$. In [6] an approach has been proposed that achieves an $O(N)$ complexity using Poisson Binomial Recurrence. Note that $O(N)$ time is asymptotically optimal in general, since the computation involves at least $O(N)$ computations, namely $P(X \in t_i) \forall 1 \leq i \leq N$. In the following, we propose a different approach that, albeit having the same linear asymptotical complexity, has other advantages.

5.1 Efficient Computation of Probabilistic Support

We apply the concept of generating functions as proposed in the context of probabilistic ranking in [16]. Consider the function: $\mathcal{F}(x) = \prod_{i=1}^{n}(a_i + b_i x)$. The coefficient of x^k in $\mathcal{F}(x)$ is given by: $\sum_{|\beta|=k} \prod_{i:\beta_i=0} a_i \prod_{i:\beta_i=1} b_i$, where $\beta = \langle \beta_1, ..., \beta_N \rangle$ is a Boolean vector, and $|\beta|$ denotes the number of 1's in β.

Now consider the following generating function:

$$\mathcal{F}^i = \prod_{t \in \{t_1, ... t_i\}} (1 - P(X \in t) + P(X \in t) \cdot x) = \sum_{j \in \{0, ..., i\}} c_j x^j$$

The coefficient c_j of x^j in the expansion of $\mathcal{F}^i$ is exactly the probability that X occurs in exactly j if the first i transactions; that is, the probability that the support of X is j in the first i transactions. Since $\mathcal{F}^i$ contains at most $i + 1$ nonzero terms and by observing that

$$\mathcal{F}^i = \mathcal{F}^{i-1} \cdot (1 - P(X \in t_i) + P(X \in t_i)x)$$

we note that $\mathcal{F}^i$ can be computed in $O(i)$ time given $\mathcal{F}^{i-1}$. Since $\mathcal{F}^0 = 1x^0 = 1$, we conclude that $\mathcal{F}^N$ can be computed in $O(N^2)$ time. To reduce the complexity to $O(N)$ we exploit that we only need to consider the coefficients c_j in the generating function $\mathcal{F}^i$ where $j < minSup$, since:

- The frequentness probability of X is defined as $P(X\ is\ frequent) = P(Sup(X) \geq minSup)) = 1 - P(Sup(X) < minSup) = 1 - \sum_{j=0}^{minSup-1} c_j$
- A coefficient c_j in $\mathcal{F}^i$ is independent of any c_k in $\mathcal{F}^{i-1}$ where $k > j$. That means in particular that the coefficients c_k, $k \geq minSup$ are not required to compute the c_i, $i < minSup$.

Thus, keeping only the coefficients c_j where $j < minSup$, $\mathcal{F}^i$ contains at most $minSup$ coefficients, leading to a total complexity of $O(minSup \cdot N)$ to compute the frequentness probability of an itemset.

Example 3. As an example, consider itemset $\{A, D\}$ in the running example database in Figure 1. Using the *ProFP-tree* (c.f. Figure 2(a)), we can efficiently extract, for each transaction t_i, the probability $P(\{A, D\} \in t_i)$, where $0 < P(\{A, D\} \in t_i) < 1$ and also the number of certain occurrences of $\{A, D\}$. Itemset $\{A, D\}$ certainly occurs in no transaction and occurs in t_2, t_3 and t_4 with a probability of 0.1, 0.4 and 0.5 respectively. Let $minSup$ be 2:

$$\mathcal{F}^1 = \mathcal{F}^0 \cdot (0.9 + 0.1x) = 0.1x^1 + 0.9x^0$$

$$\mathcal{F}^2 = \mathcal{F}^1 \cdot (0.6 + 0.4x) = 0.04x^2 + 0.42x^1 + 0.54x^0 \stackrel{*}{=} 0.42x^1 + 0.54x^0$$

$$\mathcal{F}^3 = \mathcal{F}^2 \cdot (0.5 + 0.5x) = 0.21x^2 + 0.48x^1 + 0.27x^0$$

$$\stackrel{*}{=} 0.48x^1 + 0.27x^0$$

Thus, $P(sup(\{A, D\}) = 0) = 0.27$ and $P(sup(\{A, D\}) = 1) = 0.48$. We get that $P(sup(\{A, D\}) \geq 2) = 0.25$. Thus, $\{A,D\}$ is not returned as a frequent itemset if τ is greater than 0.25. Equations marked by a * exploit that we only need to compute the c_j where $j < minSup$.

Note that at each iteration of computing $\mathcal{F}^i$, we can check whether $1 - \sum_{i<minSup} c_i \geq \tau$ and if that is the case, we can stop the computation and conclude that the respective itemset (for which $\mathcal{F}$ is the generating function) is frequent. Intuitively, the reason is that if an itemset X is already frequent considering the first i transactions only, X will still be frequent if more transactions are considered. This intuitive pruning criterion corresponds to the pruning criterion proposed in [6] for the Poisson Binomial Recurrence approach.

We remark that the generating function technique can be seen as a variant of the Poisson Binomial Recurrence. However, using generating functions instead of the complicated recursion formula gives us a much cleaner view on the problem. In addition, using generating functions, the support probability density function (sPDF) can be updated easily if a transaction t_i changes its probability of containing an itemset X. That is, if the probability $p = P(X \in t_i)$ changes to p', then we can simply obtain the expanded polynomial from the old sPDF and divide it by $px + (1 - p)$ (using polynomial division) to remove the effect of t_i and multiply $p'x + (1 - p')$ to incorporate the new probability of t_i containing X. That is, $\mathcal{F}^{i'}(x) = \mathcal{F}^i(x) : (px + 1 - p) \times (p'x + 1 - p')$, where $\mathcal{F}^{i'}$ is the generating function of the sPDF of X in the changed database containing t_i'.

6   Extracting Conditional ProFP-Trees

This section describes how conditional ProFP-trees are constructed from other (potentially conditional) ProFP-trees. The method for doing this is more involved than the analogous operation for the certain FPGrowth algorithm, since we must ensure that the information capturing the source of the uncertainty remains correct. That is, whether the uncertainty at that node comes from the prefix or from the present node. Recall from Section 4 that this is required in order to extract the correct probabilities from the tree. A conditional ProFP-tree for itemset X ($tree_X$) is equivalent to a ProFP-tree built on only those transactions in which X occurs with a non-zero probability. In order to generate a conditional ProFP-tree for itemset $X \cup i$ ($tree_{X \cup i}$) where i occurs lexicographically prior to any item in X, we first begin with the conditional ProFP-tree for X. When $X = \emptyset$, $tree_X$ is simply the complete ProFP-tree. We construct $tree_{X \cup i}$ by propagating the values at the nodes with $item = i$ upwards and accumulating these at the nodes closer to the root (cf. Algorithm 3 in the extended version of this paper [7]) . Let N_i be the set of nodes with $item = i$ (These are obtained by following the links from the header table). The values for every node n in the resulting conditional tree $tree_{X \cup i}$ are calculated as follows:

- $n.count = \sum_{n_i \in N_i} n_i.count$ since these represent certain transactions.
- $n.uft = \cup n_i.uft | n_i \in N_i$ since we are conditioning on an item that is uncertain in these transactions and hence any node in the final conditional tree will also be uncertain for these transactions.
- When collecting transactions for n that are uncertain from the prefix (i.e. $t \in ufp$), we must determine whether the item $n.item$ caused this uncertainty. If the corresponding node in $tree_X$ contained transaction t in ufp, then t is also in $n.ufp$ ($n.item$ was not uncertain in t). If $n.item$ was uncertain in t, then the corresponding node in $tree_X$ would have t listed in uft and this must also remain the case for the conditional tree. If $t \in n.ufp$ is neither in the corresponding ufp nor uft in $tree_X$, then it must be certain for $n.item$ and $n.count$ is incremented. Thus, we can avoid storing all transactions for which an item is certain. This is a key idea in our ProFP-tree.

7   ProFP-Growth Algorithm

We have now described the three fundamental operations of the ProFP-Growth Algorithm; building the ProFPTree (Section 3); efficiently extracting the certain support and uncertain transaction probabilities from it (Section 4); calculating the frequentness probability and determining whether an item(set) is a probabilistic frequent itemset (Section 5); and construction of the conditional ProF-PTrees (Section 6). Together with the fact that probabilistic frequent itemsets possess an antimonotonicity property (Lemma 17 in [6]), we can use a similar approach to the certain FPGrowth algorithm to mine all probabilistic frequent itemsets. Since, in principle, this is not substantially different from substituting the corresponding steps in FP-Growth, we omit further details.

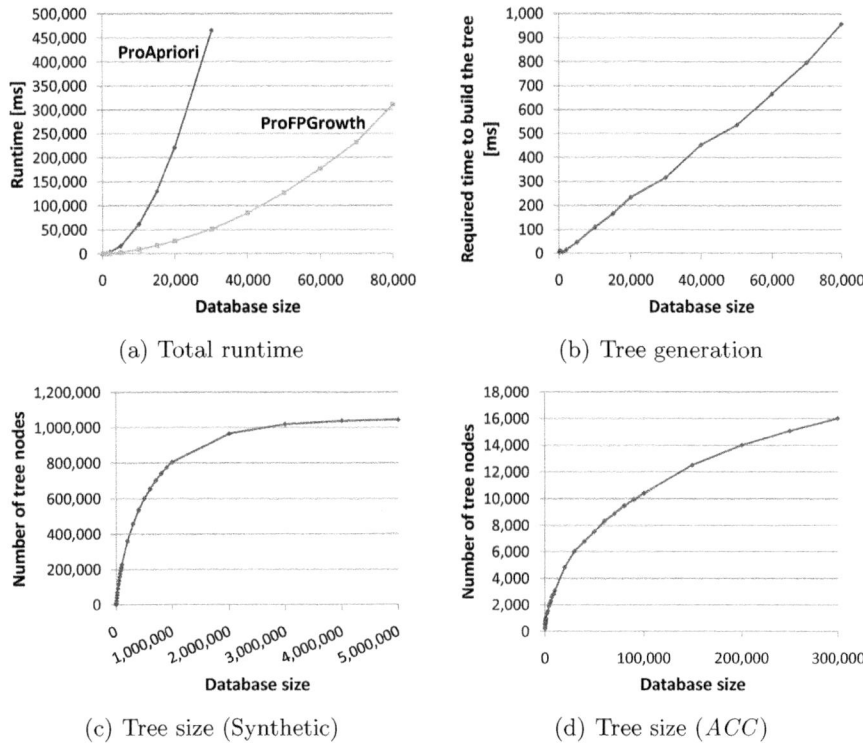

(a) Total runtime

(b) Tree generation

(c) Tree size (Synthetic)

(d) Tree size (ACC)

Fig. 4. Scalability w.r.t. the number of transactions

8 Experimental Evaluation

In this section, we present performance experiments using our proposed *ProFP-Growth* algorithm and compare the results to the Apriori-based solution (denoted as *ProApriori*) presented in [6]. We also analyze how various database characteristics and parameter settings affect the performance of *ProFP − Growth*.

All experiments were performed on an Intel Xeon with 32 GB of RAM and a 3.0 GHz processor. For the first set of experiments, we used artificial datasets with a variable number of transactions and items. Each item x has a probability $P_1(x)$ of appearing for certain in a transaction, and a probability $P_0(x)$ of not appearing at all in a transaction. With a probability $1 − P_0(x) − P_1(x)$ item x is therefore uncertain in a transaction. In this case, the probability that x exists in a transaction is picked randomly from a uniform $(0, 1)$ distribution.

For our scalability experiments, we scaled the number of items and transactions and chose $P_0(x) = 0.5$ and $P_1(x) = 0.2$ for each item. We measured the run time required to mine all probabilistic frequent itemsets that have a minimum support of 10% of the database size with a probability of a least $\tau = 0.9$.

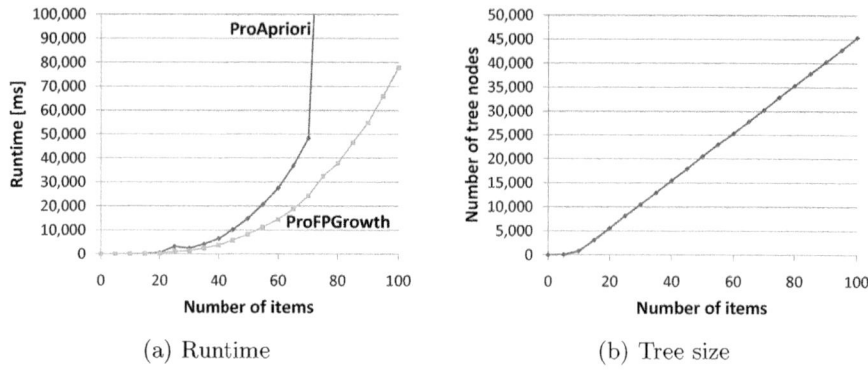

(a) Runtime   (b) Tree size

Fig. 5. Scalability with respect to the number of items

8.1 Scalability

We scaled the number of transactions and used 20 items (cf. Figure 4(a)). In this setting, our approach significantly outperforms *ProApriori* [6]. The time required to build the *ProFP-tree* w.r.t. the number of transactions is depicted in Figure 4(b). The observed linear run time indicates a constant time required to insert transactions into the tree. This is expected since the maximum height of the *ProFP-tree* is equal to the number of items. Finally, we evaluated the size of the *ProFP-tree* for this experiment, shown in Figure 4(c). The number of nodes in the *ProFP-tree* increases and then plateaus as the number of transactions increases. This is because new nodes have to be created for those transaction where a suffix of the transaction is not yet contained in the tree. As the number of transactions increases, the overlap between transaction prefixes increases, requiring fewer new nodes to be created. It is expected that this overlap increases faster if the items are correlated. Therefore, we evaluate the size of the *ProFP-tree* on subsets of the real-world dataset *accidents*[6], denoted by *ACC*. It consists of 340, 184 transactions and a reduced number of 20 items whose occurrences in transactions were randomized; with a probability of 0.5, each item appearing for certain in a transaction was assigned a value drawn from a uniform distribution in $(0, 1]$. We varied the number of transactions from *ACC* up to the first 300, 000. As can be seen in Figure 4(d), there is more overlap between transactions since the growth in the number of nodes used is slower (compared to Figure 4(c)).

Next, we scaled the number of items using 1, 000 transactions. The run times for 5 to 100 items can be seen in Figure 5(a), which shows the expected exponential runtime inherent in FIM problems. It can be clearly seen that the *ProFP-Growth* approach vastly outperforms *ProApriori*. Figure 5(b) shows the number of nodes used in the *ProFP-tree*. Except for very few items, the number of nodes in the tree grows linearly.

---

[6] The *accidents* dataset [11] was derived from the Frequent Itemset Mining Dataset Repository (http://fimi.cs.helsinki.fi/data/)

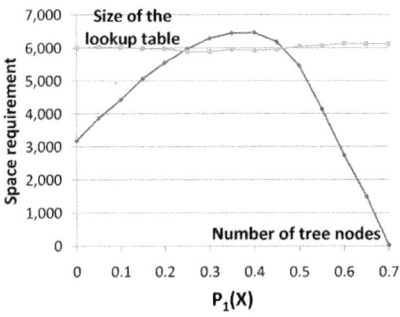
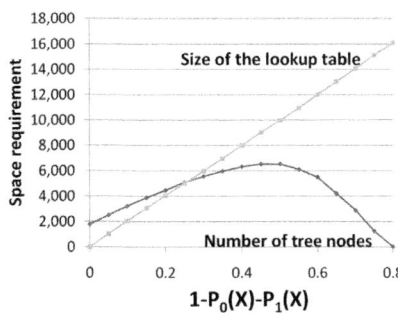

(a) Varying the probability of certain occurrences (uncertain occurrences are fixed)

(b) Varying the probability of uncertain
occurrences (certain occurrences are fixed)

Fig. 6. Effect of (un)certainty on the ProFP-tree size and uncertain item lookup table

8.2 Effect of Uncertainty and Certainty

In this experiment, we set the number of transactions to $1,000$ and the number
of items to 20 and varied the parameters $P_0(x)$ and $P_1(x)$.

First, we fixed the probability that items are uncertain $(1 - P_0(x) - P_1(x))$
at 0.3 and successively increased $P_1(x)$ from 0 (which means that no items exist
for certain) to 0.7 (cf. Figure 6(a)). It can be observed that the number of nodes
initially increases. This is what we would expect, since more items existing in
the database increases the nodes required. However, as the number of certain
items increases, an opposite effect reduces the number of nodes in the tree. This
effect is caused by the increasing overlap of the transactions – in particular, the
increased number and length of shared prefixes. When $P_1(x)$ reaches 0.7 (and
thus $P_0(x) = 0$), each item is contained in each transaction with a probability
greater than zero, and thus all transactions contain the same items with non-zero
probability. In this case, the *ProFP-tree* degenerates to a linear list containing
exactly one node for each item. Note that the size of the uncertain item lookup
table is constant, since the expected number of uncertain items is constant at
$0.3 \cdot |\mathcal{T}| \cdot |I| = 0.3 \cdot 1,000 \cdot 20 = 6,000$. In Figure 6(b) we fixed $P_1(x)$ at 0.2 and
successively decreased $P_0(x)$ from 0.8 to 0, thus increasing the probability that
items are uncertain from 0 to 0.8. We see a similar pattern as in Figure 6(a)
for the number of nodes, for similar reasons. As expected here, the size of the
lookup table increases as the number of uncertain items increases.

8.3 Effect of *minSup*

Here, we varied the minimum support threshold *minSup* using an artificial
database of $10,000$ transactions and 20 items. Figure 7 shows the results. For
low values of *minSup*, both algorithms have a high run time due to the large
number of probabilistic frequent itemsets. It can be observed that *ProFP-Growth*
significantly outperforms *ProApriori* for all settings of *minSup*.

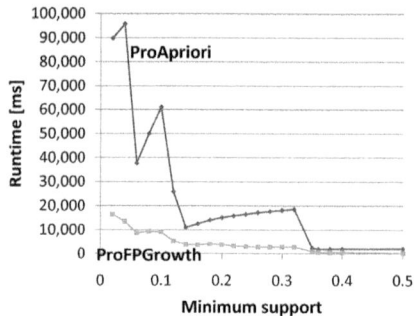

Fig. 7. Effect of *minSup*

9   Conclusion

The Probabilistic Frequent Itemset Mining (PFIM) problem is to find itemsets in an uncertain transaction database that are (highly) likely to be frequent. This problem has two components; efficiently computing the support probability distribution and frequentness probability, and efficiently mining all probabilistic frequent itemsets. To solve the first problem in linear time, we proposed a novel method based on generating functions. To solve the second problem, we proposed the first probabilistic frequent pattern tree and pattern growth algorithm. We demonstrated that this significantly outperforms the current state of the art approach to PFIM.

References

1. Aggarwal, C.C., Li, Y., Wang, J., Wang, J.: Frequent pattern mining with uncertain data. In: Proceedings of the 15th ACM International Conference on Knowledge Discovery and Data Mining (SIGKDD), Paris, France (2009)
2. Agrawal, P., Benjelloun, O., Sarma, A.D., Hayworth, C., Nabar, S., Sugihara, T., Widom, J.: Trio: A system for data, uncertainty, and lineage. In: Proceedings of the 32nd International Conference on Very Large Data Bases (VLDB), Seoul, Korea (2006)
3. Agrawal, R., Srikant, R.: Fast algorithms for mining association rules. In: Proceedings of the ACM International Conference on Management of Data (SIGMOD), Minneapolis, MN (1994)
4. Antova, L., Jansen, T., Koch, C., Olteanu, D.: Fast and simple relational processing of uncertain data. In: Proceedings of the 24th International Conference on Data Engineering (ICDE), Cancun, Mexico (2008)
5. Benjelloun, O., Sarma, A.D., Halevy, A., Widom, J.: ULDBs: Databases with uncertainty and lineage. In: Proceedings of the 32nd International Conference on Very Large Data Bases (VLDB), Seoul, Korea, pp. 1249–1264 (2006)
6. Bernecker, T., Kriegel, H.-P., Renz, M., Verhein, F., Züfle, A.: Probabilistic frequent itemset mining in uncertain databases. In: Proceedings of the 15th ACM International Conference on Knowledge Discovery and Data Mining (SIGKDD), Paris, France, pp. 119–128 (2009)

7. Bernecker, T., Kriegel, H.-P., Renz, M., Verhein, F., Züfle, A.: Probabilistic frequent pattern growth for itemset mining in uncertain databases (technical report). Computing Research Repository, abs/1008.2 (2010)
8. Chui, C.-K., Kao, B.: A Decremental Approach for Mining Frequent Itemsets from Uncertain Data. In: Washio, T., Suzuki, E., Ting, K.M., Inokuchi, A. (eds.) PAKDD 2008. LNCS (LNAI), vol. 5012, pp. 64–75. Springer, Heidelberg (2008)
9. Chui, C.-K., Kao, B., Hung, E.: Mining Frequent Itemsets from Uncertain Data. In: Zhou, Z.-H., Li, H., Yang, Q. (eds.) PAKDD 2007. LNCS (LNAI), vol. 4426, pp. 47–58. Springer, Heidelberg (2007)
10. Dalvi, N., Suciu, D.: Efficient query evaluation on probabilistic databases. The VLDB Journal 16(4), 523–544 (2007)
11. Geurts, K., Wets, G., Brijs, T., Vanhoof, K.: Profiling high frequency accident locations using association rules. In: Proceedings of the 82nd Annual Transportation Research Board, Washington, DC, USA, January 12-16, p. 18 (2003)
12. Han, J., Pei, J., Yin, Y.: Mining frequent patterns without candidate generation. SIGMOD Rec. 29(2), 1–12 (2000)
13. Leung, M.M.K., Brajczuk, D.: A Tree-Based Approach for Frequent Pattern Mining from Uncertain Data. In: Washio, T., Suzuki, E., Ting, K.M., Inokuchi, A. (eds.) PAKDD 2008. LNCS (LNAI), vol. 5012, pp. 653–661. Springer, Heidelberg (2008)
14. Kriegel, H.-P., Kunath, P., Pfeifle, M., Renz, M.: Probabilistic Similarity Join on Uncertain Data. In: Li Lee, M., Tan, K.-L., Wuwongse, V. (eds.) DASFAA 2006. LNCS, vol. 3882, pp. 295–309. Springer, Heidelberg (2006)
15. Leung, C.K.-S., Carmichael, C.L., Hao, B.: Efficient mining of frequent patterns from uncertain data. In: ICDMW 2007: Proceedings of the Seventh IEEE International Conference on Data Mining Workshops, pp. 489–494 (2007)
16. Li, J., Saha, B., Deshpande, A.: A unified approach to ranking in probabilistic databases. Proceedings of the VLDB Endowment 2(1), 502–513 (2009)
17. Ré, C., Dalvi, N., Suciu, D.: Efficient top-k query evaluation on probalistic databases. In: Proceedings of the 23rd International Conference on Data Engineering (ICDE), Istanbul, Turkey (2007)
18. Sen, R., Deshpande, A.: Representing and querying correlated tuples in probabilistic databases. In: Proceedings of the 23rd International Conference on Data Engineering (ICDE), Istanbul, Turkey (2007)
19. Soliman, M.A., Ilyas, I.F., Chang, K.C.-C.: Top-k query processing in uncertain databases. In: Proceedings of the 23rd International Conference on Data Engineering (ICDE), Istanbul, Turkey, pp. 896–905 (2007)
20. Xia, Y., Yang, Y., Chi, Y.: Mining association rules with non-uniform privacy concerns. In: DMKD 2004: Proceedings of the 9th ACM SIGMOD Workshop on Research Issues in Data Mining and Knowledge Discovery, pp. 27–34 (2004)
21. Yi, K., Li, F., Kollios, G., Srivastava, D.: Efficient processing of top-k queries in uncertain databases. In: Proceedings of the 24th International Conference on Data Engineering (ICDE), Cancun, Mexico (2008)
22. Zhang, Q., Li, F., Yi, K.: Finding frequent items in probabilistic data. In: Proceedings of the ACM International Conference on Management of Data (SIGMOD), Vancouver, BC, pp. 819–832 (2008)

Evaluating Trajectory Queries
over Imprecise Location Data

Xike Xie[1,*], Reynold Cheng[2], and Man Lung Yiu[3]

[1] Aalborg University, Denmark
xkxie@cs.aau.dk
[2] University of Hong Kong, Pokfulam Road, Hong Kong
ckcheng@cs.hku.hk
[3] Hong Kong Polytechnic University, Hung Hom, Hong Kong
csmlyiu@comp.polyu.edu.hk

Abstract. Trajectory queries, which retrieve nearby objects for every point of a given route, can be used to identify alerts of potential threats along a vessel route, or monitor the adjacent rescuers to a travel path. However, the locations of these objects (e.g., threats, succours) may not be precisely obtained due to hardware limitations of measuring devices, as well as the constantly-changing nature of the external environment. Ignoring data uncertainty can render low query quality, and cause undesirable consequences such as missing alerts of threats and poor response time in rescue operations. Also, the query is quite time-consuming, since all the points on the trajectory are considered. In this paper, we study how to efficiently evaluate trajectory queries over imprecise location data, by proposing a new concept called the u-bisector. In general, the u-bisector is an extension of bisector to handle imprecise data. Based on the u-bisector, we design several novel filters to make our solution scalable to a long trajectory and a large database size. An extensive experimental study on real datasets suggests that our proposal produces better results than traditional solutions that do not consider data imprecision.

1 Introduction

Given a set P of points, the Trajectory Nearest Neighbor Query (*TNNQ in short*) [1], retrieves the closest object in P for every query point on the given trajectory T. As an example, consider the trajectory $T = \{[q_1, q_2], [q_2, q_3], [q_3, q_4]\}$ and objects $P = \{o_1, o_2, o_3\}$, shown in Figure 1(a). The *TNNQ*'s answer is as Figure 1(b). It means for all points on $[s'_0, s'_1]$, the nearest neighbor is o_1, etc. The *TNNQ* can find applications in location-based service (*LBS in short*), such as "what is the nearest gas station along the travel route".

Unfortunately, the measured location of an object is often imprecise because of: (i) limited resolution of the measure device, (ii) infrequent measurement, (iii) environmental factors. For example, the shipping industries regard safety as their top priority. They hope to identify alerts of potential threats along the route of a vessel in advance, and take appropriate actions if necessary. People in the US and Northern Europe detect the

---

* This work was done in the University of Hong Kong.

A. Ailamaki and S. Bowers (Eds.): SSDBM 2012, LNCS 7338, pp. 56–74, 2012.

icebergs by remote sensors and satellite imaging [2], which have limited measurement accuracy and frequency. Sensors have limited battery capacity whereas satellite imaging incurs expensive deployment cost. This causes infrequent measurements, rendering the measured location of an object stale. Furthermore, as time passes by, icebergs may move according to the temperature and the ocean current / wind speed. In the *LBS* example, what if the objects being queried are not static but moving constantly (e.g. rescue vehicles positioned by GPS devices)? Again, locations obtained by GPS devices can be contaminated with measurement error, which can be further deteriorated by terrain and climate conditions [3]. Also, the positions could be tracked only periodically due to the limited battery powers [4].

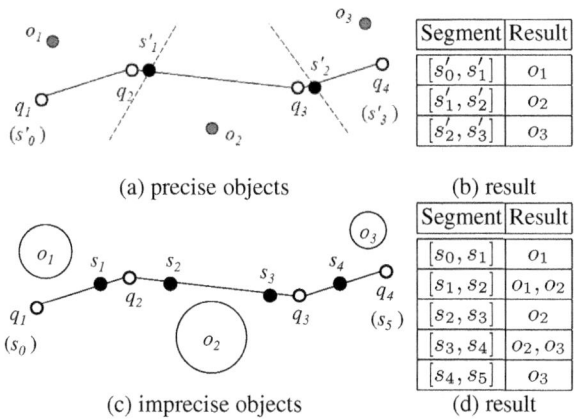

(a) precise objects

(b) result

Segment	Result
$[s_0', s_1']$	o_1
$[s_1', s_2']$	o_2
$[s_2', s_3']$	o_3

(c) imprecise objects

(d) result

Segment	Result
$[s_0, s_1]$	o_1
$[s_1, s_2]$	o_1, o_2
$[s_2, s_3]$	o_2
$[s_3, s_4]$	o_2, o_3
$[s_4, s_5]$	o_3

Fig. 1. Example Trajectory Query

A common way to represent an imprecise location or a moving object is to model the position by an area called *imprecise region* [4,5,6,7,8,9,10]. The possible location of the object is assumed to be within this region. Figure 1(c) shows a query trajectory $T = \{[q_1, q_2], [q_2, q_3], [q_3, q_4]\}$ and some imprecise objects o_1, o_2, o_3. The result (Figure 1(d)) can be represented in a compact way by partitioning the query trajectory into segments such that all locations within the same segment share the same result set. For example, o_2 is *definite nearest neighbor* to the segment $[s_2, s_3]$. On the other hand, o_1 and o_2 are *possible nearest neighbors* (**PNNs**) to the segment $[s_1, s_2]$ because both of them have potential to be the closest object. We define this query as Trajectory Possible Nearest Neighbor Query (*TPNNQ in short*). Note that [1] is a special case of our problem, where the objects being queried are precise points.

Determining the *TPNNQ* answer can be technically challenging, since the imprecise regions are considered. A simple solution is to replace the imprecise region of each object with a center point (shown as a grey dot), as illustrated in the scenario in Figure 1(a) and (b). The result consists of three segments, each associated with the closest object. For instance, the closest object to location q_2 appears to be o_1 only. The object o_2 is missing from the result. Recall from Figure 1(c) and (d) that o_2 also has possibility to become a closest object to location q_2. This "center simplification" approach causes undesirable consequences such as missing alerts of threats and poor response time in our applications. In the vessel/rescuer example, the ignorance of the imprecise region could cause potential danger. Thus, it is important to augment each threat with an imprecise region, in order to foresee the worst-case scenario. In the rescuer example, a rescue vehicle seemingly close to / far from the travel path may be actually far from / close to it. Thus, it would take longer time to respond. It is better to call up all rescuers likely to be the closest, in order to handle the emergency as soon as possible.

Another attempt to simplify the problem is to use a "sampling approach", which considers positions at every fixed length on the query trajectory, and compute the potential nearby objects at each position. However, if the sampling rate is high, it incurs a huge computation cost; on the other hand, a low sampling rate can result in many answers missing. Notice that a query trajectory consists of infinite number of possible locations, and it is not easy to determine the sampling rate. As shown in Figure 1(c), the result set changes only at a few positions (s_1, s_2, s_3, s_4). It is not clear how to determine the correct sampling rate to in order to get these answers. In fact, our experimental results show that replacing imprecise regions with points or sampling the trajectory cannot provide an accurate solution. Hence, we develop a solution that can accurately compute a trajectory query on imprecise objects.

The techniques of [1] cannot be readily applied to evaluate *TPNNQ*. [1]'s idea is to use the (perpendicular) *bisectors* of every pair of points to derive the query answer. For example, in Figure 1(a), the point s_1' is the intersection between the query trajectory and the bisector of objects o_1 and o_2, which are shown as dashed lines. Similarly, s_2' is derived by o_2 and o_3's bisector. However, the bisector, which forms the basis of [1], is limited to precise points.

We extend the concept of "bisectors" to support imprecise objects, called u-**bisector**. Figure 2 illustrates the corresponding u-bisectors for circular and rectangular imprecise regions. From this figure, we can see that the u-bisector is not a straight line anymore. It becomes a pair of lines, which partition the domain space into three parts: (1) the left area, containing points q where O_i is absolutely closer to q than O_j; (2) the right partition, consisting of points q' where O_j is absolutely closer to q' than O_i; and (3) the middle part, having points q'' where both O_i and O_j can be the nearest to q''. We

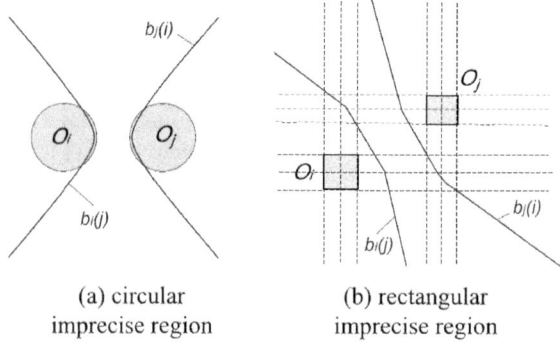

(a) circular
imprecise region

(b) rectangular
imprecise region

Fig. 2. u-bisector for imprecise regions

demonstrate how to use conceptually the intersection points of the query trajectory and the u-bisectors to answer a trajectory query.

In practice, it is expensive to compute the intersections points between the query trajectory and the u-bisectors. As shown in Figure 2, these u-bisectors can be hyperbolic curves (Figure 2(a)), or segments of straight lines/curves (Figure 2(b)). Even for one u-bisectors, they can intersect the query trajectory at more than one point. We first design a *Basic* solution, which answers the query in $O(ln^2 log n)$ (*n:database size; l:trajectory length*). To make our solution scalable to large datasets and long trajectories, we design a filter-refinement framework. In the filtering phase, *candidate objects* that may be the closest to each answer line-segment are obtained. In the refinement phase, we develop a novel technique called *ternary decomposition*, which can derive the final answers accurately. We show theoretically and experimentally that our solution is efficient and scalable. It is also more accurate than the center simplification and the sampling approaches. We assume the imprecise regions are of circular shapes for simplicity. Actually, our method is also general to other shaped objects. It would not be discussed due to page limitations.

The rest of this paper is as follows. In Section 2 we discuss the related work. Section 3 defines the problem and a basic solution based on the u-bisectors. We present our solution framework in Section 4. The filtering and refinement phases are described in Sections 5 and 6. In Section 7 we present our experiment results. Section 8 concludes.

2   Related Work

Nearest neighbor (NN) query for moving query points is a well studied topic [11] [12] [13] [1]. These works focus on reducing the computational cost at the server. Among these works, there are two major categories.

The first category does not require the user's entire trajectory in advance [11] [12] [13], but processes the query online (multiple times) based on the user's moving location. In [11], the authors propose sampling techniques to answer the moving *NN* query. They study how to calculate the upper-bound distance within which the moving point does not issue a new query to the server. [12] [13] use validity region and validity time for the query answer of moving points. They use Voronoi cells to represent the validity region. The query answer becomes invalid if the validity time is expired or the user leaves the validity region.

The second category assumes that the user's trajectory is known in advance. It evaluates the query once only [1] [4]. In our application, the trajectory, such as sailing routes, is known in advance. Thus, we elaborate the second category in details. In [1], the route of the query point is split into sub-line-segments, such that the NN answer within the same sub-line-segment remains unchanged. A perpendicular bisector $\perp(p_i, p_j)$ between two points p_i and p_j is used to partition the trajectory query into two sub-trajectories, one being definitely closer to p_i and the other being definitely closer to p_j. However, this technique is not applicable to our problem on imprecise location data. As shown in Figure 1, some segments like $[s_1, s_2]$ can have multiple *PNN*s and it is challenging to derive them.

The bisector for imprecise objects has been addressed by a few works recently [5] [6] [7]. They use bisectors to determine the dominance relationship between objects. Our work is different because we consider a query trajectory, but not a query object. For the trajectory, our solution is capable of answering the query for every point it.

The paper [4] is closely related to our problem. It also uses an imprecise region to model the location of an object and compute the object closest to a given query segment. Unlike our work, [4] only computes the answer for segments with the definite nearest neighbor, such as $[s_0, s_1]$ in Figure 1. It did not study how to compute objects that might be the closest, for some segment like $[s_1, s_2]$ in Figure 1. Furthermore, their method scans the entire database to answer the query, thus it is not very scalable to data volumes.

3   Problem and Preliminaries

In this section, we describe the query semantics in Section 3.1. We introduce the u-bisector in Section 3.2 and propose a basic method in Section 3.3.

3.1   Problem Setting

We first introduce the definition of *PNNQ* (studied in [14]), which is used to define *TPNNQ*, the query studied in this paper. Let q be a point, and let O_i be an imprecise object from an object set O. We use $dist_{min}(q, O_i)$ and $dist_{max}(q, O_j)$ to denote the minimum and maximum distances of object O_i from q, respectively.

Definition 1.   Possible Nearest Neighbor Query (PNNQ): *Given a set of imprecise objects O and a query point q, $O_i \in PNNQ(q)$, if $\nexists O_j \in O$, such that $dist_{max}(q, O_j) < dist_{min}(q, O_i)$.*

In Figure 1(c), $PNNQ(q_2) = \{O_1, O_2\}$ implies that either O_1 or O_2 could be the *NN* of the query point q_2. A query trajectory $\mathcal{T}$ can be represented by a set of query line-segments $\mathcal{T} = \{L_1, ..., L_l\}$, where L_i is a query line-segment. For a query point q, whose trajectory is $\mathcal{T}$, the *trajectory possible nearest neighbor query (TPNNQ)* returns *PNNs* for all the points in $\mathcal{T}$. In other words, the query returns $\{\langle q, PNNQ(q) \rangle\}_{q \in \mathcal{T}}$. If the connected points on the trajectory have the same *PNNs*, we could merge them into a segment.

Definition 2.   Trajectory Possible Nearest Neighbor Query (TPNNQ): *Given a set of imprecise objects O and a query trajectory $\mathcal{T}$, the answer for the TPNNQ query is a set of tuples $R = \{\langle T_i, R_i \rangle | T_i \subseteq \mathcal{T}, R_i \subseteq O\}$, where $PNNQ(q) = R_i(\forall q \in T_i)$.*

In other words, the *TPNNQ* splits $\mathcal{T}$ into a set of consecutive segments $\{T_1, T_2, ..., T_t\}$. T_i is a sub-trajectory of $\mathcal{T}$. For $\forall q \in T_i$, q has the same possible nearest neighbors (*PNNs*), then we call each T_i a **validity interval**. The connection point of two consecutive segments, say T_i and T_{i+1}, is called **turning point**, which indicates the change of *PNNQ* answers. An example for a *TPNNQ* over three imprecise objects $\{O_1, O_2, O_3\}$ is shown in Figure 1(c). The trajectory query $\mathcal{T}(s, e)$ is split into 5 pieces of segments. Also, point s_1 is the turning point for $T(s_0, s_1)$ and $T(s_1, s_2)$.

Observe that there are two major differences between the results on imprecise objects and precise objects. Comparing Figures 1 (c) and (a): (1) the *imprecise case* could have more tuples (5 compared to 3); (2) a query point in *imprecise case* might return a set of *PNN*s instead of a single answer.

Thus, the $TPNNQ$ can be answered by finding the *turning points*. Then, how to derive the *turning points* on a trajectory, given a set of imprecise objects? To address that, we first investigate the *u-bisector*, for imprecise objects. In general, the *u-bisector* splits the domain into several parts, such that query points on different parts could have different PNNs. Then, the *turning points* can be evaluated by finding the intersections of the *u-bisectors* and the query trajectory. Next, we discuss the *u*-bisector.

3.2   *u*-bisector

Definition 3. *Given two imprecise objects O_i and O_j, their u-**bisector** consists of two lines: $b_i(j)$ and $b_j(i)$. The u-**bisector half** $b_i(j)$ is a set of points satisfying*

$$b_i(j) = \{z : dist_{max}(z, O_i) = dist_{min}(z, O_j)\} \tag{1}$$

The curve $b_i(j)$ splits the domain into two *half-spaces*: $H_i(j)$ and $\overline{H_i(j)}$, where $H_i(j)$ is the half closer to O_i and $\overline{H_i(j)}$ is the complementary half. An example is shown in Figure 3. Thus we have:

$$H_i(j) = \{z : dist_{max}(z, O_i) \leq dist_{min}(z, O_j)\} \tag{2}$$

$$\overline{H_i(j)} = \{z : dist_{max}(z, O_i) > dist_{min}(z, O_j)\} \tag{3}$$

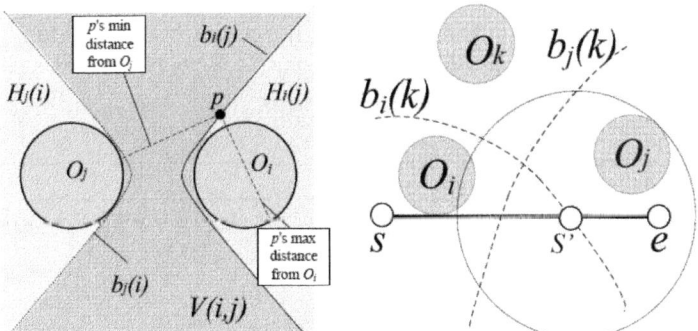

Fig. 3. *u*-bisector   **Fig. 4.** Verification

Generally speaking, the *u*-bisector half $b_i(j)$ is a curve in the domain space. If a query point $q \in H_i(j)$, q must take O_i as its nearest neighbor certainly. The *u*-bisector halves $b_i(j)$ and $b_j(i)$ separate the domain into three parts: $H_i(j)$, $H_j(i)$, and $V(i,j)$, where

$$V(i,j) = \overline{H_i(j)} \cap \overline{H_j(i)} \tag{4}$$

Notice that $V(i,j) = V(j,i)$. If O_i and O_j are degenerated into precise points, $V(i,j)$ becomes 0. Also, $b_i(j) = b_j(i)$, which becomes a straight line.

If a query line-segment is totally covered by $V(i,j)$ or $H_i(j)$, it does not intersect with $b_i(j)$. Otherwise, the intersections split the line-segment into several parts. Different parts might correspond to different PNNs, as they are located on different sides of $b_i(j)$.

For circular imprecise objects, it is easy to derive the closed form equations of the u-bisector and evaluate the analytical solution for the intersection points. The number of intersections is at most 2, since the quadratic equation has at most 2 roots (see Appendix A.1). Next, we present the basic method based on the analysis of the u-bisector's intersections. We focus our discussion on two dimension location data.

3.3   Basic Method

From Definition 2, the *TPNNQ* could be answered by deriving the *turning points*, which are intersections of the query trajectory and the u-bisectors. A u-bisector is constructed by a pair of objects. Given a set O of n objects, there can be C_2^n u-bisectors. The *Basic* method is to check the intersections of the query trajectory with the C_2^n u-bisectors. The intersections can be found by evaluating the equation's roots in Appendix A.1. Here we use *FindIntersection(.)* (in *Step 5*) to represent the process.

However, not all of the bisectors intersect with the trajectory. Even if they intersect, not all of the intersections are qualified as *turning points*. Thus, we need a "*verification*" process to exclude those unqualified intersectinos. For example, in Figure 4, the u-bisector half $b_i(k)$ intersect with $[s, e]$ at s'. For an arbitrary point $q \in [s, e]$, either O_i or O_j is closer to q than O_k, since $[s, s'] \in H_i(k)$ and $[s', e] \in H_j(k)$. Then, O_k is not PNN for $p \in [s, e]$, and s' is not a qualified *turning point*.

We use the $s_{i \vdash j}$ to represent an intersection created by $b_i(j)$ ($s_{i \vdash j} = b_i(j) \cap L$), and $s_{i \dashv j} = b_j(i) \cap L$. In other words, $s_{i \vdash j}$ can be understood as $PNNQ(q)$ answer that turns from containing O_i to both O_i and O_j, if q moves from $H_i(j)$ to $\overline{H_i(j)}$ So, O_i should definitely be $s_{i \vdash j}$'s PNN, while O_j is not. This can be implemented by issuing a $PNNQ$. Thus, we can use this for verification.

Algorithm 1. Basic

1: **function** BASIC(Trajectory $\mathbb{T}$)
2:    **for all** line-segment $L \in \mathcal{T}$ **do**
3:       **for** $i = 1 \ldots n$ **do** $\triangleright$ consider object O_i
4:          **for** $j = i + 1 \ldots n$ **do** $\triangleright$ consider object O_j
5:             $\mathcal{I}$ = FindIntersection(L, O_i, O_j);
6:             Verify $\mathcal{I}$ and delete unqualified elements;
7:    Evaluate PNNs for each Interval and Merge two successive ones if they have same PNNs;

In Algorithm 1, suppose *Step 5* can be done in time β, which is a constant if we call Appendix A.1. *Step 6* can be finished in $O(\log n)$. Suppose $\mathcal{T}$ contains l line-segments, *Basic*'s total query time is $O(l \, n^2(\log n + \beta))$. In later sections, we study several filters which can effectively prune those unqualified objects, which cannot be PNN for any point on the trajectory, in order to reduce the complexity.

4   Solution Framework

In this section, we propose the framework of the $TPNNQ$ algorithm, which follows the *filter-refinement* framework. We assume an R-tree $\mathbb{R}$ is built on the imprecise objects O and it can be stored in the main memory, as the storage capabilities increase fast in recent years.

Framework. In implementation, we organize the trajectory $\mathcal{T} = \{L_1, L_2, ..., L_l\}$ by constructing a binary tree $\mathbb{T}(\mathcal{T})$. Each binary tree node $T_i = \{L_1, ..., L_{l'}\}$ has two children: $T_i.left = \{L_1, ..., L_{\lfloor \frac{l'}{2} \rfloor}\}$ and $T_i.right = \{L_{\lceil \frac{l'}{2} \rceil}, ..., L_{l'}\}$. We show an example of $\mathcal{T} = \{L_1, L_2, L_3\}$'s trajectory tree in Figure 5(a).

The data structure for each binary tree node T_i is a triple: $T_i = \langle L, MBC, Guard \rangle$. L is a line-segment if T_i is a leaf-node and $NULL$ otherwise. MBC is the minimum bounded circle covering T_i; it is $NULL$ for leaf-nodes. $Guard$ is an entry which has minimum maximum distance to T_i. As we describe later, the entry can be either an R-tree node or an imprecise object. The $Guard$s are not initialized until the processing of $TPNNQ$. Since $\mathcal{T}$ contains l line-segments, the trajectory tree $\mathbb{T}(\mathcal{T})$ could be constructed in $O(l \log l)$ time.

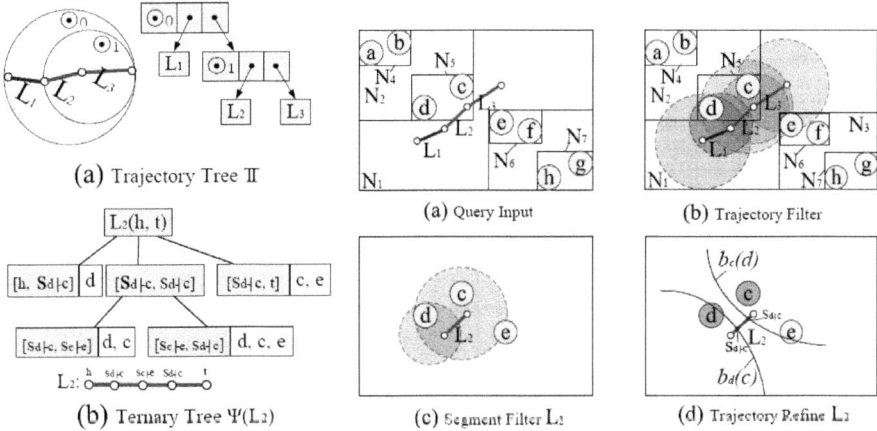

(a) Trajectory Tree $\mathbb{T}$

(b) Ternary Tree $\Psi(L_2)$

(a) Query Input

(b) Trajectory Filter

(c) Segment Filter L_2

(d) Trajectory Refine L_2

Fig. 5. Trajectory Tree $\mathbb{T}(\mathcal{T})$ and Ternary Tree $\Psi(L_2)$

Fig. 6. TPNNQ

Given a constructed trajectory tree $\mathbb{T}$ and an R-tree $\mathbb{R}$, the $TPNNQ$ algorithm, shown in Algorithm 2, consists of two phases. Phase I is the filtering phase, which includes two filters: *Trajectory Filter* and *Segment Filter*. *Trajectory Filter* is to retrieve a set of candidates from the database (*Step 3*). *Segment Filter* prunes away unqualified objects for each $L_i \in \mathcal{T}$ (*Step 4*). Phase II is to evaluate all the *validity intervals* and *turning points* for each line-segment of the trajectory (*Step 5*). Then, we scan the derived *validity intervals* once and merge two consecutive *validity intervals* if they belong to different line-segments but have the same set of PNNs (*Step 7*).

Example of TPNNQ. Suppose an R-tree built on objects $O = \{a, b, c, d, e, f\}$ and a trajectory $\mathcal{T} = \{L_1, L_2, L_3\}$, as shown in Figure 6(a). We use *Trajectory Filter* to

---

Algorithm 2. TPNNQ

1: **function** TPNNQ(Trajectory $\mathbb{T}$, R-tree $\mathbb{R}$)
2:     let ϕ be a list (of candidate objects);
3:     $\phi \leftarrow$ TrajectoryFilter($\mathbb{T}, \mathbb{R}$);   ▷ Section 5.1
4:     **for all** line-segment $L_i \in \mathcal{T}$ **do**   ▷ $\mathcal{T} = \{L_i\}_{i \leq l}$
5:         $\phi_i \leftarrow$ SegmentFilter(L_i, ϕ);   ▷ Section 5.2
6:         $\{\langle L, R \rangle\}_i \leftarrow$ TernaryDecomposition(L_i, ϕ_i);   ▷ Section 6
7:     $\{\langle T_i, R_i \rangle\}_{i=1}^t \leftarrow$ Merge($\cup_{i=1}^l \{\langle L, R \rangle\}_i$);

---

derive $\mathcal{T}$'s trajectory filtering bound, as shown by shaded areas in Figure 6(b). The objects $\{c, d, e, f\}$ overlapping with the trajectory filtering bound are taken as candidates. During the process, object d is set to be L_2's *Guard*, and stored in the trajectory tree. The *segment filter* is applied for each line-segment in $\mathcal{T}$. Taking $L_2(h, t)$ as an example, the segment filtering bound is shown as Figure 6(c), where f is excluded from L_2's candidates. Because f does not overlap with the filter bound.

In the refinement phase, we call the routine *TernaryDecomposition* to derive the *turning points*. We find the u-bisector halves $b_d(c)$ and $b_c(d)$ intersects with L_2 at $s_{d \vdash c}$ and $s_{d \dashv c}$, respectively. L_2 is split into three sub-line-segments $[h, s_{d \vdash c}]$, $[s_{d \vdash c}, s_{d \dashv c}]$, and $[s_{d \dashv c}, t]$. Meanwhile, the construction of a ternary tree $\Psi(L_2)$ starts, in Figure 5(b). The root node of $\Psi(L_2)$ derives three children correspondingly.

Then, we repeat the above process for each of the three, recursively. Finally, the process stops and we get a ternary tree $\Psi(L_2)$, in Figure 5(b). Observed from $\Psi(L_2)$, the degree of a ternary tree node is at most 3, since a line-segment is split into at most 3 sub-line-segments, as shown in Section 3. The query result on L_2 can be fetched by traversing the leaf-nodes of $\Psi(L_2)$. Then, we have: $TPNNQ(L_2) = \{\langle [h, s_{d \vdash c}], \{d\} \rangle, \langle [s_{d \vdash c}, s_{c \vdash e}], \{c, d\} \rangle, \langle [s_{c \vdash e}, s_{d \dashv c}], \{c, d, e\} \rangle, \langle [s_{d \dashv c}, t], \{c, e\} \rangle\}$. Similarly, the results of L_1 and L_3 can also be evaluated. By merging them we get the answer of $TPNNQ(\mathcal{T})$. After we get the query answer, $\mathbb{T}$ and Ψ are deleted.

In the following sections, we study the *Trajectory Filter* in Section 5.1 and *Segment Filter* in Section 5.2. The refinement step is shown in Section 6.

5   Filtering Phase

The trajectory query consists of a set of consecutive query line-segments. An intuitive way is to: (1) decompose the trajectory into several line-segments; (2) for each line-segment L_i, access R-tree to fetch candidates. Then, apply TernaryDecomposition to construct a ternary tree $\Psi(L_i)$ to evaluate its *validity intervals* and *turning points*. We call this method TP-S, which incurs multiple R-tree traversals.

Meanwhile, two consecutive line-segments might share similar $PNNs$. Also, if two line-segments are short, they could even be located within the same *validity interval*. So, considerable efficiency would be saved if the R-tree traversal for each line-segment inside the trajectory could be shared.

Algorithm 3. TrajectoryFilter

1: **function** TRAJECTORYFILTER(Trajectory tree $\mathbb{T}$, R-tree $\mathbb{R}$)
2:     let ϕ be a list (of candidate objects);
3:     Construct a min-dist Heap $\mathcal{H}$;
4:     $push\_heap(\mathcal{H}, root(\mathbb{R}), 0)$;
5:     **while** $\mathcal{H}$ is not empty **do**
6:         $E \leftarrow deheap(\mathcal{H})$;
7:         **if** E is a non-leafnode of $\mathbb{R}$ **then**
8:             **for** E's each child e **do**
9:                 **if** isProbable($\mathbb{T}, e$) **then**
10:                     $push\_heap(\mathcal{H}, e, dist_{max}(e, \mathbb{T}))$;
11:         **else**
12:             **if** isProbable($\mathbb{T}, E$) **then**
13:                 insert E into ϕ;
14:     return ϕ;

5.1 Trajectory Filter

To save the number of R-tree node access, we design Algorithm 3 as the *Trajectory Filter* to retrieve the candidates for the entire trajectory. We start Algorithm 3 by maintaining a heap in the ascending order of maximum distance between an entry to a trajectory tree node T_i's center. If T_i is leaf-node, it is a line-segment. T_i's center is its mid-point. Otherwise, the center is $MBC(T_i)$'s circle center. Then, the top element is popped to test if it/its children could be qualified to be the candidate objects. The process is repeated until the heap is empty.

To determine if an entry is qualified or not, we use Algorithm 4. Let T_i be a $\mathbb{T}$'s node and G be $T_i.Guard$. G is initialized in *Step 4*. Given an R-tree node E, if $T_i \subseteq H_G(E)$, then $\forall O_j \in E$, O_j cannot be T_i's $PNNs$. Thus, $\forall O_j \in E$ can be rejected. This helps pruning those unqualified objects in a higher index level.

Algorithm 4. isProbable

1: **function** ISPROBABLE(Trajectory tree node T_i, R-tree node E)
2:     Let G be $MBC(T_i).Guard$;   ▷ Initialize Guard Obj.
3:     **if** G is $NULL$ **then**
4:         $G \leftarrow E$;
5:     **elseif** T_i is leaf-node **and** $T_i \in H_G(E)$ **then**
6:         return $false$;   ▷ If E can be rejected, return *false*
7:     **elseif** T_i is non-leaf-node **and** $MBC(T_i) \in H_G(E)$ **then**
8:         return $false$;   ▷ If E can be rejected, return *false*
9:     **elseif** $dist_{max}(E, MBC(T_i).c) < dist_{max}(G, MBC(T_i).c)$ **then**
10:         $G \leftarrow E$;   ▷ If G can be updated
11:     **if** T_i is not leaf-node **then**
12:         return isProbable($T_i.left, E$)$\|$isProbable($T_i.right, E$);
13:     **else** return $true$;

In order to check whether E can be rejected, we consider two cases: (i) if T_i is a leaf-node; (ii) if T_i is a non-leaf-node.

(i) When T_i is a leaf-node, we can draw a pruning bound to test whether E is qualified. If we denote $\odot(c, G)$ as a circle centered at c and internally tangent with object G, and $\odot(c, r)$ as a circle centered at c with radius r, then:

$$\odot(c, G) = \odot(c, dist_{max}(c, G)) \tag{5}$$

The pruning bound is written as: $\odot(s, G) \cup \odot(e, G)$. The correctness is guaranteed by Lemma 1.

Lemma 1. *Given two imprecise objects O_i, O_j and a line-segment $L(s, e)$, O_j can not be $p \in L$'s PNN if O_j does not overlap with $\odot(s, O_i) \cup \odot(e, O_i)$.*

Proof. To judge if O_j is L's PNN, we first prove it is sufficient to check L's two end points s and e. Then, we show how the pruning bound can be derived by s and e.

Since $b_i(j)$ is a hyperbola half, it has at most two intersections with an arbitrary line. Thus, $H_i(j)$ is convex [15]. So, if s and e are in $H_i(j)$, $p \in L$ must be in $H_i(j)$. It means if O_j is not s and e's PNN given O_i, it is not a PNN for all the points on L.

Next,
$$s \in H_i(j) \Leftrightarrow dist_{max}(s, O_i) < dist_{min}(s, O_j)$$
$$\left. \begin{array}{l} \Leftrightarrow \odot(s, O_i) \cap O_j = \emptyset \\ \odot(e, O_i) \cap O_j = \emptyset \end{array} \right\} \Leftrightarrow O_j \bigcap \odot(s, O_i) \cup \odot(e, O_i) = \emptyset$$

So, the lemma is proved.

If another object O_j does not overlap with the pruning bound defined by Lemma 1, it can not be the PNN of any $p \in L$, since O_i will be always be closer. We also use Lemma 1 as the base to derive other pruning bounds in Section 6.

(ii) When T_i is a non-leaf-node, if $MBC(T_i) \in H_G(E)$, then E can be rejected from candidates. Since $MBC(T_i) \in H_G(E)$, T_i must be in $H_G(E)$. In other words, Equation 6 is satisfied when the condition below is true:

$$dist_{max}(MBC(T_i), G) \leq dist_{min}(MBC(T_i), E) \tag{6}$$

Since $MBC(T_i)$ is a circle, Equation 6 can be rewritten as ($\odot.c$ and $\odot.r$ are $\odot$'s center and radius):

$$dist_{max}(MBC(T_i).c, G) + MBC(T_i).r \leq dist_{min}(MBC(T_i).c, E) - MBC(T_i).r$$

5.2 Segment Filter

After the *trajectory filtering* step of $TPNNQ$, we get a set ϕ of candidates. Before passing ϕ to each line-segment L_i in the *refinement* phase, we perform a simple filtering process to shrink ϕ into a smaller set ϕ_i for L_i. Notice that while deriving the trajectory tree $\mathbb{T}(\mathcal{T})$, we also derive an object called "Guard" for each node T_i. Then, for a $\mathcal{T}$'s line-segment $L_i(s_i, e_i)$, we can reuse the "Guard" O_g to build the pruning bound. According to Lemma 1, the pruning bound is set to $\odot(s_i, O_g) \cup \odot(e_i, O_g)$. An example is shown in Figure 2(c), where the pruning bound for $L_2(h, t)$ is $\odot(h, d) \cup \odot(t, d)$. After that, we get ϕ_i. Empirically, the pruned candidates set ϕ_i is much smaller than ϕ.

6   Trajectory Refinement Phase

For trajectory $\mathcal{T}$, the refinement is done by applying Algorithm 5 *Ternary Decomposition* for each line-segment $L_i \in \mathcal{T}$. Essentially, Algorithm 5 is to construct a *ternary tree* $\Psi(L_i)$ for L_i.

6.1   Trajectory Refinement

Ψ is constructed in an iterative manner. At each iteration, we select two objects from the current candidate set ϕ_{cur} as seeds to divide the current line-segment L_{cur} into two/three pieces.

To split L_i, we have to evaluate a feasible u-bisector, whose intersections with L_i are *turning points*. Then, to find the u-bisector, we might have to try $\frac{C(C-1)}{2}$ pairs of objects, $C = |\phi_i|$. In fact, the object with minimum maximum distance to L_i, say O_1, must be one PNN. The correctness is shown in Lemma 2. Thus, it is often that the *turning points* on L_i is derived by O_1 and another object among the C candidates. So, in Algorithm 5, the candidates are sorted first.

Lemma 2. *If $S = \{O_1, O_2, ...\}$ are sorted in the ascending order of the maximum distance to the line-segment L, then $O_1 \in TPNNQ(L)$.*

Proof. Suppose p is a point of L, such that $dist_{max}(p, O_1) = dist_{max}(L, O_1)$. If O_1 is definitely one PNN of $p \in L$, O_1 must be one PNN of L. Thus, it is sufficient to show $O_1 \in PNNQ(p)$.

To show $O_1 \in PNNQ(p)$ is equivalent to prove $dist_{min}(p, O_1) < dist_{max}(p, O_i)$ $(O_i \in S)$. Then, it is sufficient to show $dist_{max}(p, O_1) < dist_{max}(p, O_i)(O_i \in S)$, as $dist_{min}(p, O_1) < dist_{max}(p, O_1)$.

Notice that $dist_{max}(p, O_i)$ must be no less than $dist_{max}(L, O_i)$. Then,

$$dist_{max}(p, O_1) = dist_{max}(L, O_1) \leq dist_{max}(L, O_i)(O_i \in S)$$
$$\leq dist_{max}(p, O_i)(O_i \in S)$$

So, O_1 definitely belongs to $TPNNQ(L)$.

Then, L_{cur} is split into 2 (or 3) pieces (or children). For L_{cur}'s children L^i, we derive a pruning bound B_i for L^i and select a subset of candidates from ϕ_{cur} , as shown in *Step 9* to *Step 12*.

Notice that for each leaf-node L^i of the ternary tree $\Psi(L(s, e))$, L^i's two end points must be s, e, or the *turning points* on L. If we traverse Ψ in the pre-order manner, any two successively visited leaf-nodes are the successively connected *validity intervals* in L. Suppose we have m *turning points*, we would have $m + 1$ *validity intervals*, which corresponds to $m + 1$ Ψ's leaf-nodes.

Algorithm 5 stops when any pair of objects in $\phi_{cur}^{[L]}$ does not further split L. The complexity depends on the size of the *turning points* in the final answer. Recall the splitting process of *Ternary Decomposition*, a ternary tree node T_i splits only if one or two intersections are found in T_i's line-segment. If no intersections found in its line-segment, T_i becomes a leaf-node. Given the final answer containing m *turning points*, there would

---

Algorithm 5. TernaryDecomposition

1: **function** TERNARYDECOMPOSITION(Segment $L(s,e)$, Candidates set $\phi_{cur}^{[L]}$)
2:     Sort $\phi_{cur}^{[L]}$ in the ascending of maximum distance to L
3:     **for** $i = 1 \ldots |\phi_{cur}|$ **do**   ▷ consider object O_i
4:         **for** $j = i + 1 \ldots |\phi_{cur}|$ **do**   ▷ consider object O_j
5:             $\mathcal{I} = \texttt{FindIntersection}(L, O_i, O_j)$;
6:             Verify $\mathcal{I}$ and delete unqualified elements;
7:             **if** $|\mathcal{I}| \neq 0$ **then**
8:                 Use $\mathcal{I}$ to split $L(s,e)$ into $|\mathcal{I}| + 1$ pieces
9:                 **for** each piece of line segment L^i **do**
10:                     Use Lemma 3, 4, and 5 to derive pruning bound B_i
11:                     $\phi_{cur}^{[L^i]} \leftarrow B_i(\phi_{cur}^{[L]})$
12:     release $\phi_{cur}^{[L]}$
13:     **for** each piece of line segment L^i **do**
14:         TernaryDecomposition($L^i, \phi_{cur}^{[L^i]}$)

---

be at most $2m$ nodes in the ternary tree $\Psi(\mathcal{T})$. At least, there are $\lceil 1.5m \rceil$ nodes. So, Algorithm 5 will be called $(1.5m, 2m)$ times. Step 5 is done in β and Step 6 is in $O(\log C)$. If the candidate answers returned by *Phase I* contains C objects, the complexity of *Phase II* is $O(mC^2(\log C + \beta))$. Next, we study how to derive the pruning bound B_i mentioned in *Step 11*.

6.2   Pruning Bounds for Three Cases

By a u-bisectors, a query line-segment could be divided into at most 3 sub-line-segments. The sub-line-segments fall into 3 categories according to their positions in half spaces. There are three types of sub-line-segments: *Open Case*, *Pair Case*, and *Close Case*, For example, in Figure 5, $[s_{d \vdash c}, s_{d \dashv c}]$ belongs to the *pair case*. Two examples of *open case* are $[h, s_{d \vdash c}]$ and $[s_{d \dashv c}, t]$. The *Close Case* means the line-segment is totally covered by a half-space. The three cases are formally described in Table 1.

Table 1. Three cases for a line segment

Case	Form	Position
pair	$[s_{i \vdash j}, s_{i \dashv j}]$	$l \in V(i, j)$
open	$[s, s_{i \vdash j}]$ or $[s_{i \dashv j}, e]$	$l \in H_i(j)$ (or $l \in H_j(i)$) ($s(e)$ is the line-segment's start(end) point)
close	$[s_{i \vdash j}, s'_{i \vdash j}]$	$l \in H_i(j)$ and $s_{i \vdash j}, s'_{i \vdash j} \in b_i(j)$

For *Pair Case* and *Open Case*, we can derive two types of pruning bounds. Suppose the u-bisector between O_1 and O_2 split the query line-segment $[s, e]$ into sub-line-segments: $[s, s_{1 \vdash 2}]$, $[s_{1 \vdash 2} s_{1 \dashv 2}]$, and $[s_{1 \dashv 2}, e]$, which are of *Open Case*, *Pair Case*, and *Open Case*, respectively. We shown the pruning bound derived for $[s, s_{1 \vdash 2}]$ and $[s_{1 \vdash 2} s_{1 \dashv 2}]$ in Figure 7 (a) and (b). The bounds are highlighted by shaded areas. The pruning bound of $[s_{1 \dashv 2}, e]$ is similar to Figure 7(a), so it is omitted.

Close Case is a special case, when a line-segment has two intersections and totally inside one half-space, say $H_i(j)$. It could be represented by $[s_{i\vdash j}, s'_{i\vdash j}]$, which means the two end-points are on the same u-bisector half $b_i(j)$. In this example, we known $[s_{i\vdash j}, s'_{i\dashv j}]$ must be in $H_i(j)$, so O_j cannot be the PNN for each point inside. Next, we design their pruning bounds.

Lemma 3. (Pair Case) *Suppose two imprecise objects O_i and O_j, whose u-bisector $b_i(j)$ and $b_j(i)$ intersect with a straight line at $s_{i\vdash j}$ and $s_{i\dashv j}$. For another object $\forall O_N \in O$, it cannot be $q \in [s_{i\vdash j} s_{i\dashv j}]$'s PNN, if O_N has no overlap with the pruning bound $\odot(s_{i\vdash j}, O_i) \cup \odot(s_{i\dashv j}, O_j) \bigcap \odot(s_{i\vdash j}, O_j) \cup \odot(s_{i\dashv j}, O_i)$.*

Proof. $\forall p \in [s_{i\vdash j} s_{i\dashv j}]$, both O_i and O_j have chances to be p's PNN. According to Lemma 1, a new object O_N cannot be O_i or O_j's nearest neighbor if

$$O_N \bigcap (\odot(s_{i\vdash j}, O_i) \cup \odot(s_{i\dashv j}, O_i)) = \emptyset, or O_N \bigcap (\odot(s_{i\vdash j}, O_j) \cup \odot(s_{i\dashv j}, O_j)) = \emptyset$$

So, the pruning bound is:

$$\odot(s_{i\vdash j}, O_i) \cup \odot(s_{i\dashv j}, O_i) \bigcap \odot(s_{i\vdash j}, O_j) \cup \odot(s_{i\dashv j}, O_j) \tag{7}$$

Lemma 4. (Open Case) *Given a line-segment $[s, s_{i\vdash j}]$, for other objects $\forall O_N \in O$, it cannot be query point $q \in [s, s_{i\vdash j}]$'s nearest neighbor, if O_N has no overlap with the $\odot(s, O_i) \cup \odot(s_{i\vdash j}, O_i)$.*

Lemma 5. (Close Case) *Given two split points $s_{i\vdash j}$ and $s'_{i\vdash j}$, the pruning bound for $[s_{i\vdash j}, s'_{i\vdash j}]$ is $\odot(s_{i\vdash j}, O_i) \cup \odot(s'_{i\vdash j}, O_i)$.*

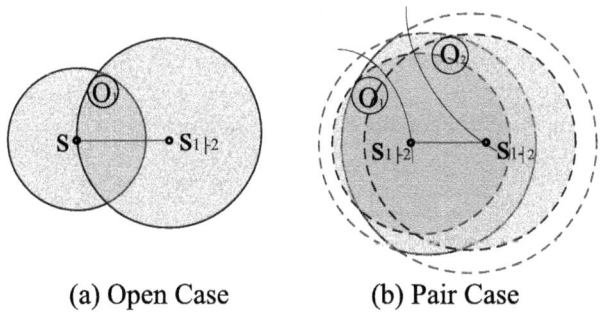

(a) Open Case   (b) Pair Case

Fig. 7. Open Case and Pair Case

Since the proofs of Lemma 5 and Lemma 4 can be easily derived from Lemma 1, they are omitted due to page limitation. The *Pair Case* could also be considered as the overlap of two *Open Cases*. For example, a *Pair Case* $[s_{i\vdash j}, s_{i\dashv j}]$ is equivalent to the overlap part of $[s, s_{i\dashv j}]$ and $[s_{i\vdash j}, e]$. Also, the *Close Case* could be viewed as the overlap of $[s, s'_{i\dashv j}]$ and $[s_{i\vdash j}, e]$. The three cases and their combinations could cover all the cases for each piece(*validity interval*) of the line segment. After Ψ's construction is done, we can view the pruning bound of a *validity interval*. It is the intersection of all its ascender nodes' pruning bounds in the ternary tree Ψ.

7   Experimental Results

Section 7.1 describes settings. We adopt a metric to measure to quality of results in Section 7.2. Section 7.3 discusses the results.

7.1   Setup

Queries. The query trajectories are generated by Brinkhoff's network-based mobile data generator [1]. The trajectory represents movements over the road-network of Oldenburg city in Germany. We normalize them into 10k × 10k space. By default, the length of trajectory is 500 units. Each reported value is the average of 20 trajectory query runs.

Imprecise Objects. We use four real datasets of geographical objects in Germany and US[2], namely *germany*, *LB*, *stream* and *block* with 30k, 50K, 199K, 550k spatial objects, respectively. We use *stream* as the default dataset. We construct the MBC for each object thus get 4 datasets with circular imprecise regions. Datasets are normalized to the same domain as queries. To index imprecise regions, we use a packed R*-tree [16]. The page size of R-tree is set to 4k-byte, and the fanout is 50. The entire R-tree is accommodated in the main memory.

For the *turning points* calculation, we call GSL Library [3] to get the analytical solution. All our programs were implemented in C++ and tested on a Core2 Duo 2.83GHz PC.

7.2   Quality Metric

To measure the accuracy of a query result, we adopt a *Error* function based on the *Jaccard Distance* [17], which is used in comparing the similarity between two sets. Recall the definition of TPNNQ the query result is a set of tuples $\{\langle T_i, R_i \rangle\}$. It can be transformed into the $PNNs$ for every point on the query trajectory. Formally, the result is $\{\langle q, PNNQ(q) \rangle\}_{q \in \mathcal{T}}$. Let $R^*(q)$ be the optimal solution for the point q, where $R^*(q) = PNNQ(q)$. We use $R^A(q)$ to represent the $PNNs$ derived for the point q in algorithm A. Then, the *Error* for algorithm A on query $\mathcal{T}$ is:

$$Error(\mathcal{T}, A) = \frac{1}{|\mathcal{T}|} \int_{q \in \mathcal{T}} 1 - \frac{R^*(q) \cap R^A(q)}{R^*(q) \cup R^A(q)} dq \qquad (8)$$

$|\mathcal{T}|$ is the total length of trajectory $\mathcal{T}$. If $\mathcal{T}$ is represented by a set of line-segments $\mathcal{T} = \{L_i\}_{i=1}^t$, the total length $|\mathcal{T}| = \sum_{i=1}^t |L_i|$.

Equation 8 captures the effect of false positives and false negatives as well. There is a *false positive* when $R^A(q)$ contains an extra item not found in $R^*(q)$. There is a *false negative* when an item of $R^*(q)$ is missing from $R^A(q)$. For a perfect method with no false positives and false negatives, the two terms $R^*(q)$ and $R^A(q)$ are the same, so the integration value is 0.

In summary, the error score is a value between 0 and 1. The smaller an *Error* score is, the more accurate the result is. On the other hand, if a method has many extra or missing results, then it obtains a high *Error*.

---

[1] http://iapg.jade-hs.de/personen/brinkhoff/generator/
[2] http://www.rtreeportal.org/
[3] http://www.gnu.org/software/gsl/

7.3   Performance Evaluation

The query performance is evaluated by two metrics: efficiency and quality. The efficiency is measured by counting the clock time. The quality is measured by the *error score*. We compare four methods: *Basic*, *Sample*, TP-S, and TP-TS. The suffixes T and S refer to *Trajectory Filter* and *Segment Filter*, respectively. *Basic* does not use any filter; TP-S does not use *Trajectory Filter*; TP-TS (Algorithm 2) uses all the filtering and refinement techniques. *Sample* draws a set of uniform sampling points $\{q\}$ from $\mathcal{T}$. Then, for all q, $PNNQ(q)$ is evaluated. The sampling interval, denoted by ϵ, is set to 0.1 unit.[4]

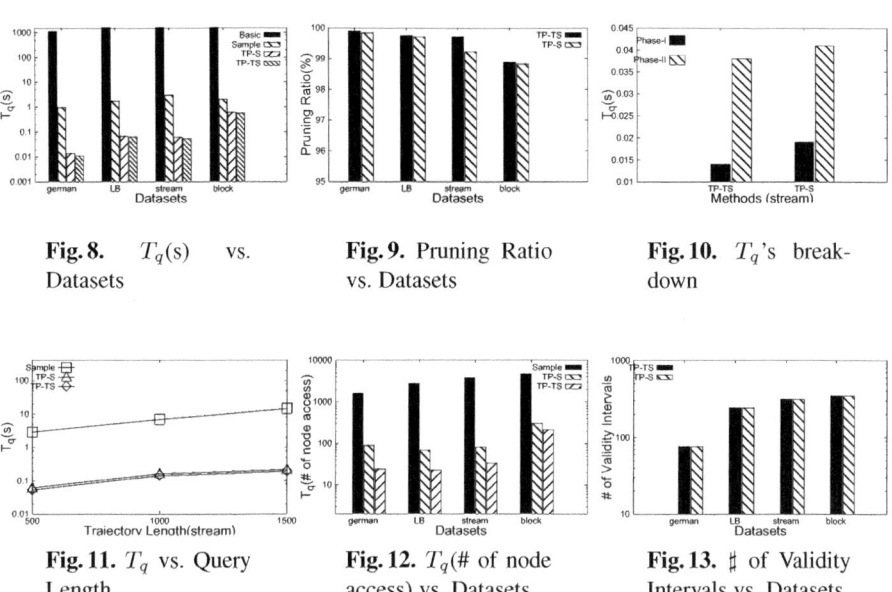

Fig. 8. $T_q(\mathrm{s})$ vs. Datasets

Fig. 9. Pruning Ratio vs. Datasets

Fig. 10. T_q's break-down

Fig. 11. T_q vs. Query Length

Fig. 12. T_q(# of node access) vs. Datasets

Fig. 13. ♯ of Validity Intervals vs. Datasets

Query Efficiency T_q. From Figure 8, the *Basic* method is the slowest method among all the four, since it elaborates all the possible pairs of objects for *turning points* (but most of them do not contribute to validity intervals). For the second slowest *Sample*, we analyze it later.

The other two methods have significant improvement over *Sample* and *Basic*. One reason is because of the effectiveness of the pruning techniques, as shown in Figure 9. For all the real datasets, the pruning ratio are as high as 98.8%. TP-S is less efficient, because some candidates shared by different line-segments in trajectory will be fetched multiple times. This drawback is overcome by TP-TS.

To get a clearer picture about the efficiency of our framework, we measure the time costs for *Phase I* and *Phase II* in Figure 10. TP-TS is faster in both phases. In *Phase I*, the combined R-tree traversal in TP-TS saves plenty of extra node access, compared to

---

[4] The sampling rate is reasonably high regarding to the trajectory's default length. More details about sampling rates are discussed later.

TP-S. The number of node access is shown in Figure 12. In *Phase II*, TP-TS is faster, since it has fewer candidates to handle. This observation is also consistent with the fact that TP-TS has a higher pruning ratio, shown in Figure 9.

We also test the query efficiency by varying the query length in Figure 11. The *Sample* method is slower than others at least one order of magnitude. The costs of other two methods increase slowly w.r.t. the query length.

TP-TS vs. Sample. *Sample* method is a straightforward solution to approximate the $TPNNQ$ answer. However, this solution suffers from the extensive R-tree traversals, since every sampling point q requires accessing of R-tree. As shown in Figure 12, *Sample* incurs at least more than one order of magnitude node access than our method.

On the other hand, *Sample* could incur *false negatives*, even with a large sampling rate. Because *Sample* only considers query points sampled on the trajectory, whereas $TPNNQ$ is for all the points in $\mathcal{T}$. To calculate *Sample*'s error score, we have to infer the $PNNs$ for a point $q \in \mathcal{T}$ not being sampled, as required by Equation 8. With limited sampled answers, q's $PNNs$ can only be "guessed" by using its closest sampling point p. In other words, $PNNQ(q)$ has to be substituted with $PNNQ(p)$.

The efficiency is reflected in Figure 14, where the sampling interval ϵ is varied from 0.01 to 10. We can observe that TP-TS outperforms *Sample* in most of the cases. *Sample* is faster only when ϵ is very large (e.g. equal to 10 units). Then, is it good if large ϵ is used? The answer is **NO**. In Table 15, when "*Sample, $\epsilon = 10$, block*", the error score of *Sample* is as high as 0.443!

We demonstrate the error score of *Sample* and TP-TS in Table 15. Since TP-TS evaluate the exact answer, the error is always 0. The error of *Sample* is small when ϵ is small, (e.g. equal to 0.01, block). However, the query time of that case is 100 times slower than TP-TS. We would like to emphasize that even the *error score* is empirically tested to be 0 over large sampling rates, there is no theoretical guarantee for the *Sample* to contain 0 false negative.

We also test the *error score* of simplifying the imprecise regions into precise points, as mentioned in the introduction. For *german* dataset, the error is as high as 0.76! Thus, the simplified solution could be harmful for applications such as safety sailing.

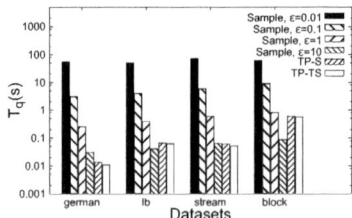

Datasets	Sample				TP-TS
	$\epsilon = 0.01$	$\epsilon = 0.1$	$\epsilon = 1$	$\epsilon = 10$	
german	0.00340	0.00457	0.01528	0.12310	0
LB	0.00005	0.00029	0.00257	0.02672	0
stream	0.00059	0.00090	0.00298	0.03962	0
block	0.01872	0.02541	0.08516	0.44310	0

Fig. 14. TP-TS vs. Sample (T_q)   **Fig. 15.** TP-TS vs. Sample(Error)

Analysis of TPNN. Observed from Figure 13, the number of validity intervals increases with the size of the datasets. TP-S and TP-TS have the same number of validity intervals, which is as expected.

In summary, we have shown that TP-TS is much more efficient than *Basic*, *Sample*, and TP-S methods. It also achieves much better quality than *Sample* method.

8 Conclusion

In this paper, we study the problem of trajectory query over imprecise data. To tackle the low quality and inefficiency in simplified methods, we study the geometric properties of the u-bisector. Based on that, we design several novel filters to support our algorithm. Extensive experiments show that our method can efficiently evaluate the $TPNNQ$ with high quality.

The geometric theories studied in this paper has no limitations in the dimensionality and shape of imprecise regions. In future, we would like to evaluate the algorithm's performance in multi-dimensional space with different shaped imprecise regions. We would also extend our work to support variants queries like k-PNN query, etc.

Acknowledgment. Reynold Cheng and Xike Xie were supported by the Research Grants Council of Hong Kong (GRF Projects 513508, 711309E, 711110). Man Lung Yiu was supported by ICRG grants A-PJ79 and G-U807 from the Hong Kong Polytechnic University. We would like to thank the anonymous reviewers for their insightful comments.

References

1. Tao, Y., Papadias, D., Shen, Q.: Continuous nearest neighbor search. In: VLDB (2002)
2. U. S. C. Guard, Announcement of 2011 international ice patrol services (2011),
   http://www.uscg.mil/lantarea/iip/docs/AOS_2011.pdf
3. Jesse, L., Janet, R., Edward, G., Lee, V.: Effects of habitat on gps collar performance: using data screening to reduce location error. Journal of Applied Ecology (2007)
4. Park, K., Choo, H., Valduriez, P.: A scalable energy-efficient continuous nearest neighbor search in wireless broadcast systems. In: Wireless Networks (2010)
5. Cheng, R., Xie, X., Yiu, M.L., Chen, J., Sun, L.: Uv-diagram: A voronoi diagram for uncertain data. In: ICDE (2010)
6. Lian, X., Chen, L.: Efficient processing of probabilistic reverse nearest neighbor queries over uncertain data. VLDBJ (2009)
7. Cheema, M.A., Lin, X., Wang, W., Zhang, W., Pei, J.: Probabilistic reverse nearest neighbor queries on uncertain data. TKDE (2010)
8. Chen, J., Cheng, R., Mokbel, M., Chow, C.: Scalable processing of snapshot and continuous nearest-neighbor queries over one-dimensional uncertain data. VLDBJ (2009)
9. Trajcevski, G., Tamassia, R., Ding, H., Scheuermann, P., Cruz, I.F.: Continuous probabilistic nearest-neighbor queries for uncertain trajectories. In: EDBT, pp. 874–885 (2009)
10. Zheng, K., Fung, G.P.C., Zhou, X.: K-nearest neighbor search for fuzzy objects. In: SIGMOD (2010)
11. Song, Z., Roussopoulos, N.: K-Nearest Neighbor Search for Moving Query Point. In: Jensen, C.S., Schneider, M., Seeger, B., Tsotras, V.J. (eds.) SSTD 2001. LNCS, vol. 2121, pp. 79–96. Springer, Heidelberg (2001)
12. Zheng, B., Lee, D.-L.: Semantic Caching in Location-Dependent Query Processing. In: Jensen, C.S., Schneider, M., Seeger, B., Tsotras, V.J. (eds.) SSTD 2001. LNCS, vol. 2121, pp. 97–113. Springer, Heidelberg (2001)
13. Zhang, J., Zhu, M., Papadias, D., Tao, Y., Lee, D.L.: Location-based spatial queries. In: SIGMOD (2003)

14. Cheng, R., Kalashnikov, D.V., Prabhakar, S.: Querying imprecise data in moving object environments. TKDE 16(9) (2004)
15. Boyd, S., Vandenberghe, L.: Convex Optimization. Cambridge University Press (2004)
16. Hadjieleftheriou, M.: Spatial index library version 0.44.2b,
    http://u-foria.org/marioh/spatialindex/index.html
17. Tan, P.-N., Steinbach, M., Kumar, V.: Introduction to data mining (2006)

A Appendix

A.1 Intersection of a Hyperbola and a Straight Line

Given a hyperbola h_1 and a straight line l_1, they could have 0, 1, or 2 intersection points, which is the roots of the following:

$$\begin{cases} h_1 : \frac{x^2}{a_1^2} - \frac{y^2}{b_1^2} = 1 \\ l_1 : a_2 x + b_2 y + c_2 = 0 \end{cases} \tag{9}$$

By solving Equation 9, we can have:

$$\begin{cases} x = \frac{-a_1^2 a_2 c_2 \pm \sqrt{-a_1^2 b_1^2 b_2^2 (a_1^2 a_2^2 - b_1^2 b_2^2 - c_2^2)}}{a_1^2 a_2^2 - b_1^2 b_2^2} \\ y = \frac{-a_2 \pm \sqrt{a_1^2 b_1^2 b_2^2 (-a_1^2 a_2^2 + b_1^2 b_2^2 + c_2^2)} - b_1^2 b_2^2 c_2}{b_2 (b_1^2 b_2^2 - a_1^2 a_2^2)} \end{cases} \tag{10}$$

where

$$\begin{cases} a_1^2 a_2^2 - b_1^2 b_2^2 \neq 0 \\ b_2 \neq 0 \\ a_1 b_1 \neq 0 \end{cases} \tag{11}$$

Notice that if any of the three pre-conditions in Equation 11 is not satisfied, there should be no intersection point for the given curve and line.

Efficient Range Queries over Uncertain Strings

Dongbo Dai[1], Jiang Xie[1,3], Huiran Zhang[1], and Jiaqi Dong[2]

[1] School of Computer Engineering and Science, Shanghai University, Shanghai, China
{dbdai,jiangx,hrzhangsh}@shu.edu.cn
[2] School of Computer Science, Fudan University, Shanghai, China
dong_jiaqi@fudan.edu.cn
[3] Department of Mathematics, University of California Irvine, CA, USA

Abstract. Edit distance based string range query is used extensively in the data integration, keyword search, biological function prediction and many others. In the presence of uncertainty, however, answering range queries is more challenging than those in deterministic scenarios since there are exponentially many possible worlds to be considered. This work extends existing filtering techniques tailored for deterministic strings to uncertain settings. We first design probabilistic q-gram filtering method that can work both efficiently and effectively. Another filtering technique, frequency distance based filtering, is also adapted to work with uncertain strings. To achieve further speed-up, we combined two state-of-the-art approaches based on cumulative distribution functions and local perturbation to improve lower bounds and upper bounds. Comprehensive experiment results show that our filter-based scheme, in the uncertain settings, is more efficient than existing methods only leveraging cumulative distribution functions or local perturbation.

Keywords: uncertain strings, range query, filtering.

1 Introduction

String data is ubiquitous in many important applications, such as customer information management, text-rich information systems, and bioinformatics [1,2,3,4]. Range query is an essential operation on string data. Given a set D of strings, a query string t, a distance function $dist(\cdot, \cdot)$ which measures the similarity between two strings, and a similarity threshold δ, the **Range query** returns all strings s from D such that $dist(s, t) \leq \delta$. Range queries have extensive applications, such as data cleaning, spell checking, plagiarism identification and bioinformatics [5,6,7,8,9].

Edit distance (or Levenshtein distance) is a widely used similarity measure for strings. The edit distance between two strings s_1 and s_2, denoted by $ed(s_1, s_2)$, is the minimum number of single-character edit operations (insertion, deletion, and substitution) that are needed to transform s_1 to s_2. Edit distance $ed(s_1, s_2)$ can be computed in $O(|s_1| \cdot |s_2|)$ time and $O(\min\{|s_1|, |s_2|\})$ space using dynamic programming [1]. In this paper, we focus on edit distance based range query.

In many cases, the given string data often contains errors or may be incomplete, due to entry typos, inaccurate information extraction from unstructured documents, or inherent limitations of the high-throughput sequencing. For example, in DNA-sequencing, it

A. Ailamaki and S. Bowers (Eds.): SSDBM 2012, LNCS 7338, pp. 75–95, 2012.

is common that scientists have identified the values (one of letters in {'A','G','C','T'}) for 90% positions in a genome sequence, while there is some degree of uncertainty for the other 10%. In this case, traditional range query performed on any one of all possible strings can not capture the genuine similarity, hence it is necessary to extend this query semantics to uncertain settings.

Uncertainty Model. There are two commonly used ways of modeling such uncertain information hidden in a string s [10]: *string-level model* and *character-level model*. Let Σ be an alphabet. In the string-level model, s has n choices to instantiate into different strings (termed as *possible worlds*) with corresponding probabilities, i.e., $s = \{(s_1, p_1), \cdots, (s_n, p_n)\}$, where $s_i \in \Sigma^*$, the associated probabilities $p_1, \cdots, p_n$ satisfying the condition: $\sum_{i=1}^{n} p_i = 1$. While the character-level model is to represent uncertain string via its uncertain positions. Specifically, for $s = s[1] \cdots s[m]$, any uncertain position $s[i] = \{(\tau_{i,1}, p_{i,1}), \cdots, (\tau_{i,m_i}, p_{i,m_i})\}$, where each character $\tau_{i,j}$ from Σ appears at position $s[i]$ with probability $p_{i,j}$ independently and $\sum_{j=1}^{m_i} p_{i,j} = 1$. String-level model may pose highly redundant information when the string has few uncertain positions, which is often the case in many applications such as bioinformatics and text-rich information systems. Thus, as a more realistic model, character-level model can concisely represent the uncertainty in strings and will be used as the working model in this paper.

Two questions should be considered in case of uncertainty which makes the query answering much complicated. One is how to define reasonable query semantics when taking probabilities into account and the other is how to answer range queries efficiently. For the first question, it may take a probability for an uncertain string to be within the given range of query string. Consequently, a range query can be formulated in multiple ways. For instance, we can output only those strings which are absolutely (i.e., with probability 100%) falling into the query range. Alternatively, we can also output those strings which have a non-zero probability to fall into the query range. Jestes et al. [10] proposed the notion of *Expected Edit Distance*(EED) over all possible worlds of two uncertain strings to perform similarity joins. The value of *EED* is computed as: $\sum_{(s_1,s_2)} p_1 \cdot p_2 \cdot ed(s_1, s_2)$, where s_1 and s_2 are two instances from two uncertain strings with the probabilities p_1 and p_2, respectively. As a similarity measure, EED appears to be natural and intuitive at first sight. However, EED is a biased measure as the following simple example shows:

query string: $t = abc$

uncertain strings: $s_1 = \{(abd, 0.45), (abe, 0.45), (s', 0.1)\}$

$s_2 = \{(uvw, 0.5), (xyz, 0.5)\}$

s_1 and s_2 are two uncertain strings in the string-level model and a possible world s' in s_1 has sufficiently large distance to the query string t. With quite large possibility value 0.9, s_1 has two possible worlds abd and abe with small edit distance 1 to t, while s_2 is totally dissimilar to t. However, if we take EED as a similarity measure, s_1 may be less similar to t than s_2 because of the enough large distance between t and s'. This example shows that EED is inappropriate for range query in the string-level model.

Ge et al. [11] further pointed out that, for substring matching over uncertain strings in the character-level model, EED may either miss real matches or has an unduly big threshold so that many false positives may mix in. To remedy this problem, the authors in [11] proposed (δ, γ)-*matching semantics*: given a pattern string t, a set D of uncertain text strings $\{s_i\}$ ($1 \le i \le |D|$) and threshold parameters δ, γ, asking for all substrings s' of s_i's such that $Pr[ed(t, s') \le \delta] > \gamma$. The semantics of (δ, γ)-matching has been justified in [11], so we also apply a similar query semantics to our range query, i.e., we extend conventional query answering on deterministic strings to uncertain threshold queries ,i.e., only those strings whose probability falling into the query range passing the threshold are returned. In the following, we use "uncertain string " for brevity to mean that it is generated by the character-level model.

Problem Statement: Given a set of uncertain strings D, a deterministic query string t, and two parameters (δ, γ) ($\gamma > 0$), a (δ, γ)-**range query** returns all strings s from D such that $Pr(ed(s, t) \le \delta) \ge \gamma$.

Note that, as illustrated in [11], we only consider deterministic query strings as it is more common in real applications.

Another fundamental challenge is how to answer range queries efficiently. It is well known that in a large collection of deterministic strings, direct edit distance computation is an expensive operation, so various filters have been designed to speed up the query processing [12,13,14,15,16,17]. For a large set of uncertain strings, even a single uncertain string may correspond to as many as exponentially possible instances. This huge size implies a prohibitively large processing cost if we enumerate all possible instances and perform range queries one by one. Thus, efficient and effective filters are urgently needed to avoid so many edit distance calculations.

In [11], to efficiently evaluate the (δ, γ)-matching queries, the authors first proposed a multilevel signature filtering structure to locate a set of positions in the text strings to be verified. Then, during the verification phase, they presented two algorithms: *Cumulative Distribution Function*(CDF) and *local perturbation* to give upper bound and lower bound of the probability of all candidate substring s' satisfying $ed(t, s') \le \delta$ for a given pattern t. Although we take the similar (δ, γ)-range query semantics as discussed in [11], the following two issues pose grand challenges in evaluating range queries over uncertain strings:

1. Although filtering techniques based on CDF and local perturbation presented in [11] can be applied to our range queries in a straightforward way, the reduction in time cost is limited and needed to be enhanced further. The reason is that CDF and local perturbation are not "light" filters in terms of time cost. Computing bounds based on CDF will incur $O(\delta^2 \times |t|)$ time complexity in order to bound the cumulative distribution functions at each cell of dynamic programming(DP) table. For large error rate δ of $O(|t|)$, the complexity is unfavorably cubic in the size of $|t|$. For local perturbation method, a complete DP algorithm computation is needed to obtain an initial *adjacent possible world*(the string s satisfying $ed(s, t) \le \delta$). Moreover, to get an initial *remote world* (the string s satisfying $ed(s, t) > \delta$), randomized testing algorithm is used, which needs several times of DP algorithm computation. Note that all of these computations are performed for every single uncertain string. Thus, for a large collection of uncertain strings, only employing these two filtering methods will incur huge time overhead from the realistic perspective. We need some "light" filtering approaches to avoid expensive filtering cost.

2. Since enumerating all possible worlds and performing range queries one by one are very time-consuming, we should reduce the number of candidate strings to be

Table 1. The symbols frequently used in the paper.

Symbol	Meaning
s_1, s_2, s, t	different strings
D	a set of strings
δ, γ	edit distance threshold and probability threshold in range queries
q	the length of q-gram
g, g', g''	different q-grams
$ed(\cdot, \cdot)$	edit distance function
$f(\cdot), FD(\cdot, \cdot)$	frequency vector and frequency distance

verified as possible as we can. Filtering only based on CDF or local perturbation is not sufficient for efficiently performing range queries, so we need to design diverse filtering schemes to achieve high efficiency.

To overcome the above challenges, we propose a novel filter-based algorithm RQUS (for Range Queries over Uncertain Strings) and make the following contributions:

- We propose a novel problem: (δ, γ)-range queries over uncertain strings. To the best of our knowledge, this problem has not been explored systematically before.
- We propose two effective filtering methods, probabilistic q-gram based filtering and frequency distance filtering, which can bound the answering probability from above and can also be implemented efficiently compared to CDF and local perturbation techniques.
- After above filtering, we subtly combine CDF and local perturbation to improve lower bound and upper bound, and use the combined filtering method to further prune away the unnecessary candidates.
- We conduct comprehensive experiments which demonstrate that, in the situations of uncertain string queries, RQUS is more efficient than the baseline algorithm only leveraging CDF or local perturbation techniques.

The rest of paper is organized as follows. Section 2 discusses the related work. Section 3 introduces the background and some commonly used filtering methods for deterministic strings. Our proposed two filtering methods and combined pruning technique are developed in section 4 and section 5, respectively. Experimental results are presented in section 6 and section 7 concludes the paper.

We summarize the symbols frequently used in the paper in Table 1.

2 Related Work

String range query is an important research issue and has been studied extensively in algorithm and database communities. In the literature, *approximate string search (queries)* [16,17], *string selection queries* [14] refer to the problem of range queries over deterministic strings.

Many algorithms have been proposed for answering string range queries efficiently. Their main strategies are to use various filtering techniques to improve the performance. Kahveci et al. [12] mapped the substrings of the data into an integer space and defined a distance function called frequency distance which is a lower bound to the actual edit distance between strings. Based on this, they built an index structure using MBRs(Minimum Bound Rectangles) to efficiently filter out non-result strings. Based on the intuition that strings with a small edit distance will share large number of common q-grams, most algorithms employed fixed or variable length q-gram filtering schemes to speed up the processing [18,16,17,5].

Since range queries are closely related to the similarity joins, some filtering techniques designed in the context of similarity joins can also be applied to the range queries such as count filtering [5,19], position filtering [5,20], length filtering [5,19], mismatching q-gram based filtering and prefix filtering [19] etc.. Recently, some extensions to traditional query type have been studied. Zhang et al. [21] developed a novel all-purpose index B^{ed}-tree that can support range queries, top-k queries and similarity joins simultbaneously. Unlike many existing algorithms with in-memory indexes, Behm et al. [22] explored external memory algorithm for range queries by contriving a new storage layout for the inverted index. Combining approximate string queries and spatial range queries have also been proposed [23]. Note that all of these algorithms focused on deterministic strings.

Recently, there have been great interests in the query processing and mining over uncertain data(see [24,25] for excellent surveys on research related to these problems). However, most of these works mainly focus on range or top-k queries over numerical data, or similarity joins over spatial datasets. To the best of our knowledge, no work has been done in the area of range queries over uncertain strings. [10] and [11] are two articles closely related to our work in this paper. Jestes et al. defined a new similarity measure EED and designed q-gram based filtering techniques in relational databases to perform similarity joins over uncertain strings. Due to aforementioned deficiencies of EED, Ge et al. [11] proposed (δ, γ)-matching semantics and applied this to approximate substring matching over uncertain strings. As we adopt (δ, γ)-range query semantics instead of EED, the q-gram based filters in [10] are of no use to our problem. In fact, in the settings of (δ, γ)-range qeury semantics, how to extend various filters on deterministic strings to uncertain settings is a non-trivial task.

3   Background on Q-grams and Frequency Distance

Let $\Sigma = \{\tau_i\}$ $(1 \le i \le |\Sigma|)$ be a finite alphabet of symbols. A string s is an ordered list of symbols drawn from Σ. The length of string s and the i-th element in s are denoted as $|s|$ and $s[i]$, respectively.

Definition 1 (positional q-gram). *Given a string s, a q-gram g is a contiguous substring of length q and all its q-grams can be obtained by sliding a window of length q over its constituent characters. The starting position of g in s is called its **position**. A **positional q-gram** is a q-gram together with its position, usually represented in the form of (g, pos). We use G_s to denote the set of all positional q-grams in s.*

Definition 2. *[possible worlds] Given an uncertain string s in the character-level model, the possible worlds Ω_s of s is the set of all possible instances from s.*

Since q-grams have fewer than q characters from Σ at the beginning and end of the string s, we introduce new characters '#' and '*' not in Σ to prefix it with $(q-1)$ '#' and suffix it with $(q-1)$ '*'. For simplicity, when there is no ambiguity, we use "positional q-gram" and "q-gram" interchangeably.

For two q-grams (g_1, p_1) and (g_2, p_2), we define $g_1 =_k g_2$ if $g_1 = g_2$ and $|p_1 - p_2| \leq k$, i.e., two q-grams are identical and their position value differ at most k. We further define $G_{s_1} \cap_k G_{s_2} = \{(g_1, g_2) \mid g_1 =_k g_2, g_1 \in G_{s_1}, g_2 \in G_{s_2}\}$

Clearly, there are $|s| + q - 1$ q-grams in the string s. Based on the observation that a single-character edit operation destroys at most q q-grams, the following lemma proposed by [5,26] captures this intuition that a few edit operations can only have limited destructive effect on a string.

Lemma 1. *([5,26]) Given two strings s_1 and s_2, if their edit distance is within δ, i.e., $ed(s_1, s_2) \leq \delta$, then:*

$$|G_{s_1} \cap_\delta G_{s_2}| \geq max(|s_1|, |s_2|) - 1 - q(\delta - 1)$$

The frequency distance is first proposed in [12]. Let s be a string from the alphabet $\Sigma = \{\tau_1, \tau_2, \cdots, \tau_{|\Sigma|}\}$. Let n_i be the number of occurrences of the character τ_i in string s for $1 \leq i \leq |\Sigma|$. The **frequency vector** $f(s)$ of s is defined as $f(s) = [n_1, n_2, \cdots, n_{|\Sigma|}]$.

By analyzing the relationship between the edit operations and the frequency vectors, Kahveci and Singh [12] present an efficient algorithm to compute the frequency distance $FD(f(s_1), f(s_2))$ between two strings s_1 and s_2. For each dimension τ_i, let $f(s_1)_i$ and $f(s_2)_i$ be the values of $f(s_1)$ and $f(s_2)$ on τ_i, respectively. We compute

$$posDistance = \sum_{\tau_i, f(s_1)_i > f(s_2)_i} (f(s_1)_i - f(s_2)_i) \qquad negDistance = \sum_{\tau_i, f(s_1)_i < f(s_2)_i} (f(s_2)_i - f(s_1)_i).$$

Then, the **frequency distance**

$$FD(f(s_1), f(s_2)) = max\{posDistance, negDistance\}.$$

It was proved that $FD(f(s_1), f(s_2))$ is a lower bound of the edit distance $ed(s_1, s_2)$:

Lemma 2. *[12] Let s_1 and s_2 be two strings from alphabet Σ, then:*

$$FD(f(s_1), f(s_2)) \leq ed(s_1, s_2)$$

Although Lemma 1 and Lemma 2 are exploited extensively to design filters in deterministic string queries, they can't directly used for uncertain strings due to uncertainty. We discuss their variant for processing uncertain string queries in the next section.

Finally, we introduce a length filtering technique:

Lemma 3. *([5,26]) For two strings s_1 and s_2, we have $ed(s_1, s_2) \geq ||s_1| - |s_2||$.*

Length filtering is used in many deterministic string queries. It can also be applied straightforwardly to uncertain strings since the length of an uncertain string is independent of its uncertainty.

4   Pruning Techniques

4.1   Probabilistic Q-gram based Filtering

For a q-gram in an uncertain string s, it may still contain uncertain characters, and we can consider it as an uncertain substring and call it *probabilistic q-gram*. Thus, in the (δ, γ)-range query, we have

$$Pr(g =_\delta g') = \sum_{g'' \in \Omega_g} Pr(g'') \cdot I_{g''=_\delta g'} \tag{1}$$

where g is a probabilistic q-gram from s and g' is a deterministic q-gram from the query string t; I_b is an indicator function, i.e., $I_b = 1$ if b is true, $I_b = 0$ otherwise.

For every uncertain string from D, a naive method by using Lemma 1 for filtering is to enumerate all its possible strings and check whether the condition stated is satisfied. Clearly, this approach is very expensive, since there are exponentially many possible worlds to be checked. Thus, we propose an uncertain version of Lemma 1 to circumvent this problem, which can perform filtering more efficiently.

Theorem 1. *In the (δ, γ)-range query, for any uncertain string s from D and query string t, let $Z = max(|s|, |t|) - 1 - q(\delta - 1)$, if $(\sum_{\substack{g \in G_s \\ g' \in G_t}} Pr(g =_\delta g')) < Z$, we have:*

$$Pr(ed(s,t) \leq \delta) < \frac{1}{Z} \sum_{\substack{g \in G_s \\ g' \in G_t}} Pr(g =_\delta g')$$

Proof. By Definition 2, let pw be any possible world from Ω_s, i.e., $pw \in \Omega_s$ and in pw, g'' be any instance of a probabilistic q-gram g in s. We first calculate the expected value of $|G_s \cap_\delta G_t|$ as follows:

$$E(|G_s \cap_\delta G_t|) = \sum_{pw \in \Omega_s} Pr(pw) \cdot |G_{pw} \cap_\delta G_t| = \sum_{pw \in \Omega_s} Pr(pw) \sum_{\substack{g \in G_{pw} \\ g' \in G_t}} I_{g=_\delta g'} = \sum_{pw \in \Omega_s} \sum_{\substack{g \in G_{pw} \\ g' \in G_t}} Pr(pw) \cdot I_{g=_\delta g'}$$

$$= \sum_{\substack{g \in G_t \\ g' \in G_t}} \sum_{pw \in \Omega_s} Pr(pw) \cdot I_{g''=_\delta g'} = \sum_{\substack{g \in G_s \\ g' \in G_t}} \sum_{g'' \in \Omega_g} Pr(g'') \cdot I_{g''=_\delta g'} = \sum_{\substack{g \in G_s \\ g' \in G_t}} Pr(g =_\delta g')$$

Note that the last step use the equation (1).
Thus, by Markov Inequality, when $(\sum_{\substack{g \in G_s \\ g' \in G_t}} Pr(g =_\delta g')) < Z$ we have:

$$Pr(|G_s \cap_\delta G_t| \geq Z) \leq \frac{1}{Z} E(|G_s \cap_\delta G_t|) = \frac{1}{Z} \sum_{\substack{g \in G_s \\ g' \in G_t}} Pr(g =_\delta g')$$

i.e.,

$$1 - \frac{1}{Z} \sum_{\substack{g \in G_s \\ g' \in G_t}} Pr(g =_\delta g') \leq Pr(|G_s \cap_\delta G_t| < Z) \tag{2}$$

Further, based on Lemma 1, it follows that for any possible world $pw \in \Omega_s$, if $|G_{pw} \cap_\delta G_t| < Z$, then $ed(pw, t) > \delta$. So,

$$Pr(|G_s \cap_\delta G_t| < Z) \le Pr(ed(s, t) > \delta) \qquad (3)$$

From (2)and (3), we have:

$$Pr(ed(s, t) \le \delta) \le \frac{1}{Z} \sum_{\substack{g \in G_s \\ g' \in G_t}} Pr(g =_\delta g')$$

$\square$

Theorem 1 provides an upper bound on the probability of those strings falling into query range. By comparing it with parameter γ, we can safely discard those candidates if their upper bound is less than γ. The upper bound in theorem 1 can be computed efficiently. First, for a given string s, the value of Z can be obtained in a constant time since the length of s is independent of its uncertainty. Second, we can also get the value of $\sum_{\substack{g \in G_s \\ g' \in G_t}} Pr(g =_\delta g')$ efficiently in $O(\delta \cdot |t|)$ time.

Example 1. Consider a query string $t = TGCTA$ and an uncertain string $s = A\{(G, 0.8), (C, 0.2)\}\{(A, 0.9), (C, 0.1)\}\{(T, 0.7), (G, 0.3)\}C$ from D. The parameters are set as $\delta = 2$ and $q = 3$. Then based on Theorem 1, the upper bound on probability is 0.056, the same as that given by CDF method, while local perturbation method outcomes the result of 0.902. This example demonstrates that, with less time cost, probabilistic q-gram based filtering works as well as CDF but provides tighter bound than local perturbation method.

4.2 Frequency-Distance Based Pruning

Another pruning technique is to apply frequency distance to uncertain strings. With the increase of uncertain positions appearing in a string, its frequency distance has great variation. As in the proof of theorem 1, we compute the expected value of all possible frequency distances, and correlate it with the probability of the corresponding edit distance falling into the query range.

First, from the definition of *posDistance* and *negDistance* in the Section 3, it is easy to derive the following facts for two strings s_1 and s_2:

$$|posDistance - negDistance| = \sum_{\tau_i} [f(s_1)_i - f(s_2)_i] = ||s_1| - |s_2|| \qquad (4)$$

$$posDistance \le |s_1|, \quad negDistance \le |s_2| \qquad (5)$$

Next, we introduce a well-known inequality: *One-Sided Chebyshev Inequality*, which will be useful for calculating upper bound on the probability of those strings falling into the query range.

One-Sided Chebyshev Inequality: Let X be a random variable with mean $E(X) = \mu$ and variance $Var(X) = \sigma^2$. For any positive number $a > 0$, the following inequality hold:

$$Pr(X \le \mu - a) \le \frac{\sigma^2}{\sigma^2 + a^2} \qquad (6)$$

Equipped with these facts and inequality, we can derive the following theorem:

Theorem 2. *Let $\Sigma = \{\tau_i\}$ $(1 \le i \le |\Sigma|)$ be a alphabet. In the (δ, γ)-range query, for any uncertain string s from D and query string t, let frequency vectors of s and t be $f(s)$ and $f(t)$, respectively, and $R_1 = \sum_{\tau_i, f(s)_i > f(t)_i} [f(s)_i - f(t)_i]$, $R_2 = \sum_{\tau_i, f(s)_i < f(t)_i} [f(t)_i - f(s)_i]$, $A = \frac{1}{2}[||t| - |s|| + E(R_1) + E(R_2)]$, $B^2 = \frac{1}{2}(|t| - |s|)^2 + \frac{1}{2}||t| - |s|| \cdot [E(R_1) + E(R_2)] + min[|s| \cdot E(R_2), |t| \cdot E(R_1)] - A^2$, if $\delta < A$, then:*

$$Pr[ed(t, s) < \delta] \le \frac{B^2}{B^2 + (A - \delta)^2}$$

Proof. By the definition of frequency distance, the expected value of $FD(s, t))$ is:

$$E[FD(f(s), f(t))] = E[max(R_1, R_2)] = E[\frac{R_1 + R_2 + |R_1 - R_2|}{2}] \quad (7)$$

Since R_1 and R_2 correspond to posDistance and negDistance respectively, by equation (4), the RHS of (7) equals to:

$$E[\frac{R_1 + R_2 + ||t| - |s||}{2}] = \frac{||t| - |s||}{2} + \frac{1}{2}[E(R_1) + E(R_2)] = A \quad (8)$$

by the linearity of expectation, we have

$$E[R_1] = \sum_{\tau_i, f(s)_i > f(t)_i} E(f(s)_i - f(t)_i), \quad E[R_2] = \sum_{\tau_i, f(s)_i < f(t)_i} E(f(t)_i - f(s)_i) \quad (9)$$

Next, we compute the variance of $FD(f(s), f(t))$:

$$Var(FD(f(s), f(t))) = E[(FD(f(s), f(t)))^2] - A^2 = E[(\frac{R_1 + R_2 + ||t| - |s||}{2})^2] - A^2$$

$$= \frac{1}{4}E[(R_1 + R_2)^2 + 2||t| - |s||(R_1 + R_2) + (||t| - |s||)^2] - A^2$$

$$= \frac{1}{4}E[|R_1 - R_2|^2 + 4R_1R_2 + 2||t| - |s||(R_1 + R_2) + (||t| - |s||)^2] - A^2 \quad (10)$$

by (4) and (5), the RHS of (10) equals to:

$$\frac{1}{2}(||t| - |s||)^2 + \frac{1}{2}||t| - |s||[E(R_1) + E(R_2)] + E(R_1R_2) - A^2$$

$$\le \frac{1}{2}(|t| - |s|)^2 + \frac{1}{2}||t| - |s||[E(R_1) + E(R_2)] + min[|s| \cdot E(R_2), |t| \cdot E(R_1)] - A^2 = B^2$$

By Lemma 2, if $ed(s, t) \le \delta$, then $FD(f(s), f(t)) \le \delta$, so if $\delta < A$, we have:

$$Pr[ed(s, t) \le \delta] \le Pr[FD(f(s), f(t)) \le \delta] = Pr[FD(f(s), f(t)) \le A - (A - \delta)] \le \frac{B^2}{B^2 + (A - \delta)^2},$$

where the last step follows from the One-Sided Chebyshev Inequality and from the monotone increase of the function $\frac{x}{x+a}$. □

Like probabilistic q-gram based filtering, Theorem 2 provides another upper-bound approach for pruning out unnecessary candidates: if the probability upper-bound $\frac{B^2}{B^2+(A-\delta)^2}$ is less than γ, the corresponding uncertain string s can be filtered out immediately. The remaining issue is how to obtain the value of A and B^2 efficiently. In the proof of theorem 2, we observe that A and B^2 are functions of $E(R_1)$ and $E(R_2)$. From (9), it is very important for us to efficiently calculate $E_{f(s)_i>f(t)_i}(f(s)_i - f(t)_i)$ and $E_{f(s)_i<f(t)_i}(f(t)_i - f(s)_i)$ for each character τ_i.

For each character τ_i from the alphabet Σ, let the maximum possible number of its occurrences from all the uncertain positions in s be M_i, and assume that the number of its occurrences with probability 1 is N_i. We use $Pr_s(\tau_i, j)$ to denote the probability that τ_i appears exactly j number of times from uncertain positions in s, then

$$\begin{cases} E_{f(s)_i>f(t)_i}(f(s)_i - f(t)_i) = \sum_{m=f(t)_i-N_i+1}^{M_i} Pr_s(\tau_i, m) \cdot (m + N_i - f(t)_i), \ if (f(t)_i - N_i + 1 \le M_i) \\ E_{f(s)_i>f(t)_i}(f(s)_i - f(t)_i) = 0, \qquad\qquad\qquad\qquad\qquad otherwise. \end{cases}$$

(11)

Note that in (11), $f(t)_i$ is a constant for a deterministic query string t. $E_{f(s)_i<f(t)_i}(f(t)_i - f(s)_i)$ can be calculated in the same way.

To evaluate $Pr_s(\tau_i, j)$, we reduce it to the following problem: *Suppose we are given r events $E_i(i = 1, \cdots, r)$, and their existence probabilities are known: $Pr[E_i] = p_i$. What is the probability that at least m events (among those r events) happen?* In this paper, we call this problem as At Least m out of r Events Happen (ALEH). This problem can be solved in time $O(r \cdot m)$ using dynamic programming method [11]. The basic idea is to first define the probability that within the first i events, at least j of them happen as $Pr(i, j)$. Thus, $Pr(r, m)$ is the answer of the ALEH problem. Then, a dynamic programming algorithm is used based on the following recursive equation:

$$Pr(i, j) = p_i \cdot Pr(i - 1, j - 1) + (1 - p_i) \cdot Pr(i - 1, j)$$

Correspondingly, if we use the notation $Pr_s^{\tau_i}(k, j)$ to denote the probability that within the first k character τ_i, at least j of them happen in s. Using dynamic programming approach, we can get the value of $Pr_s^{\tau_i}(M_i, j)$ ($j = 0, 1, \cdots, M_i$) in the last line of dynamic programming table. With these values, we have

$$Pr_s(\tau_i, j) = Pr_s^{\tau_i}(M_i, j) - Pr_s^{\tau_i}(M_i, j + 1)(j = 0, 1, \cdots, M_i - 1), \qquad Pr_s(\tau_i, M_i) = Pr_s^{\tau_i}(M_i, M_i)$$

Clearly, to perform frequency distance pruning, the running time is dominated by the computation of all $E_{f(s)_i>f(t)_i}(f(s)_i - f(t)_i)$ and $E_{f(s)_i<f(t)_i}(f(t)_i - f(s)_i)$, which in turn is mainly determined by all $Pr_s(\tau_i, m)$'s from (11). Note that the computation of $Pr_s(\tau_i, m)$ is independent of any query string t. Thus, to avoid repeated computation of $Pr_s(\tau_i, m)$ for different query strings, we adopt a trade-space-for-time strategy. It is that we store only the last line of dynamic programming table for every possible character τ_i appearing at all uncertain positions:

$$[Pr_s^{\tau_i}(M_i, 1), Pr_s^{\tau_i}(M_i, 2), \cdots, Pr_s^{\tau_i}(M_i, M_i)].$$

With this stored information , we can obtain $E(R_1)$ and $E(R_2)$ easily by just looking up the corresponding entries therein. In general, we compute the required information by preprocessing D and perform queries many times; thus query performance is of primary concern. If we assume that the size of the alphabet is constant and let the fraction of uncertain character in the strings be θ, then by frequency distance filtering for every s, both the time cost and space cost are $O(\theta \cdot |s|)$. In all practical applications, θ is typically small. Therefore, we can execute frequency distance filtering efficiently with small storage.

Example 2. If we take a given query string $t = TCCTT$, an uncertain string $s = A\{(A, 0.9), (C, 0.1)\}$
$\{(A, 0.9), (C, 0.1)\}\{(T, 0.1), (A, 0.9)\}G$ and the distance threshold $\delta = 2$, then according to Theorem 2, we obtain the upper bound on probability equal to 0.162, much smaller than that (0.991) given by local perturbation filtering. Though CDF method can provide tighter bound 0.001, we can still prune away s quickly under the practical probability constraints, say $\gamma = 0.2$.

5   Combined Pruning

For those uncertain strings that have survived the "light" filters in Section 4, we can exploit two state-of-the-art filtering techniques: cumulative distribution function (CDF) and local perturbation [11] to perform further pruning. In substring matching task [11], CDF and local perturbation have been used independent of each other for verification. However, by our observation, these two techniques can be combined naturally in (δ, γ)-range query to improve the upper bound and lower bound, which in turn can reduce the number of candidates for final verification.

Key idea for CDF CDF method changes a standard dynamic programming(DP) algorithm for two deterministic strings to accommodate uncertain characters. Specifically, for a query string t and a candidate uncertain string s, CDF is to compute (at most) $\delta + 1$ pairs of values in each cell of DP table, i.e., $\{(F_l[j], F_u[j]) \mid 0 \le j \le \delta\}$, where $F_l[j]$ and $F_u[j]$ are the lower and upper bounds of $Pr(ED \le j)$ at the cell, respectively, where ED is the edit distance at the cell.

To fill in the DP table, the following formula give methods on how to get the cumulative probability bounds of a cell $D_0 = \{(F_l[j], F_u[j]) | 0 \le j \le \delta\}$ from those of its three neighbor cells $D_1 = \{(F_l^{(1)}[j], F_u^{(1)}[j]) | 0 \le j \le \delta\}$ (upper left), $D_2 = \{(F_l^{(2)}[j], F_u^{(2)}[j]) | 0 \le j \le \delta\}$ (upper), and $D_3 = \{(F_l^{(3)}[j], F_u^{(3)}[j]) | 0 \le j \le \delta\}$ (left):

$$F_l[j] = pF_l^{(1)}[j] + (1 - p)F_l^{(argmin_i D_i)}[j - 1] \qquad (12)$$

$$F_u[j] = pF_u^{(1)}[j] + (1 - p)min(\sum_{i=1}^{3} F_u^{(i)}[j - 1], 1) \qquad (13)$$

where $p = Pr(C = c)$ is the probability of a match at cell D_0 for uncertain characters C from s and c from t, and $argmin_i D_i$ returns the index i that enables $D_i (1 \le i \le 3)$ to has the greatest probability of being small(see [11] for details). Ge and Li [11] has proved that the value $(F_l[j], F_u[j])(0 \le j \le \delta)$ thus evaluated are the lower bound and upper bound of $Pr(ED \le j)$ at the cell, respectively. Clearly, when CDF applied to our (δ, γ)-range query, the last values computed in the DP table corresponding to s

and t: $(F_l[\delta], F_u[\delta])$ can be used for early acceptance(if $F_l[\delta] > \gamma$) or early rejection (if $F_u[\delta] < \gamma$).

From equations (12) and (13), we can see that the previous values(D_1, D_2 and D_3) have great effect on the current value D_0. If the bounds in $D_i (1 \leq i \leq 3)$ are relatively loose, so are those in D_0. Similarly, these bounds in D_0 will loosen the later bounds in a cascaded way. In the end, the final result $(F_l[\delta], F_u[\delta])$ will worsen the filtering effect. Thus, to improve the filtering capability, we should tighten the bounds as possible as we can at every step of filling in the DP table.

Key Idea for Local Perturbation For a query string t and an uncertain string s, assume that s contains an *adjacent world* (a possible world pw such that $ed(t, pw) \leq \delta$). By changing a subset of the existing *bindings* (an assignment of an uncertain character to a fixed value in a possible world), if the resulting possible worlds are still adjacent to t, the sum of their possibilities is a lower bound of $Pr[ed(t, s) \leq \delta]$. Similarly, after various local perturbations are performed from a *remote world* (a possible world pw such that $ed(t, pw) > \delta$), if the resulting possible worlds are still beyond the distance δ to t, one minus the sum of their possibilities p_r, i.e., $(1 - p_r)$ is an upper bound of $Pr[ed(s, t) \leq \delta]$.

To fill in each DP table cell with the edit distance value $ed[i, j]$, the standard algorithms is based on:

$$ed[i, j] = min\{ed[i, j - 1] + 1, ed[i - 1, j] + 1, ed[i - 1, j - 1] + c(t[i], s[j])\}$$

where

$$c(t[i], s[j] = \begin{cases} 0, if \ t[i] = s[j] \\ 1, if \ t[i] \neq s[j] \end{cases}$$

In the presence of uncertain character, one of the possible ways of computing $c(t[i], s[j])$ is:

$$c(t[i], s[j] = \begin{cases} 0, if \ Pr(t[i] = s[j]) > 0 \\ 1, if \ Pr(t[i] = s[j]) = 0 \end{cases} \tag{14}$$

In [11], the authors have proved that, if the rule (14) is adopted, then the value $ed(t, s)$ thus computed is the *closet* possible world distance (pw is a closet possible world if $ed(pw, t)$ is a minimum one amongst all possible worlds). Clearly, this closet possible world can be served as an adjacent world (if there is one). To get an initial remote world, a process of binding each uncertain character randomly based on its distribution is repeated until a remote world is generated. With these adjacent and remote possible worlds, local perturbations over *crucial variable* and uncertain characters are performed to get the lower bound and upper bound of $Pr(ed(t, s) \leq \delta)$, respectively. For each cell indexed by (i, j) in the optimal path, a crucial variable is an uncertain character $s[i]$ such that one of its alternatives can match $t[j]$. The following two theorems give the computation details:

Theorem 3. *[11] Given a query string t and an uncertain string s in the range-(δ, γ) query, let $\Delta(\geq 0)$ be the difference between δ and the edit distance in an adjacent world, which has c crucial variables. Let $lp = Pr(at least \ c - \Delta \ crucial variables have the same values as in the optimal path). Then $Pr[ed(s, t) \leq \delta] \geq lp$.*

Theorem 4. *[11] Given a query string t and an uncertain string s in the range-(δ, γ) query, Consider a remote world that has a distance $\delta' > \delta$ and let $\Delta = \delta' - \delta - 1$. If there are u uncertain characters in s and up = Pr(at least $u - \Delta$ crucial variables have the same values as in the optimal path), then $Pr[ed(s,t) \leq \delta] \leq 1 - up$.*

In Theorem 3 and 4, the issue of how to evaluate lp and up is equivalent to the ALEH problem mentioned in 4.2.

CDF and local perturbation are used separately in [11]. As argued above, improving the bounds is a very important factor in terms of processing efficiency. As we will see soon, achieving bound improvement by combining CDF and local perturbation is a natural and feasible scheme.

First, we discuss how to tighten the lower bound of $Pr(ed(t, s) \leq \delta)$. The basic idea for combining goes as follows: to fill each DP table cell indexed by (x, y), besides the lower bound value $F_l[j]$ calculated by (12), we also consider the lower bound $lp[j]$ by local perturbation. Note that , like $F_l[j]$, the value $lp[j]$ is also corresponding to substring $s[1 \ldots x]$ and $t[1 \ldots y]$, i.e., $Pr(ed(s[1 \ldots x], t[1 \ldots y]) \leq j) \geq lp[j]$. Clearly, we can obtain an improved lower bound by:

$$F_l[j] = max\{F_l[j], lp[j]\} \tag{15}$$

Repeating this combining processing at each step of CDF computation, the lower bounds can be continuously improved in a cascaded way. In the end, the value $F_l[\delta]$ corresponding to the whole strings s and t is possible to be tighter than either value derived by use of CDF or local perturbation method.

As discussed above, the key issue of improving lower bound by combining is how to get the lower bound of any value $Pr(ed(s[1 \ldots x], t[1 \ldots y]) \leq j)$ $(0 \leq j \leq \delta, 1 \leq x \leq |s|, 1 \leq y \leq |t|)$ by local perturbation. Obviously, there exists a naive method to achieve this: for any value x and y, we compute the closet possible world distance between two substrings $s[1 \ldots x]$ and $t[1 \ldots y]$, along with the associated crucial variables. Then, for any given distance threshold j, we can obtain the lower bound of $Pr(ed(s[1 \ldots x], t[1 \ldots y]) \leq j)$ according to the Theorem 3. This naive method is of very low efficiency due to many times of computation for edit distances and for crucial variables. Below we design an efficient scheme to tackle this problem.

By observation, we find that once a local perturbation for s and t is performed, its associated DP table for closet possible world (denoted as *CPW table*) and evaluation table for the ALEH problem may be exploited for the evaluation of lower bound on $Pr(ed(s[1 \ldots x], t[1 \ldots y]) \leq j)$. Specifically, for each cell in CPW table indexed by (x, y), $ed(s[1 \ldots x], t[1 \ldots y])$ is also the closet possible world distance between $s[1 \ldots x]$ and $t[1 \ldots y]$. Recall that in the evaluation of ALEH problem, we use $Pr(i, j)$ to denote the probability that within the first i events, at least j of them happen. When we get the value $Pr(i, j)$ using dynamic programming method , all $Pr(x, y)(1 \leq x \leq i, 1 \leq y \leq j)$ have also been computed as long as the first x events is a *prefix* of the first i events. Let the crucial variables for CPW table be $\{s[i_1], s[i_2], \cdots, s[i_k]\}(1 \leq i_1 \leq \cdots \leq i_k \leq \cdots \leq |s|)$. According to Theorem 3, with distance threshold j and closet world distance $ed(s[1 \ldots x], t[1 \ldots y])$, if its corresponding

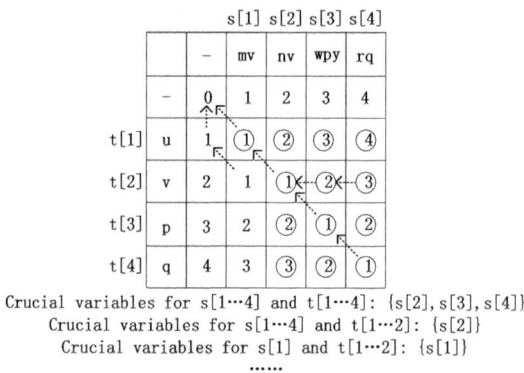

Crucial variables for s[1···4] and t[1···4]: {s[2],s[3],s[4]}
Crucial variables for s[1···4] and t[1···2]: {s[2]}
Crucial variables for s[1] and t[1···2]: {s[1]}
······

Fig. 1. Closet possible world DP table and crucial variables

crucial variables is a prefix of $\{s[i_1], s[i_2], \cdots, s[i_k]\}$, we can efficiently obtain the lower bound of $Pr(ed(s[1 \ldots x], t[1 \ldots y]) \leq j)$ by just looking up the corresponding entries in ALEH table.

Fig 1 gives an example of the CPW table and corresponding variables for string s and t. In each cell, a value in the solid circle indicates the corresponding closet distance with prefix crucial variables. For example, the closet world distance between $s[1 \ldots 4]$ and $t[1 \ldots 2]$ is 3, and the associated crucial variables $\{s[2]\}$ is a prefix of crucial variables $\{s[2], s[3], s[4]\}$ for $s[1 \ldots 4]$ and $t[1 \ldots 4]$. Thus, the lower bound of $Pr(ed(s[1 \ldots 4], t[1 \ldots 2]) \leq j)$ can be immediately obtained by looking up some entry in the ALEH table. Crucial variables can be obtained by keeping track of optimal path for each cell, which is a trivial work when constructing CPW table. When crucial variables at some cell is not a prefix of $\{s[2], s[3], s[4]\}$, like the crucial variables $\{s[1]\}$ corresponding to $s[1]$ and $t[1 \ldots 2]$, we do not care about improving the lower bound from CDF at that cell. Note that for improving the final low bound of $Pr(ed(s, t) \leq j)$, we don't need to improve each cell's low bound. Maybe a single cell's improved low bound can achieve this due to cascaded effect by (15). For instance, if we have improved the low bound of $Pr((ed(s[1 \ldots 2], t[1 \ldots 2]) \leq j)$, the low bound of $Pr((ed(s, t) \leq j)$ can be also improved by it through the tighter low bound of $Pr((ed(s[1 \ldots 3], t[1 \ldots 3]) \leq j)$.

For the upper bound $F_u[j]$ from CDF, we can improve it whenever $ed(s[1 \ldots x], t[1 \ldots y]) > j$ as required in Theorem 4, i.e., the improved upper bound can be computed as:

$$F_u[j] = min\{F_u[j], up[j]\}$$

where $up[j]$ is the upper bound derived by local perturbation if $ed(s[1 \ldots i], t[1 \ldots j]) > k$. Similarly, the evaluation table of corresponding ALEH problem is useful for any $up[j]$ computation once it is created.

Based on the proposed filtering techniques, we now present our algorithm RQUS:

We sort the dataset D by string length. This preprocessing is just executed for one-time and the amortized cost can be ignored for continuously arriving queries. In RQUS, we first use length filtering to get an initial candidate string set D' based on Lemma 3(lines 2-3). Then, probabilistic q-gram filtering and frequency distance filtering are

---

Algorithm. RQUS(D, t, δ, γ)

Input: uncertain string set D sorted by length, query string t, range parameter δ,
        probability parameter $\gamma(> 0)$
Output: subset of D: $RS = \{s | s \in D, Pr(ed(s,t) \leq \delta) \geq \gamma\}$

1 $RS \leftarrow \phi$;
2 $D' \leftarrow \{s | s \in D, max(0, |t| - \delta) \leq |s| \leq |t| + \delta\}$;
3 **for** *each* $s \in D'$ **do**
4   **if** *s survives probabilistic q-gram filtering* **then**
5   **if** *s survives frequency distance filtering* **then**
6   **if** *improved combining lower-bound* $F_l[\delta] \geq \gamma$ **then**
7                  $RS \leftarrow RS \cup s$;
8   **else if** *improved combining upper-bound* $F_u[\delta] \geq \gamma$ **then**
9   **if** $Pr(ed(t, s) \leq \delta) \geq \gamma$ **then**
10                  $RS \leftarrow RS \cup s$;

11 Return RS;

---

employed in turn (lines 4-5). If s survives these filtering, combined filtering is used to check wether it can be early accepted or rejected (lines 6-8). Otherwise, we have to resort to the exact verification (lines 9-10).

We should note that as a metric, the edit distance function satisfies the property of triangle inequality, so there exists an alternative way for us to do filtering based on this metric property in the (δ, γ)-range queries. However, in probabilistic similarity search, obtaining the bounds on probability of edit distance falling into query range instead of edit distance itself is of primary concern. For example, the assertion "*if* $ed(s_1, t) > \gamma$, *then* $ed(s_1, s_2) > \frac{1}{2}\gamma$ *or* $ed(s_2, t) > \frac{1}{2}\gamma$" follows from the property of triangle inequality. Then ,we have $Pr(ed(s_1, t) > \gamma) \leq Pr(ed(s_1, s_2) > \frac{1}{2}\gamma) + Pr(ed(s_2, t) > \frac{1}{2}\gamma)$. If we pre-build index for the range of $Pr(ed(s_1, s_2) > \frac{1}{2}\gamma)$ and $Pr(ed(s_2, t) > \frac{1}{2}\gamma)$, we can obtain the bound on $Pr(ed(s_1, t) > \gamma)$(i.e., $Pr(ed(s_1, t) \leq \gamma)$). We will study this filtering technique in the future work.

6 Experiments

6.1 Experiment Setup

The following algorithms are implemented in the experiments.

CDF and local perturbation [11] are two filtering techniques and work originally in uncertain substring-matching task. However, they can also serve as filers in range queries straightforwardly. Note that, to compare performance fairly, we also incorporate the length filtering into CDF and local perturbation, respectively.

RQUS is proposed in this paper. RQUS uses all filtering schemes: probabilistic q-gram filtering, frequency distance filtering and combined filtering.

RQUS1, RQUS2, RQUS3 To measure the effects of all filtering techniques, we bypass the probabilistic q-gram filtering, frequency distance filtering, and both of them in turn from RQUS and name the resulted algorithm **RQUS1**, **RQUS2** and **RQUS3**, respectively.

All experiments are carried out on a PC with a 2.4GHz CPU and 2GB RAM. All algorithms are implemented in C.

We generate a few synthetic datasets based on two real datasets:

DBLP is a snapshot of the bibliography records from the DBLP Web site (http://www.informatik.uni-trier.de/~ley/db/). Each record is a string of the author name(s) and the title of a publication. An uncertain version of DBLP dataset have been generated in similarity join research [10]. We adopt similar approach to generate uncertain strings. Specifically, for a name χ in the DBLP database, we find its similar set $A(\chi)$ that consists of all names in DBLP within edit distance 3 to χ. Then, we create a character-level uncertain string s from χ as follows. We randomly pick a few characters in χ, and for each such letter at position i, the pdf of $s[i]$ is generated based on the normalized frequencies of the letters in the i-th position of all strings in $A(\chi)$. The other positions of s are deterministic and the same as in χ. In addition, we also generate different size of synthetic datasets based on this real datasets by varying the parameter values of data (such as uncertain ratio θ) or the size of the data. The average length of the generated uncertain strings is $10 \sim 30$.

UNIREF is the UniRef90 protein sequence data from the UniProt project (http://www.uniprot.org/). Each string is a sequence of amino acids coded as uppercase letters. We broke long sequences into shorter strings of controlled length and generated uncertain strings with average length $50 \sim 100$ in the same way as for DBLP dataset.

The default values for some parameters are: $q=3$, $\theta = 30\%$, and the average number σ of choices that each probabilistic character $s[i]$ may have is set as 5.

6.2   Performance Comparison

In all implemented algorithms, two important measures reflecting the performance of their filtering techniques are used: (1) the size of candidate uncertain strings whose edit distances of all possible worlds are evaluated (denoted by **SCAND**); and (2) running time. On each dataset, we measured average results by running the algorithms 10 times using different query strings.

Figures 2 shows the SCAND and the running time for RQUS, RQUS1, RQUS2, RQUS3, and CDF on different sizes of input-sequences DBLP with parameter $\delta=3$, $\gamma=0.1$. As shown in Figures 2(a), although RUQS1, RUQS2, and RUQS3 do not use the probabilistic q-gram filtering, frequency distance filtering, and both of them in turn, they still get smaller SCAND than CDF on different datasets for two reasons. The first is that both of probabilistic q-gram filtering and frequency distance filtering can prune away some additional uncertain strings that CDF fails to filter out, which demonstrates that CDF can't provide absolutely tighter upper bound than probabilistic q-gram filtering or frequency distance filtering. Another reason is that, due to tighter bounds obtained by combined filtering, some strings have greater opportunity to be removed from the candidate set. In addition, by comparing the SCAND measure of RQUS1 and RQUS2, we find that although our algorithm can benefit from both probabilistic q-gram filtering and frequency distance filtering, the former is slightly better in filtering capability than the latter .

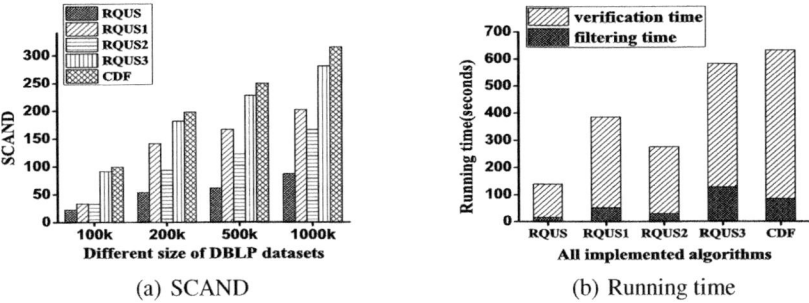

(a) SCAND   (b) Running time

Fig. 2. Performance comparison against CDF: on DBLP datasets (parameter δ =3, γ=0.1)

Figure 2(b) shows that, in terms of running time, RQUS is the most efficient algorithm, followed in turn by RQUS2, RQUS1, RQUS3 and CDF. Running time consists of two parts: filtering time and verification time. Filtering time is the processing time spent by different filtering techniques. On 500k DBLP dataset, CDF consumes more filtering time than RQUS, RQUS1, and RQUS2 because it needs to test every input string that maybe filtered out by probabilistic q-gram filter or frequency distance filter and the latter two filters spends less time than CDF. In addition, since CDF has more candidate strings to be verified, it need to spend much more verification time to enumerate all possible worlds for computing exact probability of edit distance falling into the query range, which indicate that it is very time-consuming during the verification phase. On other DBLP datasets with different size, we also get the similar results.

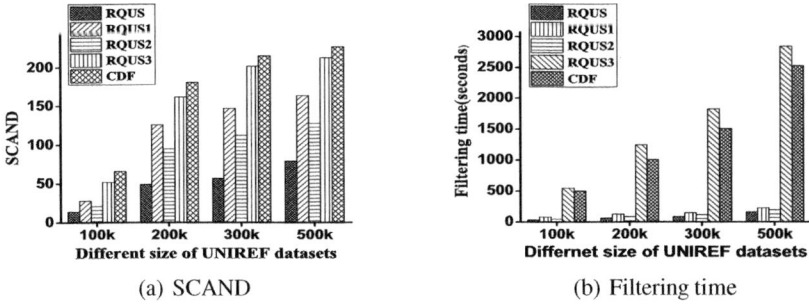

(a) SCAND   (b) Filtering time

Fig. 3. Performance comparison against CDF: on UNIREF datasets (parameter δ=8, γ=0.01)

Figure 3 shows the result for UNIREF datasets with parameter δ=8, γ=0.01. The contrast of SCAND between different algorithms is similar to that for DBLP datasets.

However, RQUS (also RQUS1 and RQUS2) shows big advantage over CDF in terms of filtering time. As analyzed in the Introduction section, the time complexity of CDF tends to be cubic in the size of query string, while probabilistic q-gram filter and frequency distance filter in RQUS work at most in linear time. Thus, for long strings, these two filters in RQUS is much more efficient than CDF filter. For instance, on the DBLP dataset with average length 20 and UNIRF dataset with average length 60 (with the same size of 500k), RQUS's filtering time increases from 16s to 164s, while CDF's filtering time increases dramatically from 86s to 2529s. In addition, the two filters in RQUS are competitive in filtering capability and thus most of strings can be first pruned away by them instead of CDF filter. These two factors contribute to the high efficiency of RQUS. Note that we do not record the verification time because more uncertain positions in long UNIREF strings incur extremely expensive calculations of edit distances.

We also compare the performance of RQUS together with its three variants against that of local perturbation method. The results are shown in Figures 4. As opposed to comparison with CDF, we find that: 1. local perturbation takes less time than CDF to perform filtering. On average, CDF spends five-fold more filtering time than local perturbation on different size of DBLP datasets. 2. CDF based bounds is better than local perturbation based bounds. So, the improvement over local perturbation is more notable since local perturbation method has to spend more time for verification. For example, on DBLP dataset with size of 500k, the SCAND of RQUS is 151 less than that of CDF, while it is 359 less than that of local perturbation, which results in 242 seconds difference in terms of running time improvement.

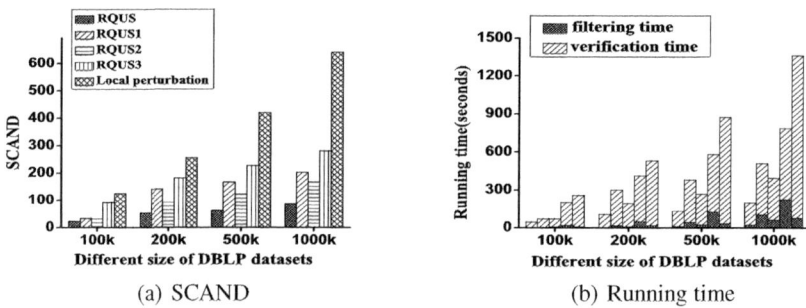

(a) SCAND   (b) Running time

Fig. 4. Performance comparison against local perturbation: on DBLP datasets (parameter $\delta = 3$, $\gamma = 0.1$)

6.3   Effect of Parameters

As Figure 5 shows, we vary the value of δ or γ to test the effect of the similarity threshold and probability threshold on the performance of the algorithms. As expected, for all algorithms, when we fix $\gamma = 0.1$ and increase δ, the total number of strings accepted as results by lower bound also increases. This number is almost the same for all algorithms as shown in Figure 5(a), which indicates that the lower bounds provided by CDF and local perturbation are equally matched. In addition, with SCAND measure, we find that the number of strings removed by upper bound of all algorithms decreases with δ.

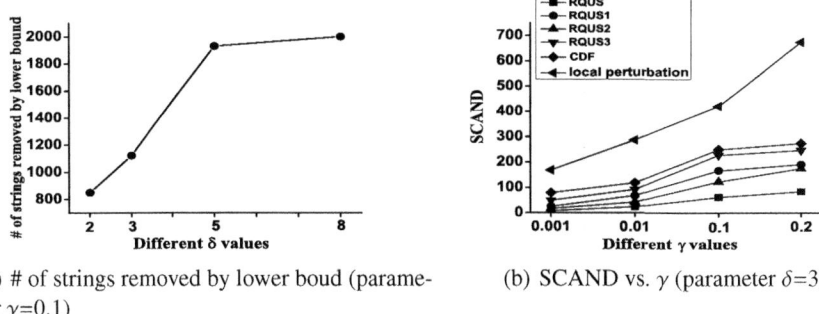

(a) # of strings removed by lower boud (parameter $\gamma=0.1$)

(b) SCAND vs. γ (parameter $\delta=3$)

Fig. 5. Effect of δ and γ on performance on 500k DBLP dataset

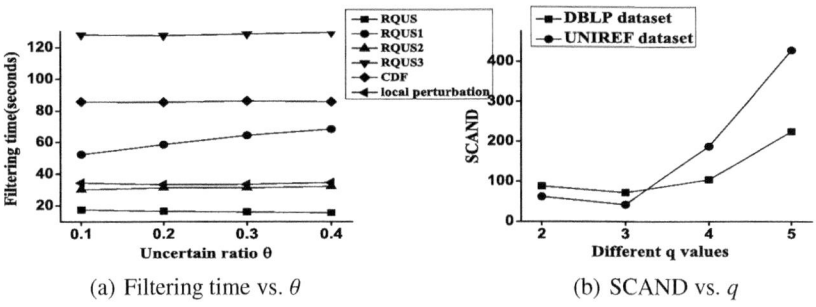

(a) Filtering time vs. θ

(b) SCAND vs. q

Fig. 6. Effect of θ and q on performance

In Figure 5(b), when we fix $\delta=3$, the general trend is that the SCAND of all algorithms increases with γ going up from 0.001 to 0.2. This is because, as γ increases, the decreased number of the accepted strings by lower bound exceeds the increased number of the removed strings by upper bound. In addition, our proposed algorithms outperform local perturbation by an increasingly larger margin when γ increases. In all scenarios, RQUS and its three variants are faster than CDF or local perturbation in terms of both SCAND and running time.

Figure 6(a) shows that another parameter, the fraction θ of uncertain positions in an uncertain string, also has an obvious impact on the filtering time of algorithm RQUS1. This is because only the performance of frequency distance filter is affected by θ directly and the frequency distance filter in RQUS1 processes much more input-strings than in RQUS. Obviously, bigger value of θ will incur exponentially more verification time for all algorithms.

As for q, the length of q-gram, only probabilistic q-gram filtering is affected by this parameter. In Figure 6(b), we run the algorithm RQUS on both DBLP and UNIREF datasets and present its result of SCAND by different value of q. It indicates that too small or too big value of q is not suitable for probabilistic q-gram filtering, so we take $q=3$ in all experiments.

7   Conclusion

In this paper, we study an unexplored problem of range queries over uncertain strings. We first discuss the range query semantics and formulate an appropriate problem on range query. With this problem, we propose two novel filtering techniques: probabilistic q-gram filtering and frequency distance filtering to speed up the query processing. To further improve the performance, we combine the two existing filtering schemes used in substring matching task: CDF and local perturbation, to make the lower bound and upper bound tighter. Extensive experiments show our proposed algorithm is more efficient than CDF or local perturbation. As future work, we plan to investigate the problem of top-k query over uncertain strings.

Acknowledgement. The research was supported by Key Project of Science and Technology Commission of Shanghai Municipality [No.11510500300], Shanghai Leading Academic Discipline Project [No.J50103], Innovation Program of Shanghai Municipal Education Commission [No.11YZ03] and Ph.D. Programs Fund of Ministry of Education of China [No. 20113108120022].

References

1. Gusfield, D.: Algorithms on strings, trees, and sequences. Cambridge University Press (1999)
2. Sarawagi, S.: Sequence Data Mining (Advanced Methods for Knowledge Discovery from Complex Data). Spinger (2005)
3. Dong, G., Pei, J.: Sequence Data Mining (Advances in Database Systems). Springer (2007)
4. Hadjieleftheriou, M., Li, C.: Efficient approximate search on string collections. In: ICDE Tutorial (2009)
5. Gravano, L., Ipirotis, P.G., Jagadish, H.V., Koudas, N., Muthukrishnan, S., Srivastava, D.: Approximate string joins in a database (almost) forfree. In: VLDB, pp. 491–500 (2001)
6. Chaudhuri, S., Ganjam, K., Ganti, V., Motwani, R.: Robust and efficient fuzzy match for online data cleaning. In: SIGMOD, pp. 313–324 (2003)
7. Bayardo, R.J., Ma, Y., Srikant, R.: Scaling up all pairs similarity search. In: WWW, pp. 131–140 (2007)
8. Henzinger, M.: Finding near-duplicate web pages: a large-scale evaluation of algorithms. In: SIGIR, pp. 284–291 (2006)
9. Buhler, J.: Efficient large-scale sequence comparison by locality-sensitive hashing. Bioinformatics 17(5), 419–428 (2001)
10. Jestes, J., Li, F., Yan, Z., Yi, K.: Probabilistic string similarity joins. In: SIGMOD, pp. 327–338 (2010)
11. Ge, T., Li, Z.: Approximate substring matching over uncertain strings. In: VLDB, pp. 772–782 (2011)
12. Kahveci, T., Singh, A.: An efficient index structure for string databases. In: VLDB, pp. 351–360 (2001)
13. Venkateswaran, J., Kahveci, T., Jermaine, C., Lachwani, D.: Reference-based indexing for metric spaces with costly distance measures. The VLDB Journal 17(5), 1231–1251 (2008)
14. Li, C., Lu, J., Lu, Y.: Efficient merging and filtering algorithms for approximate string searches. In: ICDE, pp. 257–266 (2008)
15. Xiao, C., Wang, W., Lin, X., Yu, J.: Efficient similarity joins for near duplicate detection. In: WWW, pp. 131–140 (2008)

16. Li, C., Wang, B., Yang, X.: VGRAM: Improving performance of approximate queries on string collections using variable-length grams. In: VLDB, pp. 303–314 (2007)
17. Yang, X., Wang, B., Li, C.: Cost-based variable-length-gram selection for string collections to support approximate queries efficiently. In: SIGMOD, pp. 353–364 (2008)
18. Jokinen, P., Ukkonen, E.: Two algorithms for approximate string matching in static texts. In: FOCS, pp. 240–248 (1991)
19. Xiao, C., Wang, W., Lin, X.: Ed-Join: An efficient algorithm for similarity joins with edit distance constraints. In: VLDB, pp. 933–944 (2008)
20. Xiao, C., Wang, W., Lin, X., Yu, J.: Efficient similarity joins for near duplicate detection. In: WWW, pp. 131–140 (2008)
21. Zhang, Z., Hadjieleftheriou, M., Ooi, B.C., Srivastava, D.: B^{ed}-Tree: An all-purpose index structure for string similarity search based on edit distance. In: SIGMOD, pp. 915–926 (2010)
22. Behm, A., Li, C., Carey, M.: Answering approximate string queries on large data sets using external memory. In: ICDE, pp. 888–899 (2011)
23. Yao, B., Li, F., Hadjieleftheriou, M., Hou, K.: Approximate string search in spatial databases. In: ICDE, pp. 545–556 (2010)
24. Dalvi, N., Suciu, D.: Management of probabilistic data: foundations and challenges. In: PODS, pp. 1–12 (2007)
25. Aggarwal, C.C., Yu, P.S.: A Survey of Uncertain Data Algorithms and Applications. IEEE Transaction on Knowledge and Data Engineering (TKDE) 21(5), 609–623 (2009)
26. Sutinen, E., Tarhio, J.: On Using q-Gram Locations in Approximate String Matching. In: Spirakis, P.G. (ed.) ESA 1995. LNCS, vol. 979, pp. 327–340. Springer, Heidelberg (1995)

Continuous Probabilistic Sum Queries in Wireless Sensor Networks with Ranges

Nina Hubig[1], Andreas Züfle[1], Tobias Emrich[1], Mario A. Nascimento[2],
Matthias Renz[1], and Hans-Peter Kriegel[1]

[1] Ludwig Maximilians Universität, München, Germany
{hubig,zuefle,renz,kriegel}@dbs.ifi.lmu.de
[2] University of Alberta, Edmonton, Canada
mn@cs.ualberta.ca

Abstract. Data measured in wireless sensor networks are inherently imprecise. Aggregate queries are often used to analyze the collected data in order to alleviate the impact of such imprecision. In this paper we will deal with the imprecision in the measured values explicitly by employing a probabilistic approach and we focus on one particular type of aggregate query, namely the SUM query.[1]

1 Introduction

Recent advances in sensors and wireless communication technologies have enabled the development of small and relatively inexpensive multi-functional wireless sensor nodes. This has lead to the concept of a wireless sensor network (WSN), i.e., a set of spatially distributed autonomous sensors that cooperatively monitor physical or environmental conditions in an area of interest [1]. This paper addresses efficient processing of SUM queries in such WSNs. SUM queries are very useful for many applications, for example, if we want to observe the overall amount of traffic in a road network in order to predict potential traffic jams, or areas where the risk that an accident will happen due to large volume of incoming traffic is higher. Another example is to estimate the water level of a river by aggregating over all potential inflows observed by a sensor network, which is very important to predict and avoid flooding. However, there are a number of challenges sensor network applications have to cope with, including the large number of sensor nodes used in typical sensor networks and the fact that they are prone to failures, as well as limited in power, computational capacities, and memory. Specifically, in this paper we assume that measurements taken by the sensors are imprecise, due to fluctuations in the environment itself or due to hardware limitations [4].

Our main goal is to show how to efficiently perform SUM queries on uncertain sensor data while focusing on providing reliable results. Thus allowing us to to answer probabilistic SUM queries such as *"Report the probability that the sum of sensor values exceeds a given threshold τ"*.

Table 1 shows a sample scenario for the SUM query. Let $S = \{s_1, s_2, s_3\}$ be a WSN with three sensors that are installed on the tributaries of a river measuring the water level due to rainfall. Each sensor reading contains a set of value ranges, each associated with

---

[1] This research was partially supported by NSERC, Canada and DAAD, Germany.

A. Ailamaki and S. Bowers (Eds.): SSDBM 2012, LNCS 7338, pp. 96–105, 2012.

Table 1. Example of probabilistic sensor readings (sets of (Value Range,Probability) pairs)

Sensor-ID	Set of (Value Range, Probability) pairs
s_1	$\{([0, 0], 0.5), ([0, 10], 0.3), ([10, 20], 0.2)\}$
s_2	$\{([0, 20], 0.2), ([20, 30], 0.4), ([30, 40], 0.3), ([40, 60], 0.1)\}$
s_3	$\{([0, 0], 0.5), ([10, 20], 0.3), ([20, 50], 0.1), ([50,100], 0.1)\}$

respective probabilities. A possible query in such a scenario is: "What is the probability that the total water level rises above some critical level?".

Interestingly, not much work has been done regarding the uncertainty aspect of data when processing queries in WSNs. Some efforts have addressed top-k queries, e.g., [8] but not as much for other types of aggregations. In [3] we investigated probabilistic count queries, where the task is to find the probability that a given number of sensors satisfy a query. Note that in that case a sensor either satisfies a query with a given probability or does not with a complementary probability, i.e., there is only one value and one probability involved. It turns out that even though COUNT queries are a special case of the SUM queries we study in this paper, the techniques presented in [3] are not well suited for the latter. For (probabilistic) SUM queries we allow each sensor to generate an arbitrary non-binary set of value- or range-probability pairs. In the following, we discuss in detail the probabilistic SUM queries as well as the solutions we propose to process the same in a WSN setting.

2   Problem Definition

We consider a wireless sensor network WSN, consisting of a set $S = \{s_1, s_2, \ldots, s_n\}$ of n sensors. We assume that the network topology is fixed (i.e., it does not change over time) and is a shortest path tree with one single sink node (the tree's root). Instead of a deterministic value, a sensor value of a sensor s_i is a random variable, specified by a probabilistic density function (PDF) denoted as $pdf(s_i)$. The function $pdf(s_i)$: $V(s_i) \rightarrow (0, 1]$ maps the domain $V(s_i)$ of s_i, i.e., each possible parameter value of s_i, to a non-zero probability value. A sensor s_i producing such probabilistic values, i.e., values given by a PDF, is called a *probabilistic sensor*. Sensor values are typically spatially correlated. This work allows spatial correlation of sensor values, however, it is important to note that we assume that the measurement errors (i.e., the deviation from the true value) of different sensors are mutually independent, even if the underlying observed events are mutually dependent. Consequently, the sensor value distributions of two different sensors are assumed to be independent.

Based on the above probabilistic sensor model, queries are issued in a probabilistic way by applying the *possible worlds semantic model*. However, there have been different adaptations of the model for probabilistic databases. We use the model as proposed in [7], specifically, a *possible world* w is a set $w = \{s_1^w, \ldots, s_n^w\}$ of instances of sensor values, where each sensor $s \in S$ is assigned to a (certain) value $v \in V(s_i)$. The probability $P(w)$ denotes the probability that the world w is *true*, i.e., that the instances in w coexist in the sensor network. The set of all possible worlds is denoted by $\mathcal{W} = \{w_i, \ldots, w_{|\mathcal{W}|}\}$.

We consider a wireless sensor network *WSN* consisting of a set of probabilistic sensors S and a probability threshold τ. A *probabilistic SUM query* computes the probability, that the random variable corresponding to the sum of all sensor values in the *WSN* is at least τ.

Definition 1. *A probabilistic sum query (PSQ) is defined as:*

$$PSQ(WSN, \tau) = \sum_{w \in W, sum(S^w) \geq \tau} P(w), \tag{1}$$

where W denotes the set of possible worlds and $sum(S^w) = \sum_{s_i \in S} s_i^w$ denotes the sum of all sensor (value) instances $s \in S$ in world w.

In accordance to the above definition, we can answer probabilistic sum queries by materializing all possible worlds with the corresponding probabilities and accumulate the probabilities of all worlds w where the sum of sensor instances in w exceeds τ. However, since the number of possible worlds is exponential in the number of probabilistic sensors, this naive method is not practical. In the following, we will show how to efficiently compute a probabilistic sum query in sensor networks. A special variant of our approach that is only applicable to discrete data distributions cane be found in the extended version of this paper [5]. In addition to solutions focusing on reducing the computational overhead of probabilistic sum queries, we show how to reduce the communication overhead, in terms of number of messages, in the sensor network.

3   Probabilistic Sum Queries in Probabilistic Wireless Sensor Networks Having Discrete Data Distributions

In the following, we assume that each sensor measures a large, or infinite number of possible values. The probability density function (PDF) of these values may be very complex and not representable in a parametric form. Thus, we propose to discretize the value domain into intervals and give lower and upper bounds of the aggregate probability of these intervals. Thus, the PDF of a sensor s_i is partitioned into a (finite) set of value intervals $\overline{V_i} = \{\overline{v_i^1}, ..., \overline{v_i^{|V_i|}}\}$. Each interval $\overline{v_i^j}$ is associated with the probability $pdf(\overline{v_i^j}) = cdf(ub(\overline{v_i^j})) - cdf(lb(\overline{v_i^j}))$, where $cdf(a) = \int_{-\infty}^{a} pdf(x)dx$ and $ub(\overline{v_i^j})$ and $lb(\overline{v_i^j})$ correspond to the upper and lower bound of interval $\overline{v_i^j}$, respectively.

In summary, a sensor node is now assumed to be characterized by a set of value intervals, each associated with the probability that the actual value falls into the corresponding interval. While information is absent regarding the probabilistic distribution of the sensor within the interval, we aim at bounding, efficiently, the result of a probabilistic SUM query based on these intervals.

To achieve this, consider the following generating function:

$$GF(S_n) = \prod_{i=1}^{n} \sum_{j=1}^{|\overline{V_i}|} pdf(\overline{v_i^j}) x^{lb(\overline{v_i^j})} y^{ub(\overline{v_i^j}) - lb(\overline{v_i^j})} = \sum_{i,j} c_{i,j} x^i y^j \tag{2}$$

Each monomial represents a class of possible worlds, having a total probability of $c_{i,j}$ and a sum of at least i. Additionally, the exponent j of y corresponds to a possible additional value of j, which may or may not exist in the worlds corresponding to $c_{i,j}$, leading to the following lemma:

Lemma 1. *Let $S_n = \{s_1, ..., s_n\}$ be a set of probabilistic sensors. Each coefficient $c_{i,j}$ of the expansion of $GF(S_n)$ corresponds to the probability of all worlds, in which the sum of all sensors S_n must be between i and $i + j$.*[2]

Lemma 1 allows us to compute a lower bound of the probability that the sum of S_n exceeds τ by adding up the probabilities $c_{i,j}$ of all worlds where the lower bound i (i.e., the exponent of x) exceeds τ, as $LB(PSQ(\mathcal{W}, \tau)) = \sum_{i \geq \tau, j} c_{i,j}$. The corresponding upper bound can be computed as $UB(PSQ(\mathcal{W}, \tau)) = \sum_{i+j \geq \tau} c_{i,j}$.

To allow pruning , we can use the following observation: if the sum of S_n is greater than or equal to τ with a probability of at least (at most) p, then the probability that the sum is less than τ can be at most (must be at least) $1 - p$. This permits to rewrite the two equations above as:

$$UB(PSQ(\mathcal{W}, \tau)) = 1 - \sum_{i+j<\tau} c_{i,j} \qquad (3)$$

$$LB(PSQ(\mathcal{W}, \tau)) = 1 - \sum_{i<\tau, j} c_{i,j} \qquad (4)$$

Now, we note that in this representation, for any monomial $c_{i,j}x^i y^j$ it holds that $i \leq \tau$. This directly leads to the following corollary:

Corollary 1. *Any monomial $c_{i,j}x^i y^j$ where $i > \tau$ can be pruned without loss of information.*

Furthermore, we can combine monomials $c_{i,j}x^i, y^j, c_{m,n}x^m, y^n$ where $i = m, i + j > \tau$ and $m + n > \tau$. The rationale of this is, that for both classes of worlds represented by $c_{i,j}x^i, y^j$ and $c_{m,n}x^m, y^n$ the only difference is a different upper bound sum, since the lower bounds i and n are identical. Since both upper bounds $i + j$ and $m + n$ are greater than τ, we can treat these worlds equivalently, since we do not care how much τ can be exceeded, all we need to know is whether τ can be exceeded at all in the possible worlds. This leads to the following corollary.

Corollary 2. *Any two monomial $c_{i,j}x^i, y^j, c_{m,n}x^m, y^n$ where $i = m, i + j > \tau$ represent worlds of an equivalent class of worlds. We represent this class of possible worlds by the monomial $c_{i+m,\infty}x^{i+m}y^\infty$*

In the iterative computation of $S_n, 1 \leq n \leq |S|$, Corollary 1 allows to prune, in each iteration, monomials of $GF(S_n)$ which cannot have influence on the sum of S. Furthermore, Corollary 2 allows to combine multiple monomials of $GF(S_n)$ into a single one. Both these pruning criteria reduce the number of monomials, thus decreasing the cost of computing S_{n+1}.

---

[2] Formal proofs of all lemmas and corollaries can be found in the extended version [5].

We note that the generating function we propose for the discrete case (cf. [5]) is a special case of Equation 2, where for all coefficients $c_{i,j}x^iy^j$ it holds that $y = 0$. Thus, using Equation 2 we can handle both cases of discrete and continuous data.

Example 1. Assume a set of sensors S that estimate the expected influx of water to a common river. Let us assume that if this influx exceeds a value of $\tau = 40$ with a probability of at least 10%, a flood warning should be issued. Each sensor s_i computes a continuous PDF of the water influx, using appropriate water-flow and precipitation models (e.g. [2]). Discretization of these continuous functions yields the value ranges illustrated in the table 1, associated with respective probabilities that the true water influx is within the interval[3].

Applying Equation 2 yields the following polynomial:

$$G(S_1) = 0.5x^0y^0 + 0.3x^0y^{10} + 0.2x^{10}y^{10}$$

By definition, each monomial px^iy^j corresponds to a world having a probability of p and having a sum that is lower bounded by i, and upper bounded by $i + j$. For S_2 we obtain:

$$G(S_2) = G(S_1) \times (0.2x^0y^{20} + 0.4x^{20}y^{10} + 0.3x^{30}y^{10} + 0.1x^{40}y^{20})$$

Here, monomials in $G(S_1)$ correspond to all possible worlds of S_1, while the remaining monomials correspond to worlds of S_2. Due to the observation that the sum of two values between $[lb_1, ub_1]$ and $[lb_2, ub_2]$ must be in the interval $[lb_1 + lb_2, ub_1 + ub_2]$, we now expand the above term to get all possible intervals of the sum of S_1 and S_2:

$$G(S_2) = 0.1x^0y^{20} + 0.2x^{20}y^{10} + 0.15x^{30}y^{10} + 0.05x^{40}y^{20} + 0.06x^0y^{30} + 0.12x^{20}y^{20} +$$

$$0.09x^{30}y^{20} + 0.03x^{40}y^{30} + 0.04x^{10}y^{30} + 0.08x^{30}y^{20} + 0.06x^{40}y^{20} + 0.02x^{50}y^{30}$$

Each coefficient p of each monomial px^ix^j is the product of the probabilities of two sensor values. Due to the assumption of independent sensor errors, this product is the probability of observing both values. We can combine worlds of S_2 having the same lower and upper bounds

$$G(S_2) = 0.1x^0y^{20} + 0.06x^0y^{30} + 0.04x^{10}y^{30} + 0.2x^{20}y^{10} + 0.12x^{20}y^{20} +$$

$$0.15x^{30}y^{10} + 0.17x^{30}y^{20} + 0.11x^{40}y^{20} + 0.03x^{40}y^{30} + 0.02^{50}y^{30}$$

Now, recall that ultimately, we want to compute the probability that $P(sum(S)) > 40$, which equals $1 - P(sum(S) \leq 40)$. To compute the latter probability, we can now prune any monomial having an x-exponent of $i > 40$, since for the equivalent class of worlds represented by this monomial, we can already conclude that the sum cannot be less than 40 due to non-negativity of sensor values. We obtain:

---

[3] For illustration purpose we use a very coarse discretization. In practice, a much larger set of value ranges per sensor could be used.

$$G(S_2) = 0.1x^0y^{20} + 0.06x^0y^{30} + 0.04x^{10}y^{30} + 0.2x^{20}y^{10}+$$
$$0.12x^{20}y^{20} + 0.15x^{30}y^{10} + 0.17x^{30}y^{20}$$

which will then be used in the computation of $G(S_3)$. Furthermore, we can combine monomials $c_{i,j}x^i, y^j, c_{m,n}x^m, y^n$ where $i = m$, $i + j > \tau$ and $m + n\tau$ as discussed above, yielding:

$$G(S_2) = 0.1x^0y^{20} + 0.06x^0y^{30} + 0.04x^{10}y^{30} + 0.2x^{20}y^{10}+$$
$$0.12x^{20}y^{20} + 0.15x^{30}y^{10} + 0.17x^{30}y^{\infty}$$

Due to space limitations, let us now assume that there is only two sensors, i.e., $S = S_2$. Now, in order to lower (upper) bound the probability that the sum of S is at most 40, we simply sum up all worlds where the sum must be (can be) lower than 40. Clearly, a world must have a sum of at most 40, if its corresponding upper bound is at most 40, i.e., if for the corresponding monomial px^iy^j it holds that $i + j$ is at most 40.

$$\sum_{i+j<\tau=40} c_{i,j} = 0.1 + 0.06 + 0.04 + 0.2 + 0.12 + 0.15 = 0.67$$

A world can have a sum of 40 or less, if its corresponding lower bound is 40 or less

$$\sum_{i<\tau=40} c_{i,j} = 0.1 + 0.06 + 0.04 + 0.2 + 0.12 + 0.15 + 0.17 = 0.84$$

Note, that this sum equals the sum of coefficients of all remaining monomials, since any monomial which cannot have a sum of 40 or less has already been pruned.

Using Equations 3 and 4, we bound the probability that the sum of S exceeds τ:

$$UB(PSQ(\mathcal{W},\tau)) = 1 - \sum_{i+j<\tau} c_{i,j} = 1 - 0.67 = 0.33$$
$$LB(PSQ(\mathcal{W},\tau)) = 1 - \sum_{i<\tau} c_{i,j} = 1 - 0.84 = 0.16$$

Thus we can conclude that the event that the water influx exceeds 40 must have a probability of at least 16%, and in particular must be greater than 10%, thus we will broadcast a flood warning. Note that we were able to answer this query despite a very coarse approximation of the sensor PDFs. In the general case, the probability threshold p may be between the lower and the upper bound of $PSQ(\mathcal{W}, \tau)$. In this case we cannot say for certain whether the probability is greater than p or not. However, we may re-initiate the whole query, asking each sensor for a PDF having a more refined granularity. This will reduce the uncertainty of the query result but will come at an additional network traffic, since for the new query, sensors will be forced to send a much more refined pdf, which corresponds to more information, which corresponds to more data.

This filter refinement approach can be iterated until a definite answer can be given, or until the resulting approximation is good enough. In the example, an approximation that the probability of a flood must be between 9% and 11% may be good enough to decide whether to broadcast a warning or not.

In this section, we showed how to reduce the CPU cost for computing the count distribution. Furthermore, we show in our technical report ([5]) how to reduce the communication costs. For that purpose, we consider the typical underlying characteristics of WSNs such as network topology, routing and scheduling. We propose two algorithms which solve the problem of answering continuous count queries in a WSN by taking the local distribution of data into consideration.

4   Energy Efficient Computation of Probabilistic Sum Queries

As mentioned in Section 1, it is crucial for applications on WSNs to reduce energy cost. For that purpose, we must consider the typical underlying characteristics of WSNs such as network topology, routing and scheduling. We assume that the nodes in S are connected together via a logical tree where the sink node (or base-station) is the tree's root. The choice of the tree's topology does matter, but is outside the scope of this paper. For the sake of simplicity, we assume it to be a hop-based shortest path tree commonly used in other works, e.g., [6]. In the previous Section 3, we assumed that for each sensor $s_i \in S$, the distribution $pdf(s_i)$ is readily available at the sink node. This requires each $pdf(s_i)$ to be iteratively propagated along the branch of the logical tree to the root. However, we can benefit from computing intermediate results at intermediate nodes, in order to send these condensed results to their parent node. This way, we can decrease the number of messages sent and apply early stopping conditions if a subtree already satisfies the query [4].

Lemma 2. *Let s_0 be a sensor node having (direct) children $s_1, ..., s_n$. Let S_c denote the set of sensors in the subtree rooted at s_c, including s_c itself. Then*

$$GF(S_0) = GF(s_0) \cdot \prod_{c=1}^{n} GF(S_c) \tag{5}$$

Our **in-network algorithm** now works as follows: Each leaf node s_{leaf} of the logical connection tree sends its sensor readings to its parent node after using Corollary 1 to remove worlds of s_{leaf} which are not required to compute $PSQ(WSN, \tau)$. Intermediate nodes s_{dir}, upon receiving data from their children, compute the probabilistic sum of their respective subtree using Lemma 5 using polynomial multiplication. To facilitate the expansion of two possibly large polynomials, we propose to use FFT (Fast-Fourier-Transformation). It is well known that the multiplication of two polynomials of degree O(n) can be done in $O(n \log n)$ time using FFT. Unless s_{dir} is the root, s_{dir} uses Corollaries 1 and 2 to reduce the size of its polynomial, and sends the resulting polynomial to its parent node. If s_{dir} is the root, then $PSQ(S, \tau)$ is bounded using Equations 3 and Equation 4. This bound is returned and the algorithm terminates.

5   Performance Evaluation

For the experimental evaluations we used the following setup: We used a simulation of a wireless sensor network containing between 100 and 2500 sensors (default: 1000

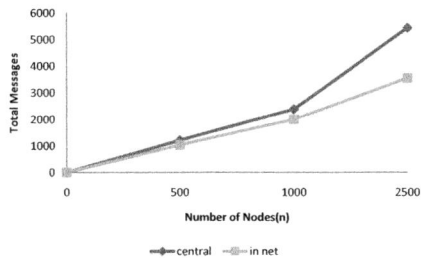

Fig. 1. Scalability w.r.t. communication cost

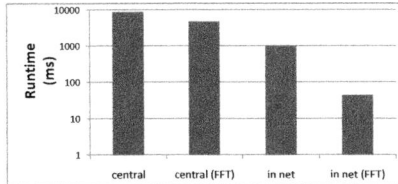 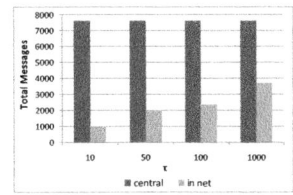

Fig. 2. a) Runtime performance.   b) Scalability w.r.t. τ.

sensors). The locations of the sensors were randomly chosen within a $100m \times 100m$ area and each sensor node was assumed to have a fixed wireless radio range of 30m. All generated sensor instances of the WSNs used a hop-wise shortest-path tree as the routing topology. We assume in all experiments that messages are delivered using a multi-hop setup. In addition, we used a synthetic sensor network tree with 1000 sensors, a height of 5 levels and each node has a branching factor around 5. The uncertain sensor values are simulated as follows: Each sensor measures a set of three intervals, each uniformly selected within the interval [0, 50]. Overlaps are allowed. The corresponding probabilities assigned to these intervals are uniformly selected and normalized such that they sum up to one. For the message size every probability value of the transmitted distributions was taken into account with a 8 bytes and the corresponding value intervals with 16 bytes. For the sake of simplicity, we assumed data packages with a header size of 0 bytes. In our experiments one message sent has a fixed number of 256 bytes. If the amount of values that has to be sent exceeds the 256 bytes the node has to sent multiple messages. The experimental results are averaged over 10 simulation runs where in each run each sensor performs one measurement. Thereby we compare the following two approaches: *central* sends all sensor signals to the central node and performs the query centrally at that node whereas *in net* performs in-network query processing as described in Section 4.

In the first experiment, we evaluate the performance of both approaches by varying the size of the network. The results are shown in Figure 1. As expected, the number of messages increases with the number n sensors in the network. As we can see, the communication cost grow super linear to the number of sensors. The reason is, that the number of sensors not only influences the number of nodes that have to send messages, but also the number of messages that have to be transmitted through the network. The in-network approach consistently improves the communication cost compared to the simple centralized approach by around 25%.

In the next experiment we evaluate the runtime performance of the two approaches, the central and the in-network approach, using different approaches to build the sum distributions. In particular, for both approaches we compare the two variants of using generating-functions for the computation of the sum probability distribution. In the first case, we simply use straight-forward polynomial multiplication. In the second case, we use fast-fourier transformation (FFT) which is well-known to speed up polynomial multiplications. The results are shown in Figure 2. As we can see, if we perform FFT instead of standard polynomial multiplication we can achieve a high performance gain for both approaches. Furthermore, we can see that the in-network approaches significantly outperform the central approach.

In the last experiment we evaluate the performance in terms of communication cost w.r.t. the query threshold τ for the discrete sensor value approach. For this experiment we used the synthetic sensor network with the fixed height of 5 and branching factor around 5. Obviously the performance of the *central* approach is not influenced by the parameter τ because all measurements (sensor value distributions) are sent to the central node as they are generated by the sensors (i.e., neither cutting them at τ nor merging the intermediate results at each non-leaf node). We can see that the in-network data processing with the ability to merge intermediate sum results significantly reduces the number of messages, in particular for lower sum threshold values (τ). With increasing threshold τ, the number of messages increases as well. The reason for this is that larger τ values require to transmit larger distributions that exceed the maximum message package size such that they have to be sent using multiple packages per distribution.

6   Conclusions

Summarizing this work, we introduced the first efficient solution to efficiently answer probabilistic count queries on continuous uncertain data by applying the concept of generating functions to build a polynomial algorithm. We adapted our solutions to wireless sensor networks which have the additional constraint of limited energy. As future work, we plan to develop an approach which is able to incrementally update the result of a probabilistic sum query when a small subset of the sensors changes, without recomputing the result from scratch. We note that the performed set of experiments is very limited, and we are currently evaluating our proposed techniques on a existing network simulators, to provide more extensive scalability experiments.

References

1. Akyildiz, I.F., Su, W., Sankarasubramaniam, Y., Cayirci, E.: Wireless sensor networks: a survey. Computer Networks 38(4), 393–422 (2002)
2. Clark, M.P., Slater, A.G.: Probabilistic quantitative precipitation estimation in complex terrain. Journal of Hydrometeorology (2006)
3. Follmann, A., Nascimento, M.A., Züfle, A., Renz, M., Kröger, P., Kriegel, H.-P.: Continuous Probabilistic Count Queries in Wireless Sensor Networks. In: Pfoser, D., Tao, Y., Mouratidis, K., Nascimento, M.A., Mokbel, M., Shekhar, S., Huang, Y. (eds.) SSTD 2011. LNCS, vol. 6849, pp. 279–296. Springer, Heidelberg (2011)

4. Hua, M., Pei, J., Zhang, W., Lin, X.: Ranking queries on uncertain data: a probabilistic threshold approach. In: Proceedings of the 2008 ACM SIGMOD International Conference on Management of Data, SIGMOD 2008, pp. 673–686 (2008)

5. Hubig, N., Züfle, A., Nascimento, M.A., Emrich, T., Renz, M., Kriegel, H.P.: Continuous Probabilistic Sum Queries in Wireless Sensor Networks with Ranges (Extended Version of this Paper). In: Ailamaki, A., Bowers, S. (eds.) SSDBM 2012. LNCS, vol. 7338, pp. 96–105. Springer, Heidelberg (2012),
http://www.dbs.ifi.lmu.de/Publikationen/Papers/Sumqueries.pdf

6. Madden, S., Franklin, M.J., Hellerstein, J.M., Hong, W.: Tag: a tiny aggregation service for ad-hoc sensor networks. SIGOPS Operating Systems Review 36, 131–146 (2002)

7. Sarma, A., Benjelloun, O., Halevy, A., Widom, J.: Working models for uncertain data. In: Proceedings of the 22nd International Conference on Data Engineering, ICDE 2006, p. 7 (2006)

8. Ye, M., Liu, X., Lee, W.C., Lee, D.L.: Probabilistic top-k query processing in distributed sensor networkss. In: Proc. of the ICDE, pp. 585–588 (2010)

Partitioning and Multi-core Parallelization of Multi-equation Forecast Models

Lars Dannecker[1], Matthias Böehm[2,*], Wolfgang Lehner[2],
and Gregor Hackenbroich[1]

[1] SAP AG, SAP Research Dresden,
Chemnitzer Str. 48, 01187 Dresden, Germany
{lars.dannecker,gregor.hackenbroich}@sap.com
[2] Technische Universität Dresden, Database Technology Group
Nöthnitzer Str. 46, 01187 Dresden, Germany
{matthias.boehm,wolfgang.lehner}@tu-dresden.de

Abstract. Forecasting is an important analysis technique used in many
application domains such as electricity management, sales and retail and,
traffic predictions. The employed statistical models already provide very
accurate predictions, but recent developments in these domains pose new
requirements on the calculation speed of the forecast models. Especially,
the often used multi-equation models tend to be very complex and their
estimation is very time consuming. To still allow the use of these highly
accurate forecast models, it is necessary to improve the data processing
capabilities of the involved data management systems. For this purpose,
we introduce a partitioning approach for multi-equation forecast models
that considers the specific data access pattern of these models to optimize
the data storage and memory access. With the help of our approach
we avoid the redundant reading of unnecessary values and improve the
utilization of the CPU cache. Furthermore, we utilize the capabilities of
modern multi-core hardware and parallelize the model estimation. Our
experimental results on real-world data show speedups of up to 73x for
the initial model estimation. Thus, our partitioning and parallelization
approach significantly increases the efficiency of multi-equation models.

Keywords: Forecasting, Multi-Equation, Partitioning, Parallelization.

1 Introduction

Forecasting is used as the basis for decisions in many application areas such as
electricity management, sales and retail, and, traffic predictions. Due to recent
developments in these domains the employed statistical models face additional
challenges and requirements. Typically the available time for estimating the models and providing accurate predictions is significantly decreasing, which requires
more efficient data processing capabilities in the employed data management
systems. In the energy domain, for example, the emerging smart grid technology
and the integration of more renewable energy sources (RES), require real-time

---

* The author is currently visiting IBM Almaden Research Center, San Jose, CA, USA.

A. Ailamaki and S. Bowers (Eds.): SSDBM 2012, LNCS 7338, pp. 106–123, 2012.

capabilities for balancing the energy demand and supply. Research projects such as MIRABEL [1], and MeRegio [2] address the issues of real-time energy balancing and improved utilization of RES by introducing new developments like dynamic price signals, special energy storage, and demand-response systems. A fundamental prerequisite for current approaches in this area including the balancing of energy in real-time is the availability of accurate forecasts at any time.

Forecasting employs mathematical models—known as forecast models—that model the behavior and development of historic time series. The most important classes of forecast models are autoregressive models [3], exponential smoothing models [4] and models that apply machine learning [5]. Most models use a number of parameters to express specific characteristics of the time series such as seasonal patterns or trends. These parameters are adapted to the specifics of a time series by estimating them on a training data set, with the goal to minimize the forecast error that is measured in terms of an error metric. Typically, each domain exhibits specific characteristics of their time series and thus, employs tailor-made forecast models that address these characteristics. With respect to energy demand time series, e.g., we observe three typical seasonal patterns for the day, the week (working days, weekend) and the year (summer, winter). While our approach can be applied to several application domains, for the remainder of this paper we use the energy domain as a running example.

A model class that is often used for forecasting time series with seasonal behavior comprises multi-equation forecast models [6,7,8,9]. In contrast to typical single-equation models that use just one equation to describe the complete time series behavior, multi-equation models apply an individual sub-model for each specific time period within a selected season, often the daily season (e.g., one model every 30 min). This partitioning of the forecast model allows each sub-model to describe a simpler behavior of the time series, compared to describing all seasonal patterns in one equation. However, the trade-off for using multi-equation forecast models is that now multiple sub-models must be estimated, with each sub-model exhibiting a multi-dimensional search space that exponentially increases with the number of parameters. This leads to higher efforts for the estimation process that typically comprises a large number of iterations per model. In addition, changing time series characteristics caused by continuously available new measurements require the adaptation of forecast models, which typically involves the re-estimation of all model parameters. The re-estimation is almost as expensive as the initial model estimation and thus, very time consuming, especially when using multi-equation models. This clearly contradicts to the requirements given by the real-time balancing.

To still allow the use of highly accurate multi-equation models in the face of real-time environments, it is necessary to optimize their calculation efficiency and thus, decrease the time needed for the model adaptation. One direction to overcome that performance bottleneck is to exploit modern hardware architectures. In this context, we observe two major characteristics. First, modern multi-core hardware systems offer a steadily increasing degree of parallelism, since the performance gain from increasing the core frequency is limited by physical constraints such as an increasing heat loss and power consumption. Second, with an increasing amount of available main memory, it is possible to store and process all considered data directly within the main memory and thus, to avoid reads from the hard disk. However, access times and bandwidth of the main memory do not increase as fast as the computational power, for what reason memory latency and bandwidth became new limiting factors for the reachable performance [10,11]. Thus, to reduce the influence of the memory latency it is important to optimize the data locality within the main memory and store the data for sequential reading instead of random access. As shown in existing work, the performance of algorithms and software greatly benefits from specifically adapting their data storage and memory access to such hardware characteristics [12]. For this purpose, we present an optimization approach that utilizes the specific time series access pattern of multi-equation forecast models to optimize the data storage with respect to modern hardware. With the help of this framework, we provide for each sub-model only the data it needs for its own calculations and avoid accessing unnecessary information. In addition, our approach optimizes the data locality and cache utilization, which greatly improves the data processing speed. In addition, we utilize the increasing parallelization capabilities of modern multi-core hardware systems and parallelize the parameter estimation of the involved sub-models. This helps us to further speed-up the parameter estimation and to meet the requirements posed by real-time environments. The paper makes the following contributions:

- First, we describe the background of multi-equation models in Section 2.
- Second, we present our partitioning approach that optimizes the data storage with respect to the access pattern of multi-equation models in Section 3.
- Third, we describe parallelization strategies that exploit model inter-similarities to estimate all sub-models in parallel in Section 4.
- Fourth, we present the results of our evaluation that show significant speed-up for the parameter estimation of multi-equation models in Section 5.
- Finally, we present related work in Section 6 and conclude the paper in Section 7.

2   Background of Multi-equation Forecast Models

There are two typical model classes used for forecasting, namely single-equation and multi-equation models. Single-equation models describe the complete time series behavior including all patterns and seasonalities within one equation. This means that they consider the most recent predecessor values from the time series as the basis for their calculation. In addition, they use further information like seasonal values or external information (e.g., weather) to increase the forecast accuracy. The example presented in Figure 1(a) considers the five most recent values from the time series plus the respective value at the same time one day and one week ago. Popular examples of single-equation models are Box-Jenkins Models (e.g., ARMA, SARIMA) [3] and adaptations of exponential smoothing (e.g., like introduced by Taylor et al. [13]).

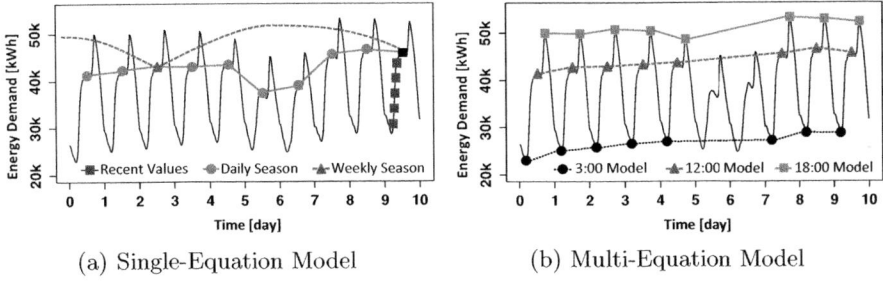

(a) Single-Equation Model   (b) Multi-Equation Model

Fig. 1. Considered Values of Single-Equation and Multi-Equation Models

In contrast to single-equation models, multi-equation models avoid the model-ing of complex seasonal patterns by decomposing the forecast model and assign-ing individual sub-models to each time period within a selected season. There, each sub-model is a separate instance of the forecast model equation, with indi-vidual values for the comprised parameters. In the energy domain, well known representatives of this model class are the EGRV forecast model [6], first order stationary vector regression from Cottet and Smith [7] and, the PCA based fore-casting method from Taylor and McSharry [8]. The reason for splitting up the forecast model with respect to a seasonal pattern is to ease the time series behav-ior a sub-model has to describe and thus, to increase the forecasting accuracy. The underlying assumption is that successive time series values corresponding to a specific time differ only slightly from season to season. Thus, the current time series value is very similar to previous time series values at the same time. In the energy domain, typically the models are divided with respect to the daily season, leading to the assignment of separate sub-models to each data point within a day. For data in hourly granularity this means that for each hour a specific sub-model is used. Some multi-equation models also consider more than one season in their partitioning. The EGRV model, for example, also considers the weekly season by assigning separate models to weekends and working days in addition to the hourly models. It is important to note that the assignment of individual sub models to specific time frames limits the use of multi-equation models to time series with equidistant data points. However, this is in line with most forecast models, which generally also require equidistant observations.

When predicting future values, each sub-model calculates a value for the next day that corresponds to its assigned time frame. Thus, to provide a complete one-day-ahead forecast, each sub-model produces exactly one value. Other than single-equation models, the sub-models base their calculation on historic values from their respective time frames; the 8:00 am sub-model, for example, considers historic values that correspond to 8:00 am. However, some additional components like for example lagged error values, might still use values from other time frames. Figure 1(b) illustrates the pattern of the considered values for the sub-models at 3:00 am, 12:00 noon and 6:00 pm. This is also the specific time series access pattern we exploit for a more efficient physical data partitioning.

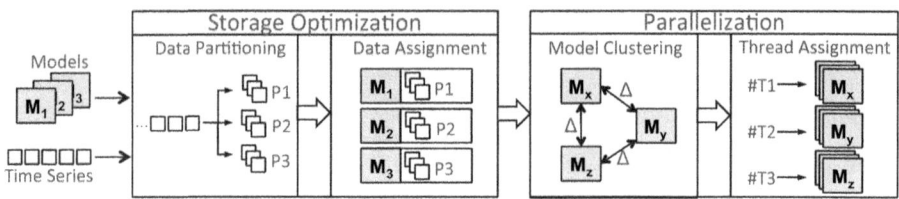

Fig. 2. Multi-Equation Model Optimization Process

An example sub-model of the EGRV forecast model is shown in Equation 1:

$$\text{Hour1} = \alpha\text{Deterministic} + \beta\text{Temperature} + \gamma\text{Load8} + \delta\text{Lags}$$
$$\text{Lags} = \delta_1 y_{t-24} + \delta_2 y_{t-48} + \delta_3 y_{t-72} + \delta_4 y_{t-96} + \delta_5 y_{t-120}. \tag{1}$$

There, the *Deterministic* variables represent additional calendar information and are included as dummy variables (value 0 or 1). Typically they do not require to read time series values. Variables from the *Temperature* category use values from an external temperature time series. In this paper, we do not separately describe the handling of this external information, since our partitioning approach works analogous for such time series. The variables named as *Load8* represent a specific aspect of the EGRV model and correspond to the load at 8:00 am on the previous day. Finally, the *Lags* variables represent the last five time series values that correspond to the specific time of each sub-model (e.g., $t_{x-24}, t_{x-48}, t_{x-72}$, etc.).

3   Partitioning for Multi-equation Forecast Models

The core idea underlying our approach is to physically partition the time series in a way that reflects the model partitioning of the multi-equation model. For this purpose we employ the process illustrated in Figure 2. There, we first partition the data and assign each data partition to its corresponding sub-model. Therefore, we ensure that each model physically accesses only the portions of the time series it needs for its own calculations. This avoids the constant scanning of unnecessary additional values and thus, significantly increases the calculation speed of the complete model. In addition, modern multi-core hardware systems offer an increasing amount of parallelism that we exploit by estimating the involved sub-models in parallel. This is done in the second part of the process. As we assume a larger number of models compared to the number of available threads, we also optimize the thread assignment of these models. The idea is to use similarities between the sub-models and exploit the parameters estimated for one model as the input for the estimation of the successive model. The goal is to reduce the number of iterations until the optimization algorithm converges. As a last step we execute the parallel parameter estimation for all sub-models.

The estimation of forecast model parameters is typically conducted using local (e.g., gradient descent, L-BFGS-B) or global search algorithms (e.g., Simulated Annealing). This optimization task involves a large number of iterations, where

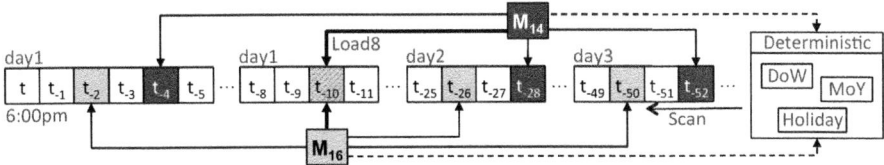

Fig. 3. Initial Sub-Model Time Series Access

each iteration requires to read all necessary time series values. With respect to single-equation models this means that the complete time series is scanned to evaluate the error of the chosen parameter combination. For multi-equation models different parameters are assigned to each sub-model and thus, each model is estimated separately. For each sub-model estimation only the values corresponding to the assigned time frame plus potentially some commonly used variables are required. Figure 3 illustrates the initial situation using the EGRV model. There, the sub-models M14 (corresponds to 2:00 pm) and M16 (corresponds to 4:00 pm) are presented. Both models only require their specific time series values, namely t_{-4}, t_{-28}, t_{-52} for M14 and t_{-2}, t_{-26}, t_{-50} for M16. In addition, both consider the time series values corresponding to the Load8 variable and some deterministic variables (that do not require time series access).

The issue in the non-partitioned case is that the time series is stored chronologically in a single, large array and reading values from cache or memory always requires to read a full cache line or—depending on the hardware platform—even larger block granularities. The number of values contained in a cache line depends on the system specification. Using the Intel Core i7, for example, a cache line contains 64Byte and a double value is 8byte, which results in 8 time series values provided per cache line. Hence, each cache line read for a required value, will also contain time series values that do not correspond to a sub-model's time frame and thus, are not needed for the sub-model estimation. In particular, each cache line will contain only a single required value. Model 16 in Figure 3 for example only needs the value t_{-26}, but the read cache line will also provide the values $t_{-25}, t_{-27}, t_{-28}$, etc. As a result, multi-equation models have a specific time series access pattern that does not correspond to the time series storage and hardware access pattern. Thus, the number of read time series values and cache lines increases with the number of sub-models, where the majority of the read values are not required for a sub-model estimation. This results in a large overhead and a poor data locality within cache and main memory, which leads to long estimation times when working with multi-equation models.

To allow multi-equation models to quickly adapt to new situations and thus, to better meet the requirements posed by real-time applications, we optimize the time series storage to reduce the number of unnecessary time series reads. To do so, we partition the time series in a way that corresponds to the time series access pattern of multi-equation models. The number of partitions directly matches the number of involved sub-models, which in most cases also reflects the granularity of the time series. Each of the used partitions represents a specific

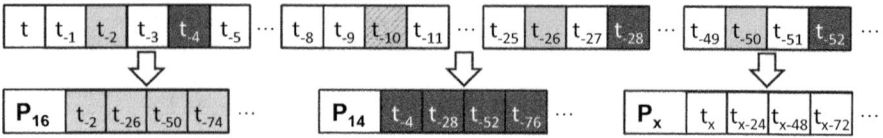

Fig. 4. Time Series Partitioning

time frame (e.g., 4:00 pm) and exclusively stores only values that pertain to this time frame. Since multi-equation models only support equidistant time series (compare Section 2), each partition comprises an equal number of values. Figure 4 illustrates an example partitioning for the EGRV model. There, Partition 14 contains only the values that correspond to 2:00 pm and Partition 16 only the values that belong to 4:00 pm. In addition, we replicate the values that are commonly accessed by all sub-models (e.g., t_{-10} - Load 8) and store the replicates in the partitions as well. This ensures the independence of the models for an optimal further parallelization. After the partitioning is finished, the partitions are assigned to the corresponding sub-model, i.e., the sub-model that describes the same hour the values from a partition belong to.

Each created partition persists in a specific area of the main memory, which means that values belonging to the same partition, i.e., values that are assigned to the same sub-model, are stored closely together within the same memory area. As a result, when reading the time series values, instead of jumping from value to value and reading cache lines that contain unnecessary values, the sub-models can sequentially process all values stored within their respective partition. This sequential reading of time series values directly increases the processing performance. In addition, the tight data storage, results in more necessary values that are contained in a single cache line. As a result, the number of cache lines and memory pages read during a sub-model estimation decreases and thus, the number of cache and memory accesses. Furthermore, for currently running estimations more necessary values can be stored in the different cache levels. This greatly increases the processing speed of the CPU and decreases the number of cache misses. Figure 5 compares the data storage within the cache for the non-partitioned and partitioned case of sub-model M16 - 4:00 pm. There, again we refer to the Intel Core i7 CPU, where each cache line contains 8 time series values. In the non-partitioned case this means that only one of the values

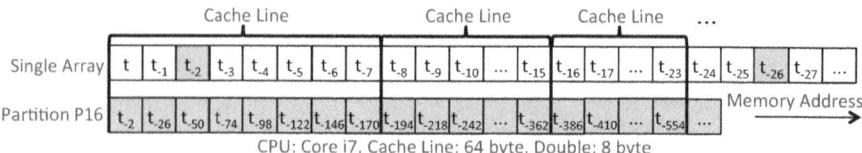

Fig. 5. Cache Organization of Non-Partitioned and Partitioned Time Series

contained in a cache line is needed for the sub-model estimation. In contrast, when using our partitioning approach, all values in a cache line are necessary for the estimation process. Thus, our partitioning approach reduces the number of cache lines processed by the CPU and therefore, reduces the calculation time for each iteration of the estimation algorithm. As a result, the memory and cache locality leads to more efficient calculations for estimating the parameters of all sub-models. Furthermore, the cache locality and the sequential storage of values within a partition also enable the usage of compression techniques like the patched frame of reference (PFOR). Hence, even more values could be stored within a single cache line, which could further increase the processing speed. In the future we will intensively evaluate the use of compression technologies in conjunction to our time series partitioning.

Altogether, our partitioning approach ensures that during all iterations of the optimization algorithm each sub-model only reads its necessary time series values and therefore, reduces the amount of redundantly accessed additional values. This also leads to an optimized memory locality of the data and thus, reduces the number of read cache lines and the amount of cache misses. This greatly improves the calculation performance of the optimization algorithms and greatly decreases the time needed for the multi-equation model parameter estimation.

4   Parallelization of Independent Forecast Models

We further optimize the multi-equation model estimation, by exploiting the parallelization capabilities of modern multi-core hardware. Therefore, we assign all sub-models including their respective partitions to a number of threads that execute the parameter estimation of the sub-models in parallel. Due to the fact that memory throughput and latency can quickly become the limiting factors when using multi-core parallelization, the parallel estimation also profits from the optimized storage and enhanced cache utilization provided by our partitioning approach. Ideally, the number of utilized threads would exactly match the number of involved sub-models, which would also bring the greatest benefit for the parallelization. However, in the real world the number of threads that can be directly executed in parallel on a specific system is limited. The number of these so called hardware threads is typically much smaller compared to the number of involved sub-models (e.g., 48, 96). A system employing a quad-core Intel Core i7 CPU, for example, has eight threads available; one hardware thread and one additional thread using Hyper-threading can be executed per core. Assigning more threads than available hardware threads creates no additional benefit in our scenario, but rather induces additional costs due to overhead for thread scheduling and cache displacement issues. As a result, for the parallel execution of the sub-model estimation, we limit the number of parallel threads to the number of hardware threads available on the executing hardware system and thus, assign multiple sub-models to each thread for sequential estimation. The assignment of the sub-models to the threads is typically conducted using a task queue, where each thread picks the next sub-model as soon as it finished the previous

Table 1. Test Results: Parameter Equality of Three Example Models

	P1	P7	P8	P9	P12	P15	P16	P20
M11	0.5250	0.9991	0.9883	0.9990	0.3798	0.4762	0.3710	0.5355
M16	0.5511	0.9974	0.9633	0.8651	0.3445	0.3161	0.3530	0.5441
M17	0.9989	0.9078	0.9928	0.8645	0.3409	0.3857	0.3012	0.5523

parameter estimation. This leads to good load balance, even if time for estimating sub-models differs significantly. However, on average when assuming, e.g, 48 sub-models and 8 hardware threads, each thread estimates six sub-models.

Given the thread-local serial estimation of a subset of models, we want to further optimize the sequential estimation for the sub-models assigned to one thread. Due to the fact that some sub-models describe similar shapes, we assume that these models should also have similar parameter combinations. In a small experiment we compared the parameters between sub-models after their initial estimation and the results supported our assumption for the most part. Table 1 presents some example parameters for three example models M11, M16 and M17. While the assumption holds for most parameters, it does not for all. Some parameters still differ for sub-models we identified as similar (marked grey in Table 1). In our example, this concerns P1 and P7 for model M17 as well as P9 and P15 for model M11. Still, both models keep their relative similarity to model M16. As a result, the parameter combinations of similar models are a much better approximation of a good starting point for the parameter estimation, compared to starting from the origin. Thus, the basic idea of our improvement is to iteratively provide the result of the preceding parameter estimation, as the input (i.e., the starting value) for the estimation of the subsequent sub-model. This *sequential start* approach reduces the number of necessary iterations for the subsequent optimization algorithms per thread, because only the first sub-model in each thread needs the full effort for the parameter estimation. All subsequent models then profit from better suited starting parameters, for what reason this should greatly reduce the time needed for the parameter estimation.

For the parameter re-estimation we can even go one step further, because the parameter values of the sub-models were already determined in the initial estimation. As described above, some sub-models exhibit a large portion of very similar parameter values. For this reason, we enhance the sequential start approach by clustering the most similar sub-models and assigning each cluster to a single thread. To do so, we measure the distances between the individual values of the parameters for all sub-models (e.g., dist(α_{M1},α_{M2}), dist(α_{M1},α_{M3})) using the *euclidean distance measure* and combine the models with the least distance to each other into one cluster. The number of used clusters directly corresponds to the number of involved threads. In detail, we follow the *k-means clustering* approach using the following process:

1. Sub-models are estimated. Maximum model number per thread calculated.
2. Sub-models are randomly assigned as centroids.
3. Distance between centroids and sub-models is computed.

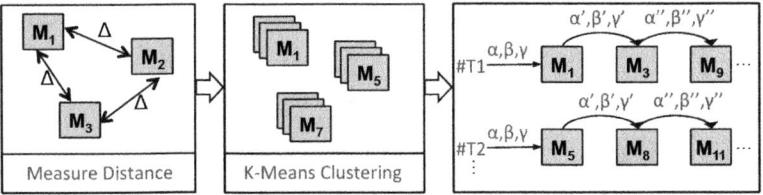

Fig. 6. Clustered Parallelization with Sequential Start

4. Models are assigned to centroid with minimal distance unless thread is full.
5. If thread is full, model is assigned to next best thread.
6. The following steps are repeated until no sub-model changes thread anymore.
   (a) Incrementally compute new centroid from all sub-models per thread.
   (b) Measure distance between new centroids and all sub-models.
   (c) Reorder models with respect to new distance measures.

As soon as our clustering process is finished, we assign the clusters to the respective threads and execute sequential start parameter estimation for each thread. To avoid degeneration in the sense that one thread estimates much more models than the other ones, we place a constraint that all threads execute the same number of models if possible. Figure 6 illustrates the parallelization process. Due to the stronger similarity between the sub-models within one thread, we can even further reduce the number of iterations conducted by the parameter estimation algorithms and thus, reduce the time needed to estimate all sub-models.

To sum up, our parallelization approach further increases the efficiency of multi-equation models. To compensate for the limited number of hardware threads on most systems, we exploited sub-model inter-similarities to optimize the parallelization. While we described our parallelization approach for a local multi-core system, it is also possible to apply the approach to a distributed setting. There, the partitioning approach would be of even more value, because it would limit the amount of transmitted data between the involved systems.

5 Experimental Evaluation

In this evaluation, we substantiate the claims of our multi-equation model optimization approach and show that with the help of our partitioning and parallelization we can greatly reduce the time needed for estimating multi-equation forecast models. Our evaluation compares the time needed for estimating and re-estimating all sub-models, the number of iterations necessary for the (re-)estimation, the amount of cache misses and, the scalability of our approach. For this purpose, we employed the EGRV forecast model as introduced in Section 2). For the parameter estimation we used the Nelder Mead Downhill Simplex approach [14] as a local optimization algorithm. The employed dataset is the energy demand data from the UK National Grid: *National Grid Demand* (publicly

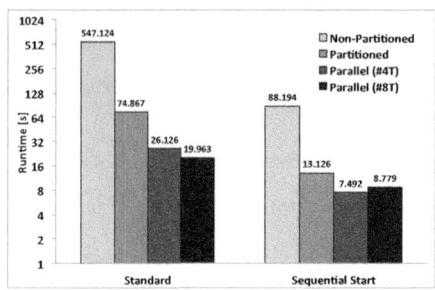
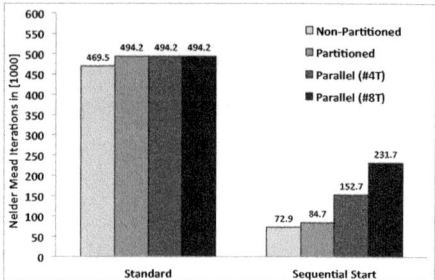

(a) Runtime per Optimization   (b) Iterations per Optimization

Fig. 7. Different Optimizations for Parameter Estimation

Table 2. Average Runtime per 1000 Iterations for used optimization approaches

Variant	Non-Partitioned	Partitioned	Parallel (#4T)	Parallel (#8T)
Avg. Runtime	1.2245s	0.1566s	0.0521s	0.0393s

available [15]): Electricity demand of the United Kingdom. Measures: INDO, January 1st 2002 to December 31st 2009, $30min$ resolution (140256 values).

For our evaluation we used the following test system: Quad-Core Intel Core i7 2635QM (2.0 GHz), 4GB RAM, 128GB SSD, Mac OSX 10.6.8. Our forecasting test suite is written in C++ using the GCC 4.2.1, with OpenMP for the parallelization. We configured OpenMP to use a pre-defined, static number of threads and dynamic thread assignment. Thus, if not explicitly specified (like for the clustering) upon finishing its job, each thread estimates the next sub-model in the queue. All presented results are the average of 20 subsequent runs.

5.1 Parameter Estimation

In the first experiment, we evaluated the runtime necessary for the estimation of a multi-equation model, using our optimization techniques. Thus, we compared the standard, non-partitioned case with the partitioned and parallelized versions. In addition, we also compared the use of the Sequential Start method. Figure 7 illustrates the results. Please note the logarithmic scale that is used in the graphs. There, in Subfigure 7(a) the most important fact is that solely the partitioning approach reduced the time needed for the estimation of all sub-models from 547.124s to 74.867s, even with slightly more iterations (compare Subfigure 7(b)). This is a significant improvement over the non-partitioned case. The parallelization then further reduces the necessary runtime. It can be seen that the runtime improvement is much larger from 1 thread to 4 threads compared to from 4 threads to 8 threads. The reason is that when using 4 threads each of them can be directly executed on one core, while when using 8 threads the additional 4 threads are subject to Hyper-threading. The sequential start method further decreases the necessary runtime for all variants. The runtime of the non-partitioned version is reduced from 547.124s to 88.194s (factor 6.2) and the runtime of the partitioned variant is reduced from 74.867s to 13.126s (factor 5.7). The reason for the reduced runtime is illustrated in Figure 7(b). There, we can see the reduction of the number of iterations conducted until the optimization algorithm converges. Thus, the sequential start provides more suitable starting points for the parameter estimation then starting from the origin for each sub-model. However, it is important to note that the benefit of the sequential start method depends on the number of forecast model parameters and the used estimator. We can further see that the number of iterations for the parallelized methods increases with the number of threads. The reason is that a full estimation is necessary for the first model that is estimated per thread; e.g., for 8 threads, 8 models cannot exploit the parameters from previous models. In addition, when using a larger number of threads, fewer models are estimated by one thread sequentially. Thus, the chance for having only non-similar sub-models assigned to a single thread is higher compared to using 4 threads, where more models are estimated in a row. As a result, the parallelization with 8 threads needs more time using the sequential start optimization then the parallelization with 4 threads. This means that the benefit of the hyper-threaded 4 additional threads is of less value, then the drawback due to the increased number of iterations. This leads us to the problem of automatically assigning an optimal degree of parallelism that we will address in the future.

Table 2 presents the average runtime for all optimizations per 1000 iterations (chosen for better readability). The sequential start and clustering approaches are not listed, because they do not influence the runtime per iteration. The results show a clear trend for the optimization approaches. While the non-partitioned variant needs more than 1.2 seconds for 1000 iterations of the Nelder Mead algorithm, all other approaches are clearly below one second, with the partitioned version marking the maximum of the optimization approaches with 0.1566s.

Overall, we can see that with the help of our optimizations the parameter estimation can be conducted in a few seconds compared to minutes needed for the non-partitioned method. Especially the partitioning optimization provides a significant speed-up. Thus, our optimizations make sure that multi-equation models can be used in the face of the challenged posed by the market dynamics.

5.2   Parameter Re-estimation

Our second experiment is similar to the first one, but we compare the runtime necessary for the parameter re-estimation rather than for the initial estimation. Thus, parameters of the sub-models are not estimated from scratch, but the local search algorithms can start from the last valid parameter combination. Typically the parameter re-estimation is triggered after additional values were added to the time series. Thus, we appended 1440 additional values (i.e., one month) to the time series used for the initial parameter estimation and triggered the re-estimation afterwards. The results are illustrated in Figure 8. There, the results of the standard execution method in Figure 8(a) are similar to the results of the initial estimation. The partitioning greatly reduces the necessary

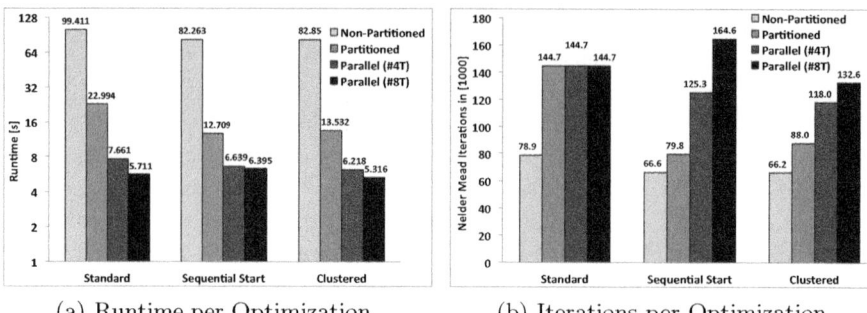

(a) Runtime per Optimization   (b) Iterations per Optimization

Fig. 8. Different Optimizations for Parameter Re-Estimation

runtime, while the parallelization distributes the models to multiple threads and thus, also speeds up the re-estimation. The sequential start reduces the time necessary for the re-estimation, but the decrease is not as significant as for the initial estimation. The reason is that the previous parameters are already a good approximation of a starting point for the local optimization algorithm. For the parallelization with 8 threads, the usage of the sequential start method even increases the necessary runtime. The reason is similar to the causes presented for the initial estimation, but additionally in some cases the parameters from the previous sub-model are worse starting points compared to the old parameters.

For the parameter re-estimation we also used our clustering method described in Section 4. There, the models are not sequentially assigned to the threads, but according to the clustering result. The clustering clearly improves the results of the parallelized version, because it provides better starting points and a more beneficial thread assignment of the sub-models than the sequential start method. Also, using our clustering approach, the calculated centroid is provided as starting point to the first models, which turns out is also a better approximation of a good start than the old parameters. Figure 8 again illustrates the results. There, the decrease/increase of the iterations correspond to the measured runtimes. Especially the increased number of iterations when using the sequential start in conjunction with the 8-thread parallelization is interesting, because it supports the assumption that sequential model assignment is worse compared to using just the old parameter combinations. The runtimes for the single-threaded, partitioned and non-partitioned versions stay roughly the same. The only slight increase is reasoned by the k-means overhead that is still conducted even for those variants. Also, increasing the number of clusters to 4 and 8 for the single threaded versions brings no benefit. The reason is similar to the simple sequential start method, meaning that an increasing number of clusters also increases the number of models that cannot benefit from the sequential start.

Overall, the re-estimation exhibits similar results like the initial estimation. However, the overall runtime for the re-estimation is lower, due to better starting points for the local search algorithm. With the help of our optimizations we also reduced the necessary runtime for adapting a multi-equation model significantly in all cases. The runtime stays always just roughly over 10s for the partitioning approach and with respect to the parallelization the time further decreases.

5.3   Cache Utilization

In this experiment we compared the cache misses for the partitioned, non-partitioned and parallelized case. For this purpose, we used the Intel Performance Counter Monitor that evaluates the values of the Performance Management Unit located directly on modern Intel CPUs. With the help of this tool we measured the number of cache misses that occurred in 20 seconds while running the estimation of an EGRV model. The used Intel Core i7-2635QM with 4 cores provides a 6 MB L3 cache and 256kB L2 cache per core. Our test data set contained 122,736 values, which resulted in a size of 737kB. This means that the complete test data set can be cached in the L3 cache and thus, we expect high L3 cache hit rates in all cases. The results are presented in Table 3. As expected, all cases exhibit a very high L3 cache hit rate. However, the non-partitioned case exhibits a far higher total number of cache misses. Due to the fact that the cache hit rate is nevertheless comparable, this means that in the non-partitioned case far more reads were executed on the L3 cache. This supports the assumption that in the partitioned case less cache lines must be read from the L3 cache. With regard to the L2 cache the result is more diverse. There we see a very low L2 hit rate of only 2% for the non-partitioned case, in comparison to almost 100% for both partitioned cases. This means that storing the data tight together leads to far more necessary values in the cache and thus to a low number of cache misses. The higher total number of L2 cache misses for the parallelized case is reasoned by the execution on 4 threads. The Intel Core i7 has one L2 cache for each CPU core and thus, the number of cache misses roughly increases with the number of threads. As a result, our partitioning approach greatly optimizes the cache utilization, which results in a very high L2 cache hit rate.

5.4   Scalability

In the last experiment we evaluate the scalability of our approach regarding data volume and number of threads. The results are presented in Figure 9. Please note the logarithmic scale of the y-axis. We first compared the behavior of our algorithm with an increasing data volume. For this evaluation we used synthetic data sets with different sizes. The drawback of synthetic data sets is that the number of iterations varies when changing the data volume, which distorts the

Table 3. Test Results: Cache Misses per Storage Approach

	# L3 Misses	% L3 Hits	# L2 Misses	% L2 Hits
Non-Partitioned	206.850 Mio.	93%	8,034.0 Mio.	2%
Partitioned	237,000	99%	7.245 Mio.	96%
Partitioned (4T)	143,000	100%	22.714 Mio.	97%

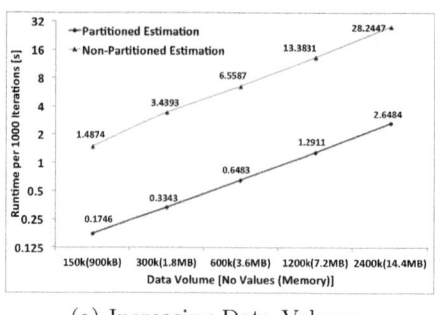

(a) Increasing Data Volume

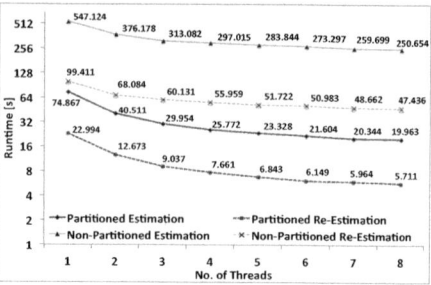

(b) Increasing Number of Threads

Fig. 9. Scalability of Our Optimization Framework

results. Thus, the runtime is for 1000 iterations to eliminate the dependency on the number of iterations. In Figure 9(a)), we see a linear increase of the necessary runtime time for the partitioned and non-partitioned estimation. Meaning that doubling the data volume also doubles the runtime of the parameter estimation. In addition, the development of our partitioning approach is constantly below the non-partitioned case and both have a similar pace of increase. Thus, there won't be a data volume where the non-partitioned case is faster than the partitioned case, which clearly renders the advantages of the time series partitioning.

The results for the scalability concerning an increased number of threads are presented in Figure 9(b). Concerning all cases, we observe the greatest runtime decrease for the parameter estimation and re-estimation, when increasing the number of threads from one to two. When adding more threads, the runtime benefit decreases. Furthermore, the start of the Hyper-threading clearly marks a specific point, after this point the runtime gain is only marginal. In the case of the used Intel Core i7, we have four cores available, meaning that the largest benefit can be observed up to a number of 4 threads. When using more than four threads the additional benefit decreases significantly. As a result, the number of used threads should at least match the number of available cores on a system, because the additional performance gain from hyper-threading is limited. Moreover, we can see that the performance gain when increasing the number of threads is higher for our partitioned case. When increasing the number of threads from one to two, the runtime decreases for the partitioned case by almost one half and for the non-partitioned case by only one third. This trend continues when further increasing the number of threads (e.g., 2 -> 3 Threads: 26% partitioned, 17% non-partitioned). Finally, the runtime difference between one and eight threads is also higher for our partitioned case. The runtime decreases by almost 75% for the partitioned case, whereas for the non-partitioned case the runtime only decreases by around 50%. As a result, our partitioning and the optimized cache utilization also increased the possible degree of parallelization and thus, the performance gain from the parallelization is significantly higher.

6   Related Work

Current research mostly focuses on increasing the accuracy of forecast models. Therefore, there exists only few related work regarding the partitioning and parallelization of forecast models, in particular multi-equation models. Some multi-equation models such as those introduced by Cottet and Smith [7], Soares and Medeiros [9], and Taylor et al. [8] directly include performance optimizations on the logical level. Taylor, for example, proposes to use Principal Component Analysis in conjunction with his multi-equation model to reduce the number of sub-models and thus, the time needed for model estimation. However, the introduced approaches directly modify the model calculations and thus, influence the achievable accuracy. In contrast, our optimizations only change the memory access of the models and do not change the model's calculations, which means that the resulting accuracy is not influenced. In addition, our proposed optimizations on the physical level can be used together with the optimizations on the logical level and thus, increase the performance of these approaches even further.

However, there is also some work that—like our approach—uses optimizations on the physical level. Ge and Zedonik [16] propose to use a skip-list with various levels to provide different data granularities and different history length for various purposes and forecast horizons. Forecasts with a long horizon (> 1 month) use a very coarse grain granularity, while forecasts with a short horizon use very fine-grained data. A similar approach is presented by Agrawal et al. [17] where they merge similar attributes into subset and calculate forecasts only on these subsets. These approaches can greatly reduce the amount of time needed for the estimation of a forecast model. However, reducing the amount of data in most cases also reduces the reachable accuracy of a forecast model. Especially in application domains, where complex patterns are described, the use of fine-grained data and a suitable history length is required for providing accurate forecasts. Our approach does not reduce the number of beneficial values considered during model estimation, but reduces the amount of redundantly read unnecessary values. This means that the accuracy is not influenced by the optimizations proposed by our approach. Also some approaches that directly involve data partitioning and parallelization of forecast models exist. Canas et al. proposed a partitioning solution for neural networks that forecast river-flows to speed up the calculations [18]. Kalaitzakis et al. propose a parallel neural network for forecasting electric load [19]. While the first approach partitions the values into river-flow-specific categories that are provided separately to the neural network, the second approach proposes a parallel calculation of hourly load, similar to multi-equation models. Overall, their partitioning is rather a model decomposition than an optimization on the physical level, especially because the publications omit details about the storage structure and parallelization. In addition, the proposed parallelization only provides the naive approach for calculating the available neurons of the neural network in parallel. In contrast, a massive parallelization approach is provided by Shimokawabe et al. [20]. They propose to distribute the calculation of a weather forecast model on a super computer that provides some thousand GPUs. However, the considered forecast model is a physical weather model, which means that their solution is specific to their approach and cannot directly be applied to multi-equation models.

Overall, our approach is the first optimization approach that exploits the specific access pattern of multi-equation models to optimize the data storage and cache utilization for less redundant and faster data processing. With the help of this technique we speed up the parameter estimation process significantly and our improvement is greatly above the current state of the art. In contrast to most related solutions, our approach does not change the calculation specifics of a forecast model and thus, does not influence the reachable accuracy. Furthermore, our approach can supplement the presented solutions and as a result, even further increase their efficiency enhancements. In addition, the proposed partitioning and parallelization approach is not limited to models from a specific application domain, but can be applied to all kinds of multi-equation forecast models.

7   Conclusion

In this paper, we presented an optimization approach for multi-equation models that greatly increases the efficiency of such models. Our time series partitioning approach ensures that each sub-model only accesses time series values that are necessary for their specific calculations, which leads to an optimized memory locality of the stored time series values. This greatly reduces the number of cache misses, read cache lines and thus, results in an increased data processing efficiency. We further increase the speed of the parameter estimation, by utilizing modern multi-core hardware systems and estimating the sub-models in parallel. There, we addressed the issue of a limited amount of threads, by presenting our sequential start execution and the clustered thread assignment technique. In our evaluation we showed that our optimization framework significantly reduces the time necessary for estimating and re-estimating an multi-equation forecast model. Especially our partitioning approach achieved a major speed up of the optimization calculations. As a result, with the help of our approaches multi-equation models can be used in the face of real-time environments.

In the future we want to enhance our approach by using compression technologies that further increase the number of values read per cache line. In addition, we want to automatically decide for which purpose to use available parallelism most beneficially. The reason is that it is either possible to use available parallelism to (1) increase the number of models estimated in parallel and thus, to potentially increase the efficiency or (2) to simultaneously start the estimation from different starting points and thus, to potentially increase the accuracy.

Acknowledgment. The work presented in this paper has been carried out in the MIRACLE project funded by the EU under the grant agreement number 248195.

References

1. MIRABEL Project (2011), http://www.mirabel-project.eu
2. MeRegio Project (2011), http://www.meregio.de/en/

3. Box, G.E.P., Jenkins, G.M., Reinsel, G.C.: Time Series Analysis: Forecasting and Control. John Wiley & Sons Inc. (1970)
4. Winters, P.R.: Forecasting sales by exponentially weighted moving averages. Management Science, 324–342 (April 1960)
5. Bunnoon, P., Chalermyanont, K., Limsakul, C.: A computing model of artificial intelligent approaches to mid-term load forecasting: a state-of-the-art- survey for the researcher. Int. Journal of Engineering and Technology 2(1), 94–100 (2010)
6. Ramanathan, R., Engle, R., Granger, C.W., Vahid-Araghi, F., Brace, C.: Short-run forecasts of electricity loads and peaks. International Journal of Forecasting 13(2), 161–174 (1997)
7. Cottet, R., Smith, M.: Bayesian modeling and forecasting of intraday electricity load. Journal of the American Statistical Association 98, 839–849 (2003)
8. Taylor, J.W., de Menezes, L.M., McSharry, P.E.: A comparison of univariate methods for forecasting electricity demand up to a day ahead. International Journal of Forecasting 22, 1–16 (2006)
9. Soares, L.J., Medeiros, M.C.: Modeling and forecasting short-term electricity load: A comparison of methods with an application to brazilian data. International Journal of Forecasting 24(4), 630–644 (2008)
10. Wulf, W.A., McKee, S.A.: Hitting the memory wall: Implications of the obvious. Computer Architecture News 23(1), 20–24 (1995)
11. Borkar, S.Y., Mulder, H., Dubey, P., Pawlowski, S.S., Kahn, K.C., Rattner, J.R., Kuck, D.J.: Platform 2015: Intel processor and platform evolution for the next decade. Technical report, Intel Corporation (2005)
12. Kim, C., Chhugani, J., Satish, N., Sedlar, E., Nguyen, A.D., Kaldewey, T., Lee, V.W., Brandt, S.A., Dubey, P.: Fast: Fast architecture sensitive tree search on modern cpus and gpus. In: Proceeding of the SIGMOD 2010 (2010)
13. Taylor, J.W.: Triple seasonal methods for short-term electricity demand forecasting. European Journal of Operational Research 204, 139–152 (2009)
14. Nelder, J., Mead, R.: A simplex method for function minimization. The Computer Journal 7(4), 308–313 (1965)
15. Nationalgrid UK: Metered half-hourly electricity demands (2010), http://www.nationalgrid.com/uk/Electricity/Data/Demand+Data/
16. Ge, T., Zdonik, S.: A skip-list approach for efficiently processing forecasting queries. In: Proceeding of the VLDB 2008 (2008)
17. Agrawal, D., Chen, D., Ji Lin, L., Shanmugasundaram, J., Vee, E.: Forecasting high-dimensional data. In: Proceeding of the SIGMOD 2010 (2010)
18. Cannas, B., Fanni, A., See, L., Sias, G.: Data preprocessing for river flow forecasting using neural networks: Wavelet transforms and data partitioning. Physics and Chemistry of the Earth 31(18), 1164–1171 (2006)
19. Kalaitzakis, K., Stavrakakis, G., Anagnostakis, E.: Short-term load forecasting based on artificial neural networks parallel implementation. Electric Power Systems Research 63, 185–196 (2002)
20. Shimokawabe, T., Aoki, T., Muroi, C., Ishida, J., Kawano, K., Endo, T., Nukada, A., Maruyama, N., Matsuoka, S.: An 80-fold speedup, 15.0 tflops full gpu acceleration of non-hydrostatic weather model asuca production code. In: Proceedings of Super Computing 2010 (2010)

Integrating GPU-Accelerated Sequence Alignment and SNP Detection for Genome Resequencing Analysis

Mian Lu, Yuwei Tan, Jiuxin Zhao, Ge Bai, and Qiong Luo

Hong Kong University of Science and Technology
{lumian,ytan,zhaojx,gbai,luo}@cse.ust.hk

Abstract. DNA sequence alignment and single-nucleotide polymorphism (SNP) detection are two important tasks in genomics research. A common genome resequencing analysis workflow is to first perform sequence alignment and then detect SNPs among the aligned sequences. In practice, the performance bottleneck in this workflow is usually the intermediate result I/O due to the separation of the two components, especially when the in-memory computation has been accelerated, e.g., by graphics processors. To address this bottleneck, we propose to integrate the two tasks tightly so as to eliminate the I/O of intermediate results in the workflow. Specifically, we make the following three changes for the tight integration: (1) we adopt a partition-based approach so that the external sorting of alignment results, which was required for SNP detection, is eliminated; (2) we perform customized compression on alignment results to reduce memory footprint; and (3) we move the computation of a global matrix from SNP detection to sequence alignment to save a file scan. We have developed a GPU-accelerated system that tightly integrates sequence alignment and SNP detection. Our results with human genome data sets show that our GPU-acceleration of individual components in the traditional workflow improves the overall performance by 18 times and that the tight integration further improves the performance of the GPU-accelerated system by 2.3 times.

Keywords: data management for e-science, GPGPU, genomic data analytics.

1 Introduction

The second-generation DNA sequencing devices have been widely used for the past few years. They produce short DNA fragments, or short *reads*, at an ultra-high throughput. For today's genomics research based on short reads, two fundamental data analysis tasks are sequence alignment and single-nucleotide polymorphism (SNP) detection. Sequence alignment matches input reads to a reference sequence. SNP detection takes the output of alignment as input, and finds genetic variation information. In practice, these two tasks are typically performed in sequence as a basic workflow for genome resequencing analysis. Furthermore,

A. Ailamaki and S. Bowers (Eds.): SSDBM 2012, LNCS 7338, pp. 124–140, 2012.

the output of this workflow is usually adopted as input for a number of higher level applications, such as minor allele frequency (MAF) computation [4][20].

Traditionally, such a genome resequencing analysis workflow is implemented using multiple software packages, each performing a task in isolation. For example, 2BWT [6], SOAP2 [11], or Bowtie [9] can be used for the sequence alignment, and SOAPsnp [10] is used to detect SNPs. Moreover, as required by the SNP detection software, additional data processing tools are adopted between the alignment and SNP detection to sort the alignment results. Due to the large amount of data to process, this workflow may take an extremely long running time, e.g., around a week for the human genome. To improve the performance, previous studies have adopted graphics processors (GPUs) to speed up individual tasks, such as SOAP3 [13] and GSNP [14] for alignment and SNP detection, respectively. With the GPU acceleration, the evaluation based on operational genomics data sets have shown that the speedup is significant, e.g., up to 50 times. However, considering the alignment and SNP detection as a workflow, few previous studies further optimize it systematically.

We observe that, with the state-of-art GPU-accelerated tools, the overall performance of the workflow is dominated by the disk I/O, especially that incurred in the intermediate data processing between the alignment and SNP detection. Therefore, it is imperative to address the intermediate data processing in order to improve the overall performance of the workflow. In this work, we develop a GPU-accelerated genome resequencing analysis system tightly integrating the alignment and SNP detection for a higher overall performance. Our focus in this paper is on the integration techniques; details about GPU-acceleration for individual tasks can be found in previous studies [14][13].

We propose three techniques for the integration of alignment and SNP detection. Note that, although our system is based on the GPU for high efficiency, these techniques are applicable to both CPU- and GPU-based implementations.

1. To avoid the external sorting for SNP detection, we propose a partition-based approach. The partitioning is integrated in the alignment component. As a result, the intermediate result processing between the alignment and SNP detection is eliminated.
2. We move the computation of a global matrix that is originally calculated in the SNP detection to the component of sequence alignment. This move saves one scan of the alignment results in the workflow.
3. To further reduce I/O, we develop customized data compression techniques for alignment results.

With these techniques, our system is optimized for the genome resequencing analysis. Compared with a traditional workflow with individual components accelerated by the GPU, our integrated, GPU-accelerated workflow achieves a speedup of 2.3X. As a result, the new system improves the overall performance of a traditional CPU-based workflow by around 43 times.

The remainder of the paper is organized as follows. We introduce the background and related work in Section 2. We describe our integration techniques in detail in Section 3. We evaluate our system in Section 4 and conclude in Section 5.

2   Background and Related Work

In this section, we first briefly introduce the sequence alignment and SNP detection. Then we present the genome resequencing analysis workflow including these two functionalities.

2.1   Sequence Alignment

The second DNA generation sequencing devices can generate short DNA fragments, or *reads*, at an ultra-high throughput. The typical length of short reads is up to around one hundred base-pair (*bp*). For a given reference sequence and a large number of short reads, sequence alignment is to match each read against the reference. Mismatches are allowed for the alignment, e.g., typically two mismatches. The output file of sequence alignment contains multiple lines, and each line has a few attributes, such as the DNA base, the aligned position on the reference, the number of mismatches, and so on. We call such a line an *alignment*. Note that, one input read may have multiple alignments, as it may be matched to multiple positions on the reference. For the whole human genome, there are typically tens of billions of input short reads, and can generate an even larger number of alignments for the SNP detection.

Short read alignment algorithms can be categorized as hashing-based and Burrows-Wheeler transform (BWT) based. A hashing-based algorithm, such as WHAM [12], constructs a hash index containing the positions of all subsequences of the reference. In comparison, a BWT index is constructed with all suffixes of the reference and stored in suffix arrays. Alignment tools employing BWT index are Bowtie [9], 2BWT [6], SOAP2 [11], and SOAP3 [13]. Overall, a hashing based algorithm is efficient when the number of alignments is small, and the disadvantage is that the memory consumption is high. Therefore, in practice, the majority of sequence alignment tasks are done through the BWT index.

To improve the performance of sequence alignment, the GPU has been studied as a hardware accelerator, such as SOAP3 [13], and shown successful to speed up the processing significantly. When building this genome resequencing analysis system, we adopt our home-made GPU-accelerated sequence alignment tool for a tight integration. Our tool adopts a BWT-based sequence alignment algorithm. Additionally, compared with SOAP3, our tool adopts GPU-CPU coprocessing and customized data compression techniques. The measured performance of our sequence alignment component is slightly better than SOAP3.

2.2   SNP Detection

SNP detection is to find DNA variations for a single nucleotide between different members of a species. It calculates the likelihood and other information to indicate whether a *site* (the position holding a base) is a SNP. For example, if the corresponding DNA fragments from two persons are ATCGGC and ACCGGC, respectively, then the second position is probably a SNP site.

A widely used SNP detection tool based on short reads is SOAPsnp [10] employing a Bayesian-based method. Due to the large size of input data, SOAPsnp reads and processes data window by window. A window is defined as a fixed number of consecutive sites on the reference. For a window of sites, the software loads the data related to the window (the corresponding alignments) from disk to memory to perform the computation and outputs SNP results.

To speed up the process of SNP detection, our previous work GSNP [14] implements the same functionality as SOAPsnp but adopts the GPU acceleration. With various optimization techniques, GSNP can achieve a speedup of around 50X over the CPU-based single-threaded SOAPsnp. This resequencing analysis system adopts GSNP as the component of SNP detection, with modifications for a tight integration.

2.3 The Workflow of Genome Resequencing Analysis

Overall, the genome resequencing analysis consists of the sequence alignment and SNP detection. Although the alignment result is the input of SNP detection, in practice, an additional data processing step is required between the alignment and SNP detection tools. This step is to sort the alignment result as well as data format conversion for the SNP detection tool. Therefore, traditionally, three separate software tools are used in the workflow, such as SOAP2 [11], *msort* [2][15], and SOAPsnp [10] for the alignment, sorting, and SNP detection, respectively. Figure 1 shows the overview of such a workflow. We describe the input and output of each software component in detail. Note that, as they are separate software packages, the input and output data are both stored on disk.

Alignment. The input for sequence alignment are the reference sequence and a large number of short reads, which are stored in plain text files. In practice, as the data size may not fit into the memory, multiple passes are performed for the alignment. The alignment result file can be very large, e.g., tens of gigabytes.

Sorting. The alignments should be sorted according to their matched positions on the reference before performing SNP detection. The purpose of sorting is to make the SNP detection tool process sites window-by-window. Figure 2 illustrates the sorting and the process of window-based SNP detection. As the data size is very large, this step is implemented using external sorting algorithms, which are expensive. The GNU *msort* [15] can be used to sort alignments. There are also other more efficient implementations, such as a dedicated alignment sorting tool [2]. By default, in this paper, the sorting program refers to this improved alignment sorting tool rather than GNU *msort*, unless otherwise specified.

SNP Detection. Overall, there are two steps in this task. The first step is to calculate a global matrix, which requires to access all alignment results. Based on the global matrix, the second step calculates likelihood for each site, which

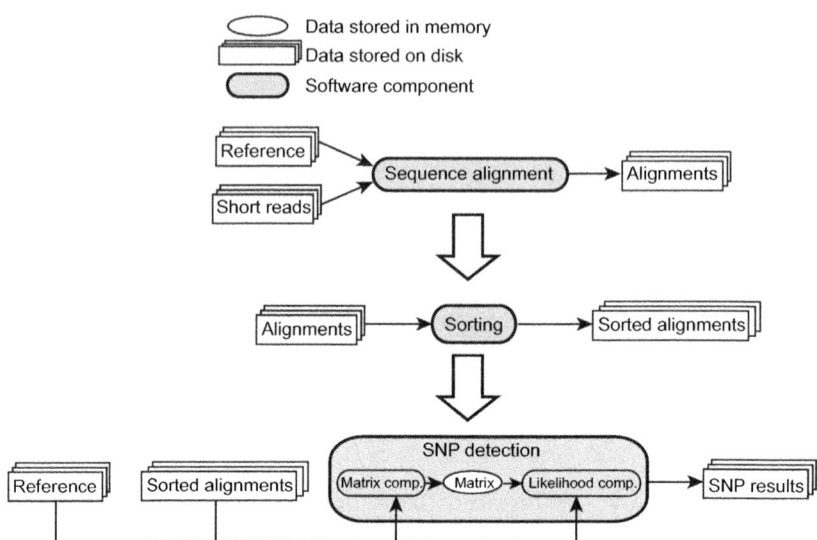

Fig. 1. The traditional workflow of genome resequencing analysis, which consists of sequence alignment, sorting, and SNP detection. These components adopt separate software packages.

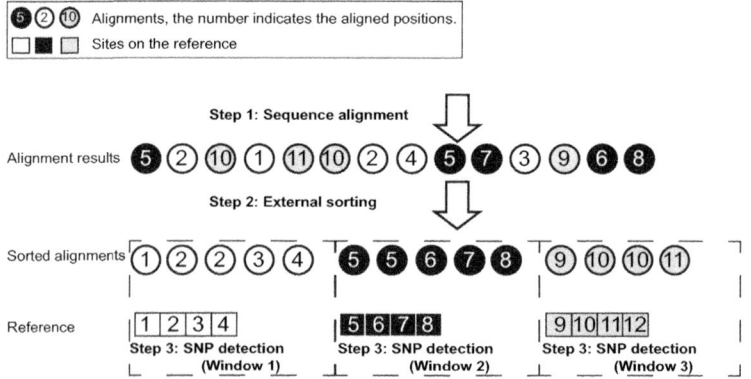

Fig. 2. Sorting and window-based SNP detection. Suppose that there are three windows, each of which contains four sites. A circle represents an alignment at a given aligned position on the reference. The color of a circle indicates which window an alignment or a site belongs to.

accesses all alignments again. Note that, in the second step, the computation for each site is independent. Therefore, with the sorted alignment results, the likelihood computation can adopt a window-based approach. This way, each window of sites and its related data that can fit into memory are loaded from disk to the main memory for the computation.

In summary, a traditional workflow employing three separate software packages contains an expensive external sorting step as well as redundant I/O accesses. Particularly, with the GPU acceleration for the in-memory computation, the I/O dominates the overall performance. We analyze the I/O cost in detail in Section 3.1. In our system, we address these issues through a tighter integration for the sequence alignment and SNP detection components. Moreover, we eliminate the external sorting through an inexpensive partition-based approach.

2.4 Related Work

Existing work on sequence alignment and SNP detection rarely considers integration techniques for a workflow. One exception is work by Wegrzyn et al. [19], which proposes a sequence alignment and SNP detection pipeline that utilizes machine learning algorithms to improve the speed and accuracy. However, their work is not based on short reads and practical algorithms used today. To the best of our knowledge, our study is the first to propose effective optimizations to tightly integrate state-of-the-art alignment and SNP detection algorithms to improve the overall performance systematically.

In addition to a single-machine solution, cloud computing solutions are investigated to improve the performance and scalability of sequence analysis. CloudBurst [16] is a parallel short sequence alignment program developed using Hadoop [1], whose running time scales near linearly with the number of nodes. Myrna [7] targets at gene expression calculation from large-scale RNA data sets, which combines Bowtie [9] and Bioconductor [3]. *Crossbow* [8] is a system that is built on Bowtie [9] and SOAPsnp [10] to perform the genome resequencing analysis in cloud computing using Hadoop. It first performs sequence alignment in the map phase on each node, then sorts the alignment result across all nodes, and finally detects SNPs on each node.

Compared with Crossbow as well as other cloud computing based systems, in addition to the GPU acceleration adopted in our system, we further consider optimizations for a tight integration on a single node. These single node optimizations can be applied on each node in the cloud computing environment. Furthermore, with our optimizations applied, the sorting phase in the MapReduce framework can be avoided.

Finally, there are a few studies for the GPU-accelerated sequence alignment, such as GPU-BLAST [18] and MUMmerGPU [17], which are designed for long reads. For the short read alignment, both SOAP3 [13] and BarraCUDA [5] implement the BWT-index based sequence alignment algorithm.

3   System Implementation

In this section, we describe the details of our integration techniques. Note that, as our system is built on the components with the GPU acceleration, by default, the sequence alignment and SNP detection components referred to in this paper are the GPU-based implementations, unless specified otherwise. Specifically, the sequence alignment is our home-made implementation, and the SNP detection is based on our previous work GSNP [14]. However, our integration techniques are applicable to both GPU- and CPU-based systems.

3.1   Analysis on the Traditional Workflow

As described in Section 2.3, the traditional workflow consists of three separate software packages to perform the sequence alignment, sorting, and SNP detection. Table 1 lists the time breakdown of a traditional workflow (with GPU acceleration) using the hardware and data sets described in Section 4.1. The three I/O intensive components, namely Output, Sorting, and Matrix Computation, take a total of 60% of the overall time. The Output component contains disk I/O operations only. In Matrix Computation, disk I/O takes around 92% of elapsed time. Since the source code of the sorting program is unavailable, we estimate the I/O in Sorting to be half of the elapsed time, assuming one read and one write for each alignment in sorting. There are two major issues through our further analysis. First, there are redundant I/O accesses. Each alignment record is accessed multiple times across the three software packages. Second, there is an expensive external sorting step. In our system, our target is to address these two issues for efficiency through a tight integration.

Table 1. Elapsed time of the traditional workflow (with GPU acceleration)

	Sequence alignment			Sorting	SNP detection	
	Input	Computation	Output		Matrix comput.	Likelihood comput.
Time (sec)	35	213	104	550	155	298
Percentage (%)	2.3	15.9	7.8	41	10.8	22.2

For the first issue, Figure 3 illustrates the multiple data reads and writes on alignment results. Note that, the sorting requires at least one read and one write for each alignment. It may incur more I/O depending on the buffer size. Additionally, as described in Section 2.3, there are two steps in the SNP detection (global matrix and likelihood computation), and each requires a full scan on the alignment results. These two scans on the same alignment results cannot be merged, as the likelihood computation relies on the result of the global matrix computation. In summary, for each alignment, there are at least five disk accesses: two reads and three writes (as shown in Figure 3). In our system, we optimize these multiple data accesses to only two necessary accesses: one write when generating the alignment, and one read when performing the SNP

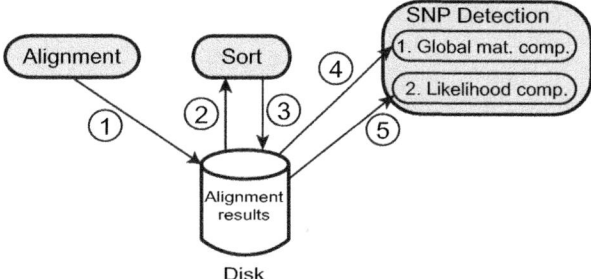

Fig. 3. Five accesses for each alignment in the traditional workflow. (1) Result output after the sequence alignment. (2) Input for the sorting. (3) Sorted result output. (4) Data input for the global matrix computation. (5) Data input for the likelihood computation.

detection. Note that, as the size of alignment file is usually large, e.g., tens of gigabytes, and some other tools may use the alignment results, we consider it necessary to store the alignment results as the intermediate data on disk. The reference sequence is also accessed twice in the workflow. However, the reference size (up to around 750 MB for the whole human genome) is much smaller than the alignment result, and it is straightforward to eliminate the second access by keeping it in memory.

For the second issue, we have presented the purpose of sorting in Section 2.3. Essentially, the sorting is used to arrange the alignments in the same SNP processing window consecutively on disk. Within a window, the order of alignments is not important for the SNP detection program, as there is a counting step to extract summary information for each alignment. Based on this observation, we propose to use range partitioning to achieve the same purpose, but with a low time cost.

3.2   System Overview

Overall, our system consists of the sequence alignment and SNP detection components, and works as follows. First, input reads are processed window-by-window for the sequence alignment. Within each window, when an alignment is produced, it is used immediately to update the global matrix that is used for detecting SNPs later. Then the partitioning function is applied to that alignment, and the alignment is stored in an in-memory buffer. When the buffer is full, its alignments will be compressed and written to the disk. After the sequence alignment for all input reads is done, we start the SNP detection component. The SNP detection component is also executed window-by-window. The window size depends on the partitioning function. Note that, there is no dependence between the window sizes of sequence alignment and SNP detection. Figure 4 illustrates the software components and the workflow in our system.

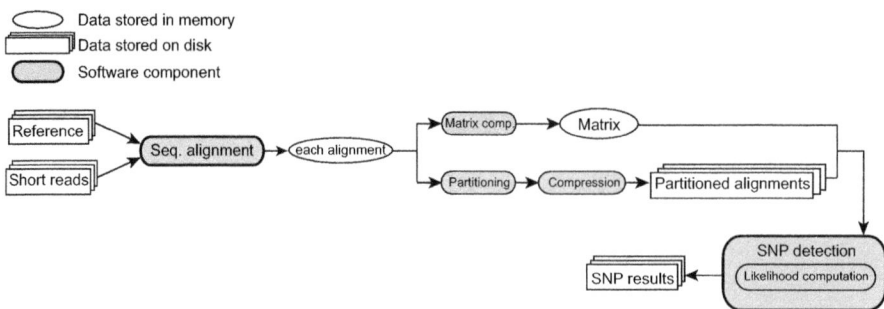

Fig. 4. Components and workflow in our genome resequencing analysis system

3.3   Range Partitioning

For range partitioning, we keep a number of buffers in memory. When an alignment is produced, it is sent into one buffer according to its aligned position in the reference. When a buffer is full, the alignments stored in that buffer are written to the disk (we call the data of each buffer a *block*). In order to facilitate the window-based processing in the SNP detection, the number of buffers and SNP processing windows are the same. Specifically, we maintain B buffers in memory, and each can hold up to m alignments. For the alignment that is matched to site i in the reference, it will be stored in the $\lfloor \frac{i}{m} \rfloor th$ buffer. If the buffer is full, m alignments in that buffer (as a data block) are written to the disk. As one window in the SNP detection may contain alignments from multiple data blocks, there is an additional data structure maintaining the block IDs for each window. Figure 5 illustrates an example of the partitioning corresponding to the sorting example (Figure 2).

The computation complexity of such a partition-based approach is $O(n)$, where n is the number of alignments. Moreover, when an alignment is generated, we can apply partitioning immediately without storing these original alignments on disk. The memory space cost of partitioning is on the in-memory alignment buffers and the block ID list. Suppose we have B buffers (or B SNP processing windows), and each can hold m alignments. Suppose each alignment requires b bytes, the total memory consumption is $(B \times m \times b)$ bytes. As the SNP detection is much more expensive than the partitioning, we mainly consider the performance of SNP detection to tune these parameters. The typical window size in the SNP detection is 256,000 sites [14] with around 1.5 GB GPU memory and 1 GB main memory consumed. A larger window size has little performance impact on SNP detection but significantly increases the memory consumption. Therefore, we set the SNP processing window size in our system as 256,000 sites by default. This way, the number of buffers is up to 11,719 ($B = 11,719$), when evaluating the whole human genome consisting of three billion sites. If the size of each buffer is 512 KB ($m \times b = 512$ KB), which saturates disk bandwidth, then each buffer can store around 2,000 alignments for 100-bp reads. As a result, the total memory consumption for the whole human genome is around 6 GB for

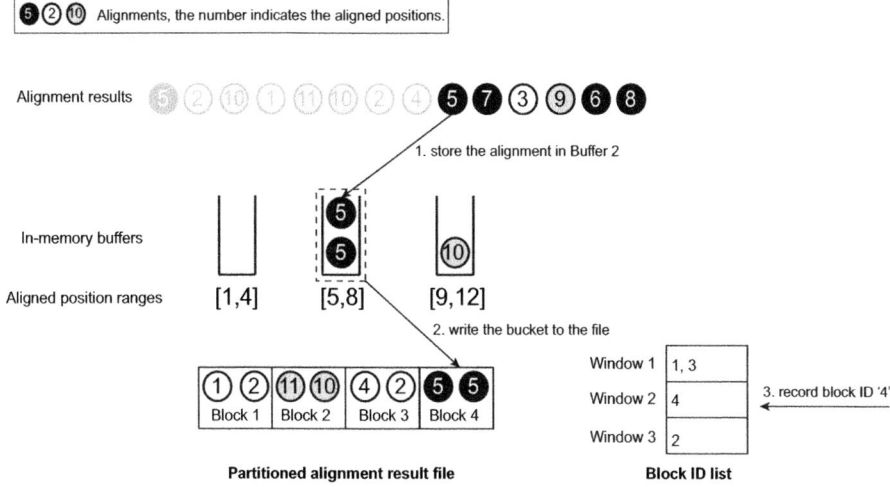

Fig. 5. An example of partitioning when handling the 9-*th* alignment with the aligned position 5. There are three buffers, and each can hold up to two alignments. The first eight alignments have been finished for the partitioning. **Step 1**: the alignment is stored in the second buffer for aligned positions from 5 to 8. **Step 2**: the second buffer is full, thus alignments are written to the disk as Block 4. **Step 3:** the block ID *4* is recorded in the list for Window 2.

in-memory buffers. This is affordable on our server. Additionally, the number of entries in the block ID list can be estimated as $\frac{n}{m}$, thus the memory consumption is around $(4 \times \frac{n}{m})$ bytes. Based on estimation, the memory consumption of the block ID list is around tens of megabytes for the whole human genome. Note that, in practice, the SNP detection is usually performed for a given individual human chromosome rather than the whole human genome, which requires less memory, allows larger buffers for efficiency.

3.4 Alignment Result Compression

Through partitioning and moving matrix computation forward, we have eliminated redundant I/O accesses, however, the size of the alignment result may still be very large. To further improve the performance of the workflow, we develop customized data compression techniques. Note that, we do not adopt general data compression algorithms or tools, such as *gzip*, as they introduce expensive computation cost for both compression and decompression. Additionally, the compression is applied before writing the alignment to the disk, and then the SNP detection component can decompress these compressed result in-memory directly, without additional disk I/O.

Recall that alignment tools output the result as multiple lines, and each line corresponds to an alignment with multiple attributes. Although various alignment tools have slightly different output formats, almost all alignment tools contain these attributes required by the SNP detection tool: the read bases, the quality scores, the number of alignments for the given read, the read length, the reference name, the aligned position, the number of mismatches, and the mismatch information (including the positions of mismatches occurred and their substitution). Specifically, the read bases and quality scores together take more than 90% of the total size. This is because each base has one quality score, and each alignment represents multiple bases. Suppose the read length is l, then the read bases and quality scores attributes both have l values for an alignment. In comparison, for other attributes, each of them only has one value for an alignment.

For the read bases, we do not store them in the alignment result file. The basic idea is that the read can be reconstructed based on the aligned position, mismatch information, and the reference. Figure 6 shows an example of such an approach. The cost of such an approach is that we need to store the reference in memory and have computation overhead when reconstructing the read bases. However, the SNP detection requires to access the reference anyway, and the reference size is much smaller than the alignment result. Additionally, the computation cost is negligible compared with the saved I/O cost.

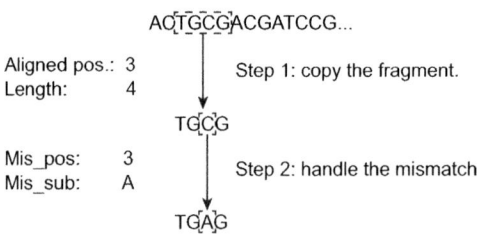

Fig. 6. An example of extracting the read bases based on the aligned position and mismatch information. Step 1: according to the aligned position and read length, we copy the fragment from the reference. Step 2: according to the position of mismatch occurred (mis_pos) and its substitution (mis_sub), we perform the mismatch on the copied fragment.

On the compression of the quality scores, the observation is that one read usually has multiple alignments. These alignments have the same quality score string. Therefore, we keep a table of all unique score strings, and append an additional ID attribute to each alignment for fetching the correct quality score string for a given alignment. To further compress this table, we apply dictionary encoding and run-length encoding.

4   Evaluation

In this section, we first study the performance impact of our integration techniques. Then we compare the end-to-end performance of our system with the original workflow, including the GPU- and CPU-based implementations.

4.1   Experimental Setup

Hardware Setup. We conduct the experiments on a server equipped with an NVIDIA Tesla C2070 and two Intel Xeon E5520, 2.27 GHz quad cores (8 cores, 16 threads in total). C2070 consists of 448 cores and 6 GB GPU memory. The server has 32 GB main memory.

Implementation Details. Our system is built using the GPU acceleration. Specifically, the alignment component is implemented by ourselves based on the Bi-BWT algorithm [6]. The alignment results are reported with up to two mismatches. The SNP detection component is modified based on our previous work GSNP [14]. By default, the window size of the alignment and SNP detection are fixed to 1,000,000,000 reads and 256,000 sites, respectively, and the buffer size in the range partitioning is 1024 KB, unless otherwise specified. Additionally, the time of loading the alignment index from the disk to the main memory is excluded from measurement, as the index can reside in memory. We compare our system with the traditional GPU-based and CPU-based workflows. Recall that the traditional workflow consists of three separate software packages. The traditional GPU-based workflow adopts our home-made GPU-accelerated alignment tool, *msort* developed by BGI-Shenzhen [2] (denoted as *msort*), and GSNP [14]. The traditional CPU-based workflow adopts 2BWT [6], *msort* [2], and SOAP-snp [10]. We use 2BWT rather than other alignment software as it outperforms other tools in our evaluation. Additionally, all CPU-based implementations are single-threaded.

Data Sets. We use a data set for human chromosome 1 (Ch. 1), which is provided by our collaborator, BGI-Shenzhen. This data set contains 15 million short reads in total, and each is 100 *bp* long. The file of short reads is around 3.3 GB. The number of alignments for this data set is 38,584,511. Without compression, the alignment result file is around 9.4 GB. The reference contains around 247 million base pairs. The final SNP detection result file is around 800 MB.

4.2   Performance Impact of Integration Techniques

We first study the performance impact of three integration techniques. For each group of experiments, we compare the implementation without a specific technique with the optimized implementation. Note that, in each group, only the investigated technique will be removed and other optimizations are still employed. Overall, for performance comparison, we divide our system into two components: sequence alignment, and SNP detection. Note that, for our system, the alignment component performs alignment, global matrix computation, partitioning, and compression, and the SNP detection component contains the likelihood calculation step of the SNP detection (as shown in Figure 4).

Range Partitioning. If we do not use partitioning, the system works as follows: we first perform alignment and store all alignments on disk; then we perform

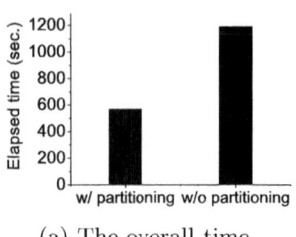
(a) The overall time.

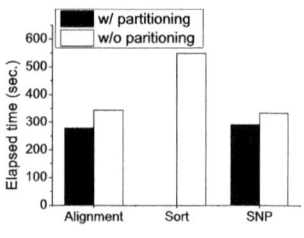
(b) Time of components.

Fig. 7. Performance comparison between using partitioning and using sorting

the external sorting; finally the SNP detection is used based on the sorted alignments stored on disk. Note that, if we use sorting, the compression becomes inapplicable, as the sorting tool can only be used to sort the file stored in specific formats. Figure 7(a) shows that with partitioning, the system is around 2X faster than that using the sorting. Figure 7(b) compares the elapsed time of three components when the system adopts the partitioning and sorting. As the data compression cannot be used when sorting is adopted, the sizes of the alignment output and SNP input both become larger, which slows down the performance of both components.

Data Compression for Alignment Results. Without the compression, the alignment result that is stored on disk as intermediate data will be larger. Figure 8(a) compares the overall elapsed time of the system with and without the data compression techniques for alignment results. The figure shows that with the compression, the overall performance is improved by around 20%. Figure 8(b) shows that, due to the reduced size of the alignment result file, the alignment and SNP detection components are around 22% and 13% faster than their counterparts without the compression, respectively. Figure 8(c) shows that the size of the compressed alignment result is only around 23.4% of that without the compression.

Move of Matrix Computation. Recall that, compared with the original workflow, our system eliminates a data scan on the alignment result when detecting

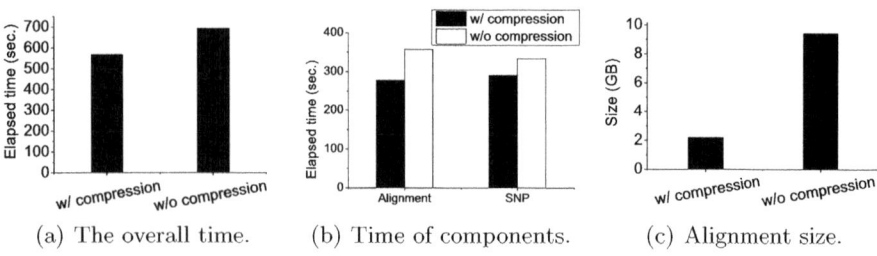

(a) The overall time.   (b) Time of components.   (c) Alignment size.

Fig. 8. Performance comparison between with and without alignment result compression

Table 2. End-to-end performance (seconds) comparison. OLD-CPU and OLD-GPU indicate the traditional workflow using the CPU- and GPU-based software, respectively.

	Alignment	Sort	SNP detection	**Overall**
OLD-CPU	1562	550	22321	24433
OLD-GPU	330	550	453	1333
Our system	278	–	290	568

SNPs due to the change of matrix computation. Figure 9(a) shows the overall elapsed time of the system with and without the move of the global matrix computation. With this technique, the overall performance is improved by around 18%. This improvement is from the elimination of additional disk I/O on the alignment result when detecting SNPs. Figure 9(b) further shows that, as the global matrix is calculated in the alignment component in our system, this component is slightly slowed down. However, the overall elapsed time is reduced due to the significant performance improvement from the SNP detection component.

In summary, the range partitioning is the most significant optimization, which can eliminate the expensive external sorting. With the range partitioning, the system is around 2X faster than that without the optimization. The data compression technique and move of matrix computation can further improve the performance by around 20% and 18%, respectively.

4.3 End-to-End Performance Comparison

We show the end-to-end performance comparison in Table 2. Note that, in our system, the partitioning, compression, and matrix computation are all included in the alignment component. This table shows that, the traditional workflow with the GPU acceleration for individual components outperforms its CPU counterpart by around 18 times. Furthermore, with the integration techniques, our system further improves the performance by around 2.3 times. This improvement is from all three components. First, for the alignment, the compression reduces the alignment size to save the I/O time. Second, the original expensive sorting is

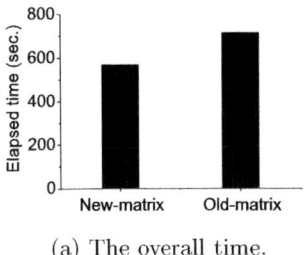

(a) The overall time.

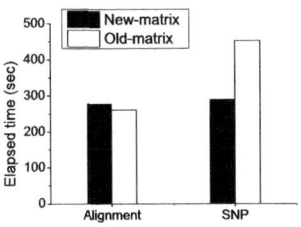

(b) Time of components.

Fig. 9. Performance comparison between the systems with the move of global matrix computation (new-matrix) and the original global matrix computation (old-matrix)

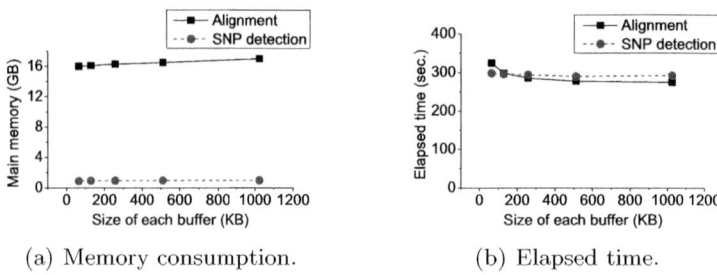

(a) Memory consumption.   (b) Elapsed time.

Fig. 10. The memory consumption and elapsed time with the buffer size in the range partitioning varied

eliminated in our system. Third, for the SNP detection, one data scan on alignment results is also eliminated in our system. Compared with the traditional CPU-based workflow (without the tight integration techniques), our system is around 43X faster.

Finally, we investigate the performance and memory consumption impact with the partitioning buffer size varied. The windows sizes of alignment and SNP detection are set according to the performance of alignment and SNP detection components. We only show the main memory consumption as the buffer size does not affect the GPU processing. Figure 10(a) shows that the memory consumption slightly increases when the buffer becomes larger for the alignment. This is because another data structure (a suffix array) dominates the overall memory usage for the alignment, which consumes around 12 GB. For the SNP detection, a larger partitioning buffer results in a smaller block ID list, which is insignificant in the overall memory consumption. Figure 10(b) shows that the alignment can benefit from a larger buffer, since the disk I/O throughput is higher when writing a larger data block. This parameter is less significant for the performance of SNP detection.

5   Conclusion

We have developed a GPU-accelerated system with a tightly integrated workflow optimized for genome resequencing analysis: the sequence alignment is used first for short reads, and then the SNPs are detected based on the alignment result. To reduce the I/O overhead in the traditional workflow, we propose three techniques for a tight integration of the alignment and SNP detection. We first use range partitioning to avoid the external sorting for alignment results. We also develop customized data compression techniques to further reduce the size of the alignment result. Finally, we calculate the global matrix computation when generating alignments, which is originally performed in the SNP detection component. As a result, compared with the traditional GPU- and CPU-based workflow consisting three separate software packages, our system can achieve a speedup of around 2.3X and 43X, respectively.

Acknowledgment. This work was supported by grants 617509 from the Hong Kong Research Grants Council and MRA11EG01 from Microsoft SQL Server China R&D. We thank our collaborator BGI Shenzhen for providing us application requirements and data sets.

References

1. Apache Hadoop, http://hadoop.apache.org/
2. Short Oligonucleotide Analysis Package, BGI-Shenzhen, China, http://soap.genomics.org.cn
3. Gentleman, R., Carey, V., Bates, D., Bolstad, B., Dettling, M., Dudoit, S., Ellis, B., Gautier, L., Ge, Y., Gentry, J., Hornik, K., Hothorn, T., Huber, W., Iacus, S., Irizarry, R., Leisch, F., Li, C., Maechler, M., Rossini, A., Sawitzki, G., Smith, C., Smyth, G., Tierney, L., Yang, J., Zhang, J.: Bioconductor: open software development for computational biology and bioinformatics. Genome Biology 5(10) (2004)
4. Kim, S.Y., Lohmueller, K.E., Albrechtsen, A., Li, Y., Korneliussen, T., Tian, G., Grarup, N., Jiang, T., Andersen, G., Witte, D., Jorgensen, T., Hansen, T., Pedersen, O., Wang, J., Nielsen, R.: Estimation of allele frequency and association mapping using next-generation sequencing data. BMC Bioinformatics 12, 231 (2011)
5. Klus, P., Lam, S., Lyberg, D., Cheung, M.S., Pullan, G., McFarlane, I., Yeo, G., Lam, B.: BarraCUDA - a fast short read sequence aligner using graphics processing units. BMC Research Notes 5(1) (2012)
6. Lam, T.W., Li, R., Tam, A., Wong, S., Wu, E., Yiu, S.M.: High throughput short read alignment via bi-directional bwt. In: IEEE International Conference on Bioinformatics and Biomedicine, pp. 31–36 (2009)
7. Langmead, B., Hansen, K., Leek, J.: Cloud-scale RNA-sequencing differential expression analysis with myrna. Genome Biology 11(8) (2010)
8. Langmead, B., Schatz, M., Lin, J., Pop, M., Salzberg, S.: Searching for SNPs with cloud computing. Genome Biology 10(11) (2009)
9. Langmead, B., Trapnell, C., Pop, M., Salzberg, S.: Ultrafast and memory-efficient alignment of short DNA sequences to the human genome. Genome Biology 10(3) (2009)
10. Li, R., Li, Y., Fang, X., Yang, H., Wang, J., Kristiansen, K., Wang, J.: SNP detection for massively parallel whole-genome resequencing. Genome Research 19(6), 1124–1132 (2009)
11. Li, R., Yu, C., Li, Y., Lam, T.-W.W., Yiu, S.-M.M., Kristiansen, K., Wang, J.: SOAP2: an improved ultrafast tool for short read alignment. Bioinformatics 25(15), 1966–1967 (2009)
12. Li, Y., Terrell, A., Patel, J.: Wham: A high-throughput sequence alignment method. In: Proceedings of the 2011 ACM SIGMOD International Conference on Management of Data (2011)
13. Liu, C.-M., Lam, T.-W., Wong, T., Wu, E., Yiu, S.-M., Li, Z., Luo, R., Wang, B., Yu, C., Chu, X., Zhao, K., Li, R.: SOAP3: GPU-based Compressed Indexing and Ultra-fast Parallel Alignment of Short Reads. In: Third Workshop on Massive Data Algorithmics (2011)
14. Lu, M., Zhao, J., Luo, Q., Wang, B., Fu, S., Lin, Z.: GSNP: A DNA Single-Nucleotide Polymorphism Detection System with GPU Acceleration. In: International Conference on Parallel Processing, ICPP (2011)

15. Poser, W.: GNU msort, http://billposer.org/Software/msort.html
16. Schatz, M.C.: CloudBurst: highly sensitive read mapping with MapReduce. Bioinformatics 25(11), 1363–1369 (2009)
17. Trapnell, C., Schatz, M.C.: Optimizing data intensive gpgpu computations for dna sequence alignment. Parallel Computing 35, 429–440 (2009)
18. Vouzis, P.D., Sahinidis, N.V.: GPU-BLAST: using graphics processors to accelerate protein sequence alignment. Bioinformatics 27(2), 182–188 (2011)
19. Wegrzyn, J.L., Lee, J.M., Liechty, J., Neale, D.B.: PineSAPsequence alignment and SNP identification pipeline. Bioinformatics 25(19), 2609–2610 (2009)
20. Yi, X., Liang, Y., et al.: Sequencing of 50 Human Exomes Reveals Adaptation to High Altitude. Science 329(5987), 75–78 (2010)

Discovering Representative Skyline Points over Distributed Data

Akrivi Vlachou[1,*], Christos Doulkeridis[1,2,**], and Maria Halkidi[2]

[1] Norwegian University of Science and Technology (NTNU), Norway
[2] University of Piraeus, Greece
vlachou@idi.ntnu.no,{cdoulk,mhalk}@unipi.gr

Abstract. Skyline queries help users make intelligent decisions over complex data. The main shortcoming of skyline queries is that the cardinality of the result set is not known a-priori. To overcome this limitation, the representative skyline query has been proposed, which retrieves a fixed set of k skyline points that best describe all skyline points. Even though the representative skyline has been studied before in centralized environments, this is the first paper that addresses efficient computation of the representative skyline in distributed systems. The distributed nature of the environment makes the task of discovering truly representative skyline points even more challenging. In this paper, we propose a novel framework for discovering the representative skyline over distributed data sources. Our experimental study demonstrates the efficiency and effectiveness of our framework.

1 Introduction

Skyline queries [1] constitute a powerful tool for multi-objective optimization, especially in the case of multiple and conflicting criteria. An important shortcoming of skyline queries is that the size of the result set is not fixed, but largely depends on various factors such as the data distribution or the dimensionality of the data space. Thus, in contrast to other popular query types, such as top-k queries [3,5] that return results of expected size, the cardinality of skyline set is unrestricted and can sometimes be comparable to the size of the complete data set. To alleviate this shortcoming, centralized approaches that select a restricted set of *representative skylines* have been proposed [7,11].

As data management becomes inherently distributed due to massive content generation at disparate locations, the importance of distributed query processing is even more evident. Lately, this is also intensified by the advent of large-scale distributed data centers and cloud computing infrastructures. In such setups, servers store portions of the data set and the objective is to support efficient and effective techniques for query processing and advanced data analysis.

In this paper, we address for the first time the problem of discovering a set of k skyline representatives over distributed data, which is even more challenging

---

* A. Vlachou was partially supported by the Greek State Scholarship Foundation.
** C. Doulkeridis was supported under the Marie-Curie IEF grant number 274063.

A. Ailamaki and S. Bowers (Eds.): SSDBM 2012, LNCS 7338, pp. 141–158, 2012.

than the centralized representative skyline query due to the lack of global knowledge. As skyline points are equivalent by definition, any representative skyline query uses an error metric that captures the loss in the expressiveness of the skyline set due to the absence of the non-representative skyline points. Thus, the representative skyline query defines an optimization problem that aims to retrieve the k skyline points that minimize the error metric. Different error metrics have been proposed for representative skyline queries, such as dominance [7] and distance-based [11] error metric. Assuming a generic distributed setup where a set of N servers store portions of the entire data set autonomously, we introduce a novel framework for distributed skyline representative algorithms which supports any error metric and retrieves a representative skyline set of low representation error by only considering a fraction of the distributed data.

More precisely, our framework encompasses a baseline approach as well as two alternative efficient algorithms. The *distributed skyline algorithm* is used as a baseline and incorporates representative skyline computation in a distributed skyline query by transferring all local skyline points to the coordinating server. In case of error metrics that are not influenced by dominated points, such as [11], the distributed skyline algorithm returns exactly the same representative points as if the query was executed on all data by a centralized algorithm. Furthermore, we prove that any algorithm that transfers fewer local data than the distributed skyline query may report non-skyline points as representative skyline points, if only a single communication phase is employed.

Motivated by this observation, we propose the *distributed skyline representative algorithm* that relies on two communication phases in order to reduce the transferred data. In the first phase, each individual server discovers its k local representative skyline points. The local representatives are transmitted to the coordinating server, which extracts an initial set of k global skyline representatives. At the second communication phase, the currently defined global representatives are forwarded back to servers, to be tested for dominance by other local skyline points. The identified set of dominating points is then sent to the coordinating server, which then re-applies the representative skyline algorithm to extract the final set of k representative skyline points. Finally, we introduce the *distributed error-based representative* algorithm that processes the query in the same spirit as the distributed skyline representative algorithm, but also exploits the information about the value of error metric at local level to reduce further the induced error of the representative skyline set.

2   Related Work

Restricting the skyline cardinality is motivated by the fact that the skyline cardinality increases with the data set dimensionality. To deal with this *dimensionality curse*, one possibility is to restrict the cardinality of the result set, by choosing k skyline points out of the entire set. Towards this goal, the authors in [2] propose the *k-dominant* skyline query. The authors relax the idea of dominance to k-dominance, in order to increase the probability of one point dominating another point, thereby restricting the skyline cardinality. *Skyline ordering* [8] is an approach that produces arbitrary size constrained skyline sets by employing skyline-based partitioning on the data set.

Selecting *representative skyline points* in centralized domains has recently attracted significant attention for retrieving exactly k points from the skyline set. In [7], the authors study the problem of selecting k skyline points, so that the number of points dominated by at least one of these k skyline points is maximized. In [11], an approach is presented for retrieving k representative skyline points, which are defined as the set of k points that minimize the maximum distance between a non-representative skyline point and its nearest representative. In [10], representative skylines are studied under the assumption that user preferences are expressed as thresholds. The thresholds indicate the worst value on each attribute that is acceptable from each user. The proposed approach relies on the probability distribution of the user's thresholds. Preference-based representative skyline queries are out of the scope of this paper.

This is the first paper that studies representative skyline queries in distributed systems. However, several approaches have been proposed for efficient skyline processing in distributed environments; a detailed survey of distributed skyline processing can be found in [6]. For example, subspace skyline computation over peer-to-peer network has been studied in [12,13]. Cui *et al.* [4] proposed the PaDSkyline algorithm for skyline query processing in a generic distributed environment. In [14], a feedback-based distributed skyline (FDS) algorithm is proposed, which aims to minimize the bandwidth consumption. However, the aforementioned papers focus on the efficient computation of the skyline query, not the representative skyline query.

3   Preliminaries and Problem Statement

Preliminaries. In our system model, a set of N servers S_i participate in the distributed skyline computation, while a coordinator server S_C is responsible for communication with the servers in order to produce the desired representative skyline set. Data is distributed in the sense of horizontal partitioning, thus each server S_i stores locally a set of points P_i. The entire data set P is the union of all sets of points P_i stored locally at any server S_i ($P = \bigcup P_i$, $P_i \bigcap P_j = \emptyset$). A representative skyline query is initiated by the coordinator. In the following, we provide the necessary definitions and preliminaries.

Given a data set P on a data space defined by a set of d dimensions $\{d_1, \ldots, d_d\}$, a point $p \in P$ is represented as $p = \{p[1], \ldots, p[d]\}$ where $p[i]$ is the value on dimension d_i. Without loss of generality, we assume that $\forall d_i : p[i] \geq 0$, and that smaller values are preferable.

Definition 1. *(Skyline set) A point $p \in P$ dominates another point $q \in P$, denoted as $p \prec q$, if (1) on every dimension d_i, $p[i] \leq q[i]$; and (2) on at least one dimension d_j, $p[j] < q[j]$. The skyline $\mathcal{S}(P)$ is a set of points that are not dominated by any other point in P.*

Consider the example in Fig. 1, where each point represents a hotel and the y-dimension represents the price of a room, while the x-dimension captures the distance of the hotel to the beach. A hotel dominates another hotel because it is cheaper and closer to the beach. Thus, the skyline points (a, i, m and k) are the best possible trade-offs between price and distance from the beach.

Problem Statement. Unfortunately, as the dimensionality of the data set grows, the skyline operator loses its discriminating power and returns a large fraction of the data. The huge size of the result set hinders decision-making and motivates the ranking of skyline points. Therefore, users prefer to retrieve k representative points instead of the whole skyline set. The representative skyline points are chosen to best describe the tradeoffs among different dimensions offered by the full skyline. As skyline points are equivalent by definition, an error metric is defined to capture the *representativeness* of a set of k skyline points.

Definition 2. *(Representative skyline set) Given an integer k, the representative skyline of a data set P is a set $\mathcal{K}$ of k skyline points of $\mathcal{S}(P)$ that minimizes the error metric $Er(\mathcal{K})$.*

As we will elaborate in the following, different definitions of the error metric for the representative skyline have been proposed: dominance-based error metric [7] and distance-based error metric [11]. Both error metrics are supported by our distributed framework. Definition 2 leads to the following problem statement of this paper.

Definition 3. *(Distributed representative skyline set) Given a distributed data set $P = \bigcup P_i$, compute its representative skyline set $\mathcal{K}$ of size k.*

Dominance-based representative skyline. In [7], the representative skyline set is defined based on the dominated points. More precisely, the authors quantify the concept of representativeness by the total number of (distinct) data points dominated by one of the k representative skyline points. In other words, the k most representative skyline points are the ones that minimize the number of the data points that are not dominated by any representative point. Thus, the error metric is defined as:

$$Er(\mathcal{K}) = \{|\{p\}| : p \in P, p \notin \mathcal{K}, \nexists p' \in \mathcal{K} : p' \prec p\}$$

For example, Fig. 2(a) depicts a data set P of hotels, along with its skyline points $\mathcal{S}(P)$. This data set contains 6 skyline points depicted with circles on a line. In addition, the representative skyline points that are derived from the dominance-based algorithm for $k=3$ are depicted with squares.

The problem is shown to be NP-hard when the dimensionality is 3 or more and it can be approximately solved by a polynomial time greedy algorithm. The proposed greedy algorithm starts by computing the skyline set and the representative error of each skyline point, i.e., the number of data points that are dominated by each skyline point. The algorithm picks as the first representative skyline point the skyline point that has the highest number of dominated points. After removing the data points that are dominated by the first representative

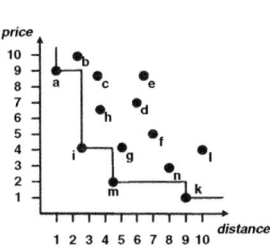

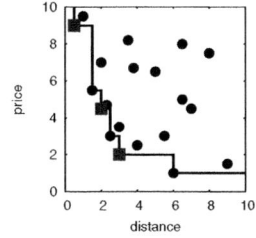   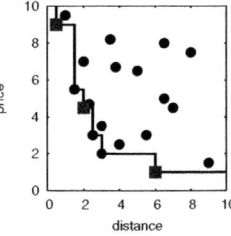

(a) Dominance-based.   (b) Distance-based.

Fig. 1. Skyline example   **Fig. 2.** Example of representative skyline

skyline point, the representative error of each skyline point is re-computed and again the skyline point with the maximum value is picked. This is repeated until k skyline points are selected. To overcome the high memory requirements of the greedy algorithm, a probabilistic counting technique can be applied for estimating the number of distinctly dominated data points. This leads to an index-based randomized algorithm for finding the representative skyline points.

Distance-based Representative Skyline. The error metric $Er(\mathcal{K})$ for the distance-based representative skyline [11] is defined as:

$$Er(\mathcal{K}) = \max_{\forall p \in \mathcal{S}(P) - \mathcal{K}} (\min_{\forall p' \in \mathcal{K}} d(p, p')),$$

where $d(p, p')$ is the Euclidean distance between points p and p'. Intuitively, a distance-based representative skyline set is good, if for every non-representative skyline point, there exists a representative skyline point nearby.

Fig. 2(b) depicts an example of the distance-based representative points $\mathcal{K}$ for illustrative purposes. The $k=3$ representative points are depicted with squares, while the skyline points depicted with circles on a line.

For dimensionality at least 3 the problem is NP-hard, thus the authors propose a greedy algorithm [11], namely I-greedy, to compute the distance-based representative. I-greedy assumes a multidimensional index on the data set and uses the concept of max-rep-dist for computing the representatives. Given a subtree in the R-tree, its max-rep-dist is a value that upper-bounds the representative distance of any potential skyline point p in this subtree. Initially, it takes as input an initial set $\mathcal{K}$ containing an arbitrary skyline point, which is used as a representative point. For example, this point can be the point with the smallest x-coordinate. I-greedy maintains the entries of the R-tree in a sorted list. In each iteration, I-greedy processes the entry E with the largest max-rep-dist. If the next entry is dominated by at least one point retrieved so far, the entry is discarded. Otherwise, I-greedy searches for the entry with the smallest L_1 distance to the origin among all entries in the sorted list whose min-corners dominate E. If such an entry exists, it must be an intermediate entry, so the entries of its child node are inserted, if they are not dominated by any point retrieved so far. If such an entry does not exist, then E is processed. If E is a point, it is inserted to $\mathcal{K}$ as the next representative skyline point. Otherwise, the entries of its child node are inserted in the list, if they are not dominated by any point retrieved so far.

Algorithm 1. *Distributed skyline algorithm DSA*

1: **INPUT:** k, Coordinator S_C, Servers S_i
2: **OUTPUT:** Representative skyline $\mathcal{K}$
3: **for** $(\forall S_i : i \in [1, N])$ **do**
4:     $\mathcal{S}(P_i) \leftarrow S_i.\underline{\text{skyline}}()$
5: **end for**
6: $\mathcal{K} \leftarrow S_C.\underline{\text{representative}}(\bigcup \mathcal{S}(P_i))$
7: **return** $\mathcal{K}$

---

4   Distributed Representative Skyline Algorithms

In this section, we present our framework that encompasses two algorithms for discovering the representative skyline points over distributed data. Our generic framework is parameterized by a centralized skyline representative algorithm that is executed locally at the participating servers. Any such algorithm for local skyline representative computation can be plugged in our framework. Currently, we have incorporated in our framework the error metrics of two existing skyline representative algorithms studied in the related work: distance-based representative [11] and dominance-based representative [7].

4.1   Distributed Skyline Algorithm (DSA)

In a generic distributed system, processing the representative skyline query can be performed by integrating the representative skyline computation in a distributed skyline algorithm. Algorithm 1, termed *DSA*, serves as a baseline and adheres to this strategy to produce a representative skyline set.

DSA is processed at S_C by first sending a skyline query to all servers S_i, which in turn process the query locally over their data P_i (line 4). Then, each server S_i reports its *local skyline set* $\mathcal{S}(P_i)$ to S_C. Similar to the case of distributed skyline query processing, a centralized algorithm for finding the representative skyline set, such as [7,11], is processed at the coordinator server S_C, in order to obtain the *representative skyline set* $\mathcal{K}$ (line 6).

An important property of the skyline operator is that the skyline set of a distributed data set P is a subset of the union of the local skyline sets of all partitions $\mathcal{S}(P) \subseteq \bigcup \mathcal{S}(P_i)$. This property of the skyline set leads to an interesting observation about the *DSA* algorithm. As long as the error metric used for defining the representative skyline is not influenced by non-skyline points of the data set, the retrieved representative skyline set of *DSA* is equivalent to the representative skyline set of the entire data set $P = \bigcup P_i$. Moreover, this is accomplished without requiring the transfer of all local data points P_i, but only the local skyline points $\mathcal{S}(P_i)$. Notice that the distance-based error metric satisfies the afore-described observation, therefore *DSA* produces the identical result of the centralized distance-based representative algorithm.

However, *DSA* has an important drawback; it needs to transfer the complete local skyline sets to the coordinator. Under certain circumstances, depending on

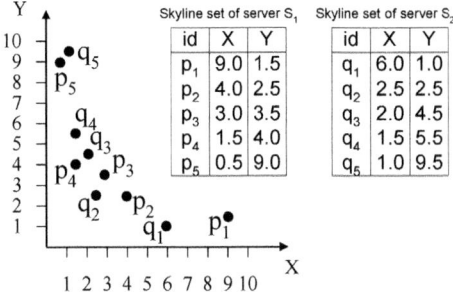

Fig. 3. Example of Lemma 1

the dimensionality or data distribution, the local skyline sets $\mathcal{S}(P_i)$ are comparable in size to the local data sets P_i. Obviously, this leads to increased network traffic, which is undesirable especially in the case of bandwidth-constrained networks. Motivated by this shortcoming, we introduce an algorithm that produces the representative skyline set $\mathcal{K}$ by transferring only a limited number of points, which is independent of the actual cardinality of local skyline sets.

4.2   Distributed Skyline Representative Algorithm (DSR)

In the following, we first show that any distributed approach that transfers fewer data points than the local skyline points requires two communication phases, in order to ensure that the representative skyline set is valid, i.e., all representative skyline points belong to the global skyline set. Then, we describe in detail the proposed algorithm, termed *DSR*.

Two Communication Phases. The design of the *DSR* algorithm is guided by the observation that we do not wish to transfer local skyline sets to the coordinator, as this would result in unrestricted size of transferred points. Consequently, our premise is to transfer to the coordinator only a fraction of the local skyline points $\mathcal{S}(P_i)$, namely only the local representative skyline points $\mathcal{K}_i$. However, the following lemma shows that this method does not guarantee that the produced representative points $\mathcal{K}$ are actually global skyline points $\mathcal{S}$.

Lemma 1. *A distributed skyline representative algorithm that produces a representative skyline set $\mathcal{K}$ over the union of local representative skyline sets $\bigcup \mathcal{K}_i$ may result in non-skyline points p, i.e., $p \in \mathcal{K} \wedge p \notin \mathcal{S}$.*

Proof. It suffices to construct an example where the algorithm will falsely report a dominated point as representative skyline point. We use the distance-based error metric, but a similar example can be constructed for the dominance error metric. Consider the example of Fig. 3, where the skyline sets of two servers S_1 and S_2 are depicted. Assume that the representative skyline set is requested for $k=3$. Applying the distance-based representative algorithm on $\mathcal{S}(P_1)$ and $\mathcal{S}(P_2)$ produces the sets $\mathcal{K}_1 = \{p_1, p_3, p_5\}$ and $\mathcal{K}_2 = \{q_1, q_3, q_5\}$ respectively. It is easy to

---

Algorithm 2. *Distributed Skyline Representative DSR*

1: **INPUT:** k, Coordinator S_C, Servers S_i
2: **OUTPUT:** Representative skyline $\mathcal{K}$
3: **for** $(\forall S_i : i \in [1, N])$ **do**
4:     $\mathcal{K}_i \leftarrow S_i.\underline{\text{representative}}(\mathcal{P}_i)$
5: **end for**
6: $\mathcal{K}' \leftarrow S_C.\underline{\text{representative}}(\bigcup \mathcal{K}_i)$
7: **for** $(\forall S_i : \overline{i \in [1, N]})$ **do**
8:     $\mathcal{D}_i \leftarrow S_i.\underline{\text{dominate}}(\mathcal{K}')$
9: **end for**
10: $\mathcal{K} \leftarrow S_C.\underline{\text{representative}}((\bigcup \mathcal{D}_i) \bigcup \mathcal{K}')$
11: **return** $\mathcal{K}$

---

see that q_1 dominates p_1, and p_5 dominates q_5, thus S_C will take as input the set $\{p_3, p_5, q_1, q_3\}$ to produce $\mathcal{K}$. Obviously, since $k{=}3$ at least one of p_3, q_3 belongs to $\mathcal{K}$. However, q_2 dominates p_3 and p_4 dominates q_3, therefore the algorithm falsely reports a dominated point as representative skyline point.

Lemma 1 practically means that no algorithm that is based solely on transfer of local representative skyline points to the coordinator can guarantee that the representative points $\mathcal{K}$ belong to the global skyline set $\mathcal{S}$, i.e., $\mathcal{K} \subset \mathcal{S}$. Consequently, we propose the *DSR* algorithm that employs two communication phases in order to guarantee that the representative skyline set consists of skyline points. In the first phase, the coordinator requests from each server the representative skyline set based on the locally stored data. Then, a centralized algorithm for representative skyline computation is applied on the union of local representative skyline sets to produce a set of representative skyline points. In the second phase, the produced representative skyline points are sent to all servers, and each server sends back a set of points $\mathcal{D}_i$ that consists of the local skyline points that dominate at least one representative skyline point. Finally, the coordinator applies again the skyline representative algorithm to produce the final set of representative skyline points $\mathcal{K}$. *DSR* improves the efficiency of *DSA* in terms of communication by requesting only the local representative skyline points.

Algorithmic Description and Correctness. *DSR* is described in Algorithm 2. First, each server S_i ($i \in [1, N]$) executes a skyline representative query[1] on the locally stored data ($\mathcal{P}_i$) to produce a set $\mathcal{K}_i$ of k local representative skyline points (line 4). The coordinator assembles the sets $\mathcal{K}_i$ ($i \in [1, N]$) and produces a new set of k representative skyline points (line 6), denoted as $\mathcal{K}'$, by applying the centralized skyline representative algorithm. Then, the coordinator sends the set $\mathcal{K}'$ to all servers S_i. Each server S_i computes all local skyline points $\mathcal{D}_i$ that dominate at least one of the points in $\mathcal{K}'$ (line 8). The coordinator merges its set $\mathcal{K}'$ together with the union of sets of local points $\mathcal{D}_i$ that dominate points of $\mathcal{K}'$, and applies the representative skyline algorithm (line 10) to produce the final set $\mathcal{K}$, which is reported to the user (line 11).

---

[1] This query is performed by using any of the algorithms proposed in [7,11].

One issue that needs further elaboration is the computation of the set $\mathcal{D}_i$ at a server S_i (line 8 of *DSR* algorithm). To support efficient processing, the data set P_i is indexed by a multidimensional index structure, such as an R-tree. Then, the sets $\mathcal{D}_i$ can be computed efficiently by applying a branch-and-bound algorithm on the R-tree similar to a constrained skyline query [9]. For each intermediate representative point $p_i \in \mathcal{K}'$, the constraint is defined by point p_i and the origin of the data space and entries of the R-tree that do not overlap with the constraint are discarded. The set $\mathcal{D}_i$ contains the union of the results for all intermediate representative points $p_i \in \mathcal{K}'$. We emphasize that the use of the R-tree is simply to increase the efficiency, and it is by no means a strict prerequisite of *DSR*. Other non-indexed techniques for computing the sets $\mathcal{D}_i$ can be used instead.

Finally, Lemma 2 ensures the correctness of *DSR* by providing guarantees that Algorithm 2 always returns representative skyline points that belong to the skyline set, i.e., the set of representative skyline points is valid.

Lemma 2. *(Correctness) Any point p of the representative skyline set $\mathcal{K}$ produced by* DSR *belongs to the skyline set, i.e., if $p \in \mathcal{K}$ then $p \in \mathcal{S}$.*

Proof. Let us assume that $p \in \mathcal{K}$ and $p \notin \mathcal{S}$. Then, there exists a point $p' \in \mathcal{S}$ such that p' dominates p. We also conclude that $p' \notin \bigcup(\mathcal{K}_i \bigcup \mathcal{D}_i)$ because otherwise $p \notin \mathcal{K}$. Let us denote as S_j the server that stores p' locally. We distinguish two cases: (a) $p \in \bigcup K_i$, then $p' \in \mathcal{D}_j$ which leads to a contradiction, or (b) $p \notin \bigcup K_i$, then $p \in \bigcup D_i$, which means that there exists a point $q \in \bigcup \mathcal{K}_i$ such that p dominates q. Due to the properties of dominations we conclude that p' dominates q, which in turn leads to $p' \in \mathcal{D}_j$ which is a contradiction.

4.3   Distributed Error-Based Representative Algorithm (DER)

As *DSA* and *DSR* transfer only a fraction of data points to the coordinator, the representation error of the produced skyline representative points may be higher than in the case of the centralized skyline representative algorithm applied on the union of the local data points ($P = \bigcup P_i$). Even though *DSA* manages to return the same representative skyline set as the centralized algorithm on P, when the error metric does not depend on dominated data points, this does not hold for all error metrics such as for example the dominance error metric. The main reason of the higher representation error is that *DSA* and *DSR* use only restricted knowledge about the underlying data due to its distribution. Thus, our premise is to additionally use the information about the error metric at each server locally (resulting from the local representative skyline query) in order to improve the representation quality of the skyline representative set $\mathcal{K}$.

In the following, we describe a generic algorithm that produces representative skyline points for any error metric by taking into account scores of the candidate representative points derived from the local query processing. Then, we demonstrate the applicability of our algorithm for both the dominance and the distance-based error metric.

Algorithm 3. *Error-based Representative Selection*

1: **INPUT:** k, Local representative skyline $\bigcup \mathcal{K}_i$
2: **OUTPUT:** Representative skyline $\mathcal{K}$
3: $p \leftarrow argmax_{\forall p \in (\bigcup(\mathcal{K}_i))}(\underline{score}(p))$
4: $\mathcal{K} = \{p\}$
5: **while** $(|\mathcal{K}| < k)$ **do**
6:     $p \leftarrow argmax_{\forall p \in (\bigcup(\mathcal{K}_i)-\mathcal{K})}(\underline{score}(p, \mathcal{K}))$
7:     $\mathcal{K} = \mathcal{K} \bigcup \{p\}$
8: **end while**
9: **return** $\mathcal{K}$

Algorithmic Description. The distributed error-based skyline representative algorithm (*DER*) processes the representative skyline query similarly to *DSR*. It consists of two phases that guarantee that the resulting set is valid. Moreover, *DER* produces candidate representative points by applying a skyline representative algorithm at local servers. The main difference to the previous algorithms is that each local representative skyline point $p \in \mathcal{K}_i$ is associated with a score of representativeness s_p. Then, the coordinator does not process a plain representative skyline query on $\bigcup \mathcal{K}_i$, but instead takes into account the score of representativeness of each point in order to minimize the error metric. In this way, an optimization problem is defined that aims to identify the k representative skyline points that minimize the error metric, given a set of candidate representative points $\bigcup \mathcal{K}_i$ each of them annotated by a score. As the representative skyline query has been shown to be NP-hard [7,11], we propose a greedy algorithm to solve our optimization problem.

The *DER* algorithm assumes that each local representative point $p \in \mathcal{K}_i$ is augmented with a numeric value (score of representativeness) that indicates its goodness. Clearly, the definition of the score depends on the selection of the representative skyline algorithm, which is applied at the local servers. After the representative points $\mathcal{K}_i$ are collected at the coordinator, Algorithm 3 is assigned with the task of selecting k representative points, i.e., by solving the optimization problem. For this purpose, the algorithm uses the *score()* function that estimates the goodness of each candidate representative point. After selecting the first representative point (line 3), in each iteration the algorithm picks as a next representative point the one that maximizes the estimated score (line 6). After the selection of k representative points $\mathcal{K}'$, the points are sent to all servers for the verification step. Local skyline points $\mathcal{D}_i$ that dominate a point in $\mathcal{K}'$ are sent to the coordinator. Finally, the coordinator produces the final k representative skyline points, by solving again the same optimization problem over the union of points in sets $\mathcal{K}'$ and $\mathcal{D}_i$. Thus, Algorithm 3 is invoked taking as input $(\bigcup \mathcal{D}_i) \bigcup \mathcal{K}'$.

The *DER* algorithm is generic and allows any error metric, i.e., any centralized representative skyline algorithm, to be plugged in our framework. *DER* is parameterized by a function *score()* that computes the error metric. For any error metric that is plugged in (or equivalently for any centralized representative skyline algorithm that should be supported), we need to define an appropriate

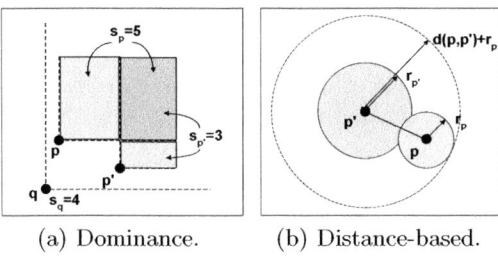

(a) Dominance.   (b) Distance-based.

Fig. 4. Example of *score()* function

score of each representative skyline point and the implementation of the abstract function *score()*. In the sequel, we demonstrate how the dominance and the distance-based error metric are easily integrated and supported by *DER*.

Dominance Error Metric. At a local level, the score s_p of a local representative skyline point $p \in P_i$ is defined as the number of points q that it dominates from the local data set P_i:

$$s_p = |\{q \in \mathcal{P}_i : p \prec q\}|$$

Notice that this definition makes the score of a representative skyline point dependent on the data points. Furthermore, we can accurately compute the aggregated score of a set of representative skyline points when all belong to different servers, as they dominate different data points.

As already mentioned, a set of representative skyline points $\bigcup \mathcal{K}_i$ is collected at the coordinator, each accompanied by its score. To use the *DER* algorithm to solve the optimization problem, we need to define the function *score()*. We use two versions of this function. The first, *score(p)*, computes the goodness for a representative skyline point p individually. This is useful in order to select the first representative skyline point. For this purpose, the most promising point is selected, therefore the function is defined as:

$$score(p) = s_p + \sum\nolimits_{\forall q \in \bigcup \mathcal{K}_i : p \prec q} s_q$$

Intuitively, we select the point p that dominates points in $\mathcal{K}_i$ with maximum number of dominated points in total.

The second, *score(p, K)*, computes the error when p is selected for inclusion in the set $\mathcal{K}$. As in each step we wish to add the next most promising point to the result set, we pick the point that maximizes the following function:

$$score(p, \mathcal{K}) = s_p + \sum\nolimits_{\forall q \in \bigcup \mathcal{K}_i : p \prec q \wedge \nexists p' \in \mathcal{K} : p' \prec q} s_q$$

Intuitively, we compute as score an upper bound of the gain in the attained representation quality, when p is added to $\mathcal{K}$. This value is an upper bound because data points dominated by two representative skyline points are double-counted, since computing the distinctly dominated points is not feasible in practice. In the example of Fig. 4(a), only the scores s_p and $s_{p'}$ are known and not the exact values of the dominated points, thus the exact number of local data points dominated by q is not known. The *score(q)* is estimated as $s_q + s'_p + s_p = 12$, which is an upper bound of the actual number of dominated points by q.

Distance-based error metric The score of a local representative skyline point $p \in \mathcal{K}_i$ is defined as the maximum distance of p to any non-representative skyline point q for which there is no other representative p' closer to q than p. Formally, the score is defined as:

$$s_p = max_{\forall q \in S(\mathcal{P}_i)}\{d(p,q) : \nexists p' \in \mathcal{K}_i, d(q,p') < d(q,p)\}$$

To select the first representative skyline point, we follow the same strategy as iGreedy [11], and define the value of the first coordinate: $score(p) = -p[1]$. Finally, the score is defined as:

$$score(p,\mathcal{K}) = \begin{cases} 0, \; if \; \exists p' \in \mathcal{K} : d(p,p') + s_p < s_{p'} \\ min_{\forall p' \in \mathcal{K}}\{d(p,p') + s_p\}, \; otherwise \end{cases}$$

The function score s_p essentially defines a covering radius $r_p = s_p$ for a hypersphere centered at p, as depicted in Fig. 4(b). This hypersphere represents the region of the space with the following property: any non-representative skyline point in this region is closer to p than to any other representative skyline p'. The score function calculates the new radius of a hypersphere centered at p that covers all skyline points that are closer to p or p' than all other representative points. The estimation of the error is an upper bound of the actual error. In worst case, the distance between a candidate point p and its closer representative point p' is $d(p,p') + s_p$. If this is smaller than $s_{p'}$, then the representation quality does not decrease by not selecting the candidate as representative, thus the estimated error is set to zero.

The algorithm proceeds as above; it first picks one representative skyline point. Then, in each iteration, the error is estimated that will be introduced if the candidate representative skyline is not selected. The point with the highest error is selected as the next representative skyline, because otherwise the error metric will become equal to the highest error.

5   Experimental Evaluation

In this section, we provide an extensive study of our framework. We developed all algorithms (the baseline *DSA*, as well as our two proposed algorithms *DSR* and *DER*) in Java and simulated the distributed aspects of our framework. We implemented the distance-based representative algorithm (I-greedy algorithm [11]), as well as the greedy algorithm proposed in [7].

We employed synthetic data sets to examine different distributions, namely uniform (UN), clustered (CL) and anti-correlated (AC). For the clustered data set (CL), each server picks 10 cluster centroids randomly and the points follow a Gaussian distribution on each axis with variance 0.05, and a mean equal to the corresponding coordinate of the centroid. The anti-correlated (AC) data set was generated as described in [1]. For our experiments on synthetic data, we report the average results over 10 different instances of the data set. In addition, we employ another synthetic data set, called *Island* (IS), which is 2-dimensional and contains 63383 points. This data set is used in [11] to demonstrate the effectiveness of distance-based representative. We also use a real data set (NBA), which consists of 17265 5-dimensional tuples representing a player's performance per year. Since our setup is distributed, we distribute IS and NBA to the N servers by choosing a server per point uniformly at random. Again, we perform this process 10 times and report average values.

To evaluate the performance of our framework, we vary N from 5 to 15, d from 2 to 5, the cardinality n from 250K to 3M (which is evenly distributed to the N servers in advance), k from 10 to 50, the network speed from 1KB/sec to 100KB/sec, and we test different data distributions (UN,AC,CL,IS,NBA). We observed that the use of larger data sets increases the total time due to increased processing time, while the networking time is not significantly affected, since the number of transferred data remains relatively stable. Unless explicitly mentioned, the default setup is $N=10$, $d=3$, $|n_i|=100$K, $k=10$, network speed 50KB/sec, and we employ the UN data set. We note that when $k < \mathcal{S}(P_i)$, then k is set to $\mathcal{S}(P_i)$. The experimental evaluation focuses on two axes; the performance of our approach and the achieved quality of results. Our main performance metrics include: (i) the amount of transferred data and (ii) the total time, which is the time until the final result is produced at S_C (including network transfer time).

To evaluate the quality of our algorithms, we employ the normalized error metric. In the case of distance-based representative, the normalized error metric is $Er(\mathcal{K})/$MAX_DIST, where MAX_DIST represents the maximum distance of the space. Assuming a d-dimensional set of points where the value of each dimension belongs to $[0, U]$, then MAX_DIST$=U\sqrt{d}$. In the case of dominance representative, the normalized error metric is $Er(\mathcal{K})/n$. In all cases the normalized error takes values that belong to the range $[0, 1]$.

5.1 Experiments with Distance-Based Representative

Evaluation for UN. In Fig. 5, we measure the amount of transferred data for various setups. In Fig. 5(a), we study the effect of increasing the cardinality of the data set from 50K to 200K points per server. *DSA* needs to transfer all local skyline points, thus resulting in much more traffic than *DSR* or *DER*. In Fig. 5(b), the number of transferred data points increases rapidly for *DSA*, due to the increase of each server's skyline cardinality as the dimensionality grows. Instead, *DSR* and *DER* show a much more stable performance, demonstrating the merits of the approaches that transfer only representative skyline points, rather than local skyline sets. In Fig. 5(c), we gradually increase the number of servers N in the system. The traffic induced by *DSA* (Fig. 5(c)) increases linearly with the number of servers. In contrast, *DSR* and *DER* scale gracefully.

Then, Fig. 6 shows the normalized error metric for different setups. Recall that in the case of distance-based representative, the error of *DSA* is equal to the error of the centralized distance-based representative algorithm and is greater than 0, unless all skyline points are reported as representative skyline points. Fig. 6(a) shows that the savings in network communication (depicted in Fig. 5(a)) cause *DSR* and *DER* to have higher error than *DSA*. It is also noteworthy that the increased cardinality does not affect the performance of the algorithms significantly. The reason is that the important factor is the skyline cardinality and

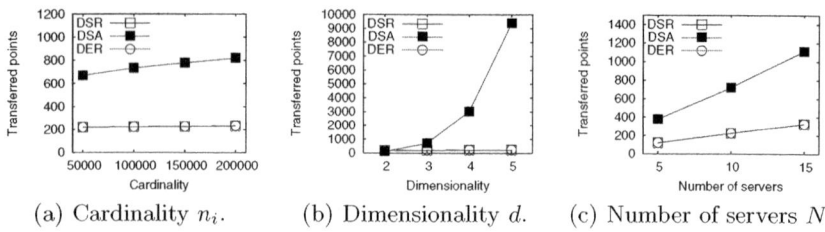

(a) Cardinality n_i.   (b) Dimensionality d.   (c) Number of servers N.

Fig. 5. Transferred data vs. cardinality, dimensionality and number of servers (UN)

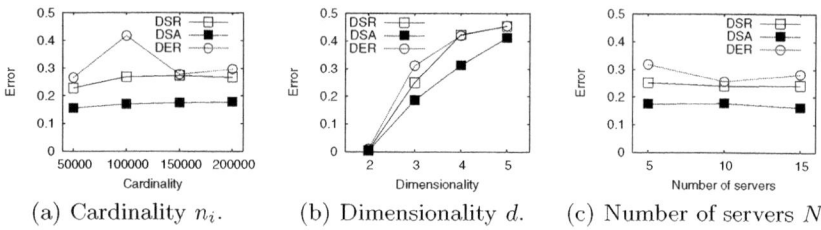

(a) Cardinality n_i.   (b) Dimensionality d.   (c) Number of servers N.

Fig. 6. Error vs. cardinality, dimensionality and number of servers (UN)

not the data cardinality. In Fig. 6(b), the induced normalized error is reported for all algorithms, which increases with dimensionality. Notice that the difference between the algorithms remains practically the same. In Fig. 6(c), the error remains relatively stable for all algorithms regardless of N.

These experimental results indicate that *DER* does not improve the performance of *DSR*. This behavior is expected because the induced error of the distance-based representative skyline algorithm depends only on the skyline points. Consequently, *DSR* achieves results of high quality even with limited knowledge. Therefore, in the remaining experimental study of the distance-based representative, we omit *DER* from the charts.

Evaluation for CL. In Fig. 7(a), the amount of transferred data is depicted for our algorithms for CL data. We emphasize that each server picks cluster centroids randomly, therefore different servers have different clusters of data. Notice that *DSR* is practically unaffected by the increased dimensionality, thus demonstrating its merits when the data set is clustered. In contrast, the traffic induced by *DSA* increases with dimensionality. Then, in Fig. 7(b), we depict the normalized error metric. As in the case of UN, *DSR* exhibits higher error values than *DSA*, however here the difference is smaller than for UN. Also, the absolute error values are smaller than in the case of UN.

In addition, we measure the total time in Fig.7(c), which increases with dimensionality for both algorithms. This is expected, as the performance of any skyline or representative skyline algorithm deteriorates with increased dimensionality. Both algorithms have similar performance in terms of total time. In our experiments, we noticed that the processing time dominates the total execution

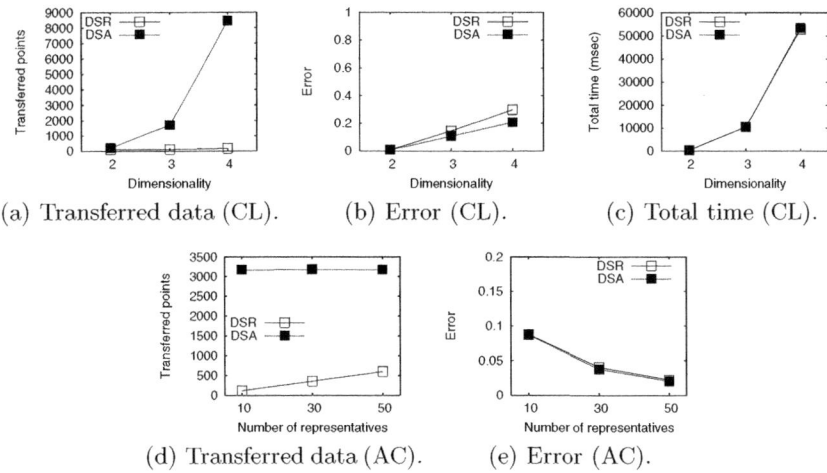

(a) Transferred data (CL).   (b) Error (CL).   (c) Total time (CL).

(d) Transferred data (AC).   (e) Error (AC).

Fig. 7. Experimental results for CL and AC data sets

time. This happens because the time required for transferring data is quite low, due to the assumed network speed of 50KB/sec. For more network-constrained networks, the transfer time is significant and affects the total time.

Evaluation for AC. In Figs. 7(d) and 7(e), we evaluate both algorithms for the 2-dimensional anti-correlated data distribution. We note that this distribution is the most challenging for skyline computation, since it results in high skyline cardinality, even for small dimensionality values. The aim of this experiment is to explore the behavior of our algorithms, when the local skyline sets at servers S_i are of high cardinality. First, in Fig. 7(d), *DSA* is unaffected by k, as it always transfers all local skyline sets regardless of k. Obviously, *DSR* needs to transfer more data as k increases. The important finding is that *DSR* requires to transfer one order of magnitude fewer data, thus demonstrating its appropriateness when the local skyline size is significant and network resources are limited. In Fig. 7(e), the normalized error is depicted. Notice that *DSR* shows marginally equal error with *DSA*, which is another strong argument in favor of *DSR*. Both algorithms exhibit a decreasing tendency with increased values of k. This is expected because as more representative skyline points are reported, the representative set describes more closely the real skyline set, thereby decreasing the error.

Evaluation for IS. For the IS data set, *DSR* is again much more communication-efficient than *DSA* as shown in Fig. 8(a), especially when the requested value k is small. Fig. 8(b) shows that the error is practically the same for both algorithms and drops for increased values of k. In Fig. 8, we also depict the data set and the representative skyline points discovered by the two algorithms. The two plots share 8 common points out of the total 10. The error is identical for both algorithms. When compared to the plot of Fig. 8(c), it is clear that both algorithms produce representative points that capture the shape of the skyline.

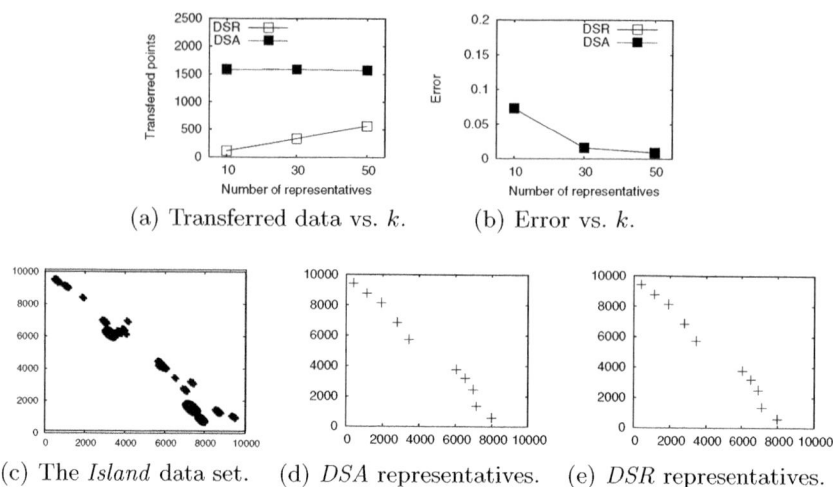

(a) Transferred data vs. k.   (b) Error vs. k.

(c) The *Island* data set.   (d) *DSA* representatives.   (e) *DSR* representatives.

Fig. 8. Experimental results for IS data set

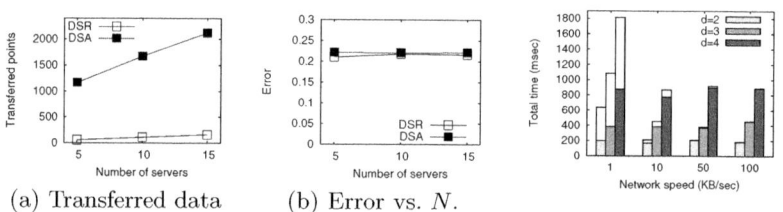

(a) Transferred data   (b) Error vs. N.

Fig. 10. Varying network

Fig. 9. Experimental results for NBA data set   speed

Evaluation for NBA. Then, in Fig. 9, we see that the conclusions drawn from the synthetic data sets are validated also in the case of the real data set. *DSR* always incurs significantly less network traffic (Fig. 9(a)), while the induced error is practically the same for both algorithms (Fig. 9(b)).

Varying Network Speed. In Fig. 10, we vary the network speed for the *DSR* algorithm. The total length of each bar corresponds to the total time, while the colored part corresponds to processing time. Larger values of network speed ($\geq$50KB/sec) do not affect performance, because the total time is dominated by the processing time, while the network transfer time is very small. In the case of smaller values of bandwidth (1KB/sec), we see that network transfer time increases and affects the total time.

5.2   Experiments with Dominance Representative

Evaluation for UN. Figs. 11(a) and 11(b) show the results of dominance representative for varying dimensionality. Both *DSR* and *DER* transfer significantly fewer data points, and the gain increases with d. An important finding is that *DER* improves the performance of *DSR* in terms of the error metric (Fig. 11(b)).

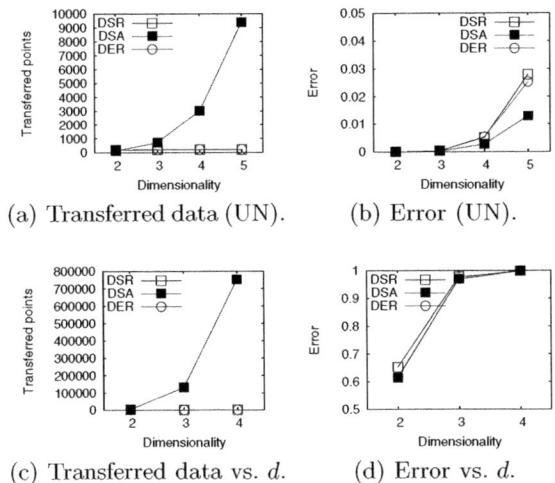

(a) Transferred data (UN).   (b) Error (UN).

(c) Transferred data vs. d.   (d) Error vs. d.

Fig. 11. Experimental results for UN and AC data sets

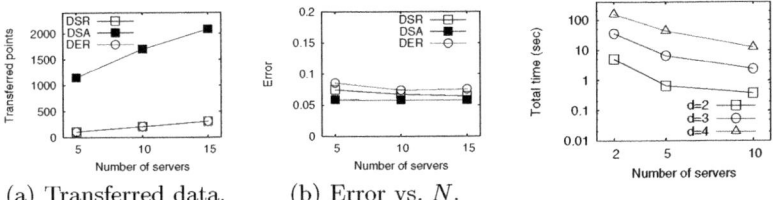

(a) Transferred data.   (b) Error vs. N.

Fig. 13. Speed-up for

Fig. 12. Experimental results for NBA data set   DSR algorithm

Evaluation for AC. In the next experiment, we test the performance of all algorithms in a hard setup (AC data distribution). As expected, when the dimensionality increases, the size of the local skyline sets increases rapidly, and DSA needs to transfer too many data points, thus becoming impractical. In contrast, both DSR and DER scale gracefully in terms of transferred data. When the error is considered (Fig. 11(d)), all algorithms induce significant error values, but DER is better than DSR. DSA exhibits lower error values because it transfers a significant part of the local data sets to the coordinator, thus easing the task of selecting representative points.

Evaluation for NBA. Then, in Fig. 12, we test the performance of all algorithms for the NBA data set. We vary the number of servers, in order to study their behavior for increased network sizes. Fig. 12(a) shows that the increase in the number transferred points as the number of servers grows is smaller for DSR and DER than DSA. Fig. 12(b) depicts the induced error as N increases. The error of all algorithms remains practically unaffected, which shows that our framework is not significantly affected when more servers are employed.

Speed-up. Finally, in Fig. 13, we perform an experiment using a data set of 1M data points and distribute it to 2, 5, and 10 servers respectively. We test the *DSR* algorithm for the dominance representative. Clearly, when a higher number of servers is used, the data set is distributed to smaller fragments, thus each server processes a smaller amount of data. In consequence, the processing cost of local computation on each server is reduced. This demonstrates that in the case of *DSR* runtime can be reduced by employing more servers.

6   Conclusions

In this paper, we addressed the challenging problem of discovering representative skyline points over distributed data, which naturally arises in various application domains, and it is mainly motivated by the unrestricted size of skyline cardinality. To address the problem effectively, we introduce a novel framework for processing the distributed skyline representative query. Our framework supports all metrics proposed for representative skyline queries in centralized settings.

References

1. Börzsönyi, S., Kossmann, D., Stocker, K.: The skyline operator. In: Proc. of ICDE (2001)
2. Chan, C.Y., Jagadish, H.V., Tan, K.L., Tung, A.K.H., Zhang, Z.: Finding k-dominant skylines in high dimensional space. In: Proc. of SIGMOD (2006)
3. Chaudhuri, S., Gravano, L.: Evaluating top-*k* selection queries. In: Proc. of VLDB (1999)
4. Cui, B., Lu, H., Xu, Q., Chen, L., Dai, Y., Zhou, Y.: Parallel distributed processing of constrained skyline queries by filtering. In: Proc. of ICDE (2008)
5. Fagin, R., Lotem, A., Naor, M.: Optimal aggregation algorithms for middleware. In: Proc. of PODS (2001)
6. Hose, K., Vlachou, A.: A survey of skyline processing in highly distributed environments. VLDBJ (2011) (to appear)
7. Lin, X., Yuan, Y., Zhang, Q., Zhang, Y.: Selecting stars: the k most representative skyline operator. In: Proc. of ICDE (2007)
8. Lu, H., Jensen, C.S., Zhang, Z.: Flexible and efficient resolution of skyline query size constraints. IEEE TKDE 23(7), 991–1005 (2011)
9. Papadias, D., Tao, Y., Fu, G., Seeger, B.: Progressive skyline computation in database systems. ACM TODS 30(1), 41–82 (2005)
10. Sarma, A.D., Lall, A., Nanongkai, D., Lipton, R.J., Xu, J.J.: Representative skylines using threshold-based preference distributions. In: Proc. of ICDE (2011)
11. Tao, Y., Ding, L., Lin, X., Pei, J.: Distance-based representative skyline. In: Proc. of ICDE (2009)
12. Vlachou, A., Doulkeridis, C., Kotidis, Y., Vazirgiannis, M.: SKYPEER: Efficient subspace skyline computation over distributed data. In: Proc. of ICDE (2007)
13. Vlachou, A., Doulkeridis, C., Kotidis, Y., Vazirgiannis, M.: Efficient routing of subspace skyline queries over highly distributed data. IEEE TKDE 22(12), 1694–1708 (2010)
14. Zhu, L., Tao, Y., Zhou, S.: Distributed skyline retrieval with low bandwidth consumption. IEEE TKDE 21(3), 384–400 (2009)

SkyQuery: An Implementation of a Parallel Probabilistic Join Engine for Cross-Identification of Multiple Astronomical Databases

László Dobos[1,2], Tamás Budavári[2], Nolan Li[2],
Alexander S. Szalay[2], and István Csabai[1]

[1] Eötvös Loránd University, Department of Physics of Complex Systems,
H-1117 Budapest, Hungary
dobos@complex.elte.hu
[2] The Johns Hopkins University, Department of Physics & Astronomy,
Baltimore, MD 21218, USA
budavari@jhu.edu

Abstract. Multi-wavelength astronomical studies require cross-identification of detections of the same celestial objects in multiple catalogs based on spherical coordinates and other properties. Because of the large data volumes and spherical geometry, the symmetric N-way association of astronomical detections is a computationally intensive problem, even when sophisticated indexing schemes are used to exclude obviously false candidates. Legacy astronomical catalogs already contain detections of more than a hundred million objects while ongoing and future surveys will produce catalogs of billions of objects with multiple detections of each at different times. One time, pair-wise cross-identification of these large catalogs is not sufficient for many astronomical scenarios. Consequently, a novel system is necessary that can cross-identify multiple catalogs on-demand, efficiently and reliably. In this paper, we present our solution based on a cluster of commodity servers and ordinary relational databases. The cross-identification problems are formulated in a language based on SQL, but extended with special clauses. These special queries are partitioned spatially by coordinate ranges and compiled into a complex workflow of ordinary SQL queries. Workflows are then executed in a parallel framework using a cluster of servers hosting identical mirrors of the same data sets.

Keywords: probabilistic join, query optimization and languages, astronomical catalogs, workflow, computational statistics.

1 Introduction

Increasingly large astronomical data warehouses are being built to support the needs of scientific collaborations. The key point in the federation of astronomical data sets is the cross-matching of detections belonging to the same physical object. The detections are usually made by using different imaging filters or entirely different instruments. Due to the exponential growth in the data volume,

A. Ailamaki and S. Bowers (Eds.): SSDBM 2012, LNCS 7338, pp. 159–167, 2012.

cross-match solutions have to be scalable. Also, since the largest data sets will be geographically distributed and data co-location might not be an option in the future, any solution will have to be optimized for slow network.

In this paper, we present a scalable solution for on-demand cross-matching of large catalogs hosted on a cluster of database servers. In Sec. 2, we explain the most important properties of astronomy catalogs. The cross-identification problem is introduced in Sec. 3. Sec. 4 describes our SQL extensions to explicitly formulate the problem of coordinate based matching in a query. In Sec. 5 we focus on the most important aspects of the implementation. Sec. 6 concludes the paper, and outlines specific future work.

2   Astronomical Surveys

Today's high-performance astronomical *imaging instruments* are mostly operated in *survey mode*, i.e. significantly large regions of the sky are mapped systematically. Every telescope is designed to work in a certain region of the electromagnetic spectrum and has filter sets to further subdivide the wavelength range. The goal of imaging sky surveys is to take snapshots of the sky in order to identify all celestial objects[1] in each imaging filter to a given faintness limit. With the advance of detector technology and growth of the mirror surface of telescopes, an exponentially growing area of the sky can be surveyed in a given amount of time. This not only helps map a larger portion of the Universe, but also to take measurements in the time domain. Multi-wavelength astronomy combines information from different instruments to investigate the physical properties and to constrain theoretical models of celestial objects. Both time-domain astronomy and multi-wavelength astronomy require reliable and fast algorithms to cross-match multiple detections of the same celestial objects.

2.1   Astronomical Catalogs

Images taken during astronomical surveys are *reduced* by the so called *photometric* or *imaging* pipe-line software. During the reduction process, individual objects are identified in the images and their readily measurable properties are determined, such as their celestial coordinates, integrated brightness, brightness profiles, morphological parameters etc. The number of measured properties is typically in the hundreds.

The numbers of objects detected by the surveys are all on scales, currently topping in the hundred million range. Ongoing and future surveys will provide information about billions of objects, about a hundred detections of each at different times. The typical data volume of current reduced catalogs tops in the 10 TB range, quickly moving toward hundreds of terabytes, reaching the petabyte range by the end of the decade. The amount of raw imaging data collected and processed during the surveys can be about ten to a hundred times more.

---

[1] Throughout the paper we will refer to discrete physical celestial objects emitting light as *objects* or *sources*. Individual observations of objects (using different filters/instruments or just different epochs) are called *detections*.

3   Coordinate-Based Cross-Identification

When we have multiple detections of the same celestial source, the measured coordinates will slightly differ. In order to cross-match the detections of two catalogs, we have to measure the distance between all detection pairs, and only accept those pairs as matches that are closer than a given threshold. The exact Bayesian probabilistic join framework is based on the statistical methods published by [Bud3].

In practice, cross-matching is done by excluding obvious false matches first. Different indexing schemes of the sphere have been invented to find matching candidates efficiently [Fek,Gor]. In our implementation, we use the so called *zone algorithm*, which is the fastest available algorithm for Microsoft SQL Server so far [Gra].

There are three different ways a catalog can be cross-matched with other catalogs. One can require that certain catalogs *must* contain good candidate detections of an object in order to accept a match. Additionally, some other catalogs *may* contain detections and should be taken into account, if possible. These are typically catalogs with brighter detection limits. In the third case, one requires that a catalog *must not* contain any candidate detections that would match with detections of other catalogs. This third case is called *drop-out detection*. The first case is similar to inner joins, the second case is to outer joins, while drop-out detection is basically an anti semi join.

Drop-out detection is particularly important in multi-wavelength astronomy because even if a certain object is not detected by an instrument due to its too high detection limits, an upper limit to the brightness of the object can still be given based on the known sensitivity of the instrument. To safely detect drop-outs, it is fundamental to know whether an object is not in the catalog because its celestial location was not observed by the survey at all, it was missed by the instrument because it was too faint or it was intentionally masked out and excluded from the catalog due to other reasons. To account for this problem, a precise description of the observed areas, the so called *footprints*, and *masks*, is necessary.

3.1   Previous Work

The first automatic cross-identification on-line service was implemented by Budavári et al. as a set of XML SOAP web services [Bud1]. As it was a prototype built to demonstrate the then new web service technology, not much attention to the performance and scaling properties was paid. Based on the idea, an open, SOAP-web-service-based standard, Open SkyQuery was developed by the National Virtual Observatory to federate geographically distributed data sets [Bud2]. Both of these versions used the SQL language with some custom functions as the main programming interface. Because Open SkyQuery could not benefit from co-located data sets, and due to performance issues, the system was limited to process only 5000 matches in a run.

The Virtual Observatory Alliance standardized the Astronomical Data Query Language (ADQL) intended to be used as the *lingua franca* of astronomical catalogs [Or1]. Though the ADQL language defines many new, astronomy-induced constructs compared to SQL, including spherical region expressions that can be used to circumscribe cross-matching problems, ADQL was not designed with query optimizability in mind.

The new CDS xMatch service implements a high performance cross-match engine that partially uses database technology, and can perform two-way joins only [BPD1]. The current version of the service features a form-based user interface only and no scripting support.

4   SQL for Astronomical Data Mining

Since the spread of relational databases in astronomy, the SQL language has become an every day tool of researchers. Although several of the typical data filtering tasks could be done with web forms or other types of custom user interfaces, SQL gives the ability to *script* the operations. This is absolutely necessary for astronomers dealing with data processing issues, as astronomical data usually have to be reprocessed many times during a research project. Also, being able to solve all problems using SQL makes it possible to process data without pulling it out from the database server.

To support analysis of astronomical data, extensive libraries of scientific functions have been developed, which can be accessed directly from SQL via user-defined functions. Functions include cosmological distance calculations [TP1] and various spherical indexing schemes for fast coordinate and region-based searches [Bud4].

4.1   SQL Language Extensions for SkyQuery

We decided to base our SkyQuery language on SQL and extend the language to support easy expression of cross-identification problems and spatial filtering. Building on the basis of SQL not only makes it easy to learn the extended syntax, but also allows for backward compatibility with traditional SQL queries.

There are some ideas that are worth considering when creating extensions to a declarative query language. First of all, we wanted to avoid any interference with the existing behavior of SQL clauses. This is why we introduced the new XMATCH clause (see Sec. 4.2) instead of incorporating its functionality into the standard FROM clause and JOIN operators.

Our language extensions were designed such a way that all queries that can be described using the language will be executable and can be optimized efficiently. This is in strong contrast, for example, with the way most GIS systems implement spatial constraints (complex boolean expression in the WHERE clause) where efficient optimization is an issue; or in contrast with the rather flexible ADQL language where even query executability is a problem. Also, simple implementation is always a main objective, especially in case of scientific projects with limited budgets.

4.2   Defining the N-way Probabilistic Join

To explain the behavior of the new clauses, we consider Query 1. The first half of the query (above the customized XMATCH clause) is in traditional SQL. Data sets listed in the FROM clause are called SDSS, TwoMASS and GALEX after three frequently used astronomical catalogs. Table names are separated from data set identifiers by colons. Each of the listed tables contains observations of galaxies. Each table has an integer field ObjID which is the primary key. Spherical coordinates are stored in the RA and Dec columns[2]. Corresponding unit vector coordinates are named Cx, Cy and Cz. Columns denoted with mag_x are brightness measurements of the objects in different imaging filters.

```
SELECT  x.RA , x.Dec ,
        s.ObjID , s.RA , s.Dec , s.mag_g , s.mag_r , s.mag_i ,
        g.ObjID , g.RA , g.Dec , g.mag_nuv , g.mag_fuv ,
        t.ObjID , t.RA , t.Dec , t.mag_J , t.mag_H , t.mag_K
INTO    MyDB : NewResults
FROM    SDSS : PhotoObjAll AS s
        CROSS JOIN GALEX : PhotoObjAll AS g
        CROSS JOIN TwoMASS : PhotoXSC AS t
WHERE   s.Galaxy = 1
XMATCH  BAYESIAN AS x
        MUST s ON POINT(s.Cx , s.Cy , s.Cz) , 0.1
        MUST g ON POINT(g.Ra , g.Dec) , 0.2
        MAY  t ON POINT(t.Ra , t.Dec) , 0.5
        HAVING LIMIT 1e6
```

Query 1. A sample cross-match query demonstrating the extended SQL syntax

The FROM clause simply produces the Cartesian product of the three catalog tables. In traditional SQL, one would write a WHERE clause which filters the Cartesian product leaving only matching detections. Obviously, tables of high cardinality cannot be matched that way. One solution would be to analyze the WHERE clause describing the cross-match criteria and optimize query execution accordingly. Such expression analysis and optimization algorithm is way too complicated. Instead, we introduce the new XMATCH clause that eliminates the need for complex expression analysis and simplifies optimization a lot.

In Query 1, the BAYESIAN keyword belongs to the XMATCH clause and defines the statistical framework of cross-identification. Currently only Bayesian is supported. The AS x alias is used to make the columns calculated by the cross-matching algorithm being able to be referenced by the rest of the query. Note the first line of Query 1 and the x.RA and x.Dec column references. These columns

---

[2] RA stands for *right ascension*, this is the angle measured around the celestial equator; the equivalent of ϕ in traditional spherical coordinates. Dec stands for declination, the angle measured from the equator toward the poles; the equivalent of θ.

will contain the best coordinate estimates computed by the Bayesian algorithm. The HAVING LIMIT clause is required and specifies the minimum value of the Bayes factor for a positive match.

5   Implementation Details

5.1   Hardware and Software Setup

The system is installed on a set of five identical eight-core database servers equipped with I/O systems able to provide a sustained sequential read speed of about 1.2 GB/s. The servers are connected with 10 Gb/s ethernet and run Microsoft SQL Server 2008 R2.

5.2   Database Setup

Database layouts are optimized for long table scans that happen when cross-matching entire catalogs. File groups contain multiple files split across the RAID volumes to benefit from the underlying hardware. Databases storing catalog data are mirrored to every cluster node to allow for parallelization and load-balancing. For every catalog, we create a so called *mini* version it to support gathering query statistics on the fly. Mini databases are uniformly sampled from the original databases at a 10^{-3} sampling rate. Sampling is done such a way that foreign key references remain intact. SkyQuery uses staging databases to store intermediate output produced by zone algorithm [Gra]. These databases are heavily used for both reading and writing.

As users interact with the system via SQL, the most convenient way to store query results is to allocate a moderately sized database for each user, and save query results there [OM1]. Users can also upload their own data tables and store them in their so called *MyDB*. The final result sets can be easily downloaded as files. In the current configuration, MyDBs are distributed among the cluster nodes; only one copy of a database per user. A future version will give the users restricted access to big staging databases on the cluster nodes. This is important in cases when the results of a computation require only limited storage but internal steps might produce larger outputs.

5.3   Query Optimization and Partitioning

The zone algorithm cross-matches catalogs pairwise. Once two catalogs have been cross-identified, the new best estimate coordinates are calculated and passed on to the next iteration. From the perspective of optimization, starting with catalogs of the least cardinality is almost always the best choice.

Since the extended SQL syntax supports filtering the data, and spatial constraints can also be applied to the queries, there is not much use to store static cardinality information about the catalogs. Instead, our system is designed to gather statistics about each query prior to optimization. In Sec. 5.2, we mentioned that random subsets of the source catalogs, the so called *mini* databases are created. We use these mini catalogs to get quick statistics about the source tables referenced by the queries. Once the referenced tables are identified in the queries, all criteria restricting the rows of those tables are also collected from the ON conditions of the JOIN expressions and from the WHERE clause. From this information we are not only able to estimate the cardinality of the source tables, but also to determine the spatial distribution of the object detections of the astronomical catalogs after all selection criteria have been applied.

Information about the spatial distribution of the data points is essential in order to be able to efficiently partition the query. Based on the histogram of coordinates, the surface of the sphere is split into disjoint partitions defined by great circles intersecting at the poles. Partition boundaries can be chosen anywhere as all the data is mirrored to every cluster node. This also eliminates the need of buffer zones along partition boundaries. Boundaries are chosen such a way that an approximately equal number of detections falls into each partition.

The number of partitions is chosen to be a multiple of the available cluster nodes. We use more partitions than the number of available physical machines to execute the cross-match tasks, because higher granularity makes error recovery much easier, only smaller parts of the job had to be redone when unexpected events happen.

5.4   Jobs as Parallel Workflows

Every partition of a cross-match query translates to a sequence of ordinary SQL queries, and these query sequences run in parallel on many machines. Because of the complexities of multi-threaded application development, we decided to implement the cross-match jobs as *workflows* written for .Net Workflow Foundation (WF) version 4. WF has extensive support for parallel execution of *activities* (atomic components of workflows), and also for exception handling and workflow cancellation logic.

Workflows make it sure that ordinary SQL queries performing the cross-matching will run in the necessary order and that partitions will be processed in parallel. All ordinary SQL queries are written such a way that they do not return any data but write all results into the staging databases of the particular cluster nodes. Also, all heavy computations are coded into these SQL queries. These design constraints make it possible to build workflows that do only basic computations and issue regular SQL queries to remote servers to do the rest. As a result, all workflows can be run on a single head node of the cluster instead of scheduling non-SQL user code on the worker nodes.

For each cross-match query, we create a new job (in the form of a WF workflow) and schedule its execution with a custom-written queueing system. The queueing system supports execution of jobs with different time-out intervals. This is particularly important in open access database systems, like SkyQuery, where queries written by the users can have any complexity. Users developing queries would submit them to the *quick* queue first to see if the queries work correctly on smaller chunks of data. Once they are satisfied with the results, they can send the queries to the long queue with much longer time-out interval for guaranteed completion.

To avoid moving large amounts of data between servers, we schedule the execution of an entire partition of a cross-identification job on the same cluster node. Cluster nodes are assigned to the partitions in a round robin fashion. We chose round robin scheduling over complex load balancing because of some problems arising from the behavior of database servers. For instance, it is hard to correctly measure the load on a particular database server as either CPU load or I/O load can vary heavily during query processing. Assigning servers to tasks in round robin seems a much simpler and reasonable way.

5.5   Performance and Scaling Considerations

Since all of the catalogs are mirrored to all nodes of the cluster, no interaction among the servers is necessary when processing independent partitions of the queries. This makes it an almost ideal scale-out situation as the only overhead coming from the partitioning is that a small amount of initial data has to be copied to the worker nodes from the users' MyDBs, and, once the query has finished, results have to be gathered from the nodes. The overall performance of the system is limited by the number of available server nodes and the intrinsic performance of the zone algorithm [Gra].

5.6   Metadata Management and Provenance from Queries

The relational data model itself only uses table names and column names to identify quantities stored in the databases. Scientific applications, on the other hand, require detailed description of the physical quantities. The Internatinal Virtual Observatory Alliance defines the ontology and metadata models to describe astronomical data. In SQL Server, extended properties can be added to every database schema object. These properties can be easily queried via special views. For Sky-Query, we use these extended properties to store meta-information. Metadata includes description of the quantities using identifiers based on the ontology but also human-readable text to display on web pages, etc.

All data in the system are manipulated with SQL scripts and results of the computations are manifested as output tables stored in the users' MyDBs. Because we parse every SQL query executed, we have complete control over the schema and metadata of the output tables as well. It will be very convenient in the future to derive metadata and provenance information about query outputs directly from the SQL scripts.

6   Summary and Future Work

In this paper we have introduced a new, scalable implementation of software for cross-identification of co-located astronomical catalogs. Compared to the earlier reincarnations, for the third version of SkyQuery, the following improvements have been made. a) Instead of single-server operation, queries are partitioned and executed on a cluster of identical database servers having identical versions of all data sets. b) An easy to optimize syntax extension to the SQL language was invented to support simple formulation of cross-match problems. c) Queries are translated into complex workflows of traditional SQL queries. Workflows are implemented using Windows Workflow Foundation to support parallel execution.

For the next versions of the system, we are working on the following additions. a) Right now, all queries are run from scratch, i.e. helper tables used to speed up cross-matching are newly created every time when needed. Certain helper tables could be cached to further speed up execution. b) We are designing a generic framework for handling metadata which will allow extract provenance information directly from the queries written by users. c) A lazy-join algorithm is being designed to allow vertically partition tables. This will make it possible to move less frequently used columns to cheaper storage. d) We will add support to reference tables from remote data sets accessible to Virtual Observatory standard protocols.

Acknowledgements. This work was supported by the following Hungarian grants: NKTH: Polányi and KCKHA005. The Project is supported by the European Union and co-financed by the European Social Fund (grant agreement no. TÁMOP 4.2.1./B-09/1/KMR-2010-0003)

References

BPD1. Boch, T.: Pineau, F.X., Derriere, S.: CDS xMatch service documentation (2011), http://cdsxmatch.u-strasbg.fr

Bud1. Budavári, T., Malik, T., Szalay, A.S., Thakar, A.R., Gray, J.: Proceedings of the Conference Astronomical Data Analysis Software and Systems XII, vol. 295, p. 31 (2003)

Bud2. Budavári, T., Szalay, A.S., Gray, J., et al.: Proceedings of the Conference Astronomical Data Analysis Software and Systems XIII, vol. 314, p. 177 (2004)

Bud3. Budavári, T., Szalay, A.S.: Astrophysical Journal 679, 301 (2008)

Bud4. Budavári, T., Szalay, A.S., Fekete, G.: Publications of the Astronomical Society of the Pacific, vol. 122, p. 1375 (2010)

Fek. Fekete, G., Szalay, A.S., Gray, J.: Proceedings of the Conference Astronomical Data Analysis Software and Systems XIII, vol. 314, p. 289 (2004)

Gor. Górski, K.M., Hivon, E., Banday, A.J., et al.: Astrophysical Journal 622, 759 (2005)

Gra. Gray, J., Szalay, A., Budavari, T., et al.: arXiv:cs/0701172 (2007)

OM1. O'Mullane, W., Gray, J., Li, N., et al.: Proceedings of the Conference Astronomical Data Analysis Software and Systems XIII, vol. 314, p. 372 (2004)

Orl. Ortiz, I., Lusted, J., Dowler, P., et al.: arXiv:1110.0503 (2011)

TP1. Taghizadeh-Popp, M.: Publications of the Astronomical Society of the Pacific, vol. 122, p. 976 (2010)

Efficient Filtering in Micro-blogging Systems: We Won't Get Flooded Again

Ryadh Dahimene, Cedric Du Mouza, and Michel Scholl

CEDRIC Laboratory - CNAM - Paris, France
firstname.lastname@cnam.fr

Abstract. In the last years, micro-blogging systems have encountered a large success. *Twitter* for instance claims more than 200 million accounts after 5 years of existence and a daily traffic of more than 200 million tweets leading to 350 billion delivered tweets. Micro-blogging systems rely on the *all-or-nothing* paradigm: a user receives all the posts from an account he follows. A consequence for a user is the risk of *flooding, i.e.*, the number of posts received implies a time-consuming scan of his list of postings to read news that match his interests. To avoid user flooding and to significantly diminish the number of posts to be delivered, we propose a filtering structure for micro-blogging systems. We present an analytical model and an experimental study on synthetical datasets and on a real Twitter dataset which consists of more than 2.1 million users, 15.7 million tweets and 148.5 million publisher-follower relationships.

Keywords: Micro-Blogging, Filtering, Indexing.

1 Introduction

Micro-blogging systems have become a major trend over the Web 2.0 as well as an important communication vector. In less than six years, *Twitter* has grown in a spectacular manner to reach more than 200 million active users in August, 2011. In such systems, the length of a published piece of news (called by post in the following) is generally limited to 140 characters (14.7 terms on average [3]), so clearly smaller than RSS items [5] or blogs (250-300 terms [9]).

In micro-blogging, a user follows other accounts to be notified whenever they publish some information. Conversely, he becomes a publisher for the accounts that follow him, which results in the existence of a large social graph. Micro-blogging is characterized by the heterogeneous nature of the accounts. In *Twitter* for instance, there exist some high update frequency accounts and others that publish less than one post a week. Moreover there exist very popular accounts and others with 0 or 1 follower. For various reasons (security, advertisement, control policy, ...) these systems rely on a centralized architecture. Each post published is received by the system that forwards it to *all* followers of the publishing account. Since most active accounts are generally the ones with the highest number of followers, the system must face a tremendous amount of posts to forward. For instance Twitter, which claims more than 200 million tweets a day,

A. Ailamaki and S. Bowers (Eds.): SSDBM 2012, LNCS 7338, pp. 168–176, 2012.

must deliver daily over 350 billion tweets. On the follower's point of view, the amount of posts received from the accounts he follows, between 30 and 200 depending on the system considered (see [8] for Twitter), lets him lost in the middle of long feed of posts. This results in poor data readability and potentially loss of valuable information. A direct consequence of this phenomenon is the high dynamicity of the graph: to avoid flooding, users follow temporary an account to cover a peculiar event, and then unsubscribe because they can't manage the continuous flow of posts [7].

In order to improve the user experience and reduce the network load, we have chosen to introduce filtering in micro-blogging systems. The main underlying idea is that a user A follows another user B for some topics, and consequently he wants to receive only a subset of B's posts that matches his interest, expressed as keywords. To scale, the structure must efficiently retrieve for an incoming post all followers of a publishing account whose filter is satisfied by the post. While designing the filtering structures, we took a particular consideration about specific aspects of micro-blogging systems which we can summarize as:

- **Short messages:** the size restriction (generally 140 characters) means that we handle short documents (*e.g.* the average length of a *tweet* is 14.7 terms [3]);
- **Graph evolution:** As observed in Twitter [7], users follow and unsubscribe often to other accounts. The filtering structure must consequently be dynamic to face this phenomenon;
- **Centralized system:** The social graph is stored by the micro-blogging system. That means the whole task is supported by the centralized system.

Therefore we must reduce the filtering process time by trying to manage the matching in central memory. This paper is, as far as we know, the first attempt to propose such a structure that combines the social graph and keyword-based filters in a micro-blogging system, considering the main characteristics of such a system.

2   Data Model

Basically a micro-blogging system like *Twitter* can be represented as a directed graph $G=(N, E)$ where the set of nodes N represent the users (accounts) of the system and the set of edges $E \subseteq N \times N$ represent the "following" relationships. More precisely a directed edge $e=(u, v)$ exists from a node u to a node v if the user whose account is u is notified whenever the user whose account is v publishes a piece of news (u receives v's updates). In the following we blur the distinction between a user, an account and a node. For a node n, we define $\Gamma^+(n)$ the set of nodes followed by n, *i.e.*, its successors in G as

$$\Gamma^+ : N \to 2^N, \Gamma^+(n) = \{n'|(n, n') \in E\}$$

We define similarly $\Gamma^-(n)$ the set of nodes that follow n (predecessors). Each node produces micro-blog piece of information, called *post* in the following. A post is defined as a sequence of terms $p = < t_0, t_1, t_2, \ldots, t_i >$. We denote by P

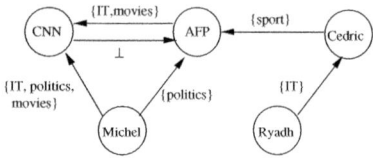

Fig. 1. A filtered social graph

the set of posts and by $\mathcal{V}_P$ the posts vocabulary. Note that we do not rely on the terms order for matching (see Section 3) and consequently we consider the bag of terms of the post $p = \{t_0, t_1, t_2, \ldots, t_i\}$ rather than the term sequence in the following.

To improve micro-blogging systems performances we propose keyword-based filters. A filter F in our system is represented as a set of distinct terms $F = \{t_1, t_2, \ldots, t_n\}$ where each term t_i belongs to the filter vocabulary denoted by $\mathcal{V}_F$. The length of F, denoted by $|F|$, is the total number of (distinct) terms it contains. Like [17], we make the common assumption that $\mathcal{V}_F \subseteq \mathcal{V}_P$. $\mathcal{F}$ denotes the set of filters, excluding the filter $\bot$ that matches all posts, *i.e.*, $\bot = \mathcal{V}_P$. A labeling function *label* associates a filter to each edge of the social graph G:

$$label : E \to \mathcal{F} \cup \bot$$

We name the social graph whose edges are labeled by filters the *filtered social graph (FSG)*. Note that the filters are associated to the edges which allows a user to express different interests (*i.e.*, filters) w.r.t. the source considered. For instance the user *Michel* wants to retrieve all posts from *CNN* concerning *IT*, *politics* and *movies* and only these ones, and from *AFP* only posts about *politics*. Thus we have $label(Michel, CNN) = \{IT, politics, movies\}$ and $label(Michel, AFP) = \{politics\}$. Figure 1 illustrates an example of *FSG* .

3   Filter Indexing

Currently the indexing scheme used in widespread micro-blogging systems like Twitter allows to efficiently retrieve for a publishing user n the set $\Gamma^-(n)$ of followers in graph G. [13] reports an average query size of 1.64 terms for the searches issued on the Twitter search engine. Assuming an average filter size similar to this value and over 300 millions of users, the problem is how to efficiently determine the set $\Gamma^-(n)$ of followers based now on the graph FSG. This issue must be especially tackled for users with a large number of followers since the notification process time largely increases due to the containment relation to be checked. We propose an index structure to manage posts filtering. To achieve notification at runtime, regarding the high incoming rate of posts, we consider structures that fit in memory. This discarded tree-based solutions. Our proposal is based on inverted lists which benefit from factorization and could be deployed on existing systems whose graph structure is already implemented as an inverted list. The set of parameters that impact our indexes constructions are:

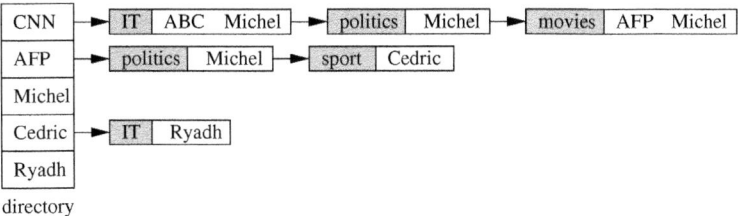

Fig. 2. The PTF-index

- N: total number of accounts
- φ: avg. number of followers for a user
- τ: avg. number of filter terms for a (publisher,follower) pair
- $\theta_{dir}, \theta_{list}$ and θ_{entry}: size of a directory entry, posting list and an entry
- $|p|$: size (distinct terms) of a post

Due to space limitation, we present only the *PTF*-Index. Our two other s proposals (*PFT & TPF*) may be found in the extended version [1].

The PTF-Index. In the *PTF*-index, (as *Publisher − Term − Follower* index), a key is an account $n \in N$, and the value is the corresponding *posting list* $Postings_{PTF}(n)$. We factorize the posting list on the terms, so each term t is associated to a list of the followers of n that choose t as a filter for the posts of n. So $Postings_{PTF}(n) = \{(t_1, \{n_{t_1}^1, n_{t_1}^2, \ldots\}), (t_2, \{n_{t_2}^1, n_{t_2}^2, \ldots\}), \ldots\}$, with $n_i \in \varGamma^-(n)$ and $t_{n_i}^j \in label(n_i, n)$. Figure 2 shows an example of PTF-index structure.

PTF-Index Memory Requirement Our index consists of N posting lists, each posting list must store the $\varphi \times \tau$ filters associated to a publisher, like in *PFT*-index, with a factorization on the different terms. If we assume that all followers of a publisher use distinct filters, the size of $Postings_{PTF}(n)$ is $\varphi \times \tau$. However we observe that followers generally express similar interests when they decide to follow a given publisher and consequently the number of distinct filters for a publisher is lower than this upper bound. We assume like in many other text-based/keyword-based application that the total number of terms in a posting list follows a *Heaps' law* [1,10], i.e., $|Postings_{PTF}(n)| = k \times T^\beta$, where k and β are constants and T is the total number of terms in the posting. Heaps' coefficients k and β depend strongly on the characteristics of the analyzed text corpora and their value in our micro-blogging system has to be determined in future work. Note that β is between 0 and 1 (generally in $[0.4, 0.6]$), so the higher the number of followers is, the better factorization is achieved. This is particularly expected in our filtering system where many users filter out on the same terms. Since the number of terms in $Postings_{PTF}(n)$ is $N \times (\varphi \times \tau)$ and the number of entries indexed is always $N \times \varphi \times \tau$, we deduce that:

$$\Delta_{memory}^{PFT}(FSG) = N \times \theta_{dir} + (N \times k \times (\varphi \times \tau)^\beta) \times \theta_{list} + (N \times \varphi \times \tau) \times \theta_{entry}$$

---

[1] http://cedric.cnam.fr/dahime_m/MicroFilterLong.pdf

PTF-Index Matching Time. Consider an incoming post p published by the user n. We access the posting list of the publisher n $Postings_{PTF}(n)$. Then for each entry $(t_i, \{n_{t_i}^1, \ldots, n_{t_i}^k\})$ we check if p contains t_i. Whenever this happens we notify each $n_{t_i}^j$ from the entry t_i. Thus the average matching time is:

$$\Delta_{time}^{PTF}(FSG, p) = |p| \times k \times (\varphi \times \tau)^\beta$$

4   Experiments

In collaboration with Forth Institute (Heraklion, Crete)[2] we gathered data on Twitter over a four month-period using the Twitter streaming API[3]. We generated a complete Twitter *graph + tweets* dataset by merging this data with graph structures from other datasets available on the Web. We obtained a dataset with 2.1 million users, 15.7 million tweets and more than 148,5 million graph edges. To generate filters we make the assumption that the average filter size (number of distinct terms that labeled an edge in the filtered social graph) corresponds to the average number of terms used on twitter querying API [13]. We decide to generate filter terms for a follower among the most frequent and significative terms (we discarded urls, terms from the common language, Web shortcuts, . . .) in the posts of the publisher he follows. Our rationale is that we usually follow a publisher because he provides some tweets that match one of our interests. Unless otherwise precised, this "realistic" filter set is used for our experiments.

4.1   Memory Requirement

All structures have different factorization criteria which lead to different memory requirements. *TPF*-index appears as the structure with the lower memory requirements (see Fig.3 and Fig.4). Many filters are shared by a significant number of users which allows a better factorization, on the terms first. Moreover we observe that many followers of a publisher filter out on the same terms. Consequently, for a term's entry in the *TPF*-index, there exists also an important factorization on publisher's id, especially for account with an important number of followers. Oppositely the *PFT*-index benefits from a poor factorization since all publishers have an entry in the directory and for each of them we have a list element for each of his followers, each of them with few filter terms. We generate synthetic datasets with a constant number of filters (τ) for each graph edge and report results in Figure 3. The memory occupancy grows linearly with τ for *PFT* and *PTF* indexes. Indeed, increasing τ does not impact the directory size that depends only on the number of publishers, *i.e.*, N. Moreover for *PFT*-index, the number of elements in the posting list remains constant and equal to the number of followers. But the number of entries follows linearly τ. For *PTF*-index the number of elements depends on the number of distinct term used as filter for a publisher. When comparing with *PFT*-index we observe the same gradient. This

_____

[2] We thank Vassilis Christophides for his support and helpful comments.
[3] http://dev.twitter.com/

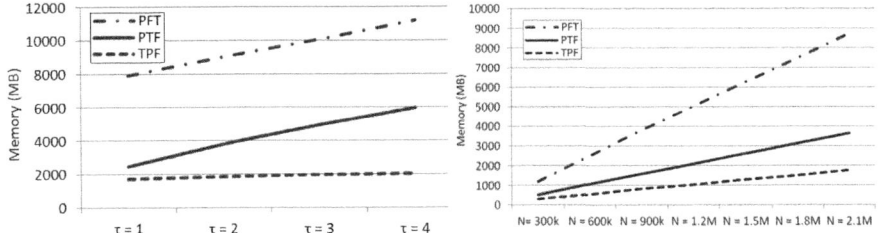

Fig. 3. Occupied memory w.r.t. τ **Fig. 4.** Occupied memory w.r.t. N

reveals that τ has a low impact on the number of elements for a posting list in *PTF*-index. The Heaps' law we propose in our model explains this result, since it assumes that the sub-vocabulary of filter terms for a given publisher increases slightly. Thus for *PTF*-index, like for *PFT*-index, the increase of τ does not impact the structure but only the number of entries. Finally we note the low impact of τ on the *TPF*-index. The rationale is (i) the directory size remains constant and equal to $\mathcal{V}_F$, (ii) since filters are generated w.r.t. the publisher's post areas, a new filter term has a high probability to be already present for the same publisher in the structure, so the number of posting elements remains low and slowly increases.

Figure 4 illustrates the impact of N on the different structures. All structures exhibit a linear growth. For *PFT* and *PTF* index, adding a new user results in adding a new directory entry, new posting list elements and new posting entries for his followers along their filters. However the factorization on terms'id in a posting list is more efficient than the one on follower's id which explains the best gradient for *PTF*. For *TPF* after a short initialization step that corresponds to the creation of the different entries of the directory and the different elements of the posting list, increasing N leads only to add new posting entries. This explains a linear growth with a lower gradient.

4.2   Matching Time

Table 1 depicts average matching times for an upcoming flow of 100,000 tweets over the different filter indexes. We observe that the *TPF*-index exhibits poor matching performances: for our realistic dataset, with around 2.5 ms a post, it can handle less than 400 posts a second, so far from being scalable (remember that in Twitter for instance there exist peaks with 8,000 posts a second). For each term of the post we retrieve a large posting list with potentially as many elements as existing publishers N. Oppositely *PTF*-index quickly retrieves the followers to be notified: it handles a post in 15 μs, so is able to manage peaks up to 66K posts a second. Here we directly access the posting list corresponding to the publisher. Then we scan all its elements that correspond to all followers of this account, and check for each of them if any filter term matches the post. Observe that the number of filter terms is low (between 1 and 3) for a follower, and that we check all terms of the post in a single scan.

Table 1. Matching time for realistic dataset

Structure	PFT	PTF	TPF
matching time (μs)	808	15	2564

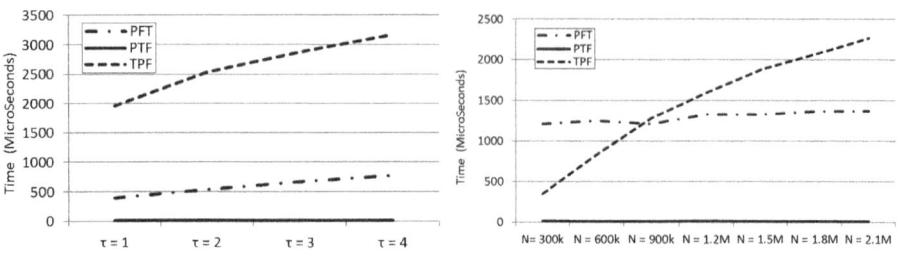

Fig. 5. Matching time w.r.t. τ **Fig. 6.** Matching time w.r.t. N

Fig 5 illustrates the impact of τ on matching time. Like in Table 1 PTF-index outperforms other proposals with 2 orders of magnitude. We observe that the matching time for PFT-index linearly increases with τ while TPF-index follows a sub-linear growth. For the former, matching implies a direct access to the posting list of a publisher and then to scan all elements in turn for this list to check if associated entries match the post terms. Increasing τ does not change neither the number of entries of the directory, nor the number of elements. So only the last step, the matching attempt against term entries requires more time. Since the number of entries is proportional to τ, this explains this linear growth. For the TPF-index, we observe the Zipf's law behavior in the term frequency distribution, so when increasing τ we generally add entries in the posting of the most frequent terms. As a consequence the more numerous filters are indexed, the higher probability we have to add an entry in an existing posting list element. Since posting lists'size has a sub-linear increase and since we scan as many lists as number of terms in the post, this justifies that the matching time increases sub-linearly w.r.t. τ.

We report also in Fig 6 the evolution of matching time w.r.t. N. We notice that both PFT and PTF-index have a constant matching time. Indeed, increasing N only impact the directory by adding new directory entries, but the posting lists keep a constant number of elements and for each element a constant number of entries. Since for an incoming post we scan a single posting list corresponding to the publisher, the matching time is constant with N. TPF-index exhibits better performance than PFT-index for N lower than $900k$. The matching time with TPF-index increases sub-linearly w.r.t. N for the same reasons as with τ.

5   Related Work

The recent success of *Twitter* and the important amount of user-generated content it handles have attracted the interest of the researchers' community. First studies attempt to capture the main characteristics of micro-blogging systems and analyzed the behavior of the users [6,8,15]. [6] presents one of the first studies that looked inside *Twitter*. It shows for instance that users with similar intentions connect with each other. [8] studied the following behavior and information diffusion patterns for a consequent snapshot of the entire Twittersphere. [15] investigates the semantics of the twitter links, and finds that the retweet relation is a strong indicator of the topical interest. We strongly rely on these works to propose structures that exploit micro-blogging characteristics, and to explain some experimental results. Some papers improve data presentation to the users and attempt to avoid user flooding [12,16,14]. [12] presents a clustering and classification of tweets based on the users profile. It relies on a trained classifier to route an upcoming *tweet* to a predefined class. [16] propose techniques to compute the user influence in *Twitter* and to rank tweets according to the user influence. [14] presents a filtering approach which takes advantage of the retweet behavior to bring more important tweets forward. All these approaches aim at avoiding the user flooding but do not diminish the number of delivered messaged by the centralized system. Oppositely our approach propose a scalable structure that handle the filtering process on the server's side reducing drastically the number of posts delivered. More generally, the problem of delivering Web 2.0 data over an underlying graph has also been treated in several papers. [11] considers high frequency update feeds. Authors propose a method to selectively materialize user events over streams in order to handle the scalability of such systems. [2] introduce the evaluation of graph constraints in content-based publish/subscribe systems. Authors suppose that the publishers and subscribers are connected by a directed graph (like in micro-blogging systems) and they implemented algorithms to efficiently evaluate constraints. [4] proposes and compares indexing schemes for a pub/sub system that scales to millions of users and high publication rates. Our context is quite different since we have very short messages (posts) and extremely short queries (filters) but with a number of users and queries much more important.

6   Conclusion

In the present paper we present inverted lists-based structures that indexes filters to decrease the number of messages delivered in micro-blogging systems. We conducted several experiments with real and synthetic datasets. PTF-index appears to achieve the best scalability since, despite memory requirements and insertion time twice more important than TPF-index, it outperformed with two orders of magnitude other proposals for matching time. As future work we intend to improve the PTF-index in several ways. First we would like to exploit the heterogeneity of the accounts as reported in [8] (*e.g.* 5% of accounts with more than 100K followers). We also envisage to rely on ontologies to perform a more "clever" filtering (for instance considering synonymy or containment).

Acknoledgement. Michel Scholl passed away Nov, 15th 2011. We would like to thank Michel for his devotion to the database research. We miss you.

References

1. Baeza-Yates, R.A., Ribeiro-Neto, B.A.: Modern Information Retrieval. ACM Press/Addison-Wesley (1999)
2. Broder, A.Z., Das, S., Fontoura, M., Ghosh, B., Josifovski, V., Shanmugasundaram, J., Vassilvitskii, S.: Efficiently Evaluating Graph Constraints in Content-Based Publish/Subscribe. In: WWW, pp. 497–506 (2011)
3. Foster, J., Çetinoğlu, Ö., Wagner, J., Roux, J.L., Hogan, S., Nivre, J., Hogan, D., van Genabith, J.: #hardtoparse: POS Tagging and Parsing the Twitterverse. In: AMW (2011)
4. Hmedeh, Z., Kourdounakis, H., Christophides, V., du Mouza, C., Scholl, M., Travers, N.: Subscription Indexes for Web Syndication Systems. In: EDBT, pp. 311–322 (2012)
5. Hmedeh, Z., Vouzoukidou, N., Travers, N., Christophides, V., du Mouza, C., Scholl, M.: Characterizing Web Syndication Behavior and Content. In: Bouguettaya, A., Hauswirth, M., Liu, L. (eds.) WISE 2011. LNCS, vol. 6997, pp. 29–42. Springer, Heidelberg (2011)
6. Java, A., Song, X., Finin, T., Tseng, B.: Why We Twitter: An Analysis of a Microblogging Community. In: Zhang, H., Spiliopoulou, M., Mobasher, B., Giles, C.L., McCallum, A., Nasraoui, O., Srivastava, J., Yen, J. (eds.) WebKDD/SNA-KDD 2007. LNCS, vol. 5439, pp. 118–138. Springer, Heidelberg (2009)
7. Kwak, H., Chun, H., Moon, S.B.: Fragile Online Relationship: A First Look at Unfollow Dynamics in Twitter. In: CHI, pp. 1091–1100 (2011)
8. Kwak, H., Lee, C., Park, H., Moon, S.B.: What Is Twitter, a Social Network or a News Media? In: WWW, pp. 591–600 (2010)
9. Ma, S., Zhang, Q.: A Study on Content and Management Style of Corporate Blogs. In: Schuler, D. (ed.) HCII 2007 and OCSC 2007. LNCS, vol. 4564, pp. 116–123. Springer, Heidelberg (2007)
10. Manning, C.D., Raghavan, P., Schütze, H.: Introduction to Information Retrieval. Cambridge University Press (2008)
11. Silberstein, A., Terrace, J., Cooper, B.F., Ramakrishnan, R.: Feeding Frenzy: Selectively Materializing Users' Event Feeds. In: SIGMOD, pp. 831–842 (2010)
12. Sriram, B., Fuhry, D., Demir, E., Ferhatosmanoglu, H., Demirbas, M.: Short Text Classification in Twitter to Improve Information Filtering. In: SIGIR, pp. 841–842 (2010)
13. Teevan, J., Ramage, D., Morris, M.R.: #TwitterSearch: a Comparison of Microblog Search and Web Search. In: WSDM, pp. 35–44 (2011)
14. Uysal, I., Croft, W.B.: User Oriented Tweet Ranking: A Filtering Approach to Microblogs. In: CIKM, pp. 2261–2264 (2011)
15. Welch, M.J., Schonfeld, U., He, D., Cho, J.: Topical Semantics of Twitter Links. In: WSDM, pp. 327–336 (2011)
16. Weng, J., Lim, E.-P., Jiang, J., He, Q.: TwitterRank: Finding Topic-sensitive Influential Twitterers. In: WSDM, pp. 261–270 (2010)
17. Yan, T.W., Garcia-Molina, H.: Index Structures for Selective Dissemination of Information Under the Boolean Model. TODS 19(2), 332–364 (1994)

Regular Path Queries on Large Graphs

André Koschmieder and Ulf Leser

Humboldt-Universität zu Berlin, Germany
Department of Computer Science
{koschmie,leser}@informatik.hu-berlin.de

Abstract. The significance of regular path queries (RPQs) on graph-like data structures has grown steadily over the past decade. RPQs are, often in restricted forms, part of graph-oriented query languages such as XQuery/XPath and SPARQL, and have applications in areas such as semantic, social, and biomedical networks. However, existing systems for evaluating RPQs are restricted either in the type of the graph (e.g., only trees), the type of regular expressions (e.g., only single steps), and/or the size of the graphs they can handle. No method has yet been developed that would be capable of efficiently evaluating general RPQs on large graphs, i.e., with millions of nodes/edges.

We present a novel approach for answering RPQs on large graphs. Our method exploits the fact that not all labels in a graph are equally frequent. We devise an algorithm which decomposes an RPQ into a series of smaller RPQs using rare labels, i.e., elements of the query with few matches, as way-points. A search thereby is decomposed into a set of smaller search problems which are tackled in a bi-directional fashion, supported by a set of graph indexes. Comparison of our algorithm with two approaches following the traditional methods for tackling such problems, i.e., the usage of automata, reveals that (a) the automata-based methods are not able to handle large graphs due to the amount of memory they require, and that (b) our algorithm outperforms the automata-based approach, often by orders of magnitude. Another advantage of our algorithm is that it can be parallelized easily.

1   Introduction

A general regular path query (RPQ) is a regular expression R over the (edge or node) labels of a graph G [28]. Its result is the set of all cycle-free paths in G whose concatenation of labels (edge or node) spells out R. Different flavors of RPQs are used in a wide range of applications. For instance, XPath supports a restricted form of RPQs on XML documents [25]. SPARQL supports a very simple form of RPQs for RDF graphs, and various proposals exist for enhancing SPARQL syntax with full RPQs (e.g., [4,5,9,20]). [11] describes a restricted form of RPQs important for graph pattern matching, and [32,33] describe languages supporting restricted forms of RPQs for studying social networks. A particularly important application domain for RPQs are the Life Sciences, where understanding the interactions of different biological entities is of great importance. Such

A. Ailamaki and S. Bowers (Eds.): SSDBM 2012, LNCS 7338, pp. 177–194, 2012.

interactions are typically modeled as graphs, and RPQs are used to find specific biochemical pathways between distant nodes [23].

None of these works support general RPQs on large graphs, but all focus on restricted languages which typically allow for more efficient evaluation. Evaluating RPQs on arbitrary graphs is an NP-hard problem [28]. We illustrate the problem and our main idea by an example. Suppose a graph of researchers (nodes), either labeled as **P**rofessors or **ST**udents, connected by directed edges such as **S**upervised or **J**oint work. In this graph, the query $P(JP)(JP)$? finds all paths between a professor and direct or indirect co-workers. $(PS)(PS) + (P|T)$ finds all paths between a professor and his doctorate descendants. Now suppose we also model research prizes as nodes (such as **N**obel Prize or Sigmod **A**ward), and connect them to researchers with edges labeled **H**onored. Then, we can find the doctorate predecessors of all Nobel Prize winners using the query $(PS) + PHN$, and those of any prize winner using the query $(PS) + PH(N|A)$.

RPQs have been studied intensively for XML, where the predominant approach is to use automata [13]. Both the graph and the query are represented as automata, whose intersection automaton is the subgraph specified by the query. In this process, the graph needs to be translated into a DFA, which can be of exponential space and may need exponential construction time. Research in XML query languages has shown that automata-based RPQ evaluation works well for trees (XML) [30], but we will show that its space consumption is enormous on general graphs (see Section 6). Furthermore, automata-based approaches completely disregard the fact that certain labels are much more frequent than others, which can be exploited for speeding up query execution. For instance, to answer the query for Nobel Laureates, it is sensible to first search for nodes labeled with N, because (a) one such node must be in any matching path and (b) there are much less Nobel Laureates than professors. Having all N nodes, complete paths can be computed easily by traversing the graph.

Such reasoning is the basis of the approach we propose in this paper; however, finding a good evaluation strategy is not always as easy as in the example, as a query usually contains various labels with different frequencies in the graph. To this end, we first gather all labels in the query that only occur a few times in the graph and use these as fix points for a series of bi-directional searches. Thus, we split the query at these rare labels into smaller queries, and answer these individually. We search all matching paths between each two adjacent rare labels as well as at the start and end of the query, and combine the results to answer the original query. The main advantage of this approach is that we do not need to consider the whole graph, but only those fractions of it that lie between adjacent rare labels. In Section 6 we will show that this strategy, over a wide range of small to large synthetic and real-life graphs, is considerably faster and especially much less space-demanding than automata-based methods ([39,14]).

Figure 1 illustrates this idea. Suppose we want to answer the RPQ $a+ b\ c+$ on a graph (Fig. 1b). Since there is only one edge labeled with b, we use it as rare label and split the query there. Now, the two smaller queries $a+$ and $c+$ have to be answered, using the b edge as end point or start point, respectively.

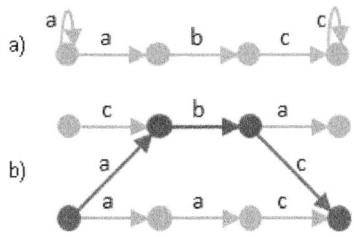

Fig. 1. a) The RPQ $a+$ b $c+$ shown as a (nondeterministic) automaton; b) a path fulfilling the RPQ in a small exemplary graph

The result of the original query is the combination of the smaller queries and the rare label.

Our idea can also be used for variations of the general RPQ problem, such as finding all shortest matching paths spelling out a regular expression, or finding all matching paths between two given nodes. Especially the latter is important if RPQs are used as predicates in a general query language (see, for instance, [20]), where variables for defining the start and the end of a path often already have been bound by other predicates before RPQs are evaluated. For space constraints, in this paper we only describe in detail the algorithm for solving the general RPQ problem (finding all acyclic paths matching a given regular expression) without bindings for the start and end nodes, as this is the most complex case. Furthermore, we only consider RPQs over edge labels; extensions to include node labels are straight-forward. Adaptations of our method targeting such variations of the RPQ problem are described in [21], which also contains a number of additional experiments we omit here for space constraints.

This paper is structured as follows. Section 2 gives an overview on related work. In Section 3, we define the basic concepts. In Section 4, we describe our novel RPQ evaluation algorithm and give implementation details in Section 5. We evaluate our method in Section 6 using various real and synthetic graphs and conclude in Section 7.

2   Related Work

The most common approach for answering RPQs is based on automata. One implementation of this idea are DataGuides by Goldman and Widom [13], based on Lorel [1]. Therein, the graph is considered as an NFA that is first converted into a DFA and then minimized. This minimized DFA (a DataGuide) is then used as an index. However, this index can become much larger than the original graph, which is a problem when dealing with arbitrary graphs (but not for regularly structured XML). Goldman and Widom therefore propose "Approximate DataGuides" [14] which reduce the index size using heuristics. As the implementation of Lorel (and DataGuides) is not supported any more since 2000, we re-implemented the algorithms and compare them to our approach in Section 6.

After the uptake of the semi-structured data model in XML, many proposals have been put forward to use automata in optimizing queries on XML (see e.g. [25,30]). However, these works mostly use tree automata [30] and are not applicable to arbitrary graphs. Additional index structures have been proposed, for instance, by Milo and Suciu [29] and Kaushnik et al. [19], but, again, they are designed to work with XML data and cannot be applied to non-tree graphs. Fernandez and Suciu present another interesting approach to speed up graph searching based on Graph Schemas [12]; however, these have to be created manually, a step that seems unfeasible for graphs with millions of nodes.

An area where RPQ queries on graphs are important is querying RDF data. However, SPARQL, the official W3C recommendation as an RDF query language, does not support regular path queries, which spurred research into extending SPARQL with RPQs. Alkhateeb et al. [4] developed the query language PSPARQL that includes RPQs, but the authors focus on formal semantics of RPQs on RDF and do only describe a proof-of-concept implementation based on backtracking. Detwiler et al. [9] present the GLEEN system, an extension to SPARQL including RPQs that is implemented as an extension to the ARQ library for SPARQL processing (see Section 6.2 for a comparison). Anyanwu et al. [5] present another extension to SPARQL that also includes RPQs, but no method for evaluating them is described. Zauner et al. [39] present a path language for RDF supporting RPQs and does provide an implementation; the system is based on automata, and we compare against it in Section 6.2. SPAR-QLeR is another RDF querying language encompassing RPQs, again based on automata, for which an implementation was described in [20], but is not publicly available. Note that the runtimes reported in the paper range in the order of seconds for queries with bound start and end nodes on moderately sized graphs, a setting in which our algorithm only needs milliseconds (see [21]).

There were also a number of proposals for general graph query languages that are not based on RDF. Leser [23] proposes a query language for querying biological pathways, which syntactically supports RPQs, but does not describe a scalable resolution technique. Graphs-at-a-time is a query language based on graph grammars that is capable of expressing RPQs, but the presented implementation does not cover such predicates [15]. The query language proposed in [10], which also includes a type of RPQs, is not accompanied by any implementation. Several systems have been designed to support extremely large graphs (hundreds of millions nodes and edges), such as Pregel [26], GRAIL [38], or DEX [27], however, none of these systems support RPQs. Finally, Sevon and Eronen [34] describe a method for querying paths in labeled graphs using context-free grammars. They traverse the graph breadth-first and use a context-free parser to find matches. While context-free grammars are more powerful than regular expressions, [34] only focus on finding paths between fix start and end nodes and do not provide optimization techniques as we do.

Another line of related work are concerned with graph pattern matching. Fan et al. [11] show that a language supporting reachability and a restricted form of RPQs allows for an evaluation in cubic time. Jin et al. [18] present algorithms for finding paths which only consist of labels from a predefined set of labels. Ronen and Shmueli [32] describe a graph query language that supports conditions on labels being contained or not contained in paths and also supports ranking (provided that the edges are weighted) and aggregation over sets of paths. In contrast to these works, our approach supports full RPQs.

Zou et al. introduced the Distance-join [40] as a technique to find subgraphs with similar connections to a query pattern. The work most related to ours is probably [7], in which Cheng et al. present algorithms for finding subgraphs that are homeomorphic to a query graph, i.e., where each edge of the query may be mapped to a path of unbounded length in the graph. In some sense, this can be seen as a generalization of the RPQ problem we study in this paper; however, a transformation would be non-trivial (mapping of edges to nodes and vice versa; introduction of special edges which must be mapped to exactly one edge for labels without '?' or '*' modifier etc.). Besides, the source code of [7] seems to be not available for comparisons.

Matching regular expressions (REs) on strings is a related problem that was recently picked up by the database community. Examples are [6,8], which both use index structures that are similar in spirit to our method, i.e., concentrating on rare characters in the query. Another line of related research is graph indexing. Here, the idea of using frequencies of labels has been used extensively, especially in mining and searching of subgraphs [36,22].

In summary, despite a large body of research around evaluating RPQs on graphs, we are aware of only two available implementations supporting full RPQs as we do [39,9]. In Section 6, we compare our approach to these methods and also to a re-implementation of the DataGuide system [13] and show that, for large graphs, they suffer from excessive memory consumption and are clearly outperformed by our method.

3   Terms and Definitions

We use labeled directed multigraphs, i.e., a graph G is a tuple $G = (V, E, f, l, \Sigma)$, where V is a finite set of nodes, E is a finite set of edges, $l : E \to \Sigma$ specifies the edge labels Σ, and $f : E \to V \times V$ is the connection function, specifying which nodes are connected by which edges.

The topological properties of a graph can be measured through node degree and label distribution. A graph is called scale-free if the number of nodes with degree k is $P(k) \sim k^{-\lambda}$ for large values of k. Scale-free graphs are the likely outcome of various random growth processes, and, indeed, many graphs discovered in biological research are scale-free [24]. The label distribution in a graph is called Zipfian if the frequency of the labels occurring in the graph follows the power law $F(k) \sim k^{-\delta}$, with $\delta \approx 1$.

A regular path query (RPQ) is a regular expression over Σ. We use the definition for regular expressions as in [16]. To evaluate regular path queries, a regular expression can be converted into an automaton that can be used to match paths. We assume definitions for deterministic (DFA) and non-deterministic automata (NFA) as in [3].

Several kinds of questions can be answered for a given regular expression R and a given graph G. (1) Does G contain any path fulfilling R? (2) Which is the shortest path in G fulfilling R? (3) Is there a path in G between two fixed nodes fulfilling R? (4) Which paths in G fulfill R? In this paper, we discuss our proposal using the latter problem as show case as it subsumes all other types of queries. In Section 6, we will show that, for instance, fixing the start and end nodes of an RPQ allows to drastically speed-up query processing compared to the more liberal problem of returning all paths in the graph.

4   Answering RPQs Using Rare Labels

In this section, we present our novel algorithm for efficiently answering RPQs. The basic idea is to search the graph while simultaneously advancing in the query automaton. Compared to an approach which converts both the graph and the query into automata, our method has several advantages: No preprocessing of the large graph is needed, it only uses space linear in the size of the graph, the search is easily parallelizable, and it can be enhanced with techniques that take label frequencies into account.

4.1   Rare Labels

Definition 1. Let G be a graph and R be a RPQ. We call a label occurrence in R *mandatory* iff it is not followed by a modifier other than $+$, i.e. it occurs in every possible result of R in G. We call a label *rare* iff it occurs at most m times in G and is mandatory in R (where m is a parameter of the method, see Section 4.3).

For example, in the regular expression $a\ b+\ c*\ d?$, a and b are mandatory, while c and d are not. Finding all rare labels in a query is very fast if a list of all labels in the graph together with their frequencies is stored. If this list is indexed by labels, finding all rare labels for a given query is linear to the size of the query pattern.

If a query contains a rare label, then any match of the query in the graph must contain an occurrence of it. Therefore, we can use the occurrences of rare labels as way-points during the search process. If we can find two or more rare labels in a query, we can use a two-way search algorithm to find all matching paths between their matches in the graph. Every additional rare label further reduces the search space.

Note that our implementation (see Section 5.1) also treats disjunctions (regular expressions of the form $(a|b|...)$) as mandatory by searching for any occurrence of the labels. A further, not yet implemented optimization would be to rewrite expressions of the form $R\ S*\ T$, where R, S, and T are regular expressions, as two expressions RT and $R\ S+\ T$. Thus, the original expression could be answered by evaluating two expressions, one of which is shorter and the other contains an additional mandatory label for our optimization.

4.2 Searching the Graph Using Rare Labels

For queries that include at least one rare label, we split the query at these rare labels and use them as fix points in the search. Searching the graph and advancing in the regular expression at the same time, we search all paths between each two adjacent rare labels, all paths from the first rare label backward to the start of the regular expression, and from the last rare label forward to the end. As shown above, the number of nodes that need to be visited during this search shrinks with every additional rare label but grows with increasing numbers of occurrences of rare labels.

Besides keeping the search space smaller, rare labels often also allow for early stops. If there is no path between any two adjacent rare labels, then there can be no path fulfilling the original query, and the search can be stopped immediately.

In the following, we use the term *first rare nodes* for all nodes that are starting point of an edge of which the label is the first rare label, according to the regular expression. Analogously, *last rare nodes* are the end nodes of all edges with the last rare label. Answering RPQs using rare labels is done in the following 6 steps.

1. Gather all rare labels for the query in the graph.
2. If more than one rare label exists, find the paths between the first and second rare label, the second and third etc. using a two-way search algorithm. If no path can be found in any of these search processes, stop the search and return an empty result for the query.
3. If more than one rare label exists: Using the results from step 2, find all paths from the *first rare nodes* to the *last rare nodes* and remove all rare nodes that are on no path, as these cannot be on a result path.
4. Beginning at all remaining *first rare nodes*, find all paths to the beginning of the regular expression, searching backward.
5. Beginning at all remaining *last rare nodes*, find all paths to the end of the regular expression (forward).
6. Using the results, enumerate all paths in the graph that fulfill the regular expression and return the result.

Figure 2 shows the principle of the algorithm. On a sample graph (edge labels and directions omitted), the query $a+\ b\ c+\ d\ e+$ is executed, assuming that b and

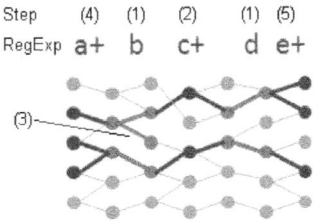

Fig. 2. Search process example for the query $a+\ b\ c+\ d\ e+$ in an arbitrary graph (edge labels and directions omitted)

d are rare labels. In step 1, rare label edges are gathered (b and d edges). In step 2, we search all paths between the end nodes of the b edges and the beginning of the d edges. These paths must fulfill the regular expression between the two rare labels, in this case $c+$. Here, two such paths can be found. For one rare edge, no path could be found, thus it is removed from further consideration in step 3.

In step 4, a one-way backward search is performed, starting at the start nodes of the b edges. The search ends once all paths have been found that fulfill the first part of the regular expression ($a+$). In step 5, we search all paths from the end of the last rare label to the end of the regular expression in forward direction. As a last step (not shown in the picture), we enumerate all paths by combining the results of the previous steps. In this case, there are 4 distinct paths. The result subgraph can be gathered by enumerating all nodes and edges of the result paths.

Our approach specifically aims at queries that include labels that do not occur often in the graph. While most queries used in Bioinformatics are interested in these rare labels, there are also queries in which no rare label is present. In such cases, our algorithm automatically switches to a brute force search, starting a search at every node in the graph (respectively at the given start/end nodes, if specified). In Section 6 we show that our implementation is faster than other approaches (in particular, the automata-based one) even for those cases.

4.3   Determining Rare Labels

The algorithm described above assumes a fixed value for m, the parameter determining which labels are considered rare. The choice of m needs to find a compromise between treating as many labels as rare as possible and keeping the number of occurrences of rare labels as small as possible. If a rare label has many occurrences in the graph, the search space increases because each occurrence needs to be included in the search (forward and backward). On the other hand, multiple different rare labels in a query speed up its execution, because partial paths that need to be searched are shorter.

We know of no simple way to determine the best value for m. It depends on the graph as well as on the query, so using a fixed value is not the best approach. We therefore use an adaptive heuristic for determining which labels to consider rare for a given graph and query. The idea is that if, for a given query, there are several possible rare labels, we set the threshold for rare labels higher than if only very few rare labels can be found. This, on the one hand, produces less queries without any rare labels. On the other hand, for queries with many potential rare labels, only labels with a small number of occurrences are included.

Our proposed heuristic works as follows. We first acquire a list of all potential rare labels for the query. We then reduce this list depending on the overall number of paths that would need to be searched in the current configuration. Labels that produce the most paths are removed first; we can compute the number of paths between any two adjacent rare labels r_1, r_2 as $|r_1| \cdot |r_2|$. The overall number of paths is the sum of all paths between all adjacent pairs. For the first and last rare labels, we also add their number of occurrences to account for the search to the beginning and to the end of the path. We repeatedly remove the rare label that produces the most paths, until the sum of all paths is below a threshold.

5   Implementation

In this section, we give a short overview of the architecture of our system and the data structures used. We describe the implementation of our algorithms in some more detail and give a complexity analysis of our search algorithms.

We use a node based storage schema, which means that the nodes in the graph are represented explicitly, while the edges only exist as attributes of the nodes in the form of adjacency lists. Edges are stored in forward and backward direction, which enables two-way searching but almost doubles memory consumption. Labels are always stored as integers; if the labels are given as strings, a global mapping table is used to map them to numbers. To be able to efficiently gather rare labels from the graph, we also use an index on the edge labels also encoding their multiplicity. Figure 3 visualizes the architecture of our system.

Regular path queries are represented as NFAs. Converting a regular expression into an NFA is straight-forward. The automaton is stored as states and transitions, which are labeled with the number representing the label for the transition. Transitions are stored in both directions for two-way search. To speed up query execution, we create a list for every state with all labels that are accepted in that state, and a list showing into which states the automaton may transition for each label. We call these *labeled follow sets* as a reference to follow sets used in compiler construction [3].

Our current implementation requires about 2 GB of memory for a graph with 10 million nodes and 20 million edges, including all additional data structures described in the following. Thus, working with graphs even much larger than the ones we use for evaluation (see Section 6.1) would, in the first place, not be a problem of memory.

5.1   Search Algorithms

Answering RPQs involves several algorithms. In this section, we give an overview of the most important ones.

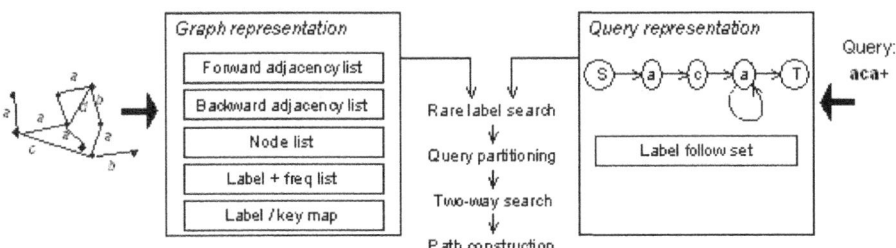

Fig. 3. Storage schema and query evaluation process of our system

RARE LABEL SEARCH. The first task in executing a query is to find the rare labels. To this end, we go through the automaton representing the query from start to end. For every state, we check if it is mandatory (i.e., does not have a modifier as * or ?). If it is mandatory, we look up the edge label the state represents in the edge label index (which requires constant time). This index gives us the number of occurrences of the label in the graph. If it is below the given threshold, the current state is added to the list of rare states. We consider alternatives (e.g., $a|b$) as rare if the sum of the number of occurrences of all alternatives is below the threshold. Items in brackets can only be mandatory if the bracket itself is mandatory.

TWO-WAY SEARCH. If two or more rare labels have been found, we use a two-way search algorithm to find paths between each two neighboring rare labels. Since rare labels can occur more than once, this is a many-to-many search, starting at the end nodes of all edges with one rare label, and ending at the start nodes of all edges with the next rare label. The search is performed breadth-first by iterating through the graph and the query automaton at the same time. A search state (one specific point during the search process) consists of the current position in the graph and the current state of the automaton. Different search states can be at the same position in the graph but in different states of the automaton, or vice versa. When traversing an edge in the graph, we check if its label is in the *labeled follow set* of the current state. In that case, new entries are added to the end of the list of search states to be processed. One search state is created for each entry in the follow set.

The aim of the search is to find all paths for each pair of fix start and end nodes. A path is found if a forward and a backward search meet at a node and are in the same state of the query automaton. We keep lists for every node where we store in which states a search passed the node. This is used to find completed paths as well as to prevent cycles in the result paths. The search ends once the list of unprocessed search states is empty.

START AND END SEARCH. Searching for the start and end of the path works much like the two-way search algorithm. The only differences are that we search one-way and that we do not end when finding specific nodes, but when hitting a finish state in the query automaton (or a start state, for the backward search).

5.2   Two-Way Search Complexity

Theoretically, all nodes might need to be searched during a two-way search, and every node could be visited in every state of the automaton, and from every start or end node, resulting in a complexity of $O(|V| \cdot |S| \cdot r)$, with S being the states of the automaton and r the number of start plus end nodes, i.e., the occurrences of the rare labels used for that search. In reality, however, the search space is limited because the labels on the path must match the automaton (a query should thus be as small as possible, e.g. not include $a*a*$ instead of $a*$). Also, the number of start and end nodes is much smaller than $|V|$ (depending on which labels are considered rare). Due to two-way search, the search space is reduced further in most cases, since complete paths are only half as long from both directions.

Additional checks have to be made during the search, but they do not add to its complexity. For finish and cycle checks, we need to check for all nodes encountered on the path whether they have already been visited in the same state in the same direction (indicating a circle) or in the other direction (indicating a completed path). This can be done in constant time. However, cycle checks might be performed more than once per node, if a node has multiple incoming edges: In the worst case, the check is performed as often as the number of edges in the graph. Thus, the overall worst-case complexity sums up to $O((|V| + |E|) \cdot |S| \cdot r)$.

5.3 Parallelization

The search process can be parallelized in different ways. For a query that contains n rare labels, $n+1$ smaller queries have to be answered. This is performed as $n+1$ independent searches which are executed in parallel. Also, the search algorithm is a many-to-many search if the rare labels occur more than once in the graph. This can also run in parallel, with different threads processing different start points.

6 Experimental Results

In this section, we present an experimental evaluation of our method. We compare our rare-label based algorithm with other implementations available and show results for different graphs and different kinds of queries. Further experiments are devoted to scalability with regard to the size and density of graphs, to the effects of parallelization, different query types, and different label distributions. All tests were executed on a Quad-Core AMD Opteron machine with 16 GB of main memory. The execution times for queries given in the following were gathered by executing 10,000 queries and building the average.

As threshold for the number of path combinations in the rare label optimization (see Section 4.3), we used a value of 100; higher values did not yield any significant changes in runtimes, while lower values lead to slower queries.

6.1 Graphs and Queries

We use real graphs (from biological research) as well as artificially created graphs for the evaluation. We present results for two real-world graphs which we call *AliBaba* and *Extracts*. *AliBaba* is a network of protein-protein-interactions extracted by text mining on all of PubMed [31]. The graph has about 50,000 nodes and 340,000 edges. *Extracts* is a graph of enzymes and their relations, also extracted by text mining from biomedical abstracts, containing about 80,000 nodes and one million edges. Note that these graphs are not toys; such networks today are used regularly in Systems Biology, for instance to improve protein function prediction [17] or disease-gene identification [2]. Their size is roughly comparable or larger to that of the largest databases of biological networks (the KEGG network currently contains approximately 45.000 nodes), but their density is considerably higher. Thus, they represent rather difficult cases for this domain. Results for other biological networks we tested were similar to those on these two graphs.

To systematically study the scalability of the algorithms with regard to various parameters, we created artificial graphs with sizes between 1000 and one million nodes, with different average degrees and different label distributions. All real and all synthetic graphs are scale-free and roughly have a Zipfian distribution of edge label frequencies. The influence of the type of graph and the label distribution is shown in [21].

For our evaluation, we used both real-life queries from the Bioinformatics domain as well as large, artificially generated sets of queries with similar properties. The biological queries on the AliBaba graph are used in a Systems Biology application analyzing networks of transcription regulation and protein-protein interactions [37]. They consist of 4-12 labels, which may or may not be rare. Artificial queries are used to evaluate the influence of the queries on the runtime (see Section 6.4). The queries have been created with special properties (e.g. number or occurrences of rare labels) and are structurally similar the the biological queries. See the full paper [21] for more details.

6.2 Comparing with other Implementations

We are aware of just two implementations of regular path queries on graphs and considered both as competitors of our algorithm.

RPL [39] evaluates RPQs on RDF graphs using an automata-based implementation. Using the original code provided by the authors, we observed that the system works well on small RDF graphs. However, it cannot handle graphs of the size we target with our work. For answering queries on a graph with approx. 10,000 nodes and 20,000 edges, the system already required 16 GB of main memory. Queries on this graph required already several seconds per query, while our algorithm answers those queries in less than 100ms. All larger graphs, including our real-world biological graphs, could not be handled anymore. Therefore, we do not include this system in the following systematic evaluation.

GLEEN [9] is implemented as an extension to the ARQ library for SPARQL processing. Thus, a comparison of runtimes would be rather unfair, as the ARQ extension mechanism is based on an iterative load-and-verify of single edges of a graph, which cannot be compared to our approach of loading the entire graph into memory at once.

This leaves the re-implementation of the DataGuides system [14] (called AUT from now on) to compare with. Automata-based methods generally work as follows. First, the graph and the query are transformed into NFAs. Both NFAs are then converted into DFAs that are minimized in a third step. Then, both minimized DFAs are intersected. The result is an automaton that is equivalent to the subgraph (of the graph) embracing all paths that match the query. Extracting all matching paths from this subgraph is straight-forward. While this approach works very well for evaluating RPQs on schema-based (i.e. sequences of labels

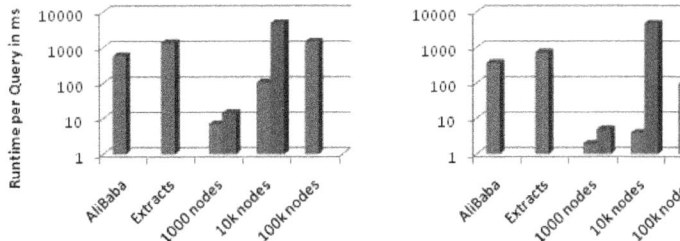

Fig. 4. Average runtime (log scale) to answer one query on different graphs. The left diagram shows the results for queries without rare labels. On the right, each query contains at least one rare label. Queries on Extracts as well as on the synthetic graph with 100K nodes could not be executed with AUT.

follow a schema) or tree-like data (e.g. XML), it has problems with general graphs and schemaless label distributions, as in our case. For such structures, the DFA generated from the NFA often grows extremely in size (recall that the DFA for an NFA can be exponential in size of the NFA in the worst case). As we show shortly, even mid-sized graphs cannot be processed by this method.

For comparing our RL method with AUT, we use sets of queries with rare labels as well as completely random queries (which may contain an arbitrary number of rare labels or none). As we are not aware of any efficient parallelization scheme for automaton minimization and determinization, we only compare single-threaded versions of both implementations; the additional speed-up that is possible with our algorithm on current (multi-core) hardware will be evaluated in Section 6.5.

Figure 4 shows the average runtime (averaged over 10,000 queries) for answering an RPQ with RL and with AUT. The former is faster in all cases. For queries without rare labels, the runtimes do not differ much for small graphs, but differences get significant for larger graphs. For a graph of 10,000 nodes and 20,000 edges, RL already performs almost two orders of magnitude faster than AUT. For queries that contain at least one rare label, again, RL is always faster, and its superiority increases with graph size; differences for those queries are larger than for queries without rare labels.

Rare labels have a considerable influence on the runtime of RL. In contrast, we found that for AUT, the runtime for different queries on the same graph is about equal. However, time is only one problem of this method; the other is space. The NFA-DFA conversion in AUT incurs an exponential increase in the number of nodes. In our implementation, the whole process requires 350 MB for the graph with 1000 nodes and 2000 edges, but already 3.8 GB for the 10 times larger graph – which still is considerably less space than required by RPL (see above). Running AUT on the AliBaba or Extracts real-life graphs or on the 100K synthetic graph failed due to memory overflow.

6.3   Scalability: Graph Size and Density

To test different scalability aspects of RL, we used artificially created graphs with varying properties. All queries contain at least one rare label. For this

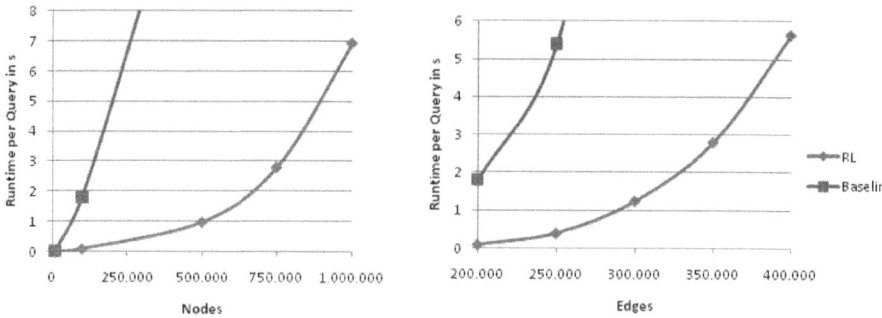

Fig. 5. Average runtime for answering one query on synthetic graphs with different numbers of nodes and edges. The left diagram shows results for a fixed node to edge ratio of 1:2, while the number of nodes is 100,000 on the right.

evaluation, we used the multi-threaded implementation with a fixed number of four threads. We evaluated the effect of our rare-label optimization by comparing it to a baseline method, which performs a brute-force search starting at every node (also in parallel) without considering label frequencies.

Figure 5 (left) shows how RL scales with the size of the graph at a fixed node/edge ratio. The smallest graph has 10K nodes and 20K edges, and the largest graph has 1 million nodes and 2 million edges. Clearly, the scaling of the implementation with the rare-label optimization is much better than that of the baseline. Even for the largest graphs we tested, RL can answer a regular path query in few seconds on average.

Figure 5 (right) shows scalability of RL with the average graph degree. Using multiple artificial graphs with 100,000 nodes, we increased the number of edges (and thus the average degree), leaving all other properties equal. Again, the increase in execution time is favorable compared to the baseline.

Both experiments also show that execution times grow super-linearly with increasing graph size and graph density. Also, the absolute times cannot be compared to those achieved for answering, for instance, reachability queries on graphs of similar size [38,35]. But one should not forget that evaluating RPQs on graphs is a NP-hard problem, whereas reachability can be answered in $O(n^3)$.

6.4   Influence of Query Types

To evaluate the influence of our optimizations on different types of queries, we created ten sets of 1000 queries each differing in the number of occurrences of rare labels. One set does not contain any rare labels (set 0). All other sets contain exactly one rare label, but with an increasing number of occurrences in the graph. The rare labels found in query set 1 appear only once in the graph, rare labels from set 2 exactly twice, and so on. The runtimes for the different query sets are shown in Figure 6.

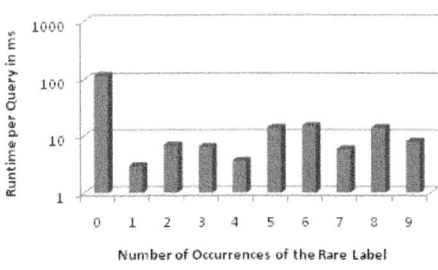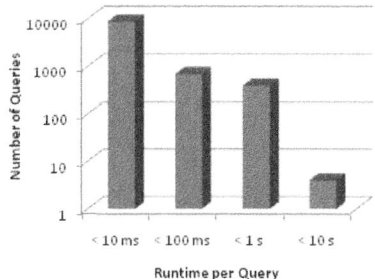

Fig. 6. Left: average runtime for answering one query from the respective query set. Right: Number of queries taking up to 10 ms, 100ms, 1s, and 10s to execute. The synthetic graph that was used has 10,000 nodes and 20,000 edges.

As expected, answering queries without rare labels is considerably slower than for queries containing rare labels. The difference between queries without rare labels and those with one rare label appearing exactly once in the entire graph is almost three orders of magnitude. Also, queries with a rare label that occurs less frequently in the graph generally are executed faster. However, this trend is partly out-weighed by noise generated through the random complexity of the queries in the workload.

To further study how runtimes change for different queries, we categorized the runtimes of all queries from our query set on a given graph. As Figure 6 shows, more than 90% of all queries are answered in less than 10ms, and more than 95% are answered in less than 100ms. But a few queries take exceptionally longer than all others. These pathological queries are queries that contain only very frequent labels, leading to exceptionally large result sets. For instance, the most difficult query from our set took about 4 seconds and generated a result set of 230 million matching paths. However, we believe that such queries rarely occur in any real application.

We also investigated the influence of query lengths, i.e., the number of labels in the regular expression of the query. Comparing the execution times of queries depending on their lengths, we found that the query length only has a minor influence on execution speed (data not shown). The reason is that, although longer queries are more complicated to answer in general, they often do not have as many matches in the graph, which reduces the search space. For execution speed, the number of occurrences of the labels and the number of rare labels are much more important than the sheer number of labels in the query.

6.5   Parallelization

Figure 7 shows the effect of using multiple threads for RL. The scale-up is very good for up to four threads, but the additional advantage of adding more threads levels out for more than 4 threads. This behavior can be explained by the fact that our current implementation uses additional threads only for additional rare

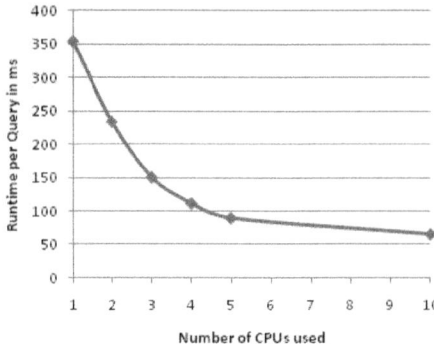

Fig. 7. Average runtime for answering a query containing rare labels on a graph with 10,000 nodes and 20,000 edges against the number of threads used

labels. For example, a query containing two rare labels uses three threads: One for searching the paths between the rare labels, one for searching to the start and one for searching to the end of the query. Since few queries contain more than three rare labels, the scaling diminishes. However, it would be possible to enhance the implementation by also running the searches between two instances of a rare label in parallel. Since most rare labels appear more than once in the graph, finding paths between them is a many-to-many search and could use different threads starting at different nodes.

7 Conclusion

We presented a novel approach for answering regular path queries on large graphs. Our main idea is to structure a graph traversal along those labels from a query that are infrequent in the graph, but guaranteed to occur in any matching path. We use these rare labels as start-, end-, and way-points during traversal, thus essentially breaking up a very large search space into many much smaller ones. We compare our novel method with a traditional approach using automata and find that the former outperforms the latter over a wide range of different graphs and queries; furthermore, it requires only linear preprocessing and is able to handle much larger graphs. We also showed that using the rare-label optimization considerably improves scalability with regard to the size of the graph and to graph density.

References

1. Abiteboul, S., Quass, D., McHugh, J., Widom, J., Wiener, J.L.: The lorel query language for semistructured data. Int. Journal on Digital Libraries 1, 68–88 (1997)
2. Aerts, S., Lambrechts, D., Maity, S., Van Loo, P., et al.: Gene prioritization through genomic data fusion. Nat. Biotechnol. 24(5), 537–544 (2006)

3. Aho, A.V., Sethi, R., Ullman, J.D.: Compilers: principles, techniques, and tools. Addison-Wesley Longman Publishing Co., Boston (1986)
4. Alkhateeb, F., Baget, J.-F., Euzenat, J.: Extending SPARQL with regular expression patterns (for querying RDF). Web Semant. 7(2), 57–73 (2009)
5. Anyanwu, K., Maduko, A., Sheth, A.: Sparq2l: towards support for subgraph extraction queries in rdf databases. In: WWW 2007, Banff, Alberta, Canada, pp. 797–806 (2007)
6. Chan, C.-Y., Garofalakis, M., Rastogi, R.: Re-tree: an efficient index structure for regular expressions. The VLDB Journal 12(2), 102–119 (2003)
7. Cheng, J., Yu, J.X., Ding, B., Yu, P.S., Wang, H.: Fast graph pattern matching. In: ICDE 2008, pp. 913–922. IEEE (2008)
8. Cho, J., Rajagopalan, S.: A fast regular expression indexing engine. In: ICDE 2002, p. 0419 (2002)
9. Detwiler, L.T., Suciu, D., Brinkley, J.F.: Regular paths in sparql: Querying the nci thesaurus. American Medical Informatics Association, 161–165 (2008)
10. Dries, A., Nijssen, S., De Raedt, L.: A query language for analyzing networks. In: CIKM 2009, New York, NY, USA, pp. 485–494 (2009)
11. Fan, W., Li, J., Ma, S., Tang, N., Wu, Y.: Adding regular expressions to graph reachability and pattern queries. In: ICDE, pp. 39–50 (2011)
12. Fernandez, M.F., Suciu, D.: Optimizing regular path expressions using graph schemas. In: ICDE 1998, pp. 14–23. IEEE, Washington, DC (1998)
13. Goldman, R., Widom, J.: Dataguides: Enabling query formulation and optimization in semistructured databases. In: VLDB 1997, pp. 436–445 (1997)
14. Goldman, R., Widom, J.: Approximate dataguides. In: Workshop on Query Processing (1999)
15. He, H., Singh, A.K.: Graphs-at-a-time: query language and access methods for graph databases. In: SIGMOD 2008, New York, USA, pp. 405–418 (2008)
16. Hopcroft, J.E., Ullman, J.D.: Introduction to Automata Theory, Languages, and Computation. Addison-Wesley, Reading (1979)
17. Jaeger, S., Gaudan, S., Leser, U., Rebholz-Schuhmann, D.: Integrating protein-protein interactions and text mining for protein function prediction. BMC Bioinformatics 9(suppl. 8), S2 (2008)
18. Jin, R., Hong, H., Wang, H., Ruan, N., Xiang, Y.: Computing label-constraint reachability in graph databases. In: Proceedings of the 2010 International Conference on Management of Data, SIGMOD 2010, New York, NY, USA, pp. 123–134 (2010)
19. Kaushik, R., Bohannon, P., Naughton, J.F., Korth, H.F.: Covering indexes for branching path queries. In: SIGMOD Conference, pp. 133–144 (2002)
20. Kochut, K.J., Janik, M.: SPARQLeR: Extended Sparql for Semantic Association Discovery. In: Franconi, E., Kifer, M., May, W. (eds.) ESWC 2007. LNCS, vol. 4519, pp. 145–159. Springer, Heidelberg (2007)
21. Koschmieder, A., Leser, U.: Regular Path Queries on Large Graphs. In: Ailamaki, A., Bowers, S. (eds.) SSDBM 2012. LNCS, vol. 7338, pp. 177–194. Springer, Heidelberg (2012)
22. Kuramochi, M., Karypis, G.: An efficient algorithm for discovering frequent subgraphs. IEEE Trans. on Knowl. and Data Eng. 16(9), 1038–1051 (2004)
23. Leser, U.: A query language for biological networks. Bioinformatics 21(2), 33–39 (2005)
24. Li, L., Alderson, D., Tanaka, R., Doyle, J.C., Willinger, W.: Towards a theory of scale-free graphs: Definition, properties, and implications (ext. version). Internet Mathematics 2(4), 431–523 (2006)

25. Li, Q., Moon, B.: Indexing and querying XML data for regular path expressions. In: VLDB 2001, Roma, Italy, pp. 361–370 (2001)
26. Malewicz, G., et al.: Pregel: a system for large-scale graph processing. In: PODC 2009, New York, NY, USA, p. 6 (2009)
27. Martínez-Bazan, et al.: Dex: high-performance exploration on large graphs for information retrieval. In: CIKM 2007, New York, NY, USA, pp. 573–582 (2007)
28. Mendelzon, A.O., Wood, P.T.: Finding regular simple paths in graph databases. SIAM Journal on Computing 24(6), 1235–1258 (1995)
29. Milo, T., Suciu, D.: Index Structures for Path Expressions. In: Beeri, C., Bruneman, P. (eds.) ICDT 1999. LNCS, vol. 1540, pp. 277–295. Springer, Heidelberg (1998)
30. Neven, F.: Automata theory for xml researchers. SIGMOD Rec. 31(3), 39–46 (2002)
31. Palaga, P., Nguyen, L., Leser, U., Hakenberg, J.: High-performance information extraction with alibaba. In: EDBT 2009, New York, USA, pp. 1140–1143 (2009)
32. Ronen, R., Shmueli, O.: SoQL: A language for querying and creating data in social networks. In: ICDE 2009, Shanghai, China, pp. 1595–1602 (2009)
33. San Martín, M., Gutierrez, C.: Representing, Querying and Transforming Social Networks with RDF/SPARQL. In: Aroyo, L., Traverso, P., Ciravegna, F., Cimiano, P., Heath, T., Hyvönen, E., Mizoguchi, R., Oren, E., Sabou, M., Simperl, E. (eds.) ESWC 2009. LNCS, vol. 5554, pp. 293–307. Springer, Heidelberg (2009)
34. Sevon, P., Eronen, L.: Subgraph queries by context-free grammars. Journal of Integrative Bioinformatics 5(2), 100 (2008)
35. Trißl, S., Leser, U.: Fast and practical indexing and querying of very large graphs. In: SIGMOD 2007, New York, NY, USA, pp. 845–856 (2007)
36. Yan, X., Yu, P.S., Han, J.: Graph indexing: a frequent structure-based approach. In: SIGMOD 2004, New York, NY, USA, pp. 335–346 (2004)
37. Yeger-Lotem, E., Sattath, S., Kashtan, N., et al.: Network motifs in integrated cellular networks of transcription-regulation and protein-protein interaction. Proc. Natl. Acad. Sci. USA 101(16), 5934–5939 (2004)
38. Yildirim, H., Chaoji, V., Zaki, M.J.: Grail: Scalable reachability index for large graphs. In: VLDB 2010. VLDB Endowment (2010)
39. Zauner, H., Linse, B., Furche, T., Bry, F.: A RPL through RDF: Expressive Navigation in RDF Graphs. In: Hitzler, P., Lukasiewicz, T. (eds.) RR 2010. LNCS, vol. 6333, pp. 251–257. Springer, Heidelberg (2010)
40. Zou, L., Chen, L., Özsu, M.T.: Distance-join: Pattern match query in a large graph database. PVLDB 2(1), 886–897 (2009)

Sampling Connected Induced Subgraphs Uniformly at Random

Xuesong Lu and Stéphane Bressan

School of Computing, National University of Singapore
{xuesong,steph}@nus.edu.sg

Abstract. A recurrent challenge for modern applications is the processing of large graphs. The ability to generate representative samples of smaller size is useful not only to circumvent scalability issues but also, per se, for statistical analysis and other data mining tasks. For such purposes adequate sampling techniques must be devised. We are interested, in this paper, in the uniform random sampling of a connected subgraph from a graph. We require that the sample contains a prescribed number of vertices. The sampled graph is the corresponding induced graph.

We devise, present and discuss several algorithms that leverage three different techniques: Rejection Sampling, Random Walk and Markov Chain Monte Carlo. We empirically evaluate and compare the performance of the algorithms. We show that they are effective and efficient but that there is a trade-off, which depends on the density of the graphs and the sample size. We propose one novel algorithm, which we call Neighbour Reservoir Sampling (NRS), that very successfully realizes the trade-off between effectiveness and efficiency.

1 Introduction

The versatility of graphs makes them an almost universal data structure in domains as varied as social network, transportation and bioinformatics, to cite a few among the obvious. The challenge for modern applications is the effective and efficient processing of very large graphs. One way to circumvent scalability issues arising from this challenge is to replace the processing of very large graphs by the processing of representative subgraphs of manageable size. This is sampling. Sampling is also useful per se for statistical analysis, data mining, and simulation as well as building block in the implementation of randomized algorithms.

We observe that the graphs of interest are often required to be connected (or that one is interested in connected components). This is the case, for instance, when one is looking for graphlets and motifs in applications such as graph pattern mining (see [10,17], for example). This is also the case when one is studying social networks where communities are characterized by their connectivity. In this domain connected induced subgraphs are the preferred basis of studies of network topology (see [2,11], for example) and evolution (see [19], for example), for instance.

A. Ailamaki and S. Bowers (Eds.): SSDBM 2012, LNCS 7338, pp. 195–212, 2012.
© Springer-Verlag Berlin Heidelberg 2012

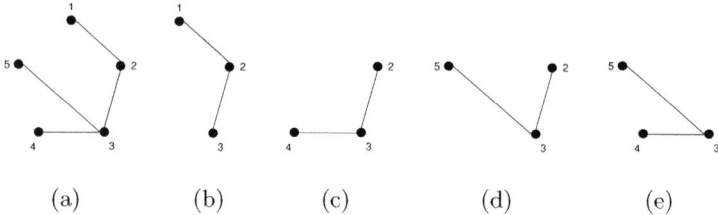

Fig. 1. A connected graph and its connected induced subgraphs

For statistical analysis, data mining and simulation applications a common requirement for a subgraph construction or sampling algorithm is that it maintains graph properties. A uniformly sampled connected induced graph has the advantage that it naturally statistically maintains local properties such as vertex degrees and clustering coefficients among others. In addition a strong guarantee on the distribution, such as uniformity, from which the graph is sampled is very useful for statistical analysis, simulation applications and randomized algorithms.

Motivated by these observations, we study the uniform (or simple) random sampling of a connected induced subgraph of a prescribed size from a graph.

For the sake of simplicity, and without loss of generality as far as the algorithms discussed are concerned, we consider *simple graphs* of the form $G =< V, E >$, where V is a set of vertices and E is a set of pairs of vertices called edges. The *size* (also called *order*) of a graph $G =< V, E >$ is the number of its vertices, $| V |$. A *subgraph* of a graph $G =< V, E >$ is a graph $G' =< V', E' >$ such that $V' \subset V$ and $E' \subset E$. A subgraph of a graph G is *induced* if it contains all the edges that appear in G between any two of its vertices.

Definition 1. *Let $G =< V, E >$ be a graph. A subgraph $G' =< V', E' >$ of G is induced if and only if:*

$$\forall x \in V' \; \forall y \in V'(\{x, y\} \in E \Rightarrow \{x, y\} \in E')$$

The problem we are studying in this paper is the **generation of connected induced subgraphs of size k uniformly at random from a connected graph G of size n, where $1 \leq k \leq n$.**

Example 1. The graph in Figure 1(a) is a simple graph of size 5. It has 4 edges. There are 4 connected induced subgraphs with 3 vertices shown in Figure 1(b), 1(c), 1(d) and 1(e), respectively. We ought to devise algorithms able to randomly generate each of this connected induced subgraphs with equi-probability, i.e. $\frac{1}{4}$ in this example.

In this paper, we devise, present and discuss several algorithms that leverage different techniques: Rejection Sampling, Random Walk and Markov Chain Monte Carlo.

The first algorithm, which we call Acceptance-Rejection Sampling (ARS) is a trivial variation of the standard Rejection Sampling [12]. It serves as a baseline reference as it is guaranteed to be a uniform random sampling.

The second algorithm, which we call Random Vertex Expansion (RVE), adapts the idea of a random walk on the original graph to the gradual construction of the sample.

The third algorithm, which we call Metropolis-Hastings Sampling (MHS), is a Markov Chain Monte Carlo algorithm. The general idea of algorithms from this known and generic family is a random walk on a graph of sample states. The effectiveness of such algorithms is guaranteed provided a sufficient duration, *mixing time*, of the random walk.

The fourth algorithm, which we call Neighbour Reservoir Sampling (NRS) is our main contribution. It operates on the same Markov Chain as MHS but tries to avoid local computation of the degrees of states and long or unbound random walks, and adjust the bias of RVE by combining the idea of expansion from RVE with the idea of reservoir from Reservoir Sampling [21].

The rest of this paper is organized as follows. In the next section, Section 2, we give a brief overview of the related work. In Section 3 we present the four algorithms and their variants. In Section 4 we comparatively and empirically evaluate the effectiveness and the efficiency of the different algorithms and evaluate the performance trade-off and its relation to graph density and sample size, and we compare with algorithms in [13] to see what properties our sampling algorithms can preserve. Finally we conclude in Section 5.

2   Related Work

Several authors [13,9,14,18] have propose algorithms that construct small (relatively to the original graph) subgraphs while trying to preserve selected metrics and properties of the original graph such as degree distribution, component distribution, average clustering coefficient and community structure. Although these algorithms have a random component, they are primarily construction algorithms and are not designed with the main concern of randomness and uniformity of the sampling. In general the distribution from which these random graphs are sampled are not known (except from the obvious distribution of the first and naïve algorithm discussed in [13].)

In their pioneering paper [13], Leskovec et al. propose several such algorithms aiming at preserving nine graphs metrics and properties.

Hübler et al. [9] propose Markov Chain Monte Carlo algorithms that produce sample subgraphs with smaller size and higher utility than Leskovecs.

Maiya et al. [14] focus on the construction of subgraphs preserving the community structure of the original graph.

Ribeiro et al. [18] are interested in properties such as in-degree distribution. They construct subgraphs by sampling edges uniformly at random. They propose a $m-$dimensional random walk algorithm. Their algorithm improves the performance of simple random walk algorithms when the sampled graphs are disconnected and loosely connected.

Several authors [8,10] are concerned with uniform sampling although they may be sampling vertices, edges or subgraphs or subgraphs with specific prescribed constraints.

Henzinger et al. [8], consider the problem of sampling connected induced subgraphs. They consider the World Wide Web as a graph. URLs of Webpages are vertices. (Undirected) edges are links from one Webpage to another. The algorithm that they propose is a random walk on this graph. The authors do not consider a prescribed size of the samples.

Kashtan et al. [10] study the estimation of network motifs concentration. Network motifs are connected subgraphs matching a prescribed pattern (and therefore a prescribed size). The proposed algorithm is not a uniform-sampling algorithm. The RVE algorithm that we discuss in Section 3.2 is a variant of their work.

Random graph generation [15,20,3,16] under specific models and with prescribed constraints can be seen as sampling from a virtual specific sample space.

Milo et al. [15] address the problem of generating uniformly at random graphs with prescribed degree sequences. Viger et al. [20] address the problem of generating uniformly at random connected graphs with prescribed degree sequences. Batagelj et al. [3] discuss efficient algorithms for generating random graphs under model such as Erdős-Rényi, Watts-Strogatz and Barabási-Albert. Nobari et at. [16] propose both sequential and parallel algorithms for the fast random generation of Erdős-Rényi graphs.

3   Algorithms

We consider the generation of connected induced subgraphs of prescribed size uniformly at random from a connected simple graph G. We first introduce a baseline algorithm based on Rejection Sampling [12]. We then revisit the algorithm of [10]. Finally, we propose two non-trivial algorithms for practically sampling the subgraphs. The two algorithms leverage the idea of traversing a Markov chain of connected induced subgraphs.

3.1   Acceptance-Rejection Sampling

The simplest algorithm to sample, uniformly at random, an induced subgraph of size k from an original graph of size n is to select k vertices uniformly at random, complete the induced graph by adding all the edges linking the vertices and then check for connectivity of the induced subgraph. If the induced subgraph is not connected then it is rejected and the selection restarts. If the induced subgraph is connected then it is accepted.

It is a simple instance of Rejection Sampling (see [12], for instance). We call this algorithm *Acceptance-Rejection Sampling* (ARS). The pseudo-code for ARS is given in Algorithm 1.

Notice that the same approach can be applied to generate uniformly at random an induced subgraph with any desired property other than connectivity.

Algorithm 1. Acceptance-Rejection Sampling

   Input: G : the original graph with n vertices, k : subgraph size
   Output: a connected induced subgraph g with k vertices

1 **do**
2 | *Select k vertices uniformly at random from G;*
3 | *Generate the induced graph g of the k vertices;*
4 **while** g is not connected;
5 *Return g*

Proposition 1. *Given a graph G of size n and an integer k where $n > k$, the ARS algorithm generates a connected induced subgraph of size k (if it exists) from G uniformly at random.*

Proof. Let S be the set of induced subgraphs of size k in G. Let C be the set of connected induced subgraphs of size k in G. Let g be the probability distribution of S. Let f be the probability distribution of C. In a uniform distribution, the probability density function (pdf) of a subgraph $c \in C$ is $f_{(c)} = 1/|C|$ and $g_{(c)} = 1/|S|$. Let $M = |S|/|C|$. For each c generated from the distribution g, we accepted it with probability $f_{(c)}/(M \times g_{(c)}) = 1$ if c is connected; otherwise, we reject c. The accepted subgraph c follows the distribution f, i.e., it is generated uniformly at random from C.

ARS is a baseline algorithm that provides a reference for effectiveness. It is a rather brute-force algorithm. We expect ARS to be efficient for dense graphs but otherwise generally inefficient when connected induced subgraphs are a small fraction of induced subgraphs.

3.2   Random Vertex Expansion

One natural method to generate connected subgraphs is to explore a graph from a starting vertex moving gradually and randomly to neighbouring vertices. This generalization of a random walk, which we call *Random Vertex Expansion* (RVE) has been independently proposed and used by many authors. In particular, Kashtan et al. in [10] use it to sample network motifs. Variants of RVE can be implemented with different random selection of the next vertex. We have experimented with several such selection functions and, for the sake of simplicity, only report here the most relevant one. Namely, in the variant we are discussing, RVE chooses the next vertex by selecting uniformly at random one of edges connecting a vertex of the current subgraph with a vertex not yet in the subgraph. The algorithm terminates when the subgraph has the desired size. The pseudo-code for RVE is given in Algorithm 2.

   The probability of sampling a $k-$vertex subgraph is then the sum of the probabilities of the permutations of its $k - 1$ edges. It is given by Equation 1

Algorithm 2. Random Vertex Expansion

Input: G : the original graph with n vertices, k : subgraph size
Output: a connected induced subgraph g with k vertices

1 $E \leftarrow \emptyset$;
2 *Select uniformly at random an edge e of G*;
3 $E \leftarrow e$;
4 **while** E *contains less than k vertices* **do**
5 | $EL \leftarrow \emptyset$;
6 | $EL \leftarrow$ *edges connecting a vertex of E with a vertex not yet in E*;
7 | *Select uniformly at random an edge e from EL*;
8 | $E \leftarrow e$;
9 **end**
10 *Return the induced graph g of E*;

---

where S_m is the set of all the valid permutations of $k - 1$ edges to sample a certain subgraph, E_j is the jth edge in an $(k - 1)$-edge permutation.

$$P = \sum_{\sigma \in S_m} \prod_{E_j \in \sigma} Pr[E_j = e_j | (E_1, E_2, \ldots, E_{j-1}) = (e_1, e_2, \ldots, e_{j-1})]. \quad (1)$$

The more edges a connected induced subgraph of size k has, the more permutations of its $k - 1$ edges, and the higher its probability to be selected. RVE has bias to those subgraphs that are denser, i.e., have higher average clustering coefficient. The main cost of RVE is computing the list of edges connecting a vertex of the current subgraph with a vertex not yet in the subgraph in each step. Suppose the maximal degree of vertices in G is D, the worst complexity is $O(Dk)$. The complexity can be reduced to $O(k^2)$ by maintaining an efficient data structure [10].

3.3   Metropolis-Hastings Sampling

Another idea for catering for the requirement of connectivity is to sample on an ergodic *Markov chain* whose states represent all the connected induced subgraphs with prescribed size of a given original graph. The approach belongs to the *Markov Chain Monte Carlo* (MCMC) family [6]. An MCMC sampling algorithm constructs and randomly traverses a Markov chain whose states are all the candidate samples, starting from any initial state of the Markov chain. After sufficiently many steps, this random walk converges and each state is visited with probability proportional to the degree of that state. The number of random walk steps needed for convergence is called the *mixing time*. The corresponding probability distribution is called *stationary distribution* and the probability is called *stationary probabilities*.

For the problem at hand, the states of the Markov chain are the connected induced subgraphs g of size k of the graph G. Two states g_i and g_j are neighbours

---

Algorithm 3. Metropolis-Hastings Sampling

Input: G : the original graph with n vertices, k : subgraph size, t : the number
of random walk steps
Output: a connected induced subgraph g with k vertices

1 *Perform any graph traverse algorithm on G to get an initial connected induced*
subgraph g of size k;
2 **while** $t > 0$ **do**
3   *Select g' uniformly at random from the neighbors of g;*
4   *Generate a random number $\alpha \in [0, 1)$;*
5   **if** $\alpha < \frac{d_g}{d_{g'}}$ **then**
6           $g = g'$
7   **end**
8       $t = t - 1$
9 **end**
10 *Return g;*

---

of each other if and only if $V(g_i) - V(g_j) = \{v_i\}$ and $V(g_j) - V(g_i) = \{v_j\}$, where
$V(g_i)$ is the vertex set of g_i, that is, one can get g_j by deleting v_i from g_i and
adding v_j to it, and vice versa. This Markov chain is constructed while it is
traversed. The starting state of the random walk can be any connected induced
subgraph g_0 of size k. In order to construct such a graph it suffices, for instance,
to start at a random vertex and to traverse, by any convenient means, the graph
G until we visit k vertices.

In each step, the random walk may select the next state uniformly at random
from the neighbours of the current state and transfers to it. We can see that
all the connected induced subgraphs can be reached from any initial state by
a sequence of replacement of the vertices, which indicates the Markov chain is
irreducible. Moreover, rejecting the transition means adding self-loops to the
states, which makes the Markov chain aperiodic. Therefore our Markov chain is
ergodic and has a stationary distribution.

However, the stationary probability of each state is proportional to its degree
in the Markov chain. Consequently for the Markov chain that we are considering,
the distribution is not necessarily uniform. One option to adjust this bias is by
using the *Metropolis-Hastings* algorithm [7]. Each step from g_i to g_j is accepted
with probability $\frac{d_i}{d_j}$ where d_i is the degree of g_i. This approach is actually balancing the probability of visiting the states with their different degrees. One may
notice that if $d_i > d_j$ the transition is accepted definitely, whereas the smaller
d_i compared to d_j the higher the chance that the random walk stays at g_i.

We call this algorithm *Metropolis-Hastings Sampling* (MHS). The pseudo-code
for MHS is given in Algorithm 3.

The Markov chain and the degree of the states are computed during the
traversal. At each step the algorithm only needs to compute the degree of the
selected neighbour of the current state and temporarily store the neighbour
information of that neighbour. If the transition is accepted, it moves to the next

state and directly uses the information computed in previous step, and then iteratively computes and stores the information of the next selected neighbour, etc. In this way the memory usage is bounded by the largest degree of a state in the Markov chain. In terms of time complexity, suppose on average each connected induced subgraph of size k has l edges and m neighbour vertices, MHS needs on average $O(km(k+l))$ time to compute its neighbours states in the Markov chain, where $O(k+l)$ is the time complexity for checking the connectivity. Suppose on average each connected induced subgraph has d neighbour states, then each local computation of the degree of neighbour states takes $O(dkm(k+l))$ time. Therefore the overall time complexity of MHS is $O(tdkm(k+l))$, where t is the number of random walk steps.

One crucial issue for Markov Chain Monte Carlo algorithms is to determine the mixing time. When the mixing time cannot be determined analytical, one can use statistical tests of convergence [4]. One simple and practical such test is the *Geweke* diagnostics [5]. The basic idea is that for a sequence of random walk iterations, Geweke diagnostics compares the distribution of some metric of interest between the beginning part and the ending part of the sequence. As the random walk iterates, the correlation between the two parts decreases and thus the distribution of the metric of interest should become identical. Geweke diagnostics defines the $z-$score such that $z = \frac{E(m_b) - E(m_e)}{\sqrt{Var(m_b) + Var(m_e)}}$, where m_b and m_e denote the metric of interest in the beginning part and the ending part of the Markov sequence, respectively. Typically the beginning part is defined as the first 10% part and the ending part is defined as the last 50% part. Then multiple chains start from different initial states. The convergence is declared when the z-scores of all the chains fall into the range $[-1, 1]$ with a mean of 0 and a variance of 1. Below we empirically evaluate the convergence of MHS using the Geweke diagnostics.

3.4   Neighbour Reservoir Sampling

In order to adjust the bias of RVE, we consider a method that decreases the probability to sample the induced subgraphs with higher average clustering coefficient. We sample the first k vertices using RVE and get a subgraph g_k. After that, we continue to choose the ith vertex v_i by selecting uniformly at random one of edges connecting a vertex of g_{i-1} with an unprocessed vertex. Then we insert v_i into g_{i-1} with probability $\frac{k}{i}$. In case of a success, one vertex of g_{i-1} is replaced by v_i uniformly at random and we get a new subgraph g_i. If g_i is connected, we keep it; otherwise, $g_i = g_{i-1}$. We iteratively select the new vertices until there is no such an edge that connects a vertex of g_i with an unprocessed vertex.

We call this algorithm *Neighbour Reservoir Sampling* (NRS) as the algorithm samples with a reservoir and always chooses the new vertices from the neighbours. The pseudo-code for NRS is given in Algorithm 4.

The algorithm begins with standard RVE and continues to select new vertices using the same strategy as RVE. However, the vertices with higher local clustering coefficient have higher probability to be replaced, because by deleting

---

Algorithm 4. Neighbour Reservoir Sampling

Input: G : the original graph with n vertices, k : subgraph size
Output: a connected induced subgraph g with k vertices

1 $V \leftarrow$ *the vertices selected using RVE*;
2 $EL \leftarrow$ *edges connecting a vertex of V with an unprocessed vertex*;
3 $i = k$;
4 **while** EL *is not empty* **do**
5     $i + +$;
6   *Select uniformly at random an edge e from EL*;
7     $v =$ *the unprocessed vertex of e*;
8   *Generate a variable $\alpha \in [0, 1)$ uniformly at random*;
9   **if** $\alpha < k/i$ **then**
10   *Select uniformly at random a vertex u from V*;
11       $V' \leftarrow V \backslash u \cup v$;
12   **if** *the induced subgraph of V' is connected* **then**
13         $V \leftarrow V'$
14   **end**
15   **end**
16   *Recompute EL of the current subgraph*
17 **end**
18 *Return the induced graph g of V*;

---

them the subgraphs have higher probability to remain connected. The bias to the subgraphs having higher average clustering coefficient is therefore adjusted by this mechanism. The new vertices enter the subgraph with a decreasing probability so that each selected vertex can be sampled with the same probability [21] despite the connectivity of the graph.

In addition, Similarly to MHS, NRS traverses a Markov chain of connected induced subgraphs of size k. However, NRS takes at most $n - k$ steps, and in each step, NRS checks the connectivity of the possible next state at most once. In each step, NRS maintains a list of edges connecting one unprocessed vertex and one vertex in the current subgraph. This process takes $O(m_e)$ time, where m_e is the number of edges connecting to the current subgraph. Therefore the overall time complexity is $O(Dk + (n - k)(m_e + k + l))$, where $O(Dk)$ is the time to compute the first subgraph of size k, $O(k + l)$ is the time to check the connectivity of each subgraph.

As a result, NRS adjusts the bias of RVE as well as avoids the local computation of the degrees of Markov chain states and the long or unbound random walks of MHS.

4   Performance Evaluation

4.1   Experimental Setup

We implement the four algorithms in C++ and run them on an Intel Core 2 Quad machine with Ubuntu 10.4 and 4GB main memory.

We conduct experiments with both synthetic datasets and real life datasets. First, we evaluate and compare the performance of the proposed algorithms with synthetic graphs generated in the Erdős-Rényi and Barabási-Albert models. We generate graphs of varying size and density. These properties can be controlled directly or indirectly by the parameters of the two models. Second, we show that induced subgraphs that are uniformly sampled can preserve significant properties of the original graph. We collect four real life graphs from *SNAP* [1] and one real life graph from *arXiv.org*. We sample a series of subgraphs with incremental sizes from these real graphs using NRS and calculate average errors on multiple graph properties between the subgraphs and the original graphs.

We preliminarily discuss the convergence of MHS using Geweke diagnostics.

We comparatively evaluate the effectiveness of the four algorithms by measuring the standard deviation from the uniform distribution and by comparing the average subgraph properties.

We comparatively evaluate the efficiency of the four algorithms by measuring their execution time.

We comparatively evaluate the overall performance of the four algorithm in terms of the efficiency versus the effectiveness.

We evaluate the property-preserving of uniform sampling by measuring the Kolmogorov-Smirnov D-statistic on different graph criteria between the subgraphs and the original graphs. The D-statistic is used to compare two distributions, even if they are of different scalings. It is defined as $D = \max_x |F'_{(x)} - F_{(x)}|$, where $F'_{(x)}$ and $F_{(x)}$ are the empirical distribution functions and x is over the range of random variable of the distribution. $F_{(x)}$ for n *iid* observations x_i is defined as $F_{(x)} = \frac{1}{n}\sum_{i=1}^{n} I_{x_i \leq x}$, where $I_{x_i \leq x}$ is equal to 1 if $x_i \leq x$ and equal to 0 otherwise. A lower value of D-statistic indicates higher similarity of two distributions.

4.2   Mixing Time

We empirically evaluate the mixing time of MHS by detecting its convergence using Geweke diagnostics.

We evaluate the convergence of MHS with an Erdős-Rényi graph with $1,000$ vertices with $p = 0.1$, and with a Barabási-Albert graph with 500 vertices and 10 new links per new vertex[1]. We sample connected induced subgraphs with prescribed size 10. For each graph, we run 10 chains from different starting vertex. We compute the z-score of Geweke diagnostics for each chain after every random walk step, using the metric of average degree and average clustering coefficient of the subgraphs.

The results are presented in Figures 2, 3, 4 and 5.

We see that for the Erdős-Rényi graph, the z-scores have a mean 0 and a variance less than 0.5 after 2000 random walk steps. For the Barabási-Albert graph, this number of random walk steps is around 1500. We then use this empirically results in the evaluation below.

---

[1] We denote by d this parameter, i.e., $d = 10$. Below we use this form.

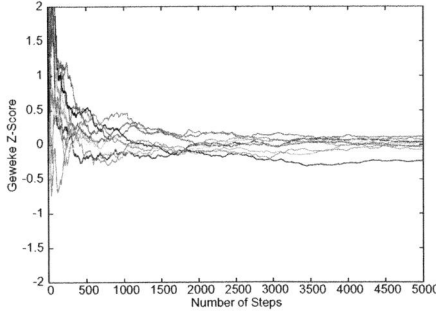

Fig. 2. Geweke diagnostics for a Erdős-Rényi graph with 1000 vertices and $p = 0.1$. The sampled subgraph size is 10. The metric of interest is average degree.

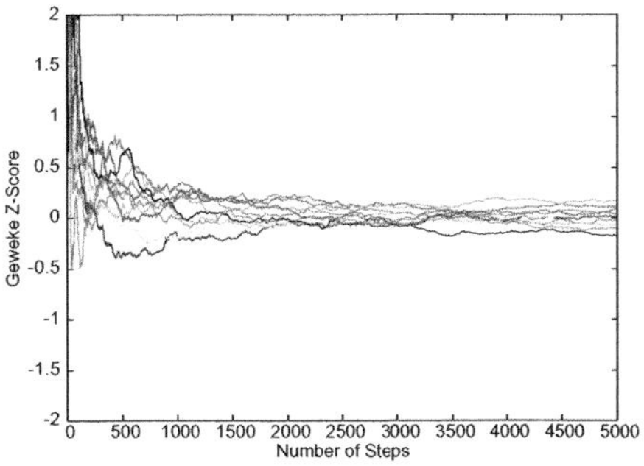

Fig. 3. Geweke diagnostics for a Erdős-Rényi graph with 1000 vertices and $p = 0.1$. The sampled subgraph size is 10. The metric of interest is average clustering coefficient.

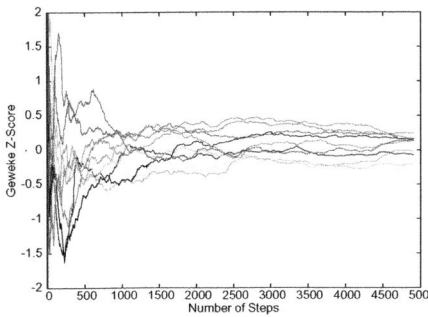
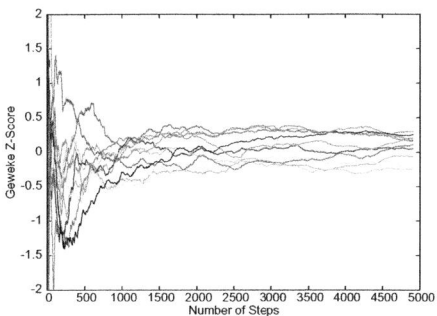

Fig. 4. Geweke diagnostics for a Barabási-Albert graph with 500 vertices and $d = 10$. The sampled subgraph size is 10. The metric of interest is average degree.

Fig. 5. Geweke diagnostics for a Barabási-Albert graph with 500 vertices and $d = 10$. The sampled subgraph size is 10. The metric of interest is average clustering coefficient.

4.3   Effectiveness

We evaluate the effectiveness of the algorithms on small and large graphs, respectively.

Small Graphs We measure the standard deviation of the four algorithms with four Barabási-Albert graphs with 15 vertices and $d = 1, 2, 3, 4$, respectively. We sample connected induced subgraphs with prescribed sizes varying from 1 to 15. For each size, we generate 10 samples on average for every distinct induced subgraphs.

The results are presented in Figures 6, 7, 8 and 9.

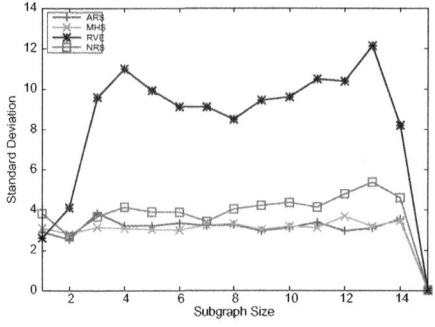

Fig. 6. Standard deviation from uniform distribution. The Barabási-Albert graph has 15 vertices and $d = 1$.

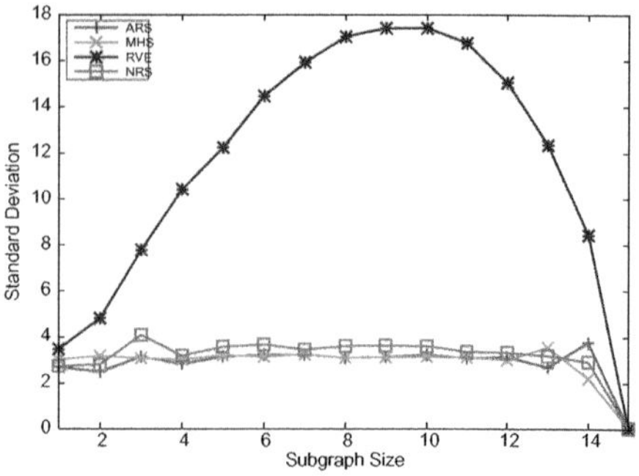

Fig. 7. Standard deviation from uniform distribution. The Barabási-Albert graph has 15 vertices and $d = 2$.

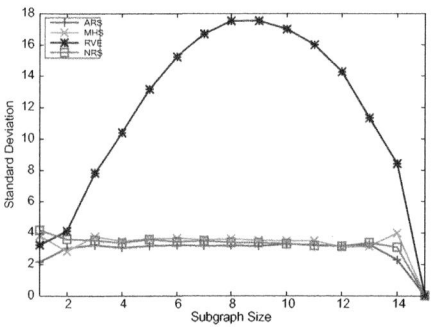
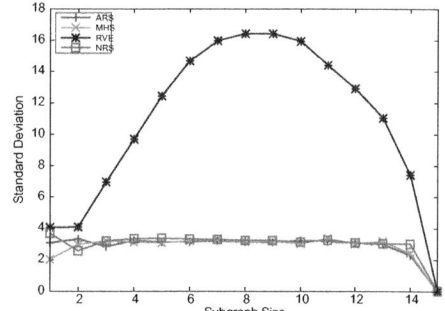

Fig. 8. Standard deviation from uniform distribution. The Barabási-Albert graph has 15 vertices and $d = 3$.

Fig. 9. Standard deviation from uniform distribution. The Barabási-Albert graph has 15 vertices and $d = 4$.

We see that ARS and MHS yield the lowest overall standard deviation while NRS remains competitive unless the graph is sparse. RVE does not perform as well as the above three algorithms.

Large Graphs. It is impractical to generate all the connected induced subgraphs for large graphs. We therefore turn to compare average graph properties of connected induced subgraphs sampled by different algorithms.

We measure the average degree and average clustering coefficient of the subgraphs generated by the four algorithms with the Erdős-Rényi graph of 1000 vertices with probability 0.1 and with the Barabási-Albert graph of 500 vertices and $d = 10$, which are used in Section 4.2. We sample connected induced subgraphs with prescribed sizes varying from 10 to 100 in increments of 10. For each size, we sample 100 connected induced subgraphs and calculate the average of average degree and average clustering coefficient of these subgraphs.

The results are presented in Figures 10, 11, 12 and 13.

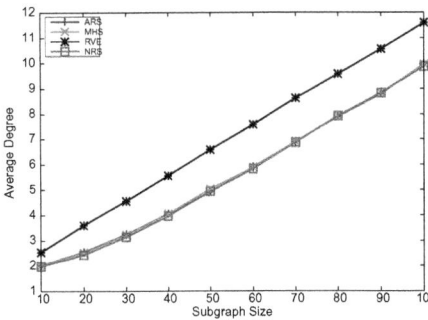

Fig. 10. Comparison of average degree of samples. The original Erdős-Rényi graph has 1000 vertices and $p = 0.1$.

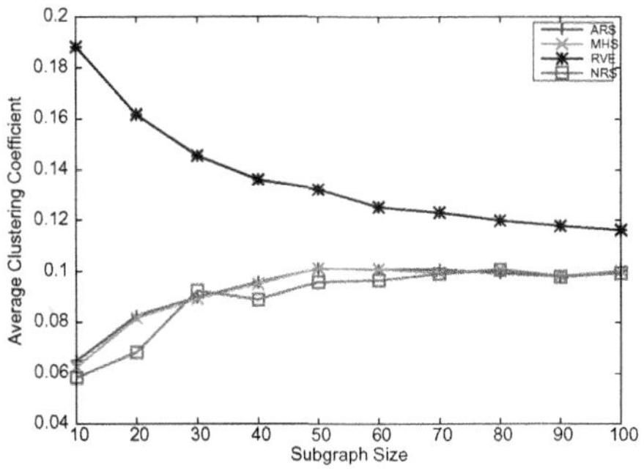

Fig. 11. Comparison of average clustering coefficient of samples. The original Erdős-Rényi graph has 1000 vertices and $p = 0.1$.

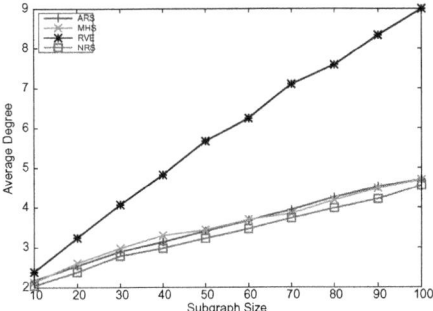
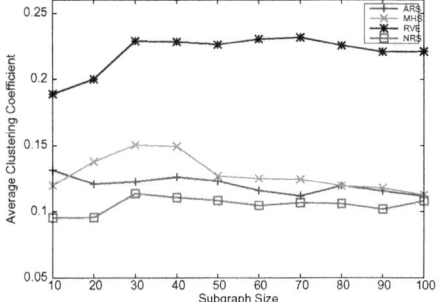

Fig. 12. Comparison of average degree of samples. The original Barabási-Albert graph has 500 vertices and $d = 10$.

Fig. 13. Comparison of average clustering coefficient of samples. The original Barabási-Albert graph has 500 vertices and $d = 10$.

We see that ARS and MHS always coincide with each other on both properties while NRS remains competitive. RVE is biased towards subgraphs with higher average degree and higher average clustering coefficient. This is because RVE tends to sample denser subgraphs.

4.4 Efficiency

We evaluate the efficiency of the algorithms on graphs of different densities and with different subgraph sizes.

Varying Density. We measure the execution time of the four algorithms with a series of Barabási-Albert graphs with 500 vertices. The number of links that each new vertex generates for each graph varies from 1 to 10 with step 1 and from 20 to 100 with step 10. For each graph we sample 10 connected induced subgraphs of size 10 and calculate the average execution time.

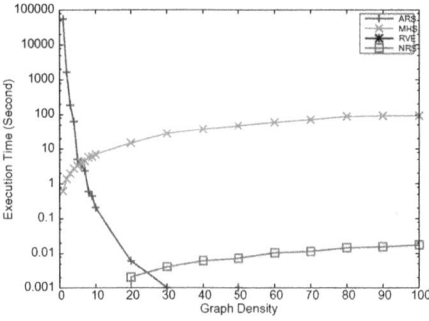 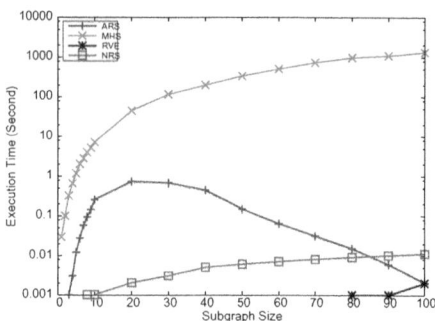

Fig. 14. Average execution times of sampling a connected induced subgraph of size 10 from Barabási-Albert graphs with 500 vertices and different densities

Fig. 15. Average execution times of sampling connected induced subgraphs of different sizes from a Barabási-Albert graph with 500 vertices and $d = 10$

The results are presented in Figures 14.

We see that the execution times of ARS and MHS are increasing rapidly as the graph density decreases and increases, respectively. When the graph density is below 5, ARS becomes unacceptably inefficient. However, real graphs often display a density less than 5 so that ARS is not practical in real applications. On the contrast, MHS caters for the sparse graph because of faster convergence. However, on dense graphs MHS is inefficient as the Markov Chain has numerous states so that the convergence is slow. Nevertheless, NRS scales well along with the graph density as its running time is bounded by the number of vertices. The execution time of NRS increases because, when the graph is denser there are more candidate vertices that can be replaced, so there are more chances to traverse a whole subgraph to confirm the connectivity. The execution time of RVE is always less than $1ms$ so that it cannot be displayed. RVE is the fastest because it terminates as long as desired number of vertices are sampled.

Varying Prescribed Size. We measure the execution time of the four algorithms with a Barabási-Albert graph with 500 vertices. The graph density is about 10. For each algorithm, the sampled subgraph size is varying from 1 to 10 with step 1 and from 20 to 100 with step 10. We compute the average execution times of 10 runs for each sample size.

The results are presented in Figures 15.

The execution time of ARS first increases because of the growing subgraph size, and then decreases because of fewer rejections of the samples. The execution time of MHS, as expected, is increasing when the subgraph size grows as there are more states in the Markov chain. NRS is slowly increasing when the subgraph size grows. This slow increase is because when the subgraph size increases, the test of connectivity consumes more time. RVE is still the fastest.

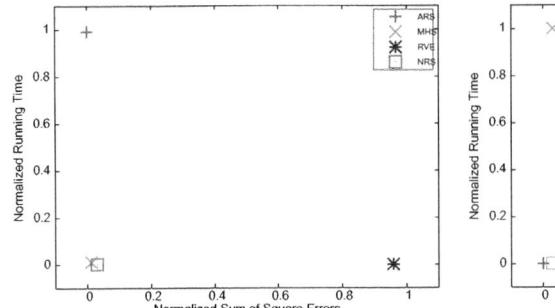 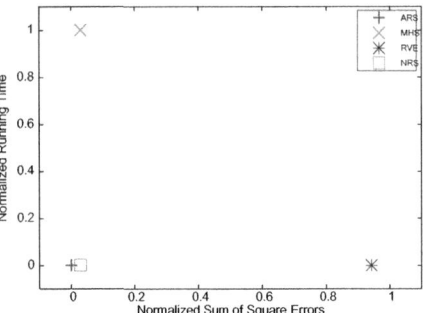

Fig. 16. Normalized efficiency versus effectiveness of sampling connected induced subgraphs of size 10 from Barabási-Albert graphs with 500 vertices and different densities

Fig. 17. Normalized efficiency versus effectiveness of sampling connected induced subgraphs of different sizes from a Barabási-Albert graph with 500 vertices and $d = 10$

4.5 Efficiency versus Effectiveness

We measure the overall performance of the four algorithms in terms of the efficiency versus the effectiveness.

For efficiency, we compute the average execution time for each algorithm of all the settings of graph density in Figure 14 and all the settings of subgraph size in Figure 15, respectively. We normalize the execution time in $[0, 1]$. A lower value corresponds to higher efficiency.

For effectiveness, we consider ARS as the most effective algorithm because it produces a true uniform distribution. For each algorithm, we compute the average value of average degree and clustering coefficient for each setting and measure the difference from ARS by computing the sum of square errors (SSE) of all the values. We normalize the SSE in $[0, 1]$. A lower value corresponds to higher effectiveness.

The results are presented in Figure 16 and Figure 17.

We see that the effectiveness of ARS, MHS, and NRS are very close to each other. However, the efficiency of ARS and MHS depends highly on the graph density and the sampled subgraph size. RVE has the best efficiency but the worst effectiveness.

4.6 Sampling Graph Properties

We measure the D-statistic for six of the nine properties used in [13] for scale-down sampling (the authors of [13] also consider back-in-time sampling). The three remaining properties concern the distribution of component size and, therefore, do not apply to connected subgraph sampling.

The first three properties, the in-degree distribution, out-degree distribution, and clustering coefficient, are properties local to vertices.

Table 1. Measuring D-statistic on six graph properties

	in-deg	out-deg	clust	sng-val	sng-vec	hops
RN	0.084	0.145	0.327	0.079	0.112	0.231
RPN	0.062	0.097	0.243	0.048	0.081	0.200
RDN	0.110	0.128	0.256	0.041	0.048	0.238
RE	0.216	0.305	0.525	0.169	0.192	0.509
RNE	0.277	0.404	0.709	0.255	0.273	0.702
HYB	0.273	0.394	0.670	0.240	0.251	0.683
RNN	0.179	*0.014*^{2}	0.398	0.060	0.255	0.252
RJ	0.132	0.151	0.235	0.076	0.143	0.264
RW	0.082	0.131	0.243	0.049	0.033	0.243
FF	0.082	0.105	0.244	0.038	0.092	0.203
NRS	**0.048**	**0.074**	**0.059**	0.060	**0.012**	0.401

[2] We suspect this number is a typo.

The next two properties, the distribution of singular values of the graph adjacency matrix versus the rank and the distribution of the first left singular vector of the graph adjacency matrix versus the rank, are spectral properties of the graph.

The sixth property, the distribution of the number of reachable pairs of nodes at distance h or less, is a global property of the sample.

We run NRS on the same datasets and using the same experimental settings as [13], and compare the results with the corresponding ones published in [13]. The results are presented in Table 1. Remember that a lower value of D-statistic indicates higher similarity of two distributions.

The results suggest that sampling connected induced subgraphs uniformly at random can preserve significant properties such as degree distribution and clustering coefficient distribution. This is because connected induced subgraphs naturally contain the information of local graph properties such as vertex degree and clustering coefficient. Also, our sampling algorithm can preserve relatively well the spectral properties such as singular values and the first left singular vector of graph adjacency matrix. The overall performance of NRS is much better than the two outperforming algorithms reported in [13], FF and RW.

4.7 Discussion

ARS, MHS and NRS are effective. They sample almost uniformly at random a connected induced subgraph with a prescribed number of vertices from a graph. NRS is slightly less effective than ARS and MHS. RVE has bias towards denser subgraphs.

RVE is more efficient than the other three algorithms. NRS is practical on all graphs but slower than RVE. MHS is efficient on sparse graphs and small prescribed sizes. ARS is only efficient for very dense graphs.

The newly proposed algorithm NRS very successfully realizes the compromise between effectiveness and efficiency for which it was designed.

5 Conclusion

In this paper, we study the uniform random sampling of connected induced subgraphs of a prescribed size from a graph.

We present and discuss four algorithms that leverage several ideas such as rejection sampling, random walk and Markov Chain Monte Carlo.

The first algorithm, Acceptance-rejection Sampling (ARS), provides a reference for effectiveness. It is generally not efficient.

The second algorithm, Random Vertex Expansion (RVE), illustrates the performance of a selection of vertices without replacement but has limited effectiveness.

The third algorithm, Metropolis-Hastings Sampling (MHS), demonstrates the practicality and the effectiveness of Markov Chain Monte Carlo.

The main contribution of this paper is the Neighbour Reservoir Sampling (NRS) algorithm that tries and finds a compromise between the effectiveness of a random walk on a Markov chain of connected induced subgraphs, as in MHS, with the performance of a sampling of vertices with no replacement, as in RVE.

Acknowledgement. This research was partially funded by the A*Star SERC project "Hippocratic Data Stream Cloud for Secure, Privacy-preserving Data Analytics Services" 102 158 0037, NUS Ref: R-702-000-005-305.

References

1. Stanford Network Analysis Project, http://snap.stanford.edu/index.html
2. Ahn, Y.-Y., Han, S., Kwak, H., Moon, S.B., Jeong, H.: Analysis of topological characteristics of huge online social networking services. In: WWW, pp. 835–844 (2007)
3. Batagelj, V., Brandes, U.: Efficient generation of large random networks. Physical Review E 71 (2005)
4. Cowles, M.K., Carlin, B.P.: Markov chain monte carlo convergence diagnostics: A comparative review. Journal of the American Statistical Association 91, 883–904 (1996)
5. Geweke, J.: Evaluating the accuracy of sampling-based approaches to the calculation of posterior moments. In: BAYESIAN STATISTICS, pp. 169–193. University Press (1992)
6. Gilks, W., Spiegelhalter, D.: Markov chain Monte Carlo in practice. Chapman & Hall/CRC (1996)
7. Hastings, W.K.: Monte carlo sampling methods using markov chains and their applications. Biometrika 57(1), 97–109 (1970)
8. Henzinger, M.R., Heydon, A., Mitzenmacher, M., Najork, M.: On near-uniform url sampling. Computer Networks 33(1-6), 295–308 (2000)

9. Hübler, C., Kriegel, H.-P., Borgwardt, K.M., Ghahramani, Z.: Metropolis algorithms for representative subgraph sampling. In: ICDM, Pisa, Italy, pp. 283–292 (December 2008)
10. Kashtan, N., Itzkovitz, S., Milo, R., Alon, U.: Efficient sampling algorithm for estimating subgraph concentrations and detecting network motifs. Bioinformatics 20(11), 1746–1758 (2004)
11. Kwak, H., Lee, C., Park, H., Moon, S.B.: What is twitter, a social network or a news media? In: WWW, pp. 591–600 (2010)
12. Leon-Garcia, A.: Probability, Statistics, and Random Processes for Electrical Engineering. Prentice Hall (2008)
13. Leskovec, J., Faloutsos, C.: Sampling from large graphs. In: KDD, Philadelphia, Pennsylvania, USA, pp. 631–636 (August 2006)
14. Maiya, A.S., Berger-Wolf, T.Y.: Sampling community structure. In: WWW, Raleigh, North Carolina, USA, pp. 701–710 (April 2010)
15. Milo, R., Kashtan, N., Itzkovitz, S., Newman, M.E.J., Alon, U.: On the uniform generation of random graphs with prescribed degree sequences (May 2004)
16. Nobari, S., Lu, X., Karras, P., Bressan, S.: Fast random graph generation. In: EDBT, pp. 331–342 (2011)
17. Przulj, N., Corneil, D.G., Jurisica, I.: Efficient estimation of graphlet frequency distributions in protein-protein interaction networks. Bioinformatics 22(8), 974–980 (2006)
18. Ribeiro, B., Towsley, D.: Estimating and sampling graphs with multidimensional random walks. In: IMC, Melbourne, Australia (November 2010)
19. Vázquez, A., Oliveira, J., Barabási, A.: Inhomogeneous evolution of subgraphs and cycles in complex networks. Physical Review E (2005)
20. Viger, F., Latapy, M.: Efficient and Simple Generation of Random Simple Connected Graphs with Prescribed Degree Sequence. In: Wang, L. (ed.) COCOON 2005. LNCS, vol. 3595, pp. 440–449. Springer, Heidelberg (2005)
21. Vitter, J.S.: Random sampling with a reservoir. ACM Trans. Math. Softw. 11(1), 37–57 (1985)

Discovery of Top-k Dense Subgraphs in Dynamic Graph Collections

Elena Valari, Maria Kontaki, and Apostolos N. Papadopoulos

Data Engineering Lab., Department of Informatics, Aristotle University
54124 Thessaloniki, Greece
{evalari,kontaki,papadopo}@csd.auth.gr

Abstract. Dense subgraph discovery is a key issue in graph mining, due to its importance in several applications, such as correlation analysis, community discovery in the Web, gene co-expression and protein-protein interactions in bioinformatics. In this work, we study the discovery of the top-k dense subgraphs in a set of graphs. After the investigation of the problem in its static case, we extend the methodology to work with dynamic graph collections, where the graph collection changes over time. Our methodology is based on lower and upper bounds of the density, resulting in a reduction of the number of exact density computations. Our algorithms do not rely on user-defined threshold values and the only input required is the number of dense subgraphs in the result (k). In addition to the exact algorithms, an approximation algorithm is provided for top-k dense subgraph discovery, which trades result accuracy for speed. We show that a significant number of exact density computations is avoided, resulting in efficient monitoring of the top-k dense subgraphs.

1 Introduction

Many modern applications require the management of large volumes of *graph data*. Graphs are very important in scientific applications such as bioinformatics, chemoinformatics, link analysis in social networks, to name a few. Dense subgraph discovery is a fundamental *graph mining* task [1,5] with increasing importance. Density is a significant property of graphs, because it is highly related to how well a graph is connected and can be used as a measure of the graph coherence. Usually, the densest the graph the more likely that the connectivity among the graph nodes will be higher.

Among the various density definitions, we adopt the one that relates the graph density to the average degree [7]. More formally, for a graph $G(V, E)$, where V is the set of nodes and E the set of edges, the density of G, denoted as $den(G)$, is given by the number of edges over the number of nodes, i.e., $den(G) = |E|/|V|$. For the rest of the work, we focus on undirected and unweighted graphs. Generalizations to other graph classes are performed easily with appropriate modifications. Figure 1 depicts some examples of density computation.

Algorithms for dense subgraph discovery that have been proposed in the literature require one or more *constraints* to be defined by the user [2]. For example,

A. Ailamaki and S. Bowers (Eds.): SSDBM 2012, LNCS 7338, pp. 213–230, 2012.

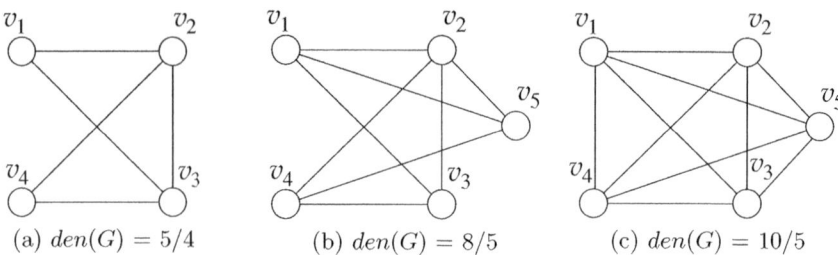

(a) $den(G) = 5/4$ (b) $den(G) = 8/5$ (c) $den(G) = 10/5$

Fig. 1. Densities of various graphs

all subgraphs having a density value higher than a threshold are reported back to the user. The major limitation of such an approach is that the determination of threshold values is difficult: i) if the threshold is too low, then we risk a cumbersome answer set and ii) if the threshold is too high then an empty answer set may be returned. A meaningful threshold value may be difficult to define without at least a limited a priori knowledge regarding the nature of the data (e.g., distribution of densities). To overcome this limitation, in this work, we focus on dense subgraph discovery by taking a top-k oriented approach. More specifically, the only input required by our methods is the number k of the dense subgraphs. The basic benefit of this approach is that the cardinality of the answer set is always known and is equal to k, thus, no surprises are expected regarding the number of answers. In addition, no "wild guesses" regarding the density threshold need to be performed by the user.

The second novelty of our research is that we focus on dynamic graph collections that take the form of a *stream of graphs*. More specifically, computation is performed on a *count-based sliding window* of size w, defined over a stream of graphs G_1, G_2, ..., G_w. New graph objects may arrive whereas old ones expire. The value of w depends on the application. The challenge is to monitor the k densest subgraphs induced by the graphs that are currently active in the sliding window. This translates in performing the necessary actions when a new graph arrives or an old one expires. Evidently, computation must be as efficient as possible to avoid significant delays during updates. Previously proposed methods working in graph streams do not consider deletions.

There are many applications that benefit from the support of top-k dense subgraph discovery. In *web usage mining*, the links followed by users form a directed graph. Many such graphs may be available in a streaming fashion, corresponding to different time instances or different geographical areas. Dense subgraphs are useful in this case because they enable community discovery. The use of the sliding window enables to center our focus at the most recent data available, rather than base the discovery process on the complete history. Mining these graphs on-the-fly is extremely important towards continuous knowledge discovery. As a second example, consider a large social network, where users communicate by means of message exchange. Each graph in this case, corresponds to the interactions among users for a specific time period, e.g., a day. By collecting these graphs for a period of time, we can monitor the evolution of interactions among users. Dense subgraphs in these graphs correspond to communities formed at certain instances. If these communities grow in size and density, then this could mean that an interesting topic emerges.

We note that, to the best of our knowledge, this is the first work that addresses the problem of top-k dense subgraph discovery and consequently, the first work that employs this concept over a stream of graphs. The rest of the paper is organized as follows. The next section, contains a brief discussion of related work in the area. Section 3 presents our methods in detail. A performance evaluation study is contained in Section 4 and finally, Section 5 concludes our work and motivates for further research in the area.

2   Related Work and Contributions

There is a significant number of research works studying the problem of dense subgraph discovery. Material in this subject can be found in [5], whereas an excellent survey chapter in the area is contained in [1]. The problem is considered important due to the numerous applications interested in dense subgraphs, such as, community discovery, genes with significant coexpression, highly correlated items in market basket data and many more.

The majority of the algorithms proposed so far in the literature, assume a fairly static case. For example, in [6] shingling is used to discover dense parts in large bipartite graphs. Dense subgraphs are also used in [8] to discover important subgraphs in a database of graphs corresponding to gene networks. Furthermore, [13] defines a generalization of the densest subgraph problem by adding an additional distance restriction (defined by a separate metric) to the nodes of the subgraphs. This method was applied to a data set of genes and their annotations.

More recently, some efforts have been performed to enable graph mining over evolving graphs [15]. In particular, to complicate things further, in many cases it is assumed that the graph (or graphs) form a data stream. In such a case, usually we are allowed to see the data only once, without the ability to perform random access. In these lines, the authors in [2] discover dense subgraphs when the graph appears in the form of an edge stream. A similar idea is studied in [4], where there are multiple streams, one for each graph. Our work differs from the previous approaches in several points:

(i) We are the first to provide algorithms for dense subgraph discovery in a top-k fashion. This is very important since the user may control the answer set and may terminate the execution at will.

(ii) There are no magic thresholds used in the algorithms. This means that there are no risks that the answer set will be too small or too big, since the k best subgraphs are returned, without any explicit reference to their density. However, our techniques may enforce constraints on the density by simply executing the top-k algorithm and reject any subgraph that does not satisfy the density constraint.

(iii) We support dense subgraph discovery in a set of graphs, rather than in a single graph. In particular, graphs are presented to the system in the form of a *stream of graphs*. Since it is natural to focus on the most recent data, we apply a count-based sliding window of size w [11,16]. This means that at any time instance there are w active graphs that are mined in a continuous manner and thus, both insertions and deletions must be supported.

3   Dense Subgraphs in Graph Collections

3.1   Preliminaries

In this section, we present some fundamental concepts necessary for top-k dense subgraph discovery. Table 1 summarizes some frequently used symbols. As mentioned previously, in this work, the density of a graph is defined as the average degree over its nodes [7]. In that work, an algorithm is given to determine the *densest subgraph* of a graph G. The densest subgraph $G^{(1)}$ of G is simply an induced subgraph with the maximum possible density among all subgraphs of G. More formally:

$$G^{(1)} = \arg\max\{den(g) : g \sqsubseteq G\}$$

The algorithm proposed by Goldberg in [7] (from now on this algorithm will be denoted as GOLD) for the computation of the densest subgraph of a graph G requires a logarithmic number of steps, where in each step a maxflow computation is performed. The maxflow computations are performed on an augmented graph G' and not on the original graph G. More specifically, G is converted to a directed graph by substituting an edge between nodes u and v by two directed edges from u to v and backwards. These edges are assigned a capacity of 1. In addition, two artificial nodes are inserted, the source s and the sink t. Node s is connected to all nodes in G by using directed arcs emanating from s, whereas t is connected by adding directed arcs emanating from each node and ending at t. The capacities of these edges are carefully selected (details in

Table 1. Basic notations used

Symbol	Description
G_i	the i-th graph in the stream
V_i, E_i	set of vertices and set of edges of G_i
n_i, m_i	order and size of G ($N_G = |V_G|$, $M_G = |E_G|$)
$d(v)$	the degree of a vertex v
$g \sqsubseteq G$	g is an induced subgraph of G
$den(G)$	density of graph G
$C(G)$	the maximum core subgraph of G
$G^{(j)}$	the j-th densest subgraph of G
w	the size of the sliding window
k	number of densest subgraphs monitored
$TOPK$	the (current) result of top-k dense subgraphs

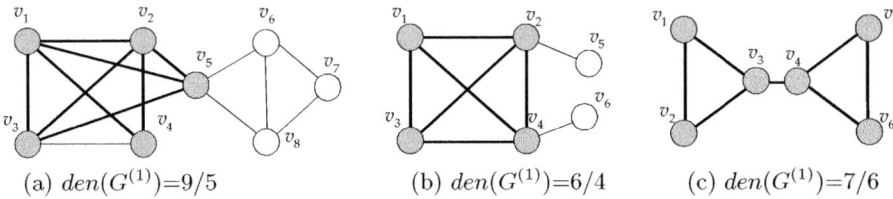

(a) $den(G^{(1)})=9/5$ (b) $den(G^{(1)})=6/4$ (c) $den(G^{(1)})=7/6$

Fig. 2. Densest subgraphs and their density values

[7]) in order to guarantee that the mincut computed after at most a logarithmic number of steps, will separate the densest subgraph from the rest of the graph. Assuming that the push-relabel maxflow algorithm is being used enhanced with dynamic trees, the complexity for the computation of the densest subgraph is $O(\log n \cdot (2n + m) \log(n^2/(n + m)))$.

Figure 2 depicts the densest subgraph for some graphs. The edges of the densest subgraph is shown bold and the nodes are gray-filled. If the density of every subgraph of G is less than the density of G then the whole graph G is its densest subgraph. Such a case is shown in Figure 2(c). Next, we state explicitly the problem investigated in this paper:

Problem Definition: Given a dynamic stream of w graphs $G_1, ..., G_w$ and an integer k, monitor the top-k most dense edge-disjoint subgraphs continuously, taking into account arrivals and expirations of graph objects.   ■

3.2   Dense Subgraph Discovery in a Set of Graphs

In this section, we study a progressive process to determine the k densest subgraphs of a graph G. The usefulness of such a mechanism raises from the fact that the user may require more dense components of the graph in a *get-next* fashion. Consequently, the search process may terminate at any time, if adequate and satisfactory results have been computed. This technique generalizes easily for a set of graphs.

To compute the k densest subgraphs we proceed as follows: *i*) first, the densest subgraph of G is computed, *ii*) the edges comprising the densest subgraph are removed from G, together with any nodes that become isolated, *iii*) if the number of answers reported is less than k we repeat the process. According to this process, two dense subgraphs cannot share edges. However, they may share some nodes. To illustrate the idea, an example is given in Figure 3, where the top-2 dense subgraphs are computed. The densest subgraph $G^{(1)}$ of G is shown in Figure 3(a) and its density is 9/5. By removing the edges contained in $G^{(1)}$ from G (along with the isolated nodes v_1, v_2, v_3, v_4) and by applying again the densest subgraph discovery algorithm in the reduced graph, we arrive at the situation depicted in Figure 3(b). Therefore, the second best (densest) subgraph is composed of the vertices v_5, v_6, v_7, v_8 and its density is 5/4.

For simplicity in the presentation, we define the operation $\ominus$ between a graph G and a subgraph $g \sqsubseteq G$. The result $G \ominus g$ is computed by removing from

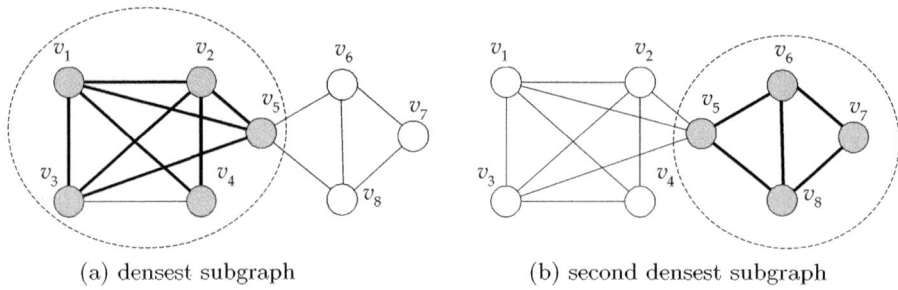

(a) densest subgraph   (b) second densest subgraph

Fig. 3. The two densest subgraphs of a graph G

G all edges in g and also the nodes that become isolated after edge removal. This process may be applied iteratively, until k dense subgraphs are reported. According to this method, the densities corresponding to the densest subgraphs are reported in a non-increasing order. This is very important towards progressive computation of dense subgraphs since it enables an early termination (e.g., before k) if the density values become too low to be of interest to the user. Note, however, that after a graph reduction process, the resulting graph may be disconnected. In such a case, the process is applied to each connected component separately, until k results are obtained.

The only available tool we have to determine the k densest subgraphs is the GOLD algorithm of [7]. This means that so far we do not have a pruning mechanism at hand in order to discard a connected component before applying the expensive sequence of maxflow computations. If this was possible, then a significant number of maxflow operations could have been avoided, resulting in a more efficient computation. To enable pruning, we will use the concept of the *maximum core* of a graph [10,14].

Definition 1. *The maximum core $C(G)$ of a graph G is a subgraph of G containing vertices with a degree at least β, where the value of β is the maximum possible.*

To illustrate the idea, an example is shown in Figure 4. A simple algorithm to compute the maximum core performs a sequence of vertex removals, starting with the vertices with the smallest degree. According to this process, vertices v_9, v_{10} and v_{11} will be removed first, since their degree is 1. The resulting graph is the 2-core of G, since all vertices have a degree at least 2. Next, we remove vertex v_7, and consequently vertices v_6 and v_8 are also removed, because their degree has been reduced due to the removal of v_7. The resulting graph, composed of the vertices v_1, v_2, v_3, v_4 and v_5 is the 3-core of G since every vertex has a degree at least 3. At this point, if we continue this process, we will result in an empty graph. Therefore, the maximum core value of G is 3.

The question is how can we use the maximum core to enable pruning during top-k dense subgraph discovery. A very interesting result has been reported in [9], stating that the density of the maximum core of a graph G is a

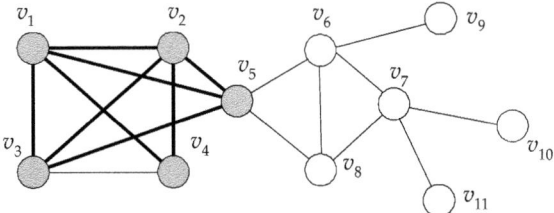

Fig. 4. The 1-core is composed of all vertices of G. The 2-core contains the vertices v_1 through v_7. Finally, the 3-core, which is also the maximum core, contains the vertices v_1, v_2, v_3, v_4 and v_5 (the vertices of maxcore are shown gray).

(1/2)-approximation of the density of the densest subgraph of G. This means that the density of the densest subgraph of G is at most twice the density of the maximum core of G. In addition, since $C(G)$ is an induced subgraph of G, its density is less than or equal to the density of the densest subgraph. More formally, we have the following inequality:

$$2 \cdot den(C(G)) \geq den(G^{(1)}) \geq den(C(G)) \tag{1}$$

According to the previous inequality, we may use the density of the maximum core to define an upper and a lower bound on the density of the densest subgraph of G. The next important issue, is how fast can we compute the maximum core. By using a binary heap enhanced with a hash table for fast decrease-key operations in the heap, the previous algorithm which is based on repetitive removal of vertices with the lowest degree requires $O(m \log n)$ comparisons, resulting in slow computation, especially for large graphs. A more efficient algorithm has been studied in [3], which is based on count-sort. The algorithm requires linear additional space but runs in $O(m)$ worst case time, resulting in a very fast maximum core computation.

Figure 5 depicts the outline of TopkDense algorithm, which computes the k densest subgraphs of an input graph G by using pruning based on the concept of maximum cores. The algorithm requires a priority queue PQ which accommodates the current connected components produced. PQ is implemented as a binary maxheap data structure which stores entries of the form $< g, C(g), den(C(g)) >$, prioritized by the density of the maxcore (the last attribute). The k best answers are stored in A. The algorithm uses Inequality 1 in Line 6 in order to decide if a subgraph is promising or not. In case where the upper bound of its density is lower than the current k-th best density, obviously it must be discarded without any further consideration. Otherwise, we should compute the density of its densest subgraph (invocation of FindDensest() algorithm in Line 7) and then proceed accordingly (Lines 8-15).

Lemma 1. *Let g_1 and g_2 be induced subgraphs of G such that $den(g_1) = den(g_2)$ $= d$. Then, if g_1 and g_2 have at least a common vertex, the density $den(g_1 \cup g_2)$ of the subgraph composed of the vertices and edges of g_1 and g_2 is strictly larger than d.*

Algorithm TopkDense (G, k)
Input: G initial input graph, k number of results
Output: A, set of k densest subgraphs of G

---

1.   initialize answer set $A \leftarrow \emptyset$;
2.   compute the maxcore $C(G)$ of G;
3.   initialize priority queue $PQ \leftarrow < G, C(G), den(C(G)) >$;
4.   **while** $(PQ$ **not** empty)
5. $< g, C(g), den > \leftarrow PQ.deheap()$; /* get the first element of the heap */
6.   **if** $(2 \cdot den > k$-th density in $A)$ **then**
7. $g^{(1)} \leftarrow$ call FindDensest(g); /* compute densest subgraph of g */
8.   **if** $(den(g^{(1)}) > k$-th density in $A)$ **then**
9.   remove subgraph with the k-th density from A;
10.   insert $g^{(1)}$ in A;
11.   remove $g^{(1)}$ from g;
12.   **for each** component h of $g \ominus g^{(1)}$
13. $C(h) \leftarrow$ maxcore subgraph of h;
14. $den(C(h))) \leftarrow$ density of $C(h)$;
15. $PQ.enheap(< h, C(h), den(C(h)) >)$;
16. **return** A;

Fig. 5. Outline of TopkDense algorithm

Proof. Let n_i and m_i denote the number of vertices and edges of g_i, where $i = 1, 2$. Based on the definition of the density, it holds that $den(g_1) = m_1/n_1$ and $den(g_2) = m_2/n_2$. Let h be the graph composed by the union of g_1 and g_2, i.e., $h = g_1 \cup g_2$. For the density of h we have that $den(h) = (m_1 + m_2)/(n_1 + n_2 - \epsilon)$, where ϵ is the number of common vertices of g_1 and g_2, which is strictly larger than zero because we have assumed that the number of common vertices is at least one. Therefore, $den(h) > d$ and this completes the proof. $\square$

Based on the previous discussion, it is not hard to verify that by using this algorithm only *maximal* dense subgraphs are returned. This is an important property, because we avoid repetitive computations to discover a large dense subgraph. It is guaranteed that the next dense subgraph returned will be maximal with respect to the number of vertices.

3.3   Dense Subgraphs in a Stream of Graphs

In this section, we extend the idea of top-k dense subgraph discovery in order to handle dynamic graph collections and more specifically, a *stream of graphs*. We center our attention in the case of a *count-based sliding window*, where we are interested only in the w most fresh graph objects. Therefore, if a new graph object is inserted in the collection, the oldest graph object must be deleted. Notice that the top-k dense subgraph set may contain subgraphs of different graphs. Without lost of generality, we assume that at each time instance only one new graph object arrives and only the oldest one expires. Arrivals and expirations of more graphs are handled similarly.

The monitoring process of top-k dense subgraphs consists of i) the *initialization phase* and ii) the *maintenance phase*. The initialization involves the computation of the top-k dense subgraphs for the first w graphs. This is performed only once, and the maintenance phase can be used to achieve this by disabling expirations. For this reason, we focus only on the maintenance phase.

Assume that at time instance t we have w graphs denoted as G_{t-w+1}, G_{t-w+2}, ..., G_t. We store the active graphs (i.e., graphs belonging to the current window) in a FIFO list with respect to their timestamps. In addition, we keep the top-k dense subgraph set separately. When an update occurs, the current time increases by one and a new graph G_{t+1} is inserted in the window. Since a count-based sliding window is used, the graph G_{t-w+1} expires. The maintenance phase should update the top-k set in order to reflect the changes of the window.

The naive approach begins with the discovery of the densest subgraph of the new graph G_{new} and if its density $den(G_{new}^{(1)})$ is greater than the current k-th best density then $G_{new}^{(1)}$ is inserted into $TOPK$ and the method continues similarly for each connected component h of $G_{new} \ominus G_{new}^{(1)}$. Next, it deletes the expired graph G_{old}. Notice that the removal of G_{old} may reduce the number of densest subgraphs. Assume that $k' \leq k$ is the cardinality of the answer set after the removal of G_{old} and its subgraphs. If $k' = k$ no more operations are required. On the other hand, if $k' < k$, the naive approach should scan all active graphs to find the remaining $k - k'$ densest subgraphs in order to retain the size of the answer set. Notice that, during this process the examined subgraphs may have been computed during the insertion of the corresponding original graph, but this does not hold for all of them. Thus, it is possible that additional densest subgraph computations may be required.

The naive approach invokes too many unnecessary densest subgraph computations. The proposed method tries to reduce the number of time consuming operations. We examine separately the insertion of the new graph and the expiration of the oldest one. Thus, our proposed method can handle different number of insertions/expirations in each update, with minimal modifications. We can modify the TopkDense algorithm to enable the insertion of G_{new}. Instead of initializing the answer set (Line 1 of Figure 5), we give the current answer set as a parameter in TopkDense. The proposed method reduces the number of densest subgraph computations, because it uses as a pruning criterion the density of the maximum core subgraph (Line 6 of Figure 5). The method is correct, i.e. if there exists a subgraph of G_{new} that has one of the highest k densities, it will be inserted in the answer set.

For the expiration of a graph, we distinguish two cases: i) the expired graph G_{old} has at least one of the k-th densest subgraphs and ii) none of its subgraphs is part of $TOPK$. For the latter case, it is sufficient to remove G_{old} from the list with the active graphs. However, if G_{old} has at least one subgraph belonging to $TOPK$, further operations are needed to update correctly the answer set.

Table 2. Last seven graphs of a stream and the density of their three densest subgraphs

	G_1	G_2	G_3	G_4	G_5	G_6	G_7
$den(G_i^{(1)})$	18.0	25.2	20.2	30.8	22.2	26.2	17.6
$den(G_i^{(2)})$	16.1	23.1	18.7	21.4	16.4	20.6	14.5
$den(G_i^{(3)})$	14.5	18.4	16.3	15.6	13.2	18.5	12.7

First, we remove the subgraphs of G_{old} from the answer set. Assume that $k - k'$ densest subgraphs are deleted. A simple approach is to scan the active graphs to find the substitute subgraphs. Remember, we reduced the number of densest subgraph discovery during the insertion of a new graph with the invocation of TopkDense algorithm. The cost to compute the densest subgraphs that are missing is prohibitive. The proposed method uses again Inequality 1 to handle this case efficiently. For a graph G, if $2 \cdot den(C(G)) < k$-th density, G can be omitted from further consideration. More specifically, the method forces the insertion of the first k' subgraphs into $TOPK$. Next, the method tries to improve the answer set by scanning the remaining graphs. For each graph G, if $G^{(1)}$ is not available, we check if $2 \cdot den(C(G)) < k$-th density, then we omit G and we proceed with the next graph. Otherwise, we compute $G^{(1)}$. If $den(G^{(1)}) \geq k$-th density, we insert $G^{(1)}$ into $TOPK$ and we proceed with the component of the residual graph.

To clarify the proposed method, we give an example. Assume the stream of graphs of Table 2. Moreover, assume that $w = 5$, $k = 4$ and the current time is 5. The current window contains the first five graphs. The $TOPK$ consists of $G_4^{(1)}$, $G_2^{(1)}$, $G_2^{(2)}$ and $G_5^{(1)}$. Assume now, that graph G_6 arrives. The densities of the three densest subgraphs of G_6 are given in Table 2. The proposed method computes the density of the maximum core subgraph $den(C(G_6)) = 20.2$. It holds that $2 \cdot den(C(G_6)) > k$-th density, therefore the method computes the densest subgraph $G_6^{(1)}$ of G_6 and its density $den(G_6^{(1)}) = 26.2$. Thus $G_6^{(1)}$ is included in $TOPK$. We proceed with the residual graph. For each component h_i, we compute the density of the maximum core subgraph. Since $2 \cdot den(C(h_1)) = 2 \cdot 10.5 < k$-th density, h_i is not further considered. The removal of G_1 does not affect the $TOPK$ set and therefore it is straight-forward.

Now, assume that graph G_7 arrives and $den(C(G_7)) = 10.0$. The method computes the maximum core subgraph and its density. Since $2 \cdot den(C(G_7)) < k$ no further action are needed. However, due to arrival of G_7, graph G_2 expires. We should update $TOPK$ because now has only two densest subgraphs. The proposed method examines the first graph G_3 and includes the first two densest subgraphs, i.e. $G_3^{(2)}$ and $G_3^{(3)}$, to $TOPK$. Next, for each available graph (initial or component of a residual) is examined the density of the maximum core subgraph. If this density is less than the half of the k-th density, the graph is omitted. On the other hand, the naive approach computes the densest subgraphs of every available graph. The proposed method is denoted as StreamTopkDense. In the sequel, we propose two enhancements in order to further improve efficiency.

Examination Order of Candidate Graphs. The most time consuming part of StreamTopkDense is the case where an element of $TOPK$ expires, because multiple densest subgraph discovery operations may be invoked. The first enhancement tries to reduce the number of examined graphs by means of a suitable examination order. In the proposed method, we favor graphs that have a maximum core subgraph with high density. We expect that during the answer set improvement, the number of examined graphs will be reduced. In order to achieve this, we use a priority queue to define the examination order of active graphs.

The maximum possible density $maxden(G)$ of a graph G is used as the key to insert G in the priority queue. We define $maxden(G)$ as $den(G^{(1)})$ if $G^{(1)}$ is available or $2 \cdot den(C(G))$ otherwise. The priority queue contains entries of the form $< G, maxden(G) >$. For two graphs G_1 and G_2, if $maxden(G_1) > maxden(G_2)$, then it is not necessary that $den(G_1^{(1)}) > den(G_2^{(1)})$. Therefore, if G is at the top of the priority queue and it holds that $maxden(G) \geq k$-th density, then we should further examine G before we insert it into $TOPK$. If $G^{(1)}$ is available (i.e., $maxden(G) = den(G^{(1)})$), we insert G to the answer set immediately. Otherwise, we compute the subgraph $G^{(1)}$ and then we check its density to determine if its inclusion in $TOPK$ is necessary. The method extracts and examines the top of the priority queue, until a graph G is found for which it holds that $maxden(G)$ is less than the current k-th best density.

Graph Pruning. The second enhancement of StreamTopkDense algorithm identifies graphs that cannot be included in $TOPK$ and discards them in order to reduce processing time and memory requirements. The key observation is that a graph can be part of top-k if it belongs to the answer set of the $(k$-1$)$-skyband query in the 2-dimensional space (time, density of densest subgraph). A similar approach has been followed in [11].

A δ-skyband query reports all the objects that are dominated by at most δ other objects [12]. In our case, the maximization of the expiration time and the maximization of the density, determine the domination relationship between graph objects, i.e., a graph H dominates another graph G if H has larger expiration time and H has a more dense subgraph than G. The use of graph pruning does not introduce false dismissals, according to the following lemma.

Lemma 2. *A graph G could not be part of $TOPK$, if there are at least k subgraphs which have greater density than that of the densest subgraph of G and their expiration times are greater than the expiration time of G.*

Proof. Assume that there are k subgraphs with density greater than that of the densest subgraph $G^{(1)}$ of a graph G and these subgraphs expire later than G. Even in the case where all the current top-k graphs are expired before G, any densest subgraph of G will never be part of $TOPK$ since during its lifetime always exist k subgraphs with greater density. □

It is evident that Lemma 2 can also be used to prune computed densest subgraphs from further consideration. In order to enable the use of δ-skyband, we keep a counter $G.c$ for each graph in the active window. When the initial graph is

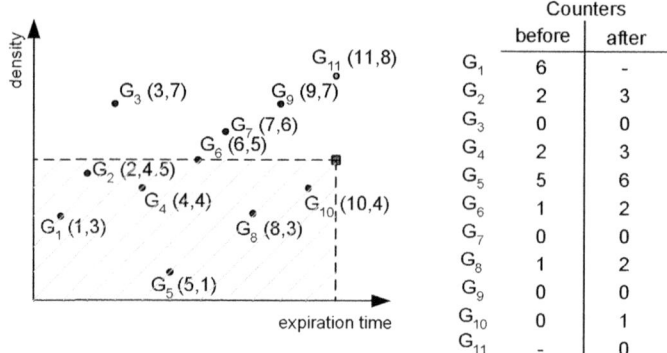

Fig. 6. Graph pruning example

inserted for the first time in the window, we initialize its counter to zero. Then, we scan all active graphs and for each graph G we increase by one $G.c$, if the density of the densest subgraph of the new graph is greater than that of G. We preserve a graph G as long as $G.c$ is less than k. If $G.c = k$ then we can safely discard G from the active window, since it does not belong to the $(k\text{-}1)$-skyband set and therefore cannot be part of the answer throughout its lifetime.

Due to the use of Inequality 1 we do not have all the densest subgraphs and therefore their corresponding densities. The question is how we can use the max-core subgraph and its density, if the densest subgraph has not been computed? There are two cases to study: the densest subgraph is not available either for the new graph or for an existing active graph. To preserve the precision of the proposed method, we use the minimum possible density of the new graph and the maximum possible density $maxden$ of the other existing graphs. The minimum possible density, $minden(G)$, is defined as $den(G^{(1)})$, if $G^{(1)}$ is available, or $den(C(G))$ otherwise.

Figure 6 gives an example of δ-skyband pruning. Assume that the current time is 10, thus the most recent graph is G_{10}. We transform each active graph, which is not part of $TOPK$, to (density, expiration time)-space by using the pair $< maxden, exptime >$. The values of these attributes are shown in parentheses in Figure 6. When a new graph arrives (graph G_{11}), we use the pair $< minden, exptime >$ to update the counters of the other active graphs. The left column shows the counters c of the graphs before update while right column shows them after update. Moreover, we discard G_1 and its counter, we set $G_{11}.c = 0$ and we use the pair $< maxden, exptime >$ to transform G_{11} to (density, expiration time)-space. For $k = 1$, assume that G_9 contains the most densest subgraph. Due to Lemma 2, it is sufficient to store graphs G_3, G_7 and G_{11} (i.e., graphs with $G.c \leq 0$). The remaining graphs can be discarded.

Recall that the density of the densest subgraph is used to prune graphs, since it affects the value of the counters of existing graphs. By using the minimum possible density of the new graph and the maximum possible density for all other existing graphs, we ensure that we do not prune graphs which are part of the

(k-1)-skyband and therefore the proposed method does not introduce any false dismissals. We integrate both enhancements in StreamTopkDense algorithm. The new algorithm is denoted as StreamTopkDense*.

4   Performance Evaluation Study

In the sequel, we report some representative performance results showing the efficiency of the proposed techniques. All algorithms have been implemented in JAVA and the experiments have been conducted on a Pentium@3.2GHz.

To study the performance of the algorithms we have used both synthetic and real-life data sets. The synthetic graphs have been generated by using the Gen-Graph tool [17]. This generator produces graphs according to power law degree distributions. In particular, GenGraph generates a set of n integers in the interval $[d_{min}, d_{max}]$ obeying a power law distribution with exponent a. Therefore, according to the degree distribution produced, a random graph is generated. The default values for the parameters of the generator are: $a \in [1.5, 2.5]$, $d_{min} = 1$, $d_{max} = 1\%$ of the number of vertices. The synthetic data set contains 10,000 graphs each with at most 1500 vertices. The real-life data set, AS-733, contains 733 daily instances (graphs) representing autonomous systems from University of Oregon. The real data set is available for download by the SNAP website at http://snap.stanford.edu/data/as.html.

We study the performance of the algorithms by varying the most important parameters, such as the window size (w) and the number of results (k). We examined several features of the algorithms, such as computational time, the time required for maxflow computations, number of maxcore computations, etc. The computational cost is represented by the running time. All measurements correspond to the total number of updates required, and this translates to w updates required to shift the whole window w times. The default values for the parameters, if not otherwise specified, are: $w = 2000$ and $k = 10$. The straight-forward approach, which performs GOLD invocations directly, is denoted as BFA (brute-force algorithm).

4.1   Performance of Exact Algorithms

Figure 7 depicts the performance of the algorithms for the synthetic data set, for several values of k. As expected, the straight-forward solution (BFA) shows the worst performance due to the excessive number of GOLD invocations. Since GOLD requires $O(\log n)$ maxflow computations, the cost is dominated by the method implementing the maxflow. In particular, it is observed that the runtime of BFA does not heavily depend on the number of results (k). BFA performs a significant number of maxflow computations, as shown in 7(b). However, many of these maxflow computations are executed on small graphs and thus, they are fast. In contrast, a maxflow computation over a larger graph is more costly. Note also, that the y axis is in logarithmic scale and thus, small changes are not easily detected. On the other hand, StreamTopkDense and StreamTopkDense*

Table 3. Evaluation of pruning

Parameter k	StreamTopkDense	StreamTopkDense$^+$	StreamTopkDense*
$k = 1$	120	91	89/2040
$k = 3$	26	6	5/1986
$k = 10$	48	14	8/1984
$k = 25$	156	32	30/1961
$k = 50$	341	240	190/1962

perform much better. In particular, StreamTopkDense* shows the best overall performance since the two optimizations applied have a significant impact in cost reduction, as shown later.

This is also depicted in Figure 8 which shows the performance of the two stream-based algorithms. Algorithm BFA is excluded from any further experiment since its performance is by orders of magnitude inferior. Again, it is evident that the optimizations applied to StreamTopkDense manage to reduce the number of maxcore computations, the number of heap operations and the number of invocations of the GOLD algorithm. Another important feature of StreamTopkDense* is that it requires significantly less storage than the other algorithms. This is illustrated in Figure 8(d) which shows the memory requirements in MBytes vs. the window size. Note that in comparison to the simple version of the stream-based algorithm, the advanced one manages to keep storage requirements low. Since many graphs are pruned due to the application of the skyband technique, the number of graphs that must be kept in memory is reduced.

In addition, Table 3 shows the performance of the pruning techniques. Specifically, we can see the number of graphs which we consider to find dense subgraphs that can participate in the Top k for each algorithm over the total number of updates. The first column shows the pruning results of the StreamTopkDense algorithm. In this algorithm, the computations are reduced by using the density

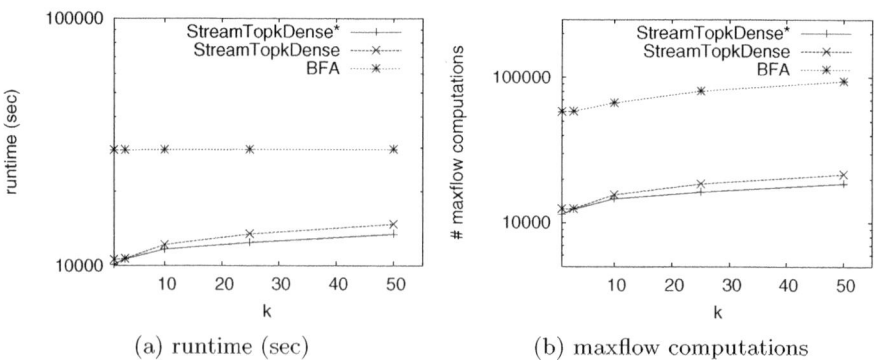

(a) runtime (sec)   (b) maxflow computations

Fig. 7. Comparison of algorithms for different values of k (synthetic data set)

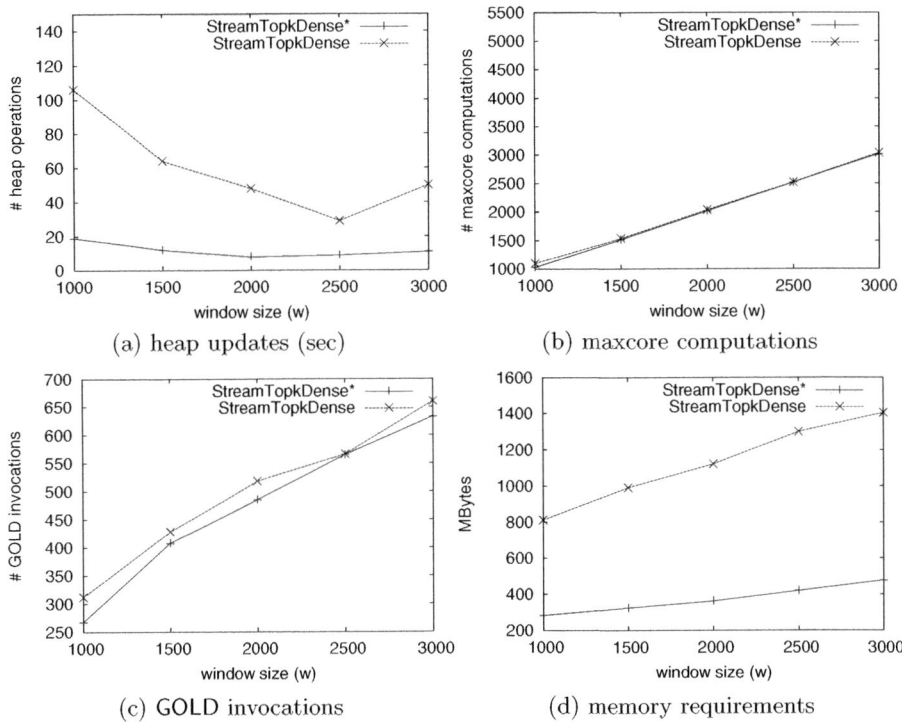

Fig. 8. Comparison of algorithms wrt window size (synthetic data set)

of the maximum core subgraph as a pruning criterion. The second column shows the pruning capabilities of StreamTopkDense[+], which is the basic algorithm enhanced by a priotity queue to determine the examination order of active graphs. As expected, StreamTopkDense[+] examines less graphs than the basic algorithm. Finally, the third column presents the pruning power of StreamTopkDense* which uses the skyband pruning technique in addition to the priority queue. As expected, the number of examined graphs (i.e., the number of graphs that we must apply the exact density computation) is smaller for StreamTopkDense* than the other two methods. In addition, the third column represents also the number of graphs which are pruned by the skyband pruning technique. A graph that is pruned is also removed from the priority queue, resulting in reduced memory requirements (as shown in Figure 8(d)).

In conclusion, StreamTopkDense* shows the best performance both in terms of running time and memory requirements. The reason for this is threefold: *i*) the reduction of the number of GOLD computations (which consequently reduces the number of maxflow computations), *ii*) the use of an appropriate examination order for the graphs resulting in a more effective pruning and *iii*) the use of the skyband to exclude graphs and subgraphs from further consideration.

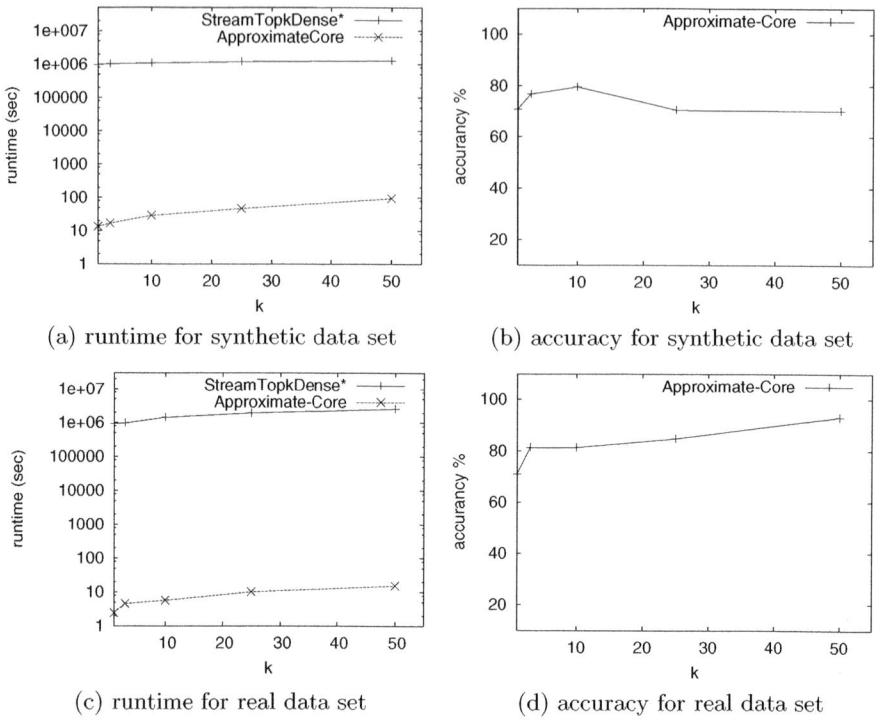

(a) runtime for synthetic data set   (b) accuracy for synthetic data set

(c) runtime for real data set   (d) accuracy for real data set

Fig. 9. Runtime-accuracy trade-off for synthetic and real data sets

4.2   Trading Accuracy for Speed

In many cases, it is important to get the answers as quickly as possible, even if the accuracy of the result is not perfect. The importance of quickly answers is more obvious when the data is from real applications.Toward this direction, we study the performance of an approximation algorithm, which uses only the core computation to determine the set of graphs containing the top-k dense subgraphs. The accuracy of the algorithm is defined as the percentage of the graphs containing the k densest subgraphs that have been reported over the set of the correct subgraphs.

The results of this study are given in Figure 9, where both the runtime and the accuracy are reported for the synthetic as well as the real-life data set. As shown, the runtime of the approximation algorithm is several orders of magnitude smaller than that of the exact one, because the approximation algorithm does not perform any maxflow computations, whereas each core computation is performed in linear time with respect to the number of edges. The performance of the algorithms is similar for the synthetic and the real-life data set. Based on the runtime comparison, the approximate algorithm can be applied when the arrival rate is large, and thus, each new graph must be processed as efficiently as possible. The quality of the approximation is at least 70% for all experiments conducted, whereas it reaches 90% in the real-life data set as k grows to 50.

5   Concluding Remarks

Dense subgraph discovery is considered an important data mining task. Although similar to clustering, dense subgraph discovery has evolved as a separate problem because the requirements are fairly different than clustering. In this paper, we presented stream-based algorithms for continuous monitoring of the k densest subgraphs from a dynamic collection of graphs. The proposed algorithm, StreamTopkDense* is the most efficient variation, because it has the lowest running time and shows the lowest memory consumption. The key issues in the design of our proposal have as follows: i) the maxcore is used to define an upper bound on the density of a subgraph, ii) a priority queue is used to enforce a particular examination order of the graphs and iii) the skyband concept is applied (as in [11]) to reduce the number of graphs that should be considered.

In addition to the study of exact algorithms, we studied the performance of an approximation algorithm, which uses solely the concept of maxcore to determine the set of graphs containing the top-k dense subgraphs. This algorithm offers an accuracy of at least 70%, whereas its running time is orders of magnitude better than that of the exact algorithms.

We point out that this is the first work that performs dense subgraph discovery in a top-k fashion. This technique alleviates the requirement for posing density constraints which are sometimes difficult to provide, especially when the graph collection evolves and graph properties change over time.

There are several interesting directions for future work. The first one, involves the use of graph summaries in order to reduce the size of the graphs. This means that we are willing to sacrifice accuracy in favor of a more efficient computation. A second direction is the adaptation of the methods in [2,4] to work in a top-k scenario, without the requirement of density constraints.

References

1. Aggarwal, C., Wang, H.: Managing and mining graph data. Springer (2010)
2. Aggarwal, C., Li, Y., Yu, P.S., Jin, R.: On dense pattern mining in graph streams. In: Proceedings of the 36th VLDB Conference, pp. 975–984 (2010)
3. Batagelj, V., Zaversnik, M.: An O(m) algorithm for cores decomposition of networks. CoRR, cs.DS/0310049 (2003)
4. Chen, L., Wang, C.: Continuous subgraph pattern search over certain and uncertain graph streams. IEEE Transactions on Knowledge and Data Engineering 22(8), 1093–1109 (2010)
5. Cook, D.J., Holder, L.B. (eds.): Mining graph data. Wiley (2007)
6. Gibson, D., Kumar, R., Tomkins, A.: Discovering large dense subgraphs in massive graphs. In: Proceedings of the 31st VLDB Conference, pp. 721–732 (2005)
7. Goldberg, A.V.: Finding a maximum density subgraph. Technical Report CSD-84-171, University of Berkeley (1984)

8. Hu, H., Yan, X., Huang, Y., Han, J., Zhou, X.J.: Mining coherent dense subgraphs across massive biological networks for functional discovery. Bioinformatics 21(1), i213–i221 (2005)
9. Kortsarz, G., Peleg, D.: Generating sparse 2-spanners. Journal of Algorithms 17(2), 222–236 (1994)
10. Luczak, T.: Size and connectivity of the k-core of a random graph. Discrete Mathematics 91(1), 61–68 (1991)
11. Mouratidis, K., Bakiras, S., Papadias, D.: Continuous monitoring of top-k queries over sliding windows. In: Proceedings of the ACM SIGMOD Conference, pp. 635–646 (2006)
12. Papadias, D., Tao, Y., Fu, G., Seeger, B.: Progressive skyline computation in database systems. ACM Transactions on Database Systems 30(1), 41–82 (2005)
13. Saha, B., Hoch, A., Khuller, S., Raschid, L., Zhang, X.-N.: Dense Subgraphs with Restrictions and Applications to Gene Annotation Graphs. In: Berger, B. (ed.) RECOMB 2010. LNCS, vol. 6044, pp. 456–472. Springer, Heidelberg (2010)
14. Seidman, S.B.: Network structure and minimum degree. Social Networks 5, 269–287 (1983)
15. Sun, J., Faloutsos, C., Papadimitriou, S., Yu, P.S.: GraphScope: parameter-free mining of large time-evolving graphs. In: Proceedings of the 13th ACM SIGKDD International Conference on Knowledge Discovery and Data Mining, pp. 687–696 (2007)
16. Tao, Y., Papadias, D.: Maintaining sliding window skylines on data streams. IEEE Transactions on Knowledge and Data Engineering 18(3), 377–391 (2006)
17. Viger, F., Latapy, M.: Efficient and Simple Generation of Random Simple Connected Graphs with Prescribed Degree Sequence. In: Wang, L. (ed.) COCOON 2005. LNCS, vol. 3595, pp. 440–449. Springer, Heidelberg (2005)

On the Efficiency of Estimating Penetrating Rank
on Large Graphs

Weiren Yu[1], Jiajin Le[2], Xuemin Lin[1], and Wenjie Zhang[1]

[1] University of New South Wales & NICTA, Australia
{weirenyu,lxue,zhangw}@cse.unsw.edu.au
[2] Donghua University, China
lejiajin@dhu.edu.cn

Abstract. P-Rank (Penetrating Rank) has been suggested as a useful measure of structural similarity that takes account of both incoming and outgoing edges in ubiquitous networks. Existing work often utilizes memoization to compute P-Rank similarity in an iterative fashion, which requires *cubic* time in the worst case. Besides, previous methods mainly focus on the deterministic computation of P-Rank, but lack the probabilistic framework that scales well for large graphs. In this paper, we propose two efficient algorithms for computing P-Rank on large graphs. The first observation is that a large body of objects in a real graph usually share similar neighborhood structures. By merging such objects with an explicit low-rank factorization, we devise a *deterministic* algorithm to compute P-Rank in *quadratic* time. The second observation is that by converting the iterative form of P-Rank into a matrix power series form, we can leverage the random sampling approach to *probabilistically* compute P-Rank in *linear* time with provable accuracy guarantees. The empirical results on both real and synthetic datasets show that our approaches achieve high time efficiency with controlled error and outperform the baseline algorithms by at least one order of magnitude.

1 Introduction

Structural similarity search that ranks objects based on graph hyperlinks is a major tool in the fields of data mining. This problem is also known as link-based analysis, and it has become popularized in a plethora of applications, such as nearest neighbor search [26], graph clustering [27], and collaborative filtering [9]. For example, Figure 1 depicts a recommender system, in which person (A) and (B) purchase itemsets {egg, pancake, sugar} and {egg, pancake, flour}, respectively. We want to identify similar users and similar items.

Existing link-based approaches usually take advantage of graph structures to measure similarity between vertices. Each object (*e.g.*, person, or item) can be regarded as a vertex, and a hyperlink (*e.g.*, purchase relationship) as a directed edge in a graph. Then a scoring rule is defined to compute similarity between vertices. Consider the well-known SimRank scoring rule [18] "two vertices are similar if they are referenced (have incoming edges) from similar vertices" in Figure 1. We can see that the items sugar and egg are similar as they are purchased by the same person (A). In spite of its worldwide popularity [1,4,6,18,24,27], SimRank has the "limited information problem" — it only takes incoming edges into account while ignoring outgoing links [26]. For instance, person (A) and (B) have the SimRank score zero as they have no incoming edges. This

A. Ailamaki and S. Bowers (Eds.): SSDBM 2012, LNCS 7338, pp. 231–249, 2012.

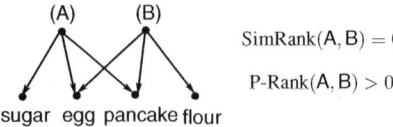

SimRank(A, B) = 0

P-Rank(A, B) > 0

Fig. 1. Purchase Relationship in a Recommender System

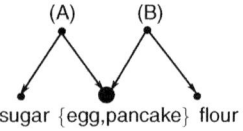

Fig. 2. Merge the nodes having the same neighborhood

is counter-intuitive since the similarity between person (A) and (B) also depends on the similarity of their purchased products. To address this issue, Zhao *et al.* [26] proposed to use P-Rank similarity to effectively incorporate both in- and out-links. Since then, P-Rank has attracted growing attention (*e.g.*, [2, 13, 17, 21]), and it can be widely used to any ubiquitous domain where SimRank is applicable, such as social graphs [2], and publication networks [13]. The intuition behind P-Rank is an improved version of Sim-Rank: "two distinct vertices are similar if (a) they are referenced by similar vertices, and (b) they reference similar vertices". In contrast with SimRank, P-Rank is a general framework for exploiting structural similarity of the Web, and has the extra benefit of taking account of both incoming and outgoing links. As an example in Figure 1, person (A) and (B) are similar in the context of P-Rank.

Nonetheless, existing studies on P-Rank have the following problems. Firstly, it is rather time-consuming to iteratively compute P-Rank on large graphs. Previous methods [17, 26] deploy a fixed-point iterative paradigm for P-Rank computation. While these methods often attain good accuracy, they do not scale well for large graphs since they need to enumerate all n^2 vertex-pairs per iteration if there are n vertices in a graph. The most efficient existing technique using memoization for SimRank computation [18] can be applied to P-Rank in a similar fashion, but still needs $O(Kn^3)$ time. The recent dramatic increase in network scale highlights the need for a new method to handle large volumes of P-Rank computation with low time complexity and high accuracy.

Secondly, it is a big challenge to estimate the error when approximation approaches are leveraged for computing P-Rank. Zhao *et al.* [26] proposed the radius- and category-based pruning techniques to improve the computation of P-Rank to $O(Kd^2n^2)$, with d being the average degree in a graph. However, this heuristic method does not warrant the accuracy of pruning results. For certain applications like ad-hoc top-k nearest neighbor search, fast speed is far more important than accuracy; it is desirable to sacrifice a little accuracy (with controlled error) for accelerating the computation.

In this paper, we address the optimization issue of P-Rank. We have an observation that many real-world graphs are low rank and sparse, such as the Web [18], bibliographic network [9], and social graph [14]. Based on this, we devise two efficient algorithms (a) to deterministically compute P-Rank in an off-line fashion, and (b) to probabilistically estimate P-Rank with controlled error in an on-line fashion. For deterministic computation, we observe that a large body of vertices in a real graph usually have the similar neighborhood structures, and some may even share the same common neighborhoods (*e.g.*, we notice in Figure 1 that the products egg and pancake are purchased by the same users—their neighborhoods are identical. Therefore, we can merge egg and pancake into one vertex, as illustrated in Figure 2). Due to these redundancy, we have an opportunity to "merge" these similar vertices into one vertex. To this end, we utilize a low-rank factorization to eliminate such redundancy. However, it is hard to develop an efficient algorithm and give an error estimate for low-rank approximation. For probabilistic computation, we notice that the iterative form of P-Rank can be characterized as a matrix power series form. In light of this, we adopt a random sampling approach to further improve the computation of P-Rank in linear time with provable guarantee.

Contributions. In summary, we make the following contributions.

(1) We characterize the P-Rank similarity as two equivalent matrix forms: The matrix inversion form of P-Rank lays a foundation for deterministic optimization, and the power series form for probabilistic computation. (Section 3).

(2) We observe that many vertices in a real graph have neighborhood structure redundancy. By eliminating the redundancy, we devise an efficient deterministic algorithm based on the matrix inversion form of P-Rank to optimize the P-Rank computation, yielding quadratic-time in the number of vertices (Section 4).

(3) We base a sampling approach on the power series form of P-Rank to further speed up the computation of P-Rank probabilistically, achieving linear-time with controlled accuracy (Section 5).

(4) Using both real and synthetic datasets, we empirically show that (a) our deterministic algorithm outperforms the baseline algorithms by almost one order of magnitude in time and (b) our probabilistic algorithm runs much faster than the deterministic method with controlled error (Section 6).

2 Preliminaries

Let $\mathcal{G} = (\mathcal{V}, \mathcal{E})$ be a directed graph with vertex set $\mathcal{V}$ and edge set $\mathcal{E}$. For a vertex $u \in \mathcal{V}$, we denote by $\mathcal{I}(u)$ and $\mathcal{O}(u)$ the in-neighbor set and out-neighbor set of u respectively, $|\mathcal{I}(u)|$ and $|\mathcal{O}(u)|$ the cardinalities of $\mathcal{I}(u)$ and $\mathcal{O}(u)$ respectively.

The *P-Rank similarity* between vertices u and v, denoted by $s(u, v)$, is defined as (a) $s(u, u) = 1$; (b) when $u \neq v$,

$$s(u, v) = \underbrace{\frac{\lambda \cdot C_{\text{in}}}{|\mathcal{I}(u)||\mathcal{I}(v)|} \sum_{i \in \mathcal{I}(u)} \sum_{j \in \mathcal{I}(v)} s(i, j)}_{\text{in-link part}} + \underbrace{\frac{(1 - \lambda) \cdot C_{\text{out}}}{|\mathcal{O}(u)||\mathcal{O}(v)|} \sum_{i \in \mathcal{O}(u)} \sum_{j \in \mathcal{O}(v)} s(i, j)}_{\text{out-link part}}, \quad (1)$$

where $\lambda \in [0, 1]$ is a *weighting factor* balancing the contribution of in- and out-links; C_{in} and $C_{\text{out}} \in (0, 1)$ are *damping factors* for in- and out-link directions, respectively.

Note that either $\mathcal{I}(\cdot)$ or $\mathcal{O}(\cdot)$ can be an empty set. To prevent division by zero, the definition in Eq.(1) also assumes that (a) in-link part= 0 if $\mathcal{I}(u)$ or $\mathcal{I}(v) = \varnothing$, and (b) out-link part= 0 if $\mathcal{O}(u)$ or $\mathcal{O}(v) = \varnothing$.

P-Rank Matrix Formula. Let $\mathbf{Q}$ be the backward transition matrix of $\mathcal{G}$, whose entry $q_{i,j} = 1/|\mathcal{I}(i)|$ if $\exists$ an edge $(j, i) \in \mathcal{E}$, and 0 otherwise; and let $\mathbf{P}$ be the forward transition matrix of $\mathcal{G}$, whose entry $p_{i,j} = 1/|\mathcal{O}(i)|$ if $\exists$ an edge $(i, j) \in \mathcal{E}$, and 0 otherwise. By virtue of our prior work [17], the P-Rank equation (1) then equivalently takes the simple form

$$\mathbf{S} = \lambda \cdot C_{\text{in}} \cdot \mathbf{Q} \cdot \mathbf{S} \cdot \mathbf{Q}^T + (1 - \lambda) \cdot C_{\text{out}} \cdot \mathbf{P} \cdot \mathbf{S} \cdot \mathbf{P}^T + \mathbf{I}_n, \quad (2)$$

where $\mathbf{S}$ is the similarity matrix whose entry $s_{i,j}$ equals the P-Rank score $s(i, j)$, and $\mathbf{I}_n$ is the $n \times n$ identity matrix [1], ensuring that each vertex is maximally similar to itself.

3   Two Forms of P-Rank Solution

In this section, we present two closed-form expressions of the P-Rank similarity matrix $\mathbf{S}$, with the aim to optimize P-Rank computation in the next sections.

Our key observation is that the P-Rank matrix formula Eq.(2) is a linear equation. This linearity can be made more explicit by utilizing the matrix-to-vector operator that converts a matrix into a vector by staking its columns one by one. This operator, denoted vec, satisfies the basic property $vec(\mathbf{A} \cdot \mathbf{X} \cdot \mathbf{B}) = (\mathbf{B}^T \otimes \mathbf{A}) \cdot vec(\mathbf{X})$ in which $\otimes$ denotes the Kronecker product. (For a proof of this property, see Theorem 13.26 in [12, p.147].) Applying this property to Eq.(2) we immediately obtain $\mathbf{x} = \mathbf{M} \cdot \mathbf{x} + \mathbf{b}$, where $\mathbf{x} = vec(\mathbf{s})$, $\mathbf{M} = \lambda \cdot C_{in} \cdot (\mathbf{Q} \otimes \mathbf{Q}) + (1 - \lambda) \cdot C_{out} \cdot (\mathbf{P} \otimes \mathbf{P})$, and $\mathbf{b} = vec(\mathbf{I}_n)$. The recursive form of $\mathbf{x}$ naturally leads itself into a power series form $\mathbf{x} = \sum_{i=0}^{\infty} \mathbf{M}^i \cdot \mathbf{b}$. Combining this observation with Eq.(2), we deduce the following lemma.

Lemma 1 (Power Series Form). *The P-Rank matrix formula Eq.(2) has the following algebraic solution*

$$vec(\mathbf{S}) = \sum_{i=0}^{\infty} [\lambda \cdot C_{in} \cdot (\mathbf{Q} \otimes \mathbf{Q}) + (1 - \lambda) \cdot C_{out} \cdot (\mathbf{P} \otimes \mathbf{P})]^i \cdot vec(\mathbf{I}_n). \qquad (3)$$

Lemma 1 describes the power series form of the P-Rank similarity. This result will be used to justify our random sampling approach for estimating P-Rank (in Section 5). One caveat is that the convergence of $\sum_i \mathbf{M}^i \cdot \mathbf{b}$ is guaranteed only if $\|\mathbf{M}\|_\infty < 1$ (see [7, p.301]), where $\| \star \|_\infty$ is the ∞-matrix norm [2]. This is true for Eq.(3) because $\lambda \in [0, 1]$ and $C_{in}, C_{out} \in (0, 1)$ imply that

$$\|\mathbf{M}\|_\infty \leq \lambda \cdot \overbrace{C_{in}}^{<1} \cdot \overbrace{\|\mathbf{Q} \otimes \mathbf{Q}\|_\infty}^{=1} + (1 - \lambda) \cdot \overbrace{C_{out}}^{<1} \cdot \overbrace{\|\mathbf{P} \otimes \mathbf{P}\|_\infty}^{=1} < \lambda + (1 - \lambda) = 1.$$

Another explicit expression for the P-Rank similarity comes from the observation that $\sum_i \mathbf{M}^i = (\mathbf{I} - \mathbf{M})^{-1}$ whenever $\|\mathbf{M}\|_\infty < 1$. Applying this observation to Lemma 1 yields the matrix inversion form of the P-Rank similarity.

Lemma 2 (Matrix Inversion Form). *The P-Rank similarity matrix $\mathbf{S}$ in Eq.(2) can be rewritten as*

$$vec(\mathbf{S}) = [\mathbf{I}_{n^2} - \lambda C_{in} (\mathbf{Q} \otimes \mathbf{Q}) - (1 - \lambda) C_{out} (\mathbf{P} \otimes \mathbf{P})]^{-1} \cdot vec(\mathbf{I}_n). \qquad (4)$$

The utility of Lemma 2 lies in the observation that computing $\mathbf{S}$ can be converted into a matrix inversion computation. Due to the huge size, the straightforward way of computing such matrix inversion is prohibitively expensive; nevertheless, optimization techniques in the next section will significantly improve the computational efficiency.

---

[1] Throughout the paper, we denote by n the number of vertices in $\mathcal{G}$.
[2] The ∞-matrix norm is simply the maximum absolute row sum of the matrix.

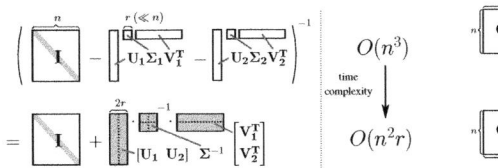

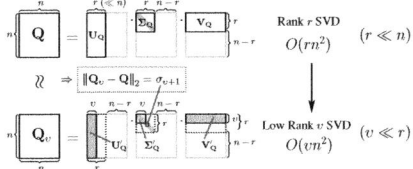

Fig. 3. Low-rank update of matrix inversion

Fig. 4. Reduced SVD Process

4   An Algorithm for P-Rank Deterministic Computation

In light of the matrix inversion form in Lemma 2, we now focus on deterministic optimization of P-Rank computation. In this section, we show the following result.

Theorem 1. *For any graph $\mathcal{G}$, given a low-rank v $(\leq r)$, it is in $O\left(vn^2 + v^6\right)$ time and $O(v \cdot \max\{v^3, n\})$ space to compute P-Rank similarity up to an additive error of $\epsilon_v \leq \frac{\lambda C_{in}\sigma_1\bar{\sigma}_{v+1}+(1-\lambda)C_{out}\bar{\sigma}_1\bar{\sigma}_{v+1}}{1-\lambda C_{in}-(1-\lambda)C_{out}}r$, where r $(\ll n)$ is the rank of the adjacency matrix, σ_i and $\bar{\sigma}_i$ $(i = 1, v + 1)$ are the i-th largest singular values of $\mathbf{Q}$ and $\mathbf{P}$ respectively.*

In particular, setting $v = r$ gives the following corollary.

Corollary 1. *The exact P-Rank similarity can be solvable in $O\left(rn^2 + r^6\right)$ time and $O(r \cdot \max\{r^3, n\})$ space.*

(A sketch proof of Theorem 1 and Corollary 1 will be provided after some discussions. See [23] for a full version of proof.)

The key observation behind P-Rank optimization is that vertices in a real graph usually have a large number of common neighborhoods (*e.g.*, many users often have the similar preferences in a recommender system). Hence, r is typically much smaller than n in practice. The main idea is (a) to devise a rank-r update formula for efficiently computing the matrix inversion in Eq.(4), and (b) to use a rank-r factorization for merging the vertices that have the same neighborhoods into one vertex.

To prove Theorem 1, we first devise a low-rank update formula of matrix inversion. We then present an algorithm for P-Rank computation with the desired properties.

Lemma 3. *Let $\mathbf{I}_n$ be an $n \times n$ identity matrix, $\mathbf{U}_i$ and $\mathbf{V}_i$ be $n \times r$ matrices $(r \ll n)$, and $\mathbf{\Sigma}_i$ be $r \times r$ matrices $(i = 1, 2)$. Then the following identity holds.*

$$\left(\mathbf{I}_n - \mathbf{U}_1\mathbf{\Sigma}_1\mathbf{V}_1^T - \mathbf{U}_2\mathbf{\Sigma}_2\mathbf{V}_2^T\right)^{-1} = \mathbf{I}_n + \left(\mathbf{U}_1\ \mathbf{U}_2\right)\begin{pmatrix}\mathbf{\Sigma}_1^{-1} - \mathbf{V}_1^T\mathbf{U}_1 & -\mathbf{V}_1^T\mathbf{U}_2 \\ -\mathbf{V}_2^T\mathbf{U}_1 & \mathbf{\Sigma}_2^{-1} - \mathbf{V}_2^T\mathbf{U}_2\end{pmatrix}^{-1}\begin{pmatrix}\mathbf{V}_1^T \\ \mathbf{V}_2^T\end{pmatrix} \quad (5)$$

(For the interest of space, please refer to [23] for a detailed proof. Lemma 3 is an extension of *the Woodbury matrix identity* [12, p.48].)

As opposed to $O\left(n^3\right)$-time of the conventional matrix inversion [7], Lemma 3 provides an efficient way of computing $\left(\mathbf{I}_n - \mathbf{U}_1\mathbf{\Sigma}_1\mathbf{V}_1^T - \mathbf{U}_2\mathbf{\Sigma}_2\mathbf{V}_2^T\right)^{-1}$ in $O(n^2r+r^2n+ r^3)$ time $(r \ll n)$ via the RHS of Eq.(5). As depicted in Figure 3, the performance gain is achieved by the observation that $\mathbf{U}_1\mathbf{\Sigma}_1\mathbf{V}_1^T$ and $\mathbf{U}_2\mathbf{\Sigma}_2\mathbf{V}_2^T$ are low rank.

One immediate consequence of Lemma 3 is the optimization of the P-Rank matrix inversion form. We have an observation that most real graphs are low rank (*e.g.*, the web graph [18], bibliographic network [9], who-trusts-whom social network [14]). By applying a *reduced singular value decomposition* [19] [3] (as depicted in Figure 4), $\lambda C_{in} (\mathbf{Q} \otimes \mathbf{Q})$ and $(1 - \lambda) C_{out} (\mathbf{P} \otimes \mathbf{P})$ in Eq.(4) can be factorized into the low-rank form of $\mathbf{U}_1 \mathbf{\Sigma}_1 \mathbf{V}_1^T$ and $\mathbf{U}_2 \mathbf{\Sigma}_2 \mathbf{V}_2^T$, respectively. Then combining Lemma 3, we have

$$vec(\mathbf{S}) = (\, \tilde{\mathbf{U}}_{\mathbf{Q}} \,\, \tilde{\mathbf{U}}_{\mathbf{P}} \,) \, \mathbf{\Sigma} \begin{pmatrix} \tilde{\mathbf{V}}_{\mathbf{Q}}^T \\ \tilde{\mathbf{V}}_{\mathbf{P}}^T \end{pmatrix} vec\,(\mathbf{I}_n) + vec\,(\mathbf{I}_n) \text{ with } \mathbf{\Sigma} = \begin{pmatrix} \frac{1}{\lambda C_{in}} \tilde{\mathbf{\Sigma}}_{\mathbf{Q}}^{-1} - \tilde{\mathbf{V}}_{\mathbf{Q}}^T \tilde{\mathbf{U}}_{\mathbf{Q}} & -\tilde{\mathbf{V}}_{\mathbf{Q}}^T \tilde{\mathbf{U}}_{\mathbf{P}} \\ -\tilde{\mathbf{V}}_{\mathbf{P}}^T \tilde{\mathbf{U}}_{\mathbf{Q}} & \frac{1}{(1-\lambda) C_{out}} \tilde{\mathbf{\Sigma}}_{\mathbf{P}}^{-1} - \tilde{\mathbf{V}}_{\mathbf{P}}^T \tilde{\mathbf{U}}_{\mathbf{P}} \end{pmatrix}^{-1},$$

where a tilde denotes the self-Kronecker product of a matrix, *e.g.*, $\tilde{\mathbf{U}}_{\mathbf{Q}} = \mathbf{U}_{\mathbf{Q}} \otimes \mathbf{U}_{\mathbf{Q}}$. Due to $\mathbf{\Sigma}$ small size, the efficiency of computing P-Rank can be greatly improved.

We next provide an algorithm for P-Rank computation, denoted by DE P-Rank.

Algorithm. In Algorithm 1, given $\mathcal{G}$, λ, C_{in}, C_{out}, and a low rank υ (an optional parameter with a default value being the rank r of adjacency matrix), DE P-Rank outputs the exact $\mathbf{S}$ if $\upsilon = r$, or the approximate $\mathbf{S}$ with an error ϵ_{υ} if $\upsilon < r$.

Some notations in the algorithm are elaborated below. (a) RowNorm $(\mathbf{A})$ returns a matrix by normalizing each nonzero row of $\mathbf{A}$. (b) Rank $(\mathbf{A})$ returns the rank of $\mathbf{A}$. (c) RSVD $(\mathbf{Q}, \upsilon)$ returns a low-rank υ factorization of $\mathbf{Q}$ (see Figure 4). (d) Reshape$(\mathbf{v}, \upsilon)$ returns an $\upsilon \times \upsilon$ matrix $\mathbf{V}$ such that $vec(\mathbf{V}) = \mathbf{v}$.

The algorithm works as follows. (a) It first initializes the adjacency matrix $\mathbf{A}$ (line 1), and computes $\mathbf{Q}$ and $\mathbf{P}$ (line 2). υ is set to Rank $(\mathbf{A})$ if the low rank υ is not specified (line 3). (b) It then utilizes RSVD () to decompose $\mathbf{Q}$ and $\mathbf{P}$ into $\mathbf{U}_{\mathbf{Q}} \mathbf{\Sigma}_{\mathbf{Q}} \mathbf{V}_{\mathbf{Q}}^T$ and $\mathbf{U}_{\mathbf{P}} \mathbf{\Sigma}_{\mathbf{P}} \mathbf{V}_{\mathbf{P}}^T$, respectively (line 4). In light of these matrices together with the self-Kronecker products, two vectors $\mathbf{v}_1$ and $\mathbf{v}_2$ can be obtained (lines 5-7). The error estimate ϵ_{υ} is also computed if $\upsilon < $ Rank$(\mathbf{A})$ (line 8). (c) Utilizing $\mathbf{v}_1$ and $\mathbf{v}_2$, the matrix $\mathbf{S}$ can be derived, which is returned as the P-Rank similarity (lines 9-11).

Example. Figures 5(a) and 5(b) show how DE P-Rank computes P-Rank in a heterogenous graph $\mathcal{G}_0$ and a homogeneous $\mathcal{G}_1$, respectively. In $\mathcal{G}_0$, there are two types of entities : person (A) and (B) purchase the items sugar, egg, flour. In $\mathcal{G}_1$, each vertex denotes a paper, and each edge a citation. For these graphs, given $C_{in} = 0.4, C_{out} = 0.6, \lambda = 0.5$, DE P-Rank first computes $\mathbf{Q}$ and $\mathbf{P}$. Since υ is not specified, it is set to Rank$(\mathbf{A})$. Then $\mathbf{Q}$ and $\mathbf{P}$ are decomposed into small matrices that can be used for computing $\begin{pmatrix} \mathbf{\Sigma}_{11} & \mathbf{\Sigma}_{12} \\ \mathbf{\Sigma}_{21} & \mathbf{\Sigma}_{22} \end{pmatrix}$ and $\mathbf{v}_1, \mathbf{v}_2$. Finally, DE P-Rank computes the exact $\mathbf{S}$.

To complete the proof of Theorem 1, we next show that the algorithm DE P-Rank (1) correctly computes the similarity values; (2) it has the time complexity bound stated in Theorem 1; (3) when $\upsilon \in [\frac{1}{2}r, r]$, the error ϵ_{υ} (line 8) is acceptable in practice.

(Due to space limitations, please refer to [23] for detailed analysis.)

(1) Correctness. The algorithm returns *exactly* the same similarity as Eq.(4) when $\upsilon = r$; and it returns the low-rank υ *approximate* similarity with an error ϵ_{υ} stated in Theorem 1 when $\upsilon < r$.

(2) Running Time. The algorithm consists of three phases: pre-processing (lines 1-3), similarity computation (lines 4-8), and result collection (lines 9-11). One can verify

---

[3] Given an matrix $\mathbf{X}$ (with its rank r) and an integer υ $(\leq r)$, the *reduced singular value decomposition* of $\mathbf{X}$ is the factorization $\mathbf{X}_{\upsilon} = \mathbf{U}_{\upsilon} \cdot \mathbf{\Sigma}_{\upsilon} \cdot \mathbf{V}_{\upsilon}^T$ *s.t.* $\|\mathbf{X} - \mathbf{X}_{\upsilon}\|_2 = \sigma_{\upsilon+1}$ is minimal, where $\mathbf{U}_{\upsilon}$ and $\mathbf{V}_{\upsilon}$ are $n \times \upsilon$ column orthonormal matrices, and $\mathbf{\Sigma} \triangleq diag\,(\sigma_1, \sigma_2, \cdots, \sigma_{\upsilon})$ is an $\upsilon \times \upsilon$ diagonal matrix whose entries are the singular values of $\mathbf{X}$.

Algorithm 1: DE P-Rank

Input : web graph $\mathcal{G} = (\mathcal{V}, \mathcal{E})$,
weighting factor λ,
damping factors C_{in} and C_{out},
low rank v.

Output: similarity matrix $\mathbf{S}$, and
approximation error ϵ_v.

1 initialize the adjacency matrix $\mathbf{A}$ of $\mathcal{G}$.

2 $\mathbf{Q} \leftarrow \text{RowNorm}(\mathbf{A}^T)$, $\mathbf{P} \leftarrow \text{RowNorm}(\mathbf{A})$.

3 **if** v *is empty* **then** $v \leftarrow \text{Rank}(\mathbf{A})$.

4 $[\mathbf{U_Q}, \Sigma_{\mathbf{Q}}, \mathbf{V_Q}; \sigma_1, \sigma_{v+1}] \leftarrow \text{RSVD}(\mathbf{Q}, v)$,
$[\mathbf{U_P}, \Sigma_{\mathbf{P}}, \mathbf{V_P}; \bar{\sigma}_1, \bar{\sigma}_{v+1}] \leftarrow \text{RSVD}(\mathbf{P}, v)$.

5 compute the small auxiliary matrices:
$\Lambda_{\mathbf{Q},\mathbf{Q}} = \mathbf{V_Q}^T \cdot \mathbf{U_Q}$, $\Lambda_{\mathbf{P},\mathbf{P}} = \mathbf{V_P}^T \cdot \mathbf{U_P}$,
$\Lambda_{\mathbf{P},\mathbf{Q}} = \mathbf{V_P}^T \cdot \mathbf{U_Q}$, $\Lambda_{\mathbf{Q},\mathbf{P}} = \mathbf{V_Q}^T \cdot \mathbf{U_P}$,
$\Lambda_{\mathbf{Q}} = \Sigma_{\mathbf{Q}}^{-1}$, $\Lambda_{\mathbf{P}} = \Sigma_{\mathbf{P}}^{-1}$.

6 compute the four blocks of the matrix Σ :
$\Sigma_{11} \leftarrow \frac{1}{\lambda C_{\text{in}}} \Lambda_{\mathbf{Q}} \otimes \Lambda_{\mathbf{Q}} - \Lambda_{\mathbf{Q},\mathbf{Q}} \otimes \Lambda_{\mathbf{Q},\mathbf{Q}}$,
$\Sigma_{12} \leftarrow -\Lambda_{\mathbf{Q},\mathbf{P}} \otimes \Lambda_{\mathbf{Q},\mathbf{P}}$,
$\Sigma_{22} \leftarrow \frac{1}{(1-\lambda)C_{\text{out}}} \Lambda_{\mathbf{P}} \otimes \Lambda_{\mathbf{P}} - \Lambda_{\mathbf{P},\mathbf{P}} \otimes \Lambda_{\mathbf{P},\mathbf{P}}$,
$\Sigma_{21} \leftarrow -\Lambda_{\mathbf{P},\mathbf{Q}} \otimes \Lambda_{\mathbf{P},\mathbf{Q}}$.

7 compute the P-Rank similarity $\mathbf{S}$:
$$\begin{pmatrix} \mathbf{v}_1 \\ \mathbf{v}_2 \end{pmatrix} \leftarrow \begin{pmatrix} \Sigma_{11} & \Sigma_{12} \\ \Sigma_{21} & \Sigma_{22} \end{pmatrix}^{-1} \begin{pmatrix} vec(\mathbf{V_Q}^T \mathbf{V_Q}) \\ vec(\mathbf{V_P}^T \mathbf{V_P}) \end{pmatrix}.$$

8 **if** $v < \text{Rank}(\mathbf{A})$ **then**
$\epsilon_v \leftarrow \frac{\lambda C_{\text{in}} \sigma_1 \sigma_{v+1} + (1-\lambda) C_{\text{out}} \bar{\sigma}_1 \bar{\sigma}_{v+1}}{1 - \lambda C_{\text{in}} - (1-\lambda) C_{\text{out}}} \text{Rank}(\mathbf{A})$

else $\epsilon_v \leftarrow 0$.

9 $\mathbf{V}_1 \leftarrow \text{Reshape}(\mathbf{v}_1, v)$,
$\mathbf{V}_2 \leftarrow \text{Reshape}(\mathbf{v}_2, v)$.

10 $\mathbf{S} \leftarrow (1 - \lambda C_{\text{in}} - (1 - \lambda) C_{\text{out}}) \cdot$
$(\mathbf{I}_n + \mathbf{U_Q} \mathbf{V}_1 \mathbf{U_Q}^T + \mathbf{U_P} \mathbf{V}_2 \mathbf{U_P}^T)$.

11 **return** $\mathbf{S}$ and ϵ_v.

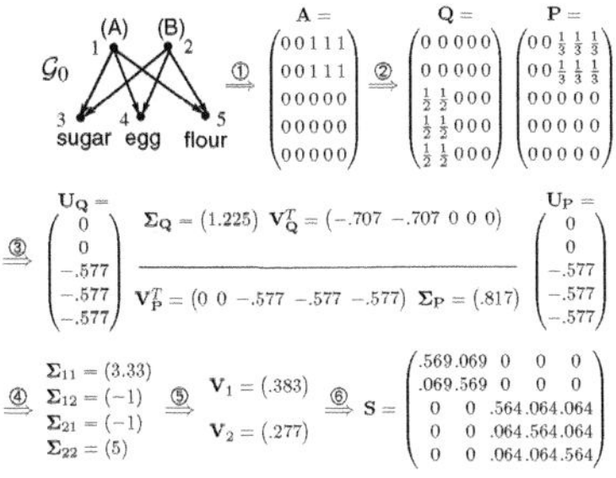

(a) Heterogenous Shopping Graph $\mathcal{G}_0$

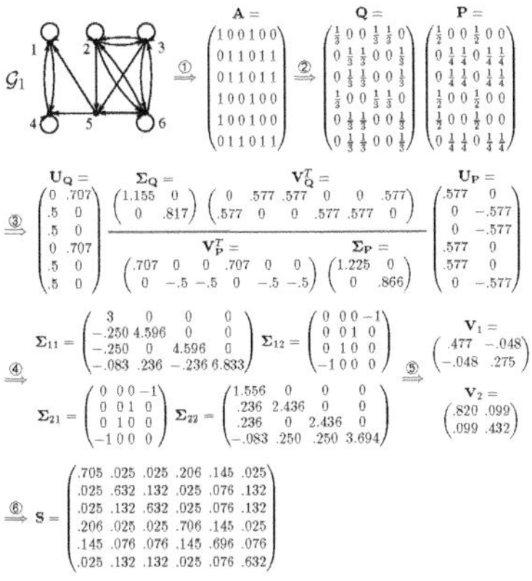

(b) Homogeneous Scientific Paper Network $\mathcal{G}_1$

Fig. 5. How DE P-Rank computes similarity

that these phases take $O(m)$, $O(v^2 n + vn^2 + v^4 + v^6)$ and $O(v)$ time, respectively. Hence, the total time is bounded by $O(vn^2 + v^6)$ with $v \leq r$.

(3) Memory Space. (a) For pre-processing, it takes $O(n)$ space to compute $\mathbf{Q}$ and $\mathbf{P}$ (line 3). (b) For similarity computation, the memory consumption is dominated by $O(v \cdot \max\{v^3, n\})$, which includes $O(vn)$ space to decompose $\mathbf{Q}$ and $\mathbf{P}$ into low-rank matrices (line 4), and $O(v^4)$ for computing Σ^{-1} (line 7). (c) The result collection requires $O(v^2)$ space (line 10). Therefore, the total space can be bounded by $O(v \cdot \max\{v^3, n\})$ with $v \leq r$. (see Table 1 for a detailed analysis)

(4) Error Bound. The error ϵ_v is reasonably small in practice when $v \in [\frac{1}{2}r, r]$. Our experimental results in Section 6 show that for such v, the singular values σ_{v+1} and $\bar{\sigma}_{v+1}$ (in line 8) are almost zero, leading to a practically acceptable NDCG$_{30}$ (see Figure 13). As an extreme case of $v = r$, $\sigma_{v+1} = \bar{\sigma}_{v+1} = 0$, which implies that $\epsilon_v = 0$.

Table 1. Running Time & Memory Space for DE P-Rank in lines 2-9

#-line	time	memory	operation
2	$O(m)$	$O(n)$	row normalization of matrices
4	$O(vn^2 + v^2n)$	$O(vn)$	low-rank v reduced SVD
5	$O(v^2n + r)$	$O(v)$	matrix multiplications and inversions
6	$O(v^4 + v^2)$	$O(v^2)$	Kronecker products
7	$O(v^6 + v^4 + v^2n)$	$O(v^4 + v^2)$	block matrix inversion
8	$O(1)$	$O(1)$	constant operations
9	$O(v)$	$O(v^2)$	reshape matrices

For instance, consider the WIKI (0715) data ($r = 15K, \sigma_1 = 1.12, \bar{\sigma}_1 = 1.08$). Setting $C_{\text{in}} = 0.8, C_{\text{out}} = 0.6$, and $\lambda = 0.5$ will yield

$$\epsilon_v \leq \frac{0.5 \times 0.8 \times 1.12 + 0.5 \times 0.6 \times 1.08}{1 - 0.5 \times 0.8 - 0.5 \times 0.6} \times 10^{-7} \times 15K = 0.0039.$$

5   Probabilistic P-Rank Similarity Estimation

Although way better than cubic, the complexity bound of DE P-Rank is still too high to compute similarity in an on-line fashion. For ad-hoc (dynamic) queries on large graphs, the execution time is one of the most crucial metrics; it is worthwhile to drastically accelerate the P-Rank computation with a little sacrifice in accuracy.

This motivates us to study *the probabilistic P-Rank computation problem*. That is, given a graph $\mathcal{G}$, a query (u, v), and a desired probabilistic accuracy, it is to estimate the P-Rank similarity $s(u, v)$ in a scalable manner (*i.e.*, in worst-case linear time).

5.1   A Probabilistic P-Rank Model

In the light of the power series form of P-Rank in Lemma 1, our key observation is that P-Rank similarity can be viewed as a geometric sum of random walks, and its score $s(u, v)$ qualifies how soon two surfers are expected to meet at the same vertex if they start from vertices u and v and do random walks on a graph backwards and forwards.

The main idea is to utilize the first hitting time $\tau(u, v)$ of coalescing walks to estimate $s_l(u, v)$ of length l. The underlying rationale is that $\tau(u, v)$ can be represented in a compact way of storing only one integer (rather than a walk of length l) for each vertex-pair. It is far less costly to estimate $\tau(u, v)$ for $s(u, v)$ than to compute the entire similarity matrix $\mathbf{S}$. Specifically, we show the following result.

Theorem 2 (Probabilistic Model). *The P-Rank similarity score between vertices u and v, with damping factors C_{in} and C_{out} for in- and out-links, is equal to the weighted mean of their expected meeting distances with uniform independent walks, i.e.,*

$$s(u, v) = \mathbb{E}(\lambda \cdot C_{in}^{\ \tau_1(u,v)} + (1 - \lambda) \cdot C_{out}^{\ \tau_2(u,v)}), \tag{6}$$

where $\mathbb{E}(\cdot)$ denotes the expectation of the random variables; and $\tau_i(u, v)$ $(i = 1, 2)$ are the first hitting time of the random surfers starting from the vertices u and v, and *following the links backwards $(i = 1)$ and forwards $(i = 2)$, respectively; $\tau_i(u, v) = \infty$ if they never hit; and $\tau_i(u, v) = 0$ if $u = v$.*

(A detailed proof of Theorem 2 will be provided after some discussions.)

Intuitively, Theorem 2 provides a stochastic model of P-Rank computation for interpreting the similarity score as the random walks of surfer pairs. From this perspective, the quality of similarity score hinges on whether the random surfers that start from two distinct vertices are close to a common "source" and meet within merely a few steps.

We first use *vertex-pair graph* $\mathcal{G}^2$ to formulate the hitting time of two surfers in $\mathcal{G}$. In $\mathcal{G}^2$, each vertex (u, v) represents a pair of vertices in $\mathcal{G}$, and each edge from (u, v) to (x, y) says that in $\mathcal{G}$, one surfer can move from u to x, and the other from v to y. Hence, in light of the power series form of P-Rank in Lemma 1, two surfers, in $\mathcal{G}$, starting from vertices u and v, following the links backwards (*resp.* forwards) and meeting within a few steps indicate that, in $\mathcal{G}^2$, there exists a path t from one singleton vertex (x, x) to (u, v) (*resp.* from (u, v) to (x, x)).

We then introduce the following notions to model the random surfers on $\mathcal{G}^2$.

(a) The *transformation* T in $\mathcal{G}^2$ is a mapping $T : t' \to t$ from one path t' into another t by adding (i) an edge $\langle (u, v), \mathcal{O}_i((u, v)) \rangle$ to the beginning of t', or (ii) an edge $\langle \mathcal{I}_i((u, v)), (u, v) \rangle$ to the end of t'.

(b) The *length* of a path t, denoted by $l(t)$ is the number of edges in t.

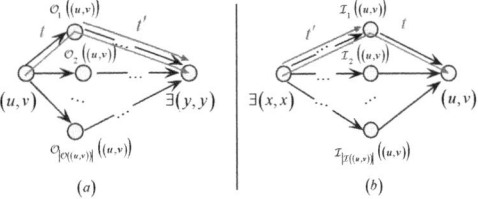

Fig. 6. Transformation from path t' into t

For one length of a random walk on $\mathcal{G}^2$, Figure 6 depicts the corresponding transformation from path t' to $t = T(t')$. Clearly,

$$l(t) = l(t') + 1. \tag{7}$$

(c) The *probability* of choosing a path $T(t')$ based on a path t', denoted by $p(T(t'))$, is defined to be

$$p(T(t')) = \begin{cases} \frac{1}{|\mathcal{I}((u,v))|} \cdot p(t'), & t' : \exists(x, x) \to (u, v); \\ \frac{1}{|\mathcal{O}((u,v))|} \cdot p(t'), & t' : (u, v) \to \exists(y, y). \end{cases} \tag{8}$$

We next complete the proof of Theorem 2 by showing that the probabilistic P-Rank model Eq.(6) is equivalent to the original model Eq.(1).

As the expectations in Eq.(6) can be rewirten as the sum *w.r.t.* probability distribution functions, it follows that

$$s(u, v) = \lambda \cdot \sum_{t:\exists(x,x)\to(u,v)} p(t) \cdot C_{\text{in}}^{\ l(t)} + (1 - \lambda) \cdot \sum_{t:(u,v)\to\exists(y,y)} p(t) \cdot C_{\text{out}}^{\ l(t)}.$$

Without loss of generality, we assume that $u \neq v$, and $\mathcal{I}(u), \mathcal{I}(v), \mathcal{O}(u), \mathcal{O}(v) \neq \varnothing$. In the above equation, we split the sums *w.r.t.* one step of the path t. Combing this with Eqs.(7) and (8), we have

$$s\left(u,v\right)=\lambda\cdot\sum_{i=1}^{|\mathcal{I}((u,v))|}\sum_{\substack{t':\exists(x,x)\\ \to\mathcal{I}_i((u,v))}}\overbrace{p\left(T\left(t'\right)\right)}^{=\frac{1}{|\mathcal{I}((u,v))|}\cdot p(t')}\cdot C_{\text{in}}\overbrace{l\left(T\left(t'\right)\right)}^{=l(t')+1}+(1-\lambda)\cdot\sum_{j=1}^{|\mathcal{O}((u,v))|}\sum_{\substack{t':\mathcal{O}_j((u,v))\\ \to\exists(y,y)}}\overbrace{p\left(T\left(t'\right)\right)}^{=\frac{1}{|\mathcal{O}((u,v))|}\cdot p(t')}\cdot C_{\text{out}}\overbrace{l\left(T\left(t'\right)\right)}^{=l(t')+1}$$

$$=\frac{\lambda\cdot C_{\text{in}}}{|\mathcal{I}(u)|\,|\mathcal{I}(v)|}\cdot\sum_{i=1}^{|\mathcal{I}((u,v))|}\sum_{t':\exists(x,x)\to\mathcal{I}_i((u,v))}p\left(t'\right)\cdot C_{\text{in}}{}^{l(t')}+\frac{(1-\lambda)\cdot C_{\text{out}}}{|\mathcal{O}(u)|\,|\mathcal{O}(v)|}\cdot\sum_{j=1}^{|\mathcal{O}((u,v))|}\sum_{t':\mathcal{O}_j((u,v))\to\exists(y,y)}p\left(t'\right)\cdot C_{\text{out}}{}^{l(t')}$$

$$=\frac{\lambda\cdot C_{\text{in}}}{|\mathcal{I}(u)|\,|\mathcal{I}(v)|}\cdot\sum_{i=1}^{|\mathcal{I}(u)|}\sum_{j=1}^{|\mathcal{I}(v)|}s\left(\mathcal{I}_i\left(u\right),\mathcal{I}_j\left(v\right)\right)+\frac{(1-\lambda)\cdot C_{\text{out}}}{|\mathcal{O}(u)|\,|\mathcal{O}(v)|}\sum_{i=1}^{|\mathcal{O}(u)|}\sum_{j=1}^{|\mathcal{O}(v)|}s\left(\mathcal{O}_i\left(u\right),\mathcal{O}_j\left(v\right)\right).$$

Hence, the probabilistic model Eq.(6) agrees with the original model Eq.(1).

5.2   A Scalable Algorithm for P-Rank Estimation

In light of Theorem 2, we next devise a probabilistic algorithm for P-Rank estimation. The main result in this subsection is the following.

Theorem 3. *For any graph $\mathcal{G}$, the probabilistic P-Rank similarity can be solvable in $O\left(N\cdot n\right)$ time and $O(n+N)$ space, where N is the sample size.*

(The proof of Theorem 3 will be provided after a few discussions.)

As will be seen shortly, N is much smaller than n and affects the accuracy of estimation. This suggests that P-Rank can be solved in *linear* time with controlled probabilistic error, as opposed to the quadratic-time of its deterministic computation.

To prove Theorem 3, we first present the general idea of the P-Rank estimation. We then devise a randomized algorithm, followed by a complexity analysis.

The central idea is to use a sampling approach to estimate s from the first hitting time τ_1 and τ_2. (i) In the pre-computation phase, we utilize a tree index structure (instead of a low-rank factorization) to represent all the first hitting time for a set of coalescing walks in a compact way. (ii) In the query phase, we use two random surfers to estimate the P-Rank similarity by following the path that is a function of the first hitting time τ_1 and τ_2, which can be justified by Theorem 2.

Algorithm. The algorithm, referred to as PR P-Rank, is shown in Algorithm 2. It takes as input a graph $\mathcal{G}$, a query vertex-pair (u,v), a sample size N, a weighting factor λ, and two damping factors C_{in} and C_{out}, it returns the approximate similarity $\hat{s}_N\left(u,v\right)$.

The algorithm maintains the following data structures to ensure estimation quality. (a) A *(reversed) fingerprint tree* FP (*resp.* RFP) for vertices in $\mathcal{G}$. For each vertex u in the i-th samples FP_i and RFP_i, $RFP_i\left(u,l\right)$ collects candidate vertices v in $\mathcal{G}$ such that each of the vertices u and v has an incoming directed path of length l that starts from some common vertex x; and $FP_i\left(u,l'\right)$ is the set of vertices w in $\mathcal{G}$ such that each of the vertices u and w has an outcoming directed path of length l' that ends to some common vertex y. (b) A *random sample list* $\hat{s}_N^{(i)}$ of size N in $\mathcal{G}$ for estimating P-Rank similarity with $\hat{s}_N$ being the mean of N independent and identically distributed (*i.i.d.*) samples $\hat{s}_N^{(i)}$. (c) The *path length*, denoted by $\text{Len}\left(v/\cdots/u\right)$, from vertex v to u in $\mathcal{G}$. The path is *nonempty* if $\text{Len}\left(v/\cdots/u\right)\geq 1$.

The algorithm works as follows. (1) It first constructs two sample lists RFP and FP for $\mathcal{G}$ by invoking Index Pre-processing procedure (lines 1-6). For each vertex

---

Algorithm 2: PR P-Rank $(\mathcal{G}, (u, v), N, \lambda, C_{\text{in}}, C_{\text{out}})$

Input : web graph $\mathcal{G} = (\mathcal{V}, \mathcal{E})$, query vertex pair $(u, v) \in \mathcal{V} \times \mathcal{V}$, sample size N,
           weighting factor λ, damping factors C_{in} and C_{out}.

Output: P-Rank similarity score $\hat{s}_N(u, v)$.

INDEX PRE-PROCESSING

1 **for** $i \leftarrow 1, \cdots, N$ **do**

2   **foreach** *vertex* $u \in \mathcal{V}$ **do**

3   **if** $\exists v \in \mathcal{V} - \{u\}$ *s.t. u and v meet at a common vertex x along a chain of l*
             in-links, and $\text{Len}(x/ \cdots /v) = \text{Len}(x/ \cdots /u) = l$ **then**

4   add v to the reversed fingerprint tree $\text{RFP}_i(u, l)$ of $\mathcal{G}$.

5   **else if** $\exists w \in \mathcal{V} - \{u\}$ *s.t. u and w meet at a common vertex y along a chain of l'*
             out-links, and $\text{Len}(v/ \cdots /y) = \text{Len}(w/ \cdots /y) = l'$ **then**

6   add w to the fingerprint tree $\text{FP}_i(u, l')$ of $\mathcal{G}$.

QUERY $\hat{s}_N(u, v)$

7 **for** $i \leftarrow 1, \cdots, N$ **do**

8   **if** *there exists a positive integer l s.t.* $\text{RFP}_i(u, l) = \text{RFP}_i(v, l)$ **then**

9          $l_0 \leftarrow \min_l \{l \in \mathbf{Z}^+ | \text{RFP}_i(u, l) = \text{RFP}_i(v, l)\}$.

10   **else if** *there exists a positive integer l' s.t.* $\text{FP}_i(u, l') = \text{FP}_i(v, l')$ **then**

11         $l'_0 \leftarrow \min_{l'} \{l' \in \mathbf{Z}^+ | \text{FP}_i(u, l') = \text{FP}_i(v, l')\}$.

12      $\hat{s}_N^{(i)} \leftarrow \lambda \cdot C_{\text{in}}^{\ l_0} + (1 - \lambda) \cdot C_{\text{out}}^{\ l'_0}$.

13 $\hat{s}_N \leftarrow \frac{1}{N} \cdot \sum_{i=1}^{N} \hat{s}_N^{(i)}$.

14 **return** $\hat{s}_N$.

---

u in $\mathcal{G}$, the i-th sample $\text{RFP}_i(u, l)$ (*resp.* $\text{FP}_i(u, l')$) collects the vertex v such that u and v have some common vertex x along a chain of l in-links (*resp.* l'out-links) in $\mathcal{G}$ (lines 4 and 6). (2) It then computes all the samples $\hat{s}_N^{(i)}$ by inspecting the vertices and path lengths collected in RFP and FP (lines 7-12). More concretely, for each sample $\hat{s}_N^{(i)}$, PR P-Rank identifies the minimum length l_0 (*resp.* l_0') of the incoming (*resp.* outgoing) directed path, along which u and v may reach a common vertex (lines 9 and 11). Furthermore, utilizing l_0 and l_0', it then computes $\hat{s}_N^{(i)}$ (line 12), which can be justified by Theorem 2. These $\hat{s}_N^{(i)}$ ($i = 1, \cdots, N$) constitute a sequence of *i.i.d.* random samples. They are averaged to produce the final score $\hat{s}_N$ (lines 13-14).

To complete the proof of Theorem 3, we next show that (1) PR P-Rank has linear time complexity bound; (2) the memory requirement is bounded by $O(N + n)$; (3) the sample size $N \ll n$ in practice; (4) the error bounds are reasonably small; (5) the relative order of PR P-Rank scores is almost preserved.

(1) Running Time. PR P-Rank consists of three phases: (a) For pre-processing (lines 1-6), PR P-Rank invokes the randomized algorithm [5] for FPT indexing (lines 4 and 6), which is in $O(N \cdot n)$ time. (b) For on-line query (lines 7-12), PR P-Rank computes l_0 and l_0' for each sample $\hat{s}_N^{(i)}$ in $O(n)$ time, being $O(N \cdot n)$ time for N samples. (c) For computing $\hat{s}_N$ (lines 13-14), it takes $O(N)$ time to collect all the samples.

Therefore, the total time of PR P-Rank is in $O(N \cdot n)$ time.

(2) Memory Space. The memory requirement is totally bounded by $O(N + n)$, comprising three phases: (a) In the precomputation phase, FPT indexing (lines 4 and 6) needs $O(n)$ space to maintain RFP_i and FP_i for every sample $\hat{s}_N^{(i)}$. (b) During the online query phase, it takes $O(n)$ space for finding the shortest meeting distance l_0 and l'_0 (lines 9 and 11), and $O(N)$ space for collecting all the similarity samples $\hat{s}_N^{(i)}$ (line 13). (c) Computing $\hat{s}_N$ on-the-fly requires $O(N)$ space.

(3) Sample Size N. (a) Choosing $N \geq -2 \lceil (\sigma/\epsilon)^2 \log \alpha \rceil$ suffices to ensure that $\Pr(|\hat{s}_N - s| \geq \epsilon) < \alpha$ (where $\hat{s}_N(u, v)$ is the sample mean, and σ^2 the variance) , given any accuracy ϵ and confidence level $1 - \alpha$ ($\alpha \in (0, 1)$). This is because applying the *Bernstein's Theorem* [11] yields $\exp(-\frac{1}{2}(\epsilon\sqrt{N}/\sigma)^2) < \alpha$. (b) N is typically much smaller than n in practice, which can be verified by our empirical results in Section 6 (see Figure 14. For instance, consider DBLP (98-07) graph with $n = 10K$ vertices. Given $\epsilon = 0.15\sigma$ and $\alpha = 0.05$, we have $N \geq -2 \lceil 0.15^{-2} \log(0.05) \rceil = 267$.

(4) Error Bound. We denoted by $Err \triangleq \sup_{N \geq 1} \Pr(|\hat{s}_N - s| \geq \epsilon)$.

(a) An upper bound can be obtained from *Bernstein's Theorem* [11], which gives

$$Err \leq \exp(-N\epsilon^2/(2\sigma^2)),$$

(b) A lower bound follows from the *Central Limit Theorem* [10], in which

$$Err \geq \Pr(|\hat{s}_N - s| \geq \epsilon) = \Pr\left(\left|\frac{1}{\sqrt{N}} \sum_{i=1}^{N} \left(\frac{\hat{s}_N^{(i)} - s}{\sigma}\right)\right| \geq \frac{\epsilon\sqrt{N}}{\sigma}\right) = 2 - 2\Phi\left(\frac{\epsilon\sqrt{N}}{\sigma}\right),$$

where $\Phi(\cdot)$ is the cumulative distribution function of normal distribution $\mathcal{N}(0, 1)$.

(c) Both bounds of Err are reasonable because (i) $\exp(-N\epsilon^2/(2\sigma^2))$ is decreasing w.r.t. N, and (ii) $\Phi(\epsilon\sqrt{N}/\sigma)$ non-decreasingly approaches 1 as N increases. Hence,

$$\lim_{N\to\infty} \exp(-N\epsilon^2/(2\sigma^2)) = \lim_{N\to\infty} 2 - 2\Phi(\epsilon\sqrt{N}/\sigma) = 0.$$

According to the *Squeeze Principle* [10], we have $\hat{s}_N(u, v) \overset{a.s.}{\to} s(u, v)$ as $N \to \infty$.

(d) Err is typically small and acceptable in practice. For instance, Setting $\epsilon = 0.15\sigma$ and $N = 300$, we have $\exp\left(-\frac{N\epsilon^2}{2\sigma^2}\right) = \exp\left(-\frac{300 \times (0.15\sigma)^2}{2\sigma^2}\right) \approx 0.034$ and $2 - 2\Phi\left(\frac{\epsilon\sqrt{N}}{\sigma}\right) = 2 - 2\Phi(0.15 \times \sqrt{300}) \approx 0.0094$. This implies that only 0.94% (at most 3.4%) of the estimated scores $\hat{s}_N$ fall outside the interval $[s - 0.15\sigma, s + 0.15\sigma]$.

(5) Relative Order. The relative order of the similarity estimated by PR P-Rank is almost preserved with the deterministic result, as shown in the following theorem.

Theorem 4. *Let $\hat{s}_N(\cdot, \cdot)$ be the estimated similarity by PR P-Rank with N being the sample size, and $s(\cdot, \cdot)$ the exact similarity. If $s(u, v) > s(u, w) + \epsilon$, then*

$$\Pr(\hat{s}_N(u, v) - \hat{s}_N(u, w) > \epsilon) \leq \exp(-N\epsilon^2/2) \quad (\forall u, v, w \in V).$$

(A detailed proof of Theorem 4 is provided in the Appendix.)

Our empirical results on DBLP will further verify that for $N \geq 350$, PR P-Rank can almost maintain the relative order of similarity (see Figure 12).

6 Experimental Evaluation

We conduct a comprehensive empirical study over several real and synthetic datasets to evaluate (1) the scalability, time and space efficiency of the proposed algorithms, (2) the approximability of DE P-Rank, and (3) the effectiveness of PR P-Rank.

6.1 Experimental Settings

Datasets. We used three real datasets (AMZN, DBLP, and WIKI) to evaluate the efficacy of our methods, and synthetic data (0.5M-3.5M RAND) to vary graph characteristics. The sizes of AMZN, WIKI and DBLP are shown in Tables 2-4.

Table 2. AMZN

	0505	0601
$|\mathcal{V}|$	410K	402K
$|\mathcal{E}|$	3,356K	3,387K

Table 3. DBLP

	98-99	98-01	98-03	98-05	98-07
$|\mathcal{V}|$	1,525	3,208	5,307	7,984	10,682
$|\mathcal{E}|$	5,929	13,441	24,762	39,399	54,844

Table 4. WIKI

	0715	0827	0919
$|\mathcal{V}|$	3,088K	3,102K	3,116K
$|\mathcal{E}|$	1,126K	1,134K	1,142K

(1) AMZN *data*[4] are based on *Customers Who Bought This Item Also Bought* feature of the Amazon website. Each node represents a product. There is a directed edge from node i to j if a product i is frequently co-purchased with product j. Two datasets were collected in May 5 2003, and June 1 2003.

(2) DBLP *data*[5] record co-authorships among scientists in the Bibliography. We extracted the 10-year (from 1998 to 2007) author-paper information, and singled out the publications on 6 conferences (ICDE, VLDB, SIGMOD, WWW, SIGIR, and KDD). Choosing every two years as a a time step, we built 5 co-authorship graphs.

(3) WIKI *data*[6] contain millions of encyclopedic articles on a vast array of topics to the latest scientific research. We built 3 graphs from the English WIKI dumps,where each vertex represents an article, and edges the relationship that "a category contains an article to be a link from the category to the article".

(4) RAND *data* were produced by C++ boost generator for digraphs, with 2 parameters: the number of vertices, and the number of edges.

Parameter Settings. To keep consistency with the experimental conditions in [26], we assigned each of the following parameters a default value.

Notation	Description	Default	Notation	Description	Default
C_{in}	in-link damping factor	0.8	λ	weighting factor	0.5
C_{out}	out-link damping factor	0.6	υ	low rank	$50\% \times \mathrm{Rank}$
ϵ	desired accuracy	0.001	N	sample size	350

Compared Algorithms. We have implemented the following algorithms. (1) DE and PR, *i.e.,* DE P-Rank and PR P-Rank; (2) Naive, a K-Medoids P-Rank iterative algorithm ($K = 10$) based on a radius-based pruning method [26]; (3) Psum, a variant

---

[4] http://snap.stanford.edu/data/index.html
[5] http://www.informatik.uni-trier.de/~ley/db/
[6] http://en.wikipedia.org/

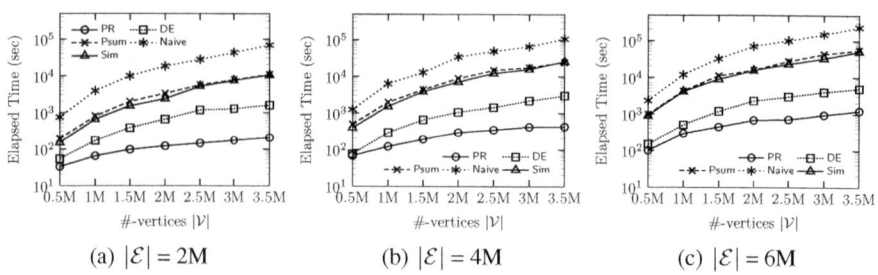

Fig. 7. Scalability on Synthetic Datasets

of P-Rank, leveraging a partial sum function [18] to compute similarity; (4) Sim, an enhanced version of SimRank algorithm [1], which takes account of the evidence factor for incident vertices. These algorithms were implemented in C++, except that the MATLAB implementation [3] for calculating RSVD () and Rank ().

All experiments were run on a machine with a Pentium(R) Dual-Core (2.00GHz) CPU and 4GB RAM, using Windows Vista. Each experiment was repeated over 5 times, and the average is reported here.

6.2   Experimental Results

Scalability. We first evaluate the scalability of the five ranking algorithms, using synthetic data. In these experiments, PR fingerprint tree indexing is precomputed and shared by all vertex pairs in a given graph, and thus its cost is counted only once.

We randomly generate 7 graphs $\mathcal{G} = (\mathcal{V}, \mathcal{E})$, with the edge size $|\mathcal{E}|$ varying from 2M to 6M. The results are reported in Figures 7(a), 7(b) and 7(c). We can notice that (1) DE is almost one order of magnitude faster than the other algorithms when $|\mathcal{V}|$ is increased from 0.5M to 3.5M. (2) Execution time for PR increases linearly with $|\mathcal{V}|$ due to the use of finger printed trees. Varying $|\mathcal{E}|$, we also see that the CPU time of DE is less sensitive to $|\mathcal{E}|$. This is because the time of DE mainly depends on the number of vertices having the similar neighbor structures. Hence, graph sparsity has not a large impact on DE. In light of this, DE scales well with $|\mathcal{E}|$, as expected.

Time & Space Efficiency. We next compare the CPU time and memory space of the five ranking algorithms on real datasets. The results are depicted in Figures 8 and 9. It can be seen that the time and space of PR clearly outperform the other approaches on AMZN, WIKI, and DBLP, *i.e.,* the use of Monte Carlo sampling approach is effective. In all the cases, DE runs faster than Psum, Naive and Sim with moderate memory requirements. This is because DE uses a singular value decomposition to cluster a large body of vertices having the similarity neighbor structures, and a low-rank approximation to eliminate the vertices of tiny singular values, which can save large amounts of memory space, and avoid repetitive calculations of "less important" vertices. Besides, with the increasing number of vertices on DBLP data, the upward trends of the time and space for DE and PR match our analysis in Sections 4 and 5.

Figure 10 depicts how the total computational time and memory space are amortized on the different phases of DE and PR, respectively, over AMZN data. We see from the results that the similarity calculation phase of DE is far more time and space consuming

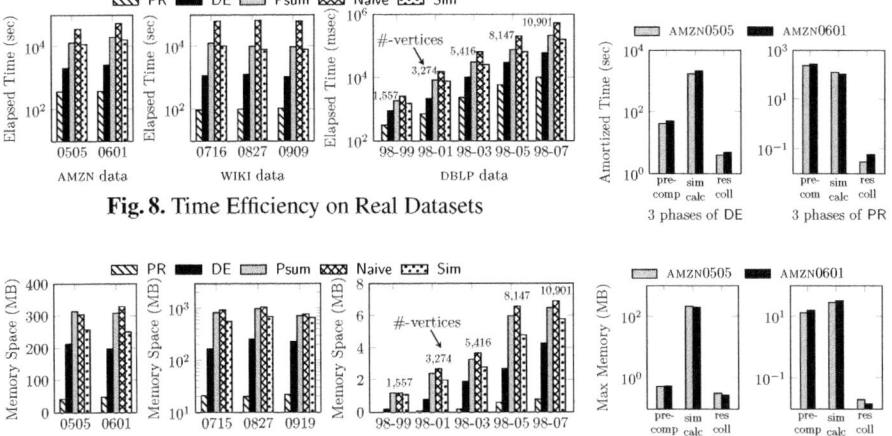

Fig. 8. Time Efficiency on Real Datasets

Fig. 9. Memory Space on Real Datasets   **Fig. 10.** Amortized Costs

(97.4% time and 99.59% space) than the other two phases (2.3% time and 0.27% space for preprocessing, and 0.3% time and 0.14% space for result collection), which is expected because factorizing $\mathbf{Q}$ and $\mathbf{P}$, and computing Σ^{-1} yield a considerable amount of complexity. We also notice that the total cost of PR is well balanced between offline pre-indexing and on-line query phases, both of which take high proportions of CPU time and memory usage, *i.e.*, almost 71.6% total time and 32.6% space are leveraged on indexing phase, and 28.3% time and 67.3% space on query phase. This tells that the use of finger printed trees can effectively reduce the overhead costs of PR.

Accuracy. We now evaluate the accuracy of the five algorithms on real data. The *Normalized Discounted Cumulative Gain* (NDCG) measure [8] is adopted. The NDCG at a rank position p is defined as $\text{NDCG}_p = \frac{1}{\text{IDCG}_p} \sum_{i=1}^{p} (2^{\text{rank}_i} - 1)/(\log_2(1+i))$, where rank_i is the average similarity at rank position i judged by the human experts, and IDCG_p is the normalization factor to guarantee that NDCG of a perfect ranking at position p equals 1.

Figure 11 compares the accuracy of DE and PR with that of Naive, Psum and Sim returned by NDCG_{30} on AMZN, WIKI and DBLP, respectively. It can be seen that DE always achieves higher accuracy than PR. The accuracy of PR is not that good because some valid finger printed trees may be neglected with certain probability by PR sampling. The results on DBLP also show that the accuracy of DE and PR is insensitive to $|\mathcal{V}|$. Hence, adding vertices does not affect the error in estimation, as expected.

We further evaluate the ground truth calculated by DE and PR on DBLP (98-07) dataset to retrieve the top-k most similar authors for a given query u. Interestingly, Figure 12 depicts the top-10 ranked results for the query "Jennifer Widom" according to the similarity scores returned by PR, DE and Naive, respectively. These members were frequent co-authors of the 6 major conference papers with "Jennifer Widom" from 1998 to 2007. It can be noticed that the ranked results for different ranking algorithms on DBLP (98-07) are practically acceptable and obey our common sense pretty well.

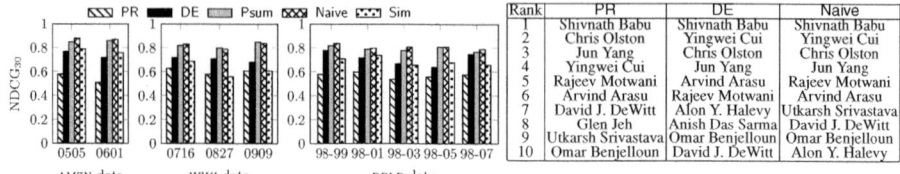

Fig. 11. Accuracy on Real Datasets

Fig. 12. Top-10 Co-authors of Jennifer Widom on DBLP

Rank	PR	DE	Naive
1	Shivnath Babu	Shivnath Babu	Shivnath Babu
2	Chris Olston	Yingwei Cui	Yingwei Cui
3	Jun Yang	Chris Olston	Chris Olston
4	Yingwei Cui	Jun Yang	Jun Yang
5	Rajeev Motwani	Arvind Arasu	Rajeev Motwani
6	Arvind Arasu	Rajeev Motwani	Arvind Arasu
7	David J. DeWitt	Alon Y. Halevy	Utkarsh Srivastava
8	Glen Jeh	Anish Das Sarma	David J. DeWitt
9	Utkarsh Srivastava	Omar Benjelloun	Omar Benjelloun
10	Omar Benjelloun	David J. DeWitt	Alon Y. Halevy

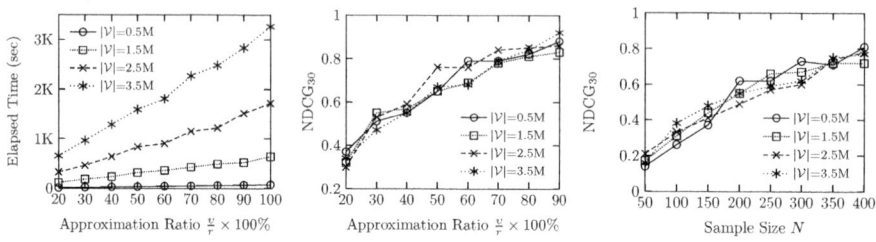

Fig. 13. Effects of v for DE

Fig. 14. Effects of N for PR

The similarities calculated by DE and PR almost preserve the relative order of Naive. Hence, both DE and PR can be effectively used for P-Rank similarity estimation in top-k nearest neighbor search on real networks.

Effects of v. For DE algorithm, we next investigate the impact of approximation rank v and adjacency matrix rank r on similarity estimation, using synthetic data.

We use 4 web graphs with the size $|\mathcal{V}|$ of the set of vertices ranging from 0.5M to 3.5M, $2|\mathcal{V}|$ edges. We consider various approximation ranks v for a given graph $\mathcal{G}$. We fix $|\mathcal{V}|$ while varying v from $10\% \times r$ to $90\% \times r$ with r being the rank of adjacency matrix for $\mathcal{G}$. The results are reported in Figure 13, which visualizes the low-rank v as a speed-accuracy trade-off. When v becomes increasingly close to r (*i.e.,* the radio $\frac{v}{r}$ approaches 1), high accuracy ($NDCG_{30}$) could be expected, but more running time needs to be consumed. This tells that adding approximation rank v will induce smaller errors for similarity estimation, but it will increase the complexity of computation up to a point of the rank r when no extra approximation errors can be reduced.

Effects of N. For PR algorithm, we evaluate the impact of the sample size N of the finger printed trees on similarity quality.

We consider 4 web graphs $\mathcal{G}$ with the size $|\mathcal{V}|$ $(= \frac{1}{2}|\mathcal{E}|)$ ranged from 0.5M to 3.5M. In Figure 14, we fixed $|\mathcal{V}|$ while varying N from 50 to 400. In all the cases, when the sample size is larger ($N > 300$), higher accuracy could be attained ($NDCG_{30} > 0.6$), irrelevant to the size $|\mathcal{V}|$ of graph. The result reveals that adding samples of finger printed trees reduces errors in estimation, and hence improves the effectiveness of PR.

Summary. We find the following. (1) DE and PR can scale well with the large size of graphs, whereas Naive, Psum, and Sim fail to run with an acceptable time. (2) DE significantly outperforms Psum and Sim by almost one order of magnitude with error guarantees (a drop 10% in NDCG). (3) PR may run an order of magnitude faster than DE with a little sacrifice in accuracy (5% relative error), which is practically acceptable for ad-hoc query performed in an on-line fashion.

7   Related Work

P-Rank has become an appealing measure of similarity used in a variety of areas, such as publication network [26], top-k nearest neighbor search [13], and social graph [2,17]. The traditional method leverages a fixed-point iteration to compute P-Rank, yielding $O(Kn^4)$ time in the worst case. Due to the high time complexity, Zhao *et al.* [26] further propose the radius- and category-based pruning techniques to improve the computation of P-Rank to $O(Kd^2n^2)$ at the expense of reduced accuracy, where d is the average degree in a graph. However, their heuristic methods can not guarantee the similarity accuracy. In contrast, our methods are based on two matrix forms for optimizing P-Rank computation with fast speed and provable accuracy.

There has also been work on other similarity optimization (*e.g.,* [5,6,15,18,20,22,24, 25]). Lizorkin *et al.* [18] proposed an interesting memoization approach to improve the computation of SimRank from $O(Kn^4)$ to $O(Kn^3)$. The idea of memoization can be applied to P-Rank computation in the same way. A notion of the weighted and evidence-based SimRank is proposed by Antonellis *et al.* [1], yielding better query rewrites for sponsored search. He *et al.* [6] and Yu *et al.* [24] show interesting approaches to paralleling the computation of SimRank.

Closer to this work are [15, 17, 27]. Our prior work [17] focuses on P-Rank computations on undirected graphs by showing an $O(n^3)$-time deterministic algorithm; however the optimization techniques in [17] rely mainly on the symmetry of the adjacency matrix. In comparison, this work further studies the general approaches to optimizing P-Rank on directed graphs, achieving quadratic-time for deterministic computation, and linear-time for probabilistic estimation. Extensions of SimRank are studied in [27] for structure- and attribute-based graph clustering, but the time complexity is still cubic in the number of vertices. Recent work by Li *et al.* shows an incremental algorithm for dynamically computing SimRank; however it is not clear that extending to the P-Rank model is possible. Besides, it seems hard to obtain an error bound for computing Sim-Rank on directed graphs as the error bound in [15] is only limited to undirected graphs. In contrast, the error bounds in our work may well suit digraphs.

In comparison to the work on deterministic SimRank computation, the work on probabilistic computation is limited. Li *et al.* [16] exploit the block structure of linkage patterns for SimRank estimation, which is in $O(n^{4/3})$ time. Fogaras *et al.* [4, 5] utilize a random permutation method in conjunction with Monte Carlo Simulation to estimate SimRank in linear time. As opposed to our probabilistic methods, (a) these algorithms are merely based on ingoing links; it seems hard to observe global structural connectivity while maintaining linear time, by using only a finger printed tree structure. (b) The theoretical guarantee of choosing a moderate sample size is not mentioned in [4, 5] as these studies ignore the central limit property of the finger printed tree by and large.

8 Conclusion

In this paper, we have studied the optimization problem of P-Rank computation. We proposed two equivalent matrix forms to characterize the P-Rank similarity. (i) Based on the matrix inversion form of P-Rank, a deterministic algorithm was devised to reduce the computational time of P-Rank from cubic to quadratic in the number of vertices; the error estimate was given as a by-product when the low rank approximation was deployed. (ii) Based on the matrix power series form of P-Rank, a probabilistic algorithm was also suggested for further speeding up the computation of P-Rank in linear time with controlled accuracy. The experimental results on both real and synthetic datasets have demonstrated the efficiency and effectiveness of our methods.

Acknowledgements. We greatly appreciate the constructive comments from the anonymous reviewers. We also thank National ICT Australia Ltd (NICTA) for their sponsorship.

References

1. Antonellis, I., Garcia-Molina, H., Chang, C.-C.: SimRank++: query rewriting through link analysis of the click graph. PVLDB 1(1) (2008)
2. Cai, Y., Zhang, M., Ding, C.H.Q., Chakravarthy, S.: Closed form solution of similarity algorithms. In: SIGIR, pp. 709–710 (2010)
3. Cowell, W.R. (ed.): Sources and Development of Mathematical Software. Prentice-Hall Series in Computational Mathematics, Cleve Moler, Advisor (1984)
4. Fogaras, D., Rácz, B.: A Scalable Randomized Method to Compute Link-Based Similarity Rank on the Web Graph. In: Lindner, W., Fischer, F., Türker, C., Tzitzikas, Y., Vakali, A.I. (eds.) EDBT 2004. LNCS, vol. 3268, pp. 557–567. Springer, Heidelberg (2004)
5. Fogaras, D., Rácz, B.: Scaling link-based similarity search. In: WWW (2005)
6. He, G., Feng, H., Li, C., Chen, H.: Parallel SimRank computation on large graphs with iterative aggregation. In: KDD (2010)
7. Horn, R.A., Johnson, C.R.: Matrix Analysis. Cambridge University Press (February 1990)
8. Järvelin, K., Kekäläinen, J.: Cumulated gain-based evaluation of IR techniques. ACM Trans. Inf. Syst. 20, 422–446 (2002)
9. Jeh, G., Widom, J.: SimRank: a measure of structural-context similarity. In: KDD, pp. 538–543 (2002)
10. Kallenberg, O.: Foundations of Modern Probability. Springer (January 2002)
11. Latuszynski, K., Miasojedow, B., Niemiro, W.: Nonasymptotic bounds on the estimation error for regenerative MCMC algorithms. Technical report (2009)
12. Laub, A.J.: Matrix Analysis For Scientists And Engineers. Society for Industrial and Applied Mathematics, Philadelphia (2004)
13. Lee, P., Lakshmanan, L.V.S., Yu, J.X.: On top-k structural similarity search. In: ICDE (2012)
14. Leskovec, J., Huttenlocher, D.P., Kleinberg, J.M.: Signed networks in social media. In: CHI, pp. 1361–1370 (2010)
15. Li, C., Han, J., He, G., Jin, X., Sun, Y., Yu, Y., Wu, T.: Fast computation of SimRank for static and dynamic information networks. In: EDBT (2010)
16. Li, P., Cai, Y., Liu, H., He, J., Du, X.: Exploiting the Block Structure of Link Graph for Efficient Similarity Computation. In: Theeramunkong, T., Kijsirikul, B., Cercone, N., Ho, T.-B. (eds.) PAKDD 2009. LNCS, vol. 5476, pp. 389–400. Springer, Heidelberg (2009)

17. Li, X., Yu, W., Yang, B., Le, J.: ASAP: Towards Accurate, Stable and Accelerative Penetrating-Rank Estimation on Large Graphs. In: Wang, H., Li, S., Oyama, S., Hu, X., Qian, T. (eds.) WAIM 2011. LNCS, vol. 6897, pp. 415–429. Springer, Heidelberg (2011)
18. Lizorkin, D., Velikhov, P., Grinev, M.N., Turdakov, D.: Accuracy estimate and optimization techniques for SimRank computation. VLDB J. 19(1) (2010)
19. Saad, Y.: Iterative Methods for Sparse Linear Systems, 2nd edn. Society for Industrial and Applied Mathematics (April 2003)
20. Sarma, A.D., Gollapudi, S., Panigrahy, R.: Estimating PageRank on graph streams. In: PODS, pp. 69–78 (2008)
21. Tsatsaronis, G., Varlamis, I., Nørvåg, K.: An Experimental Study on Unsupervised Graph-based Word Sense Disambiguation. In: Gelbukh, A. (ed.) CICLing 2010. LNCS, vol. 6008, pp. 184–198. Springer, Heidelberg (2010)
22. Xi, W., Fox, E.A., Fan, W., Zhang, B., Chen, Z., Yan, J., Zhuang, D.: SimFusion: measuring similarity using unified relationship matrix. In: SIGIR (2005)
23. Yu, W., Le, J., Lin, X., Zhang, W.: On the Efficiency of Estimating Penetrating-Rank on Large Graphs. In: Ailamaki, A., Bowers, S. (eds.) SSDBM 2012. LNCS, vol. 7338, pp. 231–249. Springer, Heidelberg (2012), http://www.cse.unsw.edu.au/~weirenyu/yu-tr-ssdbm2012.pdf
24. Yu, W., Lin, X., Le, J.: Taming Computational Complexity: Efficient and Parallel SimRank Optimizations on Undirected Graphs. In: Chen, L., Tang, C., Yang, J., Gao, Y. (eds.) WAIM 2010. LNCS, vol. 6184, pp. 280–296. Springer, Heidelberg (2010)
25. Yu, W., Zhang, W., Lin, X., Zhang, Q., Le, J.: A space and time efficient algorithm for SimRank computation. World Wide Web 15(3), 327–353 (2012)
26. Zhao, P., Han, J., Sun, Y.: P-Rank: a comprehensive structural similarity measure over information networks. In: CIKM (2009)
27. Zhou, Y., Cheng, H., Yu, J.X.: Graph clustering based on structural/attribute similarities. PVLDB 2(1) (2009)

Appendix: Proof of Theorem 4

Proof. Let $A \triangleq \{|\hat{s}_N(u,v) - s(u,v)| \geq \epsilon\}$, and $B \triangleq \{|\hat{s}_N(u,w) - s(u,w)| \geq \epsilon\}$. We first find an upper bound of variance σ^2 for any sample $\hat{s}_N^{(i)} \in [0,1]$.

$$\sigma^2 = \mathbb{E}[(\dot{s}_N^{(i)})^2] - \mathbb{E}[\dot{s}_N^{(i)}]^2 \leq \mathbb{E}[\dot{s}_N^{(i)}] - \mathbb{E}[\dot{s}_N^{(i)}]^2 = -(\mathbb{E}[\dot{s}_N^{(i)}] - 1/2)^2 + 1/4 \leq 1/4.$$

Then, using the Bernstein's inequality, we have

$$\Pr(A \cap B) \leq \Pr(A) \leq \exp(-N\epsilon^2/(2\sigma^2)) \leq \exp(-2N\epsilon^2).$$

Since $s(u,v) > s(u,w) + \epsilon$, we have

$$A \cap B \supseteq \{\hat{s}_N(u,v) - \hat{s}_N(u,w) > \hat{s}_N(u,v) - \hat{s}_N(u,w) - \overbrace{(s(u,v) - s(u,w))}^{>\epsilon(>0)}$$
$$= \underbrace{(\hat{s}_N(u,v) - s(u,v))}_{\geq \epsilon} - \underbrace{(\hat{s}_N(u,w) - s(u,w))}_{<-\epsilon} > 2\epsilon\}$$

Hence, $\Pr\{\hat{s}_N(u,v) - \hat{s}_N(u,w) > \epsilon\} \leq \exp(-N\epsilon^2/2)$.

Towards Efficient Join Processing over Large RDF Graph Using MapReduce

Xiaofei Zhang[1], Lei Chen[1], and Min Wang[2]

[1] Hong Kong University of Science and Technology, Hong Kong
[2] HP Labs China, Beijing
{zhangxf,leichen}@cse.ust.hk, min.wang6@hp.com

Abstract. Existing solutions for answering SPARQL queries in a shared-nothing environment using MapReduce failed to fully explore the substantial scalability and parallelism of the computing framework. In this paper, we propose a cost model based RDF join processing solution using MapReduce to minimize the query responding time as much as possible. After transforming a SPARQL query into a sequence of MapReduce jobs, we propose a novel index structure, called *All Possible Join* tree (APJ-tree), to reduce the searching space for the optimal execution plan of a set of MapReduce jobs. To speed up the join processing, we employ hybrid join and bloom filter for performance optimization. Extensive experiments on real data sets proved the effectiveness of our cost model. Our solution has as much as an order of magnitude time saving compared with the state of art solutions.

1   Introduction

As a well supported computing paradigm on Cloud, MapReduce substantially fits for large scale data-intensive parallel computations. The exploration of RDF query processing with MapReduce has drawn great research interests. There are a number of challenges to fit RDF query processing, especially join processing, directly into the MapReduce framework. Though attempts were made in [1][2][3], the following problems are not well solved: 1) How to map the implied join operations inside a SPARQL query to a number of MapReduce jobs? 2) Given a set of MapReduce jobs and their dependencies, how to make the best use of computing and network resources to maximize the job execution parallelism, such that we can achieve the shortest execution time? 3) How to organize and manage RDF data on Cloud such that MapReduce jobs can scale along with the data volumes involved in different queries?

To solve the above challenges, in this paper we propose a cost model based RDF join processing solution on Cloud. We develop a deterministic algorithm for optimal solution instead of the heuristic algorithm employed in [1]. To elaborate, we first decompose RDF data into *Predicate* files and organize them according to data contents. Then, we map a SPARQL query directly to a sequence of MapReduce jobs that may employ hybrid join strategies (combination of Map-side join, Reduce-side join and memory backed join). Finally, based on our cost model

A. Ailamaki and S. Bowers (Eds.): SSDBM 2012, LNCS 7338, pp. 250–259, 2012.

of MapReduce jobs for join processing, we present a *All Possible Join* (APJ) tree based technique to schedule these jobs to be executed in the most extended parallelism style. Experiments over real datasets in a real Cloud environment justified that our method has as much as an order of magnitude time saving compared with the "variable-grouping" strategy adopted in [1]. To summarize, we made the following contributions:

- We present a compact solution for RDF data management and join processing on the Cloud platform.
- We develop a cost model based deterministic solution for the optimal query processing efficiency.
- We propose a novel approach to model a MapReduce job's behavior and justify its effectiveness with extensive experiments on real data sets.

The rest of paper is structured as follows. We briefly introduce RDF and SPARQL query in Section 2. We formally define the problem in Section 3 and present the cost model in Section 4. We elaborate solution details in Section 5 and 6. Experiments on real dataset are presented in Section 7. We discuss related works in Section 8 and make the conclusion in Section 9.

2 Preliminaries

RDF, known as *Resource Definition Framework*, is originally defined to describe conceptual procedures and meta data models. A RDF triplet consists of three components: *Subject*, *Predicate* (*Property*) and *Object* (*Value*), which represent that two entities connected by a relationship specified by the *Predicate* or an entity has certain value on some property. Thus, RDF data can be visualized as a directed graph by treating entities as vertices and relationships as edges between entities.

SPARQL language is the W3C standard for RDF query. It is implemented to query RDF data in a SQL-style. The syntax of SPARQL is quite simple, with limited support on aggregation queries and no support on range queries. A SPARQL query usually involves several triple patterns, each of which is known as a BQP (*Basic Query Pattern*) and contains one or two variables. Given a SPARQL query, data that satisfy different BQPs will be joined on the same variable. In this work, we consider queries with the *Predicate* rarely appearing as a variable as suggested in [4]. In addition, the queries we that study do not support the "optional" join semantics [5].

3 Problem Definition

In this work, we only focus on join processing of a SPARQL query using MapReduce. The "*construct*" and "*optional*" semantics are not considered. Therefore, a SPARQL query can be simply modeled as "SELECT *variable(s)* WHERE {BQP(s)}". Intuitively, since we assume that the *Predicate* of BQPs is not a variable [4], the above query can be answered by the following steps: 1) select RDF triplets that satisfy a BQP; 2) join RDF triplets of different BQPs on shared variables. Essentially, we study how to map the original query to one or several join operations implemented by MRJs and schedule the execution of these jobs. To address this problem, we first partition RDF data according to the *Predicate*, each partition is called a *Predicate* file. Then for given BQP_i, we denote its corresponding *Predicate* file as $\mathbf{PF}(\text{BQP}_i)$. The statistics of *Subject* and *Object* for each *Predicate* file are also computed to serve the join order selection.

Given a SPARQL query, we can derive a query pattern graph. Each vertex represents a BQP. Two vertices are connected if the two BQPs share the same variable. A formal definition on the query pattern graph is given as follows.

Definition 1. *Graph $G\langle V, E, L_v, L_e \rangle$ is a query pattern graph, where $V = \{v | v$ is a BQP$\}$, $E = \{e | e = (v_i, v_j), v_i \in V, v_j \in V\}$, $L_v = \{l_v | l_v = (S(v), Sel(var)), v \in V$, $S(v)$ is the size of $\mathbf{PF}(v)$, var is the variable contained in v, $Sel(var)$ is the selectivity of var in $\mathbf{PF}(v)\}$ and $L_e = \{l_e | l_e$ is the variable shared by v_i and v_j, $v_i \in V$, $v_j \in V$, $(v_i, v_j) \in E\}$.*

A MRJ can be identified by selecting any subgraph of G, e.g., a subgraph containing BQP_i and BQP_j. Then the MRJ performs a join of *Predicate* file $\mathbf{PF}(\text{BQP}_i)$ and $\mathbf{PF}(\text{BQP}_j)$. Clearly, there are many possible ways to select MRJs. Our goal is to obtain an optimal MRJ selection strategy based on a cost model to achieve the minimum time span of query processing. For clear illustration, we first classify MRJs into two types, *PJoin (Pairwise Join)* and *MJoin (Multiway Join)*. We define them as follows:

Definition 2. *Given $G\langle V, E, L_v, L_e \rangle$, $PJoin(V')$ is a join operation on a set of Predicate files $\mathbf{PF}(v_i)$, $v_i \in V'$, $V' \subseteq V$, where V' in G are connected by edges labeling with the same variable(s).*

Definition 3. *Given $G\langle V, E, L_v, L_e \rangle$, $MJoin(V')$ is a join operation on a set of Predicate files $\mathbf{PF}(v_i)$, $v_i \in V'$, $V' \subseteq V$, where V' in G are connected by edges labeling with more than one variables.*

Apparently, there are many possible combinations of *PJoin* and *MJoin* to answer the query, while each combination implies an execution graph defined as follows.

Definition 4. *An execution graph P is a directed graph with form $P(V, E)$, where $V = \{v | v$ is a MRJ$\}$, $E = \{e | e = \langle \text{MRJ}_i, \text{MRJ}_j \rangle, \text{MRJ}_i, \text{MRJ}_j \in V\}$.*

Given P, we say MRJ_i depends on MRJ_j if and only if MRJ_j's output is a direct input of MRJ_i, i.e., a directed edge is added from MRJ_j to MRJ_i in P. MRJs are considered independent if they do not incident to the same edge in P, which can be executed in parallel as long as their direct inputs are ready. MRJ that has predecessor(s) must wait until its predecessor(s) is finished. We want to find such a P that guarantees a minimum query processing time.

Problem Definition: Given a platform configuration Δ, RDF data statistics S, a query pattern graph G obtained from a SPARQL query Q, find a function $F : (\Delta, S, G) \to P(V, E)$ such that,

1) Let MRJ_i's execution time be t_i. Let $\langle i, j \rangle$ denote a path from MRJ_i to MRJ_j in P, and the weight of this path to be $W_{i,j} = \sum_{k \in \langle i,j \rangle} t_k$;
3) the $Max\{W_{i,j}\}$ is minimized.

In the problem definition, $W_{i,j}$ indicates the possible minimal execution time from MRJ_i to MRJ_j. Therefore, by minimizing the maximum of $W_{i,j}$, the overall minimum execution time of P is achieved.

4   Cost Model

For MRJs conducting join operations, heavy costs on large scale sequential disk scan and frequent I/O of intermediate results dominate the execution time. Therefore, we build a model for MRJ's execution time based on the analysis of I/O and network cost.

Assume the total input size of a MRJ is S_I, the total intermediate data copied from Map to Reduce is of size S_{CP}, the number of Map tasks and Reduce tasks are m and n, respectively. Assume S_I is evenly partitioned among m Map tasks. Let J_M, J_R and J_{CP} denote the total time cost of three phases respectively, T be the total execution time of a MRJ. Then $T \leq J_M + J_{CP} + J_R$ holds due to the overlapping between J_M and J_{CP}. Let the processing cost for each Map task be t_M, which is dominated by the disk I/O of sequential reading and data spilling,

$$t_M = (C_1 + p \times \alpha) \times \frac{S_I}{m} \tag{1}$$

, where C_1 is a constant factor regarding disk I/O capability, p is a random variable denoting the cost of spilling intermediate data, which subjects to intermediate data size, and α denotes the output ratio of a Map task, which is query dependent and can be computed with selectivity estimation. Assume m' is the current number of Map tasks running in parallel in the system, then $J_M = t_M \times \frac{m}{m'}$. Let t_{CP} be the time cost for copying the output of single Map task to n Reduce tasks. It includes data copying over network as well as the overhead of serving network protocols,

$$t_{CP} = C_2 \times \frac{\alpha \times S_I}{n \times m} + q \times n \tag{2}$$

where C_2 is a constant number denoting the efficiency of data copying over network, q is a random variable which represents the cost of a Map task serving n connections from n Reduce tasks. Intuitively, there is a rapid growth of q while n gets larger. Thus, $J_{CP} = \frac{m}{m'} \times t_{CP}$. J_R is dominated by the Reduce task with the largest input size. We consider the key distribution in the input file is random. Let S_r^i denote the input size of Reduce task i, then according to the *Central Limit Theorem*[6], for $i = 1, ..., n$, S_r^i follows a normal distribution $N \sim (\mu, \sigma)$, where μ is determined by $\alpha \times S_I$ and σ subjects to data set properties, which can be learned from history query logs. By employing the rule of "*three sigmas*"[6], we make $S_r^* = \alpha \times S_I \times n^{-1} + 3\sigma$ the biggest input size to a Reduce task, then

$$J_R = (p + \beta \times C_1) \times S_r^* \tag{3}$$

where β is a query dependent variable denoting output ratio, which could be precomputed based on the selectivity estimation. Thus, the execution time T of a MRJ is:

$$T = \begin{cases} J_M + t_{CP} + J_R \text{ if } t_M \geq t_{CP} \\ t_M + J_{CP} + J_R \text{ if } t_M \leq t_{CP} \end{cases} \tag{4}$$

In our cost model, parameters C_1, C_2, p and q are system dependent and need to be derived from observations on real job execution. This model favors MRJs that have I/O cost dominate the execution time. Due to limited space, more details and experiments on cost model validation can be found in [7].

5   Query Processing

We solve the MRJ selection and ordering problem by introducing a novel tree structure, namely *All Possible Join* (APJ) tree, which implies all possible join plans to examine. We first introduce the *Var*-BQP entity concept, which shall derive generalized MRJ types (covers both *PJoin* and *MJoin*), and further help build the APJ-tree.

Definition 5. $e^{|\{var\}|} = (\{var\} : \{BQP\})$ *is a Var-BQP entity, where* $\{var\}$ *represents a set of edges labeled with elements of* $\{var\}$ *in G.* $\{BQP\}$ *represents the set of all the vertices incident to* $\{var\}$ *in G. If* $BQP_i \in \{BQP\}$ *does not only incident to edges labeled with elements from* $\{var\}$, BQP_i *is marked as optional in* $\{BQP\}$, *denoted by capping* BQP_i *with a wave symbol.*

Definition 6. *Two Var-BQP entities* $e_i^{|\{var_i\}|} = (\{var_i\}:\{BQP_i\})$ *and* $e_j^{|\{var_j\}|} = (\{var_j\}:\{BQP_j\})$ *can be joined together if and only if* $\{var_i\} \cap \{var_j\} \neq \emptyset$ *or* $\{BQP_i\} \cap \{BQP_j\} \neq \emptyset$. *Join result of two joinable Var-BQP entities is defined as follows:*

$$e_i^{|\{var_i\}|} \bowtie e_j^{|\{var_j\}|} = (\{var_i\} \cup \{var_j\} : \{BQP_i\} \cup \{BQP_j\}) \tag{5}$$

Based on the join semantic of *Var*-BQP entities, we describe a top-down approach to build an APJ-tree, as presented in Algorithm 1. By traversing G and grouping BQPs on different variables, we can easily obtain $\{e_i^1\}$. Based on the join semantics defined above, we can further obtain $\{e_i^2\}...\{e_i^n\}$. By making each entity an node, and drawing directed edges from e_i and e_j to $e_i \bowtie e_j$, we can obtain a tree structure representing all possible join semantics among *Var*-BQP entities, i.e., APJ-tree.

Algorithm 1. Bottom–up algorithm for APJ-tree generation

Input: Query pattern graph G of m vertices and n distinct edge labels; $V \leftarrow \emptyset$ and $E \leftarrow \emptyset$;
Output: APJ-tree's vertex set V and edge set E;
1: Traverse G to find e_i^1 for each label
2: Add each e_i^1 to V
3: **for** $k = 1$ to $n - 1$ **do**
4:      **if** $\exists e_i^k$ and e_j^k are joinable **then**
5:          **if** $\exists BPQ_x$ is only optional in e_i^k and e_j^k among all e^k **then**
6:              Make BPQ_x deterministic in $e_i^k \bowtie e_j^k$
7:          **end if**
8:          $V \leftarrow V \bigcup \{e_i^k \bowtie e_j^k\}$
9:          $E \leftarrow E \bigcup \{e_i^k \rightarrow e_i^k \bowtie e_j^k, e_j^k \rightarrow e_i^k \bowtie e_j^k\}$
10:     **end if**
11: **end for**

Lemma 1. *An APJ-tree obtained from query pattern graph G implies all possible query execution plans.*

Proof. Basing on the generation of APJ-tree, $\{e_i^n\}$ gives all the possible combinations that could be obtained from G. Each e_i^n for sure contains all the variables and all BQPs, which is exactly the final state in a query execution graph P. e_i^n differentiates from e_j^n as they could be obtained from different join plans. Obviously, the join of e_i^k and e_j^k, $1 \leq k < n$ (if they are *joinable*), is exactly a *PJoin*; while e_i^k itself, $1 < k \leq n$, is a *MJoin*.

For each entity e_i^k in the APJ-tree, we define its weight as the smaller one of the two cost variables, direct cost $DiC(e_i^k)$ and derived cost $DeC(e_i^k)$. $DiC(e_i^k)$ implies the cost of directly join $e_i^k.\{\text{BQP}_i\}$ on $e_i^k.\{var_i\}$. $DeC(e_i^k)$ implies the accumulative cost of obtaining e_i^k from its ancestors.

Lemma 2. *The minimum weight of e_i^n indicates the minimum total cost of joining all BQPs to answer the query.*

Lemma 2 can be easily proved by definition. We find that it is sufficient to generate and check only part of the APJ-tree to obtain the optimal solution. Our top-down search Algorithm 2 can effectively prune certain parts of APJ-tree that do not contain the optimal solution. Since we assume each BQP only has at most two variables involved, it ensures the complexity of Algorithm 4 is no worse than $O(n^2)$.

Algorithm 2. MRJ Identification for G with m vertices and n edges

---

Input: Set E for MRJ identification, $\text{E} \leftarrow \{e_i^1\}$;
    Query execution plan $P \leftarrow \emptyset$; $VT \leftarrow \emptyset$;
Output: P
1: $DeC(e_i^1) \leftarrow DiC(e_i^1)$
2: $VT \leftarrow \bigcup e.BQP, \forall e \in P$
3: $k \leftarrow 1$
4: **repeat**
5:     sort $e_i^k \in E$ on weight in ascending order
6:     **while** $\bigcup e_i^k.\{BQP\} = m$ and $\bigcup e_i^k.\{var\} = n$, $e_i^k \in \text{E}\backslash\{e_j^k\}$, e_j^k has the heaviest weight in E **do**
7:         $\text{E} \leftarrow \text{E} \backslash \{e_j^k\}$
8:     **end while**
9:     **repeat**
10:       **for any** $e_i^k \in \text{E}$
11:       **while** $\exists e_j^k \in \text{E}\backslash\{e_i^k\}$ that can be joined with e_i^k **do**
12:         **if** $(e_i^k.BQP \cup e_j^k.BQP) \not\subseteq VT$ **then**
13:             **if** $DiC(e_i^k \bowtie e_j^k) \geq DeC(e_i^k \bowtie e_j^k)$ **then**
14:                $P \leftarrow P \cup \{e_i^k, e_j^k, e_i^k \bowtie e_j^k\}$
15:             **end if**
16:             **if** $DiC(e_i^k \bowtie e_j^k) < DeC(e_i^k \bowtie e_j^k)$ **then**
17:                $P \leftarrow P \backslash \{e_i^k, e_j^k\}$
18:                $P \leftarrow P \cup \{e_i^k \bowtie e_j^k\}$
19:             **end if**
20:         **end if**
21:         $\text{E} \leftarrow \text{E} \cup \{e_i^k \bowtie e_j^k\}$
22:         update VT
23:       **end while**
24:       $\text{E} \leftarrow \text{E} \backslash \{e_i^k\}$
25:     **until** $\nexists e_i^k \in \text{E}$
26:     $k \leftarrow k + 1$
27: **until** $k = n$

Theorem 1. *P computed with Algorithm 2 is optimal.*

Proof. First we prove that Algorithm 2 finds the entity e_{opt}^n with the minimal weight. Line 5 to 8 in the algorithm guarantees this property. E only contains entities of minimal weight, which are sufficient to cover all the m BQPs and n variables. Thus, when k increases to n, the first e_i^n in E is the entity with the minimum weight. Since we already find the optimal entity e_{opt}^n, if e_{opt}^n's weight is $DiC(e_{opt}^n)$, P would only contain one MJoin that join all BQPs in one step (line

16-19); otherwise, e_{opt}^n's weight is derived from his parents, which would have been added to P (line 13-15). Iteratively, P contains all the necessary entities (equivalent to MRJs) to compute e_{opt}^n, and the longest path weight of P is just e_{opt}^n's weight, which is already the optimal.

We assume that Cloud computing system can provide as much computing resources as required (the similar claim was made by Amazon EC2 service [8]). Thus, we can make MRJs not having ancestor-descendant relationship in P be executed in parallel. Moreover, we adopt two techniques to improve the query efficiency: 1) **Hybrid join strategy**. Reduce-side join is used as the default join strategy. When a *Predicate* file is small enough to be load in memory, we load this file in several Map tasks to perform in-memory join [9]. Map-side join is adopted on the condition that ancestor MRJs' outputs are well partitioned on the same number of reducers. 2). **Bloom filter**. If the query contains a BQP which refers to a small *Predicate* file, we can always read in this file completely into main memory and generate the bloom filter of *Subject* or *Object* variables, which can be done quite efficiently with one Map task. Since the generated bloom filter file is much smaller, it can be loaded in memory later on for each MRJ and help filter out large number of irrelevant RDF triples at the Map phase.

6   Implementations

We use HDFS[10] to set up a repository of large scale RDF dataset.Figure 1 presents an system overview of our solution to RDF query processing. The whole system is backed with well organized RDF data storage on HDFS, which offers block level management that promises efficient data retrieval. The query engine accepts users' queries, and decides corresponding MRJs and an optimal execution plan for each query. Noticing the SPARQL query engine can be deployed to as many clients as desired. Therefore, the query engine will not become a performance bottleneck.

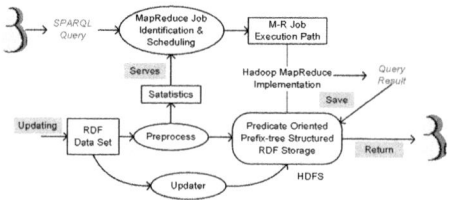

Fig. 1. System Design

The preprocessing of RDF data involves four steps: 1) Group by *Predicate*; 2) Sort on both *Subject* and *Object* for each *Predicate* file, the similar strategy was used in [11] and [12]; 3) Block-wise partition for each *Predicate* file; 4) Build a B+ tree to manage all the *Predicate* files. More details are elaborate in [7].

7   Experiments

We run all the experiments on a test bed of 16 VMs managed by xen-3.0-x86_64, running linux kernel 2.6.32.13. Each VM is configured to have 4 cores (2.26GHz), 4GB memory and 400GB HD. We use Hadoop-0.20.0 to build up the system. We test the system I/O performance with *TestDFSIO*, finding that the system performance is stable, with average writing rate 2.38Mb/sec and reading rate 20.09Mb/sec. System administrator reports the connection bandwidth of 32.96MB/s. We run each job 50 times and report the average execution time. We demonstrate the improvement of our solution over the join strategy given in [1] with benchmark queries used in [13], as shown in Fig.2.

Q2, Q3, Q5 and Q6 only have small number of *Predicate* files and variables involved, therefore, different strategies for MRJ identification and scheduling will obtain about the same result. The APJ-tree helps a lot when then query is more complex, e.g., Q4 and Q7. With bloom filter employed, our solution saves much time cost since large amount of network traffic is avoid as well as the workload of Reduce tasks is reduced. We also run benchmark queries on both enlarged (by duplication) and shrunk (uniformly sampling) datasets as shown in Fig.3 and Fig.4. The system demonstrates a scalable performance for different input size.

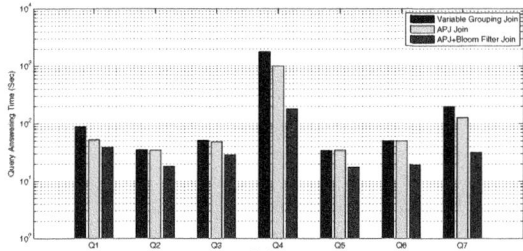

Fig. 2. Bench query evaluation with different MRJ selection strategies

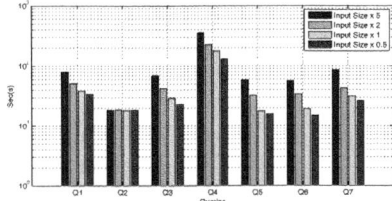

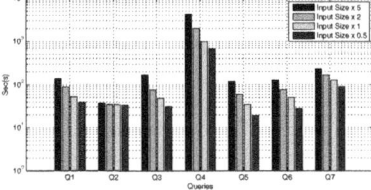

Fig. 3. APJ+Bloom Filter Join   **Fig. 4.** APJ Join

8   Related Work

Research efforts for RDF management and query processing from a traditional RDBMS perspective focus on RDF decomposition (SW-Store [14]) or composition (property table [15]), index building and searching (Hexastore [12], RDF-3X

[11]) and query optimization [13]. However, due to the limitation of RDBMS's scalability, the above solutions cannot meet the demands for the management of extremely large scale RDF data in the coming future. Researchers are trying to incorporate NoSQL database to address the scalability and flexibility issues in the first place. Many works, like [16][17][18][19][4][2][1][3], adopt the Cloud platform to solve the RDF data management problem. However, many of them focus on utilizing high-level definitive languages to create simplified user interface for RDF query processing, which omit all the underlying optimization opportunities and have no guarantees on efficiency. There are a few works directly conducting RDF query processing within the MapReduce framework. [20] provides a plug-in parser and simple aggregation processing of RDF data. Husain et al. [1] presents a greedy strategy that always picks a MRJ which may produce the smallest size of intermediate results. However, this strategy has no guarantee on the overall efficiency. [21] studies the general strategy of replicating data from Map to Reduce to conduct join operations with one MapReduce job.

9   Conclusion

In this paper we study the problem of efficiently answering SPARQL queries over large RDF data set with MapReduce. We define a transmission scheme from an original SPARQL query to MRJs and introduce a APJ-tree based MRJ scheduling technique that guarantees an optimal query processing time. Evaluations over very large real datasets in a real cloud test bed demonstrate the effectiveness of our solution.

Acknowledgment. The work described in this paper was partially supported by HP IRP Project Grant 2011, National Grand Fundamental Research 973 Program of China under Grant 2012CB316200, Microsoft Research Asia Theme Grant MRA11EG0.

References

1. Husain, M.F., et al.: Data intensive query processing for large RDF graphs using cloud computing tools. In: CLOUD 2010 (2010)
2. Farhan Husain, M., Doshi, P., Khan, L., Thuraisingham, B.: Storage and Retrieval of Large RDF Graph Using Hadoop and MapReduce. In: Jaatun, M.G., Zhao, G., Rong, C. (eds.) CloudCom 2009. LNCS, vol. 5931, pp. 680–686. Springer, Heidelberg (2009)
3. Myung, J., et al.: Sparql basic graph pattern processing with iterative mapreduce. In: MDAC 2010 (2010)
4. Tanimura, Y., et al.: Extensions to the pig data processing platform for scalable RDF data processing using hadoop. In: 22nd International Conference on Data Engineering Workshops, pp. 251–256 (2010)
5. Chebotko, A., Atay, M., Lu, S., Fotouhi, F.: Relational Nested Optional Join for Efficient Semantic Web Query Processing. In: Dong, G., Lin, X., Wang, W., Yang, Y., Yu, J.X. (eds.) APWeb/WAIM 2007. LNCS, vol. 4505, pp. 428–439. Springer, Heidelberg (2007)

6. Jaynes, E.T.: Probability theory: The logic of science. Cambridge University Press, Cambridge (2003)
7. Zhang, X., et al.: Towards efficient join processing over large RDF graph using mapreduce. Technical Report (2011)
8. http://aws.amazon.com/ec2/
9. Blanas, S., et al.: A comparison of join algorithms for log processing in mapreduce. In: SIGMOD 2010 (2010)
10. http://hadoop.apache.org/
11. Thomas, N., et al.: The RDF-3x engine for scalable management of RDF data. VLDB J. 19(1), 91–113 (2010)
12. Weiss, C., et al.: Hexastore: sextuple indexing for semantic web data management. Proc. VLDB Endow. (2008)
13. Neumann, T., et al.: Scalable join processing on very large RDF graphs. In: SIGMOD Conference, pp. 627–640 (2009)
14. Abadi, D.J., et al.: Sw-store: a vertically partitioned dbms for semantic web data management. The VLDB Journal 18, 385–406 (2009)
15. http://jena.sourceforge.net/
16. Newman, A., et al.: A scale-out RDF molecule store for distributed processing of biomedical data. In: Semantic Web for Health Care and Life Sciences Workshop (2008)
17. Newman, A., et al.: Scalable semantics - the silver lining of cloud computing. In: ESCIENCE 2008 (2008)
18. Urbani, J., Kotoulas, S., Oren, E., van Harmelen, F.: Scalable Distributed Reasoning Using MapReduce. In: Bernstein, A., Karger, D.R., Heath, T., Feigenbaum, L., Maynard, D., Motta, E., Thirunarayan, K. (eds.) ISWC 2009. LNCS, vol. 5823, pp. 634–649. Springer, Heidelberg (2009)
19. McGlothlin, J.P., et al.: Rdfkb: efficient support for RDF inference queries and knowledge management. In: IDEAS 2009 (2009)
20. http://rdfgrid.rubyforge.org/
21. Afrati, F.N., et al.: Optimizing joins in a map-reduce environment. In: EDBT 2010 (2010)

Panel on "Data Infrastructures and Data Management Research: Close Relatives or Total Strangers?"

Yannis Ioannidis[1,2]

[1] University of Athens, Dept. of Informatics & Telecommunications,
MaDgIK Lab, Athens, Greece
yannis@di.uoa.gr
[2] "Athena" Research Center, Athens, Greece
yannis@athena-innovation.gr

In 1981 the *"1st LBL Workshop on Statistical Database Management"* was held in Berkeley, CA. It was essentially the first step towards establishing the new at the time and very important branch of the data management field that deals with scientific data. A few years later, the third edition of the event already had another 'S' added in its acronym and was named the *"3rd Int'l Workshop on Statistical and Scientific Database Management"*. Eventually, the series became the well-known annual SSDBM conference. For more than 30 years, the requirements for managing and analyzing the data created in the context of research and other activities in many domain sciences have brought out several major challenging problems that have motivated and inspired much data management research. Through the years there have been a very large number of related research papers and keynote presentations in all major database conferences and journals, many significant contributions that have pushed the state of the art in various directions, several dedicated funding programs around the world, and numerous specialized and generic software systems that have been developed targeting scientific data management. Scientists from the biological, medical, physical, natural, and other sciences, as well as the arts and the humanities have worked closely together with data management researchers to obtain solutions to critical problems. All these activities have created a solid body of work that is now considered part of the data management research mainstream, on topics ranging, for example, from the traditional indexing and query processing to the more specialized data mining, real-time streaming, and provenance. Scientific data management is an area of great importance with a long history that will continue to be at the forefront of many interesting developments in the field.

On the contrary, the concept of *"data infrastructures"* is relatively new. Roughly in 2008 or 2009, it came out of several diverse efforts across the Atlantic (and elsewhere) whose main motivation was again scientific data, e.g., the Strategy Reports of ESFRI in Europe and the DataNet funding program in the US. Data infrastructures follow on the footsteps of other, lower-level digital infrastructures, i.e., networks and distributed computation (grids or clouds), and transfer the key ideas of those to data, promoting and facilitating data sharing

A. Ailamaki and S. Bowers (Eds.): SSDBM 2012, LNCS 7338, pp. 260–261, 2012.
© Springer-Verlag Berlin Heidelberg 2012

and use. They represent common platforms that transparently offer important functionality required by the creators and users of large amounts of scientific data, e.g., services related to data storage, maintenance, preservation, discovery, access, and others. There are several ongoing projects that try to build data infrastructures of various forms, such as DataONE and Data Conservancy in the US and OpenAIRE and more recently EUDAT in Europe.

This panel investigates the relationship between the two areas: scientific data management and scientific data infrastructures. Is the latter a brand new area with new challenges requiring new fundamental research and innovative solutions? Or is it just an evolution of some aspects of the former, simply calling for adaptation and good engineering of existing solutions? Would the data infrastructures community benefit from the involvement of the data management research community in addressing the issues before them? Are there data infrastructure problems that are attractive to data management researchers and/or easier solvable based on earlier research work? Do the policies that are necessary for data infrastructure governance present interesting data management research problems? These and other questions are laid before the panel, comprising experts of both backgrounds, whose answers and subsequent discussion should shed some light on how the two areas are related.

Efficient Similarity Search
in Very Large String Sets

Dandy Fenz[1], Dustin Lange[1], Astrid Rheinländer[2], Felix Naumann[1],
and Ulf Leser[2]

[1] Hasso Plattner Institute, Potsdam, Germany
[2] Humboldt-Universität zu Berlin, Department of Computer Science, Berlin,
Germany

Abstract. String similarity search is required by many real-life applications, such as spell checking, data cleansing, fuzzy keyword search, or
comparison of DNA sequences. Given a very large string set and a query
string, the string similarity search problem is to efficiently find all strings
in the string set that are similar to the query string. Similarity is defined
using a similarity (or distance) measure, such as edit distance or Ham-
ming distance. In this paper, we introduce the State Set Index (SSI) as
an efficient solution for this search problem.

SSI is based on a trie (prefix index) that is interpreted as a nondeterministic finite automaton. SSI implements a novel state labeling strategy making the index highly space-efficient. Furthermore, SSI's space
consumption can be gracefully traded against search time.

We evaluated SSI on different sets of person names with up to 170 million strings from a social network and compared it to other state-of-the-
art methods. We show that in the majority of cases, SSI is significantly
faster than other tools and requires less index space.

1 Introduction

Many applications require error-tolerant string search. For example, consider a
search application for customer support in a company. While search queries may
contain an incorrect spelling of a name, the search application should nevertheless find the matching entry of the customer in the database. Another application
arises in a biomedical context. To find and compare genomic regions in the human genome, search applications need to account for individual variations or
mutations in the genes. In these and many other scenarios, the data set consists
of millions of strings, while the search application is required to answer similarity
queries in subseconds.

In this paper, we tackle the *string similarity search problem*, which returns
for a given query all strings from a given bag of strings that are similar to the
query with respect to a previously defined string distance measure and a given
distance threshold. This problem has been covered since the 1960s [26], and is
also known as approximate string matching [19], string proximity search [24], or
error-tolerant search [4]. Much effort has been spent by the research community

A. Ailamaki and S. Bowers (Eds.): SSDBM 2012, LNCS 7338, pp. 262–279, 2012.

to develop filtering techniques, indexing strategies, or fast string similarity algorithms that improve the query execution time of similarity-based string searches. However, a fast query execution is often accomplished by storing a wealth of information in huge indexes in main memory. For very large string collections with hundreds of millions of strings, this approach often fails since indexes grow too large.

We propose the State Set Index (SSI) as a solution for this problem. The main advantage of SSI is that it has a very small memory footprint while providing fast query execution times on small distance thresholds at the same time. In particular, we extend and improve TITAN [15], a trie index that is interpreted as a nondeterministic finite automaton (NFA), developed by Liu et al. The contributions of this paper are:

- We introduce a novel state labeling approach, where only information on the existence of states is stored. Different from previous approaches, state transitions do not need to be stored and can be calculated on-the-fly.
- Using this highly space-efficient labeling strategy, SSI is capable of indexing very large string sets with low memory consumption on commodity hardware.
- SSI allows a graceful trade-off between index size and search performance by parameter adjustment. These parameters, namely labeling alphabet size and index length, determine the trade-off between index size and query runtime. We comprehensively evaluate these parameter settings and derive favorable settings such that index sizes remain small and the search performance is still competitive.
- We evaluate SSI on several data sets of person names coming from a social network. Our evaluation reveals that SSI outperforms other state-of-the-art approaches in the majority of cases in terms of index size and query response time. In particular, on a data set with more than 170 million strings and a distance threshold of 1, SSI outperforms all other methods we compared to.

The remainder of this paper is structured as follows: Section 2 describes related work. In Sec. 3, we cover basic definitions that are necessary for the following sections. We describe our approach in Sec. 4 by defining the index structure and algorithms for building the index and searching with it. We show evaluation results in Sec. 5 and conclude the paper in Sec. 6.

2   Related Work

In the past years, the research community has spent much effort on accelerating similarity-based string matching. Prominent approaches use prefiltering techniques, indices, refined algorithms for computing string similarity, or all in combination [19,20].

Filter methods are known to reduce the search space early using significantly less computational effort than computing the edit distance (or another similarity measure) directly. As a result of this so-called *filter-and-verify* approach [28], only a few candidate string pairs need to be compared using edit distance. Prominent pre-filtering approaches are based on q-grams [8,9,14], character frequencies [1], or length filtering [2].

Tries as an index structure for strings and exact string matching in tries were first introduced by Morrison [16] and were later extended by Shang et al. [25] with pruning and dynamic programming techniques to enable similarity-based string matching. Next to similarity-based string searches, tries and trie-based NFAs are also known to perform well in other areas, such as exact pattern matching [10], set joins [12], or frequent item set mining [7,11].

The Peter index structure [22] was designed for near-duplicate detection in DNA data and combines tries with filtering techniques to enable similarity-based string searches and joins. It stores additional information at each trie node for early search space pruning. Pearl [23] is a follow-up where restrictions on small alphabets were removed and a strategy for parallelizing similarity searches and joins was introduced. Closely related to SSI is TITAN [15], an index structure based on prefix trees that are converted into non-deterministic automata A, such that the initial state of A corresponds to the root node of the originating prefix tree, and leaf nodes correspond to accept states in A. Additionally, further state transitions are introduced in order to enable delete, insert, and replacement operations on edit distance-based queries.

Algorithms based on neighborhood generation were first used for similarity-based string matching by Myers [17]. One drawback of the original algorithm by Myers is its space requirements, that makes it feasible only for small distance thresholds and small alphabets. The FastSS index [3] captures neighborhood relations by recursively deleting individual characters and reduces space requirements by creating a so-called k-deletion neighborhood. Similar to filtering approaches, FastSS performs search space restriction by analyzing the k-deletion neighborhood of two strings. By adding partitioning and prefix pruning, Wang et al. [27] significantly improved the runtime of similarity search algorithms based on neighborhood generation.

The Flamingo package [2] provides an inverted-list index that is enriched with a charsum and a length filter. The filter techniques are organized in a tree structure, where each level corresponds to one filter.

We empirically compare the State Set Index to FastSS, Flamingo, Pearl, and TITAN, and show that it often outperforms these tools both in terms of query execution time and with respect to index size (see Sec. 5). We could not compare to Wang et al. since no reference implementation was available.

3   Basic Concepts and Definitions

In this section, we define basic terms and concepts that we use in the subsequent sections.

3.1   Similarity Search and Measures

Let Σ be an alphabet. Let s be a string in Σ^*. A substring of s, denoted by $s[i \ldots j]$, starts at position i and ends at position j. We call $s[1 \ldots j]$ *prefix*,

$s[i \ldots |s|]$ *suffix* and $s[i \ldots j], (1 \leq i \leq j \leq |s|)$, *infix* of s. Any infix of length $q \in \mathbb{N}$ is called q-*gram*. Conceptually, we ground our index structure and associated algorithms on a similarity search operator defined as follows:

Definition 1 (Similarity search). *Given a string s, a bag S of strings, a distance function d and a threshold k, the similarity search operator $sSearch(s, S)$ returns all $s_i \in S$ for which $d(s, s_i) \leq k$.*

All similarity-based search operations must be based on a specific similarity measure. Though there exist several techniques to measure the similarity of two strings, we focus on *edit distance* for the scope of this paper.

Definition 2 (Edit distance [13]). *The edit distance $d_{ed}(s_1, s_2)$ of two strings s_1, s_2 is the minimal number of insertions, deletions, or replacements of single characters needed to transform s_1 into s_2. Two strings are within edit distance k, if and only if $d_{ed}(s_1, s_2) \leq k$.*

Variations of edit distance apply different costs for the three edit operations. While SSI is applicable to all edit distance-based measures with integer costs for the different edit operations, we only consider the standard definition with equal weights in this paper.

The edit distance $d_{ed}(s_1, s_2)$ can be computed by dynamic programming in $\Theta(|s_1| * |s_2|)$. Apart from the dynamic programming algorithm with quadratic complexity in time and space, there exist various improvements for the edit distance computation. Bit-parallel algorithms [18] achieve a complexity of $O(\frac{|s_1| * |s_2|}{w})$, where w is the size of the computer word. If a maximum allowed distance threshold is defined in advance, the k-banded alignment algorithm [5] computes the edit distance of two strings in $\Theta(k \cdot max\{|s1|, |s2|\})$.

3.2 Tries and NFAs

The construction of *prefix or suffix trees* is a common technique for string search. In the literature, such a tree is also called *trie* [6].

Definition 3 (Trie [6]). *A trie is a tree structure (V, E, v_r, Σ, L), where V is the set of nodes, E is the set of edges, v_r is the root node, Σ is the alphabet, and $L : V \to \Sigma^*$ is the labeling function that assigns strings to nodes. For every node v_c with $L(v_c) = s[1 \ldots n]$ that is a child node of v_p, it holds $L(v_p) = s[1 \ldots n-1]$, i.e., any parent node is labeled with the prefix of its children.*

The trie root represents the empty string. The descendants of a node represent strings with a common prefix and an additional symbol from the alphabet. A trie is processed from the root to the leaf nodes. Indexed strings are attached to the node that can be reached by processing the complete string. Tries are an efficient method for exact string search.

For efficient similarity search, a trie can also be interpreted as a *nondeterministic finite automaton* [15].

Definition 4 (Nondeterministic finite automaton (NFA) [21]). *A non-deterministic finite automaton is defined as a tuple $(Q, \Sigma, \delta, q_0, F)$, where Q is the set of states, Σ is the input alphabet, $\delta : Q \times (\Sigma \cup \{\varepsilon\}) \to \mathcal{P}(Q)$ is the state transition function (with ε referring to the empty word), q_0 is the start state, and F is the set of accepting states.*

The NFA begins processing in the start state q_0. The input is processed character-wise with the state transition function. An NFA is allowed to have several active states at the same time. If, after processing the entire string, the NFA is in at least one accepting state, the string is accepted, otherwise rejected.

For similarity search, we interpret a trie as an NFA. In the NFA version of the trie, the trie root node is the start state of the NFA. The nodes with associated result strings are marked as accepting states. The trie's edges are interpreted as state transitions with reading symbols. In addition, the NFA version contains for each state transition one additional state transition for reading ε as well as one ε-transition from each state to itself. These ε-transitions allow state transitions that simulate deletion, insertion, and replacement of symbols as necessary for edit distance calculation. To do similarity search with the NFA, the query string is processed as input symbol sequence by the NFA. After the processing step, the NFA is in zero, one, or more accepting states. The query result contains all strings that are attached to the reached accepting states.

The NFA idea described so far generates for a large amount of indexed strings a large automaton with many states (but there can be no false positives in the result string set). In the next section, we describe our approach that restricts the number of NFA states and checks the result string set for false positives.

4 State Set Index

The State Set Index (SSI) is an efficient and configurable index structure for similarity search in very large string sets. In this section, we first describe the key ideas of SSI before giving details on the indexing and searching algorithms.

4.1 Index Structure

SSI is based on a trie that is interpreted as an NFA. In the following, we describe the key ideas behind SSI that go beyond the basic trie and NFA concepts described above.

State Labeling. The SSI states are labeled with numbers. Each label is calculated from the history of read symbols. For this purpose, the original input alphabet, in the following referred to as Σ_I, is mapped to a labeling alphabet $\Sigma_L = \{1, 2, \ldots, c_{max}\} \subset \mathbb{N}$ with $c_{max} \leq |\Sigma_I|$. A mapping function $m : \Sigma_I \to \Sigma_L$ defines the mapping of characters from the two alphabets. A label for a state with read symbols $s_1 \ldots s_{n-1}s_n \in \Sigma_I^n$ can be calculated as follows:

$$l(s_1 \ldots s_{n-1}s_n) = l(s_1 \ldots s_{n-1}) \cdot |\Sigma_L| + m(s_n)$$
$$l(\varepsilon) = 0$$

with ε referring to the empty word.

Restriction of Labeling Alphabet Size. SSI allows to restrict the size of the labeling alphabet Σ_L. When choosing a labeling alphabet with $c_{max} < |\Sigma_I|$ (note the strict "less than" sign), at least two symbols from the input alphabet are mapped to the same symbol from the labeling alphabet.

This can reduce the number of existing states in the resulting NFA. For any two prefixes p_1, p_2 of two indexed strings, the states $l(p_1), l(p_2)$ are merged iff $l(p_1) = l(p_2)$. This is the case iff for at least one character position pos in p_1 and p_2, it holds $p_1[1 : pos - 1] = p_2[1 : pos - 1]$ and $p_1[pos] \neq p_2[pos]$ and $m(p_1[pos]) = m(p_2[pos])$, i.e., two different characters at the same position are mapped to the same symbol in the labeling alphabet and the prefixes of the strings before this character match.

Depending on the chosen mapping, a state may contain several different strings. With $c_{max} < |\Sigma_I|$, it is not possible to reconstruct a string from a state label, as there are several different possibilities for that. Thus, we need to store which strings are stored at which state. In addition, it is possible that the accepting states may contain false positives, i.e., strings that are not query-relevant. This makes it necessary to check all resulting strings by calculating the exact distance to the query string before returning results.

Choosing a labeling alphabet size is thus an important parameter for tuning the SSI. A too large labeling alphabet size results in a large NFA with many states and thus large storage requirement, but few false positives. In contrast, a too small alphabet size leads to an NFA with only few states that does not restrict the number of result strings enough; the consequence is a large number of false positives.

Restriction of Index Length. SSI allows to restrict the number of indexed symbols. From each string $s_1 \ldots s_n \in \Sigma_I^n$, only a prefix with a maximum length of ind_{max} is indexed. The "leaf" states contain all strings with a common prefix.

Restricting the index length can reduce the number of existing states. For any two strings g_1, g_2, the two states $l(g_1)$ and $l(g_2)$ are equal iff $l(g_1[1 : ind_{max}]) = l(g_2[1 : ind_{max}])$.

Similar to choosing the labeling alphabet size, we face the challenge of handling possible false positives in the result string sets also for restricted index length. The index length is thus a second parameter to tune the trade-off between a large index (large index length, high memory consumption) and a large number of false positives to be handled (small index length, low memory consumption). In our analysis of a large string data set with the Latin alphabet (and some special characters) as input alphabet, we observed optimal results with a labeling alphabet size of 4 and an index length of 14 (see Sec. 5.1 for a discussion of this experiment).

Restricting the labeling alphabet size as well as the index length can significantly decrease the number of existing states. For example, in a data set with 1 million names, the mapping from the Latin alphabet to a labeling alphabet with 4 and restricting the index length to 14 results in a state count reduction from 5,958,916 states to 2,298,209 states (a reduction ratio of 61 %).

Storing States. Due to the history-preserving state labels, all potential successors of a state can be calculated. With a calculated state label, it is easy to check whether such a state exists: For any state ϕ, the state transition with the character $c \in \Sigma_L$ by definition exists iff $\phi_c = \phi \cdot |\Sigma_L| + c$ exists. This is because for any character $c' \in \Sigma_L \setminus \{c\}$, it holds $c \neq c'$ and thus $\phi_{c'} = \phi \cdot |\Sigma_L| + c' \neq \phi_c$.

To benefit from this observation, SSI only stores which states that actually exist, i.e., only states ϕ for which there is at least one prefix p of a string in the indexed string data set with $l(p) = \phi$. This reduces the necessary storage capacity, because it is not necessary to store state transitions. Also, during query answering, checking the existence of state transitions is not required.

Because SSI state labels are numbers, a simple storage format can be defined. A bitmap, where each bit combination represents a label of an existing or non-existing state, is sufficient to store which states do exist.

Storing Data. Due to the introduced restrictions, an accepting state may refer to multiple, different strings – the strings cannot be completely reproduced from the state labels. Thus, it is necessary to store the strings behind the states. The required data store has a rather simple interface: A set of keys (the accepting states), each with a set of values (the strings referred to by the states) needs to be stored. Any key/multi-value store is suitable for this task. Since the data store is decoupled from the state store, the data store can be held separately. Thus, while the state store can be configured to be small or large enough to fit into main memory, the data store can be held in secondary memory.

4.2   Algorithms

In the following, we describe the details for indexing a large string set with SSI and searching with the created index.

Indexing. The indexing process is shown in Algorithm 1. All strings to be indexed are processed one after another. Each string is read character-by-character. After reading a character, the current state is calculated and stored. Finally, after reading the entire string, the last state is marked as accepting state and the string is stored at this state's entry in the data store. After the initial indexing process, it is also possible to index additional strings using the same steps.

Example. We illustrate the SSI index with an example. Consider the strings Müller, Mueller, Muentner, Muster, and Mustermann and the alphabet mapping shown in Table 1. In this example, we chose a labeling alphabet size of $c_{max} = 4$ and an index length of $ind_{max} = 6$.

Table 1. Example for alphabet mapping function $m : \Sigma_L \to \Sigma_I$

Σ_L	M	u	e	l	r	ü	n	t	s	m	a
Σ_I	1	2	3	4	1	2	3	4	1	2	3

Algorithm 1. Indexing with SSI

Input: set of strings to be indexed $stringSet$,
    labeling alphabet size c_{max},
    index length ind_{max},
    mapping function $m : \Sigma_I \rightarrow \Sigma_L$
Output: set of existing states $stateSet$,
    set of accepting states $acceptingStateSet$,
    map of states with indexed strings $dataStore$
1: $stateSet := \{\}$
2: $acceptingStateSet := \{\}$
3: $dataStore = \{\}$
4: **for all** $str \in stringSet$ **do**
5:     $state := 0$
6:     **for** $pos := 1 \rightarrow min(ind_{max}, |str|)$ **do**
7:        $state := state \cdot c_{max} + m(str[pos])$
8:        $stateSet.add(state)$
9:        $acceptingStateSet.add(state)$
10:     $dataStore.add(state, str)$
11: **return** $stateSet, acceptingStateSet, dataStore$

Figure 1 shows all existing states of the resulting index. The accepting states point to the indexed strings as follows:
$1869 \rightarrow \{$Müller$\}$, $1811 \rightarrow \{$Mueller$\}$, $1795 \rightarrow \{$Muenter$\}$, $1677 \rightarrow \{$Muster, Mustermann$\}$ □

Searching. We now describe how to process a query string $q \in \Sigma_I^*$ with edit distance $k \in \mathbb{N}$. The search process is shown in Algorithm 2.

First, a set S of cost-annotated states s with state ϕ_s and associated costs λ_s (the number of edit distance operations required so far) is created. We write $s := \langle \phi_s, \lambda_s \rangle$. Initially, all states that can be reached from the start state with at most k ε-transitions are added to S. To determine these states, the labels of the successors of the start state are calculated and their existence is validated. If a state s in S is associated with several different costs λ_s, only the record with the lowest λ_s is kept; all other records are dismissed. This selection is done for all state calculations and is not stated again in the following.

Next, the query string q_I is translated into the labeling alphabet with $q := l(q_I)$. The characters of q are processed one-by-one. The following steps are processed for each character c in q.

Another empty set S^* of current cost-annotated states is created. For each cost-annotated state $\langle \phi_s, \lambda_s \rangle$ in S, a set S_c^* is created and processed with the following steps:

– To simulate deletion of characters, the cost-annotated state $\langle \phi_s, \lambda_s + 1 \rangle$ is added to S_c^* if $\lambda_s + 1 \leq k$.

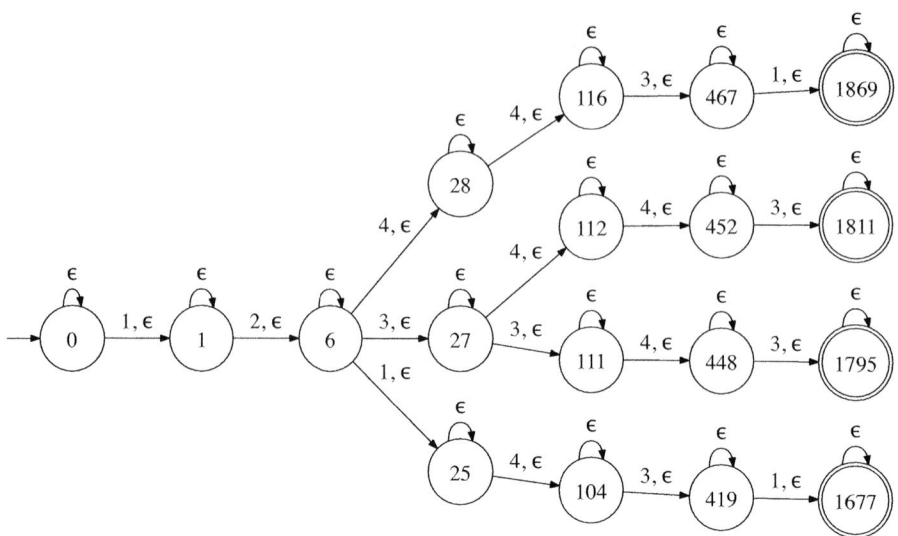

Fig. 1. Example for states created by SSI with $c_{max} = 4$ and $ind_{max} = 6$

- To simulate matching of characters, it is checked whether the state $\phi_s^* := \phi_s \cdot |\Sigma_L| + i$ exists. This state exists if and only if there is a transition from ϕ_s with the character c to ϕ_s^*. If the state exists, then $\langle \phi_s^*, \lambda_s \rangle$ is added to S_c^*.
- Next, the insertion of characters other than c is simulated. If $\lambda_s + 1 \leq k$, then for each $\phi_s^* := \phi_s \cdot |\Sigma_L| + m(c^*)$ with $c^* \in \Sigma_L \setminus \{c\}$, a new cost-annotated state $\langle \phi_s^*, \lambda_s + 1 \rangle$ is added to S_c^*.
- Inserting characters is simulated using ε-transitions. For each cost-annotated state $\langle \phi_s, \lambda_s \rangle$ in S_c^*, all states ϕ_s^* are determined that can be reached from ϕ_s with k ε-transitions. For each such state, the annotated states $\langle \phi_s^*, \lambda_s + i \rangle$ with $\lambda_s \leq i \leq k$ are added to S_c^*.

Then, S is replaced by S^* and all steps are repeated with the next character. After processing all characters, the state set S represents the final state set. For all states from S that are accepting, all strings stored at those states are retrieved. This set of strings is filtered by calculating the actual edit distance to the query string as it may contain false positives. This set of filtered strings is the result of the search.

Example. Consider the index in Fig. 1 and the example query Mustre with a maximum distance of $k = 2$. The initial state set $S = \{\langle 0, 0 \rangle, \langle 1, 1 \rangle, \langle 6, 2 \rangle\}$ contains all states reachable from the start state with at most $k = 2$ ε-transitions. Next, the first character $c = 1$ (M) is processed. The state sets $S^* = S_c^* = \emptyset$ are created. For all entries in S, the five above-described steps are executed. After processing the first character, we have:

$$S = \{\langle 0, 1 \rangle, \langle 1, 0 \rangle, \langle 6, 1 \rangle, \langle 28, 2 \rangle, \langle 27, 2 \rangle, \langle 25, 2 \rangle\}$$

Algorithm 2. Searching with SSI

Input: query string q, maximum edit distance k
Output: result string set R

```
 1: S := {⟨i · c, i⟩ | 0 ≤ i ≤ k, c ∈ ΣL} ∩ stateSet               Initial ε-transitions
 2: for pos := 1 → min(indmax, |q|) do
 3:     S* := {}
 4:     for all ⟨φs, λs⟩ ∈ S do
 5:         Sc* := {}
 6:         if λs + 1 ≤ k then                                                Deletion
 7:             Sc* := Sc* ∪ ⟨φs, λs + 1⟩
 8:         for i := 1 → |ΣL| do                                    Match & Substitution
 9:             φs* := φs · |ΣL| + i
10:             if φs* ∈ stateSet then
11:                 if i = m(q[pos]) then
12:                     Sc* := Sc* ∪ ⟨φs*, λs⟩
13:                 else if λs + 1 ≤ k then
14:                     Sc* := Sc* ∪ ⟨φs*, λs + 1⟩
15:         Sc* := Sc* ∪ ({⟨φs + i · c, i⟩ | s ∈ Sc*, λs ≤ i ≤ k, c ∈ ΣL} ∩ stateSet)  Insertion
16:         S* := S* ∪ Sc*
17:     S := S*
18: R := {}                                       Retrieve strings and filter by distance
19: for all ⟨φs, λs⟩ ∈ S do
20:     if φs ∈ acceptingStateSet then
21:         R := R ∪ {s ∈ dataStore.get(φs) | ded(s, q) ≤ k}
22: return  R
```

After that, the character $c = 2$ (u) is processed. The state set after this step is:

$$S = \{\langle 0, 2\rangle, \langle 1, 1\rangle, \langle 6, 0\rangle, \langle 28, 1\rangle, \langle 27, 1\rangle, \langle 25, 1\rangle, \langle 116, 2\rangle,$$
$$\langle 112, 2\rangle, \langle 111, 2\rangle, \langle 104, 2\rangle\}$$

After processing the third character $c = 1$ (s), we have:

$$S = \{\langle 1, 2\rangle, \langle 6, 1\rangle, \langle 28, 1\rangle, \langle 27, 1\rangle, \langle 25, 0\rangle, \langle 116, 2\rangle, \langle 112, 2\rangle,$$
$$\langle 111, 2\rangle, \langle 104, 1\rangle, \langle 419, 2\rangle\}$$

The next character $c = 4$ (t) results in:

$$S = \{\langle 6, 2\rangle, \langle 28, 1\rangle, \langle 27, 2\rangle, \langle 25, 1\rangle, \langle 104, 0\rangle, \langle 116, 1\rangle, \langle 112, 1\rangle,$$
$$\langle 111, 2\rangle, \langle 419, 1\rangle, \langle 1677, 2\rangle, \langle 467, 2\rangle, \langle 452, 2\rangle\}$$

The character $c = 1$ (r) is processed as follows:

$$S = \{\langle 28, 2\rangle, \langle 25, 2\rangle, \langle 104, 1\rangle, \langle 116, 2\rangle, \langle 112, 2\rangle, \langle 419, 1\rangle,$$
$$\langle 1677, 1\rangle, \langle 1869, 2\rangle, \langle 467, 2\rangle, \langle 452, 2\rangle\}$$

With the last character $c = 3$ (e), we finally have:

$$S = \{\langle 104, 2\rangle, \langle 419, 1\rangle, \langle 1677, 2\rangle, \langle 467, 2\rangle, \langle 1811, 2\rangle\}$$

From the set of states in S, only the accepting states 1677 and 1811 are further processed. The strings stored at these states are Muster, Mustermann, and Mueller. After filtering false positives, we finally have the result string set {Muster}. □

Complexity. To index n strings with a maximum index length ind_{max}, at most ind_{max} states need to be calculated for each string. Thus, we have an indexing complexity of $\mathcal{O}(n \cdot ind_{max})$.

The most important size factor of SSI is the number of created states. For an index length ind_{max} and an indexing alphabet Σ_L, the number of possible states is $|\Sigma_L|^{ind_{max}}$. The index size depends on the chosen parameters where ind_{max} is the dominant exponential parameter.

The search algorithm of SSI mainly depends on c_{max}, ind_{max}, and the search distance k. In the first step, $k \cdot |c_{max}|$ potential states are checked. For each existing state, its successor states are created. These consist of up to one state created by deletion, c_{max} states created by match or substitution, and $k \cdot |c_{max}|$ states created by insertion of a character. This process is repeated up to ind_{max} times. Overall, we have up to $(k \cdot |c_{max}|) \cdot (1 + k \cdot |c_{max}| + k \cdot |c_{max}|)^{ind_{max}}$ steps and thus a worst-case complexity of $\mathcal{O}((k \cdot |c_{max}|)^{ind_{max}})$. Similar to the indexing process, the complexity is bound by the parameters c_{max} and ind_{max} where ind_{max} is the dominant exponential factor. By evaluating the existence of states during the search process and proceeding only with existing states, we typically can significantly decrease the number of states that are actually evaluated.

5 Evaluation

We use a set of person names crawled from the public directory of a social network website to evaluate the performance of SSI for parameter selection, index creation, and for search operations. Table 2 shows some properties of our data set. The set D_{full} contains all person names we retrieved, whereas the sets D_i consist of i randomly chosen strings taken from D_{full}. First, we evaluate the impact of different parameter configurations on the performance of SSI and then choose the best setting to compare SSI against four competitors. In particular, we compare SSI to FastSS [3], TITAN [15], Flamingo [2], and Pearl [23], which are all main memory-based tools for index-based similarity string operations (see Sec. 2 for details). For Flamingo and Pearl, we use the original implementations provided by the authors. For FastSS and TITAN, we use our own implementations of the respective algorithms. Our evaluation comprises experiments both for indexing time and space as well as experiments on exact and similarity-based search queries.

All experiments were performed on an Intel Xeon E5430 processor with 48 GB RAM available using only a single thread. For each experiment, we report the average of three runs.

Table 2. Evaluation data sets

Set	# strings	avg./min./max. string length	input alphabet size	# exact duplicates
D_{full}	170,879,859	13.99 / 1 / 100	38	70,751,399
D_{200k}	200,000	14.02 / 1 / 61	29	5,462
D_{400k}	400,000	14.02 / 1 / 54	32	17,604
D_{600k}	600,000	14.01 / 1 / 55	35	35,626
D_{800k}	800,000	14.02 / 1 / 61	33	54,331
D_{1000k}	1,000,000	14.01 / 1 / 64	35	77,049

5.1 Evaluation of SSI Parameters

We exemplarily used the set D_{1000k} to evaluate the impact of different parameter configurations on the performance of SSI on small string sets. Since the maximum index length ind_{max} and labeling alphabet size $|\Sigma_L|$ have a large influence on the performance of SSI, we varied both ind_{max} and $|\Sigma_L|$ in the range of 2 to 15. We could not perform experiments on larger parameter ranges due to memory constraints of our evaluation platform.

As displayed in Fig. 2(a), the average query execution time drastically decreases with increased labeling alphabet size and maximum index length. In particular, a configuration of SSI with $ind_{max} = 12$ and $|\Sigma_L| = 8$ outperforms a configuration using $ind_{max} = 2$ and $|\Sigma_L| = 2$ by three orders of magnitude (factor 1521). On the other hand, when increasing ind_{max} and $|\Sigma_L|$, we observed that the index size grows significantly (see Fig. 2(b)). For example, changing the configuration from $ind_{max} = 2$ and $|\Sigma_L| = 2$ to $ind_{max} = 6$ and $|\Sigma_L| = 13$ increases memory requirements by a factor of 60. We also observed that the number of false positives and the number of accessed keys per query decreases both with increasing ind_{max} and $|\Sigma_L|$ size (data not shown) and conclude that this is the main reason for the positive outcome of a large labeling alphabet and a large index length. However, we could not increase both parameters further due to memory limitations of our platform, but we expect a further decrease of query execution time.

We also evaluated the influence of varying parameters on query execution time and index size on D_{full}. Results are shown in Fig. 3 for selected configurations. Similar to the experiments on D_{1000k}, the index size grows heavily while increasing ind_{max} and $|\Sigma_L|$. Particularly, choosing $|\Sigma_L| = 3$ and $ind_{max} = 15$ yields in an index size of approximately 12 GB, whereas a configuration with $|\Sigma_L| = 5$ and $ind_{max} = 15$ needs a bit vector of 28 GB. On the other hand, the query execution time decreases with elongating ind_{max} and $|\Sigma_L|$. Using $|\Sigma_L| = 3$ and $ind_{max} = 15$, the query execution time averages to 8 milliseconds, whereas with $|\Sigma_L| = 5$ and $ind_{max} = 15$ the query execution time diminishes to 2.6 milliseconds on average at the expense of a very large index. We also experimented with other settings of ind_{max} and $|\Sigma_L|$ in the range of 2 to 15, but these configurations either did not finish the indexing process in a reasonable amount of time or ran out of memory on our evaluation platform. Therefore, we did not consider these settings for parameter configuration on large string sets.

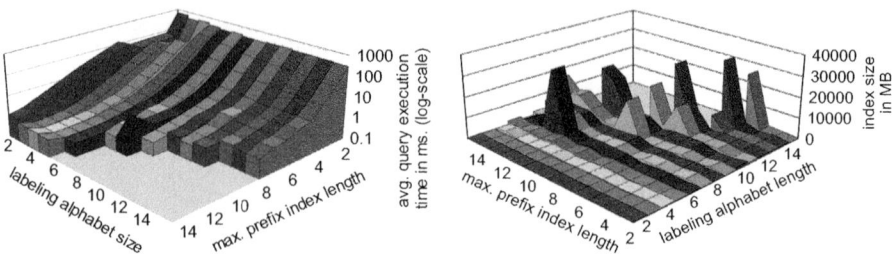

(a) Average query execution time (log-scale)   (b) Average index size in MBytes

Fig. 2. Evaluation of parameters ind_{max} and $|\Sigma_L|$ for D_{1000k} and $k = 1$

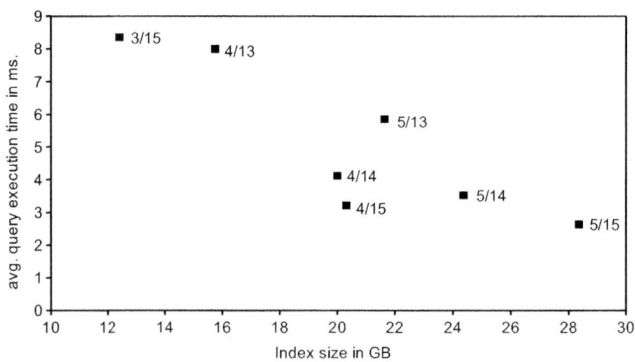

Fig. 3. Trade-off between index size and query execution time on D_{full} and $k = 1$ on varying configurations of $ind_{max} \in \{3, 4, 5\}$ and $|\Sigma_L| \in \{13, 14, 15\}$

In summary, both parameter variations of $|\Sigma_L|$ and ind_{max} have a large impact on the performance of SSI. While increasing $|\Sigma_L|$ or ind_{max}, the number of false positive results that need to be verified decreases, which yields in a considerably fast query response time. However, our experiments also revealed that at some point, no further improvements on query response time can be achieved by increasing $|\Sigma_L|$ and ind_{max}. This is caused by an increased effort for calculating involved final states that outweighs the decreased amount of false-positive and the number of lookups in this setting.

Therefore, a beneficial configuration for indexing up to one million person names is to fix $|\Sigma_L| = 8$ and $ind_{max} = 12$. Using this configuration leads to a fast query execution time with on a moderate index size. When indexing larger string collections, a shift in favor of index length is reasonable, since an increasing length yields larger performance enhancements with respect to query response time. However, elongating the index length and the labeling alphabet yields also in a vast growth of the index size, but we strive for an index structure that is efficient both in terms of space and time. Thus, we decided to configure SSI with

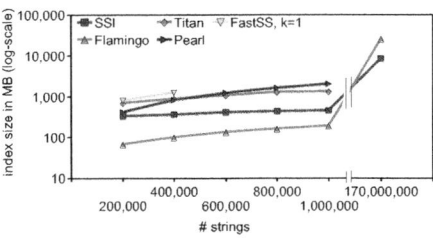
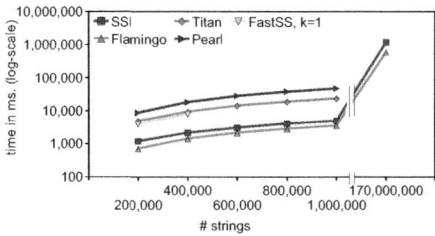

(a) Average index size in MBytes (log-scale)   (b) Average index creation time in milliseconds (log-scale)

Fig. 4. Index creation

$|\Sigma_L| = 4$ and $ind_{max} = 14$ for all following experiments using D_{full}, since this configuration gives us the best query execution time with an index size of at most 20 GB.

5.2   Index Creation Time and Memory Consumption

We evaluated SSI in terms of index creation time and memory consumption and compared it to other main-memory index structures, namely FastSS, TITAN, Pearl, and Flamingo on all available data sets. For all evaluated tools, we observed that both index sizes and indexing time grow at the same scale as the data sets.

Many of the tools we compared to are not able to handle very large string collections. Figure 4(a) displays the memory consumption of each created index in main memory. We were able to index D_{full} only with SSI and Flamingo; FastSS, Pearl, and TITAN ran out of memory during index creation. In particular, FastSS even failed to create indexes with more than 400,000 strings. Another severe drawback of FastSS is that it needs to create a separate index for each edit distance threshold k – in contrast to all other evaluated tools.

Clearly, SSI outperforms all other trie- or NFA-based tools in terms of memory consumption and outperforms FastSS, Pearl, and TITAN with factors in the range of 1.4 (Pearl on D_{200k}) to 4.5 (Pearl on D_{1000k}). Compared to Flamingo, which is based on indexing strings by their lengths and char-sums, we observed that SSI is advantageous for indexing large data sets. When indexing D_{full}, SSI needs 3.0 times less memory than Flamingo. For small data sets with up to one million strings, Flamingo outperforms SSI with factors in the range of 2.4 (D_{1000k}) to 5.0 (D_{200k}).

We also evaluated SSI on the time spent for index creation. As displayed in Fig. 4(b), SSI indexes all data sets significantly faster than the other trie- or NFA-based methods. It outperforms FastSS with factors 3.2 to 3.7 on $k = 1$, TITAN with factors 4.0 to 4.7, and Pearl with factors 7.4 to 9.6. Similar to the memory consumption, SSI is the more superior the larger the data sets grow. Compared to Flamingo, SSI is only slightly slower (factors in the range of 1.4 to 2.0).

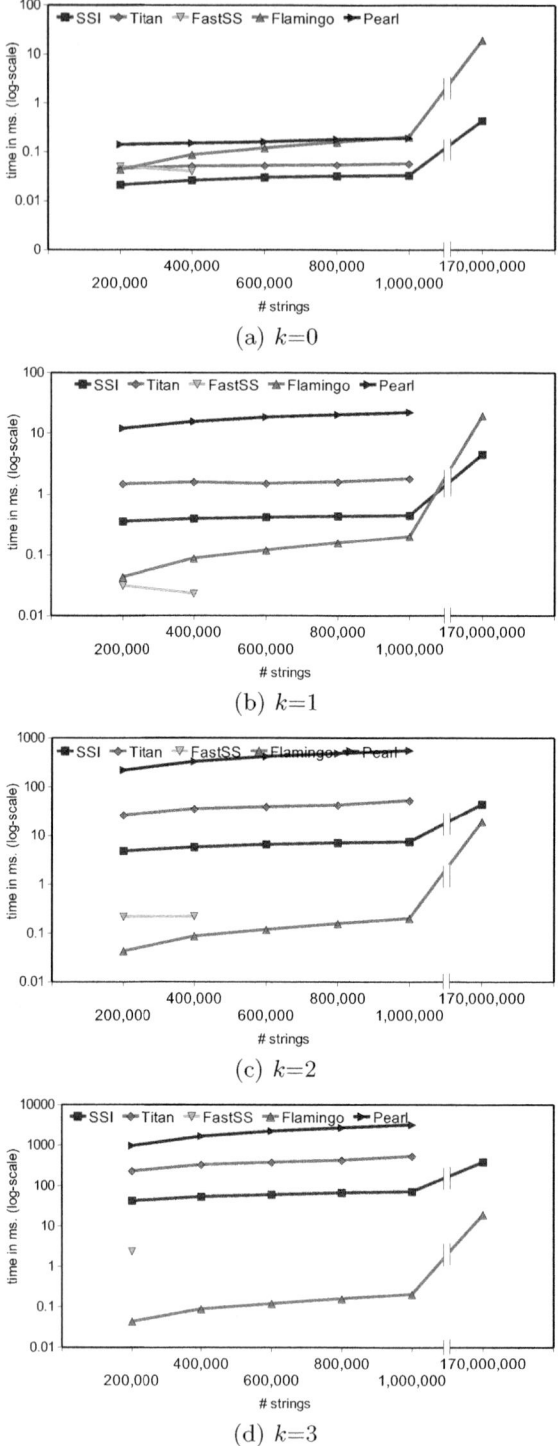

Fig. 5. Average query execution time in milliseconds (log-scale)

5.3   Query Answering

To evaluate the performance of SSI in query answering, we assembled a set of 1,000 example queries separately for each data set as follows: First, we randomly selected 950 strings from the respective data set and kept 500 of these strings unchanged. On the remaining 450 strings, we introduced errors by randomly changing or deleting one character per string. Additionally, we generated 50 random strings and added them to the set of queries. For each query, we measured the execution time and report the average of all 1,000 queries. We compared SSI to all above-mentioned tools both for exact and similarity-based queries with varying edit distance thresholds $k \in \{0, 1, 2, 3\}$. For all search experiments, indexing was performed in advance and is not included in the measured times.

For exact queries, SSI outperformed all competitors independent of the data set size (see Fig. 5(a)). Specifically, SSI outperformed FastSS with factor 2.3 on D_{200k} and factor 1.5 on D_{400k}, TITAN with factors varying between 1.6 on D_{800k} and 2.1 on D_{200k}, Pearl with factors varying between 5.3 on D_{600k} and 6.6 on D_{200k}, and Flamingo with factors from 2.0 on D_{200k} to 44.1 on D_{full}.

As displayed in Fig. 5(b – d), SSI significantly outperforms the trie- and NFA-based tools TITAN and Pearl on edit distance based queries. Using an edit distance threshold of $k = 1$, SSI outperforms TITAN with a factor of 4, using $k = 3$, SSI is 5.4 to 7.4 times faster than TITAN depending on the data set. Compared to Pearl, SSI is faster by more than one order of magnitude, independent of the data set and the edit distance thresholds. However, on comparatively small data sets (D_{200k}, D_{400k}), FastSS is by an order of magnitude faster than SSI. This observation needs to be put into perspective, since FastSS on the one hand needs to create a separate index for each k, and creating indexes with more than $400,000$ strings was not possible using FastSS. In contrast, SSI does not have these limitations.

Furthermore, we acknowledge that Flamingo, which has a different indexing and search approach (cf. Sec. 2), is significantly faster than SSI in many situations. For searches in D_{full} with $k = 0$ and $k = 1$, SSI was faster by a factor of 4.2, in all other situations, Flamingo outperformed SSI. Recall that Flamingo uses considerably more memory than SSI for indexing D_{full} to achieve this (cf. Fig. 4(a)). We also clearly observe that the advantages of Flamingo grow the larger edit distance thresholds get. However, future improvements of SSI could directly address this issue, e.g., by integrating bit-parallel edit distance computation methods which provide a fast edit distance computation that is independent of the chosen threshold k.

6   Conclusion

In this paper, we presented the State Set Index (SSI), a solution for fast similarity search in very large string sets. By configuring SSI's parameters, we can scale the index size allowing best search performance given memory requirements. Our experiments on a very large real-world string data set showed that SSI significantly outperforms current state-of-the-art approaches for string similarity search with small distance thresholds.

References

1. Aghili, S.A., Agrawal, D.P., El Abbadi, A.: BFT: Bit Filtration Technique for Approximate String Join in Biological Databases. In: Nascimento, M.A., de Moura, E.S., Oliveira, A.L. (eds.) SPIRE 2003. LNCS, vol. 2857, pp. 326–340. Springer, Heidelberg (2003)
2. Behm, A., Vernica, R., Alsubaiee, S., Ji, S., Lu, J., Jin, L., Lu, Y., Li, C.: UCI Flamingo Package 4.0 (2011)
3. Bocek, T., Hunt, E., Stiller, B.: Fast Similarity Search in Large Dictionaries. Technical report, Department of Informatics, University of Zurich (2007)
4. Celikik, M., Bast, H.: Fast error-tolerant search on very large texts. In: Proc. of the ACM Symposium on Applied Computing (SAC), pp. 1724–1731 (2009)
5. Fickett, J.W.: Fast optimal alignment. Nucleic Acids Research 12(1), 175–179 (1984)
6. Fredkin, E.: Trie memory. Commun. of the ACM 3, 490–499 (1960)
7. Grahne, G., Zhu, J.: Efficiently using prefix-trees in mining frequent itemsets. In: Proc. of the ICDM Workshop on Frequent Itemset Mining Implementations (2003)
8. Gravano, L., Ipeirotis, P.G., Jagadish, H.V., Koudas, N., Muthukrishnan, S., Srivastava, D.: Approximate string joins in a database (Almost) for free. In: Proc. of the Intl. Conf. on Very Large Databases (VLDB), pp. 491–500. Morgan Kaufmann (2001)
9. Gravano, L., Ipeirotis, P.G., Koudas, N., Srivastava, D.: Text joins in an RDBMS for web data integration. In: Proc. of the Intl. World Wide Web Conf. (WWW), pp. 90–101 (2003)
10. Gusfield, D.: Algorithms on Strings, Trees and Sequences: Computer Science and Computational Biology. Cambridge University Press (1997)
11. Han, J., Pei, J., Yin, Y., Mao, R.: Mining frequent patterns without candidate generation: A Frequent-Pattern tree approach. Data Mining and Knowledge Discovery 8(1) (2004)
12. Jampani, R., Pudi, V.: Using Prefix-Trees for Efficiently Computing Set Joins. In: Zhou, L., Ooi, B.-C., Meng, X. (eds.) DASFAA 2005. LNCS, vol. 3453, pp. 761–772. Springer, Heidelberg (2005)
13. Levenshtein, V.I.: Binary codes capable of correcting deletions, insertions and reversals. Soviet Physics Doklady (1966)
14. Li, C., Lu, J., Lu, Y.: Efficient merging and filtering algorithms for approximate string searches. In: Proc. of the Intl. Conf. on Data Engineering (ICDE), pp. 257–266. IEEE Computer Society (2008)
15. Liu, X., Li, G., Feng, J., Zhou, L.: Effective indices for efficient approximate string search and similarity join. In: Proc. of the Intl. Conf. on Web-Age Information Management, pp. 127–134. IEEE Computer Society (2008)
16. Morrison, D.R.: PATRICIA – practical algorithm to retrieve information coded in alphanumeric. Journal of the ACM 15(4), 514–534 (1968)
17. Myers, E.: A sublinear algorithm for approximate keyword searching. Algorithmica 12, 345–374 (1994)
18. Myers, G.: A fast bit-vector algorithm for approximate string matching based on dynamic programming. Journal of the ACM 46(3), 395–415 (1999)
19. Navarro, G.: A guided tour to approximate string matching. ACM Computing Surveys 33(1) (2001)
20. Navarro, G., Baeza-Yates, R., Sutinen, E., Tarhio, J.: Indexing methods for approximate string matching. IEEE Data Engineering Bulletin 24, 2001 (2000)

21. Rabin, M.O., Scott, D.: Finite automata and their decision problems. IBM J. Res. Dev. 3, 114–125 (1959)
22. Rheinländer, A., Knobloch, M., Hochmuth, N., Leser, U.: Prefix Tree Indexing for Similarity Search and Similarity Joins on Genomic Data. In: Gertz, M., Ludäscher, B. (eds.) SSDBM 2010. LNCS, vol. 6187, pp. 519–536. Springer, Heidelberg (2010)
23. Rheinländer, A., Leser, U.: Scalable sequence similarity search in main memory on multicores. In: International Workshop on High Performance in Bioinformatics and Biomedicine, HiBB (2011)
24. Sahinalp, S.C., Tasan, M., Macker, J., Ozsoyoglu, Z.M.: Distance based indexing for string proximity search. In: Proc. of the Intl. Conf. on Data Engineering (ICDE), pp. 125–136 (2003)
25. Shang, H., Merrett, T.: Tries for approximate string matching. IEEE Transactions on Knowledge and Data Engineering (TKDE) 8, 540–547 (1996)
26. Vintsyuk, T.K.: Speech discrimination by dynamic programming. Cybernetics and Systems Analysis 4, 52–57 (1968)
27. Wang, W., Xiao, C., Lin, X., Zhang, C.: Efficient approximate entity extraction with edit distance constraints. In: Proc. of the ACM Intl. Conf. on Management of Data (SIGMOD), pp. 759–770 (2009)
28. Xiao, C., Wang, W., Lin, X.: Ed-join: an efficient algorithm for similarity joins with edit distance constraints. Proc. of the VLDB Endowment 1, 933–944 (2008)

Substructure Clustering: A Novel Mining Paradigm for Arbitrary Data Types

Stephan Günnemann, Brigitte Boden, and Thomas Seidl

RWTH Aachen University, Germany
{guennemann,boden,seidl}@cs.rwth-aachen.de

Abstract. Subspace clustering is an established mining task for grouping objects that are represented by vector data. By considering subspace projections of the data, the problem of full-space clustering is avoided: objects show no similarity w.r.t. all of their attributes but only w.r.t. subsets of their characteristics. This effect is not limited to vector data but can be observed in several other scientific domains including graphs, where we just find similar subgraphs, or time series, where only shorter subsequences show the same behavior. In each scenario, using the whole representation of the objects for clustering is futile. We need to find *clusters of similar substructures*. However, none of the existing substructure mining paradigms as subspace clustering, frequent subgraph mining, or motif discovery is able to solve this task entirely since they tackle only a few challenges and are restricted to a specific type of data.

In this work, we unify and generalize existing substructure mining tasks to the novel paradigm of substructure clustering that is applicable to data of an arbitrary type. As a proof of concept showing the feasibility of our novel paradigm, we present a specific instantiation for the task of *subgraph clustering*. By integrating the ideas of different research areas into a novel paradigm, the aim of our paper is to inspire future research directions in the individual areas.

1   Introduction

Clustering – the grouping of similar objects and separation of dissimilar objects – is one of the fundamental data mining tasks for the analysis of scientific data. For decades a multitude of clustering algorithms were introduced to handle different types of data including traditional vector data, graph data, time series and many more. Considering the domain of vector data, it is well known that in many cases not all dimensions are relevant for grouping the objects [12]: the similarity of objects within a cluster is restricted to subsets of the dimensions. Ignoring this fact and using all dimensions for clustering, so-called full space clustering approaches cannot identify the hidden clusters in such data. As a solution, subspace clustering methods were developed [12], which identify for each group of objects an individual relevant subspace in which the objects are similar. In Figure 1 two subspace clusters, which do overlap in some objects, are shown.

A. Ailamaki and S. Bowers (Eds.): SSDBM 2012, LNCS 7338, pp. 280–297, 2012.

The above mentioned phenomenon is not restricted to the domain of vector data but can be observed in many scientific databases: often, objects show dissimilarity with respect to their whole characteristics,

dim. 1		$\lceil 7 \rceil$	$\lceil 8 \rceil$	$\lceil 6 \rceil$	$\lceil 6 \rceil$	$\lceil 6 \rceil$	$\lceil 6 \rceil$	C_2
dim. 2		1	5	4	4	4	4	
dim. 3	C_1	2	2	2	2	7	9	
dim. 4		3	3	3	3	8	5	

⎫ substructures in
⎭ subspace {1,2}

⎫ substructures in
⎭ subspace {3,4}

Fig. 1. 4-d database with 2 subspace clusters

thus leading to no clusters, but high similarity with respect to their substructures, resulting in meaningful *substructure* clusters. An example for the domain of graph databases is illustrated in Figure 2. The given graphs are not similar to each other. However, we can find multiple meaningful clusters that contain *similar subgraphs*. These subgraphs do not need to be isomorphic but they are clustered according to their similarity. This problem, however, cannot be solved by any of the existing methods known in the literature.

In this work, we unify and generalize existing substructure mining tasks for special data types to the novel paradigm of substructure clustering able to handle data of an arbitrary type. Thus, we develop a model to find *meaningful clusters of similar substructures*. This is useful for a broad range of scientific applications: Clustering similar subgraphs, as e.g. shown in Figure 2, can be used for the analysis of bio-chemical compounds and protein structure databases. Homology detection or structural alignments can benefit from clusters of similar subgraphs. The use of similar subgraphs for subsequent tasks, including efficient indexing [26] and the storage in semi-structured databases [4], is also possible. In network log data, similar subpatterns representing typical intrusions can be used for future anomaly detection. Also for time series, substructure clustering is beneficial. Mining similar recurrent subsequences can be used as a stand-alone tool for the analysis of climate data and stock-data, or as a subroutine for tasks including summarization [18] and rule discovery [11]. A cluster of similar subsequences is shown e.g. in Figure 3. Five similar subsequences are detected in the database: in different times series and at different points in time.

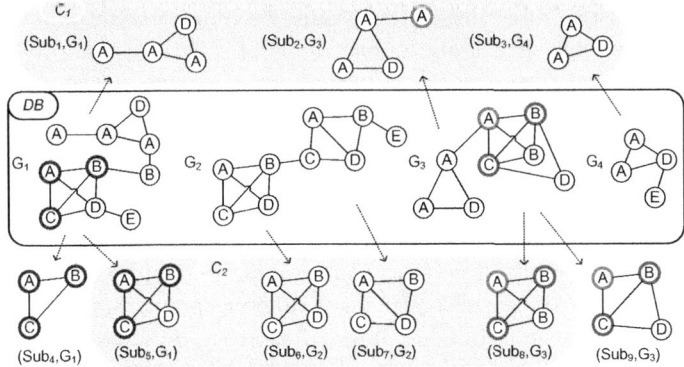

Fig. 2. Database with four graphs and two valid substructure clusters

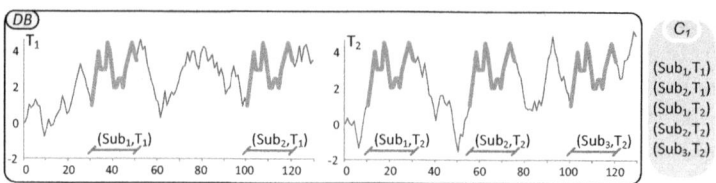

Fig. 3. Substructure clustering for time series data

Overall, clustering similar substructures – not just identical, isomorphic, or equally sized substructures – is highly relevant. Therefore, we study this problem in this work. By joining the principles of different research areas into a novel paradigm, we try to encourage future works in the individual areas.

1.1 Challenges in Substructure Clustering

Substructure clustering is inspired by subspace clustering, which, however, is restricted to vector data. Besides this difference w.r.t. the data type, we face several novel challenges for the task of substructure clustering.

From Subspace Projections to Substructures. We first have to analyze the notion of 'substructure' on its own. In traditional subspace clustering, substructures are obtained by selecting specific dimensions. Given a subspace, *each* object can be projected to it. In Figure 1 for example we can select the subspace $\{1, 2\}$ and for each object we can determine the corresponding substructure. Since the dimension 1 of one object refers to *the same characteristic/attribute* as dimension 1 of another object, this is a meaningful semantics. The notion of subspaces, however, is only valid in vector spaces and thus cannot be directly transferred to other data spaces. Naively, in graph databases we could extract from each graph e.g. the vertices $\{1, 2, 3\}$ to determine the substructures. This extraction, however, is usually not meaningful since the numbering of vertices within each graph can often be chosen arbitrarily. In general, vertex 1 of one graph *does not* refer to the same characteristic as vertex 1 of another graph.

Thus, for substructure clustering we must be able to *select and compare* arbitrary substructures. In Figure 2, subgraphs of different structure and different size are extracted and potentially grouped together in the clusters C_1 or C_2. Consequently, we need a distance measure able to determine the similarity between arbitrary substructures since clustering is based on similarity values. While in subspace clustering only substructures of the same subspace are compared – measuring similarity between objects projected to different subspaces is not possible –, such a restriction is not suitable for substructure clustering: For example, a subgraph with 4 vertices can be compared with one containing just 3 vertices. For time series, subsequences at different points in time and of different length can be clustered. Thus, novel challenges for substructure clustering include: the definition of valid substructures and the specification of distance measures between these substructures.

Redundancy within the Clusters. Assuming that we are able to measure the similarity between arbitrary substructures (and since we do not require isomorphism), several further challenges are posed. If two substructures x and y are similar to each other (as, e.g., the subgraphs Sub_5 and Sub_6 in Figure 2), we will often also get a high similarity between x and y's substructures. For example, the subgraphs Sub_4, Sub_5, and Sub_6 are pairwise similar. As another example, Sub_8 and Sub_9 are very similar. For time series, this effect is related to the 'trivial matches' [27]. Due to the general definition of clusters, i.e. similar objects are grouped together, in Figure 2 we should group the *three* substructures (Sub_4, Sub_5, Sub_6) in a single cluster. However, the subgraphs Sub_4 and Sub_5 represent nearly the same information from the original graph G_1; the cluster would contain highly redundant information. This redundancy *within* a single cluster should be avoided since it hinders the interpretation of the clustering result. We should remove either Sub_4 or Sub_5 from the cluster. This redundancy holds also between Sub_8 and Sub_9; only one of these subgraphs should be included in the cluster.

Nevertheless, it should be possible to include several substructures of a single original object within a single cluster. In Figure 3, the recurrent subsequence occurs twice in the left time series. In Figure 2, both subgraphs Sub_6 and Sub_7 belong to the graph G_2 and are included in cluster C_2. Since both subgraphs cover completely different vertices of G_2, they represent novel information and thus do not indicate redundancy. Thus, for substructure clustering *we have to distinguish between similarity/isomorphism and redundancy* of substructures. If two substructures are similar to each other, they are not necessarily redundant w.r.t. each other.

The redundancy *within clusters* is an entirely new aspect substructure clustering models have to cope with. This problem cannot occur in subspace clustering since different subspace projections of the same object never belong to the same cluster. However, in other domains such as graphs or time series we have to account for this problem. Thus, novel challenges for substructure clustering include: the handling of redundancy within clusters and the definition of appropriate cluster models.

Redundancy between the Clusters. In substructure clustering we face a further redundancy problem: different clusters can represent similar information. In Figure 4 e.g. we can get the clusters C_1 and C_2. Considering each cluster individually we do not have redundancy. *Between* both clusters, however, a high redundancy becomes apparent: the cluster C_1 contains just substructures of the cluster C_2. Such a redundancy between different clusters should be avoided.

This phenomenon is known in traditional subspace clustering where *clusters in similar subspaces* contain similar object sets. A frequently used solution is to consider the intersection of the (original) objects which appear in both clusters. If this intersection is too high, one of the clusters will be redundant [15,8,6].

This solution, however, fails for substructure clustering. In Figure 4 for example each cluster would be classified as redundant w.r.t. every other cluster as all of them represent the same original graphs. However, for the cluster C_4 this

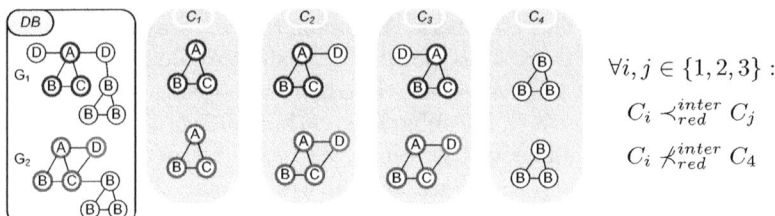

Fig. 4. Redundancy between different clusters

would be incorrect. Though the cluster also represents the same original graphs, it identifies a completely novel pattern in the data and hence is not redundant. This misclassification occurs since in substructure clustering we do not have information like similar subspaces that can be utilized between different clusters for meaningful redundancy removal. Thus, novel challenges for substructure clustering include: the redundancy removal between clusters and the identification of properties the final clustering has to fulfill.

In the next section we develop a general model for substructure clustering that accounts for the introduced challenges and is applicable for arbitrary data types. A comparison with existing paradigms is done in Section 3. We introduce an algorithm for the instantiation of our model for subgraph clustering in Section 4 and evaluate it in Section 5.

2   Substructure Clustering

Our general model for substructure clustering is introduced in three steps: In Section 2.1 we define the actual substructures to be clustered, in Section 2.2 the definition of single clusters is presented, and Section 2.3 introduces the overall clustering. In addition to these general definitions, we define a specific model for the task of *subgraph* clustering which acts as a proof-of-concept. More advanced subgraph clustering approaches are left for future work.

2.1   Substructure Definition

Defining valid (sub-)structures is highly application dependent. To introduce a unifying approach handling arbitrary data types, we therefore use the general notion of 'structures' as analyzed, e.g, in the field of model theory and type theory [10]. A (single-typed) structure is a triple $\mathcal{A} = (A, \sigma, I)$ consisting of a universe A, a signature σ, and an interpretation function I. The signature σ defines a set of function symbols and relations symbols (together with their arity) that can be used in this structure. The interpretation function I assigns actual functions and relations to the symbols of the signature. Let us consider the example of a graph database: Each graph $G \in DB$ can be described by a structure $G = (V, \sigma, I)$, where σ consists of the binary edge relation symbol R_E. The actual edges of a graph are specified by the interpretation function. A complete graph, for example, is given by the interpretation $I(R_E) = V \times V$, while a graph with only self-loops is given by $I(R_E) = \{(v,v) \mid v \in V\}$. As seen, two graphs (of the same type) differ only in their universe and their interpretation function but the signature is identical.

Since the above definition of structures uses just a single universe, they are known as single-typed structures. In real world applications, however, it is often useful to distinguish between multiple universes (also denoted as types): a labeled graph, for example, is defined based on the universe V of vertices and the universe Σ of node labels. To handle such scenarios, single-typed structures were extended to many-typed structures[1] that use a *set* of universes [10]. By using many-typed structures, we have the potential to define functions and relations over different types of elements. For example, a node labeling function is a function $l : V \to \Sigma$. Formally, we have:

Definition 1 (Many-typed signature and many-typed structure).
A many-typed signature σ is a 4-tuple $\sigma = (f, R, T, type)$ consisting of a set of function symbols f, a set of relation symbols R, a set of type symbols T, and a function $type : f \cup R \to (T \cup \{\times, \to\})^*$ that maps each function and relation symbol to a specific type (encoded as words over the alphabet $T \cup \{\times, \to\}$). As abbreviation, we denote with $type_i(x)$ the type of the ith argument of function/relation x. The sets f, R, T, and $\{\times, \to\}$ are pairwise disjoint.

A many-typed structure S is a triple $S = (U, \sigma, I)$ consisting of a set of universes $U = \{U_t\}_{t \in T}$, a many-typed signature σ, and an interpretation function I for the functions and relations symbols given in σ.[2]

Consider the example of labeled graphs: The signature σ might be given by $\sigma = (\{f_l\}, \{R_E\}, \{N, L\}, type)$. Here, $\{N, L\}$ denotes the different types 'nodes' and 'labels'. With f_l we describe the node labeling function; thus, $type(f_l) = N \to L$ (and $type_1(f_l) = N$). With R_E we describe the edge relation; thus, we have $type(R_E) = N \times N$ (and $type_1(R_E) = type_2(R_E) = N$). A labeled graph is now given by the structure $G = (\{U_N, U_L\}, \sigma, I)$ with vertex set $V = U_N$ and label alphabet $\Sigma = U_L$. Assuming that all vertices of the graph are labeled with $a \in \Sigma$, we have the interpretation $I(f_l) : V \to \Sigma, v \mapsto a$. Again, two different labeled graphs differ only in their used universes and/or their interpretation function but they are based on the same signature σ.

Using many-typed structures allows us to represent a multitude of data types analyzed in scientific domains including graphs, time series, vector spaces and strings. Based on this general definition of structures we are now also able to formalize the notion of substructures:

Definition 2 (Substructure).
Given two many-typed structures $S = (\{U_t\}_{t \in T}, \sigma, I_S)$ and $S' = (\{U'_t\}_{t \in T}, \sigma, I_{S'})$ over the same signature $\sigma = (f, R, T, type)$. S is an induced substructure of S' (denoted as $S \subseteq S'$) if

---

[1] Alternatively, also denoted as many-sorted structures.

[2] Note: We omit the formal definition of the interpretation function since it is straightforward to derive based on the function *type*.

- $\forall t \in T : U_t \subseteq U'_t$
- $\forall f_i \in f : I_{\mathcal{S}}(f_i) = I_{\mathcal{S}'}(f_i)|_X$ *where* $X = Dom(I_{\mathcal{S}}(f_i))$
- $\forall R_i \in R : I_{\mathcal{S}}(R_i) = I_{\mathcal{S}'}(R_i) \cap X$ *where* $X = Ground(I_{\mathcal{S}}(R_i))$

With $Dom(I_{\mathcal{S}}(f_i)) = U_{type_1(f_i)} \times \ldots \times U_{type_n(f_i)}$ *for every n-ary function symbol and* $Ground(I_{\mathcal{S}}(R_i)) = U_{type_1(R_i)} \times \ldots \times U_{type_n(R_i)}$ *for every n-ary relation symbol.*

For the example of graph data, the induced substructures correspond to the induced subgraphs: By considering a subset of vertices $V_A \subseteq V_B$, for the edge relation it has to hold $I_A(R_E) = I_B(R_E) \cap (V_A \times V_A)$ and the labels are simply taken over, i.e. $I_A(f_l) = I_B(f_l)|_{V_A}$. For practical applications, the substructure relation can further be restricted by additional constraints. Based on the definition of substructures, we formalize the substructure database. To enable the redundancy elimination later on, we have to store (besides the actual substructure) also the original structure from which the substructure is extracted. Thus, the substructures to be clustered in our case consist of tuples (Sub, O).

Definition 3 (Substructure database). *Given a database DB of structures over the same many-typed signature σ, the substructure database is defined by* $SubDB = \{(Sub, O) \mid Sub \subseteq O \in DB\}.$

Using this definition, two *isomorphic but not identical* substructures Sub_1 and Sub_2 are two different tuples in the database; even if both originate from the same original structure. In Figure 5 the three isomorphic subgraphs are three different tuples in the substructure database. This is meaningful since the *universes* of the substructures, e.g. their vertices, are not identical. For correctly removing redundancy, this distinction is highly relevant.

2.2 Cluster Definition

A substructure cluster C is defined as a subset $C \subseteq SubDB$, fulfilling certain characteristics. First, we have to ensure the similarity of the clustered elements. Formally, we use a distance function $dist : SubDB \times SubDB \to R_0^+$ that measures the dissimilarity between two arbitrary substructures. For our instantiation in subgraph clustering, we use the graph edit distance [19], which is an established similarity measure for graphs.

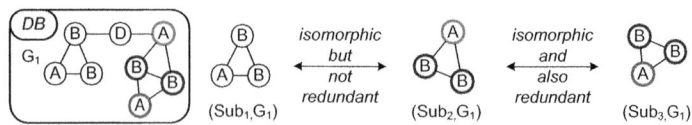

Fig. 5. Isomorphism/similarity vs. redundancy

A novel aspect is the redundancy within clusters, which has to be avoided. Keep in mind that two substructures are *not redundant* w.r.t. each other if they originate from different original structures or if they were extracted from different 'regions' of the same structure. In both cases, they represent different patterns. Only if two substructures were extracted from 'similar regions' of the same original structure, they should be denoted as redundant. In Figure 5, the subgraphs Sub_2 and Sub_3 can be denoted as redundant w.r.t. each other while Sub_2 and Sub_1 are not redundant to each other. The informal notion of 'similar regions' is represented by the structures' universes. For graphs the universe corresponds to their vertex set, whereas e.g. for time series we have the individual points in time. If two substructures S_1 and S_2 are similar w.r.t. their universes, we will get redundancy. We denote this with $S_1 \prec_{red}^{intra} S_2$ and define:

Definition 4 (Intra-cluster redundancy relation).
Let $(\mathcal{S}, G), (\mathcal{S}', G') \in SubDB$ with $\mathcal{S} = (U, \sigma, I)$ and $\mathcal{S}' = (U', \sigma, I')$, and sim be a similarity function on the set of universes U and U'.
The relation $\prec_{red}^{intra} \subseteq SubDB \times SubDB$ describes the redundancy between substructures of a single cluster and is defined by: $(\mathcal{S}, G) \prec_{red}^{intra} (\mathcal{S}', G') \Leftrightarrow$
$G = G' \wedge sim(U, U') \geq r_{intra}$ with redundancy parameter $r_{intra} \in [0, 1]$
***For subgraph clustering** we use:*
$sim(U, U') = \frac{|V \cap V'|}{|V \cup V'|}$ where $U = \{V, \Sigma\}$ and $U' = \{V', \Sigma\}$.

Since we consider the Jaccard similarity coefficient between the actual vertex sets, two very similar subgraphs (w.r.t. *dist*) need not to be redundant: for substructure clustering we have to distinguish between similarity/isomorphism and redundancy. Sub_1 and Sub_2 in Figure 5 are isomorphic but non-redundant. Only if the fraction of overlapping vertices exceeds a critical value, two structures will be redundant w.r.t. each other. Thus, in general we permit overlapping substructures, guided by the parameter r_{intra}. In our instantiation we just use the first universe, i.e. the vertex sets, to compute the substructure similarity. In general, however, all universes of the given structures might be used. The overall cluster definition can now be formalized by:

Definition 5 (Substructure cluster). *Let $SubDB$ be a substructure database and dist a substructure distance function. A valid substructure cluster $C \subseteq SubDB$ fulfills*

- *similarity: $\forall (\mathcal{S}, G), (\mathcal{S}', G') \in C : dist(\mathcal{S}, \mathcal{S}') \leq \alpha$*
- *redundancy-freeness: $\neg \exists (\mathcal{S}, G), (\mathcal{S}', G') \in C : (\mathcal{S}, G) \prec_{red}^{intra} (\mathcal{S}', G')$*
- *minimal support: $|C| \geq minSupport$*

Besides being redundancy free, a cluster should only contain similar substructures. Thus, the parameter α constrains the maximal pairwise distance between the clustered substructures. Furthermore, a cluster must be sufficiently large, indicated by the minimal support. The group C_2 in Figure 2 is a valid substructure cluster since the subgraphs are pairwise similar and non-redundant. If we added

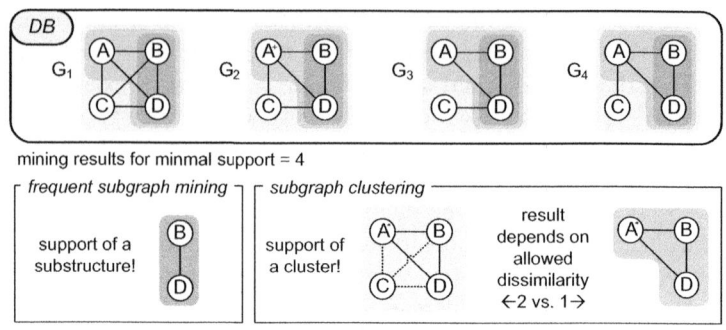

Fig. 6. Different notions of minimal support

Sub_9, however, the redundancy-free property would be violated due to Sub_8 (if, e.g., $r_{intra} \leq \frac{3}{5}$). Note that C_2 contains two non-redundant substructures of the graph G_2.

In the following, we highlight the minimal support: For the (existing) tasks of frequent substructure mining (e.g. frequent or representative subgraph mining [28] for graphs) the minimal support *of a substructure* is considered. *Each* substructure has to exceed a certain minimal support, i.e. several isomorphic substructures have to exist. In our approach, however, none of the substructures need to be frequent – there do not need to exist isomorphic substructures –; only the *whole collection* of similar substructures has to be large enough. We consider the minimal support *of a cluster*. For the example of graphs, in Figure 6 four very similar graphs are illustrated (please note the different vertex label A^+ in G_2). If we select a minimal support of 4 and $\alpha = 2$, our approach will successfully group all graphs into a single cluster. The support of the cluster is large enough. Frequent subgraph mining, however, cannot detect this group since the graphs are not isomorphic. The only frequent and isomorphic substructure is the illustrated subgraph with just 2 vertices. This pattern does not describe the clustering structure of the data very well. Even if we lower α to 1, the detected cluster by our approach will contain more meaningful patterns. By measuring similarity we achieve a further advantage: If we consider continuous valued data, the equality/isomorphism of two substructures will be virtually impossible. While previous methods are not able to detect any (frequent) pattern, our model handles such data.

2.3   Clustering Definition

Using Def. 5 we can determine the set of all valid clusters $Clusters \subseteq \mathcal{P}(SubDB)$. This set, however, contains clusters highly redundant w.r.t. each other. The next challenge we have to solve is to avoid redundancy between different clusters by selecting a clustering result $Result \subseteq Clusters$ that does not contain redundant information. Considering the cluster C_1 in Figure 4, its grouped substructures occur similarly also in the cluster C_2. The same holds between C_2 and C_3. Both clusters cover nearly the same information. Their substructures, however, are not identical; thus, intersecting the clusters is not reasonable. Instead we again have to consider the universes of the structures to also detect redundancy between clusters. We resort to the relation $\prec_{red}^{intra}$ to identify those substructures of C_3 that are redundant to at least one substructure of C_2.

Definition 6 (Cover). *Given two substructure clusters C_1, C_2, the covered substructures of C_2 due to C_1 are defined by:*
$$covered(C_2|C_1) = \{(\mathcal{S}, G) \in C_2 \mid \exists (\mathcal{S}', G') \in C_1 : (\mathcal{S}, G) \prec_{red}^{intra} (\mathcal{S}', G')\}$$

The *covered* set of a cluster denotes the whole set of redundant substructures w.r.t. another cluster without enforcing the equality of the substructures. Based on this set, the redundancy relation between two clusters C_1 and C_2 can be defined, i.e. $C_1 \prec_{red}^{inter} C_2$ if a large fraction of substructures occurs (similarly) in both clusters.

Definition 7 (Inter-cluster redundancy relation). *Let $C_1, C_2 \in Clusters$ and sim_2 be a similarity function on $Clusters$. The redundancy relation $\prec_{red}^{inter} \subseteq Clusters \times Clusters$ describes the redundancy between different clusters and is defined by: $C_1 \prec_{red}^{inter} C_2 \Leftrightarrow sim_2(C_1, C_2) \geq r_{inter}$ with redundancy parameter $r_{inter} \in [0,1]$*
For subgraph clustering we use: $sim_2(C_1, C_2) = \frac{|covered(C_2|C_1)| + |covered(C_1|C_2)|}{|C_1| + |C_2|}$

In Figure 4, the clusters C_1, C_2 and C_3 are pairwise redundant. The cluster C_4, however, is not redundant to any of the other clusters although substructures from the same original graphs are grouped.

Using this relation we can select a set $Result \subseteq Clusters$ containing only pairwise non-redundant clusters. Moreover, we want to select the *most interesting* redundancy-free clustering. For example, in Figure 4 we could prefer clustering $\{C_2, C_4\}$ to $\{C_1, C_4\}$ since larger subgraphs are grouped. Thus, in our general model we need to judge the quality of each clustering:

Definition 8 (Quality of a substructure clustering).
Given a substructure clustering $Result \subseteq Clusters$, the quality is defined by the function: $q : \mathcal{P}(Clusters) \to \mathbb{R}$
For subgraph clustering we use:
$$q(Result) = \sum_{C \in Result} \frac{1}{|C|} \sum_{(\mathcal{S} = (\{V_S, \Sigma\}, \sigma, I), G_i) \in C} |V_S|$$

The quality function enables us to determine the interestingness of a clustering in a flexible way. In our instantiation we sum over the average subgraph sizes for each cluster. Thus, we prefer clusters containing large substructures, which often allow better interpretation. Overall, we are now able to define the optimal substructure clustering that accounts for all aforementioned challenges:

Definition 9 (Optimal substructure clustering). *Given the set $Clusters$ of all substructure clusters, the optimal substructure clustering $Result \subseteq Clusters$ fulfills*

(1) redundancy-freeness: $\neg \exists C_1, C_2 \in Result : C_1 \prec_{red}^{inter} C_2$
(2) maximal quality: $\forall Result' \subseteq Clusters$ fulfilling (1): $q(Result') \leq q(Result)$.

3   Related Work

Substructure clustering generalizes existing substructure mining tasks that were introduced for specific data types.

Vector Data. In traditional clustering, the objects are grouped based on the similarity of *all* their attributes. As in high-dimensional data it is very unlikely that the objects of a cluster are similar w.r.t. all attributes [2], subspace clustering approaches [12,15,8,7] were introduced which cluster objects based on *subsets of their attributes*. This can also be seen as clustering 'sub-objects', where a sub-object consists of a subset of the attributes of an original object. However, the sub-objects within a single cluster have to lie in exactly the same subspace. In our approach this can be realized by restricting the distance function to objects of the same subspace, i.e. $dist(.,.)=\infty$ for substructures located in different subspaces. Our substructure clustering is more general than subspace clustering, as it is able to cluster objects from different subspaces together. This is especially useful if objects show missing values in some dimensions or erroneous values are present. These errors can be suppressed by grouping based on different subspaces.

Graph Data. An overview of the various graph mining tasks is given in [1]. The term 'graph clustering' is somewhat ambiguous: Besides finding *clusters of similar graphs* where the input consists of a set of graphs [23], it also refers to clustering the vertices of a single large graph based on their density [20]. Our paradigm varies from both approaches as it finds clusters of similar *sub*structures.

Another graph mining task is *frequent subgraph mining* (FSM) that aims at finding subgraphs in a graph database whose frequencies (number of isomorphic subgraphs) exceed a given threshold. Since the whole set of freq. subgraphs is often too large to enumerate, mining closed [25] or maximal [22] frequent subgraphs was introduced. These results still contain high redundancy whereas substructure clustering avoids this. To get smaller result sets, representative FSM [28] chooses a set of frequent subgraphs such that each frequent subgraph is 'represented' by one of these. The FSM methods use graphs with categorical labels for finding frequent subgraphs. However, if some errors in the data occur, e.g. edges are missing, or if the data is continuous valued, it is very unlikely to find a large number of isomorphic subgraphs. However, if no frequent subgraphs exist, no representative subgraphs can be found. In substructure clustering we first cluster similar subgraphs together, and then retain a cluster if its support is high enough. Therefore, substructure clustering can find meaningful clusters even if there are no single frequent subgraphs in the data. Consequently, we cannot use (representative) FSM to solve substructure clustering; but substructure clustering subsumes (representative) FSM.

Transaction Data. Frequent Itemset Mining [9] is the task of determining *subsets of items* that occur together frequently in a database of transactions. To reduce the huge output if all frequent itemsets are determined, closed and maximal mining approaches were introduced [5]. The result, however, still contains high redundancy between the itemsets; substructure clustering avoids this redundant information. Summarizing itemsets [24] can be seen as clustering frequent itemsets, to reduce the number of generated patterns and thus the redundancy. However, these approaches still use the strict notion of frequent itemsets which correspond to isomorphic substructures. In contrast, we group based on similarity values. Approaches for *error-tolerant* itemset mining generalize the definition of frequent itemsets to tolerate errors in the data [17]. These approaches consider an itemset as frequent if there exist many transactions in the database that contain 'most' of the items in the set; this corresponds to an oversimplified definition of similarity between itemsets. For substructure clustering we can use arbitrary distance functions between the substructures.

Time Series Data. Another related task is the *subsequence* clustering, which aims at finding sets of similar subsequences in a database of time series or within a single streaming time series [14,3]. As stated in [14], naive subsequence clustering is meaningless if not for certain conditions: trivial matches have to be eliminated and only subsets of the data should be grouped, i.e. not any subsequence should belong to a cluster. Both properties are fulfilled by substructure clustering: trivial matches are avoided with the intra-cluster redundancy and clusters are only generated by similar substructures. Thus, substructure clustering can be used to find meaningful subsequence clusters. Related to subsequence clustering is *motif discovery* [27], which detects approximately repeated subsequences. Both paradigms, however, are restrained to group subsequences with the same length together. Even more, the user has to set the length w as a parameter. In our substructure clustering model, subsequences of different length can be grouped together if they are similar (and non-redundant).

Specialized Data. Aside from the previous data types, substructure clustering for a very specialized data type is introduced by [21]. The sCluster approach finds clusters of similar substructures in sequential 3d objects, which are sets of points located in a three dimensional space and forming a sequence. sCluster only groups substructures of the same length together and redundancy is not removed from the clustering result. Substructure clustering overcomes these drawbacks and is applicable to arbitrary data types.

Overall, substructure clustering generalizes all presented paradigms and unifies them into a single consistent model.

4   An Algorithm for Subgraph Clustering

In the following we present an overview of our algorithm SGC that implements the instantiation of our substructure clustering model for the domain of subgraph mining. Again, our aim is to demonstrate the principle of substructure clustering by implementing a proof-of-concept.

Determining the exact optimal subgraph clustering is not efficiently possible. Obviously, our model is NP-hard since it subsumes (cf. Sec. 3) the NP-hard paradigm of (repr.) FSM. We have to tackle several challenges: Given a graph database DB, the size of the subgraph database grows exponentially w.r.t. the

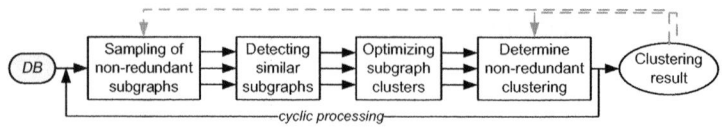

Fig. 7. Processing scheme of the SGC algorithm

number of vertices in the original graphs, i.e. $|SubDB| \in \mathcal{O}(\sum_i(2^{|V_i|}))$. Furthermore, the number of possible clusters exponentially depends on the number of subgraphs; $|Clusters| \in \mathcal{O}(2^{|SubDB|})$. Last, we have to select the final result $Result \subseteq Clusters$, which again implies exponentially many combinations.

Clearly, enumerating all these combinations or even just all possible subgraphs is not manageable. Thus, we propose an approximative algorithm that *directly* extracts valid, high-quality clusters of non-redundant subgraphs from the original graphs. Moreover, the extracted clusters should be non-redundant to each other. Based on these considerations, the general processing of our algorithm is illustrated in Figure 7. In a cyclic process, consisting of four successive phases, potential clusters are generated and based on their redundancy characteristics rejected or added to the result.

In phase 1, we randomly draw subgraph samples from the input graphs that act as cluster centers. Therefore, any subgraph sampling method [13] can be integrated in SGC; in this version we simply use random sampling of connected subgraphs. Since not necessarily each sample leads to a valid cluster, we draw several samples in each iteration hence increasing the probability to identify the hidden clusters.

In phase 2, we identify for each sample a cluster of similar subgraphs. For each graph of the database we determine the subgraphs whose edit-distance to the sample is small enough. Besides similarity, we have to take care of redundancy. Since several subgraphs of the same graph could be similar to the considered sample, we potentially induce intra-cluster redundancy. Thus, while computing the cluster we also remove redundancy. If the cluster's support is large enough, it will correspond to a valid substructure cluster. Otherwise, the sample did not lead to a valid cluster: We store the sample to prevent its repetitive drawing and hence ensure the termination of the algorithm.

In phase 3, we optimize the quality of the clusters. Since the final clusters should primarily contain large subgraphs, we enlarge the clustered subgraphs by adding further vertices. We randomly select a subgraph from the cluster, enlarge it by a neighboring vertex, and recalculate the edit-distance to the remaining subgraphs of the cluster. If the distance is still small enough, the subgraph will remain in the cluster. Otherwise, it is removed. By this optimization step, the support of a cluster decreases while the subgraph sizes increase. We enlarge the elements until the minimal support would be violated.

In phase 4, we choose from the set of clusters a valid substructure clustering without inter-cluster redundancy. Starting with an empty solution, we iteratively select the cluster with the highest quality. If the cluster is redundant w.r.t. an already selected cluster, we will reject it, else we add it to the new clustering. After all clusters have been processed, we have the new redundancy-free clustering.

This result is used as background knowledge in the next iteration to prevent the generation of redundant samples in the sampling phase and it corresponds to the final clustering if no further clusters are generated. The effectiveness of our SGC is analyzed in the next section.

5   Experiments

Setup. We compare SGC with four different subgraph mining paradigms: 1. We use closed frequent subgraph mining (CFSM), considering each pattern as a single cluster. 2. We cluster the frequent subgraphs based on their graph edit-distances using k-medoid and thus get a kind of representative subgraph mining. 3. We use full-graph k-medoid clustering that groups the graphs based on their whole characteristics. 4. We randomly sample subgraphs of certain sizes from the graphs such that on average each vertex is contained in one subgraph and we cluster the resulting subgraphs using k-medoid and graph edit-distance. *We are aware that the competitors 1-3 try to solve different (but similar) problems. However, as there exists no method that exactly solves the same task as our approach (besides the baseline competitor 4), we think a comparison is beneficial.* We use synthetic data, by default with 1000 graphs of size 15 in the database and with 30 hidden clusters each containing 50 subgraphs of size 6. In average, the clustered subgraphs differ by a distance of 2. Furthermore, we use real world data containing structures for chemical substances: the TOXCST data[3] , the NCI database[4] and the roadmap dataset[5] (cf. PubChem DB). Efficiency is measured by the approaches' runtime; clustering quality via the CE measure [16] by considering each vertex as a data object. For comparability all experiments were run on Opteron 2.3GHz CPUs using Java6 64 bit. As the competing approaches require the number of clusters as input, we provide them with the true number of clusters. The clustering based on random samples is provided with the true size of the subgraphs. The minimal support of the closed/representative mining approaches is optimized such that the highest clustering quality is achieved. *Thus, these paradigms already get a huge advantage for finding the clusters.* For SGC we use $r_{intra}=r_{inter}=0.75$, $\alpha=5$, $minSupport=40$.

Graph Sizes. The primary objective of our method is to achieve high clustering quality. Nonetheless, we also depict runtimes. We start by varying the size of the database graphs, i.e. we increase their number of vertices. As shown in Figure 8 (left) SGC outperforms all competing approaches and constantly gets high clustering quality. Clustering based on random samples achieves good quality only for small graphs; for larger graphs the quality drops since many samples do not

---

[3] www.epa.gov/ncct/dsstox/sdf_toxcst.html, TOXCST_v3a_320_12Feb2009.sdf

[4] http://cactus.nci.nih.gov/download/nci/, NCI-Open_09-03.sdf.gz

[5] http://cactus.nci.nih.gov/download/roadmap/, roadmap_2008-01-28.sdf.gz

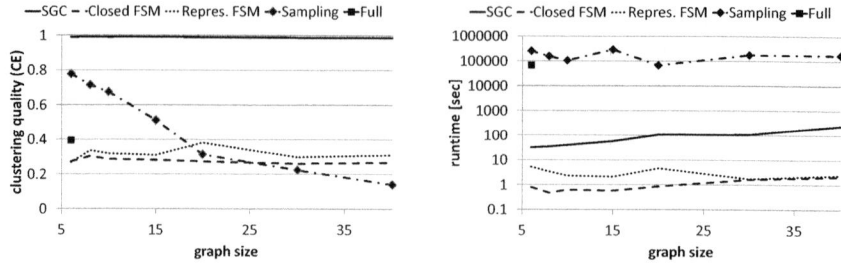

Fig. 8. Varying graph size

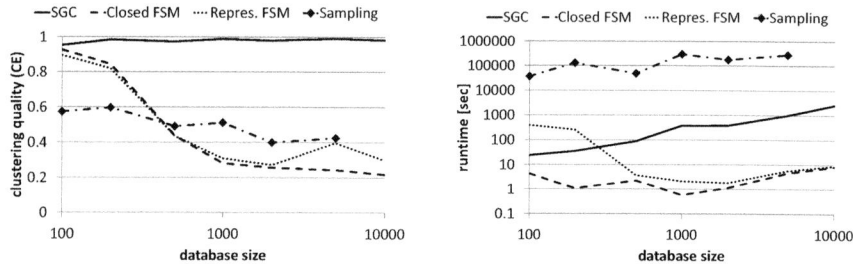

Fig. 9. Varying database size

belong to meaningful clusters. Random sampling is not an option for performing subgraph clustering. Mining based on closed or representative subgraphs fails for these data too. Since the hidden subgraphs are not isomorphic, these approaches detect only subsets of the clusters. Representative FSM is slightly better than CFSM. Full-graph clustering also achieves low quality. The whole graphs do not show similarity; good clusters cannot be detected.

Another problem of full-graph clustering is its high runtime (Figure 8 (right)). Since large graphs have to be compared, an efficient execution is not possible. Only for graphs up to a size of 6, this method was applicable. Thus, we do not include full-graph clustering in the following. Clustering based on random samples is orders of magnitude slower than SGC since many - but not interesting - samples are drawn. The runtime of SGC increases only slightly since we just have to determine the similarity between subgraphs. Although some algorithms have smaller runtime, SGC is still efficient and, as we believe, this aspect is compensated by its higher clustering quality.

Database Size. Next we increase the database size, i.e. the number of graphs. As shown in Figure 9 (left), SGC again constantly outperforms the other paradigms. Only for very small datasets, closed and representative mining obtain comparable quality but they drop heavily afterwards. With increasing database size, a larger number of redundant patterns is generated leading to lower quality. Considering the runtime - Figure 9 (right) - we observe similar behavior as in the previous experiment. SGC is the only method with acceptable runtime that also gets high quality.

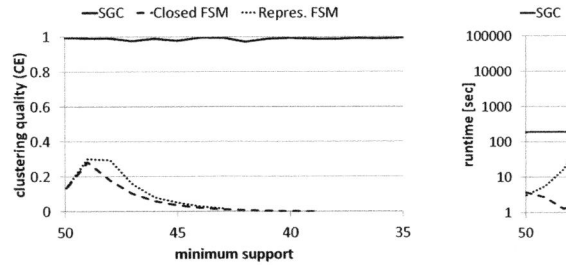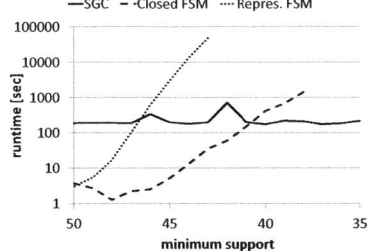

Fig. 10. Varying minimal support

Minimal Support. A large difference between substructure clustering and the other paradigms is the notion of the minimal support. Figure 10 analyzes this difference by varying the minimal support parameter of the algorithms from the true cluster size (50) to 35. SGC is robust w.r.t. this parameter: quality and runtime are stable. If the minimal support is smaller than the true cluster size, SGC still collects all similar subgraphs. A different effect is observed for the other methods: If we set the minimal support to 50 only very small subgraphs fulfill this support: the methods get very low quality. However, by decreasing the support, too many subgraphs are frequent, leading to highly redundant results and poor quality. Generating many redundant subgraphs also affects the runtime, which heavily increases with decreasing support.

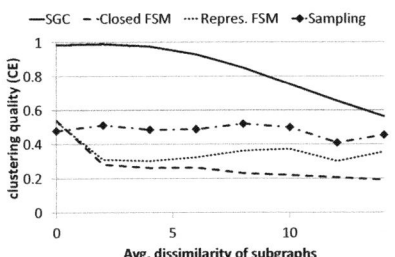

Fig. 11. Varying dissimilarity

Dissimilarity. Next, we highlight the strength of our method to group similar and not just isomorphic subgraphs. In Figure 11 we increase the dissimilarity between the clustered subgraphs; the clusters get obfuscated. This effect is weakened by our approach: SGC still detects the clusters even if the subgraphs are not perfectly similar. Of course, if the dissimilarity is too high, the clusters cannot and should not be detected anymore. Overall, SGC is able to detect clusters under the condition of non-isomorphism, which especially occurs for continuous valued data. The other algorithms get low quality even if the subgraphs differ only slightly.

Real World Data. In Figure 12 we use real world data of different sizes (as shown by the adjacent numbers). Since no hidden clusters are given, we cannot determine clustering quality. Besides the runtime we thus analyze

	Toxcst 320		NCI 500		NCI 5000	Road 1000		Road 10000
	SGC	Clos.	SGC	Clos.	SGC	SGC	Clos.	SGC
runtime [s]	207	11	322	956	5840	465	1224	14943
# cluster	20	503	13	2165	23	19	2232	19
intra-redun.	1.7	3.9	2.3	9.9	1.2	1.7	10.4	1.2
inter-redun.	1.2	48.7	1.2	430.5	1.3	1.2	253.9	1.3
∅ subgr. size	9.3	6.9	11.8	10.3	8.9	12.2	10.1	10.7

Fig. 12. Clustering properties for real world data

different properties of the clustering results. We just compare SGC and CFSM since the other methods require the number of clusters as input, which is not known for the data. In any case, we set the minimum support to 10% of the whole database size. SGC is applicable on all datasets; CFSM only runs for small databases (≤ 1000 graphs). The number of clusters generated by SGC is reasonable while CFSM generates an overwhelming result of redundant clusters. SGC avoids redundancy as shown by two facts: First, we measure in how many subgraphs of the same cluster a vertex appears (in avg.), corresponding to intra-cluster redundancy. Second, we measure in how many different clusters a vertex appears, corresponding to inter-cluster redundancy. For SGC both measures are around 1.2–2.3. Thus, we permit overlapping clusters but avoid high redundancy. CFSM has high redundancy in the result, especially between different clusters. Redundancy removal as e.g. performed by our approach is essential for inter-pretable results and incorporated in our novel paradigm. Finally, SGC groups larger subgraphs since it is not restricted to isomorphism.

Overall, all experiments indicate that SGC achieves the highest clustering quality and acceptable runtimes. Even though our model and algorithm for subgraph clustering are proofs-of-concept, they are already superior to related methods. Thus, SGC is the only method of choice for analyzing the clustering structure of subgraph patterns and confirms the strength of our substructure clustering paradigm.

6   Conclusion

We introduced the novel mining paradigm of substructure clustering that defines clusters of similar substructures. This paradigm unifies and generalizes existing substructure mining tasks: the substructures of our clusters do not need to be isomorphic or of the same size. To ensure interpretability, our model avoids redundant information in the clustering result – between different clusters as well as within single clusters. As a proof-of-concept we presented an instantiation of our model for subgraph clustering. In our experiments we demonstrate that our novel method outperforms existing approaches for subgraph mining. As future work, we want to apply our novel paradigm to time series subsequence clustering and to vector data containing missing values.

Acknowledgments. This work has been supported by the UMIC Research Centre, RWTH Aachen University, Germany, and the B-IT Research School.

References

1. Aggarwal, C., Wang, H.: Managing and Mining Graph Data. Springer, New York (2010)
2. Beyer, K., Goldstein, J., Ramakrishnan, R., Shaft, U.: When Is "Nearest Neighbor" Meaningful? In: Beeri, C., Bruneman, P. (eds.) ICDT 1999. LNCS, vol. 1540, pp. 217–235. Springer, Heidelberg (1998)
3. Chen, J.: Making subsequence time series clustering meaningful. In: ICDM, pp. 114–121 (2005)
4. Deutsch, A., Fernández, M.F., Suciu, D.: Storing semistructured data with stored. In: SIGMOD, pp. 431–442 (1999)

5. Gouda, K., Zaki, M.J.: Genmax: An efficient algorithm for mining maximal frequent itemsets. DMKD 11(3), 223–242 (2005)
6. Günnemann, S., Färber, I., Boden, B., Seidl, T.: Subspace clustering meets dense subgraph mining: A synthesis of two paradigms. In: ICDM, pp. 845–850 (2010)
7. Günnemann, S., Kremer, H., Seidl, T.: Subspace clustering for uncertain data. In: SDM, pp. 385–396 (2010)
8. Günnemann, S., Müller, E., Färber, I., Seidl, T.: Detection of orthogonal concepts in subspaces of high dimensional data. In: CIKM, pp. 1317–1326 (2009)
9. Han, J., Pei, J., Yin, Y., Mao, R.: Mining frequent patterns without candidate generation: A frequent-pattern tree approach. DMKD 8(1), 53–87 (2004)
10. Jacobs, B.: Categorical Logic and Type Theory. Studies in Logic and the Foundations of Mathematics, vol. 141. North Holland, Amsterdam (1999)
11. Jin, X., Lu, Y., Shi, C.: Distribution Discovery: Local Analysis of Temporal Rules. In: Chen, M.-S., Yu, P.S., Liu, B. (eds.) PAKDD 2002. LNCS (LNAI), vol. 2336, pp. 469–480. Springer, Heidelberg (2002)
12. Kriegel, H.-P., Kröger, P., Zimek, A.: Clustering high-dimensional data: A survey on subspace clustering, pattern-based clustering, and correlation clustering. TKDD 33(1), 1–58 (2009)
13. Leskovec, J., Faloutsos, C.: Sampling from large graphs. In: KDD, pp. 631–636 (2006)
14. Lin, J., Keogh, E., Truppel, W.: Clustering of streaming time series is meaningless. In: SIGMOD, pp. 56–65 (2003)
15. Müller, E., Assent, I., Günnemann, S., Krieger, R., Seidl, T.: Relevant subspace clustering: Mining the most interesting non-redundant concepts in high dimensional data. In: ICDM, pp. 377–386 (2009)
16. Patrikainen, A., Meila, M.: Comparing subspace clusterings. TKDE 18(7), 902–916 (2006)
17. Poernomo, A.K., Gopalkrishnan, V.: Towards efficient mining of proportional fault-tolerant frequent itemsets. In: KDD, pp. 697–706 (2009)
18. Rombo, S.E., Terracina, G.: Discovering Representative Models in Large Time Series Databases. In: Christiansen, H., Hacid, M.-S., Andreasen, T., Larsen, H.L. (eds.) FQAS 2004. LNCS (LNAI), vol. 3055, pp. 84–97. Springer, Heidelberg (2004)
19. Sanfeliu, A., Fu, K.S.: A distance measure between attributed relational graphs for pattern recognition. IEEE Transactions on Systems, Man, and Cybernetics 13, 353–362 (1983)
20. Shi, J., Malik, J.: Normalized cuts and image segmentation. PAMI 22(8), 888–905 (2000)
21. Tan, Z., Tung, A.: Substructure clustering on sequential 3D object datasets. In: ICDE, pp. 634–645 (2004)
22. Thomas, L., Valluri, S., Karlapalem, K.: Margin: Maximal frequent subgraph mining. In: ICDM, pp. 1097–1101 (2006)
23. Tsuda, K., Kudo, T.: Clustering graphs by weighted substructure mining. In: ICML, pp. 953–960 (2006)
24. Wang, C., Parthasarathy, S.: Summarizing itemset patterns using probabilistic models. In: KDD, pp. 730–735 (2006)
25. Yan, X., Han, J.: CloseGraph: mining closed frequent graph patterns. In: KDD, pp. 286–295 (2003)
26. Yan, X., Yu, P.S., Han, J.: Graph indexing: A frequent structure-based approach. In: SIGMOD, pp. 335–346 (2004)
27. Yankov, D., Keogh, E.J., Medina, J., Chiu, B.Y., Zordan, V.B.: Detecting time series motifs under uniform scaling. In: KDD, pp. 844–853 (2007)
28. Zhang, S., Yang, J., Li, S.: RING: An Integrated Method for Frequent Representative Subgraph Mining. In: ICDM, pp. 1082–1087 (2009)

BT* – An Advanced Algorithm for Anytime Classification

Philipp Kranen, Marwan Hassani, and Thomas Seidl

RWTH Aachen University, Germany
{lastname}@cs.rwth-aachen.de

Abstract. In many scientific disciplines experimental data is generated at high rates resulting in a continuous stream of data. Data bases of previous measurements can be used to train classifiers that categorize newly incoming data. However, the large size of the training set can yield high classification times, e.g. for approaches that rely on nearest neighbors or kernel density estimation. Anytime algorithms circumvent this problem since they can be interrupted at will while their performance increases with additional computation time. Two important quality criteria for anytime classifiers are high accuracies for arbitrary time allowances and monotonic increase of the accuracy over time. The Bayes tree has been proposed as a naive Bayesian approach to anytime classification based on kernel density estimation. However, the employed decision process often results in an oscillating accuracy performance over time. In this paper we propose the BT* method and show in extensive experiments that it outperforms previous methods in both monotonicity and anytime accuracy and yields near perfect results on a wide range of domains.

1 Introduction

Continuous experimental data in scientific laboratories constitutes a stream of data items that must be processed as they arrive. Many other real world applications can be associated with data streams as large amounts of data must be processed every day, hour, minute or even second. Examples include traffic/network data at web hosts or telecommunications companies, medical data in hospitals, statistical data in governmental institutions, sensor networks, etc. Major tasks in mining data streams are classification as well as clustering and outlier detection. Optimally, an algorithm should be able to process an object in a very short time and use any additional computation time to improve its outcome. The idea of being able to provide a result regardless of the amount of available computation time led to the development of anytime algorithms. Anytime algorithms have been proposed e.g. for Bayesian classification [19,23] or support vector machines [5], but also for anytime clustering [13] or top-k queries [2].

The Bayes tree proposed in [19] constitutes a statistical approach for stream classification. It can handle large amounts of data through its secondary storage index structure, allows for incremental learning of new training data and is

A. Ailamaki and S. Bowers (Eds.): SSDBM 2012, LNCS 7338, pp. 298–315, 2012.

capable of anytime classification. It uses a hierarchy of Gaussian mixture models, which are individually refined with respect to the object to be classified. In [14] we proposed in the MC-Tree a top down construction of the mixture hierarchy that led to an improved anytime classification performance. However, the employed decision process yields strong oscillations of the accuracy over time in several domains, which contradicts the assumption that the performance increases monotonically.

In this paper we propose BT* as an advanced algorithm for anytime classification. We investigate three methods to improve the parameters in a given Bayes tree and propose two alternative approaches for the decision design. The goals of the research presented in this paper are: *maintain the advantages* of the Bayes tree, including the applicability to large data sets (index structure) and the individual query dependent refinement, *overcome the oscillating behavior* of the anytime accuracy and achieve a monotonically increasing accuracy over time, and *increase the accuracy* of the classifier both for the ultimate decision and for arbitrary time allowances. The BT* algorithm is one approach to Bayesian anytime classification that we investigate in this paper and whose effectiveness is clearly shown in the experiments. Other solutions can be investigated, such as SPODEs (see Section 2) for categorical data, which are beyond the scope of this paper. In the following section we review related work on anytime algorithms. In Section 3 we provide details on the BT* algorithm, Section 4 contains the experimental evaluation and Section 5 concludes the paper.

2   Related Work

Anytime algorithms have first been discussed in the AI community by Thomas Dean and Mark Boddy in [4] and have thereafter been an active field of research. Recent work includes an anytime A* algorithm [16] and anytime algorithms for graph search [15]. In the data base community anytime measures for top-k algorithms have been proposed [2], in data mining anytime algorithms have been discussed for clustering [13] and other mining tasks.

Anytime classification is real time classification up to a point of interruption. In addition to high classification accuracy as in traditional classifiers, anytime classifiers have to make best use of the limited time available, and, most notably, they have to be interruptible at any given point in time. This point in time is usually not known in advance and may vary greatly. Anytime classification has for example been discussed for support vector machines [5], nearest neighbor classification [21], or Bayesian classification on categorical attributes [23]. Bayes classifiers using kernel density estimation [11] constitute a statistical approach that has been successfully used in numerous application domains. Especially for huge data sets the estimation error using kernel densities is known to be very low and even asymptotically optimal [3].

For Bayesian classification based on kernel densities an anytime algorithm called Bayes tree has been proposed in [19]. The Bayes tree maintains a hierarchy of mixture densities that are adaptively refined during query processing to allow for anytime density estimation. An improved construction methods has been discussed in [14] that generates the hierarchy top down using expectation maximization clustering. Our proposed BT* algorithm builds upon this work.

An important topic in learning from data streams is the handling of evolving data distributions such as concept drift or novelty. A general approach to building classifiers for evolving data streams has been proposed in [22]. The main idea is to maintain a weighted ensemble of several classifiers that are build on consecutive chunks of data and are then weighted by their performance on the most recent test data. The approach is applicable to any classifier and can hence be combined with our proposed method in case of concept drift and novelty.

A different line of research focuses on anytime learning of classifiers, e.g. for Bayesian networks [17] or decision tree induction [6]. BT* allows for incremental insertion and can thereby be interrupted at will during training. Our focus in this paper is on varying time allowances during classification to allow processing newly incoming data at varying rates.

3   BT*

We start by describing the structure and workings of the Bayes tree in the following section. We recapitulate the top down build up strategy proposed in [14] along with its performance. In Section 3.2 we develop three approaches to improve the parameters in a given model hierarchy and Section 3.3 introduces two alternative decision processes for anytime classification. Finally we evaluate the proposed improvements in Section 4, individually as well as combined, to find BT* as the best performing alternative.

3.1   Anytime Bayesian classification

Let $\mathcal{L} = \{l_1, \ldots, l_{|\mathcal{L}|}\}$ be a set of class labels, $\mathcal{T}$ a training set of labeled objects and $\mathcal{T}_l \subseteq \mathcal{T}$ the set of objects with label l. A classifier assigns a label $l \in \mathcal{L}$ to an unseen object x based on its features and a set of parameter values Θ. The decision function of a Bayes classifier is generally

$$f_{Bayes}(\Theta, x) = \arg\max_{l \in \mathcal{L}} \{P(l) \cdot p(x|\Theta, l)\} \tag{1}$$

where $P(l) = |\mathcal{T}_l|/|\mathcal{T}|$ is the a priori probability of label l and $p(x|\Theta, l)$ is the class conditional density for x given label l and Θ. The class conditional density can for example be estimated using per class a unimodal distribution or a mixture of distributions. The Bayes tree, referred to as BT in the following, maintains a hierarchy of Gaussian mixture models for each label $l \in \mathcal{L}$ (see Figure 1).

Definition 1. *The model $M = \{\mathcal{N}_1, \ldots, \mathcal{N}_r\}$ of a Bayes tree is a set of connected nodes that build a hierarchy. Each node $\mathcal{N} = \{e_1, \ldots, e_s\}$ stores a set of entries with $2 \leq s \leq maxFanout$. An entry $e = \{p_e, n_e, \boldsymbol{LS}_e, \boldsymbol{SS}_e\}$ stores a pointer p_e to the first node $\mathcal{N}_e$ of its subtree, the number n_e of objects in the*

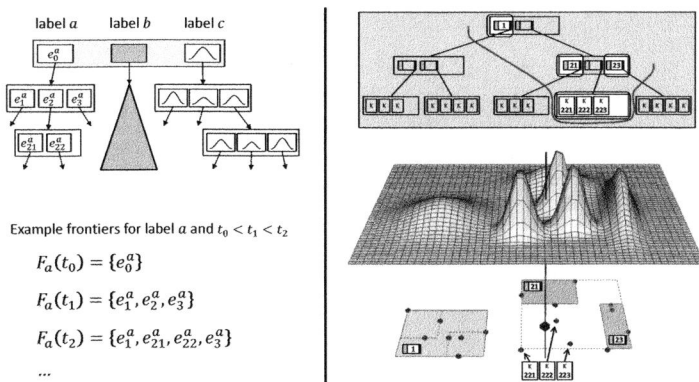

Fig. 1. The Bayes tree uses per class a hierarchy of entries that represent Gaussians (top left). For classification, the class conditional density is computed using the entries in the corresponding frontier (right). The initial frontier contains only the root entry; in each refinement one frontier entry is replaced by its child entries (bottom left).

subtree and their linear and quadratic sums per dimension. The root node $\mathcal{N}_{root}$ stores exactly one entry e_o^l for each label $l \in \mathcal{L}$ that summarizes all objects with label l. Entries in leaf nodes correspond to d-dimensional Gaussian kernels.

We refer to the set of objects stored in the subtree corresponding to entry e as $\mathcal{T}_{|e}$. Figure 1 illustrates the structure of a Bayes tree. Each entry e is associated with a Gaussian distribution

$$g(x, \mu_e, \Sigma_e) = \frac{1}{(2\Pi)^{d/2} \cdot |\Sigma_e|} \cdot e^{-\frac{1}{2}(x-\mu_e)\Sigma_e^{-1}(x-\mu_e)^T} \tag{2}$$

where μ_e is the mean, Σ_e the covariance matrix and $|\Sigma_e|$ its determinant. The parameters of the Bayes tree are

$$\Theta = \{(\mu_e, \Sigma_e) \,|\forall e \in \bigcup_{i=1}^{r} \mathcal{N}_i \} \tag{3}$$

i.e. the set of parameters for all Gaussian distributions in the tree structure. These can be easily computed from the information that is stored in the entries. The mean values can be computed as

$$\mu_e = \boldsymbol{LS}_e / n_e \tag{4}$$

Since the Bayes tree constitutes a naive Bayes approach, the covariance matrix $\Sigma_e = [\sigma_{e,ij}]$ is a diagonal matrix with $\sigma_{e,ij} = 0 \; \forall_{i \neq j}$ and

$$\sigma_{ii} = \boldsymbol{SS}_e[i]/n_e - (\boldsymbol{LS}_e[i]/n_e)^2 \tag{5}$$

where $\boldsymbol{LS}_e[i]$ is the i-th component in $\boldsymbol{LS}_e$, $i, j \in \{1, \dots, d\}$.

To estimate the class conditional density $p(x|\Theta, l)$, the Bayes tree maintains at each time t one mixture model for each label l. The mixture model is composed

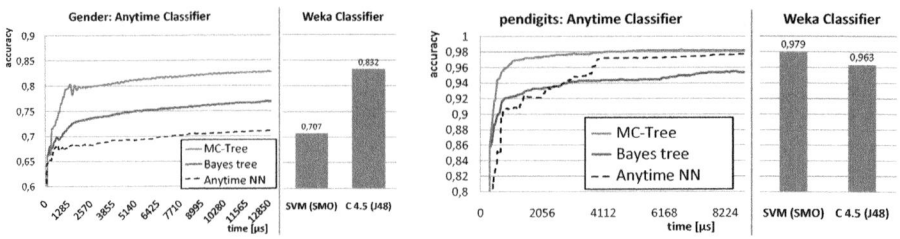

Fig. 2. MC-Tree [14] shows constantly best anytime accuracy performance against the other Bayes tree classifier variant [19], the anytime nearest neighbor [21] and the Weka classifier implementations of SVM and decision tree

of the most detailed entries that have been read up to that point in time such that each training object is represented exactly once. This set of entries is called a frontier $\mathcal{F}_l(t)$. Initially $\mathcal{F}_l(0) = \{e_0^l\}$ (see Figure 1), i.e. the initial frontier for label l contains only the root node entry $\{e_0^l\}$ corresponding to a unimodal Gaussian distribution for that class. In each improvement step one frontier is refined by

$$\mathcal{F}_l(t+1) = \mathcal{F}_l(t) \setminus \{\hat{e}\} \cup \mathcal{N}_{\hat{e}} \tag{6}$$

where $\hat{e} = \arg\max_{e' \in \mathcal{F}_l(t)} \{g(x, \mu_{e'}, \sigma_{e'})\}$ is the entry in $\mathcal{F}_l(t)$ that yields the highest density for x. To decide which frontier is refined in the next improvement, the labels are sorted according to the posterior probability with respect to the current query. In this order the top $k = \log(|\mathcal{L}|)$ frontiers are consecutively refined before resorting the labels. The decision rule of the Bayes tree at time t is then

$$f_{BT}(\Theta, x, t) = \arg\max_{l \in \mathcal{L}} \left\{ P(l) \cdot \sum_{e \in \mathcal{F}_l(t)} \frac{n_e}{n_l} g(x, \mu_e, \sigma_e) \right\} \tag{7}$$

where n_e/n_l is the fraction of objects from class l in the subtree corresponding to entry e. Hence, the Bayes tree can provide a classification decision at any time t and has an individual accuracy after each refinement.

The original Bayes tree builds separate hierarchies for each class label which are created using the incremental insertion from the R-tree [9]. In [14] we investigated combining multiple classes within a single distribution and exploiting the entropy information for the refinement decisions. While it turned out that separating the classes remained advisable, the novel construction method EM-TopDown proposed in [14] yielded constantly the best performance (see Fig. 2).

However, despite the improved anytime accuracy, the resulting anytime curves exhibit strong oscillations on several domains (see Figures 2 or 9), which does not constitute a robust performance. The oscillation results from alternating decisions between the individual refinement steps. With our proposed BT* algorithm we achieve near perfect results in terms of monotonicity and at the same time successfully improve the anytime accuracy performance. In our experiments we will use both the original incremental insertion (denoted as R) as well as the EM-TopDown construction (denoted as EM) as baselines for comparison. We use the

$\diamond$ operator to denote combinations of the baselines with different optimizations, e.g. $EM \diamond BN$ (see below).

3.2 Parameter Optimization

The Bayes tree determines the parameters (μ_e, Σ_e) for the Gaussians corresponding to the entries according to Equations 4 and 5. In this section we develop both discriminative and generative approaches to optimize the parameters of the Bayes tree. The first approach works on a single inner entry of the tree, the second approach processes an entire mixture model per class, and the third approach considers only leaf node entries.

BN. Our first strategy constitutes a generative approach. The goal is to fit the Gaussian distribution of an entry e better to the data in its corresponding subtree. So far the Bayes tree only considers variances and sets all covariances to zero, i.e. Σ constitutes a diagonal matrix of a naive Bayesian approach. Hence, the resulting distributions functions can only reflect axis-aligned spread of the training data. The advantage is that the space demand with respect to the dimensionality d is $O(d)$ compared to $O(d^2)$ for a full covariance matrix. Using full covariance matrices would mean that $O(d^2)$ covariances have to be stored for every single entry. Moreover, the time complexity for the evaluation of the Gaussian density function (see Equation 2) increases from $O(d)$ to $O(d^2)$. However, not all covariances might be useful or necessary. Small and insignificant correlations and corresponding rotations of the Gaussians can be neglected.

Our goal is to add important correlations at low time and space complexity. To this end we fix a maximal block size $\mathbf{s}$ and constrain Σ to have a block structure, i.e.

$$\Sigma = \text{diag}(B_1, \ldots, B_u) \tag{8}$$

where $B_i \in \mathbb{R}^{\mathbf{s}_i \times \mathbf{s}_i}$ are quadratic matrices of block size $\mathbf{s}_i$ and $\mathbf{s}_i \leq \mathbf{s} \; \forall i = 1 \ldots u$. The resulting space demand is in $O(d \cdot s)$. So is the time complexity, since the exponent in the Gaussian distribution (see Equation 2) can be factorized as

$$z\Sigma^{-1}z^T = \sum_{i=1}^{u} z[L_i..U_i]B_i^{-1}z[L_i..U_i]^T \tag{9}$$

with $z = (x - \mu)$, $L_i = 1 + \sum_{j=1}^{i-1}\mathbf{s}_j$, $U_i = L_i + \mathbf{s}_i - 1$ and $z[L_i..U_i]$ selects dimensions L_i to U_i from z.

To decide which covariances to consider and which to ignore we adapt a hill climbing method for Bayesian network learning (as e.g. proposed in [12]) that finds the most important correlations. We first describe how we derive a block structured matrix from a Bayesian network and detail the learning algorithm thereafter.

A Bayesian network $\mathcal{B} = \langle G, \Theta \rangle$ is characterized by a directed acyclic graph G and a set of parameter values Θ. In our case $G = (V, E)$ contains one vertex for the class label and one vertex $i \in V$ for each attribute. The class vertex is connected to every attribute vertex. An edge $(i, j) \in E$ between attribute i and j induces a dependency between the corresponding dimensions. We derive an undirected graph G' from G by simply removing the orientation of the edges.

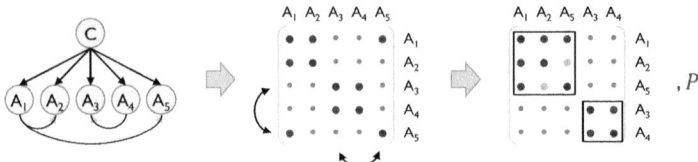

Fig. 3. An example for deriving a block structured covariance matrix (right) from a Bayesian network (left) performed on each inner entry

Edges $(i, j) \in E'$ between two attributes i and j are then transferred to a co-variance σ_{ij} and its symmetric counter part σ_{ji} in Σ (see Figure 3). Since there are no constraints on the dependencies in the Bayesian network, the resulting matrix is unlikely to contain non-zero entries only in blocks along the diagonal. To ensure a block structure, we apply a permutation P to Σ by $P\Sigma P^{-1}$ that groups dimensions, which are connected in G', as blocks along the diagonal. In the resulting blocks, covariances that have not been set before are added, since these harm neither the space nor the time complexity. During classification, P is then also applied to z before computing Equation 9. Since P is just a reordering of the dimensions, it can be stored in an array of size d and its application to z is in $O(1)$ per feature.

To create the block covariance matrix for an entry e we start with a naive Bayes, i.e. initially $(i, j) \notin E' \; \forall i, j$ and $\Sigma_e = \text{diag}(\sigma_{e,11}, \ldots, \sigma_{e,dd})$. From the edges that can be added to G' without violating the maximal block constraint in the resulting block matrix, we iteratively select the one that maximizes the likelihood of e given $\mathcal{T}_{|e}$. Since all objects $x \in \mathcal{T}_{|e}$ have the same label l, the log likelihood is

$$LL(e|\mathcal{T}_{|e}) = \sum_{x \in \mathcal{T}_{|e}} \log \left(p(l, x|(\mu_e, \Sigma_e)) \right) \tag{10}$$

We stop when either the resulting matrix does not allow for further additions or no additional edge improves Equation 10.

In general, we determine a block covariance matrix for each entry in the Bayes tree. However, on lower levels of the Bayes tree the combination of single components is likely to capture already the main *directions* of the data distribution. This might render the additional degrees of freedom given by the covariances useless or even harmful, since they can lead to overfitting. We therefore evaluate in Section 4 in addition to s the influence of restricting the BN optimization to the upper m levels of the Bayes tree.

MM. In the previously described approach we fitted a single Gaussians to the underlying training data using a generative approach. The approach we propose next considers an entire mixture model per class and tries to optimize the mixture parameters simultaneously in a discriminative way. To this and we adapt an approach for margin maximization that has been proposed in [18] (referred to as MM). It seeks to improve the classification performance of Bayesian classifiers based on Gaussian mixtures and is therefore a good starting point for the hierarchical mixtures in the Bayes tree. In the following we first describe how we derive the mixture models per class from the tree structure and then explain the optimization procedure.

We need a heuristic to extract one mixture for each label $l \in \mathcal{L}$ from the Bayes tree. The mixture models are described by a set of entries $\mathcal{E} \subset \bigcup_{\mathcal{N}_i \in M} \mathcal{N}_i$ and we define

$$\Theta(\mathcal{E}) = \bigcup_{e \in \mathcal{E}} \{\mu_e, \Sigma_e\} \tag{11}$$

as the corresponding set of parameter values. Initially we set $\mathcal{E}_0 = \mathcal{N}_{root}$, i.e. for each label $l \in \mathcal{L}$ the unimodal model describing $\mathcal{T}_l$ is represented by $\Theta(\mathcal{E}_0)$. After optimizing the parameters in $\Theta(\mathcal{E}_0)$ we update the corresponding parameters in the Bayes tree (see Figure 4). In subsequent steps we descend the Bayes tree and set

$$\mathcal{E}_{i+1} = \bigcup_{e \in \mathcal{E}_i} \begin{cases} \mathcal{N}_e & \text{if } p_e \neq \text{null} \\ \{e\} & \text{otherwise.} \end{cases} \tag{12}$$

As above, the sets $\mathcal{E}_i$ are optimized and the Bayes tree parameters are updated. Similar to the previously described BN approach we test as an additional parameter the maximal number m of steps taken in Equation 12 in our evaluation in Section 4. For example, for $m = 2$ we only optimize the upper two levels of the Bayes tree.

We explain the MM approach for a given set of entries $\mathcal{E}$ representing one mixture model for each label $l \in \mathcal{L}$. Let $\Theta = \Theta(\mathcal{E})$ be the corresponding set of parameter values and $l_x \in \mathcal{L}$ the label of an object x. The goal is to find parameters such that for each object $x \in \mathcal{T}$ the class conditional density $p(l_x, x|\Theta)$ of its own class is larger than the maximal class conditional density among the other labels. The ratio between the two is denoted as the multi class margin $d_\Theta(x)$:

$$d_\Theta(x) = \frac{p(l_x, x|\Theta)}{\max_{l \neq l_x} p(l, x|\Theta)} \tag{13}$$

If $d_\Theta(x) > 1$ the object is correctly classified. The optimization then strives to maximize the following global objective

$$D(\mathcal{T}|\Theta) = \prod_{x \in \mathcal{T}} \bar{h}((d_\Theta(x))^\lambda) \tag{14}$$

where the hinge function $\bar{h}(y) = \min\{2, y\}$ puts emphasis on samples with a margin $d_\Theta(x) < 2$, and samples with a large positive margin have no impact on the optimization. The optimization steps require the global objective to be differentiable. To this end the multi class margin $d_\Theta(x)$ is approximated by

$$d_\Theta(x) \approx \frac{p(l_x, x|\Theta)}{\left[\sum_{l \neq l_x} p(l, x|\Theta)^\kappa\right]^{1/\kappa}} \tag{15}$$

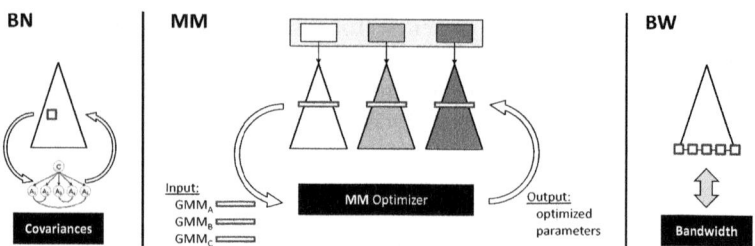

Fig. 4. Left: adding selected covariances individually to inner entries. Center: discriminative parameter optimization for entire mixture models using the margin maximization concept (MM). Right: changing the bandwidth parameter for leaf entries.

using $\kappa \geq 1$. In the global objective the hinge function is approximated by a smooth hinge function $h(y)$ that allows to compute the derivative $\partial \log D(\mathcal{T}|\Theta)/\partial\Theta$:

$$h(y) = \begin{cases} y + \frac{1}{2} & \text{if } y \leq 1 \\ 2 - \frac{1}{2}(y-2)^2 & \text{if } 1 < y < 2 \\ 2 & \text{if } y \geq 2 \end{cases} \qquad (16)$$

The derivatives with respect to the single model parameters $\theta_i \in \Theta$ are then used in an extended Baum-Welch algorithm [8] to iteratively adjust the weights, means and variances of the Gaussians. Details for the single derivations as well as the implementation of the extended Baum Welch can be found in [18].

BW. In its leaves the Bayes tree stores d-dimensional Gaussian kernels (see Definition 1). The kernel bandwidth h_i, $i = 1 \ldots d$, is a parameter that is chosen per dimension and that can be optimized by different methods for bandwidth estimation. A method discussed in [20,10] uses

$$h_i = \sqrt[d+4]{\frac{4}{(d+2) \cdot |\mathcal{T}|}} \cdot \widehat{\sigma}_i \qquad (17)$$

where $\widehat{\sigma}_i$ is the variance of the training data $\mathcal{T}$ in dimension i. A second method [11] determines the bandwidth as

$$h_i = \frac{max_i - min_i}{\sqrt{|\mathcal{T}|}} \qquad (18)$$

where max_i and min_i are the maximal and minimal values occuring in $\mathcal{T}$ in dimension i. Additionally we test a family of bandwidths

$$h_i = \alpha \cdot \widehat{\sigma}_i \qquad (19)$$

in Section 4, where a factor α is multiplied to $\widehat{\sigma}_i$. We refer to the three methods in Equations 17, 18 and 19 as *haerdle*, *langley* and *fα* respectively.

3.3 Decision Design

In the previous section we discussed different approaches to optimize distribution parameters in a given Bayes Tree. In this section we investigate alternatives for making decisions over time given an optimized tree structure. The original decision function for the Bayes tree is given in Equation 7. To estimate the class conditional density for a class l at time t the entries that are stored in the current frontier $\mathcal{F}_l(t)$ are evaluated and summed up according to their weight. We propose two approaches that both change the set of entries that are taken into consideration for the classification decision.

ENS. The first approach constitutes a special kind of ensemble methods. Ensembles are frequently used both for a single paradigm, as e.g. in Random Forests where multiple decision trees are created and employed, or for several different paradigms. The basic concept of ensembles is simple and can easily be transferred to any classification method. A straightforward way for the Bayes tree would be to build several tree structures, e.g. using different samples of the training data, and combine the individual outcomes to achieve a classification decision. The method we propose here uses a single Bayes tree and builds an ensemble over time.

In the Bayes tree so far only the most recent frontiers $\mathcal{F}_l(t)$ were used in the decision function. To create an ensemble over time using the Bayes tree, we combine all previous frontiers in the modified decision function

$$f_{BT\diamond ENS}(\Theta, x, t) = \arg\max_{l \in \mathcal{L}} \left\{ P(l) \cdot \sum_{s=0}^{t} \sum_{e \in \mathcal{F}_l(t-s)} \frac{n_e}{n_l} g(x, \mu_e, \sigma_e) \right\} \quad (20)$$

The additional computational cost when using the ensemble decision from Equation 20 compared to the original decision from Equation 7 is only a single operation per class. More precisely, we just have to add the most recent density, which we compute also in the original Bayes tree, to an aggregate of the previous densities. Since we sum up the same amount of frontiers for each label, we can skip the normalization without changing the decision and do not have to account for additional operations. The ensemble approach widens the basis on which we make our decision in the sense that it takes mixture densities of different granularities into account for the classification decision. The approach we propose next takes the opposite direction in the sense that it narrows the set of Gaussian components that are used in the decision process.

NN. The NN approach is inspired by the nearest neighbor classifier. The nearest neighbor classifier finds a decision based on the object that is the closest to the query object x, i.e. it selects only the most promising object from $\mathcal{T}$. This concept can be transferred to the Bayes tree in a straightforward way by using the modified decision function

$$f_{BT\diamond NN}(\Theta, x, t) = \arg\max_{l \in \mathcal{L}} \left\{ P(l) \cdot \max_{e \in \mathcal{F}_l(t)} \left\{ \frac{n_e}{n_l} g(x, \mu_e, \sigma_e) \right\} \right\} \quad (21)$$

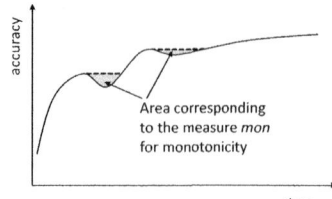

name	#obj	d	$|\mathcal{L}|$	name	#obj	d	$|\mathcal{L}|$
page-blocks	5473	10	5	pendigits	10992	16	10
optdigits	5620	64	10	vowel	990	10	11
letter	20000	16	26	spambase	4601	57	2
segment	2310	19	7	gender	189961	9	2
kr-vs-kp	3196	36	2	covtype	581012	10	7

Fig. 5. Left: Illustration of the monotonicity measure. The larger the area resulting from decreasing accuracy over time, the worse the monotonicity performance. Right: Data sets used for evaluation and their corresponding number of objects (#obj), dimensions (d), and classes ($|\mathcal{L}|$).

The NN approach comes at no additional cost since we replace the addition by a comparison in Equation 21. Variants of the nearest neighbor classifier use k closest objects. The actual label can then be assigned based on a simple majority voting among the neighbors or taking their distance or the prior probability of the labels into account. We test the standard nearest neighbor concept with $k = 1$ for $f_{BT \diamond NN}$ in our experiments.

4   Experiments

We evaluate our improvements over both [19] (called R) and [14] (called EM). Comparisons of the Bayes tree to anytime nearest neighbor, decision tree and SVM can be found in [14] (see Section 3.1 and Figure 2).

In classifier evaluation mostly the accuracy acc or the error rate, i.e. $1 - acc$, is used as a measure. Since the Bayes tree is an anytime classifier that incrementally refines its decision, we get an individual accuracy $acc(n)$ for each number n of refinements (see for example Figure 7 right). In all experiments we report the results for the first $r = 200$ refinements. To compare the different approaches we use the average accuracy avg as well as the maximal accuracy max over all refinements. As a third objective, which penalizes descending or oscillating anytime curves, we use the monotonicity

$$mon = 1 - \frac{1}{r} \sum_{n=1}^{r} \widehat{acc}(n) - \min\{\widehat{acc}(n), acc(n)\}$$

where $\widehat{acc}(n) = \max_{1 \leq n' < n} acc(n')$ is the maximal accuracy over all $n' < n$. Figure 5 illustrates the monotonicity measure. The larger the sum of all areas resulting from decreasing anytime accuracy, the worse the monotonicity.

When choosing best results we select according to a linear combination of all three measures with equal weights. For the Bayes tree we set $maxFanout = 7$ and use the bandwidth estimation from [11] (*langley*) for the baselines. All experiments use 10-fold cross validation. For all objects in the test set we evaluate the classification decision after each improvement and report the average accuracy

	Bandwidth Estimation (BT ◊ BW)						Bayesian Network (BT ◊ BN)						MaxMargin (BT ◊ MM)					
	diff. to R baseline			diff. to EM baseline			diff. to R baseline			diff. to EM baseline			diff. to R baseline			diff. to EM baseline		
	avg	max	mon	avg	max	mon	avg	max	mon	avg	max	mon	avg	max	mon	avg	max	mon
page-blocks	1.5%	1.1%	5.2%	0.6%	-0.3%	9.8%	0.2%	0.3%	-18.0%	0.1%	-0.2%	-8.9%	-0.9%	-0.5%	-1.1%	-0.5%	-0.4%	4.7%
optdigits	0.7%	0.8%	2.4%	0.0%	0.0%	0.0%	0.0%	0.0%	0.0%	0.0%	0.0%	0.0%	-1.0%	-0.3%	8.3%	-0.6%	-0.3%	0.2%
letter	7.3%	9.5%	2.9%	5.5%	4.7%	3.7%	3.6%	4.0%	-1.7%	1.5%	1.5%	-1.1%	-1.3%	0.1%	0.4%	-1.5%	-0.6%	1.3%
segment	3.9%	2.1%	24.8%	6.2%	3.1%	23.5%	0.0%	-0.3%	10.8%	-0.1%	-0.1%	6.4%	-5.8%	-2.6%	23.9%	-1.3%	-0.7%	20.2%
kr-vs-kp	0.0%	0.0%	0.0%	1.1%	0.5%	-1.4%	0.0%	0.0%	-0.5%	0.0%	0.0%	-0.1%	-0.5%	-1.1%	-2.3%	-18.7%	-13.1%	-17.4%
pendigits	2.7%	2.9%	11.4%	1.1%	0.8%	3.0%	1.8%	2.0%	-3.4%	0.2%	0.1%	-2.5%	0.0%	0.3%	3.3%	-0.6%	-0.5%	1.1%
vowel	6.0%	4.1%	8.0%	4.9%	3.8%	6.4%	0.2%	0.9%	-5.1%	0.3%	0.1%	-1.9%	-0.8%	-1.1%	4.7%	-1.8%	-1.5%	4.4%
spambase	0.8%	-0.3%	1.2%	0.0%	0.0%	0.0%	0.0%	0.0%	0.7%	0.0%	0.0%	0.1%	-7.1%	-5.1%	-0.2%	-6.3%	-3.1%	-1.9%
gender	2.6%	3.9%	1.2%	3.5%	3.9%	1.3%	0.0%	0.0%	0.0%	0.0%	0.0%	-0.2%	1.1%	1.3%	5.9%	-4.7%	-3.3%	0.7%
covtype	3.4%	5.6%	3.0%	14.3%	12.3%	5.4%	-2.4%	-2.4%	2.3%	-0.3%	-0.3%	1.7%	-	-	-	-	-	-
averages	2.9%	3.0%	6.0%	3.7%	2.9%	5.1%	0.3%	0.5%	-1.5%	0.2%	0.1%	-0.7%	-1.8%	-1.0%	4.8%	-4.0%	-2.6%	1.5%

Fig. 6. Approaches for parameter optimization.

over all folds per improvement. The employed data sets and their characteristics are listed in Figure 5 (right). They are available at [7] (and [1] for *gender*) where further details and background information can be found. We summarize the results and our findings in Section 4.5.

4.1 Parameter Optimization

Figure 6 shows the improvements in all three measures for the three proposed parameter optimization approaches. The numbers are absolute differences to the corresponding baseline method, highlighted cells indicate improved performance. The additional row contains the average values over all domains. We report the actual values of the measures for the baselines and the final BT* in Figure 9 below. In Figure 7 (right) we show the resulting anytime accuracy plots for the single approaches using the *letter* data set as an example.

The results shown for the bandwidth estimation in Figure 6 are the best results over the three heuristics from Equations 17 to 19, where we tested for the latter $\alpha \in \{0.001, 0.005, 0.01, 0.05, 0.1, 0.5\}$. In the 20 results (*EM* and *R* on 10 data sets) *haerdle* and *langley* were both chosen three times, twice *f*0.01 was the best choice, and the remaining results were achieved by *f*0.05. By the optimized bandwidth estimation all accuracy values, i.e. *avg* and *max* on both *R* and *EM*, could be improved with the exception of *max* for *EM* on *page-blocks* and *max* for *R* on *spambase*. The largest improvement is achieved for the monotonicity with the single exception of *EM* on *kr-vs-kp*. The anytime plot in Figure 7 illustrates the good performance of *BT ◊ BW* in all three measures.

The performance of using block covariance matrices in the *BT ◊ BN* approach is hardly better than any of the two baselines. As above, the results shown in Figure 6 are the best among all parameter settings for *BT ◊ BN*, i.e. over all block sizes **s** and numbers of levels *m* (see Section 3.2). The largest improvements are achieved on the *letter* data set with **s** = *d* and *m* = 1. The performance gain was less for smaller block sizes (results not shown). As mentioned in Section 3.2, the additional degrees of freedom gained through the covariances seem to be useless or even harmful on lower levels of the tree: the displayed results, which

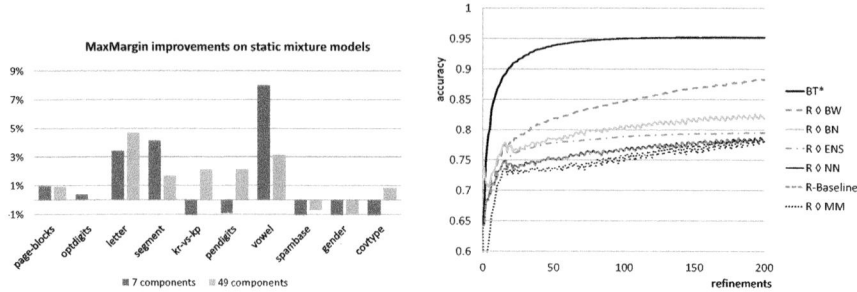

Fig. 7. Left: Results for MM on static mixture models of size 7 and 49. Right: Effects of single approaches on *letter*.

are the best among all m and s, all use $m = 1$ and $s = d$. Nonetheless, as stated above, the performance gain is only marginal on 9 of 10 data sets.

The margin maximization approach $BT \diamond MM$ does not add covariances but only seeks to improve the weights, means and variances of the Gaussians in the Bayes tree. The results on the 10 tested data sets are shown in Figure 6 (The results for *covtype* could not be achieved with 4GB RAM). As above we show the best results over all parameter settings where $\kappa \in \{1..10\}$, $\lambda \in [0, 10]$ and m from 1 to the maximal tree height. On average $BT \diamond MM$ improves the monotonicity over the baselines, but neither of the accuracy measures. The detailed results show slight improvements over the R baseline on three data sets. This result is surprising at first sight, since the original concept from [18] is designed for improving Gaussian mixture models. We discuss possible reasons below.

To exclude the possibility that the poor performance of $BT \diamond MM$ is solely due to the data set characteristics, we tested our implementation of MM on static Gaussian mixture models. Using the same expectation maximization clustering that we use for constructing the Bayes tree in the EM baseline, we created for each data set two mixtures, each contains one model per class. In the first mixture each model has 7 components, in the second 49 components per class were used. We chose multiples of 7 since it corresponds to the chosen fanout of the Bayes tree. We report the resulting absolute gain in classification accuracy from the optimized over the initial mixtures in Figure 7. MM improves the accuracy for at least one mixture on 8 data sets and for both mixtures on 5 of 10 data sets in our experiments. For the *vowel* data set the improvement is 3% for the 49 components and nearly 8% for the 7 components. However, neither *avg* nor *max* are improved by $BT \diamond MM$ on *vowel* (see Figure 6). One reason for this is the fact that $BT \diamond MM$ optimizes the mixtures in the Bayes tree level by level, but the decision f_{BT} uses arbitrary mixtures that can contain components from many different levels of the tree. These components, or rather these mixtures, were never optimized together. Optimizing all possible mixtures is not feasible, since on the one hand the sheer number of possible mixtures makes the optimization computationally infeasible, and on the other hand such an approach does not yield a single set of parameters values per Gaussian.

	Ensemble (BT ◊ ENS)						Nearest Neighbor (BT ◊ NN)					
	diff. to R baseline			diff. to EM baseline			diff. to R baseline			diff. to EM baseline		
	avg	max	mon	avg	max	mon	avg	max	mon	avg	max	mon
page-blocks	1.2%	0.4%	10.7%	0.8%	-0.2%	9.7%	0.7%	0.5%	3.8%	-46.3%	-0.8%	-47.2%
optdigits	-1.1%	-2.0%	11.4%	-3.2%	-3.5%	1.1%	0.0%	0.0%	0.1%	0.0%	0.0%	0.0%
letter	1.7%	0.7%	3.1%	1.6%	0.0%	3.8%	0.1%	0.2%	-0.3%	-0.2%	0.3%	1.5%
segment	5.9%	1.5%	37.7%	4.2%	0.5%	24.3%	0.0%	0.0%	-0.1%	4.4%	2.4%	18.6%
kr-vs-kp	-1.7%	-3.4%	2.0%	-2.7%	-1.6%	0.7%	-0.1%	-0.1%	-1.1%	0.1%	0.1%	-0.2%
pendigits	2.0%	1.1%	12.5%	0.3%	-0.1%	3.1%	0.1%	0.0%	0.8%	0.6%	0.3%	2.7%
vowel	2.4%	0.7%	8.1%	2.8%	1.1%	4.8%	4.9%	4.0%	7.4%	3.6%	3.9%	5.4%
spambase	2.1%	-0.2%	12.9%	-4.0%	-5.7%	-3.8%	0.0%	0.0%	0.0%	-0.2%	-0.2%	-0.6%
gender	0.4%	1.5%	1.3%	2.3%	1.7%	1.4%	-2.1%	-1.7%	-4.4%	-3.4%	-0.8%	-17.3%
covtype	2.3%	3.4%	3.0%	6.2%	1.8%	5.4%	-0.5%	-0.1%	-3.2%	-11.9%	-6.9%	-37.9%
averages	1.5%	0.4%	10.3%	0.8%	-0.6%	5.1%	0.3%	0.3%	0.3%	-5.3%	-0.2%	-7.5%

Fig. 8. Decision design approaches

4.2   Decision Design

For the decision design we tested $f_{BT \diamond ENS}$ and $f_{BT \diamond NN}$ and show the absolute improvements for the three objectives in Figure 8. The ensemble over time in the Bayes tree (see Equation 20) yields on average a slight increase in the accuracy measures avg and max for the R baseline and is rather neutral on the EM baseline. The monotonicity, however, is drastically increased by $BT \diamond ENS$, on average by more than 10% compared to the R baseline and more than 5% compared to the EM baseline. This is underlined by the anytime accuracy plot of $BT \diamond ENS$ in Figure 7 which shows a smooth and monotonically increasing behaviour.

In contrast, the results of $BT \diamond NN$ hardly show any improvement over the R baseline except for *vowel*. The anytime plot for $BT \diamond NN$ is hardly visible in Figure 7, since it coincides with the curve of the R baseline. This is another surprising result: it indicates that taking per label only the one single Gaussian, which yields the highest class conditional density for the query object, results in almost exactly the same decisions as taking the entire mixture models. As can be seen in Figure 8, this strongly holds for 7 of the 10 tested data sets on the R baseline. Compared to the EM baseline the performance of $BT \diamond NN$ is worse on average. Summarizing the evaluation of the single approaches we can conclude that BW, ENS and BN successfully improve the anytime accuracy (see Figure 7 (right)). The first two additionally drastically improve the monotonicity on all domains.

4.3   Combining Approaches

Next we study the cumulative performance gain when improving BT by more than one concept. Figure 9 shows the anytime accuracy plots on *letter* for the EM baseline and the combined versions using ENS and/or BN and/or BW. The curve of the EM baseline exhibits a strong oscillation. $EM \diamond BN$ improves

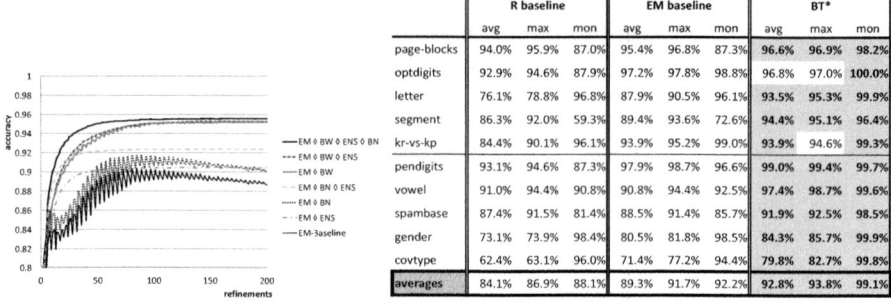

Fig. 9. Left: Anytime plots for combined approaches. Right: Baseline results for R and EM and the results for the proposed BT*.

the accuracy throughout on this data set, but cannot diminish the oscillation. Near perfect monotonicity is reached when using ENS, either alone ($EM\diamond ENS$) or additionally ($EM\diamond BN\diamond ENS$). The cumulation of the two positive effects, i.e. increased accuracy and monotonicity, is clearly expressed by the corresponding anytime plots. $EM \diamond BW$ pushes up the accuracy and can also improve the monotonicity. Adding ENS yields again near perfect monotonicity (see $EM \diamond BW \diamond ENS$). Finally, combining the three concepts with EM yields the best results on this domain.

To find the globally best results we allowed all combinations of the proposed improvements and selected per data set the best setting with respect to the linear combination of all three measures. In the resulting settings ENS was used on all data sets and $f0.05$ was used eight times for BW, while other parameter optimizations were rarely employed (once BN and twice MM).

For the final setting that is used on all data sets we chose BT* $= EM \diamond f0.05 \diamond ENS$. The results are shown in Figure 9 (right) along with the two baselines R and EM. The values shown are the absolute values for the measures. Highlighted cells indicate an improvement over both baselines. On all data sets all three measures are improved by the BT* except for avg and max on optdigits and max on kr-vs-kp, where it shows slightly worse performance compared to EM. Overall, improving BT to BT* yields very good results on all tested data sets. Figure 10 (left) shows the anytime accuracy plots for BT* which illustrate the great performance with respect to all three measures. Figure 10 visually summarizes the average results of the single concepts and BT*, underlining the superior performance of BT*.

4.4   Scalability

To investigate the scalability of the BT* classifier we have to consider both large training data sets and large test data sets. The former affect the construction time and the storage of the data structure. For the latter, the classification time is of interest. Before we discuss these issues, we introduce three final results that are important for both training and testing. The anytime accuracy plots in

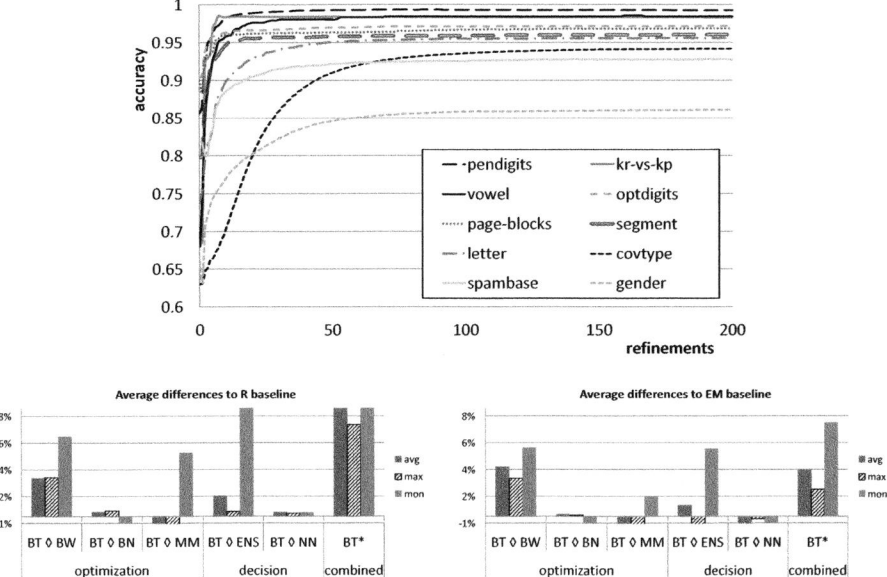

Fig. 10. Top: Anytime accuracy results for BT^*. Bottom: Average absolute differences over all 10 data sets for single and combined approaches.

Figures 7, 9 and 10 show the accuracy values for the first 200 refinements. If even more refinements are performed, the classification decision barely changes, since the density estimates change only marginally with remote mixture components being refined. Figure 11 shows for three data sets the accuracy performance for all possible refinements.

As described in the previous section, BT* uses the EMTopDown construction. For very large training data sets the complexity of the EM clustering algorithm can yield high training times. However, if the data is collected over time, BT* can be trained using the EMTopDown strategy on an initial data base and new training data can be incrementally learned in addition. This strategy can also be applied for large training data sets that are readily available. Another option in this scenario is to sample the data base and perform the EM construction on the sample. The results from Figures 10 and 11 suggest that accurate classification decisions can be achieved based on relatively small parts of the training data.

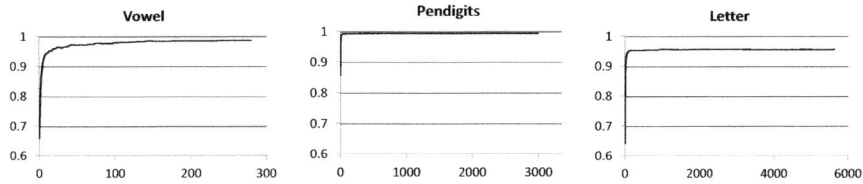

Fig. 11. Computing all possible refinements for *vowel*, *pendigits*, and *letter*

If the size of the resulting data structure exceeds the main memory, it can be stored and accessed from secondary storage. One page hosts in this scenario one node of the tree structure (as in common index structures), experiments for page accesses and different page sizes can be found in [19].

During classification, the time that is needed for a single refinement depends on the dimensionality d and on the number of previous refinements r. For every entry in the newly read node a Gaussian has to be evaluated, which is in $O(d)$. After that the entry has to be sorted into the frontier, which can be done in $O(\log_2^2(r \cdot maxFanout))$ using a heap (after r refinements the frontier contains maximally $r \cdot maxFanout$ entries). The actual times for *gender* and *pendigits* shown in Figure 2 correspond to 64 and 82 μs per refinement, respectively. From the results shown in Figures 10 and 11 we can derive that a rather small number of refinement (around 200) most often suffices to achieve a classification accuracy that is comparable to the ultimate performance. Hence, the classification time for a single object is very low even for large training data sets, which renders the BT* algorithm well suited for applications with very large test data sets.

4.5 Summary

We investigated three approaches for parameter optimization and two novel methods for the decision design. The margin maximization concept MM and the nearest neighbor like decision method NN both yielded only marginal improvements, if any. Adding covariances using the Bayesian network approach BN improved the performance only on very few domains and left it unchanged in most other cases. Two approaches that together significantly improved both accuracy and monotonicity are the bandwidth estimation BW and the ensemble over time ENS. Therefore we evaluated in Section 4.3 BT* $= EM \diamond f0.05 \diamond ENS$ (see Figure 10) that we suggest as the final variant of our anytime Bayesian classifier.

5   Conclusion

Applications for stream classification are numerous and anytime classifiers are well suited for this task since they flexibly use all available time and can provide a result after any time. Two important properties of an anytime classifier are high accuracies regardless of the available time and monotonic increase of the accuracy with additional time allowance. In this paper we have significantly improved both the monotonicity and the anytime accuracy of a recent anytime classifier. The proposed BT* algorithm achieved near perfect results on all tested data sets. It uses a special kind of ensemble that combines mixtures of different granularities resulting from the same classifier over time. The improved performance is achieved without sacrificing the time complexity, which is the same as in previous approaches. In summary, the BT* algorithm constitutes an efficient and consistent solution for anytime classification.

Acknowledgments. This work has been supported by the UMIC Research Centre, RWTH Aachen University, Germany.

References

1. Andre, D., Stone, P.: Physiological data modeling contest, ICML 2004 (2004), http://www.cs.utexas.edu/sherstov/pdmc/
2. Arai, B., Das, G., Gunopulos, D., Koudas, N.: Anytime measures for top-k algorithms on exact and fuzzy data sets. VLDB Journal 18(2), 407–427 (2009)
3. Bouckaert, R.R.: Naive Bayes Classifiers That Perform Well with Continuous Variables. In: Webb, G.I., Yu, X. (eds.) AI 2004. LNCS (LNAI), vol. 3339, pp. 1089–1094. Springer, Heidelberg (2004)
4. Dean, T., Boddy, M.S.: An analysis of time-dependent planning. In: AAAI, pp. 49–54 (1988)
5. DeCoste, D.: Anytime query-tuned kernel machines via cholesky factorization. In: Proc. of the 3rd SIAM SDM (2003)
6. Esmeir, S., Markovitch, S.: Anytime learning of anycost classifiers. Machine Learning, 25th Anniversary 82(3), 445–473 (2011)
7. Frank, A., Asuncion, A.: UCI machine learning repository (2010)
8. Gopalakrishnan, P.S., Kanevsky, D., Nadas, A., Nahamoo, D.: An inequality for rational functions with applications to some statistical estimation problems. IEEE Transactions on Information Theory 37(1), 107–113 (1991)
9. Guttman, A.: R-trees: A dynamic index structure for spatial searching. In: ACM SIGMOD, pp. 47–57 (1984)
10. Härdle, W., Müller, M.: Multivariate and semiparametric kernel regression. In: Smoothing and Regression. Wiley Interscience (1997)
11. John, G., Langley, P.: Estimating continuous distributions in bayesian classifiers. In: UAI. Morgan Kaufmann (1995)
12. Keogh, E.J., Pazzani, M.J.: Learning the structure of augmented bayesian classifiers. Intl. Journal on AI Tools 11(4), 587–601 (2002)
13. Kranen, P., Assent, I., Baldauf, C., Seidl, T.: Self-adaptive anytime stream clustering. In: ICDM, pp. 249–258 (2009)
14. Kranen, P., Günnemann, S., Fries, S., Seidl, T.: MC-Tree: Improving Bayesian Anytime Classification. In: Gertz, M., Ludäscher, B. (eds.) SSDBM 2010. LNCS, vol. 6187, pp. 252–269. Springer, Heidelberg (2010)
15. Likhachev, M., Ferguson, D., Gordon, G.J., Stentz, A., Thrun, S.: Anytime search in dynamic graphs. Artificial Intelligence 172(14), 1613–1643 (2008)
16. Likhachev, M., Gordon, G.J., Thrun, S.: ARA*: Anytime A* with provable bounds on sub-optimality. In: NIPS (2003)
17. Liu, C.-L., Wellman, M.P.: On state-space abstraction for anytime evaluation of bayesian networks. SIGART Bulletin 7(2), 50–57 (1996)
18. Pernkopf, F., Wohlmayr, M.: Large Margin Learning of Bayesian Classifiers Based on Gaussian Mixture Models. In: Balcázar, J.L., Bonchi, F., Gionis, A., Sebag, M. (eds.) ECML PKDD 2010. LNCS, vol. 6323, pp. 50–66. Springer, Heidelberg (2010)
19. Seidl, T., Assent, I., Kranen, P., Krieger, R., Herrmann, J.: Indexing density models for incremental learning and anytime classification on data streams. In: EDBT/ICDT, pp. 311–322 (2009)
20. Silverman, B.W.: Density Estimation for Statistics and Data Analysis. Chapman & Hall/CRC (1986)
21. Ueno, K., Xi, X., Keogh, E.J., Lee, D.-Y.: Anytime classification using the nearest neighbor algorithm with applications to stream mining. In: ICDM, pp. 623–632 (2006)
22. Wang, H., Fan, W., Yu, P.S., Han, J.: Mining concept-drifting data streams using ensemble classifiers. In: KDD, pp. 226–235 (2003)
23. Yang, Y., Webb, G.I., Korb, K.B., Ting, K.M.: Classifying under computational resource constraints: anytime classification using probabilistic estimators. Machine Learning 69(1) (2007)

Finding the Largest Empty Rectangle
Containing Only a Query Point
in Large Multidimensional Databases[*]

Gilberto Gutiérrez[1] and José R. Paramá[2]

[1] Universidad del Bío-Bío, Departamento de Ciencias de la Computación y
Tecnologías de la Información, Chillán, Chile
ggutierr@ubiobio.cl
[2] University of A Coruña, Department of Computer Science, España
jose.parama@udc.es

Abstract. Given a two-dimensional space, let S be a set of points stored
in an R-tree, let R be the minimum rectangle containing the elements
of S, and let q be a query point such that $q \notin S$ and $R \cap q \neq \emptyset$. In this
paper, we present an algorithm for finding the empty rectangle with the
largest area, sides parallel to the axes of the space, and containing only
the query point q. The idea behind algorithm is to use the points that
define the minimum bounding rectangles (MBRs) of some internal nodes
of the R-tree to avoid reading as many nodes of the R-tree as possible,
given that a naive algorithm must access all of them. We present several
experiments considering synthetic and real data. The results show that
our algorithm uses around 0.71–38% of the time and around 3–4% of the
main storage needed by previous computational geometry algorithms.
Furthermore, to the best of our knowledge, this is the first work that
solves this problem considering that the points are stored in an R-tree.

1 Introduction

In computational geometry, there is a research line that is aimed at finding empty
geometric figures in a space that contains a set of points. For example, one of
them is to find the largest empty axis-parallel rectangle in a space containing a
set of points (see Figure 1(a)). A variant of the previous problem is to find the
largest rectangle that only contains a given query point, assuming that the query
point does not belong to the set of points in the space (see Figure 1(b)). More
variants of this problem are those that find a circumference, a square, or a convex
hull. In addition, in the case of rectangles and squares, another alternative is to
consider figures with sides that are not parallel to the axes.

The search for empty geometric figures with the largest area, or any other
metric, has applications in several fields. Among them, we can cite data mining

---

[*] This work was supported in part by the project MECESUP UBB0704 (Chile) in
the context of a postdoctoral stay of the first author at the University of A Coruña
(Spain); and (for second author) by the Spanish Ministerio de Educación y Ciencia
[TIN2010-21246-C02-01].

A. Ailamaki and S. Bowers (Eds.): SSDBM 2012, LNCS 7338, pp. 316–333, 2012.

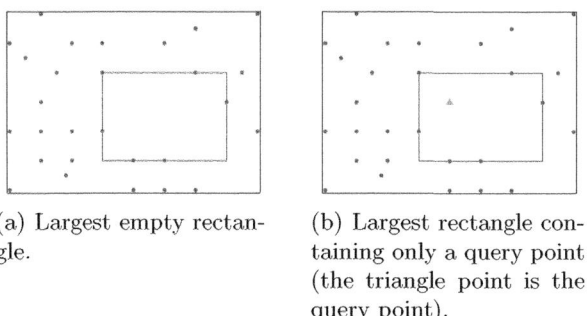

(a) Largest empty rectangle.

(b) Largest rectangle containing only a query point (the triangle point is the query point).

Fig. 1. The two variants of the problem

[EGLM03], geographical information systems (GIS), and very-large-scale integration design [ADM+10a].

In the case of data mining, Edmonds, et al. [EGLM03] propose the search for empty spaces as a complementary technique for the characterization of data patterns. More precisely, it is interesting to discover if there are certain ranges of values that never appear together. For example, suppose a database that stores the amounts and dates of bank deposits. Consider a graph where we have the time in the x axis and the amount in the y axis. An empty space indicates that during a given period of time there were not deposits within a certain range of amounts. For example, if we find that during years 2007 and 2008, there were no deposits of more than one million dollars, this could a symptom of a new economic crisis that is arising. In this scenario, the query point could be defined by a time point and a minimum deposit amount. Apart from the discovered knowledge, the empty spaces have value by their own [EGLM03]. Another example could be a database of a Hospital or Social Security system. Considering data about surgery operations, it is possible to discover that there are not face transplants in the database before 2008. This knowledge indicates that such procedure was not possible before that year, and it can be introduced as a integrity restriction of the database, in order to perform semantic query optimization [Kin81].

As an example of GIS application, suppose that we want to build a park in a region, and that we have a database that stores the buildings/houses, electric towers, and other important facilities of that region. The following queries can be interesting: *which is the largest empty rectangular space?* (this gives the space where it is cheaper to place the park) or *which is the largest empty rectangular space around a certain position q?*, if we have a restriction in the position of the park.

The spatial databases (SDB) represent an important aid for GIS to manage large amounts of data. Yet, SBDs require the development of efficient algorithms and data structures in order to address several query types, among others, the window query, the intersection query, the nearest neighbor, or the pair of nearest neighbors [GG98, SC03]. Many of those query types are problems that were first tackled in the field of the computational geometry, where it is assumed that all spatial objects can be fit into main memory, and later, those problems were faced in the field of the SBDs. Following this path, several algorithms have been proposed considering that objects are stored in a multidimensional structure, in most cases an R-tree [Gut84], for example; [HS98, Cor02, CMTV04] present several algorithms that solve the k-pairs ($k \geq 1$) of nearest neighbors between two sets, [RKV95] shows an algorithm to find the nearest neighbor to a given point, and more recently, [BK01] presents an algorithm to obtain the convex hull of a set of points stored in an R-tree.

Given a space with a set of points stored in an R-tree, the main contribution of this paper is an algorithm, called $q-$MER, that finds the largest axis-paralell rectangle that only contains a given query point. The algorithm computes a set of candidate maximum empty rectangles (CERs). At least one of these CERs is either the solution or a higher bound of the solution. Instead of inspecting the complete R-tree, we only inspect the blocks/nodes of the R-tree that intersect with those CERs. Our approach takes advantage of the extensive use of the R-tree in real database management systems (Oracle, PostgreSQL, etc.), extending the usefulness of that data structure.

The results show that our algorithm requires between 0.71% and 38% of the running time and between 3% and 4% of the storage required by the naive approach of reading all the points from disk, storing them in main memory, and finally running a computational geometry algorithm with those points. The improvement of our approach comes, in part, by its filtering capability; in our experiments, $q-$MER only needed to access between 10% and 55% of the blocks of R-tree

The outline of the paper is as follows: Section 2 presents some previous related work. Section 3 presents our algorithm. Section 4 shows the results of our experiments. Finally, Section 5 shows our conclusions and directions for future work.

2   Related Work

Given a space with a set of n points, the search for the largest empty geometric figure (circumference, square, rectangle, or convex hull) has been an active research field in last decades. Focusing on the problem of finding the largest rectangle with sides parallel to the axes of the space, two variants have been considered: (i) no information about the position of the figure is provided, and (ii) information about the position is provided, typically by means of a point.

The first variant has been extensively studied. The first work is [NLH84], where two algorithms are described: the first one takes $O(n^2)$ time and $O(n)$ space, the second one takes $O(n \log^2 n)$ expected time considering that the points are randomly arranged into the space. Later, in [CDL86], it is presented a divide-and-conquer algorithm with $O(n \log^3 n)$ time complexity using $O(n \log n)$ space. An algorithm with similar time complexity is discussed in [AS87], this one using $O(n)$ space. In [Orl90], it is shown an algorithm that takes $O(s \log n)$ time, where s is the number of restricted rectangles [Orl90]. Moreover, that algorithm has an expected time $O(n \log n)$. A more recent approach [DN11] takes $O(n \log^2 n + s)$ time and $O(\log n)$ space by using a priority search tree.

For the second variant, [ADM+10a], [ADM+10b], and [KS11] present algorithms to find the largest empty rectangle that contains a query point. The algorithm presented in [ADM+10a] and [ADM+10b] performs a preprocessing step where the space is divided into a set of cells such that all points that fall in the same cell produce the same maximum empty rectangle (MER). These cells are stored in main memory organized into a data structure for objects in two dimensions called range tree. The preprocessing stage takes time and storage $O(n^2 \log n)$ and, to retrieve the MER corresponding to a query point q, an additional $O(\log n)$ time is needed. The algorithm in [KS11] corresponds to a significant improvement in terms of time and space with respect to those in [ADM+10a] and [ADM+10b]. Specifically, this algorithm requires $O(n\alpha(n) \log^3 n)$ storage to maintain the data structure (a segment tree), $O(n\alpha(n) \log^4 n)$ time to build the structure, and $O(\log^4 n)$ time to find the MER that only contains q, where the term $\alpha(n)$ is the slowly increasing inverse Ackermann function.

All the algorithms commented so far assume that the objects can be fit into main memory. Edmonds, et al. [EGLM03] face the problem of finding all the empty spaces left by a set of objects, assuming that the main memory does not have enough space to store all the objects. That algorithm takes $O(|X||Y|)$, where X e Y are the distinct values of the coordinates of the data set. Yet, this work does not consider the case where the objects are stored in a multidimensional structure.

3   Empty Rectangle with Largest Area That Contains only a Query Point

Given a set of points stored in an R-tree, a naive solution could be to read all of them from the R-tree and then find the largest empty rectangle using some of the algorithms described in Section 2. Instead of reading all nodes (disk blocks) of the R-tree, $q-$MER uses the query point q and the properties of the MBRs of the R-tree to avoid inspecting as many blocks as possible. It requires two steps:

1. First, it computes a set of CERs, where at least one of them is either the solution or a higher bound of the solution. In order to obtain those CERs, another two steps are needed:

   (a) A set of points, called C, is generated from the query point q and the MBRs of the R-tree. Specifically, for each MBR in parent nodes of the leaves of the R-tree, the algorithm might add one or two points to C. Those points correspond to the most distant points to the query point q that could be located in the considered MBR. Therefore, it is likely that most of the points in C do not exist in reality. Figure 2 displays an example. From the MBRs in parent nodes of leaves (the rectangles drawn with solid lines) and the query point q, $q-$MER produces the set of points $C = \{p_1, p_2, p_3, p_4, p_5, p_6\}$. The MBRs in parent nodes of leaves are processed sequentially. The process of the MBR R_1 produces the point p_1, since it is the farthest point with respect to q that could be located in that MBR, but, as explained, p_1 is probably not part of the set of points actually stored in the R-tree. In the same way, the process of R_2 produces the point p_2, R_3 generates p_5 and p_6, and finally R_4 adds p_3 and p_4.

(b) A computational geometry algorithm is run using the set C as input to finally obtain the CERs. A CER is similar to a MER, which is a rectangle that cannot be enlarged if we want to keep only the query point inside it. The difference between a CER and a MER is that while MERs are computed with real points, CERs are computed using C.

Observe that a MER (CER) is not necessarily the largest empty rectangle containing q. Figure 2 displays two CERs, labeled as A and B, which are the rectangles with dotted lines. There are others, but they are not shown to simplify the illustration. For example, observe that the CER B can not be enlarged in any direction. To the south, the space ends; to the east, the CER can not be enlarged, otherwise B would contain the point p_6; to the north, the CER founds a barrel in points p_3 and p_4; and finally to the west, p_5 represents an obstacle to the growth of B. This does not mean that B is the largest empty rectangle containing q, for instance, A is larger.

2. In the second step, the CERs are processed according to their area, from largest to smallest. For each CER, our algorithm accesses the leaves of the R-tree that contain the real points that intersect with such a CER. Those real points, that we call C', are used to obtain a candidate solution (by means of the same computational geometry algorithm used to obtain the CERs). This candidate solution is the real MER (since it is computed using real points) that is equal to or contained into the processed CER. As the process of CERs progresses, the candidate solutions may improve previous ones. For example, when the CER B of Figure 2 is processed, it is necessary to access the children of the entries containing the MBRs R_3 and R_4. Those nodes are leaves of the R-tree, and contain the real points that caused the creation of R_3 and R_4. Then $q-$MER inserts in C' the points that intersect with B and processes C' with the computational geometry algorithm. Finally, if the obtained candidate solution is better than the previous ones, then it passes to be considered the best candidate solution so far.

3.1   Basic Definitions

First of all, let us present the problem more formally and some definitions that will be used later.

Given a set of points S in a space $R \subseteq \mathcal{R}^2$, which is stored in an R-tree, and a point $q \notin S$ that is in R, find the rectangle, also in R, with the largest area that only includes q.

Let p be a point, v_p is the vertical line that covers R from north to south and passes through p. h_p is the horizontal line that covers R from west to east and passes through p. These two lines define four regions in R: (i) $NW(p)$ is the

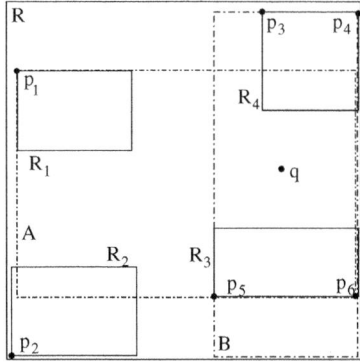

Fig. 2. An example of the elements involved in the first step of $q-$MER

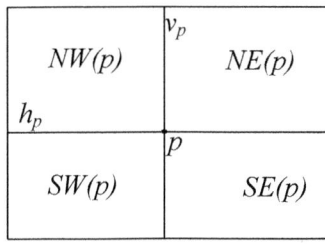

(a) Elements defined by a point.

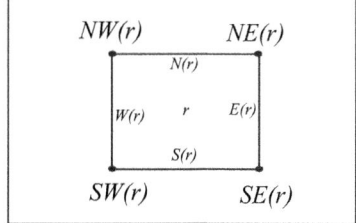

(b) Elements defined by a rectangle.

Fig. 3. Definitions

northwestern region of p, (ii) $NE(p)$ is the northeast region of p, (iii) $SW(p)$ is the southwestern region of p, and (iv) $SE(p)$ is the southeast region of p. Figure 3(a) displays all these elements.

The four corners of a rectangle r are denoted as: (i) $NW(r)$, the northwestern corner, (ii) $NE(r)$, the northeast corner, (iii) $SW(r)$, the southwest corner, and (iv) $SE(r)$ the southeast corner. In addition, a rectangle r defines four lines: (i) $W(r)$ is the line that connects $NW(r)$ with $SW(r)$, (ii) $E(r)$ is the line that connects $NE(r)$ with $SE(r)$, (iii) $N(r)$ is the line that connects $NW(r)$ with $NE(r)$, and (iv) $S(r)$ is the line that connects $SW(r)$ with $SE(r)$. Figure 3(b) shows these elements.

3.2 Obtaining the CERs

As explained, $q-$MER starts by computing a set of CERs. To obtain them, we use a variant of a computational geometry algorithm that obtains the rectangle with the largest empty area. This variant, that we call *ComputeCER*, produces the set of MERs, instead of obtaining only the largest one.

The key idea is to use *ComputeCER* with much fewer points than in the case of using the computational geometry algorithm over the whole set of points.

Algorithm 1. Algorithm that processes an R-tree to obtain the CERs.

1: step1(q, T)
2: INPUT: q {the query point and T the $R - tree$}
3: OUTPUT: L_{CER} {a set of CERs}
4: Let $C = \emptyset$ {a set of points}
5: **for each** node n parent of the leaves of T **do**
6:     **for each** MBR MBR_i in n **do**
7:         **if** q is not inside MBR_i **then**
8:             **if** h_q and v_q do not cross $N(MBR_i)$, $S(MBR_i)$, $E(MBR_i)$, and $W(MBR_i)$ **then**
9:                 **if** MBR_i is completely inside $NW(q)$ **then**
10:                     **add** $NW(MBR_i)$ to C
11:                 **else if** MBR_i is completely inside $SW(q)$ **then**
12:                     **add** $SW(MBR_i)$ to C
13:                 **else if** MBR_i is completely inside $NE(q)$ **then**
14:                     **add** $NE(MBR_i)$ to C
15:                 **else if** MBR_i is completely inside $SE(q)$ **then**
16:                     **add** $SE(MBR_i)$ to C
17:                 **end if**
18:             **else**
19:                 **if** h_q or v_q crosses exactly two of the lines $N(MBR_i)$, $S(MBR_i)$, $E(MBR_i)$, and $W(MBR_i)$ **then**
20:                     **if** MBR_i intersects with $NW(q)$ and $SW(q)$ **then**
21:                         **add** $NW(MBR_i)$ and $SW(MBR_i)$ to C
22:                     **else if** MBR_i intersects with $SW(q)$ and $SE(q)$ **then**
23:                         **add** $SW(MBR_i)$ and $SE(MBR_i)$ to C
24:                     **else if** MBR_i intersects with $NE(q)$ and $NW(q)$ **then**
25:                         **add** $NE(MBR_i)$ and $NW(MBR_i)$ to C
26:                     **else if** MBR_i intersects with $SE(q)$ and $NE(q)$ **then**
27:                         **add** $SE(MBR_i)$ and $NE(MBR_i)$ to C
28:                     **end if**
29:                 **end if**
30:             **end if**
31:         **end if**
32:     **end for**
33: **end for**
34: $L_{CER} = ComputeCER(C, q)$

As it can be seen in Algorithm 1, the first step of $q-$MER obtains zero, one, or two points from each processed MBR, depending on three cases.

The first case is when the *if* of line 8 is true. This means that the considered MBR_i is completely inside one of the regions defined by the query point (see Figure 4(a)). In this case, the algorithm produces the point of the farthest corner of MBR_i with respect to the query point. In Figure 4(a), it is supposed that MBR_i is in $SE(q)$, and therefore the point $SE(MBR_i)$ is added to the set of points C.

Another treated case is when the *if* of line 19 is true. This means that MBR_i intersects with two of the regions defined by the query point (see Figure 4(b)). In this case, two points are added to the set C, those in the farthest corners of MBR_i with respect to the query point. In Figure 4(b), MBR_i intersects with regions $NE(q)$ and $SE(q)$, and therefore the algorithm adds $NE(MBR_i)$ and $SE(MBR_i)$ to C.

The last case appears when the query point is inside the considered MBR_i. For this situation, we have three options:

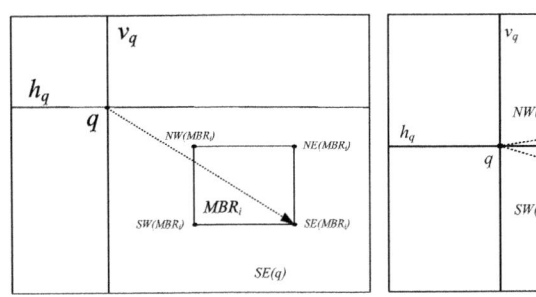 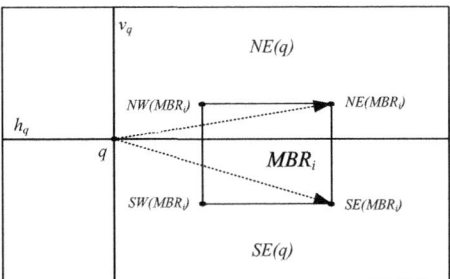

(a) Example of first case.   (b) Example of the second case.

Fig. 4. The two cases tackled by the first step of $q-$MER

1. The first option is to split C in two sets of points $C_1 = C \cup \{NE(MBR_i), SW(MBR_i)\}$ and $C_2 = C \cup \{NW(MBR_i), SE(MBR_i)\}$. Now each set should continue the whole process independently. This apparently does not represent a big issue. The problem arises when the query point is in more than one MBR. In this case the number of set of points increases rapidly, since for C_1 two new sets should be created C_{11} and C_{12}, and the same for C_2. Furthermore, since for each set of points C_i, several CERs should be created, if the number of sets of points grows fast, the same will happen with the number of CERs.
2. To access the leaf node corresponding to MBR_i and add all the real points it contains to C. This significantly increases the number of points in C and thus, the number of CERs to be processed. Observe again, that the query point might be inside several MBRs, and then all the points in the leaves corresponding to the entries of those MBRs should be added to C.
3. No point is added to C.

We chose the third option since the other two options increase the computation time, whereas the benefits we found in the filtering capability were not significant.

As explained, once we have the set of points C, the algorithm runs the *ComputeCER* to obtain the CERs. Next we prove that the solution cannot be larger than, at least, one of these CERs.

Theorem 1. *Let L_{CER} the list of CERs obtained by the first step of q-MER. The solution or solutions can not be larger than one of the CERs in L_{CER}.*

Proof:
It is clear that the solution should be one (or more) of the real MERs. For each real MER, we are going to prove that the first step of $q-$MER produces at least one CER that is either equal or a higher bound of that MER. We prove this by showing that the points computed by the first step of $q-$MER using one MBR can not shorten a CER with respect to the corresponding real MER. Once we prove this for the points obtained from one MBR, the proof trivially extends for any number of MBRs.

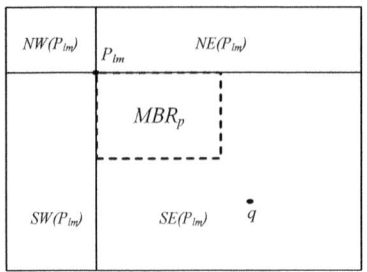
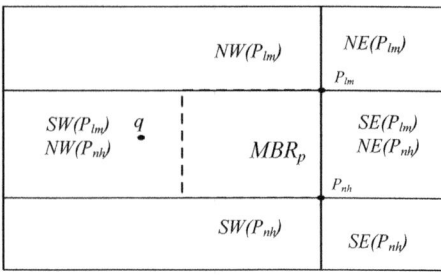

(a) Example of Case 1.   (b) Example of Case 2.

Fig. 5. The two cases of the proof

Let MBR_p a MBR in an entry of a node of the R-tree, which is parent of leaves:

Case 1. Assume that MBR_p is fully inside one of the regions defined by q, then the first step of $q-$MER adds only one point to the set C. Let P_{lm} be that point and let us suppose without loss of generality that P_{lm} is in $NW(q)$. Figure 5(a) shows an example of the four regions defined by P_{lm} and the MBR responsible of its creation.

Assuming no other MBR, the algorithm *ComputeCER* creates two CERs: (1) the union of $SW(P_{lm})$ and $SE(P_{lm})$, and (2) the union of $NE(P_{lm})$ and $SE(P_{lm})$. The region $NW(P_{lm})$ can not be part of a CER that contains q and does not contain P_{lm}, by construction.

If by chance, there is a point in S (the set of real points) with the same coordinates as P_{lm}, we are going to show that the CERs (1) and (2) are larger than corresponding the real MERs. It is clear that the existence of MBR_p requires the presence of more points than P_{lm}, at least one more. One of these two cases should occur. (i) At least the point $SE(MBR_p)$ exists: this point would not allow the existence of MERs with the same size as the CERs (1) and (2), given that $SE(MBR_p)$ is closer to q than P_{lm}, since it is more to the right and in a lower position with respect to P_{lm}. Thus, it would represent an obstacle that would produce MERs shorter than (1) and (2). (ii) The existence of, at least, another two points, one that intersects with $N(MBR_p)$ and another one that intersects with $W(MBR_p)$. These points would be placed with respect P_{lm} more to the right (that lying on $N(MBR_p)$) and in a lower position (that lying on $W(MBR_p)$). Therefore, again the real MERs would be shorter than the CERs (1) and (2).

If P_{lm} does not exist in reality, by construction, MBR_p requires the existence of at least two points; one intersecting $N(MBR_p)$ and another one intersecting $W(MBR_p)$. Those points, once again, must be placed more to the right and in a lower position with respect to P_{lm}, respectively. Therefore, the real MERs would be shorter than the created CERs, as well.

Thus, we can conclude that the CERs (1) and (2) are higher bounds of the real MERs.

Case 2. Now we consider that MBR_p is between two regions of those defined by q. Let P_{lm} and P_{nh} be the two points produced by the first step of q−MER. Without loss of generality, let us suppose that those points are to the east of q (see an example in Figure 5(b)). As it can be seen in the figure, there are only two CERs that only contain q: (1) The area resulting from $(SW(P_{lm}) \cap NW(P_{nh})) \cup (SE(P_{lm}) \cap NE(P_{nh}))$, and (2) the area resulting from $NW(P_{lm}) \cup SW(P_{lm})$, or which is the same, $NW(P_{nh}) \cup SW(P_{nh})$.

For the CER (1), if the real points in MBR_p are placed only intersecting with $N(MBR_p)$ or $S(MBR_p)$, the obtained CER is equal to the real MER. In any other case, the MER that contains q would be, at least, a rectangle less high, and therefore it would have a shorter area.

In the case of CER (2), the existence of MBR_p requires the existence of, at least, one point lying on $W(MBR_p)$. That point would be more to the west than P_{lm} and P_{nh} and, at the same time, more to the east of q (otherwise MBR_p would include q), therefore that point would shorten the real MER with respect to the CER (2).

Observe in Figure 2 that, for example, any point that do not lie on the east and south lines of the MBR R_3 would shorten the CER A. Indeed, those points should exist, otherwise R_3 would not be created. Therefore A is a higher bound for the real MER.

Figure 6 displays a possible arrangement of the real points. As seen, the real MER A' is shortened with respect to the CER A, due to, among other points, the existence of a point intersecting with $N(R_3)$. The case of the CER B is even worse, as its corresponding MER B' is not a MER anymore, as now B' is included in the MER A'.

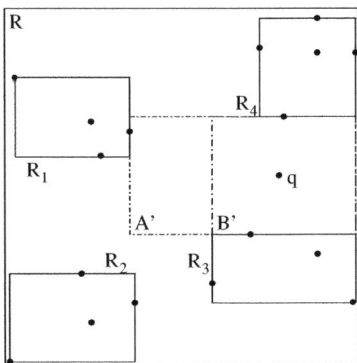

Fig. 6. An example of a real MER corresponding to Figure 2

3.3   Computing the Rectangle with the Largest Area Containing q

Algorithm 2 shows the second step of $q-$MER, which obtains the largest MER containing q. The set of CERs obtained from the first step are stored in a heap where the largest CER is at the top.

The algorithm starts by checking the largest CER. The function *Get* extracts the top of the heap. Now the corresponding real MER is computed by accessing the real points stored in the leaves of the R-tree. We run the computational geometry algorithm *computeER*[1] with the real points that intersect the considered CER. To obtain them, we check all the MBRs of parent nodes of leaves that intersect with the current CER. Moreover, from the points inside those MBRs, we only consider those that actually intersect with the considered CER.

The real MER is stored in a temporary object (TMPMER). The function *area* computes the area of that MER, and if its area is greater than that of the current MER_{max}, then TMPMER becomes MER_{max}. The process ends when the heap becomes empty or the area of the CER at the top of the heap is shorter than that of the current MER_{max}.

Algorithm 2. Algorithm that computes the rectangle with the largest area containing a query point

---

```
1:  Step2(H_CER, q, T)
2:  INPUT: {H_CER a heap with the CERs obtained by ComputeCER, q the query point, and T
        the R − tree}
3:  OUTPUT:MER_max {the largest rectangle containing only q}
4:  Let a = 0 {The area of the MER currently stored at MER_max}
5:  repeat
6:      Let C' = ∅ {A set of points}
7:      Let CER = Get(H_CER) {Extracts the first CER of the heap H_CER}
8:      for  each node n parent of leaves with MBRs intersecting with CER do
9:          Let C' = C'∪ all the points stored in the children of n that intersect with CER
10:     end for
11:     Let TMPMER = computeER(C', q) {The computational geometry algorithm computes the
        largest rectangle only containing q considering the points in C'}
12:     if area(TMPMER) > a then
13:         Let MER_max = TMPMER
14:         Let a=area(TMPMER)
15:     end if
16: until (H_CER is empty) OR (area(CER)< a)
```

---

4   Experimental Results

We compared $q-$MER against a naive algorithm that retrieves all the points stored in the R-tree by reading all the blocks and then solving the problem in main memory with the computational geometry algorithm (*computeER*). We used the algorithm of Naamad, et al. [NLH84] modified to meet our requirements, that is, the computation of the largest rectangle containing only q. *computeCER*

---

[1] *computeER* is similar to *computeCER*, with the difference that *computeER* only returns the largest MER.

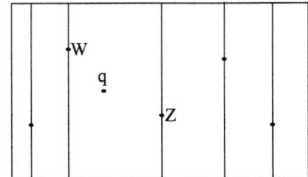

Fig. 7. Naamad's Type A MERs

is also a modification of the same algorithm, which obtains all the possible MERs containing q.

The restriction of the query point allows some improvements in the algorithm that speed up its execution times. Naamad's algorithm compute three types of MERs (called A, B, and C). The MERs of Type A are obtained by drawing a vertical line from the top to the bottom of the space passing through each point (see Figure 7). In our case, we only have to compute the MER delimited by the lines that intersect with points W and Z, since that MER is the only one (of this type) that contains q. To compute this type of MERs, Naamad's algorithm sorts the points by the x coordinate, we take advantage of this ordering to obtain the target MER by performing a binary search that obtains the nearest points to q, in our example, the points W and Z. This reduces the cost from linear to logarithmic. Similar improvements were applied to MERs of Type B and C.

We suppose that it is possible to store all the points in main memory. This eliminates the effect of the memory over our experiments, since we provide the computational geometry with all the memory it needs that, as we will see, it is much more than $q-$MER.

The algorithms were implemented in Java and the programs were run on an isolated Intel®Xeon®-E5520@2.26GHz with 72 GB DDR3@800MHz RAM with a SATA hard disk model Seagate® ST2000DL003-9VT166. It ran Ubuntu 9.10 (kernel 2.6.31-19-server).

The performance of both algorithms was measured comparing the number of accessed blocks and the real time that each algorithm required to find the solution. The time includes the time required to read the points from disk. We recall that we assume that the R-tree is already built (that is, the time required to build it is not included in our times) since it is the structure that stores the points. Both $q-$MER and the naive approach use a stack to traverse the R-tree. This stack is used to avoid the repetitive read of internal nodes of the R-tree from disk. As a search traverses the R-tree downwards, the nodes are stored in the stack, if we need to go back upwards, we already have the parent node at the top of the stack, therefore the maximum size of the stack is the height of the R-tree. For measurement purposes, these reads of nodes in the stack are not counted. We do not use any other read buffer, therefore whenever the algorithms read a node (disk block) that is not in the stack, that read is counted regardless of whether the node comes from disk or from the operating system cache. Therefore, the measure of accessed blocks eliminates the effect of any type of buffer cache (excepting the simple stack).

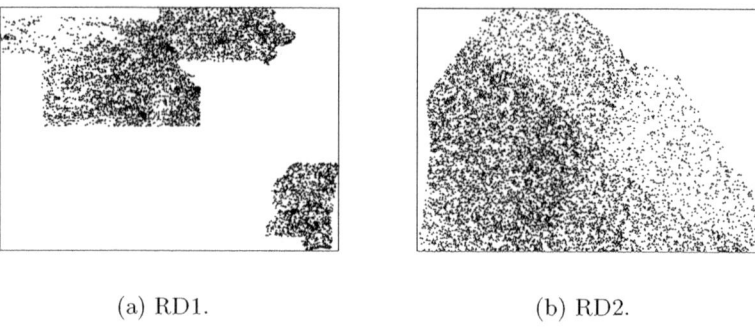

(a) RD1.   (b) RD2.

Fig. 8. Real data sets

We considered real and synthetic set of points in a two-dimensional space $[0,1] \times [0,1]$. The synthetic data sets have uniform distribution. The real data sets are the Tiger Census Blocks data set (RD1) from the web site `rtreeportal.org` and a data set (RD2) that was given by a Chilean company provided that the source were not published. Figure 8 shows some of the points of these data sets (we only plot some of them in order to simplify the graphs). The sizes of the sets were as follows: for the synthetic data, we used sets with size $200K^2$, 500K, 1,000K, 2,000K, and 5,000K; and the real data sets have size 556K (RD1) and 700K (RD2). In the case of the synthetic data sets, we considered block sizes of $1KB^3$ and 4KB. Furthermore, for all measures, we computed the average of 100 queries.

The results of the experiments using synthetic data are summarized in Figures 9–11 and Table 1. In Figures 9 and 10, we can observe the filtering capability of the q−MER algorithm. For example, q−MER only needs to access around between 10% and 55% (Figure 9) of the blocks of the R-tree, whereas the naive algorithm needs to access all of them. Furthermore, we can see that when the size of the set increases, the percentage of the blocks accessed by q−MER decreases.

Figure 10 shows the amount of blocks accessed by q−MER and the total number of blocks that the R-tree uses to store the points. When the size of the collection of points increases, we can see that the slope of the line that shows the blocks accessed by q−MER to solve the problem is less pronounced than that showing the size of the R-tree.

Figure 10 also shows the effects of the size of the blocks over the amount of accessed blocks, which, as expected, decreases inversely proportional to the size of the block.

As explained, we do not use any read buffer (except the stack). The effect of any read buffer, for example the operating system buffer cache, would benefit only the q−MER algorithm, since it might access several times the same R-tree leaf node when processing different CERs. However, the naive approach would

---

[2] $1K = 1,000$ points.
[3] $1KB = 1,024$ Bytes.

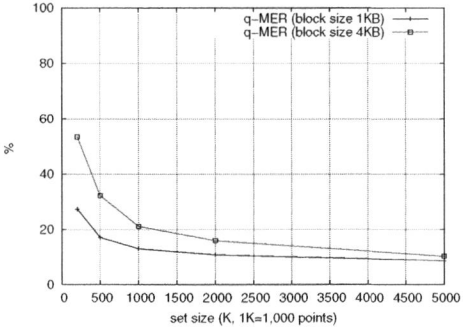

Fig. 9. Percentage accessed blocks

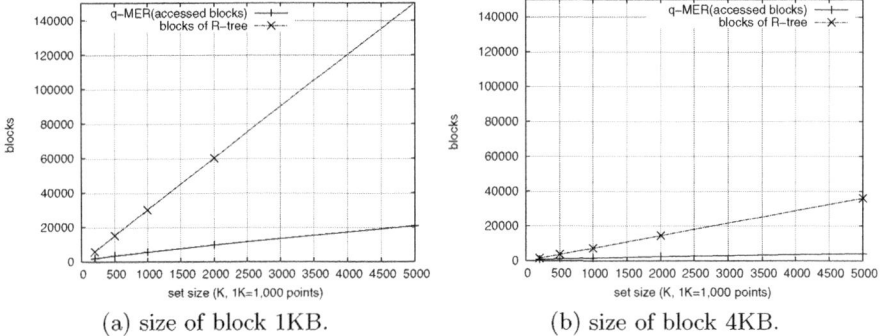

(a) size of block 1KB.   (b) size of block 4KB.

Fig. 10. Performance (accessed blocks) against total blocks of R-tree

not improve its performance as it reads from disk each leaf node once, since the repetitive reads of intermediate nodes are solved by the stack.

The second step of $q-$MER obtains several sets of points C', one for each CER computed by the first step. Each set is provided as input to the algorithm *ComputeER*, which solves the problem of finding the largest empty rectangle with those points in main memory. In Table 1, we denote each run of *ComputeER* with a set of points C' (line 11 of Algorithm 2) a *problem*. Table 1 describes each of those problems considering different sizes of original sets of points and sizes of blocks. The third column shows the amount of problems solved by $q-$MER to obtain the final solution. This amount also includes the run of the algorithm *ComputeCER* of the first step (line 33 of Algorithm 1). Note that each CER obtained by the first step of $q-$MER represents a *problem*, that is, the total number of problems is the number of CERs provided by the first step of $q-$MER, plus one. Observe that the naive algorithm only solves one *problem*, but with many more points.

Table 1 shows that the bigger problem treated by *ComputeER* in the $q-$MER algorithm includes a very short percentage of the total number of points, namely

Table 1. Description of the problems

size of set (thousand)	size of block(KB)	# problems	# points average	# points minimum	# points maximum
200	1	54	928	57	5,893
	4	48	2,206	217	8,464
500	1	60	1,307	90	14,520
	4	54	3,002	306	12,941
1,000	1	63	1,799	81	28,893
	4	58	3,601	416	17,508
2,000	1	69	2,599	89	57,555
	4	66	4,774	344	32,559
5,000	1	77	4,345	82	143,402
	4	74	6,782	365	53,101

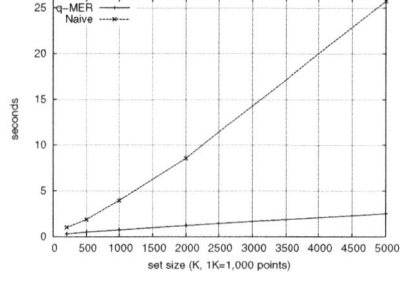   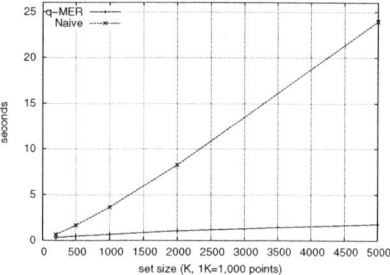

(a) size of block 1 KB   (b) size of block 4 KB

Fig. 11. Performance (real time) of algorithm $q-$MER

around 3%–4%. This means that $q-$MER only needs around the 3%–4% of the main memory needed by the naive algorithm to solve the same problem.

Figure 11 compares the average execution time of a query using $q-$MER and the naive algorithm when the points are retrieved from the R-tree. In all cases, $q-$MER overcomes the naive approach, since $q-$MER uses between 7%–38% of the time required by the naive approach. The differences get bigger as the size of the problem increases. In order to avoid any distortion due to the arrangement of the data in a R-tree that might increase the disk seek times, we stored all the points in a sequential file, in such a way that the naive approach can read the points sequentially. The results of this experiment are shown in Figure 12. As it can be seen, $q-$MER also overcomes the naive approach in this scenario. Figure 12 displays the values for disk blocks of 1 KB, since the results for 4KB block size were similar as the time required to read the points is a very small portion of the time required to solve the problem.

With regard to the main memory consumption, each *problem* (as explained each run of *ComputeER* or *ComputeCER*) needs to store in main memory the points involved in that computation. In the worst case, a problem solved by $q-$MER needs only 3%–4% of the points (or main memory), which needs the naive approach. If the available memory is not enough for the naive approach,

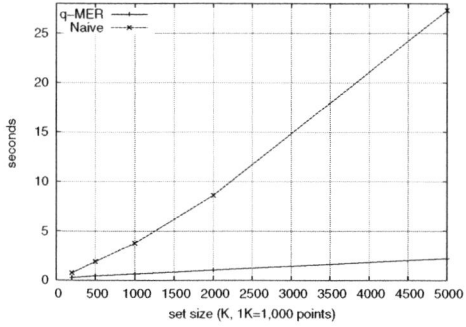

Fig. 12. Experiment where the naive approach reads the points from a sequential file

Table 2. Blocks accessed by q−MER with real data

	RD1	RD2
Size of R-tree(# blocks)	18,509	21,066
#Accessed blocks	1,614	2,416
% accessed blocks with respect the total	8.7	11.5

Table 3. Performance with real datasets and disk block size 1KB

	RD1	RD2
Algorithm	time (seconds)	
Naive (read from R-tree)	62.155	32.939
Naive (read from sequential file)	62.085	32.807
q−MER	0.441	0.494

this implies a important increment of the time required to run the algorithm, since external sorts would be required.

Tables 2 and 3 show the performance of our algorithm and the naive approach over the real datasets. In this experiment, we used only one size of disk block (1 KB). Table 2 summarizes the disk block accesses, whereas Table 3 shows the time consumed to solve the queries.

As in previous data distributions, our algorithm overcome the naive approach. In this case, differences are even bigger; q−MER needs only 0.71% and 1.5% (for the data sets RD1 and RD2, respectively) of the time required by the naive approach (see Table 3).

5   Conclusions

Given a space with a set points stored in spatial index R-tree, we described the algorithm q−MER that obtains the largest rectangle containing just a query point q. The results of the experiments with synthetic and real data show that q−MER requires between 0.71% and 38% of the running time and between 3%

and 4% of the storage required by the naive algorithm. In part, the performance of $q-$MER can be explained by its filtering capability, since it requires to access only around between 10% and 55% of the total blocks of the R-tree.

The experiments also show the scalability of our algorithm, which obtains better improvements as the size of the collection grows. In addition, $q-$MER would benefit from any improvement in the computational geometry algorithm used by *computeCER* and *computeER*.

To the best of our knowledge, this is the first work that solves this problem considering that the points are stored in a multidimensional data structure R-tree.

As future work, we want to extend our proposal to objects with more dimensions and to rectangles with sides that are not necessarily parallel to the axes of the original space. We plan also to work in developing a cost model to predict the time and space consumed by our approach.

Acknowledgements. We would like to thank Juan Ramón López Rodríguez for his comments and suggestions.

References

[ADM⁺10a]   Augustine, J., Das, S., Maheshwari, A., Nandy, S.C., Roy, S., Sarvat-tomananda, S.: Recognizing the largest empty circle and axis-parallel rectangle in a desired location. CoRR, abs/1004.0558 (2010)

[ADM⁺10b]   Augustine, J., Das, S., Maheshwari, A., Nandy, S.C., Roy, S., Sarvat-tomananda, S.: Querying for the largest empty geometric object in a desired location. CoRR, abs/1004.0558v2 (2010)

[AS87]   Aggarwal, A., Suri, S.: Fast algorithms for computing the largest empty rectangle. In: Proceedings of the Third Annual Symposium on Computational Geometry, SCG 1987, pp. 278–290. ACM, New York (1987)

[BK01]   Böhm, C., Kriegel, H.-P.: Determining the Convex Hull in Large Multidimensional Databases. In: Kambayashi, Y., Winiwarter, W., Arikawa, M. (eds.) DaWaK 2001. LNCS, vol. 2114, pp. 294–306. Springer, Heidelberg (2001)

[CDL86]   Chazelle, B., Drysdalet, R.L., Lee, D.T.: Computing the largest empty rectangle. SIAM Journal Computing 15, 300–315 (1986)

[CMTV04]   Corral, A., Manolopoulos, Y., Theodoridis, Y., Vassilakopoulos, M.: Algorithms for processing k-closest-pair queries in spatial databases. Data & Knowledge Engineering 49(1), 67–104 (2004)

[Cor02]   Corral, A.: Algoritmos para el Procesamiento de Consultas Espaciales utilizando R-trees. La Consulta de los Pares Más Cercanos y su Aplicación en Bases de Datos Espaciales. PhD thesis, Universidad de Almería, Escuela Politécnica Superior, España, Enero (2002)

[DN11]   De, M., Nandy, S.C.: Inplace algorithm for priority search tree and its use in computing largest empty axis-parallel rectangle. CoRR, abs/1104.3076 (2011)

[EGLM03]   Edmonds, J., Gryz, J., Liang, D., Miller, R.J.: Mining for empty spaces in large data sets. Theoretical Computer Science 296, 435–452 (2003)

[GG98]   Gaede, V., Günther, O.: Multidimensional access methods. ACM Computing Surveys 30(2), 170–231 (1998)

[Gut84]   Guttman, A.: R-trees: A dynamic index structure for spatial searching. In: ACM SIGMOD Conference on Management of Data, pp. 47–57. ACM (1984)

[HS98]   Hjaltason, G.R., Samet, H.: Incremental distance join algorithms for spatial databases. In: ACM SIGMOD Conference on Management of Data, Seattle, WA, pp. 237–248 (1998)

[Kin81]   King, J.J.: Query optimization by semantic reasoning. PhD thesis, Stanford University, CA, USA (1981)

[KS11]   Kaplan, H., Sharir, M.: Finding the maximal empty rectangle containing a query point. CoRR, abs/1106.3628 (2011)

[NLH84]   Naamad, A., Lee, D.T., Hsu, W.-L.: On the maximum empty rectangle problem. Discrete Applied Mathematics 8, 267–277 (1984)

[Orl90]   Orlowski, M.: A new algorithm for the largest empty rectangle problem. Algorithmica 5, 65–73 (1990)

[RKV95]   Roussopoulos, N., Kelley, S., Vincent, F.: Nearest neighbor queries. In: SIGMOD 1995: Proceedings of the 1995 ACM SIGMOD International Conference on Management of Data, pp. 71–79. ACM Press, New York (1995)

[SC03]   Shekhar, S., Chawla, S.: Spatial databases - a tour. Prentice Hall (2003)

Sensitivity of Self-tuning Histograms: Query Order Affecting Accuracy and Robustness

Andranik Khachatryan, Emmanuel Müller, Christian Stier, and Klemens Böhm

Institute for Program Structures and Data Organization (IPD)
Karlsruhe Institute of Technology (KIT), Germany
{khachatryan,emmanuel.mueller,klemens.boehm}@kit.edu,
christian.stier2@student.kit.edu

Abstract. In scientific databases, the amount and the complexity of data calls for data summarization techniques. Such summaries are used to assist fast approximate query answering or query optimization. Histograms are a prominent class of model-free data summaries and are widely used in database systems.

So-called self-tuning histograms look at query-execution results to refine themselves. An assumption with such histograms is that they can learn the dataset from scratch. We show that this is not the case and highlight a major challenge that stems from this. Traditional self-tuning is overly sensitive to the order of queries, and reaches only local optima with high estimation errors. We show that a self-tuning method can be improved significantly if it starts with a carefully chosen initial configuration. We propose initialization by subspace clusters in projections of the data. This improves both accuracy and robustness of self-tuning histograms.

1   Introduction

Histograms are a fundamental data-summarization technique. They are used in *query optimization, approximate query answering, spatio-temporal, top-k* and *skyline query processing*. Query optimizers use histograms to obtain accurate size estimates for sub-queries. When the query predicate refers to more than one attribute, a joint distribution of attributes is needed to obtain these size estimates. There are two different paradigms of histogram construction: *Static* and *Self-Tuning* histograms.

Static Histograms And Dimensionality Reduction. Static histograms [11,5,1,16,10] are constructed by scanning the entire dataset. They need to be rebuilt regularly to reflect any changes in the dataset. For large relations, building a static multi-dimensional histogram in the full attribute space is expensive, both regarding construction time and occupied space. *Dimensionality reduction techniques* try to solve the problem by removing less relevant attributes [4].

A. Ailamaki and S. Bowers (Eds.): SSDBM 2012, LNCS 7338, pp. 334–342, 2012.

These approaches leave aside that different combinations of attributes can be correlated in different subregions of the data set.

Traditional Self-Tuning Histograms. In contrast to static histograms, self-tuning histograms [3,15,8] use query feedback to learn the dataset. They amortize the construction costs: the histogram is constructed on the fly as queries are executed. Such histograms adapt to the querying patterns of a user. As one representative, we consider the data structure of STHoles [3], which is very flexible and has been used in several other histograms [15,12]. It tries to learn bounding rectangles of regions which have close to uniform tuple density. However, similar to traditional index structures, such as R-Trees [6], STHoles fails in high dimensional data spaces (curse of dimensionality [2]) and is affected by the order of tree construction [14].

An attempt to address the latter problem is to re-schedule queries to achieve a better learning [8]. The central assumption behind the approach is that it is permissible to delay the execution of queries or switch the query order.

Self-Tuning Histograms and Subspace Clustering. We consider the common scenario where the queries are executed as they arrive. The focus is on the general assumption with self-tuning methods, i.e. that they can learn the dataset from scratch – starting with no buckets and relying only on query feedback. We show that this is not the case and highlight open challenges in self-tuning. In particular, we focus on its sensitivity w.r.t. the query order.

In general, the first few queries define the top-level bucket structure of the histogram. If this structure is bad, then, regardless of the amount of further training, the histogram is unlikely to become good.

To solve this top-level construction problem, we initialize the histogram with few buckets which define the top levels of the histogram. Such buckets are obtained using recent subspace clustering methods [9]. In contrast to dimensionality reduction techniques, which aim at one projection, subspace clustering methods aim at multiple local projections. As a first step [7], we have proposed a method to transfer arbitrary subspace clustering results into an initial histogram structure. First, we have focused on the clustering aspect and evaluated the performance of several clustering algorithms. The main result of [7] is that Mineclus performs best as an initializer. Thus, we take it as our basis for this paper. We extend the discussion to the sensitivity problem, a general challenge for self-tuning histograms. It provides the *reasons* why self-tuning methods without initialization struggle to learn the dataset from scratch.

2   Self-tuning and Its Sensitivity to Learning

We use STHoles [3] as a representative for self-tuning histograms and describe its main properties and problems. We do this in several steps. First, we describe how the histogram partitions the data space and estimates query cardinalities. Then we describe how new buckets are inserted into the histogram, and briefly mention how redundant buckets are removed to free up space. Last, we derive the open challenges in its sensitivity to learning.

2.1   Histogram Structure and Cardinality Estimation

STHoles partitions the data space into a tree of rectangular buckets. Each bucket b stores $n(b)$, the number of tuples. This number does not include the tuples in its child buckets. Similarly, $vol(b)$ is the volume of the rectangular space occupied by a bucket, without its child buckets. Figure 1 shows a histogram with 4 buckets. STHoles estimates query cardinality of query q using the *Uniformity Assumption*. This means that it assumes the tuples inside the buckets are distributed uniformly:

$$est(q, H) = \sum_{b \in H} n(b) \cdot \frac{vol(q \cap b)}{vol(b)} \tag{1}$$

where H is the histogram.

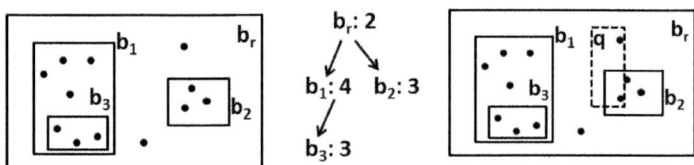

Fig. 1. STHoles with 4 buckets, corresponding tree and query (dashed rectangle)

When estimating the number of tuples in q (cf. Fig. 1), STHoles computes the intersections with histogram buckets. q intersects with b_r and b_2. Using (1), we estimate the number of tuples in $b_r \cap q$ to be ≈ 0. The estimated number of tuples in $b_2 \cap q$ is ≈ 1.5. So the overall approximated number of tuples will be a little over 1.5. We can see that the real number of tuples is 2.

Adding Buckets. After the query is executed, the real numbers of tuples falling into $b_r \cap q$ and $b_2 \cap q$ become known. The histogram *refines* itself by *drilling* new buckets. The process consists of adding new buckets and updating the frequencies of existing buckets. Figure 2 shows the histogram with two buckets added. Because the intersection of $b_r \cap q$ is not a rectangle, it is shrunk across dimension(s) to become rectangular. Note that the frequencies of b_r and b_2 are also updated to reflect the new information about the tuple placement in the histogram.

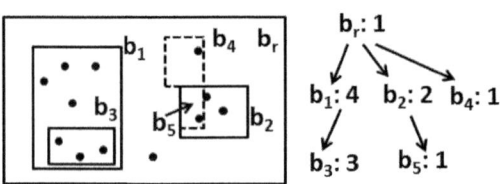

Fig. 2. The histogram with newly added buckets b_4 and b_5

Removing Buckets. The histogram constantly adds buckets as new queries are executed. When the number of histogram buckets exceeds a certain number, some buckets need to be removed. STHoles merges buckets with close densities to free up space. Only sibling-sibling or parent-child merges are allowed. In case of a sibling-sibling merge, the bounding box of the new bucket is the minimal rectangle that includes the initial buckets and does not partially intersect with any other bucket. For a detailed discussion of the merging procedure see [3].

2.2   The Problem with Self-tuning: Sensitivity to Learning

Informally, *Sensitivity to Learning* is determined by the significant impact of the query order on the estimation accuracy of self-tuning histograms.

First, let us define the underlying workload for the histogram construction.

Definition 1. *A workload is a sequence of queries:* $W = (q_1, \ldots, q_n)$

We will call the workloads W_1 and W_2 permutations of each other if they consist of the same queries, but in different order. We will write $W_2 = \pi(W_1)$, where π is some permutation. Given a histogram H and a workload W, we will write $H|W$ to indicate the histogram which results from H after it learns the query feedback from W.

At first sight, histograms resulting from two workloads where one is a permutation of the other one, $H|W$ and $H|\pi(W)$, should produce very close estimates. However, we can show in a counter example how permutation of queries can result in histograms which differ in structure considerably.

Example 1. Figure 3 demonstrates what happens when we change the order of the queries. On Figure 3(a), left, we see two queries (numbered 1 and 2 by the order at which they arrive). The histogram bucket capacity is 2 buckets. The queries are executed and the resulting histogram is depicted on the right. Figure 3(b), left shows the same queries, but now the insertion order is reversed. On the right side of Figure 3(b) we see the histogram resulting from this query order. Clearly, the histogram on Figure 3(a) is the better one. It captures the data distribution well, while the one on Figure 3(b) misses out some tuples and has one bucket which encloses regions of different densities. The reason why this happens is that STHoles shrinks incoming query rectangles if the intersection with existing buckets is non-rectangular. On Figure 3(b) we see that important information is being discarded because of such shrinking.

Let ε be some quality measure for the histogram. For instance, we can take

$$\varepsilon = \int_{u \in D} |real(u) - est(u)| \, du$$

where $real(u)$ is the real cardinality of the point-query u, and D is the attribute-value domain. For an error measure ε we define δ-sensitivity to learning.

Definition 2. *We call a histogram H δ-sensitive to learning w.r.t workload W if, for some permutation π exists, with:* $|\varepsilon(H|W) - \varepsilon(H|\pi(W))| > \delta$

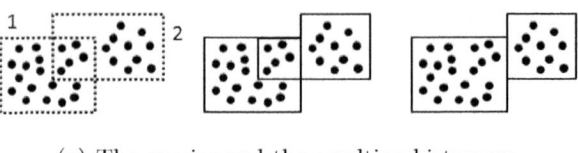

(a) The queries and the resulting histogram

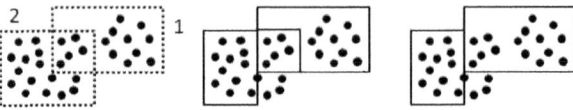

(b) The queries in the reverse order and the resulting histogram

Fig. 3. The queries and resulting histograms for two queries

δ-sensitivity means that changing the order of learning queries changes the histogram estimation quality by more than δ. Example 1 demonstrates *Sensitivity to Learning*, an effect of histogram construction, which shows significantly different histogram structures and resulting selectivity estimates for transposed workloads. Our intuition would be: "A good histogram should not be sensitive to workload permutations." We discuss in the following how we can achieve this intuitive goal.

3   Histogram Initialization by Subspace Clustering

Self-tuning is able to *refine* the structure of the histogram. If started with no buckets at all, the histogram has to rely on the first few queries to determine the top-level partitioning of the attribute-value space. If this partitioning is bad, the later tuning is unlikely to "correct" it. The solution is to provide the histogram with a good initial configuration. This configuration should:

1. provide a top-level bucketing for the dataset, which can be later tuned using feedback;
2. capture the data regions in relevant dimensions, that is, it should exclude irrelevant attributes for each bucket.

We now describe how to initialize the histogram with subspace buckets. The subspace clustering algorithm finds dense clusters together with the set of relevant attributes. Then these clusters are transformed into histogram buckets.

Generally, clustering algorithms output clusters as a set of points. We need to transform this set of points into a rectangular representation. Cell-based clustering algorithms such as Mineclus [18] look for rectangular clusters. We could take these rectangles as the STHoles buckets. However, we have found out in preliminary experiments that this has a drawback, which is illustrated in Figure 4. Although the cluster found is one-dimensional (left), the MBR is two-dimensional (dashed rectangle on the right). The two-dimensional MBR would

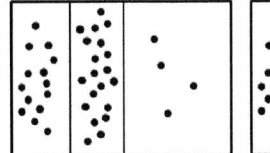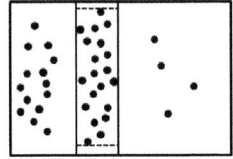

Fig. 4. On the left, the cluster found. On the right, the dashed rectangle is the MBR of the cluster. The solid rectangle on the right is the extended BR.

introduce additional intersections with incoming query rectangles without measurable difference in estimation quality. This is undesirable. We can bypass this problem using the information produced by Mineclus. Mineclus outputs clusters as sets of tuples together with the relevant dimensions. This means that the cluster spans $[min, max]$ on any unused dimension. To preserve subspace information, we introduce *extended BRs*.

Definition 3. *Extended BR*. *Let cluster C consist of tuples $\{t_1, \ldots, t_n\}$ and dimensions $d_1, \ldots, d_k$. The extended BR of C is the minimal rectangle that contains the points $\{t_1, \ldots, t_n\}$ and spans $[min, max]$ for every dimension not in $d_1, \ldots, d_k$.*

Another characteristic of Mineclus is that is assigns importance to clusters. The algorithm has a score function which decides whether a set of points is a cluster or not. The clusters themselves are then sorted according to this score. We found out that, if we use the important clusters as first queries in the initialization, we have a better estimation quality.

4 Experiments

We have used one synthetic and one real-world data set. The synthetic dataset (*Cross*) is a 2-dimensional database, which contains two one-dimensional clusters. Each cluster contains 10,000 tuples. Another 2,000 tuples are random noise. The *Cross* dataset is very simple, in the sense that it is possible to perfectly describe it using 5 buckets. There can still be an estimation error due to randomly generated data, but the error should be very low. As real-world dataset (*Sky*), we use one of the datasets published by the Sloan Digital Sky Survey (SDSS) [13]. It contains approximately 1,7 million tuples of observations in a 7-dimensional data space (two coordinates in the sky and five attributes with brightness values w.r.t. different filters). Complex data correlations exist in the *Sky* dataset, which make it a hard evaluation scenario for histogram construction.

In our evaluation, we generate queries which span a certain volume in the data space. The query centers are generated randomly. We vary the number of histogram buckets from 50 to 250 like most others [15,12,3,17]. The quality of estimations is measured by the error the histogram produces over a series of queries. Given a workload W and histogram H, the *Mean Absolute Error* is:

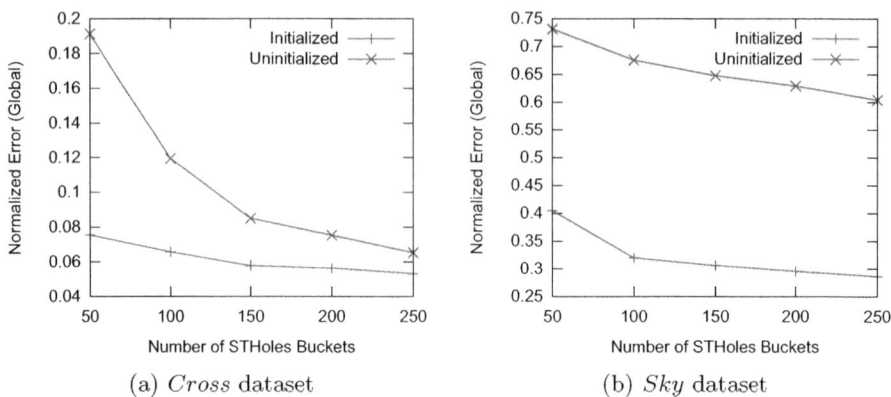

(a) *Cross* dataset   (b) *Sky* dataset

Fig. 5. Errors of initialized vs uninitialized histograms

$$E(H, W) = \frac{1}{|W|} \sum_{q \in W} |est(H, q) - real(q)| \tag{2}$$

where $real(q)$ is the real cardinality of the query. In order for the results to be somewhat comparable across datasets, we normalize this error by dividing it by the error of a trivial histogram H_0 [3]. H_0 contains only one bucket which simply stores the number of tuples in the database:

$$NAE(H, W) = \frac{E(H, W)}{E(H_0, W)} \tag{3}$$

Unless stated otherwise, the workload is the same for all histograms and contains 1,000 training and 1,000 simulation queries. The error computation starts with with the simulation queries.

4.1   Accuracy

In the first set of experiments we show that initialization improves estimation quality. Figures 5(a) and 5(b) show the error comparison for the *Cross* and the *Sky* datasets. For all datasets, the initialized histogram outperforms the uninitialized version.

As mentioned, the *Cross* dataset is simple and can easily be described with 5 to 6 buckets. Nevertheless, Figure 5(a) shows that initialization has a significant effect in improving the estimation accuracy. Initialization finds the 5-6 buckets which are essential for the good histogram structure, while a random workload of even 1,000 training queries is not enough for the uninitialized histogram to find this simple bucket layout. This shows that uninitialized histograms have trouble even with the simplest datasets.

Figure 6 shows the comparison on the *Sky* dataset. Here, the errors are higher than both for *Cross* and *Gauss* datasets. The benefit of initialization is again clear: The initialized version has about half the error rate compared to the uninitialized version.

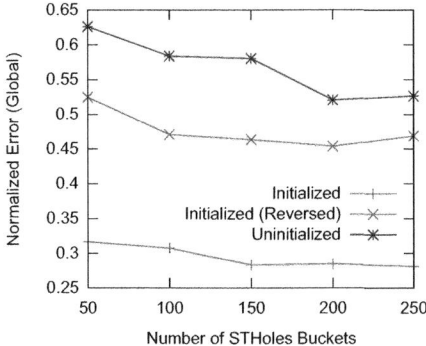

Fig. 6. Error comparison for $Sky[1\%]$ setting

In both cases, the initialized histogram outperforms the uninitialized version. Moreover, for the *Sky* dataset, the initialized histogram with only 50 buckets is significantly better than the uninitialized histogram with 250 buckets. On the simple *Cross* dataset the uninitialized histogram with 250 buckets reaches the quality of the initialized histogram with 50 buckets.

4.2   Robustness

We revisit the challenges described in Section 2.2. To show that STHoles is sensitive to learning, we conducted experiments using permuted workloads. To show the effect of changing the order of queries, recall how we initialize the histogram. We generate rectangles with frequencies from the clustering output and feed this to the histogram in the order of importance. This importance is an additional output of the clustering algorithm. In the experiment in Figure 6, we use the same set of clusters to initialize the histogram, but in a reverse order of importance. Clearly, there is a significant difference between the normal initialization and the reverse one. This shows two things. First, it is clear that permuting a workload changes the histogram error significantly (Sensitivity to Learning). Second, it shows the the importance of the order of initialization, as the "correct order" has a noticeably lower error compared to the reversed order.

5   Conclusions

A central assumption with self-tuning histograms is that they can learn a dataset relying solely on query feedback. We show that this is not the case. Self-tuning methods are overly sensitive to the query order. A "bad" order can negatively influence the histogram quality. We show that initialization by subspace clusters can make the histogram less sensitive to learning and significantly improve the estimation quality.

References

1. Baltrunas, L., Mazeika, A., Böhlen, M.H.: Multi-dimensional histograms with tight bounds for the error. In: IDEAS, pp. 105–112 (2006)
2. Beyer, K., Goldstein, J., Ramakrishnan, R., Shaft, U.: When Is "Nearest Neighbor" Meaningful? In: Beeri, C., Bruneman, P. (eds.) ICDT 1999. LNCS, vol. 1540, pp. 217–235. Springer, Heidelberg (1998)
3. Bruno, N., Chaudhuri, S., Gravano, L.: Stholes: a multidimensional workload-aware histogram. SIGMOD Rec. 30, 211–222 (2001)
4. Deshpande, A., Garofalakis, M., Rastogi, R.: Independence is good: dependency-based histogram synopses for high-dimensional data. SIGMOD Rec. 30, 199–210 (2001)
5. Gunopulos, D., Kollios, G., Tsotras, V.J., Domeniconi, C.: Approximating multidimensional aggregate range queries over real attributes. SIGMOD Rec. 29, 463–474 (2000)
6. Guttman, A.: R-trees: A dynamic index structure for spatial searching. In: Yormark, B. (ed.) SIGMOD 1984, Proceedings of Annual Meeting, Boston, Massachusetts, June 18-21, pp. 47–57. ACM Press (1984)
7. Khachatryan, A., Müller, E., Böhm, K., Kopper, J.: Efficient Selectivity Estimation by Histogram Construction Based on Subspace Clustering. In: Bayard Cushing, J., French, J., Bowers, S. (eds.) SSDBM 2011. LNCS, vol. 6809, pp. 351–368. Springer, Heidelberg (2011)
8. Luo, J., Zhou, X., Zhang, Y., Shen, H.T., Li, J.: Selectivity estimation by batch-query based histogram and parametric method. In: ADC 2007, pp. 93–102. Australian Computer Society, Inc., Darlinghurst (2007)
9. Müller, E., Günnemann, S., Assent, I., Seidl, T.: Evaluating clustering in subspace projections of high dimensional data. PVLDB 2(1), 1270–1281 (2009)
10. Muthukrishnan, S., Poosala, V., Suel, T.: On Rectangular Partitionings in Two Dimensions: Algorithms, Complexity, and Applications. In: Beeri, C., Bruneman, P. (eds.) ICDT 1999. LNCS, vol. 1540, pp. 236–256. Springer, Heidelberg (1998)
11. Poosala, V., Ioannidis, Y.E.: Selectivity estimation without the attribute value independence assumption. In: VLDB 1997, pp. 486–495. Morgan Kaufmann Publishers Inc., San Francisco (1997)
12. Roh, Y.J., Kim, J.H., Chung, Y.D., Son, J.H., Kim, M.H.: Hierarchically organized skew-tolerant histograms for geographic data objects. In: SIGMOD 2010, pp. 627–638. ACM, New York (2010)
13. SDSS Collaboration. Sloan Digital Sky Survey (August 27, 2011)
14. Sellis, T.K., Roussopoulos, N., Faloutsos, C.: The r+-tree: A dynamic index for multi-dimensional objects. In: VLDB, pp. 507–518 (1987)
15. Srivastava, U., Haas, P.J., Markl, V., Kutsch, M., Tran, T.M.: Isomer: Consistent histogram construction using query feedback. In: ICDE 2006, p. 39. IEEE Computer Society, Washington, DC (2006)
16. Wang, H., Sevcik, K.C.: A multi-dimensional histogram for selectivity estimation and fast approximate query answering. In: CASCON 2003, pp. 328–342. IBM Press (2003)
17. Wang, H., Sevcik, K.C.: Histograms based on the minimum description length principle. The VLDB Journal 17 (May 2008)
18. Yiu, M.L., Mamoulis, N.: Frequent-pattern based iterative projected clustering. In: ICDM 2003, pp. 689–692 (2003)

Database Support for Exploring Scientific Workflow Provenance Graphs

Manish Kumar Anand[1], Shawn Bowers[2], and Bertram Ludäscher[3]

[1] Microsoft Corporation, Redmond, WA, USA
[2] Dept. of Computer Science, Gonzaga University, Spokane, WA, USA
[3] Dept. of Computer Science, University of California, Davis, CA, USA

Abstract. Provenance graphs generated from real-world scientific workflows often contain large numbers of nodes and edges denoting various types of provenance information. A standard approach used by workflow systems is to visually present provenance information by displaying an entire (static) provenance graph. This approach makes it difficult for users to find relevant information and to explore and analyze data and process dependencies. We address these issues through a set of abstractions that allow users to construct specialized views of provenance graphs. Our model provides operations that allow users to expand, collapse, filter, group, and summarize all or portions of provenance graphs to construct tailored provenance views. A unique feature of the model is that it can be implemented using standard relational database technology, which has a number of advantages in terms of supporting existing provenance frameworks and efficiency and scalability of the model. We present and formalize the operations within the model as a set of relational queries expressed against an underlying provenance schema. We also present a detailed experimental evaluation that demonstrates the feasibility and efficiency of our approach against provenance graphs generated from a number of scientific workflows.

1 Introduction

Most scientific workflow systems record provenance information, i.e., the details of a workflow run that includes data and process dependencies [11,20]. Provenance information is often displayed to users as a (static) dependency graph [16,13,7]. However, many real-world scientific workflows result in provenance graphs that are large (e.g., with upwards of thousands of nodes and edges) and complex due to the nature of the workflows, the number of input data sets, and the number of intermediate data sets produced during a workflow run [10,11], making them inconvenient to explore visually.

The goal of the work described here is to help users more easily explore and analyze provenance information by allowing them to specify and navigate between different abstractions (or *views*) of complex provenance graphs. Specifically, we describe a set of abstraction mechanisms and operators for scientific workflow provenance graphs that allow users to create, refine, and navigate between different views of the same underlying provenance information. We consider the following levels of granularity: (1) A workflow *run* represents the highest level of abstraction; (2) An *actor dependency graph* consists of the types of processes (actors) used in a workflow run and the general

A. Ailamaki and S. Bowers (Eds.): SSDBM 2012, LNCS 7338, pp. 343–360, 2012.

flow of data between them; (3) An *invocation dependency graph* consists of individual processes (invocations) within a run and their implicit data dependencies; (4) A *structure flow graph* consists of the actual data structures input to and output by each invocation; and (5) A *data dependency graph* consists of the detailed data dependencies of individual data items. In addition, we consider navigation operations that allow all or a portion of each provenance view to be *expanded* or *collapsed*, *grouped* into composite structures, *filtered* using a high-level provenance query language, and *summarized* through aggregation operators.

Contributions. We present a general provenance model that enables users to explore and analyze provenance information through novel graph-based summarization, navigation, and query operators. We also show how this model can be implemented using standard relational database technology. We adopt a relational implementation for two reasons. First, many existing workflow systems that record provenance store this information within relational databases [11], making it relatively straightforward to adopt the implementation described here. Second, the use of standard database technology can provide advantages in terms of efficiency and scalability over other general-purpose graph-based approaches (e.g., [12]) that provide only in-memory implementations. Finally, we demonstrate the feasibility of our implementation through an experimental evaluation of the operators over real and synthetic scientific-workflow traces.

Organization. The basic provenance model and query language that our view abstractions and navigation operators are based on is presented in Section 2. The different levels of abstraction and navigation operators supported by our model is defined in Section 3. We show how our model can be implemented within a relational framework in Section 4, which describes the relational schemas used to store provenance information and the queries used to execute each of the navigation operations and views. Our experimental results are presented in Section 5, which demonstrates the feasibility and scalability of the implementation. Related work is discussed in Section 6 and we conclude in Section 7.

2   Preliminaries: Provenance Model and Query Language

In our provenance model, we assume that workflows execute according to standard dataflow-based computation models (e.g., [15,17]). In addition, the provenance model supports processes that can be executed (i.e., *invoked*) multiple times in a workflow run and can receive and produce data products that are *structured* according to labeled, nested data collections (e.g., as XML). To help illustrate, consider the simple workflow definition shown in Fig. 1a. This workflow consists of three actors a, b, and c; distinguished input and output nodes; and four data-flow channels. The channels constrain how data is passed between actors within a workflow run. In paticular, the input to the run is passed to invocations of a, the results of a's invocations are passed to invocations of b, and so on.

Fig. 1b shows a high-level view of an example run of this workflow, called a *structure flow graph* (SFG). As shown, each actor invocation receives a nested-collection data structure, performs an update on a portion of the structure (by either adding or

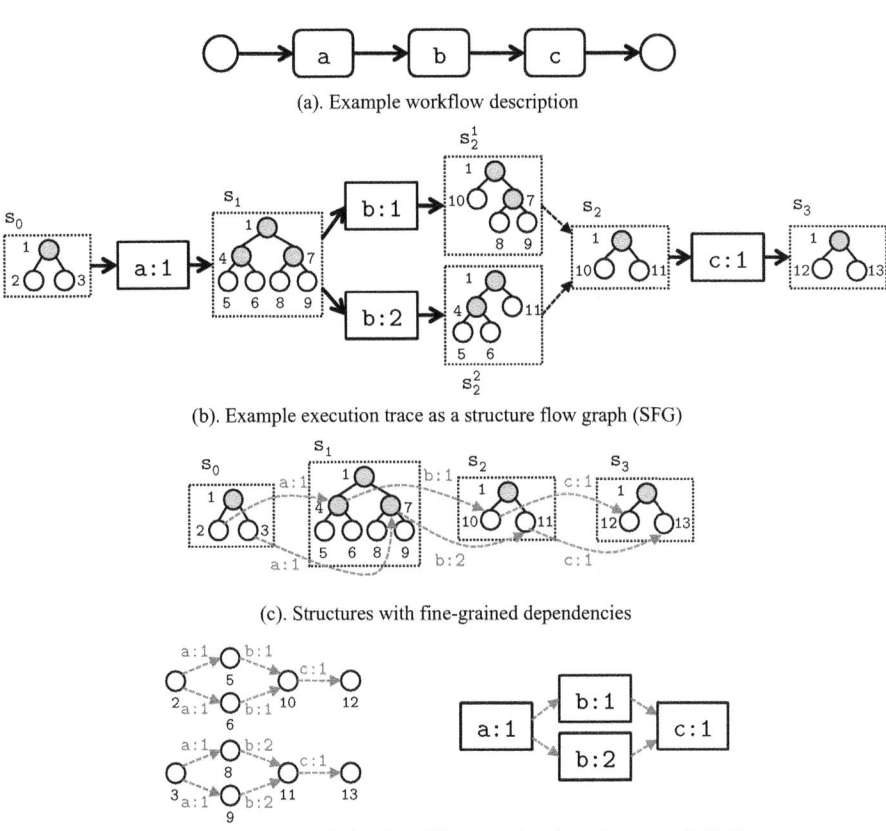

(a). Example workflow description

(b). Example execution trace as a structure flow graph (SFG)

(c). Structures with fine-grained dependencies

(d). Data dependency graph (DDG)   (e). Invocation dependency graph (IDG)

Fig. 1. An example workflow graph (a) with a corresponding structure flow (b) and fine-grained dependency (c) graph together with associated data (d) and invocation (e) dependency graphs

removing items), and then passes the updated version to downstream actors. An SFG consists of the intermediate data structures s_i that were input to and output by actor invocations, where edges denote "coarse-grained" dependencies [2]. In this example, the first invocation a:1 of actor a takes as input s_0 and produces the updated version s_1. The actor b is invoked twice to produce the modified structure s_2. The first invocation b:1 removes item 6 in s_1 and adds item 10, and similarly, the second invocation b:2 removes item 7 and adds item 11. The structure s_2 is the result (union) of the two independent modifications s_2^1 and s_2^2. Finally, invocation c:1 modifies s_2 to produce the output of the run s_3. This use of nested data collections within scientific workflows is supported within both Kepler [7] and Taverna [18] as well as more recent approaches such as [2]. These systems also often support independent invocations as in Fig. 1, where actor b is "mapped" over its input structure such that each invocation of b is applied independently to a specific sub-collection of b's input. Workflow systems that support these and other types of iterative operations (e.g., [17,5]) typically require each of the independent invocations to process a non-overlapping portion of the input to avoid downstream structure conflicts.

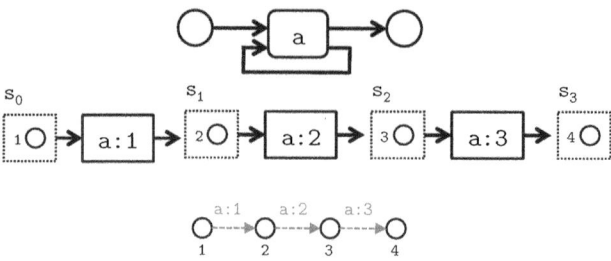

Fig. 2. A cyclic workflow graph (a) with a corresponding SFG (b) and data dependency graph (c)

In addition to coarse-grained dependency information, many applications of provenance [2,17,4] also require "fine-grained" dependencies. For instance, invocation a:1 in Fig. 1b resulted in two new collections (items 4 and 7). However, without also capturing fine-grained dependencies, it is unclear which items in the input of a:1 were used to derive these new collections. Fig. 1c shows the fine-grained (or explicit) dependencies for each structure in our example run. For example, the arrow from data item 2 in s_0 to collection item 4 in s_1 states that 4 (including its containing items) was created from (i.e., depended on) 2 via invocation a:1, whereas item 7 was introduced by a:1 using item 3. As another example, the dependency from item 4 in s_1 to item 10 in s_2 states that item 10 was created by invocation b:1 using item 4 and its containing items 5 and 6. Note that from Fig. 1c it is possible to recreate the SFG in Fig. 1b as well as the other views including the standard data and invocation dependency graphs of Fig. 1d and 1e.

A workflow execution *trace* is represented as a fine-grained dependency graph, i.e., the trace stores the information shown in Fig. 1c. Each trace stores the data structures input to and produced by the workflow run and the corresponding fine-grained dependencies among structures. The parameter values supplied for each invocation are also stored within a trace (not shown in Fig. 1). A trace can be represented in a more condensed form by only storing data and collection items shared by intermediate structures once together with special provenance annotations for item insertions (including dependency information) and deletions [5]. This approach is similar to those for XML-based version management (based on storing "diffs"). The provenance model is able to represent a large number of workflow patterns and constructs, including iteration and looping. Fig. 2 gives a simple example of a trivial iterative calculation in which an actor a is invoked repeatedly until it reaches a fixed point (the output value computed is the same as the input value).

The provenance model supports queries expressed using the *Query Language for Provenance* (QLP) [4]. In QLP, queries are used to filter traces based on specific data or collection items, fine-grained dependencies, and the inputs and outputs of invocations. QLP is similar in spirit to tree and graph-based languages such as XPath and generalized path expressions [1]. Unlike these approaches, however, QLP queries are closed under dependency edges. That is, given a set of dependency edges (defining a fine-grained dependency graph), a QLP query selects and returns a subset of dependencies denoting the corresponding subgraph. This approach provides advantages in terms of supporting incremental querying (e.g., by treating queries as views) and for query optimization [4].

As an example of QLP, the query " $*$ .. 10 " returns the set of dependency edges within a fine-grained dependency graph that define paths starting from any item in a data structure and ending at item 10.[1] Expressed over the trace graph in Fig. 1c, this query returns the dependencies $(2, a:1, 4)$, $(4, b:1, 10)$, $(5, b:1, 10)$, and $(6, b:1, 10)$. Similarly, the query " $*$ .. 4 .. $*$ " returns all dependencies defining paths that start at any data-structure item, pass through item 4, and end at any item in the trace. For Fig. 1, this query returns the same dependencies as the previous query plus the additional dependency $(10, c:1, 12)$. In addition to items in data structures, QLP allows paths to be filtered by invocations. For example, the query "#a .. #b:1 ..$*$" returns dependencies that define paths starting at input items of any invocation of actor a, contain a dependency edge labeled by invocation b:1, and end at any data-structure item. Applied to the example in Fig. 1, this query returns the same set of dependencies as the previous query. QLP uses " @in " and " @out " to obtain the inputs and outputs, respectively, of invocations or runs. For example, the query " $*$ ..@in b:1 " returns dependencies defining paths from any item in a structure to an input item of the first invocation of actor b. Similarly, the query " @in .. 10 " returns the dependencies defining paths that start at any item within an input data structure of the run and that end at item 10.

3 Operators for Exploring Workflow Provenance Graphs

While languages such as QLP can help users quickly access and view relevant parts of large provenance graphs, doing so requires knowledge of the graph prior to issuing a query. When a user does not know ahead of time the parts of the graph that are of interest, or would like to create summarized views of only certain parts of a provenance graph to place the relevant portions in context, additional techniques beyond basic query languages are required. Here we describe extensions to the provenance model of the previous section to help support users as they explore provenance graphs. We consider both specific operations for transforming a provenance graph as well as a set of default views. Using these extensions, a user can switch (or *navigate*) to different views of all or a portion of the provenance graph by applying the transformation operators, or by navigating directly to any of the default views (bringing the current view to the same level of granularity). Given a transformation operator, a new view is constructed using the navigate function. If v_i is the current provenance view, t is the underlying trace, and op is a transformation operator, $\text{navigate}(t, v_i, op) = v_{i+1}$ returns the new provenance view v_{i+1} that results from applying op to v_i under t.

Fig. 3 shows the default views and their relationship to the transformation operators. In addition to the operators shown, we also consider operations for filtering views using QLP queries and for accessing summary data on current views. The rest of this section defines the default views of Fig. 3 and the various operations supported by the provenance model.

---

[1] In QLP, "..." specifies a fine-grained dependency path of one or more edges, " $*$ " specifies any item in the trace, "#" specifies invocations, and " @ " specifies data structures.

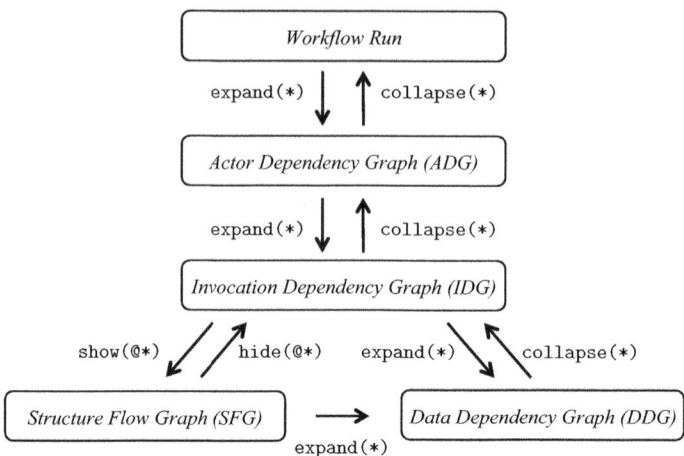

Fig. 3. Default views supported by the provenance model and corresponding navigation operators

Default Provenance Views. Navigation begins at a workflow run (e.g., see the top of Fig. 4). Expanding the run gives an *actor dependency graph* (ADG), which is similar in structure to a workflow graph, but where only actors that were invoked within the run are shown. Expanding an ADG produces an invocation dependency graph (IDG). An IDG consists of invocations and the implicit data dependencies between them, e.g., as in Fig. 1e. A structure flow graph (SFG) can be obtained from an IDG by showing all input and output structures of the trace, e.g., Fig. 1b. Expanding either an SFG or IDG results in a data dependency graph (DDG), e.g., Fig. 1d. It is also possible to navigate from a DDG to an IDG, an IDG to an ADG, and so on, by collapsing the current view (or in the case of an SFG, by hiding all structures). We provide transformation operations that allow a user to directly navigate to any of the graphs in Fig. 3 from any other graph. In particular, the operators ADG, IDG, SFG, and DDG can be used to go directly to the ADG, IDG, SFG, or DDG view, respectively.

Expand and Collapse. In general, the expand and collapse operators allow users to explore specific portions of a provenance view at different levels of detail. We consider three versions of the expand operator based on the type of entity being expanded. For a run r, $expand(r) = \{a_1, a_2, \dots\}$ returns the set of actors $a_1, a_2, \dots$ that were invoked as part of the run. Similarly, given an actor a, $expand(a) = \{i_1, i_2, \dots\}$ returns the set of invocations $i_1, i_2, \dots$ of a. For an invocation i, $expand(i) = \{d_1, d_2, \dots\}$ returns the set of fine-grain dependencies $d_1, d_2, \dots$ introduced by i, where each d_j is a dependency edge of the form (x, i, y) for data items x, y. The collapse operator acts as the inverse of expand. Given a set of of dependencies $\{d_1, d_2, \dots\}$ generated by an invocation i, $collapse(\{d_1, d_2, \dots\}) = i$. Note that a user may select a single dependency to collapse, which will result in all such dependencies of the same invocation to also collapse. Given a set of invocations $\{i_1, i_2, \dots\}$ of an actor a, $collapse(\{i_1, i_2, \dots\}) = a$. In a similar way, if a user selects only a single invocation to collapse, this operation will cause all invocations of the corresponding actor to also collapse. Finally, for a set

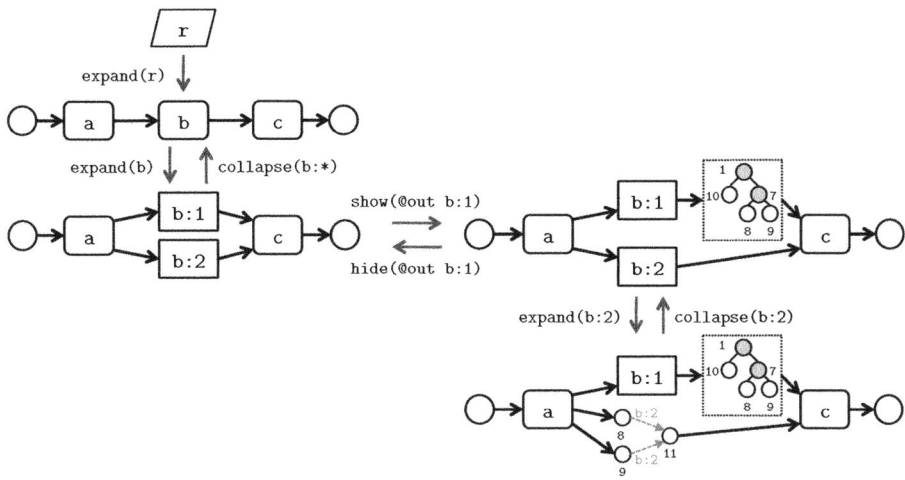

Fig. 4. Applying `expand`, `collapse`, `show`, and `hide` operators to only a part of each view

of actors $\{a_1, a_2, \dots\}$ of run r, $\text{collapse}(\{a_1, a_2, \dots\}) = r$. Again, collapsing any one actor will result in setting the current view to the run view.

Example 1. The top left of Fig. 4 shows an initial expand step to the corresponding actor dependency graph for the example run of Fig. 1 (labeled r in the figure). The second navigation step expands only actor b in the ADG. Similarly, the right of the figure shows invocation b:2 being expanded, resulting in the portion of the data-dependency graph associated with invocation b:2 (i.e., with dependency edges labeled by b:2). The final view shown contains each level of granularity, namely, actors, invocations, data structures, and fine-grain dependencies. Fig. 4 also shows each of the corresponding `collapse` operations. Note that the actor dependency graph is reconstructed by calling `collapse` on the expression b:* which denotes all invocations of b in the current view.

Show and Hide. The `show` operator displays data structures within the current view. The structures displayed depend on the type of entity selected (either a run, actor, or invocation) and whether the input or output of the entity is chosen. We use the QLP "@" construct within `show` to select both the entity and whether the input or output is desired. The expression "@*" denotes all inputs and outputs of each entity in the view. The `hide` operator acts as the inverse of `show` by removing the specified structures. As an example, the `show` operator is used in the third navigation step of Fig. 4 to display the output structure of invocation b:1 (s_2^1 in Fig. 1b). Note that in this example, `show(@in c:1)` would additionally display structure s_2 from Fig. 1b.

Group and Ungroup. The `group` and `ungroup` operators allow actors and invocations to be combined into composite structures. The `group` operator explicitly allows users to control which items should be grouped and supports both actor and invocation granularity. We consider two versions of `group` and `ungroup`. Given a set of actors $\{a_1, a_2, \dots\}$, $\text{group}(\{a_1, a_2, \dots\}) = g_{\{a_1, a_2, \dots\}}$ returns a composite actor $g_{\{a_1, a_2, \dots\}}$ over the given set.

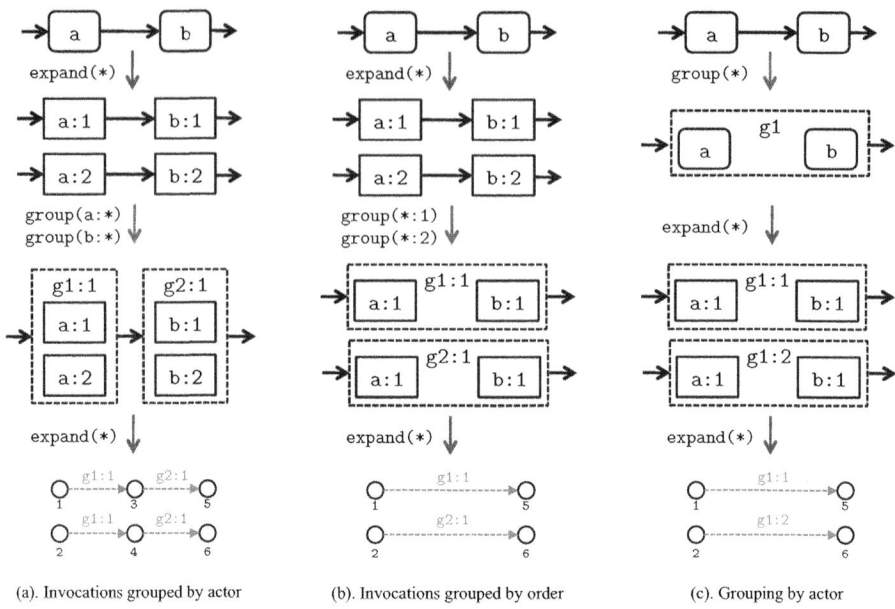

(a). Invocations grouped by actor   (b). Invocations grouped by order   (c). Grouping by actor

Fig. 5. Actors and invocations combined into composite structures

For a composite actor $g_{\{a_1,a_2,\dots\}}$, $\mathrm{ungroup}(g_{\{a_1,a_2,\dots\}}) = \{a_1,a_2,\dots\}$ simply returns the set of actors corresponding to the group (i.e., ungroup is the inverse of group). Similarly, for a set of invocations $\{i_1,i_2,\dots\}$, $\mathrm{group}(\{i_1,i_2,\dots\}) = g_{\{i_1,i_2,\dots\}}$ returns a composite invocation $g_{\{i_1,i_2,\dots\}}$, and $\mathrm{ungroup}(g_{\{i_1,i_2,\dots\}}) = \{i_1,i_2,\dots\}$ returns the original set.

Example 2. Fig. 5 shows three examples of using the group operator. Here we consider only a portion of a workflow showing two actors a and b such that both were invoked twice resulting in the actor dependency graph shown in the second navigation step of Fig. 5a. Each invocation of a and b consume and produce one data item, where the output of invocation a : i is used by invocation b : i. In Fig. 5a, invocations of the same actor are grouped such that a:1 and a:2 form one group and b:1 and b:2 form a different group. The result of the grouping is shown in the third view of Fig. 5a, and the corresponding data dependency graph is shown as the fourth view. In Fig. 5b, invocations with the same invocation number are grouped such that a:1 and b:1 form one group, and a:2 and b:2 form a different group. Finally, in Fig. 5c, actors a and b are first grouped, resulting in a composite actor with two distinct invocations. Unlike in Fig. 5b, these invocations are of the same actor group and have different invocation numbers, whereas in Fig. 5b two distinct groups are created. In general, forming invocation groups explicitly, as opposed to first forming actor groups and then expanding actor groups, supports grouping at a finer-level of granularity by allowing various patterns of composite invocations that are not possible to express at the actor level.

As shown in Fig. 5, composites created by the group operator are assigned special identifiers. In addition, the inputs, outputs, and dependencies associated with grouped items are inferred from the underlying inputs, outputs, and dependencies of the invocations of the groups. For showing dependencies in particular, this often requires computing the transitive dependency closure associated with invocations of the group, e.g., as in Fig. 5b and 5c (since items 3 and 4 are "hidden" by the grouped invocation). When a group is created at the actor level, expanding the group results in a correspondingly grouped set of invocations (as in Fig. 5c). These invocations are constructed based on the invocation dependency graph. In particular, each invocation group of the actor group contains a set of connected invocations, and no invocation within an invocation group is connected to any other invocation in a different invocation group. Thus, the portion of the invocation graph associated with the actor group is partitioned into connected subgraphs, and each such subgraph forms a distinct invocation group of the actor group. When an invocation group is expanded, this composite invocation is used in structure flow and data dependency graphs, resulting in provenance views where dependencies are established between input and output items, without intermediate data in between. This approach allows scientists to continue to explore dependencies for grouped invocations (since fine-grain dependencies are *maintained* through groups, unlike, e.g., the approach in [11]).

Filter. The `filter` operation allows provenance views to be refined using QLP query expressions. Issuing a query using `filter` results in only the portion of the current provenance view corresponding to the query answer to be displayed. Given a QLP query q, $\texttt{filter}(q) = \{d_1, d_2, \ldots\}$ returns the set of fine-grain dependencies that result from applying the query to the trace graph. These dependencies are used to remove or add entities to the current view. In general, items can be added to a view when the current view is based on a more selective query.

Aggregation. It is often convenient to see summary information about entities (actors, invocations, etc.) when exploring provenance information. We provide standard aggregate operators (`count`, `min`, `max`, `avg`) to obtain statistics for a user's current provenance view. The "`count` *entity_type* of *scope*" operator returns the number of entities of *entity_type* within a given *scope* expression. Entity types are either `actors`, `invocations`, or `data`, which count the number of actors, invocations, or data items, respectively. For the `actors` entity type, the scope is either $*$, denoting the entire view, or a group identifier. For example, "`count actors of` $*$" returns two for each view in Fig. 5, whereas "`count actors of g1 : 1`" returns one for the third view in Fig. 5. For the `invocations` entity type, the scope is either $*$ (the entire view) or an actor or group identifier. For instance, in the first view of Fig. 5a "`count invocations of` $*$" returns four whereas "`count invocations of a`" returns two. The `count` operation is also useful for exploring workflow loops, e.g., the expression "`count invocations of a`" can be used to obtain the number of iterations of a in Fig. 2. For the `data` entity type, the QLP "`@`" syntax is used to define the scope. For instance, for the invocation dependency graph of Fig. 1e, "`count data of @in`" returns two (since two data items are input to the workflow), and "`count data of @out a:1`" returns four (the number of data items output by invocation a:1). The `min`, `max`, and `avg` operations return the minimum, maximum, and average number of entity types within a view, respectively. These operations can be used to compute the number of actors by group (e.g., "min actor by group"), the number of invocations by actors or groups (e.g., "min invocations by actor"), and the number of input or output data items by actor, group, or invocation (e.g., "min input data by actor"). Both the min and max operators return the entity with the minimum or maxiximum count value, respectively, as well as the number of corresponding entities. Finally, the params operation returns the set of parameter values used in invocations of the current view. This operation can also be restricted to specific actors, invocations, or groups, e.g., params(a:1) returns the parameter settings for invocation a:1 whereas params(*) returns the parameter settings for all invocations in the current view.

4   Implementation

In this section we describe an implementation using standard relational database technology for the operators presented in Section 3. As mentioned in the introduction, this approach has benefits for efficiency and scalability (see Section 5). We first describe a set of relational schemas for representing traces and default views, and then show how the navigation operators can be implemented as relational queries over these schemas.

We consider three distinct schemas: (1) a *trace schema* $\mathbf{T}$ for representing the trace information corresponding to a workflow run; (2) a *dependency schema* $\mathbf{D}$ for representing the result of executing QLP queries expressed through filter operations; and (3) a *view schema* $\mathbf{V}$ for representing the user's current provenance view. The trace schema $\mathbf{T}$ consists of the following relations: $\mathrm{Run}(r,w)$ denotes that r was a run of workflow w; $\mathrm{Invoc}(r,i,a,j)$ denotes that invocation i was the j-th invocation of actor a in run r; $\mathrm{Node}(r,n,p,t,l,i_{ins},i_{del})$ denotes that n was an item (node) in run r such that p is the parent collection of n, t is the type of the item (either data or collection), l is the label of n (e.g., XML tag name), and that n was inserted by invocation i_{ins} and deleted by i_{del}; $\mathrm{DDep}(r,n,n_{dep})$ denotes a dependency from item n_{dep} to item n in run r; $\mathrm{DDepc}(r,n,n_{dep})$ stores the transitive closure of DDep; $\mathrm{IDep}(r,i,i_{dep})$ denotes an invocation dependency from invocation i to i_{dep}; and $\mathrm{IDepc}(r,i,i_{dep})$ stores the transitive closure of IDepc. In general, storing the transitive closure improves query time for more complex path-expression queries and simplifies a number of the operations presented here. We show in [5] an approach for efficiently compressing the transitive closure relations that only marginally affect the response time of basic queries (adding only an additional join in many cases), and adopt this approach in Section 5 when describing our experimental results. In [4] we extend this approach for efficiently answering QLP queries, which we assume here for implementing filter operations. We also assume the following views (expressed using Datalog) as part of T.

$$\mathrm{ADep}(r,a,a_{dep}) :\!\!- \mathrm{IDep}(r,i,i_{dep}), \mathrm{Invoc}(r,i,a,j_1), \mathrm{Invoc}(r,i_{dep},a_{dep},j_2).$$
$$\mathrm{ADepc}(r,a,a_{dep}) :\!\!- \mathrm{IDepc}(r,i,i_{dep}), \mathrm{Invoc}(r,i,a,j_1), \mathrm{Invoc}(r,i_{dep},a_{dep},j_2).$$

The dependency graph schema $\mathbf{D}$ consists of the single relation $\mathrm{DepView}(n_{from},i,n_{to})$ denoting a dependency edge from item n_{from} to n_{to} labelled by invocation i. Similarly, the view schema $\mathbf{V}$ consists of the single relation $\mathrm{CurrView}(e_{from},t_{from},l,e_{to},t_{to})$ for e_{from} and e_{to} entites connected via an edge label l such that each entity's type is denoted by t_{to} and t_{from}, respectively. The entity type can be either a run, actor, invoc (invocation), data, coll (collection), or struct (structure). A structure is denoted by a QLP expression of the form @in $[i \mid a]$ or @out $[i \mid a]$ where invocation i and actor a are optional. The following are examples of possible tuples stored within the current view relation: $\text{CurrView}(r, \text{run}, \perp, \perp, \perp)$ stores a run view ($\perp$ denotes a null value); $\text{CurrView}(a_1, \text{actor}, \perp, a_2, \text{actor})$ stores an edge in an actor dependency graph; $\text{CurrView}(i_1, \text{invoc}, \perp, i_2, \text{invoc})$ stores an edge in an invocation dependency graph; $\text{CurrView}(\text{@out } i_1, \text{struct}, \perp, i_2, \text{invoc})$ stores an edge in a structure flow graph; and $\text{CurrView}(n_1, \text{data}, \text{a:1}, n_2, \text{data})$ stores an edge in a data dependency graph.

When a user begins navigating a trace, the instance D of the dependency schema **D** consists of the entire set of dependencies. After applying a navigation operation, the instance D of the dependency schema **D** and V of the current view schema **V** are updated as needed. For example, after applying an initial filter operation, the dependencies of D are updated (denoted as D_1) to store the result of the given QLP query over the instances T of the trace schema **T**. The instance V of the view schema **V** is also updated (denoted as V_1) based on D_1. Similarly, after applying an expand, collapse, show, or hide operator, a new view V_2 is created from D_1, V_1, and T. This process continues for each navigation step performed by the user, where only the current view and dependency graph is stored together with the initial trace.

Queries to Generate Default Views. We can implement the default view operators as queries over a trace instance T and dependency instance D as follows. First, we define the following notation as shorthand for filtering relations in T by D. Given a trace relation R, we write $R^{(D)}$ to denote the filtered version of R with respect to the dependencies in D. For instance, $\text{IDep}^{(D)}$ is the invocation dependency relation (IDep) containing only invocations that participate in dependency edges within D. Given a dependency relation D, we define the following:

$$\text{IDep}^{(D)}(r, i, i_{dep}) :\text{-IDep}(r, i, i_{dep}), \text{DepView}(n_1, i, n_2), \text{DepView}(n_2, i_{dep}, n_3).$$
$$\text{ADep}^{(D)}(r, a, a_{dep}) :\text{-IDep}^{(D)}(r, i, i_{dep}), \text{Invoc}(r, i, a, j_1), \text{Invoc}(r, i_{dep}, a_{dep}, j_2).$$
$$\text{Node}^{(D)}(r, n, p, t, l, i_{ins}, i_{del}) :\text{-Node}(r, n, p, t, l, i_{ins}, i_{del}), \text{DepView}(n, i, n_2).$$
$$\text{Node}^{(D)}(r, n, p, t, l, i_{ins}, i_{del}) :\text{-Node}(r, n, p, t, l, i_{ins}, i_{del}), \text{DepView}(n_1, i, n).$$

These relations are used to compute the default view operators for a run r (each of which follow the same relation structure as CurrView):

$$\text{ADG}(a, \text{actor}, \perp, a_{dep}, \text{actor}) :- \text{ADep}^{(D)}(r, a, a_{dep}).$$
$$\text{IDG}(i, \text{invoc}, \perp, i_{dep}, \text{invoc}) :- \text{IDep}^{(D)}(r, i, i_{dep}).$$
$$\text{DDG}(n_1, t_1, i, n_2, t_2) :- \text{DepView}(n_1, i, n_2), \text{Node}(r, n_1, p_1, t_1, i_{ins_1}, i_{del_1}),$$
$$\text{Node}(r, n_2, p_2, t_2, i_{ins_2}, i_{del_2}).$$

Each of these operators returns a new view that replaces the current view. In a similar way, we can define the structure flow graph for run r, which is similar to computing the invocation dependency graph.

$$\text{SFG}(i, \text{invoc}, \perp, \text{@out } i, \text{struct}) :- \text{IDep}^{(D)}(r, i, i_{dep}).$$
$$\text{SFG}(\text{@out } i, \text{struct}, \perp, \text{@in } i_{dep}, \text{struct}) :- \text{IDep}^{(D)} r, i, i_{dep}).$$
$$\text{SFG}(\text{@in } i_{dep}, \text{struct}, \perp, i_{dep}, \text{invoc}) :- \text{IDep}^{(D)}(r, i, i_{dep}).$$

Queries to Generate Input-Output Structures. Given a trace T and a dependency D, we compute the input and output structures of runs r, actors a, and invocations i as follows. The input of a run r includes all data and collection items in the trace that were not inserted by any invocation:

$$\texttt{RunInput}(n,p) \ :- \ \texttt{Node}^{(D)}(r,n,p,t,l,\bot,i_{del}).$$

The output items of r include those that (1) were either input to the run or inserted by an invocation, and (2) were not deleted by an invocation:

$$\texttt{RunOutput}(n,p) \ :- \ \texttt{Node}^{(D)}(r,n,p,t,l,i_{ins},\bot).$$

The input of an invocation i includes all nodes not deleted by an invocation that i depends on such that either the item (1) was inserted by an invocation that i depended on or (2) was not inserted by an invocation and thus was an input to the run. The output of i is computed similarly, i.e., by removing from the input of i the nodes deleted by i and adding the nodes inserted by i. The following rules compute the input and output of a given invocation i and a run r.

$$\texttt{InvocInput}(n,p)\text{:-}\texttt{Node}^{(D)}(r,n,p,t,l,\bot,\bot).$$
$$\texttt{InvocInput}(n,p)\text{:-}\texttt{Node}^{(D)}(r,n,p,t,l,i_{ins},\texttt{i}).$$
$$\texttt{InvocInput}(n,p)\text{:-}\texttt{Node}^{(D)}(r,n,p,t,l,\bot,i_{del}),\texttt{Invoc}(r,i_{del},a,j),\neg\texttt{IDepc}(r,i_{del},\texttt{i}).$$
$$\texttt{InvocInput}(n,p)\text{:-}\texttt{Node}^{(D)}(r,n,p,t,l,i_{ins},i_{del}),\texttt{IDepc}(r,i_{ins},\texttt{i}),\texttt{Invoc}(r,i_{del},a,j),$$
$$\neg\texttt{IDepc}(r,i_{del},\texttt{i}).$$

Note that the first rule selects input items that were neither inserted or deleted within a run, and the second rule selects input items that were deleted by the given invocation i. The last two rules ensure the item was not deleted by an invocation that i depended on. The output of invocations are defined similarly:

$$\texttt{InvocOutput}(n,p) \ :- \ \texttt{Node}^{(D)}(r,n,p,t,l,\bot,\bot).$$
$$\texttt{InvocOutput}(n,p) \ :- \ \texttt{Node}^{(D)}(r,n,p,t,l,\texttt{i},i_{del}).$$
$$\texttt{InvocOutput}(n,p) \ :- \ \texttt{Node}^{(D)}(r,n,p,t,l,\bot,i_{del}),\texttt{Invoc}(r,i_{del},a,j),$$
$$\neg\texttt{IDepc}(r,i_{del},\texttt{i}),i_{del}\neq\texttt{i}.$$
$$\texttt{InvocOutput}(n,p) \ :- \ \texttt{Node}^{(D)}(r,n,p,t,l,i_{ins},i_{del}),\texttt{IDepc}(r,i_{ins},\texttt{i}),$$
$$\texttt{Invoc}(r,i_{del},a,j),\neg\texttt{IDepc}(r,i_{del},\texttt{i}),i_{del}\neq\texttt{i}.$$

The input and output structures of actors (as opposed to invocations) are computed by first retrieving the invocations that are in the current view, and then for each such invocation, unioning the corresponding structures.

Queries to Group and Ungroup Actors and Invocations. To implement the group and ungroup operators, we store the set of entities supplied to the group operator in a temporary relation $\texttt{Group}(e,t,n,g)$ where e is one of the entities being grouped, t is the entity type (either actor or invoc), n is the group identifier, and g is the grouping type. For example, to group invocations b : 1 and b : 2 into a group g1 : 1 we store the tuples $\texttt{Group}(b:1,\texttt{invoc},\texttt{g1}:1,\texttt{invoc\_group})$ and $\texttt{Group}(b:2,\texttt{invoc},\texttt{g1}:1,$

invoc_group). The new view with grouped entities is generated as follows: (1) retrieve those tuples from the current view relation $\mathtt{CurrView}_i$ where e_{from} and e_{to} are not entities to be grouped and store these in the new view $\mathtt{CurrView}_{i+1}$; (2) if e_{from} is an entity to be grouped, then retrieve the tuple from $\mathtt{CurrView}_i$ and modify e_{from} and t_{from} with the group identifier n and group type g, respectively; and similarly (3) if e_{to} is an entity to be grouped, perform a similar operation as in (2). These steps are performed by the following queries. Note that we assume for each entity e of type t that is not involved in a group within the current view, there exists a tuple $\mathtt{Group}(e, t, \perp, \perp)$.

$$\mathtt{CurrView}_{i+1}(e_{from}, t_{from}, l, e_{to}, t_{to}) \; \text{:-} \; \mathtt{CurrView}_i(e_{from}, t_{from}, l, e_{to}, t_{to}),$$
$$\mathtt{Group}(e_{from}, t_{from}, \perp, \perp),$$
$$\mathtt{Group}(e_{to}, t_{to}, \perp, \perp),$$
$$\mathtt{CurrView}_{i+1}(n, g, l, e_{to}, t_{to}) \; \text{:-} \; \mathtt{CurrView}_i(e_{from}, t_{from}, l, e_{to}, t_{to}),$$
$$\mathtt{Group}(e_{from}, t_{from}, n, g),$$
$$\mathtt{Group}(e_{to}, t_{to}, \perp, \perp).$$
$$\mathtt{CurrView}_{i+1}(e_{from}, t_{from}, l, n, g) \; \text{:-} \; \mathtt{CurrView}_i(e_{from}, t_{from}, l, e_{to}, t_{to}),$$
$$\mathtt{Group}(t_{to}, e_{to}, n, g),$$
$$\mathtt{Group}(e_{from}, t_{from}, \perp, \perp).$$
$$\mathtt{CurrView}_{i+1}(n_1, g_1, l, n_2, g_2) \; \text{:-} \; \mathtt{CurrView}_i(e_{from}, t_{from}, l, e_{to}, t_{to}),$$
$$\mathtt{Group}(e_{from}, t_{from}, n_1, g_1),$$
$$\mathtt{Group}(t_{to}, e_{to}, n_2, g_2).$$

Grouping has implications on how inputs, outputs, and data dependencies across grouped entities are displayed. Inputs of grouped entities are computed by performing the union of all the inputs of those entities that are a source node in the graph with respect to entities to be grouped. Similarly, the outputs of grouped entities are computed by performing the union of all the outputs of those entities that are sink nodes with respect to the entities to be grouped. Also, as discussed in the previous section, data dependency views over grouped entities require computing the transitive dependency closure of inputs and outputs of grouped entities. We denote the source entities of a group as $\mathtt{Group}_{src}$ and the sink entities of a group as $\mathtt{Group}_{sink}$. These relations are computed by first deriving the dependency relations between the entities of the group and then checking which one has no incoming edges (for $\mathtt{Group}_{src}$), and similarly, which one has no outgoing edges (for $\mathtt{Group}_{sink}$).

The ungroup operator for the current view $\mathtt{CurrView}_i$ is performed as follows: (1) for all invocations of the group, we retrieve their invocation dependencies; (2) for all source invocations of the group (with respect to the invocation dependencies), we add an edge to $\mathtt{CurrView}_{i+1}$ from the e_{to} to the source invocation; and (3) for all sink invocations of the group, we add an edge to $\mathtt{CurrView}_{i+1}$ from the sink invocation to e_{from}. Ungrouping of an actor is done in a similar way.

$$\mathtt{CurrView}_{i+1}(i, \mathtt{invoc}, l, i_{dep}, \mathtt{invoc}) \; \text{:-} \; \mathtt{IDep}^{(D)}(r, i, i_{dep}),$$
$$\mathtt{Group}(i, \mathtt{invoc}, n, \mathtt{invoc\_group}),$$
$$\mathtt{Group}(i_{dep}, \mathtt{invoc}, n, \mathtt{invoc\_group}).$$
$$\mathtt{CurrView}_{i+1}(e_{from}, t_{from}, l, e, t) \; \text{:-} \; \mathtt{CurrView}_i(e_{from}, t_{from}, l, n, g),$$
$$\mathtt{Group}_{src}(n, g, e, t).$$
$$\mathtt{CurrView}_{i+1}(e, t, l, e_{to}, t_{to}) \; \text{:-} \; \mathtt{CurrView}_i(n, g, l, e_{to}, t_{to}), \mathtt{Group}_{sink}(n, g, e, t).$$

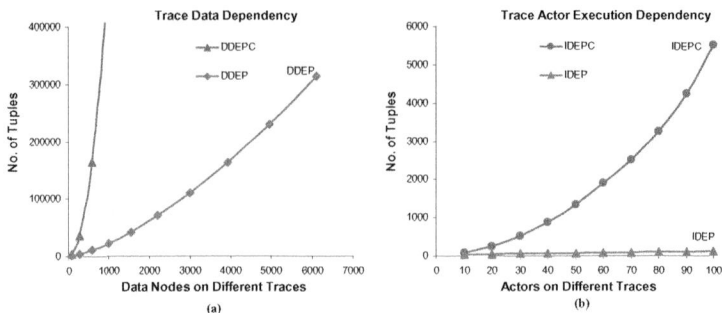

Fig. 6. (a) Data dependency and (b) actor invocation complexity of synthetic traces

Expand, Collapse, and Aggregates. We expand a given invocation i to show the dependency relationship between its inputs and outputs using the relation DepView(n_{from}, i, n_{to}) in D. We collapse an invocation i by showing the actor corresponding to the invocation via the Invoc relation in T. We expand a given actor a by obtaining the invocations of a that are in D. The current view is constructed directly from these operations in a similar way as for grouping and ungrouping (i.e., by inserting the corresponding relations in the new current view CurrView$_{i+1}$). Finally, the aggregate operations are straightforward to compute using standard relational aggregation queries over the current view, dependency, and group relations.

5   Experimental Results

Here we evaluate the feasibility, scalability, and efficiency of executing the navigation operators over the approach in Section 4 on both real and synthetic traces. Real traces were generated from existing workflows implemented within the Kepler scientific workflow system. Our experiments were performed using a 2.8GHz Intel Core 2 duo PC with 4 GB RAM and 500 GB of disk space. Navigational operators were implemented as SQL queries (views over the schema), which were executed against a PostgreSQL database where all provenance traces were stored. The QLP parser was implemented in Java using JDBC to communicate with the provenance database.

We evaluated the feasibility of executing the navigation operator queries using the following real traces from scientific workflows implemented within Kepler: the *GBL* workflow [7] infers phylogenetic trees from protein and morphological sequence data; the *PC1* workflow was used in the first provenance challenge [19]; the *STAP* and *CYC* workflows are used in characterizing microbial communities by clustering and identifying DNA sequences of 16S ribosomal RNA; the *WAT* workflow characterizes microbial populations by producing phylogenetic trees from a list of sequence libraries; and the *PC3* workflow was used within the third provenance challenge[2]. These traces ranged from 100–10,000 immediate data dependencies and 200–20,000 transitive data dependencies. We also evaluated our approaches to determine the scalability of executing navigation operators queries using synthetic traces ranging from 500–100,000 immediate data dependencies, 1,000–10^8 transitive data dependencies, and data dependency

---

[2] see http://twiki.ipaw.info/bin/view/Challenge/

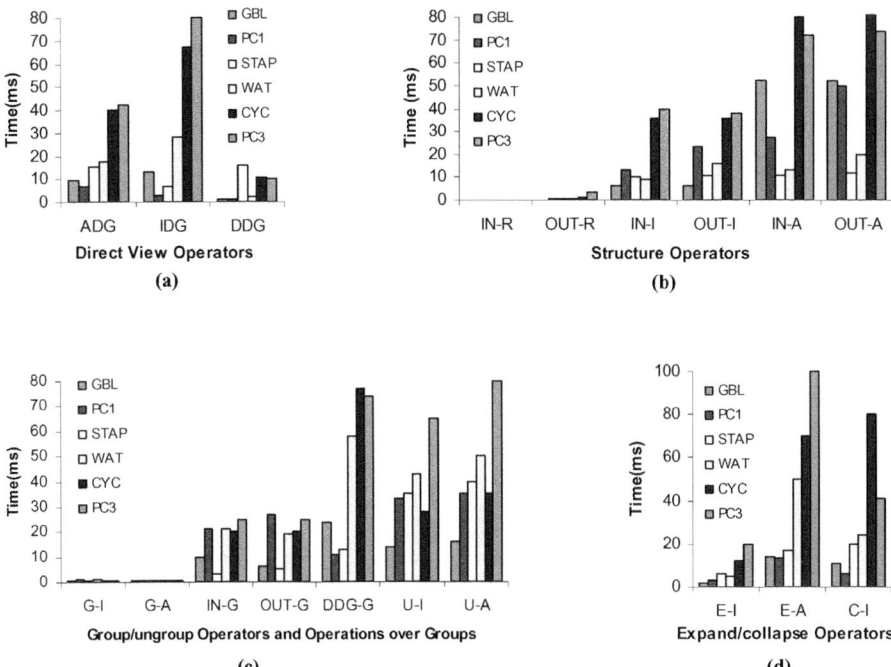

Fig. 7. Average query time for operators over real traces: (a) Q_D; (b) Q_S; (c) Q_G; and (d) Q_E

paths of length 10–100. These traces also contained 10–100 actors, 10–100 immediate invocation dependencies, and 10–6,000 transitive invocation dependencies. The synthetic traces were taken from [5], and represent typical dependency patterns of common scientific workflows [5,7,19]. Fig. 6a shows the complexity of data dependencies (immediate and transitive) as the number of nodes in the synthetic traces increase, and Fig. 6b shows the the complexity of invocation dependencies (immediate and transitive) as the number of actors in the synthetic traces increase.

We use four types of operators in our evaluation: (Q_D) queries to generate default ADG, IDG, and DDG views; (Q_S) queries to generate data structures (run, actor, and invocation inputs and outputs); (Q_G) queries to group and ungroup actors and invocations, and to retrieve their inputs, outputs, and fine-grained dependencies; and (Q_E) queries to expand and collapse actors and invocations. Section 4 details the underlying datalog queries for each operator type.

Feasibility and Efficiency Results. Fig. 7 shows timing results of the navigation operators over real traces. The time to execute DDG operations (Fig. 7a) is smaller since we store the result of previous QLP queries in the dependency instance D, which is used to generate the result of DDG calls, whereas ADG and IDG operations need additional queries over provenance schema tables. The time to retrieve the input and output structures of a run is less expensive than for an invocation (Fig. 7b). However, the time to retrieve input and output structures for an invocation is less expensive for an actor. This is because an actor can be invoked many times, which requires computing the union of structures for all such invocations. Grouping invocations and actors is also less

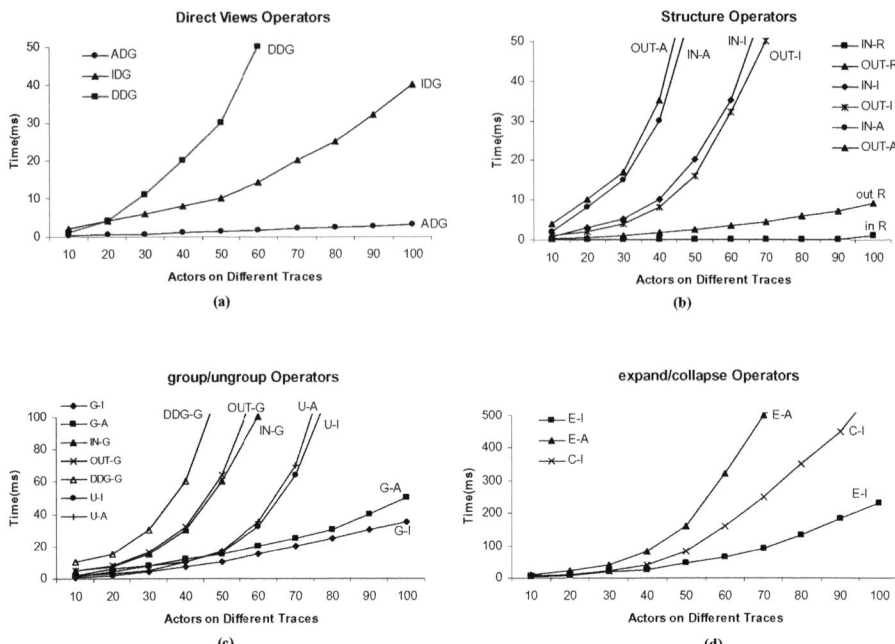

Fig. 8. Average query time for operators over synthetic traces: (a) Q_D; (b) Q_S; (c) Q_G; and (d) Q_E

expensive than their inverse ungrouping operations (see Fig. 7c), since ungrouping has to perform many conditional joins with the current view to reconstruct the ungrouped invocations, actors, and their relations to other items within the current view. Expanding an actor is more expensive than expanding an invocation (see Fig. 7d) since expanding an invocation involves dependencies that are already materialized as QLP query results, but expansion of an actor must execute queries against schema tables through additional conditional joins. Despite the complexity involved in executing such queries, our experimental results show that each type of operation takes *less then* 1 sec, demonstrating the feasibility and efficiency that can be obtained using a purely relational approach.

Scalability Results. Fig. 8 shows the results of executing navigation queries over the synthetic traces. As shown, most of the queries are still executed in less than 1 sec (100 ms) for larger trace sizes, suggesting that a purely relational approach can scale to larger trace sizes (compared with those obtained from the real traces used above). Note that queries for implementing the expand and collapse operators (i.e., of type Q_E) take more time than the other operator types, which is due to the number of conditional joins (as discussed earlier) that are required. Overall, however, the results for synthetic traces confirm those discussed in the case for real traces above.

6   Related Work

Current approaches for exploring workflow provenance are based on statically visualizing entire provenance graphs [16,14,13,7]. In these approaches, provenance graphs are typically displayed at the lowest level of granularity (e.g., fine-grain dependencies). In the case of [14], query results are viewed independently of the rest of a provenance trace. Some systems divide provenance information into distinct layers, e.g., VisTrails divides provenance information into workflow evolution, workflow, and execution layers [8], and PASS divides provenance into data and process layers [21]. In all these approaches, however, these levels are largely either orthogonal or hierarchical, whereas the provenance views supported by our model (i) combine both hierarchical abstractions (i.e., ADGs, IDGs, and SFGs) with (ii) the ability to seamlessly navigate between these different levels of granularity, while (iii) allowing users to summarize, group, and filter portions of these views to create new views for further exploration of relevant provenance information. The Zoom*UserViews system [6] (extended in [2] to work with a fine-grained database provenance model for Pig Latin) provides a mechanism for defining composite actors to abstract away non-relevant provenance information. Composites are constructed over "relevant" actors to maintain certain dataflow connections, thereby generating a view over the composites that is similar to the original. However, unlike in our approach, users of the Zoom*UserViews system cannot explicitly define their own composites, and composition is defined only at the actor level (where each actor is assumed to have at most one invocation). Our approach also maintains grouping across views, maintains the original data dependencies for composites (unlike in the Zoom*UserViews approach, which switches to coarse-grain dependencies), and we support a more general provenance model that explicitly handles structured data.

Our navigation approach is inspired by and has similarities to those proposed previously for exploring object-oriented [9] and XML databases, where graphical environments allow users to "drill-down" from schema to instances and navigate relationships among data. For example, [9] provides an integrated browsing and querying environment that allows users to employ a "query-in-place" paradigm where navigation and query can be mixed. In contrast, provenance information is largely schema-free, i.e., the information contained within an ADG, IDG, and SFG is not constrained by an explicit schema, and queries in our model are posed directly against the items contained within these views (or generally, the fine-grain dependency graph).

7   Conclusion

We have presented a general model to help users explore provenance information through novel graph-based summarization, navigation, and query operators. The work presented here extends our prior work [7,3] by providing a significantly richer model for navigation (adding additional views, the `show` and `hide` constructs, and aggregation) as well as an implementation using standard relational database technology. We also provide experimental results demonstrating the feasibility, scalability, and efficiency of our approach. Because of the size and complexity of real-world scientific workflow provenance traces [10,4], providing users with high-level navigation operations and views for abstracting and summarizing provenance information can provide a powerful environment for scientists to explore and validate the results of scientific workflows.

Acknowledgements. Work supported through NSF grants IIS-1118088, DBI-0743429, DBI-0753144, DBI-0960535, and OCI-0722079.

References

1. Abiteboul, S., Quass, D., McHugh, J., Widom, J., Wiener, J.L.: The Lorel query language for semistructured data. IJDL (1997)
2. Amsterdamer, Y., Davidson, S.B., Deutch, D., Milo, T., Stoyanovich, J., Tannen, V.: Putting lipstick on pig: Enabling database-style workflow provenance. PVLDB 5(4) (2011)
3. Anand, M.K., Bowers, S., Ludäscher, B.: A navigation model for exploring scientific workflow provenance graphs. In: Proc. of the Workshop on Workflows in Support of Large-Scale Science, WORKS (2009)
4. Anand, M.K., Bowers, S., Ludäscher, B.: Techniques for efficiently querying scientific workflow provenance graphs. In: EDBT, pp. 287–298 (2010)
5. Anand, M.K., Bowers, S., McPhillips, T.M., Ludäscher, B.: Efficient provenance storage over nested data collections. In: EDBT (2009)
6. Biton, O., Boulakia, S.C., Davidson, S.B., Hara, C.S.: Querying and managing provenance through user views in scientific workflows. In: ICDE (2008)
7. Bowers, S., McPhillips, T., Riddle, S., Anand, M.K., Ludäscher, B.: Kepler/pPOD: Scientific Workflow and Provenance Support for Assembling the Tree of Life. In: Freire, J., Koop, D., Moreau, L. (eds.) IPAW 2008. LNCS, vol. 5272, pp. 70–77. Springer, Heidelberg (2008)
8. Callahan, S., Freire, J., Santos, E., Scheidegger, C., Silva, C., Vo, H.: VisTrails: Visualization meets data management. In: SIGMOD (2006)
9. Carey, M.J., Haas, L.M., Maganty, V., Williams, J.H.: PESTO: An integrated query/browser for object databases. In: VLDB (1996)
10. Chapman, A., Jagadish, H.V., Ramanan, P.: Efficient provenance storage. In: SIGMOD (2008)
11. Davidson, S.B., Freire, J.: Provenance and scientific workflows: challenges and opportunities. In: SIGMOD (2008)
12. He, H., Singh, A.K.: Graphs-at-a-time: Query language and access methods for graph databases. In: SIGMOD, pp. 405–418 (2008)
13. Hunter, J., Cheung, K.: Provenance explorer-a graphical interface for constructing scientific publication packages from provenance trails. Int. J. Digit. Libr. 7(1) (2007)
14. Lim, C., Lu, S., Chebotko, A., Fotouhi, F.: Opql: A first opm-level query language for scientific workflow provenance. In: IEEE SCC, pp. 136–143 (2011)
15. Ludäscher, B., et al.: Scientific workflow management and the Kepler system. Concurr. Comput.: Pract. Exper. 18(10) (2006)
16. Macko, P., Seltzer, M.: Provenance map orbiter: Interactive exploration of large provenance graphs. In: TAPP (2011)
17. Missier, P., Paton, N.W., Belhajjame, K.: Fine-grained and efficient lineage querying of collection-based workflow provenance. In: EDBT, pp. 299–310 (2010)
18. Missier, P., Soiland-Reyes, S., Owen, S., Tan, W., Nenadic, A., Dunlop, I., Williams, A., Oinn, T., Goble, C.: Taverna, Reloaded. In: Gertz, M., Ludäscher, B. (eds.) SSDBM 2010. LNCS, vol. 6187, pp. 471–481. Springer, Heidelberg (2010)
19. Moreau, L., et al.: The first provenance challenge. Concurr. Comput.: Pract. Exper. 20(5) (2008)
20. Moreau, L., et al.: The open provenance model core specification (v1.1). Future Generation Computer Systems 27(6), 743–756 (2011)
21. Muniswamy-Reddy, K.K., et al.: Layering in provenance systems. In: USENIX Annual Technical Conference (2009)

(Re)Use in Public Scientific Workflow Repositories

Johannes Starlinger[1], Sarah Cohen-Boulakia[2], and Ulf Leser[1]

[1] Humboldt-Universität zu Berlin, Department of Computer Science,
Unter den Linden 6, 10099 Berlin, Germany
{starling,leser}@informatik.hu-berlin.de
[2] Université Paris-Sud, Laboratoire de Recherche en Informatique,
CNRS UMR 8623 and INRIA AMIB, France
cohen@lri.fr

Abstract. Scientific workflows help in designing, managing, monitoring, and executing in-silico experiments. Since scientific workflows often are complex, sharing them by means of public workflow repositories has become an important issue for the community. However, due to the increasing numbers of workflows available in such repositories, users have a crucial need for assistance in discovering the right workflow for a given task. To this end, identification of functional elements shared between workflows as a first step to derive meaningful similarity measures for workflows is a key point. In this paper, we present the results of a study we performed on the probably largest open workflow repository, myExperiment.org. Our contributions are threefold: (i) We discuss the critical problem of identifying same or similar (sub-)workflows and workflow elements, (ii) We study, for the first time, the problem of cross-author reuse and (iii) We provide a detailed analysis on the frequency of re-use of elements between workflows and authors, and identify characteristics of shared elements.

Keywords: scientific workflows, similarity measures, workflow reuse.

1 Introduction

Scientific workflow management systems (SWFM) recently gained increasing attention as valuable tools for scientists to create and manage reproducible in-silico experiments. Nowadays, several SWFM, each with its particular strengths and weaknesses, are freely available, such as Taverna [1], Kepler [2], VisTrails [3], or Galaxy [4]. Yet, creating scientific workflows using an SWFM is still a laborious task and complex enough to prevent non computer-savvy researchers from using these tools [5]. On the other hand, once designed, a scientific workflow is a valuable piece of knowledge, encoding a (usually proven) method for analysing a given data set. Sharing such workflows between groups is a natural next step towards increasing collaboration in research which may have a large impact on research, in the same manner as the increased sharing of data sets across

A. Ailamaki and S. Bowers (Eds.): SSDBM 2012, LNCS 7338, pp. 361–378, 2012.

the Internet was important for advancing science [6]. As a consequence, online repositories, such as myExperiment.org [7] or CrowdLabs [8], have emerged to facilitate reusing and repurposing of scientific workflows. Such repositories, together with the increasing number of workflows uploaded to them, raise several new research questions [5,9,10]. Among them, the question of how to find the workflow(s) that best match a given analysis need is of particular interest due to the rapidly growing size of the repositories.

Since one cannot expect that workflows perfectly match such needs, searching best matches must be based on meaningful similarity measures. In turn, similarity measures between complex objects, such as workflows, typically first require a method for identifying shared elements, e.g., shared analysis tools [11]. Only then we can identify similar workflows in a repository, making similarity search worth while. Therefore, an important step towards the investigation of workflow similarity searching is the analysis of the workflows contained in a repository, in particular to find elements that are shared across workflows. For web-services such an analysis is presented in [12], who found that only few web-services are used in more than one workflow. Yet, another analysis of processors contained in the workflows stored in myExperiment [13], revealed that the majority of basic tasks are local processors and not web services. In [14], functional properties of several types of processors, including local ones, are used as features to classify workflows. Yet, no differentiation is made between single types of processors after extraction, hindering fine-grained analysis of shared elements. Processor labels are used for identification in [15] in order to match a very limited set of workflows using subgraph-isomorphism. From our experience, this label-based approach is not generally applicable to large public collections of scientific workflows: The broad author base results in substantially heterogeneous labels.

We believe that all the above studies failed to give a comprehensive account on the degree of sharing between workflows in a repository. In particular, three key aspects were not adequately considered.

(1) To determine elements shared by different workflows, one needs to establish a method to test the identity of (or similarity between) two elements from different workflows. While the identity of web-services is determined by service name and operation, this approach is not generalizable to arbitrary processors. In [14], the authors build such a method by extracting features from processors, yet omit important information; for instance, local processors are only identified by their respective class, disregarding their actual instantiation parameters.

(2) From a structural perspective, workflows can be described at three levels of functionality: single processors (typically treated as black boxes), groups of processors organized as dataflows, and whole workflows. A workflow always consists of one top-level dataflow which contains processors and potentially includes subworkflows, that is, further, nested dataflows. However, previous work only considered sharing at the level of processors, although sharing of more coarse-grained functional units such as dataflows is equally important.

(3) Another important aspect of studying workflow connectivity is authorship. We argue that true sharing of workflow elements is only achieved if elements

(processors or dataflows) are shared between workflows authored by different persons; however, authorship has, to the best of our knowledge, not been considered in any of the prior repository analyses. Mere statistics for web-service usage across workflows, for instance, leave the question unanswered, whether preferences in service usage vary between authors, or which services are used by the broadest author base.

This paper presents results on all of these three issues. We describe the method and the results of performing a comprehensive study on reuse in the largest open scientific workflow repository to-date, myExperiment.org. We discuss and evaluate different methods to establish element identity at all three levels of workflows, i.e., at the processor level, the dataflow level, and the workflow level, and further refine the results based on classes of processors. Overall, we find that elements are shared at all levels, but that most sharing only affects "trivial" elements, e.g., processors for providing input parameters or type conversions. Furthermore, we provide a detailed analysis of cross-author reuse and show that much of the previously identified reuse is by single authors. Still, a significant number of non-trivial elements are shared cross-author, and we present first results on how these can be used to identify clusters of related workflows.

The remainder of this paper is structured as follows. We first describe in Section 2 the data sets and the methods used for identifying workflow elements. In Section 3 we present the results of our analysis on each of the aforementioned levels. Section 4 discusses our findings, and provides an outlook on future research.

2   Materials and Methods

2.1   Data Sets

We study the reuse of elements in myExperiment.org, which, to our knowledge, is the public scientific workflow repository containing the highest number of publicly available workflows. myExperiment allows upload of workflows from several systems, but approximately 85% of its content are workflows for Taverna [1]. Taverna workflows can appear in two different formats, namely *scufl* (used by Taverna 1) and *t2flow* (used by Taverna 2). In this work, we only consider Taverna workflows, yet, in either format.

For these two types, figure 1 shows the number of workflows submitted to myExperiment per month. The first Taverna 2 workflows appeared in January 2009 when the new version of Taverna was released. But, even in recent times, still a sizable number of Taverna 1 workflows are being uploaded. Note that Taverna 2 can load and process both *scufl* and *t2flow*. Figure 1 also shows a steady overall growth in total available workflows. This growth of the repository is accompanied by an increase in duplicate workflows. Most but not all of these apparent redundancies are caused by the format change (the same workflow uploaded once as a *scufl* and once as a *t2flow* workflow). We get back to this point in section 3.3.

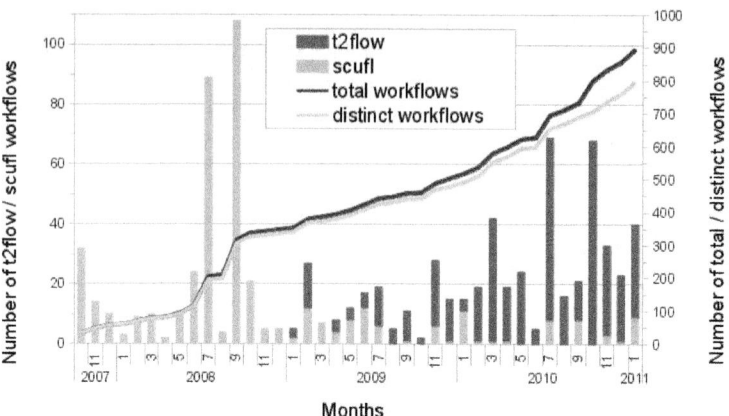

Fig. 1. Taverna workflows uploaded to myExperiment by month (left hand scale), and the numbers of overall and distinct workflows in the myExperiment repository the uploads amount to (right hand scale)

498 workflows in *scufl* format, and 449 in *t2flow* format were downloaded from myExperiment.org. Of the 498 *scufl* files, 449 could be converted to *t2flow* by manually loading them into the Taverna 2 Workbench and storing them in the new format. 49 files showed format inconsistencies and were removed from further analysis. Altogether, our analysis is comprised of 898 workflows.

Objects of Study. Figure 2 provides an example workflow from our study set. This workflow has two global inputs and four global outputs; it is composed of one top-level dataflow and one nested dataflow named *EBI_InterProScan_poll_job*. In general, a Taverna workflow is defined by one top-level dataflow and may contain nested dataflows. Each dataflow has its own main inputs and outputs and consists of one or more processors representing activities operating on data. Processors are connected by datalinks, describing the flow of data from one processor's output(s) to the next processor's input(s). Each processor has a type and a configuration, where the type denotes the executable class and the configuration contains the parameters passed to the class for instantiation. For example, the configuration of a *WSDLActivity* processor denotes the url of the WSDL document to enact and the operation to call with each piece of data received by the processor via its input ports. Taverna's workflow model is presented in more detail in [1].

The 898 workflows used in our analysis contain a total of 1,431 dataflows, including both 898 top-level and 533 nested dataflows, and a total of 10,242 processors. On average, each workflow uses 1.6 dataflows and 11.4 processors, where the largest workflow has 455 processors and the smallest has 1 processor. In terms of dataflows, the largest workflow contains 19 nested dataflows.

In myExperiment, each workflow is further accompanied by metadata provided in RDF-format including title, description, author and version information,

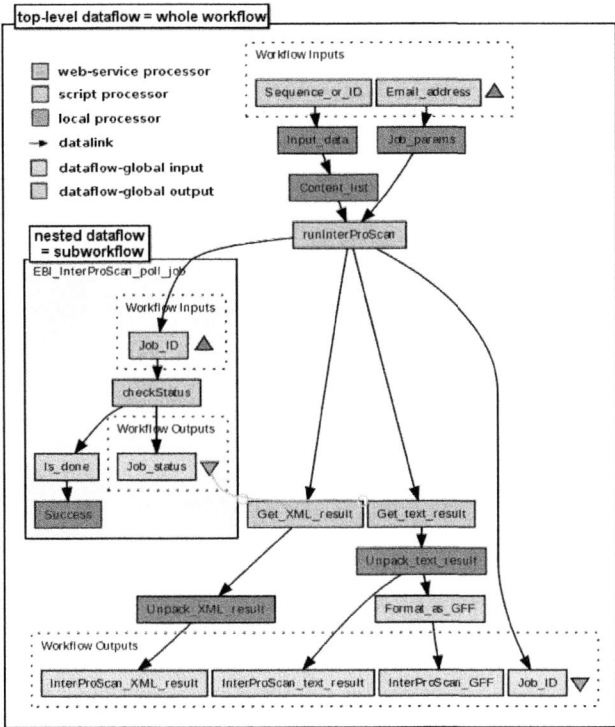

Fig. 2. Example of workflow (myExperiment id 204)

and creation and modification dates. For each workflow in our set, we downloaded its accompanying metadata and specifically extracted the workflows author.

2.2   Identifying Shared Workflow Elements

Investigating interconnections between (parts of) workflows and cross-author reuse necessitates precise methods for identifying elements. At first glance, elements can be deemed identical if they are functionally equivalent, i.e, if they, for every possible input, always produce the same output. However, such a stringent definition is not possible as it is fundamentally undecidable if two programs are identical in this sense. Also, this definition of identity is not advisable for us because we ultimately work in a retrieval setting: Imagine a user has chosen a workflow X for analyzing her data set and is interested in exploring workflows that perform similar analyses to learn about alternatives; such a user is naturally interested in elements (and workflows) implementing a similar method, not an identical one [5]. In the following, we discuss various methods to account for this fuzzyness in comparing workflow elements at each of the three levels, i.e., at processor level, at dataflow level, and at workflow level.

Identifying Processors. The configuration of each processor specified in the *t2flow* workflow definition file contains both the type of processor and the code it is to be instantiated with. After having cleansed from whitespace all the configurations, we used them as the way to identify identical processors. We thus assume processors to be identical if their cleansed configurations match exactly. Clearly, this is a very strict definition of identity: however, please note that it still leads to a meaningful definition of workflow similarity as workflows typically contain multiple (identical or not) processors. We shall explore the impact of relaxing this definition in Section 4.

Identifying Dataflows. Both myExperiment and Taverna provide identifiers for dataflows: while myExperiment assigns an integer to each workflow uploaded, Taverna assigns a 36 character, alphanumeric string value to each dataflow created. When a dataflow is included in a workflow as a subworkflow, the Taverna id is used to reference the dataflow. The way this id is determined is not defined by Taverna; although it seems to be meant to be able to uniquely identify a dataflow across systems (like a checksum), we found this id to be not sufficient for this task. We both found the same Taverna id to be assigned to different dataflows (e.g., workflows 1702 and 1703) as well as identical dataflows having different Taverna ids (e.g., workflows 359 and 506). Furthermore, using an id would restrict our analysis to identity of dataflows. Not relying on identifiers provided by Taverna has the additional advantage of making our methods transferable to other workflow systems. Therefore, we devised three alternative ways to compare two dataflows:

- Method 1: We consider two dataflows as shared if the sets of processors are identical;
- Method 2: We consider two dataflows as shared if the sets of processors and the numbers of datalinks are identical;
- Method 3: We consider two dataflows as shared if their sets of processors, the numbers of datalinks, and the numbers of global inputs and global outputs are identical.

Note that we use multisets here: if two processors contained in a dataflow's processor set have identical cleansed configurations, they are still both included in the set. Figure 3 gives an overview of the dataflow occurrences found using Taverna ids and each of our three methods. A first observation is that the number of dataflows occurring (i.e. used) only once is just above 80% of the total number of distinct dataflows when identification is based on method 1, 2, or 3 while it is 95% when the identification is based on Taverna ids. Assuming that two dataflows must be highly similar already if they consist of the same set of processors, we conclude that any of the three methods appear more suitable than Taverna ids to study reuse across workflows. Second, while method 1 assimilates several dataflows which do not share a Taverna id, the addition of datalinks by method 2 - not surprisingly - re-adds some distinction. However, the fact that the total number of distinct dataflows only increases slightly (from 1,071 to 1,081)

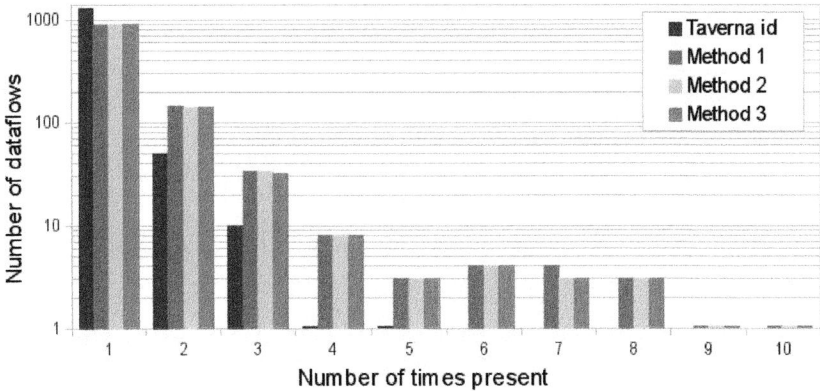

Fig. 3. Occurrences of distinct dataflows in the analysis set for several methods of identification

when going from method 1 to method 2 shows that almost every identical set of processors is connected by the same number of datalinks.

To gain more insight into this finding, we looked more closely into the dataflows which had the same set of processors but were connected by different numbers of links. We found that those only differ in the way they deal with outputs (use of additional processor outputs to propagate results to the final output). We further studied the difference in results between methods 2 and 3 which affect only a group of three dataflows (contained in workflows 1214, 1512 and 1513) which method 3 splits into two groups of one and two dataflows, respectively. However, the differences in these dataflows, again, only affects the number of inputs and outputs, but not the analysis performed by the workflow itself.

Based on these observations, we decided to continue our study using only method 1. In the following, when we speak of 'distinct' dataflows, we thus mean them to differ by their respective sets of processors.

Identifying Workflows. We follow the same thoughts for comparing workflows as for comparing dataflows. We consider two workflows as identical if they are built from the exact same set of dataflows and processors.

Note that we did not include the textual descriptions of processors and dataflows in the process of identification. We did find workflows that are deemed related when using the identification scheme established above but differ in the names assigned to some of their processors. This difference does, however, not cause any structural or functional divergence.

3 Results

Using method 1 outlined before, we found 3,598 of 10,242 processors (30%), 1,071 of 1,431 dataflows (75%), and 792 of 898 workflows (88%) to be distinct.

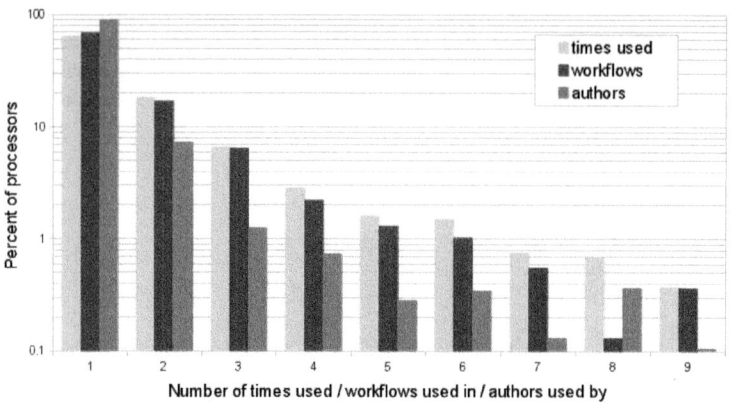

Fig. 4. Usage of processors overall, in workflows and by authors

In the following, we look at workflow interconnections and cross-author reuse on each of these structural levels separately and also break up our analysis by different processor categories.

3.1 Processors

We first looked at the general usage of processors by comparing the total numbers of processor occurrences with the number of distinct processors, and their use across workflows and authors. On average, each processor is used 2.85 times in 2.24 workflows by 1.31 authors. Figure 4 shows the relative usage frequencies for all processors in our set. Overall reuse of processors and cross-workflow reuse of processors closely correlate (Pearson Correlation Coefficient > 0.99, p-value $\ll 0.01$). The slight difference is caused by single processors being used more than once in single workflows. Cross-author reuse, on the other hand, is much lower, indicating that workflow authors reuse their own workflow components more often than those created by others.

Maxima[1] of 314 usages, 177 workflows and 39 authors and minima of 1 for single processor usage call for a more detailed investigation. To this end, we used the categorization established in [13], organizing Taverna processors into the four main categories *local, script, subworkflow*, and *web-service*. Note that subworkflows are nested dataflows which will be discussed in more detail in Section 3.2. Within these categories, processors are further divided into subcategories based on functional or technological characteristics. The authors of [13] showed that overall usage of processors varies greatly between these categories and subcategories. Here, we extend their analysis by considering the reuse frequencies of processors within each category.

Usage statistics by category and by subcategory are shown in Table 1, listing total and distinct numbers of processors for each subcategory, together with the numbers of workflows and authors using processors from these subcategories.

---

[1] Usage maxima not shown in Figure 4 for better visualization.

Table 1. Statistics on processor usage by subcategory

(Sub)category	Processors			Times used			Workflows			Authors		
	total	distinct	reused	avg	stddev	max	avg	stddev	max	avg	stddev	max
local	6518	1786	777	3.65	13.91	314	2.71	7.30	177	1.47	2.16	39
cdk	63	56	7	1.12	0.33	2	1.08	0.28	2	1.00	0.00	1
conditional	109	2	2	54.50	19.50	74	38.50	13.50	52	10.50	3.50	14
string constant	2817	844	387	3.33	9.83	173	2.71	6.21	114	1.44	1.81	28
data conversion	2979	771	307	3.86	17.84	314	2.56	8.27	177	1.43	2.34	39
user interaction	62	9	8	6.88	6.40	22	5.55	4.59	16	2.22	1.54	6
operation	5	2	2	2.50	0.50	3	2.50	0.50	3	1.50	0.50	2
database access	81	13	10	6.23	7.94	28	4.69	5.01	20	2.92	4.21	16
testing	13	6	3	2.16	1.21	4	2.16	1.21	4	1.33	0.47	2
util	389	83	51	4.68	11.25	84	3.73	8.49	63	1.79	3.16	25
script	1393	753	227	1.85	3.69	83	1.60	2.03	30	1.07	0.46	8
beanshell	1333	701	220	1.90	3.81	83	1.63	2.09	30	1.07	0.47	8
r	60	52	7	1.15	0.41	3	1.03	0.19	2	1.01	0.13	2
subworkflow	533	358	91	1.48	1.23	11	1.38	1.01	9	1.03	0.19	2
subwf only	380	279	54	1.36	0.95	9	1.27	0.85	9	1.03	0.17	2
subwf + top	153	79	32	1.93	1.82	11	1.75	1.38	7	1.05	0.21	2
web-service	1798	701	309	2.56	4.27	47	2.14	3.32	36	1.29	1.28	21
biomart	79	49	12	1.61	1.49	7	1.61	1.49	7	1.16	0.50	3
biomoby	58	44	9	1.31	0.66	3	1.31	0.66	3	1.11	0.31	2
cagrid	9	8	1	1.12	0.33	2	1.12	0.33	2	1.00	0.00	1
rest	6	6	0	1.00	0.00	1	1.00	0.00	1	1.00	0.00	1
sadi	5	5	0	1.00	0.00	1	1.00	0.00	1	1.00	0.00	1
soaplab	332	81	46	4.09	5.50	31	3.41	4.69	22	1.53	1.34	8
wsdl	1304	503	241	2.59	4.41	47	2.11	3.33	36	1.29	1.39	21
xmpp	5	5	0	1.00	0.00	1	1.00	0.00	1	1.00	0.00	1

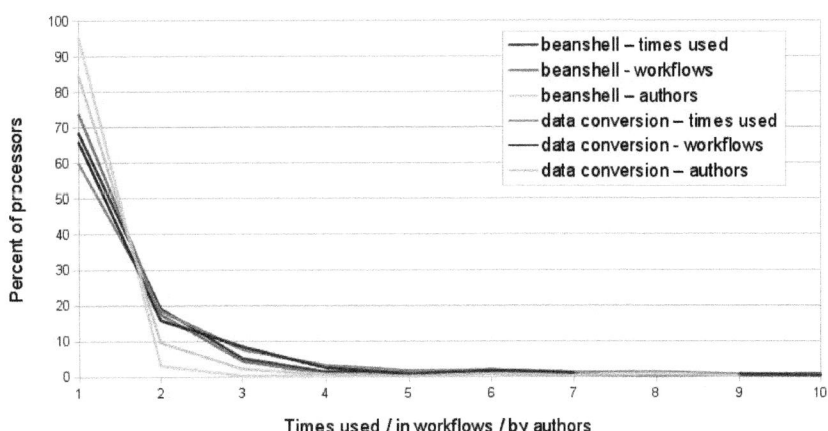

Fig. 5. Exemplary usage distributions for beanshell and data conversion processors showing relative overall, cross-workflow and cross-author usage frequencies

Figure 5 exemplifies total, workflow-based, and author-based relative usage distributions in subcategories *beanshell* and *data conversion*.

For the **local** category 27% of processors are distinct with a reuse rate of 44%. Processors from this category access functionality provided and executed by the SWFM. This category is particularly interesting for workflow interconnections, as all of its subcategories show comparably high reuse rates, overall as well as across workflows and authors. This is especially true for *conditional, user interaction, operation, database access, testing* and *util*: These subcategories exhibit a much broader usage distribution than the other processor subcategories whose distributions are similar to the distributions shown in Figure 5. The reason for this fact is that many of the respective processors are from the standard set that Taverna provides for building workflows, and used without further modification by the author. Table 2, listing the top five most frequently used processors, underpins this finding: These processors provide very common functionality which is likely to be widely used.

54% of all **script** processors are distinct, of which 30% are reused. Within this categroy, *R* scripts are far less often used than *beanshell* scripts and hardly used more than once. *Beanshell* scripts, on the other hand, are the third most popular type of processor with only 53% of its instances being distinct. 31% of them are reused. This seems remarkable, as we would expect these processors to contain user-created functionality for data processing. Yet, looking at the author-based distribution of these processors reveals that almost 96% of them are used only by single authors. It appears that users of beanshell processors have personal libraries of such custom-made tools which they reuse quite frequently, while usage of others' tools is rare.

Web-service processors show 39% distinctiveness, and 44% reuse. By far the most popular types of web-service invocations are *soaplab* and *wsdl* processors. 24% of *soaploab* processors and 39% of *wsdl* processors are distinct, and reuse is at 57% and 48%, respectively. As for scripts, reuse across authors is low, with single author usage rates of 78% and 87%, respectively. This gap between overall and cross-author reuse shows quite clearly that authors use and reuse certain web-services preferentially, while these preferences are not too widely shared between workflow authors. An exception to this are some rather popular, well-known web-services, such as *Blasts SimpleSearch*.

Returning from the categorized to the global view, Figure 6 shows the 300 most frequently used processors and their cross-workflow and cross-author reuse. Overall usage counts clearly follow a Zipf-like distribution. Zipf's law [16] states that when ranking words in some corpus of natural language by their frequency, the rank of any word in the resulting frequency table is inversely proportional to its frequency. Carrying over this distribution to processors in scientific workflows, it means that only few processors are used very often, while usage of the vast majority of processors is very sparse. The corresponding counts for reuse across workflows and authors exhibit the same trend. Yet, they show peaks of increased reuse which mostly are synchronized between the two. These peaks are caused by the aforementioned Taverna built-in processors.

Table 2. Top 5 most used processors (T: times used; W: workflows used in; A: authors used by)

Category	Subcategory	Description	T	W	A
local	data conversion	Regular expression splitter taking an optional regex as input (default ',') and splitting an input string into a list of strings.	314	177	39
local	data conversion	String list merger taking an optional string separator (default *newline character*) and merging an input list of strings into one new string with the original strings separated by the given separator	309	87	25
local	string constant	string constant *newline character*	173	114	28
local	string constant	string constant '1'	157	82	22
local	data conversion	string concatenator for two input strings	118	66	28

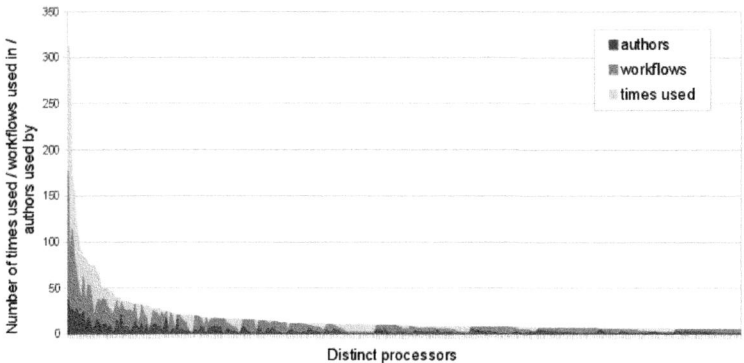

Fig. 6. Usage counts of the 300 most used processors showing a Zipf-like distribution

3.2   Dataflows

As shown in Figure 7, the pattern of usage for dataflows closely follows that of single processors. In contrast, overall reuse is lower by 20%: Over 80% of dataflows are used only once, and only 5% used more than twice. Usage across workflows is slightly lower, implying that some workflows use single dataflows multiple times. 1,038 dataflows are used by single authors, 29 by two authors, and 1 each by 3,4,6 and 7 authors, resulting in an overall of only 3% cross-author reuse for dataflows.

As described in section 2.1, dataflows can be either top-level or nested. A top-level dataflow is the same as the entire workflow, while a nested dataflow is a subworkflow. Due to this dual nature of dataflows, three different cases of reuse may occur: (a) Reuse of whole workflows as whole workflows; (b) Reuse of subworkflows as subworkflows; (c) Reuse of whole workflows as subworkflows and vice versa.

Case (a) can be deemed undetectable when looking only at a repository, as re-using a workflow does not mean re-uploading it (the reason why myExperiment still contains duplicates is investigated in the next section). To distinguish cases (b) and (c), we grouped all dataflows in our analysis by their appearance

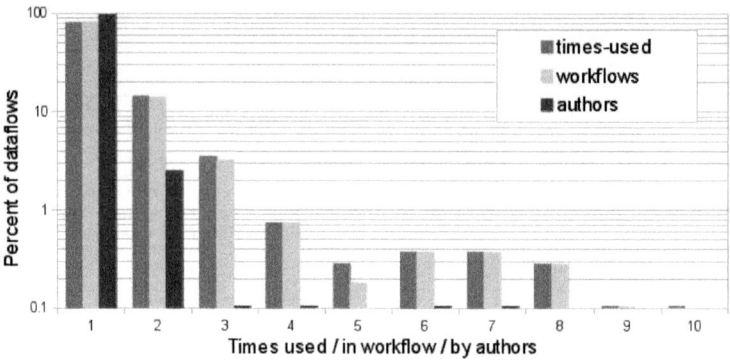

Fig. 7. Dataflow occurrences in total, in workflows and by authors

as workflows or subworkflows. 380 (75% distinct) dataflows are only present as nested, but not as top-level dataflows. 153 (55%) nested dataflows are also published as standalone top-level dataflows. For the second group we identified a total of 86 standalone workflows which are used as subworkflows in these 153 cases. Numbers of reuse for the two groups are also shown in Table 1.

We did not find significant differences in cross-author reuse between these groups. For cross-workflow and overall reuse, on the other hand, major differences exist: Numbers are as high as 40% for those subworkflows which have a corresponding standalone workflow published[2], while for those that don't they are at only 19%. This can be interpreted in two ways. First, it indicates that authors publish the dataflows they use most often as standalone workflows. This eases their inclusion as nested, functional components inside other workflows. On the other hand, the finding that such dataflows are, for the most part, not used by different authors in derivative work shows that modular extension of existing workflows created by others is still uncommon.

3.3   Workflows (Top-Level Dataflows)

Of the 898 workflows, 83 appear more than once in the repository, 19 of which where uploaded by more than one user. This indicates that there are users which upload workflows which are equal (by our definition of identity) to already existing ones. Figure 8 shows author contributions of workflows and their dataflows, both total and distinct. It reveals that a single user (the one with the highest overall number of workflows uploaded) is responsible for the majority of the cases of duplicate workflows. By looking into this in more detail, we found that all of this user's duplicates are caused by equivalent workflows being uploaded in both *scufl* and *t2flow* formats. Figure 8 also shows that this user alone has authored 23% of all workflows analyzed. Communication with the respective author, who is part of the Taverna development team, revealed that most of his workflows

---

[2] The standalone workflow itself was not included in reuse computation.

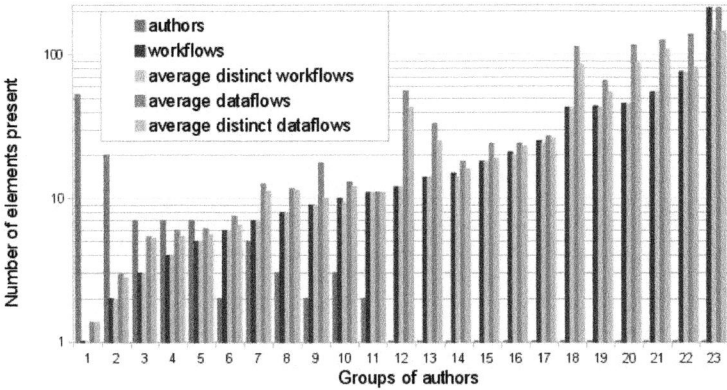

Fig. 8. Authors grouped by the total number of workflows they have created. Total amounts of workflows, and averages of distinct workflows, and total and distinct dataflows shown for each group.

serve the purpose of testing the functionality of Taverna-provided processors and giving examples for their usage. The remainder of duplicate workflows is largely due to users following tutorials including uploads of workflows to myExperiment: They upload an unmodified workflow.

As another interesting finding, the top 10 single authors (groups 14 through 23) have created 554 worfklows, i.e., app. 62% of all workflows in our analysis set. Conversely, 43% of all 124 authors have only created one workflow.

4   Discussion

We studied reuse of elements in scientific workflows using the to-date largest public scientific workflow repository. In contrast to previous work, our analysis covers all types of processors and looks into reuse not only at the general level, but also by author and by processor category. In this section, we discuss our results and some of the decisions we took when defining our methods. We also point to how our findings should influence next steps towards meaningful similarity measures for scientific workflows.

Processor identification. In this paper, we use exact matching of the processors' configurations to determine processor identity. Thus, we only identify verbatim syntactic reuse. To assess the impact of this limitation, we computed pairwise Levenshtein edit distance between all processors[3]. We compared the level of reuse using our strict definition of processor identity with one that assumes processors as identical if their edit distance is below a given threshold. Results are shown in Figure 9. Clearly, relaxing the threshold even until 20% difference does not have a significant impact on the number of distinct processors. On the other hand, relaxing

---

[3] 15 processors with configurations longer that 20,000 characters were excluded from computation. Results were normalized using the lengths of the compared processor configurations. Only processors from equal functional subcategories where compared.

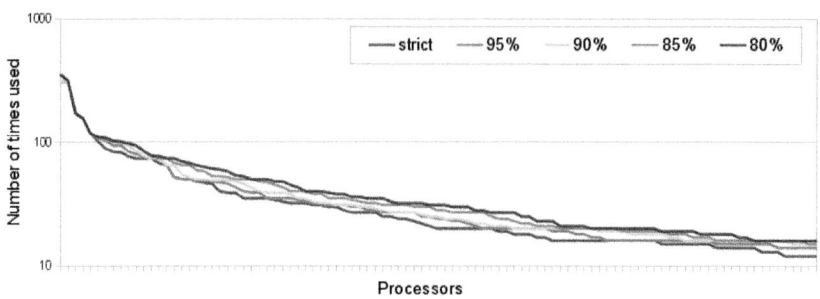

Fig. 9. Change in overall usage frequencies for processors when matched by 95, 90, 85, and 80 percent similarity regarding their Levenshtein edit distance

syntactic similarity comes with the risk of assimilating processors with different functionality, and thus, different usage intent; a problem also present in [14].

Of course, string similarity is a purely syntactic measure, while processor identity in reality is a semantic issue. Thus, more differentiated solutions could be explored. [17] additionally compares the numbers and types of input/output ports, while [18] suggested to use semantic or functional information provided by catalogues of tools. However, both approaches have their problems. Regarding the former, one must consider that the majority of processors in our data set only have a single input and a single output port, mostly of type String. On the other hand, the latter depends on the existence of well-curated ontologies for describing the function of a processor, and on authors using these ontologies to tag their processors[4]. Another option would be to exploit provenance traces to infer functional similarity between processors. If two sets of provenance data can be mapped onto each other, so might the processors responsible for the corresponding changes in the data. Unfortunately, repositories specifically targeted at collecting provenance are just starting to emerge [19], and provenance data is currently not available for large groups of workflows.

Reuse characteristics. As shown in this paper, the most commonly used processors are generic string operations and string constants. More generally, the main glue points of workflows are processors with non-specific functionality, provided by the system and used as-is. As such, these processors provide neither specialized, nor author-created functionality, limiting their usefullness for detecting both cross-workflow and cross-author reuse. Other types of processors are most often only used across workflows created by single authors, and their reuse frequencies are lower than those of unspecific ones. Yet, these processors are especially interesting: The custom-made nature of processors from the *script* category differentiates them from other processors in terms of their highly user-provided functionality. Thus, if such a processor is found more than once, it is

---

[4] Note that myExperiment allows tagging of workflows, but does not enforce a fixed vocabulary for this purpose.

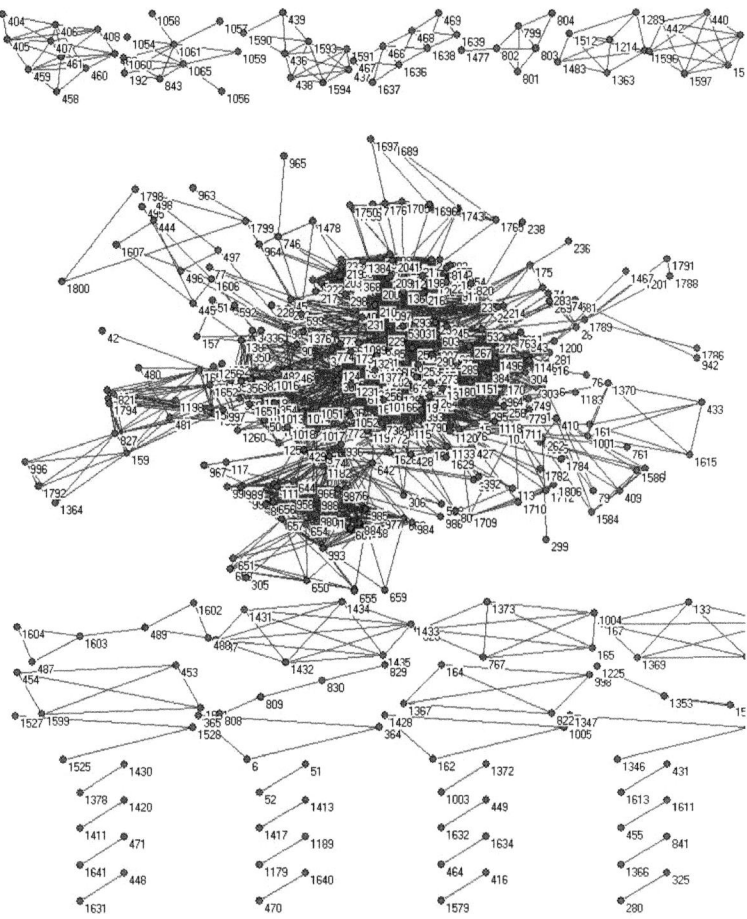

Fig. 10. Excerpt of a network of workflow-workflow interconnections by at least 3 mutual processors. Labels are myExperiment workflow ids.

an indication of non-trivial reuse. The case is similar for web-services. On the other hand, even apparently trivial operations should not be ignored completely. For instance, a string constant containing a complex XML configuration file is highly specific and its reuse a strong indicator for functional similarity.

Thus, the suitability of processors or groups of processors to determine functional reuse needs further investigation. For scalability and practicability, a general and automated approach has to be found to distinguish such cases. One solution could use TF/IDF scores [20] to assign weights to processors based on their usage frequencies. The Zipf-like distribution of processor frequencies suggests such an approach.

Workflow Interconnections. Our analysis focussed on characterizing processors that are re-used. An equally valid view is to ask how large the overlap in processors is between two workflows. 769 of our 898 workflows share one or more processors with at least one other workflow. On average, each workflow is thus connected to 92.12 other workflows by the use of at least one common processor, averaging at 7.23 distinct processors being shared between two workflows. This is remarkable given the findings from Section 3.1, as the majority of processors is shown to only be present in single workflows. Apparently, the remaining one third of processors interconnect workflows quite densely.

Figure 10 shows how workflows cluster by shared processors. In the figure, two workflows are connected if they share at least three processors. The figure clearly shows clusters of highly interconnected workflows. We manually studied a sample of these clusters and found them to be highly similar in function.

User assistance. The fact that processor reuse is uncommon across authors could be interpreted in several ways. One explanation could be that authors are simply not aware (enough) of other people's dataflows and workflows. This situation could be alleviated by using the repository for providing better support for designing scientific workflows [5]. Some work has started already in this direction; for instance, [21] presents a system which recommends web-services for use during workflow design. Accomplishing such functionality for other types of processors and even for subworkflows is highly desirable.

5 Summary

This paper introduces the first study performed on reuse of scientific workflows which has considered reuse at various levels of granularity (processor, dataflows, and workflows), at various categories of processors, and also differentiated by workflow authorship. Thereby, we provided three major contributions.

First, we introduced and compared different **methods to identify** processors, dataflows, and whole workflows which we deem suitable for detection of reuse. Our methods allows us to provide fine-grained analyses, and to distinguish functionally important cases of reuse from trivial ones.

Second, our study is the first to consider **authorship**. This allowed us to characterize different kinds of users depending on their usage of the repository, ranging from single time 'authors' uploading duplicate workflows when following a tutorial, to advanced authors creating many functionally interlinked workflows. An important observation obtained by entering this level of detail is that while 36% of workflow elements are reused, only 11% of workflow elements are used by more than one author. Cross-author reuse for dataflows is even lower at 3%. This calls for actions to make authors more aware of the repository contributions by others.

Third, our study investigated **reuse and duplication of workflow elements** in more detail than ever before. Using a categorization of processors helped to better characterize re-use in terms of the types of processors that are reused. Furthermore, it showed that the appearance of single processors in multiple workflows is not per se an indication of functional similarity, and that not all processors are equally well suited for deriving information about functional workflow similarity.

We believe that our findings are important for future work in scientific workflow similarity.

References

1. Oinn, T., Addis, M., Ferris, J., Marvin, D., Senger, M., Greenwood, R., Carver, K., Pocock, M.G., Wipat, A., Li, P.: Taverna: a tool for the composition and enactment of bioinformatics workflow. Bioinformatics 20(1), 3045–3054 (2003)
2. Bowers, S., Ludäscher, B.: Actor-oriented design of scientific workflows. In: 24th Int. Conf. on Conceptual Modeling (2005)
3. Freire, J., Silva, C.T., Callahan, S.P., Santos, E., Scheidegger, C.E., Vo, H.T.: Managing Rapidly-Evolving Scientific Workflows. In: Moreau, L., Foster, I. (eds.) IPAW 2006. LNCS, vol. 4145, pp. 10–18. Springer, Heidelberg (2006)
4. Goecks, J., Nekrutenko, A., Taylor, J.: Galaxy: a comprehensive approach for supporting accessible, reproducible, and transparent computational research in the life sciences. Genome Biology 11, R86 (2010)
5. Cohen-Boulakia, S., Leser, U.: Search, Adapt, and Reuse: The Future of Scientific Workflow Management Systems. SIGMOD Record 40(2) (2011)
6. Berners-Lee, T., Hendler, J.: Publishing on the Semantic Web. Nature, 1023–1025 (2001)
7. Roure, D.D., Goble, C.A., Stevens, R.: The design and realisation of the myexperiment virtual research environment for social sharing of workflows. Future Generation Computer Systems 25(5), 561–567 (2009)
8. Mates, P., Santos, E., Freire, J., Silva, C.T.: CrowdLabs: Social Analysis and Visualization for the Sciences. In: Bayard Cushing, J., French, J., Bowers, S. (eds.) SSDBM 2011. LNCS, vol. 6809, pp. 555–564. Springer, Heidelberg (2011)
9. Goderis, A., Sattler, U., Lord, P., Goble, C.A.: Seven Bottlenecks to Workflow Reuse and Repurposing. In: Gil, Y., Motta, E., Benjamins, V.R., Musen, M.A. (eds.) ISWC 2005. LNCS, vol. 3729, pp. 323–337. Springer, Heidelberg (2005)
10. Xiang, X., Madley, G.: Improving the Reuse of Scientific Workflows and Their By-products. In: IEEE Int. Conf. on Web Services (2007)
11. Tversky, A.: Features of Similarity. Psychological Review 84, 327–352 (1977)
12. Tan, W., Zhang, J., Foster, I.: Network Analysis of Scientific Workflows: a Gateway to Reuse. IEEE Computer 43(9), 54–61 (2010)
13. Wassink, I., Vet, P.E.V.D., Wolstencroft, K., Neerincx, P.B.T., Roos, M., Rauwerda, H., Breit, T.M.: Analysing Scientific Workflows: Why Workflows Not Only Connect Web Services. In: IEEE Congress on Services (2009)
14. Stoyanovich, J., Taskar, B., Davidson, S.: Exploring repositories of scientific workflows. In: 1st Int. Workshop on Workflow Approaches to New Data-centric Science (2010)
15. Goderis, A., Li, P., Goble, C.: Workflow discovery: the problem, a case study from e-Science and a graph-based solution. In: IEEE Int. Conf. on Web Services (2006)
16. Zipf, G.: The Psycho-Biology of Language. MIT Press, Cambridge (1935)
17. Silva, V., Chirigati, F., Maia, K., Ogasawara, E., Oliveira, D., Braganholo, V., Murta, L., Mattoso, M.: Similarity-based Workflow Clustering. J. of Computational Interdisciplinary Science (2010)

18. Gil, Y., Kim, J., Florez, G., Ratnakar, V., Gonzalez-Calero, P.A.: Workflow matching using semantic metadata. In: 5th Int. Conf. on Knowledge Capture (2009)
19. Missier, P., Ludaescher, B., Dey, S., Wang, M., McPhillips, T., Bowers, S., Agun, M.: Golden-Trail: Retrieving the Data History that Matters from a Comprehensive Provenance Repository. In: 7th Int. Digital Curation Conf. (2011)
20. Salton, G., McGill, M. (eds.): Introduction to Modern Information Retrieval. McGraw-Hill (1983)
21. Zhang, J., Tan, W., Alexander, J., Foster, I., Madduri, R.: Recommend-As-You-Go: A Novel Approach Supporting Services-Oriented Scientific Workflow Reuse. In: IEEE Int. Conf. on Services Computing (2011)

Aggregating and Disaggregating Flexibility Objects

Laurynas Šikšnys, Mohamed E. Khalefa, and Torben Bach Pedersen

Department of Computer Science, Aalborg University
{siksnys,mohamed,tbp}@cs.aau.dk

Abstract. Flexibility objects, objects with flexibilities in time and amount dimensions (e.g., energy or product amount), occur in many scientific and commercial domains. Managing such objects with existing DBMSs is infeasible due to the complexity, data volume, and complex functionality needed, so a new kind of *flexibility database* is needed. This paper is the first to consider flexibility databases. It formally defines the concept of flexibility objects (flex-objects), and provide a novel and efficient solution for aggregating and disaggregating flex-objects. This is important for a range of applications, including smart grid energy management. The paper considers the grouping of flex-objects, alternatives for computing aggregates, the disaggregation process, their associated requirements, as well as efficient incremental computation. Extensive experiments based on data from a real-world energy domain project show that the proposed solution provides good performance while still satisfying the strict requirements.

1 Introduction

Objects with inherent flexibilities in both the time dimension and one or more amount dimensions exist in both scientific and commercial domains, e.g., energy research or trading. Such objects are termed *flexibility objects* (in short, flex-objects). Capturing flexibilities explicitly is important for *smart grid* domain, in order to consume energy more flexibly. In the ongoing EU FP7 research project MIRABEL project [2], the aim is to increase the share of renewable energy sources (RES) such as wind and solar, by capturing energy demand and supply, and the associated flexibilities. Here, flex-objects facilitate planning and billing of energy. First, consumers (automatically) specify the flexible part of their energy consumption, e.g., charging an electric vehicle, by issuing flex-objects to their energy company. A flex-object defines how much energy is needed and when, and the tolerated flexibilities in time (e.g., between 9PM and 5AM) and energy amount (e.g. between 2 and 4 kWh). In order to reduce the planning complexity, similar flex-objects are aggregated into larger "macro" flex-objects. Then, the energy company tries to plan energy consumption (and production) such that the desired consumption, given the macro flex-objects plus a non-flexible base load, matches the forecasted production from RES (and other sources) as well as possible. During the process, the flex-objects are *instantiated* (their flexibilities

A. Ailamaki and S. Bowers (Eds.): SSDBM 2012, LNCS 7338, pp. 379–396, 2012.

are *fixed*), resulting in so-called *fix-objects*, where fixed concrete values are assigned for time and amounts, within the original flexibility intervals. Finally, the "macro" fix-objects are *disaggregated* to yield "micro" fix-objects corresponding to the instantiation of the original flex-objects issued by the consumers, which are then distributed back to the consumers. The consumers are rewarded according to their specified flexibility.

For typical real world and scientific applications (e.g., MIRABEL), there exists, hundreds of millions of flex-objects have to be managed efficiently. This is infeasible using existing DBMSs. First, due to the complexity of flex-objects, queries over flex-objects are also complex and cannot be formulated efficiently, if at all, using standard query languages. Second, the number of flex-objects can be very large and timing restrictions are tight. Third, incremental processing must be supported natively. In order to efficiently evaluate queries over flex-objects, a new tailor-made database for handling flex-objects is needed.

Contributions. This paper is the first to introduce the vision of flex-object databases by discussing their functionality, queries, and application. As the most important operations of a flex-object database, the paper focuses on flex-object aggregation and disaggregation, which are analogous to a *roll-up* and *drill-down* queries in an OLAP database. As will be shown later, the aggregation of even just two flex-objects is non-trivial as they can be combined in many possible ways. The paper formally defines flex-objects, how to measure the flexibility of a flex-object, aggregation and disaggregation of flex-objects. Our flex-object aggregation approach takes a set of flex-objects as input and, based on the provided aggregated parameters, partitions the set of flex-objects into disjoint groups of similar flex-objects. This partitioning is performed in two steps - grid-based grouping and bin-packing. The grouping of flex-objects ensures that flex-objects in a group are sufficiently similar (in terms of chosen flexibility attributes). The bin-packing ensures that the groups themselves conform to the given (aggregate) criteria. After bin-packing, an aggregation operator is applied to merge similar flex-objects into aggregated flex-objects. In one possible scenario, when the flexibilities of the aggregated flex-objects are fixed, e.g., in the planning phase, our disaggregation approach takes the respective fix-objects as input and disaggregate them into fix-objects that correspond to the original flex-objects. Our solution inherently supports efficient incremental aggregation, which is essential to handle continuously arriving new flex-objects. An extensive set of experiments on MIRABEL data shows that our solution scales well, handling both aggregation and disaggregation efficiently.

The remainder of the paper is structured as follows. Section 2 outlines the functionality of envisioned flex-object databases. Section 3 formally defines the concept of flex-object and the problem of aggregating and disaggregating flex-objects. Section 4 describes how to aggregate several flex-objects into one. Section 5 generalizes the approach to aggregate a set of input flex-objects into a set of output flex-objects. Section 6 considers incremental aggregation. Section 7 describes the experimental evaluation, while Section 8 discusses related work. Finally, Section 9 concludes and points to future work.

2   Flex-object Databases

Although the paper focuses on two specific operations on flex-objects, aggregation and disaggregation, in this section, we provide a broader context by outlining our vision for the functionality of *flex-object databases*. A flex-object database is a database storing flex-objects, possibly of several different types and with different types of flexibilities. Thus, flex-objects must be first-class citizens in such a database and the associated complex functionality must be supported. In some application scenarios, the flex-object database will be stand-alone and focused only on flexibility management, in other scenarios, the flex-object database will be part of a larger database storing also other kinds of objects and with a mixed query workload. Support for dimensions and hierarchies is important to be able to view flex-objects at the desired level of granularity, however, the dimension hierarchies must be more complex than in current systems to support complex real-world hierarchies such as energy distribution grids. Similarly, spatio-temporal support is essential, as many flex-object applications have strong spatial and/or temporal aspects. Several options for query languages are possible. For standalone flex-object-specific applications, a JSON-style declarative language like that used by MongoDB would be effective. For more general mixed databases, an extension of SQL with specific syntax and operators for flexibility manipulation is desirable.

The storage of flex-objects is not a trivial issue. A flex-object database should be able to store massive amounts of flex-objects while still ensuring very fast response time. Due to their complex internal structures (to be detailed in the next section), flex-offers cannot be directly stored as atomic objects in standard relational databases. Here, accessing flex-objects becomes an expensive operation requiring joins and aggregations over two or more large tables. Alternatives include nested object-relational storage and or dedicated native storage, where it is important to strike the right balance between efficiency and the ability to integrate easily with other types of data. The storage issue is beyond the scope of this paper, and will be addressed in future work. We now present at the most important types of queries to be supported by a flex-object database.

Flexibility availability queries provide an overview over the *amount flexibilities* that are available at given time intervals. For example, such a query may retrieve the *min, max, and average* amounts available, or build a time series with the (expected) distribution of amounts at every time instance. Such queries are used in feasibility/risk analysis where nominal or peak values are explored, e.g., to see how much energy consumption can be increased or decreased at a given time to counter unexpected highs or lows in the RES production.

Adjustment potential queries computes the distribution of amounts that can be potentially injected into (or extracted away from) a given time interval, taking into account the amounts which are already fixed with fix-objects. Several options for the amount injection (or extraction) are possible, including adjusting amounts within amount flexibility ranges, shifting amounts within available time flexibility ranges, or a combination.

Fixing queries alter (or create) fix-objects (the plan), based on user selected amounts to inject or extract. Fixing queries are used in the analysis and planning phase, to interactively explore flexibility potentials (the first two types of queries), followed by modifying the existing plan (fix-objects) if needed.

Flex-object aggregation queries combine a set of flex-objects into fewer, "larger" flex-objects. In some sense, this is analogous to a *roll-up* query in an OLAP database (going from finer to coarser granularities), although considerably more complex (as will be discussed in the next sections). The aggregation usually reduces flexibilities, so it is important to quantify and minimize how much of the original flexibilities are lost due to aggregation. The aggregation of flex-objects can substantially reduce the complexity of the above-mentioned analysis queries as well as the complexity of various flex-object-based planning processes. For example, a very large number of flex-objects must be scheduled in MIRABEL. Since scheduling is NP-complete problem, it is infeasible to schedule all these flex-objects individually within the (short) available time. Instead, flex-objects can be *aggregated*, then *scheduled* (not considered in this paper), and finally *disaggregated* (see below) into fix-objects.

Flex-object disaggregation queries go the opposite way, "exploding" a large "macro" fix-object into a set of smaller "micro" fix-objects corresponding to the instantiation of the original flex-objects. This yields the refinement of the "macro" plan necessary for carrying out the plan in practice. In some sense, this is like a *drill-down* query in an OLAP database (going from coarser to finer granularities), but more complex.

Flex-object aggregation and disaggregation queries are particularly important and more challenging. In the next section, we formulate the problem of aggregating and disaggregating flex-objects.

3   Problem Formulation

We now formalize our proposed problem of aggregating and disaggregating flexibility objects. Our formalization includes (1) a definition of flex-object, (2) a measure to quantify flex-object flexibility, and (3) *aggregation* and *disaggregation* functions and their associated constraints.

The introduced flex-object is a multidimensional object capturing two aspects: (1) the *time flexibility interval*, and (2) a *data profile* with a sequence of consecutive *slices* each defined by (a) its start and end time and (b) the minimum and maximum amount bounds for one or more amount dimensions. We can formally define a flex-object as follows:

DEFINITION 1: *A flex-object f is a tuple $f = (T(f), profile(f))$ where $T(f)$ is the start time flexibility interval and $profile(f)$ is the data profile. Here, $T(f) = [t_{es}, t_{ls}]$ where t_{es} and t_{ls} are the earliest start time and latest start time, respectively. The $profile(f) = s^{(1)}, \ldots, s^{(m)}$ where a slice $s^{(i)}$ is a tuple $([t_s, t_e], [a_{min}^{(1)}, a_{max}^{(1)}], \ldots, [a_{min}^{(D)}, a_{max}^{(D)}])$ where $[a_{min}^{(i)}, a_{max}^{(i)}]$ is a continuous range of the amount for dimension $i = 1..D$ and $[t_s, t_e]$ is a time interval defining the extent of $s^{(i)}$ in the time dimension. Here, $t_{es} \leq s^{(1)}.t_s \leq t_{ls}$ and $\forall j = 1..m :$*

$s^{(j)}.t_e > s^{(j)}.t_s$, $s^{(j+1)}.t_s = s^{(j)}.t_e$. *We use the terms* profile start time *to denote* $s^{(1)}.t_s$, duration of the slice *to denote* $s_{dur}(s) = s.t_e - s.t_s$, duration of profile *to denote* $p_{dur}(f) = \sum_{s \in profile(f)} s.t_e - s.t_s$, *and* latest end time *to denote* $t_{le}(f) = f.t_{ls} + p_{dur}(f)$.

For simplicity and without loss of generality, time is discretized into equal-sized units, e.g., 15 minute intervals, and we have only one amount dimension (i.e., $D = 1$). Figure 1 depicts the example of a flex-object having a profile with four slices: $s^{(1)}$, $s^{(2)}$, $s^{(3)}$, and $s^{(4)}$. Every slice is represented by a bar in the figure. The area of the light-shaded bar represents the minimum amount value (a_{min}) and the combined area of the light- and dark-shaded bars represents the maximum amount value (a_{max}). The left and the right sides of a bar represent t_e and t_s of a slice, respectively.

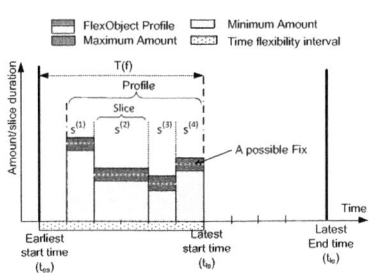

Fig. 1. A generic flex-object

We distinguish two types of flexibility associated with f. The *time flexibility*, $tf(f)$, is the difference between the latest and earliest start time. Similarly, the *amount flexibility*, $af(s)$, is the sum of amount flexibility of all slices in the profile of f, i.e.,

$$af(f) = \sum_{s \in profile(f)} (s.t_e - s.t_s) \cdot (\sum_{j=1}^{D} s.a_{max}^{(j)} - \sum_{j=1}^{D} s.a_{min}^{(j)}).$$

Based on these notations, the *total flexibility* of f is defined as follows:

DEFINITION 2: *The total flexibility of an flex-object f is the product of the time flexibility and the amount flexibility, i.e., $flex(f) = tf(f) \cdot af(s)$.*

Consider a flex-object $f=([2,7], s^{(1)}, s^{(2)})$ where $s^{(1)} = ([0,1], [10,20])$ and $s^{(2)} = ([1,4], [6,10])$. The time flexibility of f is equal to $7 - 2 = 5$. The amount flexibility $af(f)$ is equal to $(1 - 0)(20 - 10) + (4 - 1)(10 - 6) = 22$. Hence, the total flexibility of f is equal to 110.

A flex-object with time and profile flexibilities equal to zero is called a *fix-object*. In this case, the fix-object $f = ([l_{es}, l_{ls}], s^{(1)}, \ldots, s^{(m)})$ is such that $l_{es} - t_{ls}$ and $s.a_{min}^{(d)} = s.a_{max}^{(d)} \; \forall s \in profile(f), \forall d = 1..D$.

DEFINITION 3: *An* instance *(instantiation) of a flex-object $f = ([t_{es}, t_{ls}], s^{(1)}, \ldots, s^{(m)})$ is a fix-object $f_x = ([t_s, t_s], s_x^{(1)}, \ldots, s_x^{(m)})$ such that $t_{es} \leq t_s \leq t_{ls}$ and $\forall i = 1..m, d = 1..D : s^{(i)}.a_{min}^{(d)} \leq s_x^{(i)}.a_{min}^{(d)} = s_x^{(i)}.a_{max}^{(d)} \leq s^{(i)}.a_{max}^{(d)}$. We refer to t_s as the* start time *of the flex-object f.*

There is an infinite number of possible instances (fix-objects) of a flex-object. One possible instance is shown as the dotted line in Figure 1. We can now define *aggregation* and *disaggregation* as follows:

DEFINITION 4: *Let AGG be an aggregate function which takes a set of flex-objects F and produces a set of flex-objects A. Here, every $f_a \in A$ is called an* aggregated flex-object, *and $|A| \leq |F|$.*

DEFINITION 5: *Let* DISAGG *be a function which takes a set of A instances and produces a set of F instances. We denote these sets of fix-objects as A_X and F_X, respectively, and assume that $A = AGG(F)$, $\forall f \in F \Leftrightarrow \exists f^x \in F_X$ and $\forall f_a \in A \Leftrightarrow \exists f_a^x \in A_X$. Moreover, to ensure the balance of amounts at aggregated and non-aggregated levels, for all time units $T = 0, 1, 2, \cdots$ and all dimensions $d = 1..D$, the following equality must hold:*

$$\sum_{t=0}^{T} \left[s.a_{min}^{(d)} | \forall f_a^x \in A_X, \forall s \in profile(f_a^x), s.t_e \le t \right] =$$

$$\sum_{t=0}^{T} \left[s.a_{min}^{(d)} | \forall f_x \in F_X, \forall s \in profile(f_x), s.t_e \le t \right].$$

Evaluation of the functions *AGG* and *DISAGG* is called flex-object *aggregation* and *disaggregation*, respectively. Due to the amount balance requirement, disaggregation is, however, not always possible for any arbitrary *AGG* function. Depending on whether disaggregation is possible or not for all instances of aggregated flex-objects, we identify two types of flex-object aggregation: *conservative* and *greedy*, respectively. Aggregated flex-objects resulting from greedy aggregation might define more time and amount flexibilities compared to the original flex-objects. Obviously, instances of such flex-objects might not be disaggregated using *DISAGG*. Nevertheless, this type of aggregation is still important in feasibility/risk analysis where extreme amount values are explored (see *flexibility availability queries* in Section 2). On the contrary, aggregated flex-objects resulting from conservative aggregation always define less (or equal) flexibilities compared to the original flex-objects. Consequently, it is always possible to disaggregate instances of such flex-objects using *DISAGG*. Conservative aggregation is important in use-cases where planning is involved, e.g., in MIRABEL. The flex-object database has to support both types of aggregation, however, in this paper, we focus on conservative aggregation only.

The following requirements for the aggregation originate from the MIRABEL use-case, but they are also important for general flex-object aggregation:

Compression and Flexibility Trade-off Requirement. It must be possible to control the trade-off between (1) the number of aggregated flex-objects and (2) the flexibility loss, i.e., difference between the total flex-object flexibility (see Definition 2) before and after aggregation.

Aggregate Constraint Requirement. Every aggregated flex-object $f_a \in AGG(F)$ must satisfy a user-defined so-called *aggregate constraint C*, which is satisfied if the value of a certain flex-object attribute, e.g., *total maximum amount*, is within the given bounds. For example, such constraints can ensure that aggregated flex-objects are "properly shaped" to meet energy market rules and power grid constraints.

Incremental Update Requirement. Flex-object updates (addition/removal) should be processed efficiently and cause minimal changes to the set of aggregated flex-objects. Thus functionality is vital in scenarios, e.g., MIRABEL, where addition/removal of flex-objects are very frequent.

In the following sections, we present a technique to perform flex-object aggregation and disaggregation while satisfying all requirements.

4   Aggregation and Disaggregation

In this section, we first propose a basic N-to-1 flex-object aggregation algorithm and explain how to generalize it for a large set of flex-objects. Additionally, we explain how disaggregation can be performed.

According to the flex-object definition, the profile start time $(s^{(1)}.t_s)$ of a flex-object f is not pre-determined, but must be between the earliest start time $f.t_{es}$ and the latest start time $f.t_{ls}$. Hence, the aggregation of even two flex-objects is not straightforward. Consider aggregating two flex-objects f_1 and f_2 with time flexibility values equal to six and eight, respectively. Thus, we have 48 $(6*8)$ different profile start time combinations, each of them realizing a different aggregated flex-object. Three possible profile start time parameter combinations are shown in Figure 2(a-c).

In general, to aggregate a set of flex-objects F into a single aggregated flex-object f_a, we follow these three steps:

1. Choose a profile start time value $f.s^{(1)}.t_s = s_f \ \forall f \in F$ such that $f.t_{es} \le s_f \le f.t_{ls}$. Later, we will refer to this choice of profile start time as *profile alignment*.
2. Set the time flexibility interval for f_a such that $f_a.t_{es} = min_{f \in F}(s_f)$ and $f_a.t_{ls} = f_a.t_{es} + min_{f \in F}(f.t_{ls} - s_f)$.
3. Build a profile for f_a by summing the corresponding amounts for each slice across all profiles.

There are many ways to align profiles (by choosing the constants s_{f_1}, s_{f_2}, $\ldots$, $s_{f_{|F|}}$). Each of these alignments determine where amounts from individual flex-objects are concentrated within the profile of f_a. We focus only on the three most important alignment options: *start-alignment, soft left-alignment*, and *soft right-alignment*. Start-alignment spreads out amounts throughout the time extent of all individual flex-objects, making larger amounts available as early as possible. On the contrary, soft left-/right-alignment builds shorter profiles with amounts concentrated early (left) or late (right) in the profile. In the context of MIRABEL, start-alignment is suitable for the near real-time balancing of electricity, where energy has to be available as early as possible; and soft left/right alignment allows the consumption of anticipated wind production peaks with steep rises (left-alignment) or falls (right-alignment). The three alignment options are illustrated in Figure 2. Here, the crossed area in the figure represents the amount of time flexibility that is lost due to aggregation when different profile alignment options are used. The alignment option are elaborated below:

Start-alignment. We set s_{f_1}, s_{f_2}, $\ldots$, $s_{f_{|F|}}$ so that $\forall f \in F : s_f = f.t_{es}$. This ensures that profiles are aligned at their respective *earliest start time* values (see f_1 and f_2 in Figure 2(a)).

Soft Left-Alignment. We set s_{f_1}, $s_{f_2}, \ldots, s_{f_{|F|}}$ so that $\forall f \in F : s_f = min(f.t_{ls} - min_{g \in F}(g.t_{ls} - g.t_{es}), max_{g \in F}(g.t_{es}))$. Figure 2(b) illustrates the effect of soft

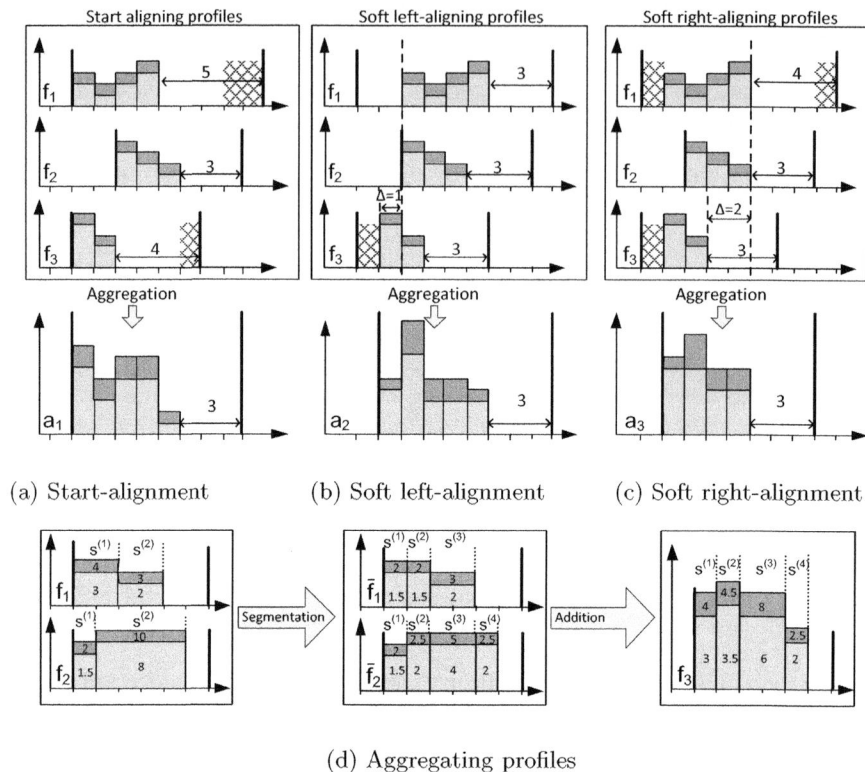

Fig. 2. Profile alignment and aggregation

left-alignment. Here, f_1 and f_2 are left-aligned, meaning that their profile start times are equal. However, the profile of f_3 cannot be left-aligned with respect to the profiles of f_1 and f_2 as that would shorten the remaining time flexibility range of the aggregate. f_3 lacks one time unit ($\Delta = 1$) for its profile to left-align.

Soft Right-Alignment. We set $s_{f_1}, s_{f_2}, ..., s_{f_{|F|}}$ so that $\forall f \in F : s_f = min(f.t_{ls} - min_{g \in F}(g.t_{ls} - g.t_{es}), max_{g \in F}(g.t_{es} + p_{dur}(g)) - p_{dur}(f))$. Figure 2(c) illustrates the effect of soft right-alignment. Here, f_1 and f_2 are right-aligned, meaning that their profiles align at the right hand side (i.e., have equal $t_{es} + p_{dur}$ values). However, the profile of f_3 cannot be right-aligned with respect to the profiles of f_1 and f_2 as this would shorted the remaining time flexibility range of the aggregate.

After the alignment, the time flexibility interval is computed for the aggregated flex-object. As illustrated in Figure 2(a-c), for all three alignment options, the time flexibility of f_a is equal to that of the flex-object with the smallest time flexibility in the set F, i.e., $f_a.t_{ls} - f_a.t_{es} = min_{f \in F}(f.t_{ls} - f.t_{es})$. However, other types of alignment, e.g., *hard left* or *hard right* where all profiles are forced to align at the left or right hand side, might reduce the time flexibility of the aggregated flex-object.

Finally, the minimum and maximum amounts of adjacent slices in the aligned profiles are summed to construct the profile of the aggregated flex-object. If adjacent slices at any time unit have different durations, those slices are partitioned to unify their durations. During the partitioning, minimum and maximum amounts are distributed proportionally to the duration of each divided slice. This step is called *segmentation*. The segmentation step reduces the amount flexibility, $af(f)$, as it imposes more restrictions on the amount for each divided segment. For example, consider the slice $s^{(1)}$ of f_1 in Figure 2(d), which illustrates the segmentation for two flex-objects. Originally, the minimum amount is 3 and the maximum amount is 4 over two time units. Thus, we can supply one amount unit in the first time unit, and three units in the second time unit. However, this supply is not acceptable after dividing the slice into two equal-sized slices $s^{(1)}$ and $s^{(2)}$ with minimum and maximum amount of 1.5 and 2, respectively. After the segmentation, the addition of profiles is performed. During the addition, $a_{min}^{(1)}$ and $a_{max}^{(1)}$ amounts are added for every corresponded profile slice.

It is always possible to disaggregate a flex-object produced by this aggregation approach. Consider the following disaggregation procedure. For a given instance f_a^x of flex-object f_a, we produce the set of fix-objects $\{f_1^x, f_2^x, ..., f_{|F|}^x\}$ such that $\forall i = 1..|F| : f_i^x.t_{es} = f_i^x.t_{ls} = s_{f_i} + (f_a^x.t_{es} - f_a.t_{es})$. It is always possible to fix the *start time* of every $f_i^x, i = 1..|F|$ because the time flexibility range of the aggregate f_a is computed conservatively, and the aligned profiles of $f \in F$ can always be shifted within this range (see Figure 2). Also, the amount values from every slice $s_a^x \in profile(f_a^x)$ are distributed proportionally to the respective slices of $f_i^x, i = 1..|F|$ so that minimum and maximum amount constraints are respected for every $f \in F$. This can always be achieved, and consequently, for any instance of f_a, it is always possible to build instances of flex-objects from F. Moreover, the newly built fix-objects will collectively define a total amount which is equal to that of the initial fix-object f_x^a.

To summarize, the N-to-1 aggregation approach can be used to aggregate flex-objects in F. However, the time (and total) flexibility loss depends on the flex-object with smallest time flexibility in the set F. Due to this issue, much of the flexibility will be lost when aggregating flex-objects with distinct time flexibilities. To address this, we will now propose an N-to-M aggregation approach.

5 N-to-M Aggregation

As discussed in Section 4, aggregating "non-similar" flex-objects results in an unnecessary loss of time flexibility. This loss can be avoided, and the profile alignments can be better enforced, by carefully grouping flex-objects and thus ensuring that their time flexibility intervals overlap substantially. We now describe an (*N-to-M*) approach to aggregate a set of flex-objects, $\mathcal{F}$, to a set of aggregated flex-objects, $\mathcal{A}$, while satisfying all requirements (see Section 3). The algorithm consists of three phases: *grouping*, *bin-packing*, and *N-to-1 aggregation*:

Grouping Phase. We partition the input set $\mathcal{F}$ into disjoint groups of similar flex-objects. Based on the application scenario, the user specifies which attributes to use in the grouping. For example, the user may choose the earliest start time, latest start time, time flexibility, and/or amount flexibility as grouping attributes. Two flex-objects are grouped together if the values of the specified attributes differ no more than a user-specified threshold. The thresholds and the associated grouping attributes are called *grouping parameters*. As shown later, the choice of grouping parameters yields a trade-off between compression and flexibility loss. In the example in Figure 3, flex-objects $f_1, f_2, ...,$ and f_5 are assembled in two groups g_1 and g_2 during the grouping phase.

Bin-Packing Phase. This phase enforces the aggregate constraint (see Section 3). Each group g produced in the *grouping* phase is either passed to the next phase (if g satisfies the constraint already) or further partitioned into the minimum number of bins (groups) such that the constraint $w_{min} \leq w(b) \leq w_{max}$ is satisfied by each bin b. Here, $w(b)$ is a weight function, e.g., $w(b) = |b|$, and w_{max} and w_{min} are the upper and lower bounds. We refer to w_{min}, w_{max}, and w as *bin-packing parameters*. By adjusting these parameters, groups with a bounded number of flex-objects or a bounded total amount can be built. Note that it may be impossible to satisfy a constraint for certain groups. For example, consider a group with a single flex-object, while we impose a lower bound of two flex-objects in all groups. These groups are discarded from the output (see g_{22} in Figure 3), and, depending on the application, these flex-objects can be either: (1) excluded from the N-to-M aggregation output, or (2) aggregated with another instance of the N-to-M aggregation with less constraining grouping or bin-packing parameters.

N-to-1 Aggregation Phase. We assemble the output set $\mathcal{A}$ by applying N-to-1 aggregation (see Section 4) for each resulting group g. The alignment option is specified as the *aggregation parameter*. Every aggregated flex-objects satisfies the aggregate constraint enforced in the bin-packing phase.

The complete N-to-M aggregation process is visualized in Figure 3. Here, given the initial flex-object set $\{f_1, f_2, ..., f_5\}$ and grouping, bin-packing, and aggregation parameters, two aggregated flex-objects, f_{a1} and f_{a2}, are produced. The grouping parameters are set so that the difference between the earliest start time (t_{es}) is at most 2. The bin-packing parameters require that the number

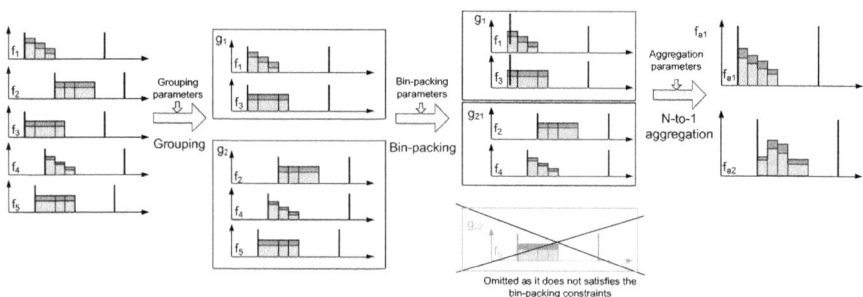

Fig. 3. Aggregation of flex-objects in the N-to-M aggregation approach

of flex-objects in resulting groups are 2, i.e., $w_{min} = w_{max} = 2$, $w(g) = |g|$. In the aggregation phase, the *start-aligned* option is used. In practice, the user will choose from a number of meaningful pre-defined parameter settings, e.g., *short/long profiles* or *amount as early as possible*.

6   Incremental N-to-M Aggregation

In this section, we present an incremental version of the N-to-M aggregation approach. The set of flex-objects F is updated with a sequence of incoming updates: u_1, u_2,...,u_k. Each update u_i is of the form (f, c_i), where f is a flex-object and $c_i \in \{+, -\}$ indicates insertion ('+') or deletion ('−') of f to/from F. The incremental approach outputs a sequence of aggregated flex-object updates which correspond to u_1, u_2,...,u_k. The approach has four phases: *grouping*, *optimization*, *bin-packing*, and *aggregation*.

Grouping Phase. We map each flex-object into a d-dimensional point. This point belongs to a cell in a d-dimensional uniform grid. Users specify the extent of each cell in each dimension using thresholds $T_1, T_2, ..., T_d$ from the grouping parameters. Every cell is identified by its coordinates in the grid. We only keep track of *populated* cells, using an in-memory hash table, denoted as the *group hash*. This table stores key-value pairs, where the key is the cell coordinates and the value is the set of flex-objects from F mapped to this cell. We combine adjacent populated cells into a *group*. A group can be either *created*, *deleted*, or *modified*. Group changes are stored in a list, denoted as the *group changes list*. Figure 4 visualizes the effect of adding a flex-object f_1. f_1 is mapped to a 2-dimensional point which lies in the grid cell c_2. The coordinates of c_2 are used to locate a group in the group hash. The found group is updated by inserting f_1 into its list of flex-objects. Finally, a change record indicating that the group was *modified* is inserted into the group changes list. In the cases when a group is not found in the group hash, a new group with an unique *id* and a single populated cell c_2 is created. Also, if the group changes list already contains a change record for a particular group, the record is updated to reflect the combination of the changes.

Optimization Phase. This phase is only executed when aggregation is triggered, either (1) periodically, (2) after a certain number of updates, or (3) when

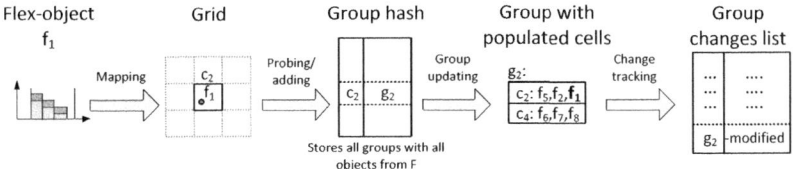

Fig. 4. Processing the addition of a flex-object in the grouping phase

the latest aggregates are requested. During this phase, we consolidate the group changes list. For each update of a group g in the list, we identify its adjacent groups g_1^a, g_2^a, ... by probing the group-hash. Then, for each adjacent group g_i^a, a minimum bounding rectangle (MBR) is computed over all points that contains flex-objects from the groups g and g_i^a. If the extent of the MBR in all dimensions are within the user-specified thresholds, we combine the groups g and g_i^a (see merge in Figure 5). Otherwise, if the MBR of g in any dimension is larger than the size of a grid cell, we perform a group split (depicted in Figure 5). Any over-sized group is partitioned into groups of a single grid cell, and, for every individual group, an MBR is computed. Then, the two groups with the closest MBRs are merged until the grouping constraint is violated. Then, g is substituted the with newly built groups. Groups changes incurred during merging and splitting are added to the group change list.

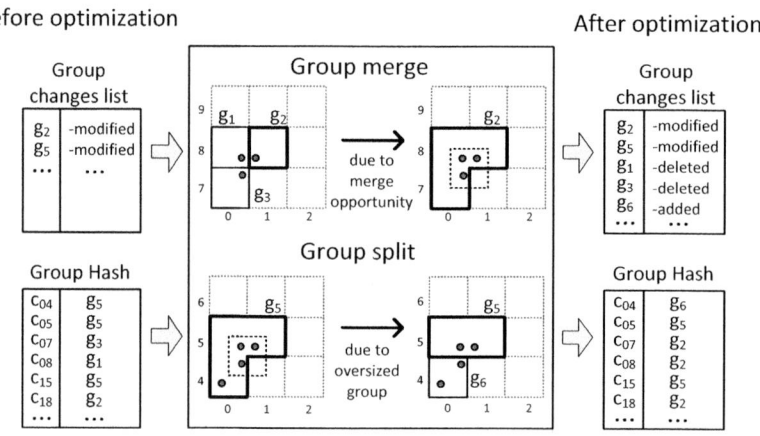

Fig. 5. Flow of data in the optimization phase

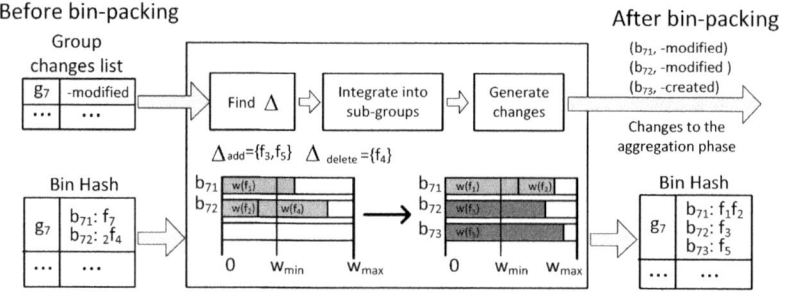

Fig. 6. Flow of data in the bin packing phase

Bin-Packing Phase. We maintain a hash table, denoted the *bin hash*, which maps from each group, produced in the grouping phase, to its bins (as described in Section 5). In this phase, we propagate updates from the group change list to bins. We first compare existing bins with an updated group to compute the deltas to obtain added and deleted flex-objects, Δ_{added} and Δ_{delete}, respectively. Then, we discard from the bins the flex-objects that are in Δ_{delete}. Groups with total weight less than w_{min} are deleted and flex-objects from these groups as well as from Δ_{added} are included into other existing bins using the first fit decreasing strategy [22]. New bins are created, if needed.

Figure 6 shows how the bins of the group g_7 are updated when lower and upper bounds w_{min} and w_{max} are set. Finally, all bins changes are pipelined to the aggregation phase. Flex-objects that did not fit to any bin (due to their weight being lower than w_{min} or higher than w_{max}) are stored in a separate list.

Aggregation Phase. We maintain a hash table, denoted as the *aggregate hash*, which maps from each individual bin to an aggregated flex-object. Each aggregated flex-object has references to the original flex-objects. Thus, for every bin change, added and deleted non-aggregated flex-objects (see Δ_{add} and Δ_{delete} in Figure 7) are found and used to incrementally update an aggregate flex-object. If there are no deletes, N-to-1 aggregation is applied for every added object. Otherwise, an aggregate is recomputed from scratch by applying N-to-1 aggregation on all object in a bin. Finally, all changed aggregated flex-objects are provided as output.

7 Experimental Evaluation

In this section, we present the experimental evaluation of the full incremental N-to-M aggregation approach. As there are no other solutions for flex-object aggregation and disaggregation, we propose two rival implementations: *Hierarchical Aggregation* and *SimGB*. In *Hierarchical Aggregation*, we use agglomerative hierarchical clustering for the grouping phase. First, the approach assigns each flex-object to individual clusters. Then, while no grouping constraints are violated, it incrementally merges the two closest clusters. The distance between two clusters is calculated based on the values of the grouping parameter flex-object attributes. For *SimGB*, we apply the *similarity group-by operator* [19] for one grouping parameter at a time, thus partitioning the input into valid

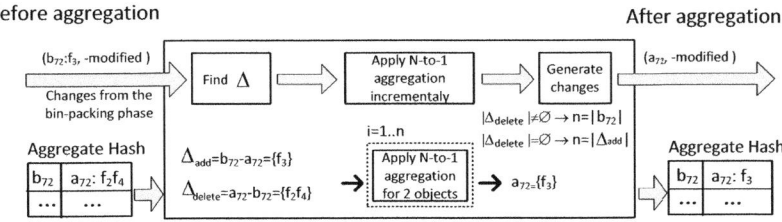

Fig. 7. Aggregation phase

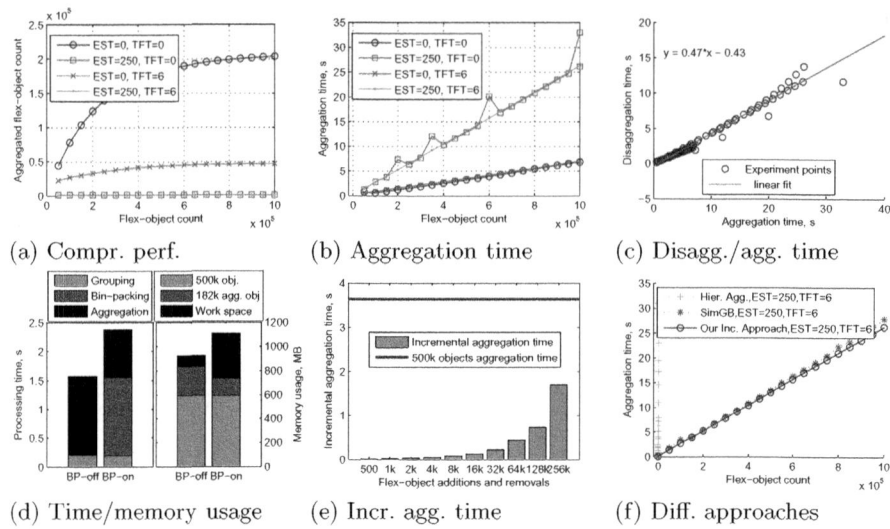

(a) Compr. perf.   (b) Aggregation time   (c) Disagg./agg. time

(d) Time/memory usage   (e) Incr. agg. time   (f) Diff. approaches

Fig. 8. Results of the scalability and incremental behavior evaluation

groups of similar flex-objects. For the evaluation, we use a synthetic flex-object dataset from the the MIRABEL project. The dataset contains one million energy consumption request flex-objects. The *earliest start time* (t_{es}) is distributed uniformly in the range $[0, 23228]$. The number of slices and the time flexibility values ($t_{ls} - t_{es}$) follow the normal distributions $\mathcal{N}(8, 4)$ and $\mathcal{N}(20, 10)$ in the ranges $[10, 30]$ and $[4, 12]$, respectively; the slice duration is fixed to 1 time unit for all flex-objects, thus profiles are from 2.5 to 7.5 hours long. Experiments were run on a PC with Quad Core Intel®Xeon®E5320 CPU, 16GB RAM, OpenSUSE 11.4 (x86_64), and Java 1.6. Unless otherwise mentioned, the default values of the experiment parameters are: (a) The number of flex-objects is $500k$. (b) $EST = 0$ (*Earliest Start Time Tolerance*) and $TFT = 0$ (*Time Flexibility Tolerance*) are used as the grouping parameters. They apply on the *Earliest Start Time* (t_{es}) and *Time Flexibility* ($t_{ls} - t_{es}$) flex-object attributes, respectively. (c) The aggregate constraint is unset (bin-packing is disabled). We also perform experiments with bin-packing enabled (explicitly stated).

Scalability. For evaluating flex-object compression performance and scalability, the number of flex-offers is gradually increased from $50k$ to $1000k$. Aggregation is performed using two different EST and TFT parameter values: EST equal to 0 or 250, and TFT equal to 0 or 6. Disaggregation is executed with randomly generated instances of aggregated flex-objects. The results are shown in Figure 8(a-d). Figure 8(a-b) shows that different aggregation parameter values lead to different compression factors and aggregation times. Disaggregation is approx. 2 times faster than aggregation (see Figure 8(c)) regardless of the flex-object count and grouping parameter values. Most of the time is spent in the bin-packing (if enabled) and N-to-1 aggregation phases (the 2 left bars in Figure 8(d)). Considering the overhead associated with incremental behavior, the

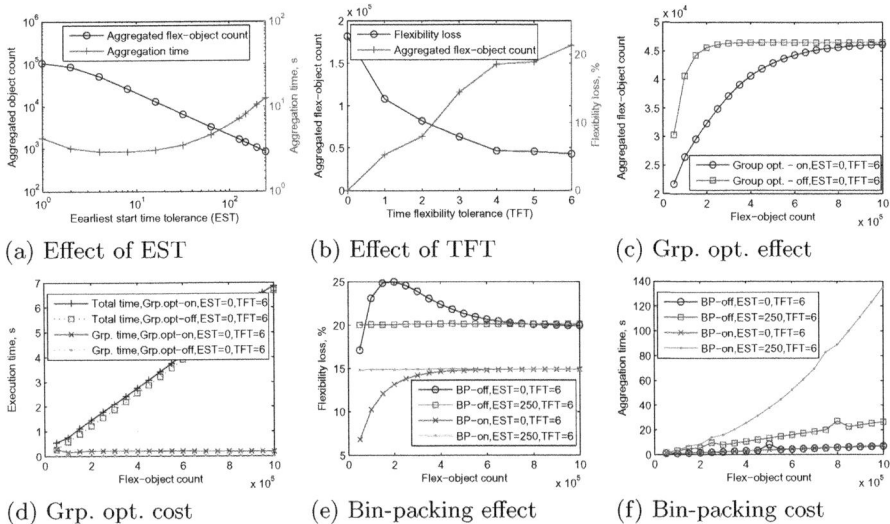

(a) Effect of EST   (b) Effect of TFT   (c) Grp. opt. effect

(d) Grp. opt. cost   (e) Bin-packing effect   (f) Bin-packing cost

Fig. 9. Results of the grouping, optimization, and bin-packing evaluation

amount of memory used by the approach is relatively small compared to the footprint of the original and aggregated flex-objects. Memory usage increases when bin-packing is enabled.

Incremental Behavior. When evaluating incremental aggregation performance, we first aggregate $500k$ flex-objects. Then, for different k values ranging from 500 to $256k$, we insert k new flex-objects and remove k randomly selected flex-objects. The total number of flex-objects stays at $500k$. For every value of k, we execute incremental aggregation. As seen from Figure 8(e), the updates can be processed efficiently so our approach offers substantial time savings compared to the case when all $500k$ flex-objects are aggregated from scratch (the line in the figure). We then compare the total time to process flex-objects with our incremental approach to the other two (inherently non-incremental) approaches - *Hierarchical Aggregation* and *SimGB*. As seen in Figure 8(f), our incremental approach is competitive to SimGB in terms of scalability. The overhead associated with the change tracking in our approach is not significant in the overall aggregation time. Additionally, the hierarchical clustering-based approach (Hier. Agg.) incurs very high processing time even for small datasets (due to a large amount of distance computations), and is thus not scalable enough for the flex-object aggregation problem.

Grouping Parameters Effect. As seen in Figure 9(a), the *EST* significantly affects the flex-object compression factor. For this dataset, increasing *EST* by a factor of two leads to a flex-object reduction by approximately the same factor. However, the use of high *EST* values results in aggregated flex-object profiles with more slices. Aggregating these requires more time (see "aggregation time" in Figure 9(a)). The *TFT* parameter has a significant impact on the flexibility loss (see "flexibility loss" in Figure 9(b)). Higher values of TFT incur higher flexibility losses. When it is set to 0, aggregation incurs no flexibility loss, but results in a larger amount of aggregated flex-objects. When the number of distinct time flexibility values in a flex-object dataset is low (as in our case), the best compression with no flexibility losses can be achieved when $TFT = 0$ and the other grouping parameters are unset (or set to high values). However, due to the long durations of profiles and high total amount values, the produced aggregated flex-object might violate the aggregate constraint.

Optimization and Bin-packing. We now study the optimization and bin-packing phases. As seen in Figure 9(c-d), the optimization phase is relatively cheap (Figure 9(d)), and it substantially contributes to the aggregated flex-object count reduction (Figure 9(c)). For bin-packing evaluation, the aggregate constraint was set so that the time flexibility of an aggregate is always at least 8 ($w_{min} = 8$, equiv. to 2 hours). By enabling this constraint, we investigate the overhead associated to bin-packing and its effect on the flexibility loss. As seen in Figure 9(e), by bounding the time flexibility for every aggregate, the overall flexibility loss can be limited. However, bin-packing introduces a substantial overhead that depends on the number of objects in flex-object groups after the optimization phase (see Figure 9(f)). When this number is small ($EST = 0, TFT = 6$), the overhead of bin-packing is insignificant. However, when groups are large ($EST = 250, TFT = 6$), bin-packing overhead becomes very significant.

In summary, we show that our incremental aggregation approach scales linearly in the number of flex-object inserts. The overhead associated with incremental behavior is insignificant. Our approach performs aggregation *incrementally* just as fast as efficient non-incremental grouping approaches ($SimGB$). The trade off-between flex-object compression factor and flexibility loss can be controlled using the grouping parameters. The compression factor can be further increased efficiently by group optimization. Disaggregation is approx. 2 times faster than aggregation.

8   Related Work

Related research fall in several categories.

Clustering. Many clustering algorithms have been proposed, including density-based (e.g., BIRCH [24]), centroid-based (e.g., K-Means [13]), hierarchical clustering (e.g., SLINK [18]), and incremental algorithms such as incremental K-means [25] and incremental BIRCH [10]. In comparison to our approach, clustering solves only the grouping part of the problem, which is a lot simpler than the whole problem. For grouping alone, the closest work is incremental grid-based clustering [16,9,12], where we, in comparison, improve the clusters across the grid boundaries and limit the number of items per each cluster.

Similarity Group By. SimDB [20] groups objects based on the similarity between tuple values, and is implemented as a DBMS operator in [19]. However, SimDB again only solves the grouping part of the problem, and is (unlike our approach) not incremental, which is essential for us.

Complex objects. Complex objects with multidimensional data exists in many real-world applications [14] and can be represented with *multidimensional data models* [17]. Several research efforts (e.g., [5] and [23]) have been proposed to aggregate complex objects. However, these efforts do not consider the specific challenges related to aggregating flex-objects.

Temporal Aggregation. Several papers have addressed aggregation for temporal and spatio-temporal data including: instantaneous temporal aggregation [3], cumulative temporal aggregation [21,1,11], histogram-based aggregation [6] and multi-dimensional temporal aggregation [4]. These techniques differ in the way how a time line is partitioned into time intervals and how an aggregation group is associated with each time instant. The efficient computation of these time intervals poses a great challenge and therefore various techniques that allow computing them efficiently are proposed [8,7,15]. Unfortunately, these techniques only deal with simple data items without flexibilities, making them unsuitable for aggregation of flex-objects.

9   Conclusion and Future Work

Objects with inherent flexibilities, so-called flexibility objects (flex-objects), occur in both scientific and commercial domains. Managing flex-objects with existing DBMSs is infeasible due to their complexity and data volume. Thus, a new tailor-made database for flex-objects is needed. This paper was the first to discuss flex-object databases, focusing on the most important operations of a flex-object database: aggregation and disaggregation. The paper formally defined the concept of flexibility objects and provided a novel and efficient grid-based solution considering the grouping of flex-objects, alternatives for computing aggregates, the disaggregation process, and the requirements associated to these. The approach allowed efficient incremental computation. Extensive experiments on data from a real-world energy domain project showed that the apporach provided very good performance while satisfying all entailed requirements.

As future work, other challenges related to the flex-object database have to be addressed. These include flex-object storage and visualization, as well as support for other types of queries (flexibility availability, adjustment potential, etc.). Another interesting topic is aggregation and disaggregation techniques for flex-objects with flexibility in the profile slices durations.

References

1. Arasu, A., Widom, J.: Resource sharing in continuous sliding-window aggregates. In: Proc. of VLDB, pp. 336–347 (2004)
2. Boehm, M., Dannecker, L., Doms, A., Dovgan, E., Filipic, B., Fischer, U., Lehner, W., Pedersen, T.B., Pitarch, Y., Siksnys, L., Tusar, T.: Data management in the mirabel smart grid system. In: Proc. of EnDM (2012)
3. Böhlen, M.H., Gamper, J., Jensen, C.S.: How would you like to aggregate your temporal data? In: Proc. of TIME, pp. 121–136 (2006)

4. Böhlen, M.H., Gamper, J., Jensen, C.S.: Multi-dimensional Aggregation for Temporal Data. In: Ioannidis, Y., Scholl, M.H., Schmidt, J.W., Matthes, F., Hatzopoulos, M., Böhm, K., Kemper, A., Grust, T., Böhm, C. (eds.) EDBT 2006. LNCS, vol. 3896, pp. 257–275. Springer, Heidelberg (2006)
5. Cabot, J., Mazón, J.-N., Pardillo, J., Trujillo, J.: Specifying Aggregation Functions in Multidimensional Models with OCL. In: Parsons, J., Saeki, M., Shoval, P., Woo, C., Wand, Y. (eds.) ER 2010. LNCS, vol. 6412, pp. 419–432. Springer, Heidelberg (2010)
6. Chow, C.Y., Mokbel, M.F., He, T.: Aggregate location monitoring for wireless sensor networks: A histogram-based approach. In: Proc. of MDM, pp. 82–91 (2009)
7. Gao, D., Gendrano, J.A.G., Moon, B., Snodgrass, R.T., Park, M., Huang, B.C., Rodrigue, J.M.: Main memory-based algorithms for efficient parallel aggregation for temporal databases. Distributed Parallel Databases 16(2), 123–163 (2004)
8. Gordevičius, J., Gamper, J., Böhlen, M.: Parsimonious temporal aggregation. In: Proc. of EDBT, pp. 1006–1017 (2009)
9. Hou, G., Yao, R., Ren, J., Hu, C.: A Clustering Algorithm Based on Matrix over High Dimensional Data Stream. In: Wang, F.L., Gong, Z., Luo, X., Lei, J. (eds.) WISM 2010. LNCS, vol. 6318, pp. 86–94. Springer, Heidelberg (2010)
10. Jensen, C.S., Lin, D., Ooi, B.C.: Continuous clustering of moving objects. IEEE Trans. Knowl. Data Eng. 19(9), 1161–1174 (2007)
11. Jin, C., Carbonell, J.G.: Incremental Aggregation on Multiple Continuous Queries. In: Esposito, F., Raś, Z.W., Malerba, D., Semeraro, G. (eds.) ISMIS 2006. LNCS (LNAI), vol. 4203, pp. 167–177. Springer, Heidelberg (2006)
12. Lei, G., Yu, X., Yang, X., Chen, S.: An incremental clustering algorithm based on grid. In: Proc. of FSKD, pp. 1099–1103. IEEE (2011)
13. Macqueen, J.B.: Some methods of classification and analysis of multivariate observations. In: Proc. of 5th Berkeley Symposium on Math. Stat. and Prob., pp. 281–297 (1967)
14. Malinowski, E., Zimnyi, E.: Advanced Data Warehouse Design: From Conventional to Spatial and Temporal Applications, 1st edn. Springer (2008)
15. Moon, B., Fernando Vega Lopez, I., Immanuel, V.: Efficient algorithms for large-scale temporal aggregation. TKDE 15(3), 744–759 (2003)
16. Park, N.H., Lee, W.S.: Statistical grid-based clustering over data streams. SIGMOD Rec. 33(1), 32–37 (2004)
17. Pedersen, T.B., Jensen, C.S., Dyreson, C.E.: A foundation for capturing and querying complex multidimensional data. Information Systems 26(5), 383–423 (2001)
18. Sibson, R.: SLINK: An optimally efficient algorithm for the single-link cluster method. The Computer Journal 16(1) (January 1973)
19. Silva, Y.N., Aly, A.M., Aref, W.G., Larson, P.A.: SimDB: A similarity-aware database system. In: Proc. of SIGMOD (2010)
20. Silva, Y.N., Aref, W.G., Ali, M.H.: Similarity group-by. In: Proc. of ICDE, pp. 904–915 (2009)
21. Yang, J., Widom, J.: Incremental computation and maintenance of temporal aggregates. VLDB 12(3), 262–283 (2003)
22. Yue, M.: A simple proof of the inequality $FFD(L) \leq \frac{11}{9}OPT(L) + 1, \forall L$ for the FFD bin-packing algorithm. Acta Mathematicae Applicatae Sinica 7(4), 321–331 (1991)
23. Zhang, D.: Aggregation computation over complex objects. Ph.D. thesis, University of California, Riverside, USA (2002)
24. Zhang, T., Ramakrishnan, R., Livny, M.: BIRCH: an efficient data clustering method for very large databases. In: Proc. of SIGMOD, pp. 103–114 (1996)
25. Zhang, Z., Yang, Y., Tung, A.K.H., Papadias, D.: Continuous k-means monitoring over moving objects. IEEE Trans. Knowl. Data Eng. 20(9), 1205–1216 (2008)

Fine-Grained Provenance Inference for a Large Processing Chain with Non-materialized Intermediate Views

Mohammad Rezwanul Huq, Peter M.G. Apers, and Andreas Wombacher

University of Twente, 7500AE, Enschede, The Netherlands
{m.r.huq,p.m.g.apers,a.wombacher}@utwente.nl

Abstract. Many applications facilitate a data processing chain, i.e. a workflow, to process data. Results of intermediate processing steps may not be persistent since reproducing these results are not costly and these are hardly re-usable. However, in stream data processing where data arrives continuously, documenting fine-grained provenance explicitly for a processing chain to reproduce results is not a feasible solution since the provenance data may become a multiple of the actual sensor data. In this paper, we propose the *multi-step provenance inference* technique that infers provenance data for the entire workflow with non-materialized intermediate views. Our solution provides high quality provenance graph.

1 Introduction

Stream data processing involves a large number of sensors and a massive amount of sensor data. To apply a transformation process over this infinite data stream, a window is defined considering a subset of tuples. The process is executed continuously over the window and output tuples are produced. Applications take decisions as well as control operations using these output tuples. In case of any wrong decision, it is important to reproduce the outcome for validation. Reproducibility refers to the ability of producing the same output after having applied the same operation on the same set of input data, irrespective of the operation execution time. To reproduce results, we need to store provenance data, a kind of metadata relevant to the process and associated input/output dataset.

Data provenance refers to the derivation history of data from its original sources [1]. It can be defined either at the tuple-level or at the relation-level [2] also known as fine-grained and coarse-grained data provenance respectively. Fine-grained data provenance can achieve reproducibility because it documents the used set of input tuples for each output tuple and the transformation process as well. On the other hand, coarse-grained data provenance cannot achieve reproducibility because of the updates and delayed arrival of tuples. However, maintaining fine-grained data provenance in stream data processing is challenging. If a window is large and subsequent windows overlap significantly, the size of provenance data becomes a multiple of the actual sensor data. Since provenance

A. Ailamaki and S. Bowers (Eds.): SSDBM 2012, LNCS 7338, pp. 397–405, 2012.

data is 'just' metadata and less often used by the end users, this approach seems to be infeasible and too expensive [3].

In addition, researchers often facilitate scientific workflows consisting of multiple processing elements to produce results. Some of these processing elements are intermediary steps and produce intermediate results. The source and output tuples are stored persistently since they are the basis for the processing and the result used for the analysis. However, the intermediate results may not be persistent due to the lack of their reuse. It is possible to document provenance information explicitly for these intermediate processing steps. However, explicit documentation is expensive in terms of storage requirements and it can be significantly reduced by inferring provenance data. Since intermediate results are transient, provenance inference in presence of non-materialized views is different than what has been proposed in [4].

In this paper, we propose the *multi-step provenance inference* technique which can infer provenance for an entire processing chain with non-materialized intermediate views. To accomplish this, we facilitate coarse-grained provenance information about the processing elements as well as reproducible states of the database enabled by a temporal data model. Moreover, the *multi-step* inference technique only needs to observe the processing delay distribution of all processing elements as a basis of inference, unlike the work reported in [5], which requires to observe more specific distributions. The *multi-step* algorithm provides an *inferred provenance graph* showing all the contributing tuples as vertices and the relationship between tuples as edges. This provenance graph is useful to researchers for analyzing the results and validating their models.

2   Motivating Scenario

RECORD[1] is one of the projects in the context of the Swiss Experiment[2], which is a platform to enable real-time environmental experiments. Some sensors measure electric conductivity of water which refers to the level of salt in the water. Controlling the operation of a nearby drinking water well by using the available sensor data is the goal.

Fig. 1 shows the workflow. This workflow is used to visualize the fluctuation of electric conductivity in the selected region of the river. Three sensors are deployed, known as: Sensor#1, Sensor#2 and Sensor#3. For each sensor, there is a corresponding source processing element named PE_1, PE_2 and PE_3 which provides data tuples in persistent *views* S_1, S_2 and S_3 respectively. Views hold data tuples and processing elements are executed over views. S_1, S_2 and S_3 are the input for the *Union* processing element which produces a view V_1 as output. Each data tuple in the view V_1 is attached with an explicit timestamp referring to the point in time when it is inserted into the database, i.e. *transaction time*. Next, V_1 is fed to the processing element P_1 which calculates the average value per window and then generates a new view V_2. V_2 is not materialized since

---

[1] http://www.swiss-experiment.ch/index.php/Record:Home
[2] http://www.swiss-experiment.ch/

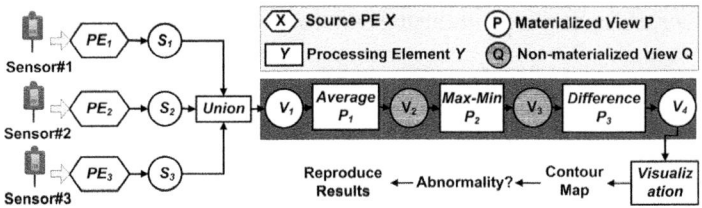

Fig. 1. Example workflow

it holds the intermediate results which are not interesting to the researchers as well as the results are easy to reproduce. The task of P_2 is to calculate the maximum and minimum value per input window of view V_2 and store the aggregated value in view V_3 which is not persistent. Next, V_3 is used by the processing element P_3, calculating the difference between the maximum and minimum electric conductivity. The view V_4 holds these output data tuples and gives significant information about the fluctuation of electric conductivity over the region. Since this view holds the output of the processing chain which will be used by users to evaluate and interpret different actions, view V_4 is materialized. Later, *Visualization* processing element facilitates V_4 to produce a contour map of the fluctuation of the electric conductivity. If the map shows any abnormality, researchers may want to reproduce results to validate their model. We consider the shaded part in Fig. 1 to explain and evaluate our solution later in this paper.

3   Proposed Multi-step Provenance Inference

3.1   Overview of the Algorithm

At first, we document coarse-grained provenance information of all processing elements which is a one-time action. Next, we observe the processing delay distributions δ of all processing elements which allow us to to make an *initial tuple boundary* on the materialized input view. This phase is known as *backward computation*. Then, for each processing step, processing windows are reconstructed, i.e. inferred windows, and we compute the probability of existence of an intermediate output tuple at a particular timestamp based on the δ distributions and other windowing constructs documented as coarse-grained provenance. Our algorithm associates the output tuple with the set of contributing input tuples and this process is continued till we reach the chosen tuple for which provenance information is requested. This phase is known as *forward computation*. It provides an *inferred provenance graph* for the chosen tuple. To explain these phases, we consider the shaded processing chain in Fig. 1 and focus on time-based windows.

3.2   Documenting Coarse-Grained Provenance

The stored provenance information is quite similar to *process provenance* reported in [7]. Inspired from this, we keep the following information of a processing element specification based on [8] as coarse-grained data provenance.

- Number of input views: indicates the total number of input views.
- View names: a set of input view names.
- Window types: a set of window types; one element for each input view. The value can be either *tuple* or *time*.
- Window Size: a set of window size; one element for each input view. The value actually represents the size of the window.
- Trigger type: specifies how the *processing element* will be triggered for execution (e.g. *tuple* or *time* based)
- Trigger rate: specifies when a *processing element* will be triggered.

3.3   Backward Computation: Calculating Initial Tuple Boundary

We apply a temporal data model on streaming data to retrieve appropriate tuples based on a given timestamp. The temporal attributes are: i) **valid time** or application timestamp represents the point in time when a tuple is created and ii) **transaction time** or system timestamp represents the point in time when a tuple is entered into the database. A view V_i contains tuples $t_k{}^i$ where k indicates the *transaction time*. We define a window $w_j{}^i$ based on tuples' transaction time over the view V_i which is an input view of processing element P_j. The window size of $w_j{}^i$ is referred to as $WS_j{}^i$. The processing element P_j is triggered after every TR_j time units defined as trigger rate. The processing delay distribution of P_j is referred to as δ_j distribution.

To calculate the *initial tuple boundary*, δ_j distributions of all processing elements and window size of all input views are considered assuming that the view V_j is the input of P_j. Fig. 2 shows a snapshot of all the associated views during the execution. It also shows the original provenance information represented by solid edges for a chosen output tuple $t_{46}{}^4$. It means that the chosen tuple is in view V_4 and the *transaction time* is 46 which is our *reference point*. To calculate the *upper bound* of the *initial tuple boundary*, the minimum delays of all processing elements are subtracted from the *reference point*. The *lower bound* is calculated by subtracting the maximum delays of all processing elements along with the associated window sizes from the reference point. Thus:

$$upper\,Bound = reference\ point - \sum_{j=1}^{n}(min(\delta_j)) \tag{4.1}$$

$$lower\,Bound = reference\ point - \sum_{j=1}^{n}(max(\delta_j)) - \sum_{j=1}^{n}(WS_j{}^j) \tag{4.2}$$

where $n = total\ number\ of\ processing\ elements$. In the formula, the *upper bound* is always exclusive and the *lower bound* is inclusive.

For the chosen tuple $t_{46}{}^4$, according to Eq. 4.1 and Eq. 4.2, *upper Bound* = $46 - 3 = 43$ and *lower Bound* = $46 - 6 - 24 = 16$ respectively based on the given parameters mentioned in Fig. 2. Therefore, the *initial tuple boundary* is $[16, 43)$. This boundary may contain some non-contributing input tuples to the chosen output tuple which will be removed during the next phase of inference.

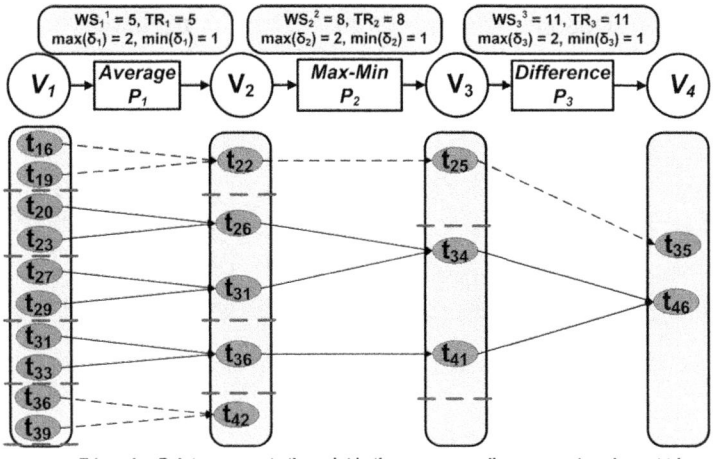

Fig. 2. Snapshot of views during the execution

3.4   Forward Computation: Building Provenance Graph

In this phase, the algorithm builds the *inferred provenance graph* for the *chosen tuple*. Our proposed algorithm starts from the materialized input view V_1. Since V_1 is materialized, all the tuples in V_1 with *transaction time* k have been assigned with *probability*, $P(t_k{}^1) = 1$. Fig. 2 shows that V_1 has 5 different triggering points in it's *initial tuple boundary* which are at time 20, 25, 30, 35 and 40 based on the trigger rate of P_1. Since the output view of P_1, V_2 is not materialized, the exact *transaction time* of the output tuple of each of these 5 executions is not known. Therefore, we calculate the probability of getting an output tuple at a specified *transaction time* k based on the δ_1 distribution. We call these output tuples as *prospective tuples*. Assume that, for all P_j, $P(\delta(w_j{}^j)) = 1) = 0.665$ and $P(\delta(w_j{}^j)) = 2) = 0.335$. For all the triggering points at l of P_1, the probability of getting a prospective tuple at k in V_2 can be calculated based on the following formula.

$$P(t_k{}^j) = P(\delta(w_{j-1}{}^{j-1}) = k - l)\quad [j = 2] \tag{4.3}$$

Therefore, based on Eq. 4.3 the probability of getting an output tuple at time 26 and at time 27 for the triggering at time 25 is 0.665 and 0.335 respectively. For the triggering point at 40, the output tuple could be observed either at 41 or 42. Since both of these timestamps fall outside the last triggering point of P_2, these tuples are not considered in the provenance graph. The same pruning rule also applies to the output tuple observed due to the triggering at time 20. In this case, the output tuple falls outside the window of the last processing element P_3. The associations among these pruned tuples are shown as dotted edges in Fig. 2.

Next, we move to view V_2 which is the input view of intermediate processing element P_2. In an intermediate processing step, the input tuples are produced

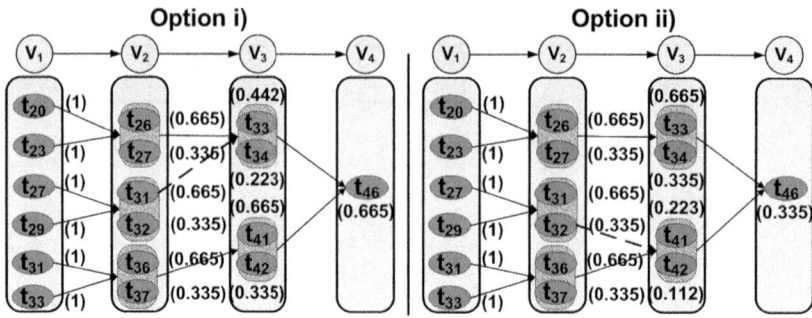

Fig. 3. The inferred provenance graph for both options

by different triggering points of the previous processing element and they might fall within the same window of the current processing element. In P_2, there is a triggering point at 32 and the window contains tuples within the range $[24, 31)$ which are the output tuples produced by triggering points at 25 and 30 of P_1. We define this as *contributing points*, *cp* and here $cp = 2$. Moreover, the possible timestamps to have the output tuple due to a particular triggering point might fall in two different input windows which results into different choice of paths to construct the provenance graph. Suppose, for the triggering point at 32 of P_2, there are two options: i) inclusion of $t_{31}{}^2$ and ii) exclusion of $t_{31}{}^2$. Fig. 3 shows provenance graph for both options. The probability of the existence of a tuple at transaction time k produced by a triggering point at time l of P_j where $j > 2$ can be calculated as:

$$P(t_k{}^j) = \prod_{x=1}^{cp} (\sum P(prospective\ tuples)) \times P(\delta(W_{j-1}{}^{j-1}) = k - l) \quad [j > 2]$$

$$(4.4)$$

Assuming aforesaid option i), the probability of getting an output tuple at time 33 due to the triggering at time 32 is:

$$P(t_{33}{}^3) = [\{P(t_{26}{}^2) + P(t_{27}{}^2)\} \times \{P(t_{31}{}^2)\}] \times P(\delta(W_2{}^2) = 1) = 0.442$$

where as in option ii) which excludes $t_{31}{}^2$:

$$P(t_{33}{}^3) = [\{P(t_{26}{}^2) + P(t_{27}{}^2)\}] \times P(\delta(W_2{}^2) = 1) = 1 \times 0.665 = 0.665$$

Eq. 4.4 is slightly modified while calculating the probability of the *chosen tuple*. Since this output view is materialized, the existence of the chosen tuple at *reference point* is certain. Therefore, δ_j distribution does not play a role in the formula. Assuming option i) which indicates the inclusion of $t_{31}{}^2$, the probability of getting an output tuple at time 46 for the execution at time 44 is:

$$P(t_{46}{}^4) = [\{P(t_{33}{}^3) + P(t_{34}{}^3)\} \times \{P(t_{41}{}^3) + P(t_{42}{}^3)\}] \times 1$$
$$= [\{0.442 + 0.223\} \times \{0.665 + 0.335\}] \times 1 = 0.665$$

Assuming option ii), $P(t_{46}{}^4)$ is 0.335. Fig. 3 shows the *inferred provenance graph* for both options. The probability of each tuple is shown by the value within parenthesis. Since the provenance graph using option i) provides maximum probability for the *chosen tuple*, our algorithm returns the corresponding provenance graph. Comparing it with the original provenance graph shown in Fig. 2 by solid edges, we conclude that the *inferred provenance graph* provides accurate provenance.

4   Evaluation

4.1   Evaluating Criteria and Test cases

We evaluate our proposed *multi-step provenance inference* algorithm using i) accuracy and ii) precision and recall. The accuracy compares the inferred *multi-step* fine-grained data provenance with explicit fine-grained provenance information, that is used as a ground truth. The precision and recall assess the quality of the provenance graph. The simulation is executed for 10000 time units for the entire processing chain. Based on queuing theory, we assume that tuples arrive into the system following Poisson distribution. The processing delay δ for each processing element also follows Poisson distribution. The δ-column for each processing element in Table 1 represents $avg(\delta_j)$ and $max(\delta_j)$. The test cases are chosen carefully based on the different types of window (e.g. overlapping/non-overlapping, sliding/tumbling) and varying processing delay. Specially, test case 2 involves longer processing delay than the others.

4.2   Accuracy

Accuracy of the proposed technique is measured by comparing the inferred provenance graph with the original provenance graph constructed from explicitly documented provenance information. For a particular output tuple, if these two graphs match exactly with each other then the accuracy of inferred provenance information for that output tuple is 1 otherwise, it is 0. We calculate the average of the accuracy for all output tuples produced by a given test case, called as *average accuracy* which can be expressed by the formula: *Average accuracy* $= (\frac{\sum_{i=1}^{n} acc_i}{n} \times 100)\%$ where $n =$ number of output tuples.

Table 1 shows the *average* and *expected* accuracy for different test cases. The avg. accuracy of test case 1 is 81%. The 100% average accuracy has been achieved

Table 1. Different test cases used for evaluation and Evaluation Results

Test case	P_1 WS	TR	δ	P_2 WS	TR	δ	P_3 WS	TR	δ	Exp. Accuracy	Avg. Accuracy	Avg. Precision	Avg. Recall
1	5	5	(1,2)	8	8	(1,2)	11	11	(1,2)	83%	81%	87%	98%
2	5	5	(2,3)	8	8	(2,3)	11	11	(2,3)	75%	61%	78%	89%
3	10	5	(1,2)	15	10	(1,2)	20	15	(1,2)	100%	100%	100%	100%
4	5	10	(1,2)	10	15	(1,2)	15	20	(1,2)	100%	100%	100%	100%
5	7	5	(1,2)	13	11	(1,2)	23	17	(1,2)	91%	90%	94%	97%

for test cases 3 and 4. However, for test case 2, we achieve only 61% accuracy due to the longer processing delay. If the processing delay is longer and new tuples arrive before finish the processing, it increases the chance of failure of the inference method.

Expected accuracy is calculated by taking the average of the probability of being accurate of an inferred provenance graph generated for all output tuples. It can be expressed as: *Expected accuracy* $= (\frac{\sum_{i=1}^{n} P(acc_i=1)}{n} \times 100)\%$. For the given test case 1 and 5, *expected* and *average accuracy* are similar. For test case 3 and 4, they are the same. However, we see a notable difference in test case 2 where *average accuracy* is smaller than the *expected* one.

4.3   Precision and Recall

To calculate precision and recall of an inferred provenance graph, we consider the edges between the vertices which represent the association between input and output tuples, i.e. provenance information and then we compare the set of edges between the inferred and original graph. Assume that, I is the set of edges in the inferred graph and O is the set of edges in the original graph. Therefore,

$$precision = (\frac{|I \cap O|}{|I|} \times 100)\% \qquad recall = (\frac{|I \cap O|}{|O|} \times 100)\%$$

We calculate *precision* and *recall* for each output tuple and then compute the *average precision* and *average recall*. In most of the cases, *recall* is higher than *precision*. It means that the inferred provenance graph may contain some extra edges which are not present in the original one. However, high values of both *precision* and *recall* in all test cases suggest that the probability of an inferred provenance graph to be meaningful to a user is high.

5   Related Work

In [10], authors described a data model to compute provenance on both relation and tuple level. However, it does not address the way of handling streaming data and associated overlapping windows. In [11], authors have presented an algorithm for lineage tracing in a data warehouse environment. They have provided data provenance on tuple level. LIVE [12] is an offshoot of this approach which supports streaming data. It is a complete DBMS which preserves explicitly the lineage of derived data items in form of boolean algebra. Since LIVE explicitly stores provenance information, it incurs extra storage overhead.

In sensornet republishing [13], the system documents the transformation of online sensor data to allow users to understand how processed results are derived and support to detect and correct anomalies. They used an annotation-based approach to represent data provenance explicitly. However, our proposed method does not store fine-grained provenance data rather infer provenance data.

A layered model to represent workflow provenance is introduced in [14]. The layers presented in the model are responsible to satisfy different types of provenance queries including queries about a specific activity in the workflow.

A relational DBMS has been used to store captured provenance data. The authors have not introduced any inference mechanism for provenance data.

6 Conclusion and Future Work

The *multi-step provenance inference* technique provides highly accurate provenance information for an entire processing chain, if the processing delay is not longer than the sampling time of input tuples. Our evaluation shows that in most cases, it achieves more than 80% accuracy. Our solution also provides an inferred provenance graph with high precision and recall. In future, we will try to extend this technique to estimate the accuracy beforehand.

References

1. Simmhan, Y.L., et al.: A survey of data provenance in e-science. SIGMOD Rec. 34(3), 31–36 (2005)
2. Buneman, P., Tan, W.C.: Provenance in databases. In: International Conference on Management of Data, pp. 1171–1173. ACM SIGMOD (2007)
3. Huq, M.R.: et al.: Facilitating fine grained data provenance using temporal data model. In: Proc. of Data Management for Sensor Networks (DMSN), pp. 8–13 (2010)
4. Huq, M.R., Wombacher, A., Apers, P.M.G.: Inferring Fine-Grained Data Provenance in Stream Data Processing: Reduced Storage Cost, High Accuracy. In: Hameurlain, A., Liddle, S.W., Schewe, K.-D., Zhou, X. (eds.) DEXA 2011, Part II. LNCS, vol. 6861, pp. 118–127. Springer, Heidelberg (2011)
5. Huq, M.R., Wombacher, A., Apers, P.M.G.: Adaptive inference of fine-grained data provenance to achieve high accuracy at lower storage costs. In: IEEE International Conference on e-Science, pp. 202–209. IEEE Computer Society Press (December 2011)
6. Bishop, C.M.: Patter Recognition and Machine Learning. Springer Science+Business Media LLC (2006)
7. Simmhan, Y.L., et al.: Karma2: Provenance management for data driven workflows. International Journal of Web Services Research, 1–23 (2008)
8. Wombacher, A.: Data workflow - a workflow model for continuous data processing. Technical Report TR-CTIT-10-12, Centre for Telematics and Information Technology University of Twente, Enschede (2010)
9. Gebali, F.: Analysis of Computer and Communication Networks. Springer Science+Business Media LLC (2008)
10. Buneman, P., Khanna, S., Tan, W.-C.: Why and Where: A Characterization of Data Provenance. In: Van den Bussche, J., Vianu, V. (eds.) ICDT 2001. LNCS, vol. 1973, pp. 316–330. Springer, Heidelberg (2000)
11. Cui, Y., Widom, J.: Lineage tracing for general data warehouse transformations. VLDB Journal 12(1), 41–58 (2003)
12. Das Sarma, A., Theobald, M., Widom, J.: LIVE: A Lineage-Supported Versioned DBMS. In: Gertz, M., Ludäscher, B. (eds.) SSDBM 2010. LNCS, vol. 6187, pp. 416–433. Springer, Heidelberg (2010)
13. Park, U., Heidemann, J.: Provenance in Sensornet Republishing. In: Freire, J., Koop, D., Moreau, L. (eds.) IPAW 2008. LNCS, vol. 5272, pp. 280–292. Springer, Heidelberg (2008)
14. Barga, R., et al.: Automatic capture and efficient storage of e-science experiment provenance. Concurrency and Computation: Practice and Experience 20(5) (2008)

Automatic Conflict Resolution in a CDSS

Fayez Khazalah[1], Zaki Malik[1], and Brahim Medjahed[2]

[1] Department of Computer Science
Wayne State University, MI 48202
{fayez,zaki}@wayne.edu
[2] Department of Computer & Information Science
University of Michigan - Dearborn, MI. 48128
brahim@umd.umich.edu

Abstract. A Collaborative Data Sharing System (CDSS) allows groups of scientists to work together and share their data in the presence of disparate database schemas and instances. Each group can extend, curate, and revise its own database instance in a disconnected mode. At some later point, the group may publish the updates it made for the benefit of others and to get updates from other groups (if any). Any conflicting updates are handled by the reconciliation operation which usually rejects such updates temporally and marks them as "deferred". The deferred set is then resolved manually according to pre-defined data reconciliation policies, priorities, etc. for the whole group. In this paper, we present an approach to resolve conflicts in an automatic manner. The focus is to resolve conflicts in the deferred set by collecting feedbacks about the quality of conflicting updates from local users, and then weighing, and aggregating the different feedbacks to assess the most "trusted" update.

1 Introduction

A collaborative data sharing system facilitates users (usually in communities) to work together on a shared data repository to accomplish their "shared" tasks. Users of such a community can add, update, and query the shared repository [1], and may thus contain inconsistent data at a particular time instance [2]. While a relational DBMS can be used to manage the shared data, RDBMSs lack the ability to handle such conflicting data [1]. Traditional integration systems (using RDBMSs) usually assume a global schema, where autonomous data sources are mapped to this global schema, and data inconsistencies are solved by applying conflict resolution strategies [3] [4]. However, these systems only support queries on the global schema and do not support update exchange. To remedy this shortcoming, peer data management systems support disparate schemas [5] [6], but they are also not flexible in terms of update propagation between different schemas, and handling data inconsistency issues. Therefore, the concept of a "multi-versioned database" has been proposed that allows for conflicting data to be stored in the same database [7] [8]. Other similar approaches allow users to annotate data items to express their positive or negative beliefs about different

A. Ailamaki and S. Bowers (Eds.): SSDBM 2012, LNCS 7338, pp. 406–415, 2012.

data items, and each user is shown a consistent instance of the shared database according to his own beliefs [1][9].

While the above mentioned collaborative data sharing approaches fulfill the need of some scientific communities for data sharing, it is not the case with others [6]. Collaborative data sharing systems (CDSSs) have been defined to support the collaboration needs in such type of communities [6]. In a CDSS, groups of scientists normally work on disparate schemas and database instances in a disconnected mode, and at some later point, may decide to publish the data updates publicly to other peers and/or get the updates from other peers. The reconciliation operation in the CDSS engine imports the updates, and then filters them based on trust policies and priorities for the current peer. It then applies the non-conflicting and accepted updates on the local database instance, while conflicting updates are grouped into individual conflicting sets of updates. Each update of a set is assigned a *priority level* according to the trust policies of the reconciling peer, and the update with the highest priority is selected to be applied on the local database instance and the rest are rejected. When multiple updates having the same preference or with no assigned preferences are found, it marks those updates as "deferred". The deferred updates are not processed and not considered in future reconciliations until a user manually resolves the deferred conflicts. We propose a conflict resolution approach that uses community feedbacks to handle the conflicts that may arise in collaborative data sharing communities, with potentially disparate schemas and data instances. The focus is to allow the CDSS engine to automatically utilize community feedback for the purpose of handling conflicting updates that are added to the deferred set during the reconciliation operation and minimize (or omit altogether) the user intervention.

2   Automated Conflict Resolution in CDSS

A CDSS only applies semi-automatic conflict resolution by accepting the highest-priority conflicting updates, but it leaves for individual users the responsibility of resolving conflicts for the updates that are deferred. In our proposed approach, after the reconciliation operation adds a new conflict group to the deferred set, the following steps are taken:

1. Local users are informed to rate the updates in the conflict group. When a predefined number of user ratings are received, then it is marked as closed.
2. When a conflict group is closed, then the credibilities of the raters are (re)computed based on the majority rating and the aggregation of the previously computed locally assessed reputations of this particular provider peer. The reported ratings are then weighted according to the new credibilities.
3. The weighted aggregated value represents the locally assessed reputation of the update's provider peer as viewed by the reconciling peer.
4. Finally, the update which is imported from the provider peer with the highest assessed reputation value is applied to the reconciling peer's instance (making sure it does not violate any local constraints).

Local Reputation of a Provider Peer ($LRPP$). When a new conflict group is added to the deferred set of a peer, it needs to resolve the conflict by choosing a single update from the group and rejecting others. The decision is based on the feedbacks of the local community.

Rating Updates. The reconciliation operation in a consumer peer p_i notifies local users when a new conflict group of updates (G_c) is inserted into the deferred set $Deferred(p_i)$, along with a *closing time* for the rating process. Local users of p_i rate the updates of unresolved G_c in $Deferred(p_i)$. A user x (p_i^x) of p_i assigns a rating in the *interval* $[0, 1]$ to each update of a provider peer p_j in G_c, where 0 identifies the rater's extreme disbelief and 1 identifies the rater's extreme belief in an update. If the number of users who rate this G_c exceeds *percentage threshold* (σ_i), the reconciliation operation marks this G_c as "closed" and users cannot rate this G_c anymore. Otherwise, the reconciliation operation extends the rating period.

Computing the Assessed $LRPP$. We assume that each participant peer keeps records of all previously computed $LRPPs$ for each provider peer. Let p_j be a provider peer and p_i^x be a rater user. $Rep(p_j, p_i^x)$ represents the rating assigned by p_i^x to the update of p_j in G_c. Formally, the $LRPP$ of a provider peer p_j as viewed by a consumer peer p_i, computed post closing a conflict group G_c, is defines as:

$$LRPP(p_j, p_i) = \frac{\sum_{x=1}^{L}(Rep(p_j, p_i^x) * C_{p_i^x})}{\sum_{x=1}^{L} C_{p_i^x}} \tag{1}$$

where L denotes the set of local users who have rated p_j's update in G_c, $Rep(p_j, p_x)$ is the last personal evaluation of p_j as viewed by p_x, and $C_{p_i^x}$ is the credibility of p_i^x as viewed by p_i. In the above equation, the trust score of the provider peer is calculated according to the credibility scores of the rater peers, such that ratings of highly credible raters are weighed more than that of raters with low credibilities. The credibility of a rater lies in the interval $[0,1]$ with 0 identifying a dishonest rater and 1 an honest one. We adopt a similar notion as in [10][11] to minimize the effects of *unfair* or inconsistent ratings. The basic idea is that if the reported rating agrees with the majority opinion, the rater's credibility is increased, and decreased otherwise. We use the k-mean clustering algorithm to compute the *majority rating* (denoted $\mathcal{MR}$) by grouping current similar feedback ratings together. The Euclidean distance between the majority rating ($\mathcal{MR}$) and the rating (R) is used to adjust the rater credibility. The change in credibility due to majority rating, denoted by $\mathcal{MR}_\Delta$ is then defined as:

$$\mathcal{MR}_\Delta = \begin{cases} 1 - \frac{\sqrt{\sum_{k=1}^{n}(\mathcal{MR} - R_k)^2}}{\sigma}, & \text{if } \sqrt{\sum_{k=1}^{n}(\mathcal{MR} - R_k)^2} < \sigma; \\ 1 - \frac{\sigma}{\sqrt{\sum_{k=1}^{n}(\mathcal{MR} - R_k)^2}}, & \text{otherwise.} \end{cases} \tag{2}$$

where σ is the standard deviation in all the reported ratings. Note that $\mathcal{MR}_\Delta$ does not denote the rater's credibility (or the weight), but only defines the effect on credibility due to aggrement/disagreement with the majority rating. We supplement the majority rating scheme by adjusting the credibility of a rater based on its past behavior as well. The historical information provides an estimate of the *trustworthiness* of the raters [11]. We believe that under controlled situations, a consumer peer's perception of a provider peer's reputation should not deviate much, but stay consistent over time. We assume the interactions take place at time t and the consumer peer already has record of the previously assessed $LRPP$ (denoted $\overline{LRPP}$), which is defined as:

$$\overline{LRPP} = \prod_{t-1}^{t-k} LRPP(p_j, p_i)^t \qquad (3)$$

where $LRPP(p_j, p_i)$ is the assessed $LRPP$ of p_j by p_i for each time instance t, $\prod$ is the aggregation operator and k is the time duration defined by p_i. It can vary from one time instance to the complete past reputation record of p_j. Note that $\overline{LRPP}$ is the "local assessed reputation" calculated by p_i at the previous time instance(s). If p_j does not change much from the previous time instances, then $\overline{LRPP}$ and the present rating R should be somewhat similar. Thus, the effect on credibility due to agreement or disagreement with the aggregation of the last k assessed $LRPP$ values (denoted $\overline{LRPP_\Delta}$) is defined in a similar manner as Equation (2). We just put $\overline{LRPP}$ instead of $\mathcal{MR}$ in the equation to compute $\overline{LRPP_\Delta}$.

In real-time situations it is difficult to determine the different factors that cause a change in the state of a provider peer. A user may rate the same provider peer differently without any malicious motive. Thus, the credibility of a rater may change in a number of ways, depending on the values of R, $\mathcal{MR}_\Delta$, and $\overline{LRPP_\Delta}$. The general formula is:

$$C_{p_i^x} = C_{p_i^x} \pm \Phi * \Psi \qquad (4)$$

where Φ is the credibility adjustment normalizing factor, while Ψ represents amount of change in credibility due to the equivalence or difference of R with $\mathcal{MR}$ and $\overline{LRPP}$. The signs $\pm$ indicate that either $+$ or $-$ can be used, i.e., the increment or decrement in the credibility depends on the situation.

We place more emphasis on the ratings received in the current time instance than the past ones, similar to previous works as [10] and [11]. Thus, equivalence or difference of R with $\mathcal{MR}$ takes a precedence over that of R with $\overline{LRPP}$. This can be seen from Equation (4), where the $+$ sign with Φ indicates $R \simeq \mathcal{MR}$ while $-$ sign with Φ means that $R \neq \mathcal{MR}$. Φ is defined as:

$$\Phi = C_{p_i^x} * (1 - |R_x - \mathcal{MR}|) \qquad (5)$$

Equation (5) states that the value of the normalizing factor Φ depends on the credibility of the rater and the absolute difference between the rater's current feedback and the majority rating calculated. Multiplying by the rater's credibility allows the honest raters to have greater influence over the ratings aggregation process and dishonest raters to lose their credibility quickly in case of a false or malicious rating. Ψ is made up of $\mathcal{MR}_\Delta$ and/or $\overline{LRPP_\Delta}$, and a "pessimism factor" (ρ). The exact value of ρ is left at the discretion of the consumer peer, with the exception that its minimum value should be 2. The lower the value of ρ, the more optimistic is the consumer peer and higher value of ρ are suitable for pessimistic consumers. We define a pessimistic consumer as one that does not trust the raters easily and reduces their credibility drastically on each false feedback. Moreover, honest rater's reputations are increased at a high rate, meaning that such consumers make friends easily. R, $\mathcal{MR}$, and $\overline{LRPP}$ can be related to each other in one of four ways, and each condition specifies how $\mathcal{MR}_\Delta$ and $\overline{LRPP_\Delta}$ are used in the model.

1. The reported reputation value is similar to both the majority rating and the aggregation of the previously computed $LRPP$ values (i.e., $R \simeq \mathcal{MR} \simeq \overline{LRPP}$). The equality $\mathcal{MR} \simeq \overline{LRPP}$ suggests that majority of the raters believe that the quality of updates imported from a provider peer p_j has not changed. The rater peers credibility is updated as:

$$C_{p_i^x} = C_{p_i^x} + \Phi * \left(\frac{|\mathcal{MR}_\Delta + \overline{LRPP_\Delta}|}{\rho} \right) \tag{6}$$

Equation (6) states that since all factors are equal, the credibility is incremented.

2. The individual reported reputation rating is similar to the majority rating but differs from the previously assessed reputation, i.e. ($R \simeq \mathcal{MR}$) and ($R \neq \overline{LRPP}$). In this case, the change in the reputation rating could be due to either of the following. First, the rater peer may be colluding with other raters to increase or decrease the reputation of a provider peer. Second, the quality of updates imported from the provider peer may have actually changed since $\overline{LRPP}$ was last calculated. The rater peer's credibility is updated as:

$$C_{p_i^x} = C_{p_i^x} + \Phi * \left(\frac{\mathcal{MR}_\Delta}{\rho} \right) \tag{7}$$

Equation (7) states that since $R \simeq \mathcal{MR}$, the credibility is incremented, but the factor $R \neq \overline{LRPP}$ limits the incremental value to ($\frac{\mathcal{MR}_\Delta}{\rho}$).

3. The individual reported reputation value is similar to the aggregation of the previously assessed $LRPP$ values but differs from the majority rating (reasons omitted here due to space restrictions), i.e. ($R \neq \mathcal{MR}$) and ($R \simeq \overline{LRPP}$). Thus, the rater peer's credibility is updated as:

$$C_{p_i^x} = C_{p_i^x} - \Phi * \left(\frac{\overline{LRPP_\Delta}}{\rho} \right) \tag{8}$$

Equation (8) states that since $R \neq \mathcal{MR}$, the credibility is decremented, but here the value that is subtracted from the previous credibility is adjusted to ($\frac{\overline{LRPP_\Delta}}{\rho}$).

4. The individual reported reputation value is not similar to both the majority rating and the calculated aggregation of assessed $LRPP$ values, i.e. ($R \neq \mathcal{MR}$) and ($R \neq \overline{LRPP}$). R may differ from the majority rating and the past aggregation of $LRPP$ values due to either of the following. First, R may be the first one to experience the provider peer's new behavior. Second, R may not know the actual quality of the provider peer's imported updates. Third, R may be lying to increase/decrease the provider peer's reputation. In this case, the rater peer's credibility is updated as:

$$C_{p_i^x} = C_{p_i^x} - \Phi * \left(\frac{|\mathcal{MR}_\Delta + \overline{LRPP_\Delta}|}{\rho} \right) \tag{9}$$

Equation (9) states that the inequality of all factors means that rater peer's credibility is decremented, where the decremented value is the combination of both the effects $\mathcal{MR}_\Delta$ and $\overline{LRPP_\Delta}$.

3   Example Scenario

Consider a CDSS of three participant peers (p_1, p_2, and p_3) that represent three bioinformatics warehouses (example adapted from [12]). The three peers share a single relation $F(organism, protein, function)$ for protein function, where the key of the relation is composed of the fields $organism$ and $protein$. Peer p_1 accepts updates from both p_2 and p_3 with the same trust priority level. For illustration, we also assume that there are 10 other participant peers (p_4 through p_{13}), and we assign different roles for the participant peers.

We consider peers p_2 and p_3 as provider peers for the rest of peers, peer p_i as a consumer peer who imports updates from the provider peers and needs to reconcile its own instance though, and the rest of peers (p_4 through p_{13}) play the role of the raters which interacted with the provider peers in the past and are willing to share their experience with other consumer peers. Similar to [12], we illustrate the reconciliation operation of this CDSS example as shown in Table 1, taking into consideration our proposed modification for the system.

In the beginning, we assume that the instance of relation F at each participant peer p_i, denoted by $I_i(F)|0$, is empty (i.e., at time 0). At time 1, p_3 conducts two transactions $T_{3:0}$ and $T_{3:1}$. It then decides to publish and reconcile its own state (to check if other peers made any changes). Since the other two participant peers have no updates, the operation is completed, $I_3(F)|1$ denotes the result. At time 2, p_2 conducts two transactions $T_{2:0}$ and $T_{2:1}$. It then publishes and reconciles its own state. Note that the resulting instance $I_2(F)|2$ of p_2 contains only its own updates. Although there is a recently published update by p_3, which is trusted by it, p_2 does not accept p_3's published update because it conflicts with its own updates. At time 3, p_3 reconciles again. It accepts the transaction $T_{2:0}$ that is published by p_2 and rejects p_2's second update $T_{2:0}$ because it conflicts with its

Table 1. Reconciliation of F(organism, protein, function)

t p_3	p_2	p_1
0 $I_3(F)\|0=\{\}$	$I_2(F)\|0 = \{\}$	$I_1(F)\|0 = \{\}$
$T_{3:0}$:{+F(rat,prot1,cell-metab;3)} $T_{3:1}$:{F(rat,prot1,cell-metab → rat,prot1,immune;3)} 1 <publish and reconcile> $I_3(F)\|1$:{(rat,prot1,immune)}		
2	$T_{2:0}$:{+F(mouse,prot2,immune;2)} $T_{2:1}$:{+F(rat,prot1,cell-resp;2)} <publish and reconcile> $I_2(F)\|2$:{(mouse,prot2,immune), (rat,prot1,cell-resp)}	
3 <reconcile> $I_3(F)\|3$:{(mouse,prot2,immune), (rat,prot1,immune)}		
4		<reconcile> $I_1(F)\|4$:{(mouse,prot2,immune)} DEFER: {$T_{3:1}$, $T_{2:1}$}

own state. At time 4, p_1 reconciles. It gives the same priority for transactions of p_2 and p_3. Thus, it accepts the non-conflicting transaction $T_{2:0}$, and it defers both the conflicting transactions $T_{2:1}$ and $T_{3:1}$. p_1's reconciliation operation forms a conflict group G_1 (shown in Table 2) that includes both deferred transactions that are added to the deferred set of p_1 during the reconciliation. p_1 then notifies its local users that a new conflict group is added to $Deferred(p_1)$, so they can start rating updates in this particular conflict group.

Computing the *LRPP*. The local peer p_1 maintains a table of all the previously assessed $LRPP$ values of provider peers that it interacts with. For instance, the last 5 $LRPP$ values for p_2 and p_3 are $\{0.41, 0.43, 0.58, 0.52, 0.38\}$ and $\{0.90, 0.89, 0.89, 0.94, 0.90\}$, respectively.

Table 2. The deferred set of peer p_1

							$Deferred(p_1)$						
G_c	Trans	p_1^1	p_1^2	p_1^3	p_1^4	p_1^5	p_1^6	p_1^7	p_1^8	p_1^9	p_1^{10}	Status	σ_i
G_1	$T_{3:1}$	0.95	0.65	1.00	0.60	0.97	1.00	0.95	0.90	0.95	1.00	Closed	100%
	$T_{2:1}$	0.45	0.80	0.45	0.75	0.40	0.40	0.45	0.40	0.45	0.43		

1. p_1 computes the values of $\mathcal{MR}$, $\mathcal{MR}_\Delta$, $\overline{LRPP}$, and $\overline{LRPP_\Delta}$ factors for each provider peer in G_1. The computed values for p_2 are $(0.50, 0.67, 0.47, .68)$ and for p_3 are $(0.90, 0.67, 0.90, 0.67)$, respectively.

Table 3. The result of computation for local users who rated the update of p_2 in G_1

p_2	Local Users									
Factor	p_1^1	p_1^2	p_1^3	p_1^4	p_1^5	p_1^6	p_1^7	p_1^8	p_1^9	p_1^{10}
$C_{p_i^x}(old)$	0.96	0.70	0.90	0.85	0.97	0.95	0.95	0.90	0.94	0.90
Φ	0.91	0.49	0.86	0.64	0.87	0.86	0.90	0.81	0.89	0.84
$V \simeq \mathcal{MR}$	0.05	0.30	0.05	0.25	0.10	0.10	0.05	0.10	0.05	0.07
$\mathcal{MR} \simeq \overline{LRPP}$	0.03	0.03	0.03	0.03	0.03	0.03	0.03	0.03	0.03	0.03
$V \simeq \overline{LRPP}$	0.02	0.34	0.02	0.29	0.07	0.07	0.02	0.07	0.02	0.04
$Case(1-4)$	1	4	1	4	1	1	1	1	1	1
$C_{p_i^x}(new)$	0.96	0.65	0.90	0.80	0.98	0.96	0.95	0.91	0.94	0.90
V_w	0.43	0.52	0.41	0.60	0.39	0.38	0.43	0.36	0.42	0.39

2. p_1 computes the new credibility values for each rater user, as shown in Table 3, who has participated in rating the update of p_2 in G_1. We provide more details about the computations done in Table 3 in the following sub-steps:
   (a) The first row of Table 3, titled $(C_{p_i^x}(old))$, shows the current credibility values for rater users $(p_1^1, p_1^2, ..., p_1^{10})$.
   (b) In the second row of Table 3, the values of Φ variable are shown after Equation (5) is applied.
   (c) The rows (3-5) show the equalities between the factor pairs $(R \simeq \mathcal{MR})$, $(\mathcal{MR} \simeq \overline{LRPP})$, and $(R \simeq \overline{LRPP})$, for each rater user. Here, we assume that the two compared factors are equal if the amount of difference between them is equal or less than 0.20. Otherwise, they are considered not to be equal. The values are then adjusted using Equations 6-9.
   (d) The rows (6-8) of Table 3 show the matched case, the value of Ψ, and the new computed credibility value $(C_{p_i^x}(new))$, for each rater user.

(e) The last row, titled (R_w), shows the weights of reputation ratings, provided by local users of p_1, as shown in Table 2.

(f) Based on the last two rows, p_1's reconciling operation computes the $LRPP$ for p_2 $(LRPP(p_2, p_1) = 0.49)$ by applying Equation (1).

3. Step 2 is repeated to compute the $LRPP$ for p_3 $(LRPP(p_3, p_1) = 0.91)$.

Conflict Resolution. After the conflict group $G1$ is closed, p_1's reconciliation operation computes $LRPP$ for each provider peer in G_1. Because p_3 has the highest reputation value (i.e., the highest $LRPP$), the transaction $T_{3:1}$ of p_3 is then accepted and applied to the local instance of peer p_1 as it does not violate its local state, while the transaction $T_{2:1}$ of p_2 is rejected.

Experimental Evaluations. We simulated (using Java code) a CDSS with three participants peers. p_1 is the reconciling peer, whereas p_2 and p_3 are the provider peers. p_1 has 100 local users. The provider peers are assigned degrees of quality or behavior randomly; between 0.1 and 0.7 for p_2, and between 0.7 and 1.0 for p_3. We conducted two sets of experiments: In the first one, 80% of users are high-quality users, and 20% are low-quality. Both groups of users randomly generate rating values with qualities in the range (0.8-1) and (0.1-0.4), respectively. In the second set, we keep the quality level of rating for both groups of users the same as in the first set, but we only increase the percentage of dishonest raters and decrease the percentage of honest ones (50% of users are high quality users, and 50% are low quality users). At the beginning of the simulation, we assume that all local users of the reconciling peer have credibility of 1. Each time during the simulation, p_2 and p_3 generate identical tuples (i.e., tuples that have the same key but differ in values of the non-key attributes) and then publish their updates. When p_1 reconciles (i.e., imports the newly published updates from both p_2 and p_3), a conflict is found in the pair of updates with the same key but imported from different providers. The conflict is resolved by either accepting the update of p_2 or p_3, according to the weighted ratings of users. The simulation ends when p_1 resolve the conflict numbered 3600. Figure 1 shows the results for the above experimental sets. The shown results represent the average of 10 rounds of experiments. In the first set, honest raters out-number dishonest ones. Fig. 1(a) shows the effect of this inequality in calculating raters' credibilities and number of accepted updates per each provider. The average credibility of each group of users is shown in Fig. 1(a.1) with increasing number of conflicts, while Fig. 1(a.2) represents the number of accepted updates from each provider. Because there are more honest raters we can see that all the accepted updates comes from the provider with the highest-quality generated updates. We also can see that the average credibility of honest raters is always high compared to that of dishonest group where it is drastically decreasing for the consecutive conflicts. The result of the second set where the number of honest and dishonest raters are equal is shown in Fig. 1(b). In this set, it may happen that the dishonest raters' ratings form the majority rating. This causes a degradation in the credibility of honest raters since their opinion now differs from the majority opinion, and

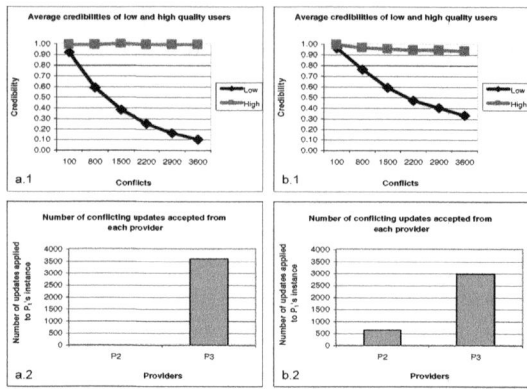

Fig. 1. Simulation results of the two experimental sets (a and b)

an increment in the dishonest raters' credibilities (Fig. 1(b.1)). Therefore, some
updates that come from the provider with the lowest-quality generated updates
are accepted (Fig. 1(b.2)).

4   Conclusion and Future Work

We presented an approach to resolve conflicts in the deferred set of a CDSS's
reconciling peer by collecting feedbacks about the quality of conflicting updates
from local community. We plan to extend our proposed work in the future, to
also deploy community feedbacks for the purpose of automatically defining trust
policies for the local peer, thereby omitting the role of the administrator.

References

1. Gatterbauer, W., Balazinska, M., Khoussainova, N., Suciu, D.: Believe it or not:
   adding belief annotations to databases. Proc. VLDB Endow. 2, 1–12 (2009)
2. Gatterbauer, W., Suciu, D.: Data conflict resolution using trust mappings. In: Proc.
   of SIGMOD 2010, pp. 219–230. ACM (2010)
3. Motro, A., Anokhin, P.: Fusionplex: resolution of data inconsistencies in the integration of heterogeneous information sources. Information Fusion 7(2), 176–196
   (2006)
4. Bleiholder, J., Draba, K., Naumann, F.: Fusem: exploring different semantics of
   data fusion. In: Proc. 33rd VLDB 2007, pp. 1350–1353. VLDB Endowment (2007)
5. Bernstein, P.A., Giunchiglia, F., Kementsietsidis, A., Mylopoulos, J., Serafini, L.,
   Zaihrayeu, I.: Data management for peer-to-peer computing: A vision. In: Proc. of
   the 5th WebDB 2002, pp. 89–94 (2002)
6. Ives, Z.G., Green, T.J., Karvounarakis, G., Taylor, N.E., Tannen, V., Talukdar,
   P.P., Jacob, M., Pereira, F.: The orchestra collaborative data sharing system. SIG-
   MOD Rec. 37, 26–32 (2008)
7. Snodgrass, R.T.: Developing time-oriented database applications in SQL. Morgan
   Kaufmann Publishers Inc., San Francisco (2000)

8. Cudre-Mauroux, P., Kimura, H., Lim, K.-T., Rogers, J., Simakov, R., Soroush, E., Velikhov, P., Wang, D.L., Balazinska, M., Becla, J., DeWitt, D., Heath, B., Maier, D., Madden, S., Patel, J., Stonebraker, M., Zdonik, S.: A demonstration of scidb: a science-oriented dbms. Proc. VLDB Endow. 2, 1534–1537 (2009)

9. Pichler, R., Savenkov, V., Skritek, S., Hong-Linh, T.: Uncertain databases in collaborative data management. In: Proc. 36st VLDB 2010. VLDB Endow. (2010)

10. Buchegger, S., Le Boudec, J.-Y.: A robust reputation system for p2p and mobile ad-hoc networks. In: Proc. of 2nd Workshop on Economics of P2P Sys. (2004)

11. Whitby, A., Josang, A., Indulska, J.: Filtering out unfair ratings in bayesian reputation systems. Science 4(2), 106–117 (2004)

12. Taylor, N.E., Ives, Z.G.: Reconciling while tolerating disagreement in collaborative data sharing. In: Proc. of the SIGMOD 2006, pp. 13–24. ACM (2006)

Tracking Distributed Aggregates
over Time-Based Sliding Windows*

Graham Cormode[1] and Ke Yi[2,**]

[1] AT&T Labs–Research
graham@research.att.com
[2] Hong Kong University of Science and Technology
yike@cse.ust.hk

Abstract. The area of distributed monitoring requires tracking the value of a function of distributed data as new observations are made. An important case is when attention is restricted to only a recent time period, such as the last hour of readings—the sliding window case. In this paper, we introduce a novel paradigm for handling such monitoring problems, which we dub the "forward/backward" approach. This view allows us to provide optimal or near-optimal solutions for several fundamental problems, such as counting, tracking frequent items, and maintaining order statistics. The resulting protocols improve on previous work or give the first solutions for some problems, and operate efficiently in terms of space and time needed. Specifically, we obtain optimal $O(\frac{k}{\varepsilon} \log(\varepsilon n/k))$ communication per window of n updates for tracking counts and heavy hitters with accuracy ε across k sites; and near-optimal communication of $O(\frac{k}{\varepsilon} \log^2(1/\varepsilon) \log(n/k))$ for quantiles. We also present solutions for problems such as tracking distinct items, entropy, and convex hull and diameter of point sets.

1 Introduction

Problems of distributed tracking involve trying to compute various aggregates over data that is distributed across multiple observing sites. Each site observes a stream of information, and aims to work together with the other sites to continuously track a function over the union of the streams. Such problems arise in a variety of modern data management and processing settings—for more details and motivating examples, see the recent survey of this area [6]. To pick one concrete example, a number of routers in a network might try to collaborate to identify the current most popular destinations. The goal is to allow a single distinguished entity, known as the "coordinator", to track the desired function. Within such settings, it is natural to only want to capture the recent behavior—say, the most popular destinations within the last 24 hours. Thus, attention is limited to a "time-based sliding window".

---

* These results were announced at PODC'11 as a 'brief announcement', with an accompanying 2 page summary.
** Ke Yi is supported by an RPC grant from HKUST and a Google Faculty Research Award.

A. Ailamaki and S. Bowers (Eds.): SSDBM 2012, LNCS 7338, pp. 416–430, 2012.
© Springer-Verlag Berlin Heidelberg 2012

For these problems, the primary goal is to minimize the (total) communication required to achieve accurate tracking. Prior work has shown that in many cases this cost is asymptotically smaller than the trivial solution of simply centralizing all the observations at the coordinator site. Secondary goals include minimizing the space required at each site to run the protocol, and the time to process each new observation. These quantities are functions of k, the number of distributed sites, n, the total size of the input data, and ε, an a user-supplied approximation parameter to tolerate some imprecision in the computed answer (typically, $0 < \varepsilon < 1$).

Within this context, there has been significant focus on the "infinite window" case, where all historic data is included. Results have been shown for monitoring functions such as counts, distinct counts, order statistics, join sizes, entropy, and others [1,4,7,14,19,20]. More recently there has been interest in only tracking a window of recent observations, defined by all those elements which arrived within the most recent w time units. Results in this model have been shown for tracking counts and frequent items [4], and for sampling [8].

The most pertinent prior work is that of Chan $et\ al.$ [4], which established protocols for several fundamental problems with sliding windows. The analysis used quickly becomes quite complicated, due to the need to study multiple cases in detail as the distributions change. Perhaps due to this difficulty, the bounds obtained are not always optimal. Three core problems are studied: basic counting, which is to maintain the count of items observed within the window; heavy hitters, which is to maintain all items whose frequency (within the window) is more than a given fraction; and to maintain the quantiles of the distribution. Each problem tolerates an error of ε, and is parametrized by k, the number of sites participating in the computation, and n, the number of items arriving in a window. [4] shows (per window) communication costs of $O(\frac{k}{\varepsilon} \log \frac{\varepsilon n}{k})$ bits for basic counting, $O(\frac{k}{\varepsilon} \log \frac{n}{k})$ words[1] for frequent items and $O(\frac{k}{\varepsilon^2} \log \frac{n}{k})$ words for quantiles. Our main contributions in this paper are natural protocols with a more direct analysis which obtain optimal or near optimal communication costs. To do this, we outline an approach for decomposing sliding windows, which also extends naturally to other problems in this setting. We call this the "forward/backward" framework, and provide a general claim, that the communication complexity for many functions in the model with a sliding window is no more than in the infinite window case (Section 2). We instantiate this to tracking counts (Section 3), heavy hitters (Section 4) and quantiles (Section 5) to obtain optimal or near optimal communication bounds, with low space and time costs. Lastly, we extend our results to functions which have not been studied in the sliding window model before, such as distinct counts, entropy, and geometric properties in Section 6.

Other Related Work. Much of the previous work relies on monotonic properties of the function being monitored to provide cost guarantees. For example, since a count (over an infinite window) is always growing, the cost of most approximate tracking algorithms grows only logarithmically with the number of updates [7]. But the adoption of a time-based sliding window can make a

---

[1] Here, words is shorthand for machine words, in the standard RAM model.

previously monotonic function non-monotonic. That is, a function which is monotonic over an infinite window (such as a count) can decrease over a time-based window, due to the implicit deletions. Sharfman *et al.* [19] gave a generic method for arbitrary functions, based on a geometric view of the input space. This approach relies on keeping full space at each monitoring site, and does not obviously extend to functions which do not map on to single values (such as heavy hitters and quantiles). Arackaparambil *et al.* [1] study (empirical) entropy, which is non-monotonic. The protocols rely on a slow changing property of entropy: a constant change in the value requires a constant factor increase in the number of observations, which keeps the communication cost logarithmic in the size of the input. This slow-changing property does not hold for general functions. Distributed sliding window computations have also received much attention in the non-continuous-tracking case [3,11], where the goal is to keep a small amount of information over the stream at each site, so that the desired aggregate can be computed upon request; here, we have the additional challenge of tracking the aggregate at all times with small communication.

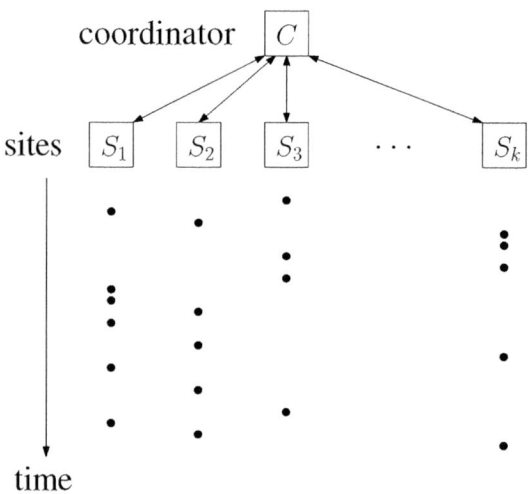

Fig. 1. Schematic of the distribute streaming model

1.1   Problem Definitions and Our Results

Now we more formally define the problems studied in this paper. Figure 1 shows the model schematically: k sites each observe a stream S_i of item arrivals, and communicate with a single distinguished coordinator node to continuously compute some function of the union of the update streams.

The *basic counting* problem is to track (approximately) the number of items which have arrived across all sites within the last w time units. More precisely, let the stream of items observed at site i be S_i, a set of $(x, t(x))$ pairs, which indicates that an item x arrives at time $t(x)$. Then the exact basic count at time t is given by

$$C(t) = \sum_{1 \leq i \leq k} |\{(x, t(x)) \in S_i \mid t - t(x) \leq w\}|.$$

Tracking $C(t)$ exactly requires alerting the coordinator every time an item arrives or expires, so the goal is to track $C(t)$ approximately within an ε-error, i.e., the coordinator should maintain a $\tilde{C}(t)$ such that $(1 - \varepsilon)C(t) \leq \tilde{C}(t) \leq (1 + \varepsilon)C(t)$ at all times t. We will assume that at each site, at most one item arrives in one time unit. This is not a restriction, since we can always subdivide the time units into smaller pieces so that at most one item arrives within one unit. This rescales w but does not fundamentally change our results, since the bounds provided do not depend on w.

The *heavy hitters* problem extends the basic counting problem, and generalizes the concept of finding the mode [15]. In the basic counting problem we count the total number of all items, while here we count the frequency of every distinct item x, i.e., the coordinator tracks the approximate value of

$$n_x(t) = \sum_{1 \leq i \leq k} |(x, t(x)) \in S_i \mid t - t(x) \leq w\}|.$$

Since it is possible that many $n_x(t)$ are small, say 0 or 1 for all x, requiring a multiplicative approximation for all x would require reporting all items to the coordinator. Consequently, the commonly adopted approximation guarantee for heavy hitters is to maintain a $\tilde{n}_x(t)$ that has an additive error of at most $\varepsilon C(t)$, where $C(t)$ is the total count of all items. This essentially makes sure that the "heavy" items are counted accurately while compromising on the accuracy for the less frequent items. In particular, all items x with $n_x(t) \leq \varepsilon C(t)$ can be ignored altogether as 0 is considered a good approximation for their counts.[2] This way, at most $1/\varepsilon$ distinct items will have nonzero approximated counts.

The *quantiles* problem is to continuously maintain approximate order statistics on the distribution of the items. That is, the items are drawn from a total order, and we wish to retain a set of items $q_1, \ldots, q_{1/\varepsilon}$ such that the rank of q_i (number of input items within the sliding window that are less than q_i) is between $(i - 1)\varepsilon C(t)$ and $(i + 1)\varepsilon C(t)$ [18]. It is known that this is equivalent to the "prefix-count" problem, where the goal is to maintain a data structure on the sliding window such that for any given x, the number of items smaller than x can be counted within an additive error of at most $\varepsilon C(t)$.

Figure 2 summarizes our main results. The communication cost is measured as the total amount of communication between all k sites and the central co-ordinator site, as a function of n, the number of observations in each window, and ε, the approximation parameter. All communication costs are optimal or near-optimal up to polylogarithmic factors. We also list the space required by each site to run the protocol.

---

[2] We may subsequently drop the (t) notation on variables when it is clear from the context.

Problem	Communication Cost	Communication lower bound	Space Cost
Basic Counting	$O(\frac{k}{\varepsilon}\log(\varepsilon n/k))$ bits	$\Omega(\frac{k}{\varepsilon}\log(\varepsilon n/k))$ bits	$O(\frac{1}{\varepsilon}\log\varepsilon n)$
Heavy Hitters	$O(\frac{k}{\varepsilon}\log(\varepsilon n/k))$	$\Omega(\frac{k}{\varepsilon}\log(\varepsilon n/k))$ bits	$O(\frac{1}{\varepsilon}\log\varepsilon n)$
Quantiles	$O(\frac{k}{\varepsilon}\log^2(1/\varepsilon)\log(n/k))$	$\Omega(\frac{k}{\varepsilon}\log(\varepsilon n/k))$ bits	$O(\frac{1}{\varepsilon}\log^2(1/\varepsilon)\log n)$

Fig. 2. Summary of Results. All bounds are in terms of words unless specified otherwise.

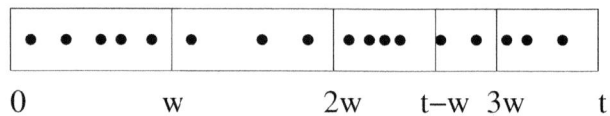

$$0 \qquad\qquad w \qquad\qquad\qquad 2w \qquad t{-}w\ \ 3w \qquad\qquad t$$

Fig. 3. Item arrivals within fixed windows

2   The Forward/Backward Framework

To introduce our framework, we observe that the problems defined in Section 1.1 all tolerate an error of $\varepsilon C(t)$, where $C(t)$ is the total number of items in the sliding window from all k sites. If we can track the desired count for every site within an error of $\varepsilon C^{(i)}(t)$, where $C^{(i)}(t)$ is the number of items at site i in the sliding window, then the total error will be $\sum_{i=1}^{k}\varepsilon C^{(i)}(t) = \varepsilon C(t)$. So we can focus on accurately tracking the data of one site, and combine the results of all sites to get the overall result.

Next, assuming the time axis starts at 0, we divide it into fixed windows of length w: $[0, w)$, $[w, 2w)$, $\ldots$, $[jw, (j + 1)w), \ldots$. Then at at time t, the sliding window $[t - w, t)$ overlaps with at most two of these fixed windows, say $[(j - 1)w, jw)$ and $[jw, (j + 1)w)$. This splits the sliding window into two smaller windows: $[t - w, jw)$ and $[jw, t)$. We call the first one the *expiring window* and the second the *active window*. Figure 3 shows this schematically: item arrivals, shown as dots, are partitioned into fixed windows. At the current time, t, which in this example is between $3w$ and $4w$, it induces the expiring window $[t - w, 3w)$ (with two items in the example) and the active window $[3w, t)$ (with a further three items). As the window $[t - w, t)$ slides, items expire from the expiring window, while new items arrive in the active window. The problem is again decomposed into tracking the desired function in these two windows, respectively. Care must be taken to ensure that the error in the approximated count is with respect to the number of items in the active (or expiring) window, not that of the fixed window. However, a key simplification has happened: now (with respect to the fixed time point jw), the counts of items in the expiring window are only decreasing, while the counts of items in the active window are only increasing. As a result we make an (informal) claim about the problem:

Claim. For tracking a function in the sliding window continuous monitoring setting, the asymptotic communication complexity per window is that of the infinite window case.

To see this, observe that, using the above simplification, we now face two sub-problems: (i) forward: tracking the active window and (ii) backward: tracking the expiring window. Tracking the active window is essentially the same as the infinite window case, hence the cost (per window) is that of running a protocol for the infinite window case. However, for the expiring window we also face essentially the same problem: we need a protocol which ensures that the coordinator always knows a good approximation to the function for the window as items expire instead of arrive. When we view this in the reverse time direction, the expirations become arrivals. If we ran the infinite window protocol on this time-reversed input, we would meet the requirements for approximating the function. Therefore, we can take the messages that are sent by this protocol, and send them all to the coordinator in one batch at the end of each fixed window. The site can send a bit to the coordinator at each time step when it would have sent the next message (in the time-reversed setting). Thus, the coordinator always has the state in the forward direction that it would have had in the time-reversed direction. □

This outline requires the site to record the stream within each window so it can perform this time-reversal trick. The problem gets more challenging if we also require small space at the site. For this we adapt small-space sliding window algorithms from the streaming literature to compactly "encode" the history. Next we show how to instantiate the forward/backward framework in a space efficient way for each of the three problems defined earlier. The forward problem (i.e., the full stream case) has been studied in prior work (for example, [20] gave results for heavy hitters and quantiles), but we are able to present simpler algorithms here. The lower bounds for the forward problem apply to the forward/backward case, and so we are able to confirm that our solutions are optimal or near-optimal.

3   Warm-Up: Basic Counting

The Forward Problem. For basic counting, the forward problem is to track the number of items that have arrived since a *landmark* t_0, up to a multiplicative error of $(1+\varepsilon)$. This is straightforward: the site simply sends a bit every time this number has increased by a $(1 + \varepsilon)$ factor. This results in a communication cost of $O(1/\varepsilon \cdot \log n^{(i)})$ bits, where $n^{(i)}$ is the number of items in the fixed window at the site i when this forward tracking problem takes place. This can be tightened to $O(1/\varepsilon \cdot \log(\varepsilon n^{(i)}))$ by observing that the site actually sends out $1/\varepsilon$ bits for the first $1/\varepsilon$ items. Summing over all k sites and using the concavity of the log function gives $O(\frac{k}{\varepsilon} \log \frac{\varepsilon n}{k})$, matching the bound of [4]. The space required is minimal, just $O(1)$ for each site to track the current count in the active window.

The Backward Problem. As noted above, if we can buffer all the input, then we can solve the problem by storing the input, and compute messages based on item expires. To solve the backward problem with small space (without buffering the current window) is more challenging. To do so, we make use of the *exponential histogram* introduced in [9]. Let the active window be $[t_0, t)$. Besides running the forward algorithm from t_0, each site also maintains an exponential histogram starting from t_0. It records the ε^{-1} most recent items (and their timestamps), then every other item for another stored ε^{-1} items, then every fourth item, and so on. This is easily maintained as new items arrive: when there are more than $\varepsilon^{-1} + 1$ items at a particular granularity, the oldest two can be "merged" into a single item at the next coarser granularity. Let t be the current time. When $t = t_0 + w$, the site freezes the exponential histogram. At this time, we set $t_0 \leftarrow t_0 + w$, and the active window becomes the expiring window while a new active window starts afresh. It follows from this description that the size of the exponential histogram is $O(\varepsilon^{-1} \log(\varepsilon n^{(i)}))$.

With an exponential histogram for the window $[t_0 - w, t_0)$, one can approximately count the items in the interval $[t - w, t_0)$, i.e., the expiring window at time t. We find in the exponential histogram two adjacent timestamps t_1, t_2 such that $t_1 < t - w \leq t_2$. Note that from the data structure we can compute the number of items in the time interval $[t_2, t_0)$ exactly, which we use as an estimate for the number of items in $[t - w, t_0)$. This in the worst case will miss all items between t_1 and t_2, and there are 2^a of them for some a. The construction of the exponential histogram ensures that $C_i(t_2, t_0) \geq \varepsilon^{-1} 2^a$, where $C_i(t_2, t_0)$ denotes the number of items that arrived between time t_2 and t_0. So the error is at most $\varepsilon C_i(t_2, t_0) \leq \varepsilon C_i(t - w, t_0)$, as desired.

There are two ways to use the exponential histogram in a protocol. Most directly, each site can send its exponential histogram summarizing $[t_0 - w, t_0)$ to the coordinator at time t_0. From these, the coordinator can approximate the total count of the expiring window accurately. However, this requires the coordinator to store all k windows, and is not communication optimal. Instead, the space and communication cost can be reduced by having each site retain its exponential histogram locally. At time t_0, each site informs the coordinator of the total number of timestamps stored in its histogram of $[t_0 - w, t_0)$. Then each site sends a bit to the coordinator whenever any timestamp recorded in the histogram expires (i.e., reaches age w). This information is sufficient for the coordinator to recreate the current approximate count of the expiring window for each site. The communication cost is the same as the forward case, i.e., $O(1/\varepsilon \cdot \log(\varepsilon n^{(i)}))$ bits.

Theorem 1. *The above protocol for basic counting has a total communication cost of $O(\frac{1}{\varepsilon} \log(\varepsilon n^{(i)}))$ bits for each site, implying a total communication cost of $O(\frac{k}{\varepsilon} \log \frac{\varepsilon n}{k})$ per window. The space required at each site is $O(\frac{1}{\varepsilon} \log(\varepsilon n^{(i)}))$ words, and $O(k)$ at the coordinator to keep track of the current estimates from each site.*

This bound is optimal: the lower bound for an infinite window is $\Omega(\frac{k}{\varepsilon} \log \frac{\varepsilon n}{k})$ bits of communication. We see the power of the forward/backward framework: the analysis matches the bound in [4], but is much more direct. The dependence on $O(\log n)$ is unavoidable, as shown by the lower bounds. However, note that we do not require explicit knowledge of n (or an upper bound on it). Rather, the communication used, and the space requires, scales automatically with $\log n$, as the stream unfolds.

Algorithm 1. HEAVYHITTERSARRIVALS

---

1 $\forall x, n_x^{(i)} = 0$;

2 $A^{(i)} = 1$;

3 **foreach** *arrival of* x **do**

4 $\quad$ $n_x^{(i)} \leftarrow n_x^{(i)} + 1$;

5 $\quad$ $n^{(i)} \leftarrow n^{(i)} + 1$;

6 $\quad$ **if** $n_x^{(i)} \bmod A^{(i)} = 0$ **then**

7 $\qquad$ $\lfloor$ Send $(x, n_x^{(i)})$ to coordinator;

8 $\quad$ **if** $n^{(i)} \geq 2\varepsilon^{-1} A^{(i)}$ **then**

9 $\qquad$ $\lfloor$ $A^{(i)} \leftarrow 2A^{(i)}$;

---

4 Heavy Hitters

The Forward Problem. For simplicity, we first present an algorithm for the forward problem that assumes that each site has sufficient space to retain all "active" (non-expired) items locally. Then we discuss how to implement the algorithm in less space below.

Starting from time t_0, each site executes HEAVYHITTERARRIVALS (Algorithm 1) on the newly arriving items until $t = t_0 + w$. This tracks counts of each item in the active window ($n_x^{(i)}$ for the count of item x at site i), and ensures that the coordinator knows the identity and approximate count of any item with an additive error of at most $A^{(i)} - 1$. Note that $A^{(i)}$ is always at most $\varepsilon n^{(i)}$, where $n^{(i)}$ is the total number of items in the active window at site i, so correctness follows easily from the algorithm.

We next bound the communication cost. While $n^{(i)}$ is between $2^a \varepsilon^{-1}$ and $2^{a+1} \varepsilon^{-1}$, $A^{(i)} = 2^a$. For each distinct x, line 7 of the algorithm is called whenever $A^{(i)}$ new copies of x have arrived, except possibly the first call. Ignoring the first call to every distinct x, line 7 is executed at most $2^a \varepsilon^{-1}/A^{(i)} = \varepsilon^{-1}$ times. Note that for an item x to trigger line 7 at least once, $n_x^{(i)}$ has to be at least $A^{(i)}$, and there are at most $2^{a+1} \varepsilon^{-1}/A^{(i)} = O(\varepsilon^{-1})$ such distinct items, so the number of first calls is at most $O(\varepsilon^{-1})$. Hence, the total amount of information sent during this phase of the algorithm is $O(\varepsilon^{-1})$ items and counts. In total, there are $O(\log(\varepsilon n^{(i)}))$ such phases corresponding to the doubling values of $A^{(i)}$, and the total communication cost is $O(\varepsilon^{-1} \log(\varepsilon n^{(i)}))$. Summed over all sites the protocol costs $O(k\varepsilon^{-1} \log(\varepsilon n/k))$ per window.

The Backward Problem. In the case where we can afford to retain all the stream arrivals during the current window, we use a similar algorithm to solve the backward problem. Each site executes HEAVYHITTEREXPIRIES in parallel on the expiring window (Algorithm 2). Conceptually, it is similar to running the previous algorithm in reverse. It maintains a parameter $B^{(i)}$ which denotes the local tolerance for error. The initial value of $B^{(i)}$ is equivalent to the final value

---

Algorithm 2. HEAVYHITTERSEXPIRIES

---

1 Send $n^{(i)}$ to the coordinator;

2 $B^{(i)} \leftarrow 2^{\lfloor \log \varepsilon n^{(i)} \rfloor}$;

3 **while** $n^{(i)} > 0$ **do**

4   **foreach** x **do**

5   **if** $n_x^{(i)} \geq B^{(i)}$ **then** send $x, n_x^{(i)}$

6   **while** $n_x^{(i)} > \varepsilon^{-1} B^{(i)}$ *or* $(B^{(i)} \leq 1$ *and* $n^{(i)} > 0)$ **do**

7   **foreach** *expiry of* x **do**

8              $n^{(i)} \leftarrow n^{(i)} - 1$;

9              $n_x^{(i)} \leftarrow n_x^{(i)} - 1$;

10   **if** $n_x^{(i)} \bmod B^{(i)} = 0$ **then**

11   Send $x, n_x^{(i)}$ to the coordinator

12      $B^{(i)} \leftarrow B^{(i)}/2$;

---

of $A^{(i)}$ from the active window which has just concluded. Letting $n^{(i)}$ denote the number of items from $[t - w, t_0)$ (i.e., those from the expiring window which have not yet expired), $B^{(i)}$ remains in the range $[\frac{1}{2}\varepsilon n^{(i)}, \varepsilon n^{(i)}]$. Whenever $B^{(i)}$ is updated by halving, the algorithm sends all items and counts where the local count is at least $B^{(i)}$. Since $B^{(i)}$ is $O(\varepsilon n)$, there are $O(\varepsilon^{-1})$ such items to send. As in the forward case, these guarantees ensure that the accuracy requirements are met.

The communication cost is bounded similarly to the forward case. There are $\log(\varepsilon n^{(i)})$ iterations of the outer loop until $B^{(i)}$ reaches 1. In each iteration, there are $O(\frac{1}{\varepsilon})$ items sent in line 5 that exceed $B^{(i)}$. Then at most $O(\frac{1}{\varepsilon})$ updates can be sent by line 11 before $B^{(i)}$ decreases. When $B^{(i)}$ reaches 1, there are only $1/\varepsilon$ unexpired items, and information on these is sent as each instance expires. This gives a total cost of $O(\frac{1}{\varepsilon} \log(\varepsilon n^{(i)}))$, which is $O(\frac{k}{\varepsilon} \log(\varepsilon n))$ when summed over all k sites.

At any time, the coordinator has information about a subset of items from each site (from both the active and expiring windows). To estimate the count of any item, it adds all the current counts for that item together. The error bounds ensure that the total error for this count is at most εn. To extract the heavy hitters, the coordinator can compare the estimated counts to the current (estimated) value of n, computed by a parallel invocation of the above basic counting protocol.

Reducing the Space Needed at Each Site. To reduce space used at the site for the forward problem, it suffices to replace the exact tracking of all arriving items with a small space algorithm to approximate the counts of items. For example, the SpaceSaving algorithm [17] tracks $O(1/\varepsilon)$ items and counts, so that item frequencies are reported with error at most $\varepsilon n^{(i)}$. This adds another $\varepsilon n^{(i)}$ to the error at the coordinator side, making it $2\varepsilon n^{(i)}$, but a rescaling of ε suffices to bring it back to $\varepsilon n^{(i)}$. The communication cost does not alter: by the guarantee of the algorithm, there can still be only $O(\varepsilon^{-1})$ items whose approximate count exceeds $A^{(i)}$. While these items exceed $A^{(i)}$, their approximate counts are monotone increasing, so the number of messages does not increase.

For the backward part, the details are slightly more involved. We require an algorithm for tracking approximate counts within a window of the last w time steps with accuracy $\varepsilon n^{(i)}$. For each site locally, the data structure of Lee and Ting can track details of the last W arrivals (for a fixed parameter W) using $O(\varepsilon^{-1})$ space [16][3]. We begin the (active) window by instantiating such a data structure for $W = 2(\varepsilon/3)^{-1}$. After we have observed $n^{(i)} = 2^a$ items, we also instantiate a data structure for $W = 2^a(\varepsilon/3)^{-1}$ items, and run it for the remainder of the window: the omitted $n^{(i)} = O(\varepsilon W)$ items can be ignored without affecting the $O(\varepsilon W)$ error guarantee of the structure. Over the life of the window, $O(\log n^{(i)}/\varepsilon) = O(\log n^{(i)})$ (since $n^{(i)} > \varepsilon^{-1}$) such data structures will be instantiated. When the window is expiring, during the phases where $n^{(i)}$ is in the range $2^a(\varepsilon/3)^{-1}\ldots2^{a-1}(\varepsilon/3)^{-1}$, the local site uses the instance of the data structure to monitor items and approximate counts, accurate to $\varepsilon n^{(i)}/3$. The structure allows the identification of a set of items which are frequent when this range begins (lines 4-5 in Algorithm 2). The structure also indicates how their approximate counts decrease as old instances of the items expire, and so when their (approximate) counts have decreased sufficiently, the coordinator can be updated (line 11). In this case, the estimated counts are monotone decreasing, so the communication cost does not alter. The space at each site is therefore dominated by the cost of these data structures, which is $O(\varepsilon^{-1}\log n^{(i)})$.

Theorem 2. *The above protocol for heavy hitters has a total communication cost of $O(\frac{k}{\varepsilon}\log\frac{\varepsilon n}{k})$ words per window. The space required at each site is $O(\frac{1}{\varepsilon}\log n^{(i)})$, and $O(\frac{k}{\varepsilon})$ at the coordinator to keep track of the current heavy hitters from each site.*

This protocol is optimal: it meets the communication lower bound for this problem stated in [4], and improves the upper bound therein. It similarly improves over the bound for the infinite window case in [20].

5   Quantiles

In this section, we study the problem of tracking the set of quantiles for a distribution in the sliding window model. Yi and Zhang [20] study this problem in the infinite window model, and provide a protocol with communication cost $O(\frac{k}{\varepsilon}\log n\log^2\frac{1}{\varepsilon})$. As a byproduct, our solution slightly improves on this. The improvement over the best known solution for the sliding window model is more substantial.

In order to achieve small space and communication, we make use of the data structure of Arasu and Manku [2], referred to as the AM structure. The AM

---

[3] The λ-counter data structure defined therein can be extended to store timestamps in addition to items, which makes it sufficient for our purpose.

structure stores the ε-approximate quantiles over a sequence of W items, for W fixed in advance. The W items are divided along the time axis into blocks of size εW, and summaries are built of the blocks. Specifically, at level 0, an ε_0-approximate quantile summary (of size $1/\varepsilon_0$) is built for each block, for some ε_0 to be determined later. An ε-approximate quantile summary for a set of m items can be computed by simply storing every tth item in the sorted order, for $t = \varepsilon m$: from this, the absolute error in the rank of any item is at most t. Similarly, summaries are built for levels $\ell = 1, \ldots, \log(1/\varepsilon)$ with parameter ε_ℓ by successively pairing blocks in a binary tree: level ℓ groups the items into blocks of $2^\ell \varepsilon W$ items. Using this block structure, any time interval can be decomposed into $O(\log(1/\varepsilon))$ blocks, at most two from each level, plus two level-0 blocks at the boundaries that partially overlap with the interval. Ignoring these two boundary level-0 blocks introduces an error of $O(\varepsilon W)$. The blocks at level ℓ contribute an uncertainty in rank of $\varepsilon_\ell 2^\ell \varepsilon W$. Hence if we choose $\varepsilon_\ell = 1/(2^\ell \log(1/\varepsilon))$, the total error summed over all $L = \log(1/\varepsilon)$ levels is $O(\varepsilon W)$.

The total size of the structure is $\sum_\ell (1/\varepsilon_\ell \cdot 1/(2^\ell \varepsilon)) = O(1/\varepsilon \cdot \log^2(1/\varepsilon))$. Rather than explicitly storing all items in each block and sorting them to extract the summary, we can instead process the items in small space and a single pass using the GK algorithm [12] to summarize the active blocks. This algorithm needs space $O(1/\varepsilon_\ell \cdot \log \varepsilon^2 2^\ell W)$ for level ℓ. When a block completes, i.e., has received $2^\ell \varepsilon W$ items, we produce a compact quantile summary of size $1/\varepsilon_\ell$ for the block using the GK summary, and discard the GK summary. The space required is dominated by the GK algorithm at level $\ell = \log(1/\varepsilon)$, which is $O(1/\varepsilon \cdot \log(1/\varepsilon) \log(\varepsilon W))$.

The Forward Problem. Recall that in the forward problem, the coordinator needs to estimate ranks of items from site i with error at most $\varepsilon n^{(i)}$, where $n^{(i)}$ is the current number of items received since time t_0. To achieve this, we build multiple instances of the above structure for different values of W. When the $n^{(i)}$-th item arrives for which $n^{(i)}$ is a power of 2, the site starts building an AM structure with $W = n^{(i)}/\varepsilon$. Whenever a block from any level of any of the AM structures completes, the site sends it to the coordinator. This causes communication $O(1/\varepsilon \cdot \log^2(1/\varepsilon) \log n^{(i)})$ for the entire active window (the communication is slightly less than the space used, since only summaries of complete blocks are sent). After the $n^{(i)}$-th item has arrived, all AM structures with $W < n^{(i)}$ can be discarded.

We argue this is sufficient for the coordinator to track the quantiles of the active window at any time. Indeed, when the $n^{(i)}$-th item arrives, the site has already started building an AM structure with some W that is between $n^{(i)}$ and $2n^{(i)}$,[4] and the completed portion has been communicated to the coordinator. This structure gives us quantiles with error $O(\varepsilon W) = O(\varepsilon n^{(i)})$, as desired. Note that the site only started building the structure after εW items passed, but ignoring these contributes error at most εW.

---

[4] The special case $n^{(i)} \le 1/\varepsilon$ is handled by simply recording the first $1/\varepsilon$ items exactly.

The Backward Problem. To solve the backward problem, we need a series of AM structures on the last W items of a fixed window so that we can extract quantiles when this fixed window becomes the expiring window. Fortunately the AM structure can be maintained easily so that it always tracks the last W items. Again for each level, new items are added to the latest (partial) block using the GK algorithm [12]; when all items of the oldest block are outside the last W, we remove the block.

For the current fixed window starting from t_0, we build a series of AM structures, as in the forward case. The difference is that after an AM structure is completed, we continue to slide it so as to track the last W items. This remains private to the site, so there is no communication until the current active window concludes. At this point, we have a collection of $O(\log(n^{(i)}))$ AM structures maintaining the last W items in the window for exponentially increasing W's. Then the site sends the summaries for windows of size between ε^{-1} and $2n^{(i)}$ to the coordinator, and the communication cost is the same as in the forward case. To maintain the quantiles at any time t, the coordinator finds the smallest AM structure (in terms of coverage of time) that covers the expiring window $[t - w, t_0)$, and queries that structure with the time interval $[t - w, t_0)$. This will give us quantiles with error $O(\varepsilon W) = O(\varepsilon C_i(t - w, t_0))$ since $W \leq 2C_i(t - w, t_0)$.

Theorem 3. *The above protocol for quantiles has a total communication cost of $O(k/\varepsilon \log^2(1/\varepsilon) \log(n/k))$ words per window. The space required at each site is $O(\frac{1}{\varepsilon} \log^2(1/\varepsilon) \log(\varepsilon n^{(i)}))$, and $O(\frac{k}{\varepsilon} \log^2(1/\varepsilon) \log(\varepsilon n/k))$ at the coordinator to keep copies of all the data structures.*

Yi and Zhang [20] show a lower bound of $\Omega(\frac{k}{\varepsilon} \log \frac{\varepsilon n}{k})$ messages of communication (for the infinite window case), so our communication cost is near-optimal up to polylogarithmic factors. Meanwhile, [4] provided an $O(k/\varepsilon^2 \cdot \log(n/k))$ cost solution, so our protocol represents an asymptotic improvement by a factor of $O(\frac{1}{\varepsilon \log^2(1/\varepsilon)})$. We leave it open to further improve this bound: removing at least one $\log(1/\varepsilon)$ term seems feasible but involved, and is beyond the scope of this paper.

6   Other Functions

The problems discussed so far have the nice property that we can separately consider the monitored function for each site, and use the additivity properties of the function to obtain the result for the overall function. We now discuss some more general functions that can also be monitored under the same model.

Distinct Counts. Given a universe of possible items, the distinct counts problem asks to find the number of items present within the sliding window (counting duplicated items only once). The summary data structure of Gibbons and Tirthapura can solve this problem for a stream of items under the infinite window semantics [10]. A hash function maps each item to a level, such that the probability of being mapped to level j is geometrically decreasing. The algorithm tracks the set of items mapped to each level, until $O(1/\varepsilon^2)$ distinct items have been seen there, at which point the level is declared "full". Then the algorithm uses the number of distinct items present in the first non-full level to estimate the overall number of distinct items.

This leads to a simple solution for the active window: each site independently maintains an instance of the data structure for the window, and runs the algorithm. Each update to the data structure is echoed to the coordinator, ensuring that the coordinator has a copy of the structure. The communication cost is bounded by $O(1/\varepsilon^2 \log n^{(i)})$. The coordinator can merge these summaries together in the natural way (by retaining the set of items mapped to the same level from all sites) to get the summary of the union of all streams. This summary therefore accurately summarizes the distinct items across all sites.

The solution for the expiring window is similar. Each site retains for each level the $O(1/\varepsilon^2)$ most recent distinct arrivals that were mapped to that level, along with their timestamps. This information enables the distinct count for any suffix of the window to be approximated. This data structure can then be shared with the coordinator, who can again merge the data structures straightforwardly. The total communication required is $O(\frac{k}{\varepsilon^2} \log \frac{n}{k})$ over all k sites.

Entropy. The (empirical) entropy of a frequency distribution is $\sum_j \frac{f_j}{n} \log \frac{n}{f_j}$, where f_j denotes the frequency of the jth token. As more items arrive, the entropy does not necessarily vary in a monotone fashion. However, the amount by which it can vary is bounded based on the number of arriving items: specifically, for m new arrivals after n current arrivals, it can change by at most $\frac{m}{n} \log(2n)$ [1]. This leads to a simple protocol for the forward window case to track the entropy up to additive error ε: given n current items, each site waits to see $\frac{\varepsilon n}{\log(2n)}$ new arrivals, then communicates its current frequency distribution to the coordinator. Within a window of n_i arrivals, there are at most $O(\frac{1}{\varepsilon} \log^2 n_i)$ communications.

For the backward case, the protocol has to send the frequency distribution when the number of items remaining reaches various values. This can be arranged by use of the exponential histogram outlined in Section 3: for each timestamp stored, it keeps the frequency distribution associated with that timestamp. When a timestamp is "merged", and dropped, the corresponding distribution is dropped also. Thus, the space required is $O(\frac{1}{\varepsilon} \log(\varepsilon n^{(i)}))$ entries in the histogram. The histogram introduces some uncertainty into the number of items remaining, but after rescaling of parameters, this does not affect the correctness.

When the domain is large, the size of the frequency distributions which must be stored and sent may dominate the costs. In this case, we replace the exact distributions with compact *sketches* of size $\tilde{O}(\frac{1}{\varepsilon^2})$ [13]. The coordinator can combine the sketches from each site to obtain a sketch of the overall distribution, from which the entropy can be approximated.

Geometric Extents: Spread, Diameter and convex Hull. Given a set of points in one dimension p_i, their *spread* is given by $(\max_i p_i - \min_i p_i)$. The forward case is easy: send the first two points to the coordinator, then every time a new point causes the spread to increase by a $1 + \varepsilon/2$ factor, send the new point to the coordinator. This ensures that spread is always maintained up to a $(1 + \varepsilon)$ factor, and the communication is $O(\frac{1}{\varepsilon} \log R)$, where R is the ratio between the closest and furthest points in the input, a standard factor in computational geometry. For the backward case, we can use the algorithm of Chan and Sadjad [5] to build a summary of size $O(\frac{1}{\varepsilon} \log R)$ as the points arrive in the active window, and communicate this to the coordinator to use for the expiring window. Lastly, observe that spread of the union of all points can be approximated accurately from the union of points defining the (approximate) spread for each site.

The diameter of a point set in two (or higher) dimensions is the maximum spread of the point set when projected onto any line. A standard technique is to pick $O(1/\varepsilon^{(d-1)/2})$ uniformly spaced directions in d dimensions, and project each input point onto all of these directions: this preserves the diameter up to a $(1 + \varepsilon)$ factor, since there is some line which is almost parallel to the line achieving the diameter. This immediately gives a protocol for diameter with cost $O(\frac{1}{\varepsilon}^{(d+1)/2} \log R)$, by running the above protocol for each of the $O(1/\varepsilon^{(d-1)/2})$ directions in parallel. A similar approach also maintains the approximate convex hull of the point set, by observing that the convex hull of the maximal points in each direction is approximately the convex hull of all points.

7   Concluding Remarks

The forward/backward framework allows a variety of functions to be monitored effectively within a sliding window, and improves over the results in prior work [4]. The underlying reason for the complexity of the the analysis of the protocols previously proposed is that they focus on the current count of items at each site. This count rises (due to new arrivals) and falls (due to expiry of old items). Capturing this behavior for heavy hitters and quantiles requires the analysis of many cases: when an item becomes frequent through arrivals; when an item becomes frequent because the local count has decreased; when an item becomes infrequent due to arrivals of other items; when an item becomes infrequent due to expiry; and so on. By separating streams into streams of only arrivals and streams of only expirations, we reduce the number of cases to consider, and remove any interactions between them. Instead, we just have to track the function for two different cases. This allowed a much cleaner treatment of this problem, and opens the door for similar analysis of other monitoring problems.

References

1. Arackaparambil, C., Brody, J., Chakrabarti, A.: Functional Monitoring without Monotonicity. In: Albers, S., Marchetti-Spaccamela, A., Matias, Y., Nikoletseas, S., Thomas, W. (eds.) ICALP 2009. LNCS, vol. 5555, pp. 95–106. Springer, Heidelberg (2009)
2. Arasu, A., Manku, G.S.: Approximate counts and quantiles over sliding windows. In: ACM Principles of Database Systems (2004)

3. Busch, C., Tirthapura, S., Xu, B.: Sketching asynchronous streams over sliding windows. In: ACM Conference on Principles of Distributed Computing (PODC) (2006)
4. Chan, H.-L., Lam, T.-W., Lee, L.-K., Ting, H.-F.: Continuous monitoring of distributed data streams over a time-based sliding window. In: Symposium on Theoretical Aspects of Computer Science, STACS (2010)
5. Chan, T.M., Sadjad, B.S.: Geometric Optimization Problems Over Sliding Windows. In: Fleischer, R., Trippen, G. (eds.) ISAAC 2004. LNCS, vol. 3341, pp. 246–258. Springer, Heidelberg (2004)
6. Cormode, G.: Continuous distributed monitoring: A short survey. In: Algorithms and Models for Distributed Event Processing, AlMoDEP (2011)
7. Cormode, G., Muthukrishnan, S., Yi, K.: Algorithms for distributed, functional monitoring. In: ACM-SIAM Symposium on Discrete Algorithms (2008)
8. Cormode, G., Muthukrishnan, S., Yi, K., Zhang, Q.: Optimal sampling from distributed streams. In: ACM Principles of Database Systems (2010)
9. Datar, M., Gionis, A., Indyk, P., Motwani, R.: Maintaining stream statistics over sliding windows. In: ACM-SIAM Symposium on Discrete Algorithms (2002)
10. Gibbons, P., Tirthapura, S.: Estimating simple functions on the union of data streams. In: ACM Symposium on Parallel Algorithms and Architectures (SPAA), pp. 281–290 (2001)
11. Gibbons, P., Tirthapura, S.: Distributed streams algorithms for sliding windows. In: ACM Symposium on Parallel Algorithms and Architectures (SPAA) (2002)
12. Greenwald, M., Khanna, S.: Space-efficient online computation of quantile summaries. In: ACM SIGMOD International Conference on Management of Data (2001)
13. Harvey, N.J.A., Nelson, J., Onak, K.: Sketching and streaming entropy via approximation theory. In: IEEE Conference on Foundations of Computer Science (2008)
14. Keralapura, R., Cormode, G., Ramamirtham, J.: Communication-efficient distributed monitoring of thresholded counts. In: ACM SIGMOD International Conference on Management of Data (2006)
15. Kuhn, F., Locher, T., Schmid, S.: Distributed computation of the mode. In: ACM Conference on Principles of Distributed Computing (PODC), pp. 15–24 (2008)
16. Lee, L., Ting, H.: A simpler and more efficient deterministic scheme for finding frequent items over sliding windows. In: ACM Principles of Database Systems (2006)
17. Metwally, A., Agrawal, D., Abbadi, A.E.: Efficient computation of frequent and top-k elements in data streams. In: International Conference on Database Theory (2005)
18. Patt-Shamir, B.: A note on efficient aggregate queries in sensor networks. In: ACM Conference on Principles of Distributed Computing (PODC), pp. 283–289 (2004)
19. Sharfman, I., Schuster, A., Keren, D.: A geometric approach to monitoring threshold functions over distributed data streams. In: ACM SIGMOD International Conference on Management of Data (2006)
20. Yi, K., Zhang, Q.: Optimal tracking of distributed heavy hitters and quantiles. In: ACM Principles of Database Systems, pp. 167–174 (2009)

Hinging Hyperplane Models
for Multiple Predicted Variables

Anca Maria Ivanescu, Philipp Kranen, and Thomas Seidl

RWTH Aachen University, Germany
{lastname}@cs.rwth-aachen.de

Abstract. Model-based learning for predicting continuous values involves building an explicit generalization of the training data. Simple linear regression and piecewise linear regression techniques are well suited for this task, because, unlike neural networks, they yield an interpretable model. The hinging hyperplane approach is a nonlinear learning technique which computes a continuous model. It consists of linear submodels over individual partitions in the regressor space. However, it is only designed for one predicted variable. In the case of r predicted variables the number of partitions grows quickly with r and the result is no longer being compact or interpretable.

We propose a generalization of the hinging hyperplane approach for several predicted variables. The algorithm considers all predicted variables simultaneously. It enforces common hinges, while at the same time restoring the continuity of the resulting functions. The model complexity no longer depends on the number of predicted variables, remaining compact and interpretable.

1 Introduction

Researchers from various fields (e.g. engineering, chemistry, biology) deal with experimental measurements in form of points. For example in sensor network applications, measurements are performed at spatially and temporally discrete positions. In engineering sciences test benches are build and measurements performed at discrete operating points. This data representation though is incomplete since it is only a discretization of the underlying process. For understanding this set of measurement points, researchers model these points by mathematical functions. These models offer a compact and intuitive representation of the underlying process. Hence sensor measurements can be predicted for spatial and temporal points where no measurements were performed, and measurements predicted at operating points which were not tested. Database systems like FunctionDB [14] and MauveDB [6] embed an algebraic query processor, which allows the user to handle and query data in form of mathematical functions.

The problem we address in this paper is that of learning a generalized mathematical model from a given set of data. The goal is to obtain a compact, intelligible description of the data and to accurately predict continuous values. Often linear models are not sufficient to accurately describe the data. In this case

A. Ailamaki and S. Bowers (Eds.): SSDBM 2012, LNCS 7338, pp. 431–448, 2012.
© Springer-Verlag Berlin Heidelberg 2012

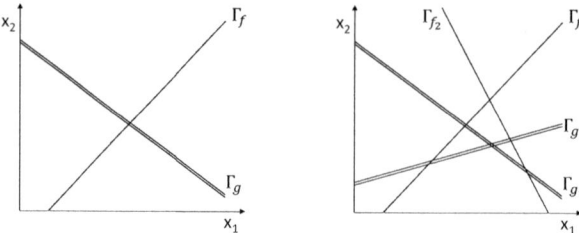

Fig. 1. Increasing number of partitions

nonlinear models are used. We focus our attention on approximating nonlinear data with several linear functions. An arbitrarily high approximation accuracy is obtained at the expense of a sufficiently large number of linear submodels.

Having the task of describing the output as a linear or piecewise linear function of the input, the main challenge is to partition the input space such that the corresponding fitted linear functions have a high approximation accuracy. Both regression trees [7,17] and correlation clustering methods [2,15,1] are well suited for clustering linearly dependent points. However, the resulting piecewise affine functions are not guaranteed to be continuous. This is a severe drawback, since most systems or processes represent continuous dynamics and require the continuity of the piecewise affine functions.

The hinging hyperplanes (HH) model introduced by Breiman in [4] is a technique for continuous approximation of nonlinear data. The HH model is defined as a sum of basis functions, called hinge functions, each consisting of two continuously joined hyperplanes. The contained hinges can also be used to determine splits in the input space for the construction of regression trees. This way an interpretable model of the data is built which offers a better understanding of the underlying process.

The hinge finding algorithm (HFA) is only designed to deal with one dimensional output variables. In the case of r output dimensions r independent models are built, one for each output. Since the input and output dimensions are correlated and describe the same process, our goal is to model their relationship in a single multivariate regression tree. While the leafs of univariate regression trees contain a linear model, the leafs of multivariate regression trees contain several linear models, one for each output. The challenge is to find a partitioning of the input space, such that in each leaf of the resulting regression tree all linear models are as accurate as possible.

The most straight-forward way to construct a multivariate regression tree is to construct for each output a univariate regression tree and combine their partitionings. Note that in this case the number of partitions grows quickly with K, the number of hinges in the HH model, and with r, the number of output variables. Figure 1 illustrates this for a 2-dimensional input space. The left image shows one partitioning generated by Γ_f, and one partitioning generated by Γ_g. Combining the two yields 4 partitions. The right image illustrates a possible result in the case of 2 output dimensions and 2 hinges for each output. For

3 predicted variables 3 hinges can amount to 46 partitions which is in strong contrast to maximally 7 for a shared partitioning. With such a high number of leafs, the resulting multivariate regression tree looses its property of being easily interpretable. Moreover, the increased model size induces performance losses during the algebraic query processing.

We propose the mHFA approach that extends the estimation of HH models to the multivariate case. The output of our approach is a multivariate regression tree that represents a single partitioning of the input space for all predicted variables. Despite the common partitions, the continuity of the hinge functions is guaranteed for all predicted variables. An intuitive and compact model of the underlying process is constructed, which can be used for both gaining a deeper understanding of the process and performing efficient algebraic queries in function-based database systems.

The structure of the paper is as follows: We discuss related work in the following section and provide preliminaries as well as the hinge finding algorithm (HFA) for a single output in Section 3. In Section 4 we present our hinge finding algorithm for multiple outputs (mHFA) which generates a shared partitioning of the input space. Section 5 contains the experimental evaluation and Section 6 concludes the paper.

2   Related Work

Different approaches for approximating a nonlinear function from a set of points by fitting several linear segments were proposed in the literature. Algebraic procedures [16] and bounded-error procedures [3] proved not to be robust against noise and parameter settings [12]. More robust are clustering based procedures [9]. However these approaches do not deliver continuous models and are only designed to deal with 1-dimensional outputs. Correlation clustering algorithms are well suited to cluster linearly correlated points, and are not bounded to any output dimension. A shared partitioning in the input space can be computed, as described in [11]. Still, the resulting models are not guaranteed to be continuous.

Methods for nonlinear regression, classification, and function approximation often use expansions into sums of basis functions. Widely adopted are neural networks [8] which use sigmoidal functions as basis functions. A major drawback is that these produce black box models, while in many applications understanding the model is as important as an accurate prediction. Regression trees, like CART [5] and MARS [10], build interpretable models but are limited to axis parallel partitionings. More flexible are the hinging hyperplanes (HH) introduced by Breiman in [4]. They allow arbitrary oriented partitions in the input space. All these approaches are restricted to deal with one output dimension.

We extend the HH model to deal with the multivariate case by applying hinge finding algorithm for each output in an interleaved manner. Each hinge is conjointly computed to fit all outputs. We thus obtain a single partitioning of the input space.

3   The Hinging Hyperplane Model

We introduce notations and preliminaries in the next section an describe the hinge finding algorithm from [4] in Section 3.2.

3.1   Preliminaries

An $m-1$-dimensional hyperplane in the m-dimensional Euclidean space $\mathbb{R}^m$ in its Hessian normal form is described by a normal vector $n = [n_1, \ldots, n_m]^T \in \mathbb{R}^m$ with $\|n\| = 1$ and the distance $d \in \mathbb{R}$ to the origin as

$$x^T n - d = 0, \tag{1}$$

Alternatively, an $m-1$-dimensional hyperplane can be described by a parametric equation:

$$x_m = a_1 x_1 + \ldots + a_{m-1} x_{m-1} + b, \tag{2}$$

where $\beta = [a_1, \ldots, a_{m-1}, b]^T$ is the parameter vector describing the hyperplane, $\beta \in \mathbb{R}^m$. This form is encountered e.g. in linear regression, where the predicted variable is described as a parametric equation of the predictor variables. One representation can easily be converted into the other:

$$(1) \Rightarrow (2): \quad a_i = -\frac{n_i}{n_m}, i = 1 \ldots m-1, \quad b = \frac{d}{n_m}$$

$$(2) \Rightarrow (1): \quad n_i = a_i, i = 1 \ldots m-1, \quad n_m = -1, \quad d = -b$$

Throughout the paper we consider a data set $\mathcal{D} \subset \mathbb{R}^{m+r}$ where an observation $(x, y) \in \mathcal{D}$ consists of an m-dimensional input vector $x = [x_1, \ldots, x_m]^T$ and an r-dimensional output $y = [y_1, \ldots, y_r]^T$. For the simple case of $r = 1$, linear regression fits a hyperplane to all observations in $\mathcal{D}$. Using the notation $\mathbf{x} = [x, 1]^T$ the hyperplane can be expressed as $\hat{y} = \mathbf{x}^T \beta$, and the residuals as $y_{[1]} = y - \mathbf{x}^T \beta$.

Let $X \in \mathbb{R}^{|\mathcal{D}| \times m}$ be the matrix that contains as rows all input vectors $\mathbf{x}$ and $Y \in \mathbb{R}^{|\mathcal{D}| \times 1}$ the vector that contains all corresponding output values of the observations in $\mathcal{D}$, then the residual sum of squares is

$$\|Y - X\beta\|_2^2 = \sum_{(x,y) \in \mathcal{D}} (y - \mathbf{x}^T \beta)^2.$$

A solution to the unconstrained least squares problem

$$\min_\beta \|Y - X\beta\|_2^2 \tag{3}$$

is found by solving the equation $\beta = (X^T X)^{-1} X^T Y$.

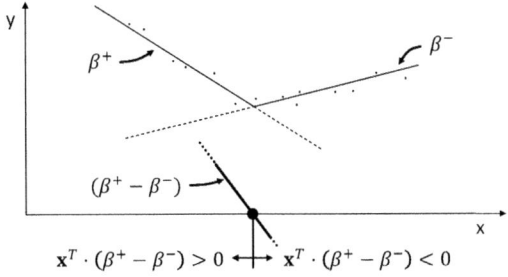

Fig. 2. A set of observations approximated by two regression models of a hinge function

3.2   The Hinge Finding Algorithm for a Single Output

Let $\mathcal{D} \subset \mathbb{R}^{m+1}$ be a data set as defined above. A hinge function $h(x)$ approximates the unknown function $\hat{y} = f(x)$ by two hyperplanes: $\mathbf{x}^T \beta^+$ and $\mathbf{x}^T \beta^-$, that are continuously joined together at their intersection $\{\mathbf{x} : \mathbf{x}^T \Delta = 0\}$, where $\Delta = (\beta^+ - \beta^-)$ is called hinge. The hinge Δ separates the data set in $S^+ = \{\mathbf{x} : \mathbf{x}^T \Delta \geq 0\}$ and $S^- = \{\mathbf{x} : \mathbf{x}^T \Delta < 0\}$. The explicit form of the hinge function $h(x)$ is either $h(x) = \min(\mathbf{x}^T \beta^+, \mathbf{x}^T \beta^-)$ in the case of concave functions, or $h(x) = \max(\mathbf{x}^T \beta^+, \mathbf{x}^T \beta^-)$ in the case of convex functions. Figure 2 shows an example for $h(x) = \max\{\mathbf{x}^T \beta^+, \mathbf{x}^T \beta^-\}$ for a set of observations with a single input.

The hinge finding algorithm (HFA) is described by Breiman in [4] as follows: start with an arbitrary initial hinge $\Delta^{(0)}$ and partition the dataset into the two sets S^+ and S^-. Use least squares regression to fit a hyperplane $\mathbf{x}^T \beta^+$ to the observations in S^+ and another hyperplane $\mathbf{x}^T \beta^-$ to the observations in S^-. The new hinge is $\Delta^{(1)} = \beta^+ - \beta^-$. Update the sets S^+ and S^- and repeat the steps until the hinge function converges.

Pucar et al. proposed in [13] a line search strategy in order to guarantee convergence to a local minimum. In each iteration i a weighted average of the previous hinge $\Delta^{(i-1)}$ and the new suggestion $\beta^+ - \beta^-$ is used. The next hinge $\Delta^{(i)}$ is set to

$$\Delta^{(i)} = \lambda \cdot \Delta^{(i-1)} + (1 - \lambda) \cdot (\beta^+ - \beta^-) \tag{4}$$

with λ being the first value that decreases the residual error in the sequence $\lambda = 2^{-\tau}$ for $\tau = 1, \ldots, \tau_{max}$.

For a better approximation of the observations in $\mathcal{D}$ several hinge functions $h_{[k]}(x)$ can be combined yielding $\hat{y} = \sum_{k=1}^{K} h_{[k]}(x)$. The first hinge $h_{[1]}(x)$ is computed as described above. To compute the second hinge $h_{[2]}(x)$ a temporary data set $\mathcal{D}_{[2]}$ is generated that contains for each observation $(x, y) \in \mathcal{D}$ an observation $(x, y_{[2]})$, where $y_{[2]} = y - h_{[1]}(x)$ is the residuum that is not fitted by $h_{[1]}(x)$. After computing $h_{[2]}(x)$ the first hinge is *refitted* on a temporary data set with output values $y_{[1]} = y - h_{[2]}(x)$. The general procedure for finding K hinges is as follows: Initially $h_{[k]}(x) = 0 \; \forall k = 1 \ldots K$. To compute the k-th hinge

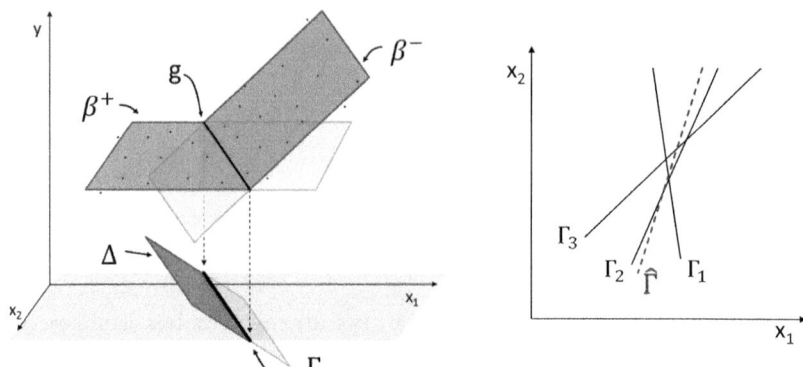

Fig. 3. Left: Two hinging hyperplanes and the corresponding separator Γ. Right: Three separators Γ_1, Γ_2, Γ_3 in the input space and the consensus separator $\hat{\Gamma}$.

a temporary data set

$$\mathcal{D}_{[k]} = \left\{ (x, y_{[k]}) \mid (x,y) \in \mathcal{D} \right\} \quad \text{with} \quad y_{[k]} = y - \sum_{i=1, i \neq k}^{K} h_{[k]}(x) \tag{5}$$

is generated and used in the hinge finding algorithm. After computing a new hinge $h_{[k]}(x)$, all hinges $h_{[i]}(x)$ from $i = 1$ to k are refitted by running the HFA on $\mathcal{D}_{[i]}$ according to Equation 5 using the current $h_{[i]}(x)$ as initialization.

4   Hinge Regression for Multiple Outputs

Before we describe our method, we first discuss the geometric interpretation of the hinges (cf. Figure 3). We denote the intersection of the two m-dimensional hyperplanes β^+ and β^- as g. Of special interest is the orthogonal projection of g onto the input space, which is the intersection of the hinge Δ with the input space obtained by setting $\mathbf{x}^T \Delta = 0$. We denote this intersection as Γ and refer to it as the *separator*, since it separates the m-dimensional input space into two half spaces.

Let $\Delta = (\delta_1, \ldots, \delta_{m+1})$ be the parameters of the hinge for an output y:

$$y = x_1 \delta_1 + \ldots + x_m \delta_m + \delta_{m+1}.$$

By setting this equation equal to 0, we obtain for Γ in the input space the following equation in Hessian form:

$$x_1 \delta_1 + \ldots + x_m \delta_m + \delta_{m+1} = 0.$$

with normal vector $[\delta_1, \ldots, \delta_m]^T$ and offset $d = -\delta_{m+1}$.

For a given data set $\mathcal{D} \subset \mathbb{R}^{m+r}$, with $r > 1$, our goal is to find hinge functions for each output with identical separators, which simultaneously minimize

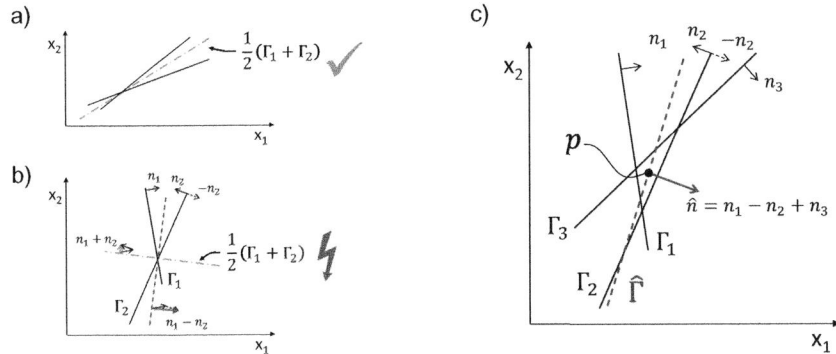

Fig. 4. Finding the best consensus separator $\hat{\Gamma}$

the residual error in all output dimensions. The basic idea is to compute one hinge per output y_j and to combine the corresponding separators Γ_j to a single *consensus separator* $\hat{\Gamma}$ (cf. Figure 3 right). Combining several separators in a meaningful manner in arbitrary dimensions is not straight forward. Moreover, by imposing a hinge on a hinge function, the continuity property of hinge functions gets lost. Hence, we additionally force the regression models β_j^+ and β_j^- for each output y_j to join continuously on $\hat{g}_j$ whose projection onto the input space is $\hat{\Gamma}$. The main steps of the procedure are

1. Generate separate data sets $\mathcal{D}_j$ per output y_j, i.e. $\forall (x, y) \in \mathcal{D} : (x, y_j) \in \mathcal{D}_j$
2. Compute one hinge Δ_j per output y_j using $\mathcal{D}_j$
3. Combine all separators Γ_j, $j = 1 \dots r$ to a consensus separator $\hat{\Gamma}$
4. Force the regression models for each output y_j to join continuously at $\hat{g}_j$

Step 1 generates r temporary data sets, Step 2 finds a hinge for each output as described in Section 3.2. Step 3 is described in the following Section, details on Step 4 are provided in Section 4.2. The entire hinge finding algorithm for multiple outputs is provided in Section 4.3.

4.1   Finding the Consensus Separator $\hat{\Gamma}$

A naive solution to combine the individual separators to a consensus separator $\hat{\Gamma} = (\hat{\gamma}_1, \dots, \hat{\gamma}_m)$ is to use a linear combination of the parameters as

$$\hat{\gamma}_i = \frac{1}{r} \sum_{j=1}^{r} \gamma_{j,i} \ , i = 1 \dots m \tag{6}$$

This solution does not necessarily find the best consensus separator as depicted in Figure 4. Part a) illustrates an example in a 2-dimensional input space where two separators Γ_1 and Γ_2 are combined to consensus separator $\hat{\Gamma}$ using Equation 6. $\hat{\Gamma}$ is the bisector of the smaller angle between Γ_1 and Γ_2. Part b) shows that

$\hat{\Gamma}$ according to Equation 6 yields the bisector of the larger angle between Γ_1 and Γ_2, which does not represent the best consensus. The normal vectors of the two different solutions result from adding the individual normal vectors with different orientations, i.e. either $n_1 + n_2$ or $n_1 - n_2$. We denote the two possible orientations of a normal vector n_j as $\xi_j n_j$ with $\xi_j \in \{-1, 1\}, j = 1 \ldots r$ and define the normal vector of the consensus separator as a sum of these oriented normals

$$\hat{n} = \frac{\sum_{j=1}^{r} \xi_j n_j}{\| \sum_{j=1}^{r} \xi_j n_j \|} \tag{7}$$

In general we search for the separator $\hat{\Gamma}$, defined by its normal $\hat{n}$ and its offset $\hat{d}$, with minimal deviation from the r separators. The normal $\hat{n}$ with minimal orientation deviation is found by minimizing the sum of angles α_j between $\hat{\Gamma}$ and each Γ_j, where $\alpha_j < \pi/2$ is the smaller angle between the two separators. This is the same as maximizing the sum of all cosines of the α_j. The offset with minimal deviation is found by constraining $\hat{\Gamma}$ to include the point with minimal distance to all Γ_j. Figure 4 c) illustrates a consensus separator in a 2-dimensional input space.

Definition 1. *Consensus separator.* *In an m-dimensional input space let Γ_j, $j = 1 \ldots r$, be separators defined by $x^T n_j - d_j = 0$. The consensus separator $\hat{\Gamma}$ is defined by the following normal vector:*

$$\sum_{j=1}^{r} |\hat{n}^T n_j| \longrightarrow \max \tag{8}$$

and contains the point p with

$$\sum_{j=1}^{r} (p^T n_j - d_j)^2 \longrightarrow \min \tag{9}$$

Lemma 1. *Let $\Xi = (\xi_1, \ldots, \xi_r)$ with $\xi_j \in \{-1, 1\}$, $j = 1 \ldots r$. The normal according to Equation 7 that maximizes Equation 8 is the one that maximizes the sum of cosines of all angles between oriented normals $\xi_i n_i$ and $\xi_j n_j$*

$$\max_{\Xi} \left(\sum_{i=1}^{r} \sum_{j=1}^{r} \xi_i n_i^T \xi_j n_j \right) \tag{10}$$

The proof is provided in appendix A. We determine Ξ using binary integer programming.

To find the point p that minimizes the sum of squared distances to all hyperplanes (cf. Equation 9) we use the notation $\mathbf{p} = (p, 1)$, $\mathbf{n}_j = (n_j, d_j)$ and define the matrix $N \in \mathbb{R}^{r \times (m+1)}$ as the matrix that contains as rows all $\mathbf{n}_j$, $j = 1 \ldots r$. Equation 9 can then be rewritten as

$$\min_{p} \left((N\mathbf{p})^2 \right) = \min_{p} \left((\mathbf{p}^T N^T)(N\mathbf{p}) \right) = \min_{p} \left(\mathbf{p}^T A \mathbf{p} \right) \tag{11}$$

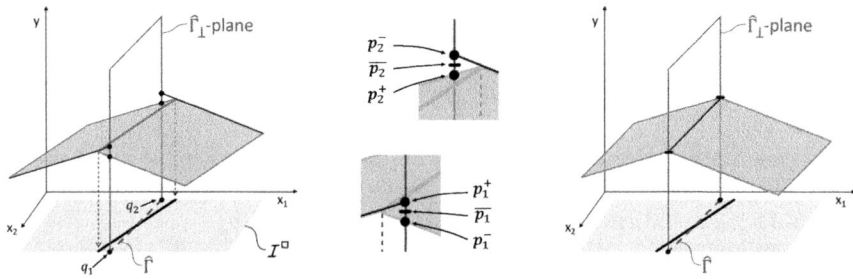

Fig. 5. Enforcing the regression planes to continuously join on $\hat{\Gamma}_\perp$

where $A = N^T N$ is an $(m + 1) \times (m + 1)$ matrix. A solution can be found by computing an eigenvalue decomposition of A or solving

$$\frac{\partial}{\partial \mathbf{p}} \mathbf{p}^T A \mathbf{p} = 0 \quad \Leftrightarrow \quad \left[A_1 \cdots A_m \right] p = \left[- A_{m+1} \right] \tag{12}$$

where A_i is the i-th column of matrix A.

4.2   Forcing Continuous Joins

Recall that the two regression models β_j^+ and β_j^- for output y_j are continuously joined on a $(m - 1)$-dimensional hyperplane g_j whose projection onto the input space is the separator Γ_j (cf. Figure 3). To enforce the consensus separator $\hat{\Gamma}$ we have to ensure that the regression planes are continuously joined on an $(m - 1)$-dimensional hyperplane $\hat{g}_j$ whose projection onto the input space is $\hat{\Gamma}$. The idea is depicted in Figure 5 and consists of two parts:

(a) Compute a consensus hyperplane $\hat{g}_j$, in which the two hyperplanes shall be continuously joined
(b) Recompute β_j^+ and β_j^- such that they both include $\hat{g}_j$ (cf. Figure 5 right)

(a) We find the consensus hyperplane $\hat{g}_j$ by computing a set of m affinely independent points $\bar{p}_i$ that lie on $\hat{g}_j$. $\hat{g}_j$ itself lies in the $\hat{\Gamma}_\perp$-plane, which is the m-dimensional hyperplane that contains $\hat{\Gamma}$ and that is perpendicular to the input space (cf. Figure 5 left). Each point $\bar{p}_i$ is the average between two points p_i^+ and p_i^-. p_i^+ lies in the intersection of β_j^+ and the $\hat{\Gamma}_\perp$-plane, analogously for p_i^-, and $p_{i,l}^+ = p_{i,l}^-$ for $l = 1 \ldots m$, i.e. they have the same input values.

To compute the points p_i^+ we choose a set $\mathcal{Q} \subset \mathbb{R}^m$ of points q_i in the input space, set $p_{i,l}^+ = q_{i,l}$, $l = 1 \ldots m$ and determine the output values as $p_{i,m+1} = q_i^T \cdot \beta^+$. The p_i^- are computed analogously and $\bar{p}_i = (p_i^+ + p_i^-)\frac{1}{2}$. Choosing $\mathcal{Q}$ is done as follows: Let $x_{i,min} = \min_{(x,y) \in \mathcal{D}} \{x_i\}$ and $x_{i,max} = \max_{(x,y) \in \mathcal{D}} \{x_i\}$ be the minimal and maximal values in $\mathcal{D}$ in dimension i. We define the input cube

$\mathcal{I}^\square \subset \mathbb{R}^m$ as the m-dimensional cube spanned by the minimum and maximum values

$$\mathcal{I}^\square = \left\{ (x_1, \ldots, x_m) \,\middle|\, \exists l \; x_{l,min} \leq x_l \leq x_{l,max} \wedge \forall i \neq l \; x_i = x_{i,min} \vee x_i = x_{i,max} \right\}$$

We choose the set $\mathcal{Q}$ to be m intersection points q_i of $\hat{\Gamma}$ with the input cube $\mathcal{I}^\square$, which can easily be computed for arbitrary dimensions. Figure 5 (left) shows an example for $\mathcal{I}^\square$ and the two intersection points q_1 and q_2.

(b) For output y_j let $\hat{S}_j^+$ and $\hat{S}_j^-$ be the two partitions of $\mathcal{D}_j$ corresponding to the consensus separator $\hat{\Gamma}$. We describe how we recompute the regression plane β_j^+ using $\hat{S}_j^+$, β_j^- is computed analogously. Let $X \in \mathbb{R}^{|\hat{S}_j^+| \times m}$ be the matrix that contains as rows all input vectors in $\hat{S}_j^+$ and $Y \in \mathbb{R}^{|\hat{S}_j^+| \times 1}$ the vector that contains all corresponding output values of the observations in $\hat{S}_j^+$. Similarly we define $Q \in \mathbb{R}^{|\mathcal{Q}| \times m}$ as the matrix that contains as rows all points $q_i \in \mathcal{Q}$ and $Z \in \mathbb{R}^{|\mathcal{Q}| \times 1}$ as the vector that contains all corresponding values $\bar{p}_{i,m+1}$. A regression plane that contains all points $\bar{p}_i$ can be approximated by assigning a high weight w to Q and Z in the unconstrained least squares problem

$$\min_\beta \left\| \begin{bmatrix} X \\ wQ \end{bmatrix} \beta - \begin{bmatrix} Y \\ wZ \end{bmatrix} \right\| \tag{13}$$

We set $w = |\mathcal{D}|$ in our experiments, i.e. the weight for each point $\bar{p}_i$ is equal to the size of the data set.

4.3   The Hinge Finding Algorithm for Multiple Outputs

The hinge finding algorithm for multiple outputs (mHFA) has a similar process flow as the HFA for a single output. It starts with a random hinge $\Delta^{(0)}$ that is iteratively improved until no significant error reduction occurs. Since the algorithm does not guarantee convergence to a global optimum, we let it run for a number of ι random initializations and retain the model yielding the smallest error. Unlike the case with a single output, each observation has several errors, one for each output. To avoid an influence of different output ranges we normalize the errors per output y_j by the corresponding range $range_j = y_{j,max} - y_{j,min}$. The approximation error for $\mathcal{D}$ over all outputs is then:

$$Err_\mathcal{D} = \sum_{(x,y) \in \mathcal{D}} \sum_{j=1}^{r} \left(\left| y_j - \sum_{k=1}^{K} h_{j[k]}(x) \right| \frac{1}{range_j} \right)^2 . \tag{14}$$

When the error is no longer reduced from on iteration to the next, mHFA is stopped.

As in the case of a single output (cf. Section 3.2), for multiple outputs the approximation error can be reduced using $K > 1$ hinges and estimating output y_j as $\hat{y}_j = \sum_{k=1}^{K} h_{j[k]}(x)$. To this end, for each hinge $h_{[k]}(x)$ a temporary data

---

Algorithm 1. computeHinges $(\mathcal{D}, K)$

---

1 **for** $k = 1$ to K **do**
2   create $\mathcal{D}[k]$;
3   compute Γ_{init};
4     $\hat{\Gamma}[k] = findConsensusSeparator(\mathcal{D}[k], \Gamma_{init})$;
5   **for** $i = 1$ to k **do**
6   create $\mathcal{D}[i]$;
7         $\hat{\Gamma}[i] = findConsensusSeparator(\mathcal{D}[i], \hat{\Gamma}[i])$;
8   **endfor**
9 **endfor**

Algorithm 2. findConsensusSeparator $(\mathcal{D}, \Gamma_{init})$

---

1 create $\mathcal{D}_j, \forall j = 1, \ldots, r$;
2 $forceJoin(\Gamma_{init})$ for each output j;
3 compute $newErr$;
4 $crtErr = \infty$; $\Gamma_{new} = \Gamma_{init}$;
5 **while** $newErr < crtErr$ **do**
6     $crtErr = newErr$;
7   call $HFA(\mathcal{D}_j, \Gamma_{new})$ for each output j;
8     $\Gamma_{new} =$ compute $\hat{\Gamma}$;
9     $forceJoin(\Gamma_{new})$ for each output j;
10   compute $newErr$;
11 **endw**
12 **return** Γ_{new};

---

set is computed that contains as output values the residuals that are not yet fitted by all other hinge functions:

$$\mathcal{D}[k] = \left\{ (x_1, \ldots, x_m, y_1[k], \ldots, y_r[k]) \middle| (x_1, \ldots, x_m, y_1, \ldots, y_r) \in \mathcal{D} \right\}, \quad (15)$$

$$\text{where} \quad y_j[k] = y_j - \sum_{i=1,\, i \neq k}^{K} h_{j[i]}(x) \quad \text{for} \quad j = 1 \ldots r,$$

and $h_{j[i]}(x)$ is initially set to 0 for all outputs and all hinges.

The main steps of the algorithm to fit K hinges to a given data set $\mathcal{D}$ are summarized in Algorithm 1. The normal form of the initial separator (cf. line 1) is computed using a random normal vector, and the mean vector of all input vectors in $\mathcal{D}$ is used to compute the distance to the origin. The method $findConsensusSeparator$ is listed in Algorithm 2. $forceJoin(\Gamma)$ (cf. line 2) is done as described in Section 4.2. $HFA(\mathcal{D}_j, \Gamma_{new})$ (cf. line 2) performs the hinge finding algorithm from Section 3.2 on $\mathcal{D}_j$ using Γ_{new} as initialization.

Complexity Analysis. The main building block of mHFA is the HFA, which iterates until the error no longer decreases. Let ι_1 and ι_2 be the number of random initialization for HFA and mHFA, respectively, and ι_{HFA} and ι_{mHFA} the

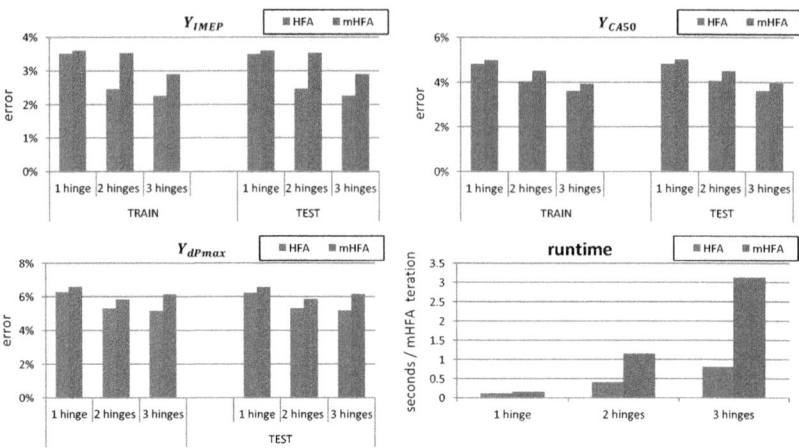

Fig. 6. Error measurements for the DIESEL dataset, and the corresponding runtimes

corresponding number of iterations until convergence. One such HFA iteration consists of fitting two regression hyperplanes and recomputing the two partitions S^+ and S^-, and has a runtime complexity of $O(|\mathcal{D}|^2 \cdot m + m^3)$. In total the runtime of HFA for one hinge is $O(\iota_1 \cdot \iota_{HFA} \cdot (|\mathcal{D}|^2 \cdot m + m^3))$. Our mHFA algorithm first performs the HFA for each output independently, then finds a consensus separator and forces it to all outputs. Hence the runtime complexity is $O(\iota_2 \cdot \iota_{mHFA} \cdot (r \cdot O(HFA) + 2^r + |\mathcal{D}|^2 \cdot m))$. The exponential term is due to the binary integer program in the computation of the consensus hinge. For the computation we employ the highly optimized Gurobi solver (www.gurobi.com). Hence, the scalability of mHFA depends on three factors: dataset size, input dimensionality, and output dimensionality. All of these aspects are investigated in the Section 5.

Trade-off between Quality, Compactness, and Runtime. There are two trade-offs which the user has to consider when intending to use HFA or mHFA. The first one is the trade-off between the quality and the compactness of a model. It generally holds, that the quality increases with more hinges at the cost of a poorer compactness. The second trade-off concerns the runtime complexity of building a model versus the quality of a model. This trade-off can be steered by the number of random initializations. The more noise or nonlinearity the data contains, the more probable it gets for the algorithms to get stuck in a local optimum. With more random initializations the probability of achieving a better result increases.

5   Experiments

In this section we investigate the mHFA algorithm compared to the HFA. First we show that by imposing the constraint of a common partitioning in the input space, the quality of the resulting models is not considerably affected. Second, we investigate the scalability of mHFA w.r.t. different aspects.

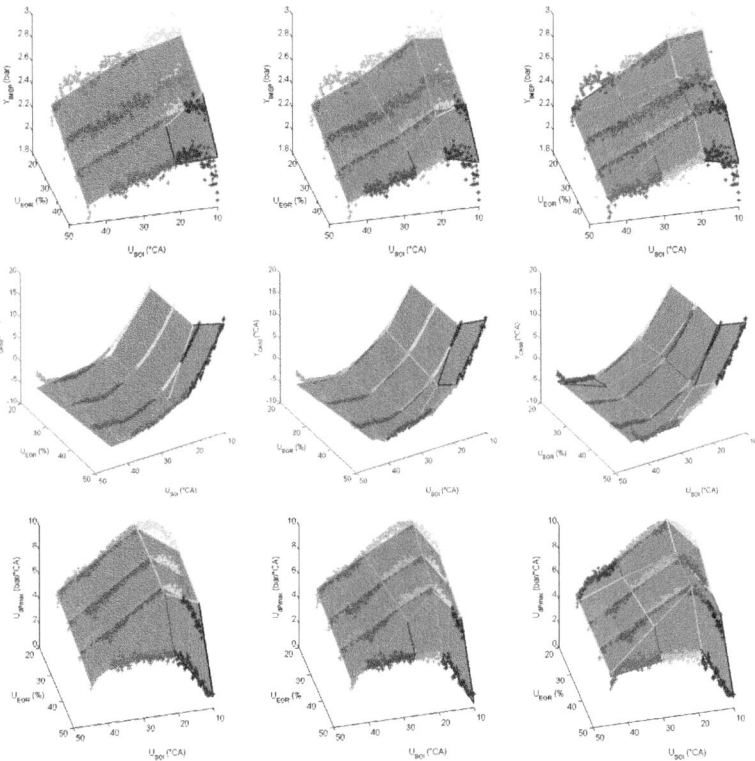

Fig. 7. Plots for the DIESEL dataset with 2 hinges ($\sum errors = 13.31\%$), 3 hinges ($\sum errors = 11.15\%$), and 4 hinges ($\sum errors = 9.79\%$)

For evaluating the model quality we consider following two datasets: the DIESEL dataset and the CONCRETE SLUMP dataset [18]. The DIESEL dataset contains 8020 measurements from a combustion process in a Diesel engine, with 3 input dimensions (U_{FMI}, U_{FMI}, and U_{EGR}), and 3 output dimensions (Y_{IMEP}, Y_{IMEP}, and Y_{dPmax}). The goal is to obtain a model describing the combustion process. The CONCRETE SLUMP dataset contains 103 measured points, with 7 input dimensions and 3 output dimensions. The aim is to obtain from these measurements a material workability behavior model to predict the concrete slump, flow, and strength. To evaluate the quality of a constructed model we use the error as defined in Equation 14. In our experiments we performed $\iota_1 = \iota_2 = 100$ random initializations and chose at the end the model with the lowest error. In all our experiments we used 10-fold cross validation.

Figure 6 illustrates the error of the different outputs for the DIESEL dataset, for the model constructed with mHFA compared to the models independently constructed with HFA. We can see that by forcing a common partitioning of the input space, mHFA obtains only a small error increase compared with HFA. The fourth diagram (in the lower right corner) illustrates the mean runtimes of the

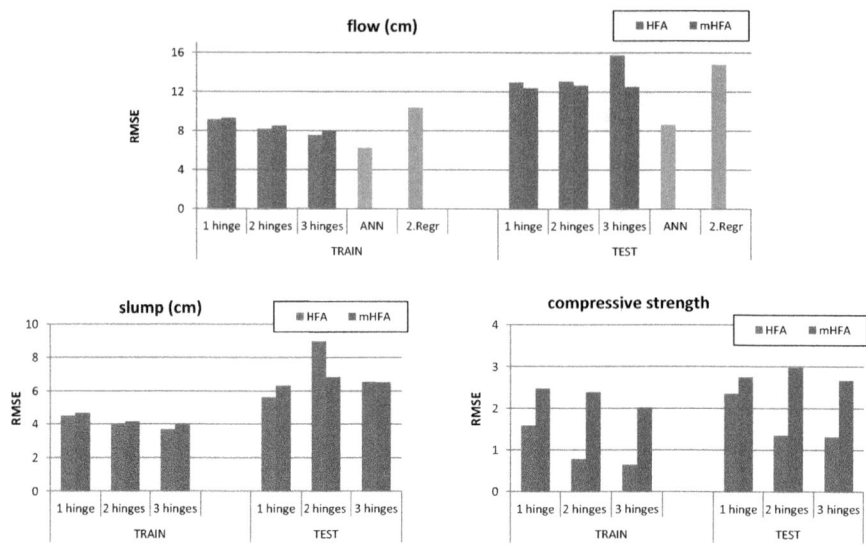

Fig. 8. Error measurements for the SLUMP dataset

two algorithms for one random initialization, ranging from 0.1 seconds for one hinge up to 3 seconds for three hinges (for mHFA).

Figure 7 illustrates the fitted models for 2 input dimenisions (for $U_{FMI} = 10$). The model for each output is plotted separately, with two, three, and four hinges. We recognize in these images, that the partitioning of the input space is the same for each model, only the linear equations for each output differ. The more hinges are used, the better the model fits the data.

Figure 8 compares the errors of the computed models for the CONCRETE SLUMP dataset. For the slump flow we additionally show the RMSE results that were reported in [18] for artificial neural networks and second order regression. For comparability, we also use the root mean square error for our experiments ($ RMSE = \frac{1}{N}\sqrt{\sum_{i=1}^{N}(y_i - \hat{y}_i)^2} $). We see that the artificial neural network has a smaller RMSE than both HFA and the mHFA . Although they have a higher prediction accuracy, they do not generate intelligible models but rather prediction functions which act as black boxes. The model built with second order regression has a lower prediction accuracy than both hinge approaches. This is because using piecewise linear functions an arbitrary high approximation accuracy is obtained with enough linear pieces. Taking a closer look at the slump output and the flow output, we see that in the test phase the RMSE is higher for the separately computed hinges. This is because of the noisy measurements, which are smoothed out when several outputs are considered.

Another interesting aspect is number ι of required initializations. To empirically investigate the convergence of mHFA, we plotted in Figure 9 the resulting RMSE values for several random initializations. In the case of the DIESEL dataset, we observe that, because of its nonlinearity, the more hinges are used,

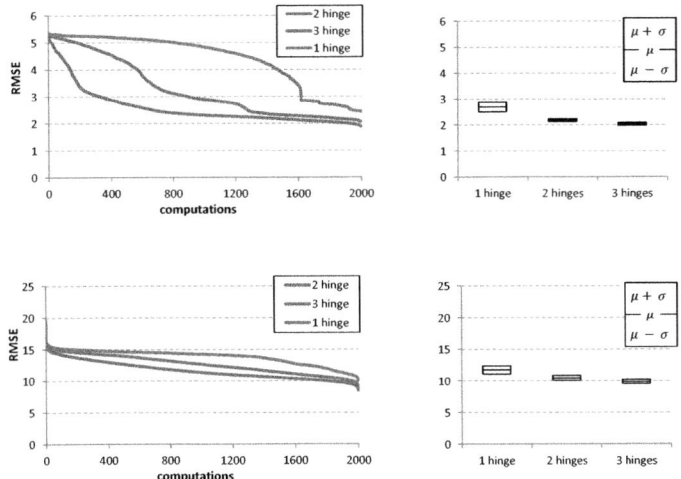

Fig. 9. Convergence of RMSE

the faster mHFA converges. In the case of the CONCRETE SLUMP dataset, mHFA converges faster even with only one hinge. Choosing ι generally depends on the linearity of the dataset, and on the number of desired hinges.

To evaluate the scalability of mHFA w.r.t. the three different aspects mentioned in Section 4.3 we used synthetically generated data. These are obtained by sampling continuous functions and adding noise. Figure 10 illustrates the results of our experiments. In the left figure we can see that the runtime of mHFA increases almost linearly with the size of the dataset, for a 3-dimensional input and 3-dimensional output. For a dataset with 1000 points, one random initialization run of mHFA requires around 1 second, and reaches 244 second on a dataset of around 280,000 points.

The figure in the middle shows the results of our experiments made to investigate the scalability of mHFA w.r.t. the input dimensionality. Together with the dimensionality, we also increased the dataset size, since sparse data does not offer much flexibility for fitting the hinges. We compared the resulting runtimes with the ones obtained on a dataset of the same size, but with 3-dimensional inputs. The numbers in the columns represent the corresponding dataset size. For each output dimensionality we tested the runtime on two different dataset sizes for a better comparability. We see that with a 4-dimensional input the runtime of mHFA is very similar to the runtime on a dataset with 3-dimensional input, while the runtime on a dataset with a 7-dimensional input is almost twice as high as on a dataset with a 3-dimensional input. In the right Figure we see the runtime performance of mHFA for a growing output dimensionality for a dataset of 8,000 points and 3-dimensional input. While the worst case runtime complexity is exponential, we do not observe this in the experiments. One reason is the highly optimized Gurobi solver. Another reason is that correlated output dimensions cause the mHFA algorithm to converge faster.

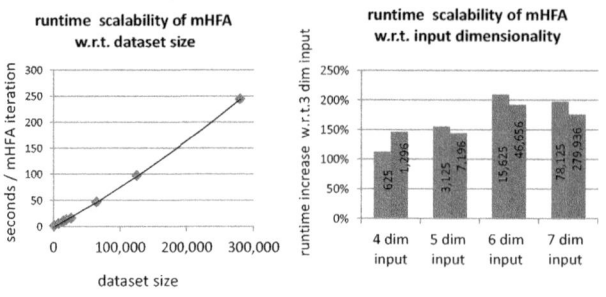
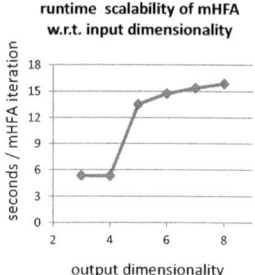

Fig. 10. Scalability of mHFA

6   Conclusion

Hinging hyperplanes yield a continuous piecewise linear approximation function for a single predicted variable. Besides prediction, the computed hinges can be used to construct a regression tree, a compact and interpretable model of the underlying data. In this paper we presented the mHFA approach that builds continuous models for multiple predicted variables and use it to build a multivariate regression tree. In contrast to previous approaches, where the number of partitions grows exponentially in the number of predicted variables, our approach maintains the low model complexity as in the case of a single output. mHFA enforces common hinges and restores the continuity of the resulting functions for each predicted variable. We evaluated the performance of our approach and discussed the convergence of the RMSE over several initializations. In summary mHFA yields compact and intelligible models with continuous functions and low approximation errors.

Acknowledgments. The authors gratefully acknowledge the financial support of the Deutsche Forschungsgemeinschaft (DFG) within the Collaborative Research Center SFB-686 "Model-Based Control of Homogenized Low-Temperature Combustion".

References

1. Achtert, E., Böhm, C., David, J., Kröger, P., Zimek, A.: Robust clustering in arbitrarily oriented subspaces. In: SDM (2008)
2. Aggarwal, C., Yu, P.S.: Finding generalized projected clusters in high dimensional spaces. In: SIGMOD, pp. 70–81 (2000)
3. Bemporad, A., Garulli, A., Paoletti, S., Vicino, A.: A bounded-error approach to piecewise affine system identification. IEEE Transactions on Automatic Control 50(10), 1567–1580 (2005)
4. Breiman, L.: Hinging hyperplanes for regression, classification, and function approximation. IEEE Transactions on Information Theory 39(3), 999–1013 (1993)
5. Breiman, L., Friedman, J.H., Stone, C.J., Olshen, R.A.: Classification and Regression Trees. Wadsworth and Brooks, Monterey (1984)
6. Deshpande, A., Madden, S.: Mauvedb: supporting model-based user views in database systems. In: SIGMOD, pp. 73–84 (2006)

7. Dobra, A., Gehrke, J.: Secret: a scalable linear regression tree algorithm. In: KDD, pp. 481–487 (2002)
8. Duda, R.O., Hart, P.E., Stork, D.G.: Pattern Classification, 2nd edn. Wiley Interscience (2000)
9. Ferrari-Trecate, G., Muselli, M., Liberati, D., Morari, M.: A clustering technique for the identification of piecewise affine systems. Automatica 39, 205–217 (2003)
10. Friedman, J.H.: Multivariate adaptive regression splines. The Annals of Statistics 19(1), 1–67 (1991)
11. Ivanescu, A.M., Albin, T., Abel, D., Seidl, T.: Employing correlation clustering for the identification of piecewise affine models. In: KDMS Workshop in Conjunction with the 17th ACM SIGKDD, pp. 7–14 (2011)
12. Paoletti, S., Juloski, A.L., Ferrari-trecate, G., Vidal, R.: Identification of hybrid systems: a tutorial. Eur. J. of Control 513(2-3), 242–260 (2007)
13. Pucar, P., Sjöberg, J.: On the hinge-finding algorithm for hinging hyperplanes. IEEE Transactions on Information Theory 44(3), 1310–1319 (1998)
14. Thiagarajan, A., Madden, S.: Querying continuous functions in a database system. In: SIGMOD, pp. 791–804 (2008)
15. Tung, A.K.H., Xu, X., Ooi, B.C.: Curler: Finding and visualizing nonlinear correlated clusters. In: SIGMOD, pp. 467–478 (2005)
16. Vidal, R., Soatto, S., Ma, Y., Sastry, S.: An algebraic geometric approach to the identification of a class of linear hybrid systems. In: IEEE CDC, pp. 167–172 (2003)
17. Vogel, D.S., Asparouhov, O., Scheffer, T.: Scalable look-ahead linear regression trees. In: KDD, pp. 757–764 (2007)
18. Yeh, I.-C.: Modeling slump flow and concrete using second-order regressions and artificial neural networks. Cement Concrete Composites 29(6), 474–480 (2007)

Appendix A

Proof. (Lemma 1). Let $\Xi = (\xi_1, ..., \xi_r)$ with $\xi_j \in \{-1, 1\}$, $j = 1, ..., r$ be the set of orientations for the normal vectors $n_1, ..., n_r$, such that following equation is maximized:

$$\sum_{j=1}^{r} \sum_{i=1}^{r} \xi_j \xi_i n_j^T n_i \qquad (16)$$

We prove that the normal $\hat{n}$ according to equation 7 that uses the above orientations Ξ is the one that maximizes the equation:

$$\sum_{j=1}^{r} |\hat{n}^T n_j| \qquad (17)$$

We provide a proof by contradiction and assume that Equation 16 is maximal for Ξ, but not Equation 17. This implies that $\exists e \in \{1, \ldots, r\}$ such that $|\hat{n}^T n_e| \neq \xi_e \hat{n}^T n_e$. Since $\xi_e \in \{-1, 1\}$, this further implies that $\hat{n}^T n_e \xi_e < 0$. Substituting $\hat{n}$ from Equation 7 in $\hat{n}^T n_e \xi_e < 0$ we obtain:

$$\left\| \sum_{j=1}^{r} \xi_j n_j \right\| \left(\sum_{j=1}^{r} \xi_j n_j \right)^T \xi_e n_e < 0 \Leftrightarrow \left(\sum_{j=1}^{r} \xi_j n_j \right)^T \xi_e n_e < 0 \qquad (18)$$

This can be further transformed to:

$$\sum_{j=1,j\neq e}^{r} \xi_j\xi_e n_j^T n_e + \xi_e\xi_e n_e^T n_e < 0 \Leftrightarrow \sum_{j=1,j\neq e}^{r} \xi_j\xi_e n_j^T n_e < -1 \qquad (19)$$

We next split the set $\Xi \setminus \{\xi_e\}$ into two sets K and L such that: $\forall l \in L :$ $\xi_l\xi_e n_l^T n_e > 0$, and $\forall k \in K : \xi_k\xi_e n_k^T n_e < 0$. We rewrite Equation 19 as:

$$\sum_{l\in L} \xi_l\xi_e n_l^T n_e + \sum_{k\in K} \xi_k\xi_e n_k^T n_e < -1 \Rightarrow \left|\sum_{l\in L} \xi_l\xi_l n_l^T n_e\right| - \left|\sum_{k\in K} \xi_k\xi_e n_k^T n_e\right| < -1$$

and conclude that:

$$\left|\sum_{l\in L} \xi_l\xi_e n_l^T n_e\right| < \left|\sum_{k\in K} \xi_k\xi_e n_k^T n_e\right|$$

Hence by inverting n_e, i.e. setting $\xi_e' = (-1)\cdot\xi_e$, we obtain:

$$\left|\sum_{l\in L} \xi_l\xi_e' n_l^T n_e\right| > \left|\sum_{k\in K} \xi_k\xi_e' n_k^T n_e\right| \Rightarrow \sum_{l\in L} \xi_l\xi_e' n_l^T n_e + \sum_{k\in K} \xi_k\xi_e' n_k^T n_e > 0$$

Since $\xi_e\xi_e n_e^T n_e = \xi_e'\xi_e' n_e^T n_e = 1$ we obtain:

$$\sum_{j=1,j\neq e}^{r} \xi_j\xi_e' n_j^T n_e + \xi_e'\xi_e' n_e^T n_e > 0 \qquad (20)$$

From Equation 18 and Equation 20 it follows that:

$$\sum_{j=1}^{r} \xi_j\xi_e' n_j^T n_e > \sum_{j=1}^{r} \xi_j\xi_e n_j^T n_e \qquad (21)$$

By switching ξ_e with ξ_e' in Eq. 16 we obtain according to Eq. 21 the following:

$$\sum_{j=1,j\neq e}^{r}\sum_{i=1}^{r} \xi_j\xi_i n_j^T n_i + \sum_{i=1}^{r} \xi_e'\xi_i n_e^T n_i > \sum_{j=1,j\neq e}^{r}\sum_{i=1}^{r} \xi_j\xi_i n_j^T n_i + \sum_{i=1}^{r} \xi_e\xi_i n_e^T n_i \quad (22)$$

Hence we obtained a higher sum than before, which is a contradiction to the assumption that Equation 16 is maximal.

$\square$

Optimizing Notifications of Subscription-Based Forecast Queries

Ulrike Fischer[1], Matthias Böhm[1,*], Wolfgang Lehner[1],
and Torben Bach Pedersen[2]

[1] Dresden University of Technology, Database Technology Group, Dresden, Germany
{ulrike.fischer,matthias.boehm,wolfgang.lehner}@tu-dresden.de
[2] Aalborg University, Center for Data-intensive Systems, Aalborg, Denmark
tbp@cs.aau.dk

Abstract. Integrating sophisticated statistical methods into database management systems is gaining more and more attention in research and industry. One important statistical method is time series forecasting, which is crucial for decision management in many domains. In this context, previous work addressed the processing of ad-hoc and recurring forecast queries. In contrast, we focus on subscription-based forecast queries that arise when an application (subscriber) continuously requires forecast values for further processing. Forecast queries exhibit the unique characteristic that the underlying forecast model is updated with each new actual value and better forecast values might be available. However, (re-)sending new forecast values to the subscriber for every new value is infeasible because this can cause significant overhead at the subscriber side. The subscriber therefore wishes to be notified only when forecast values have changed relevant to the application. In this paper, we reduce the costs of the subscriber by optimizing the notifications sent to the subscriber, i.e., by balancing the number of notifications and the notification length. We introduce a generic cost model to capture arbitrary subscriber cost functions and discuss different optimization approaches that reduce the subscriber costs while ensuring constrained forecast values deviations. Our experimental evaluation on real datasets shows the validity of our approach with low computational costs.

1 Introduction

Empirically collected data constitutes time series in many domains, e.g., sales per month or energy demand per minute. *Forecasting* is often applied on these time series in order to support important decision-making processes. Sophisticated forecasts require the specification of a stochastic model that captures the dependency of future on past observations. We will refer to such models as *forecast models*. The creation of forecast models is typically computationally expensive, often involving numerical optimization schemes to estimate model parameters. Once a model is created and parameters are estimated, it can efficiently be used to forecast future values, where the *forecast horizon* denotes the

---

* The author is currently visiting IBM Almaden Research Center, San Jose, CA, USA.

A. Ailamaki and S. Bowers (Eds.): SSDBM 2012, LNCS 7338, pp. 449–466, 2012.

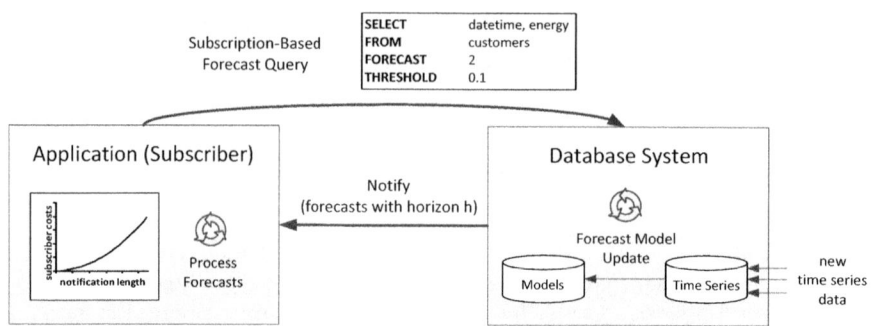

Fig. 1. Overview Subscription-Based Forecast Queries

number of requested forecast values. As new data arrives, the forecast model requires *maintenance* and improved forecast values might be obtained.

Integrating advanced statistical methods into database management systems is getting more and more attention [1]. These approaches allow for improved performance and additional functionality inside a DBMS. Research on integrating time series forecasting has mainly focused on accuracy and efficiency of ad-hoc [7] and recurring forecast queries [8,9]. However, applications might continuously require forecast values in order to do further processing. Forecast queries incorporate the unique characteristic that they can provide an arbitrary number of forecast values. However, with each new actual value the underlying model is updated and better forecasts might be available. A dependent application could obtain these values by repeatedly polling from the database. This is very inefficient if forecasts have changed only marginally, especially if the application executes a computational expensive algorithm based on the received forecasts.

In order to tackle this problem, we introduce the concept of *subscription-based forecast queries* that may be seen as continuous forecast queries associated with constraints guiding notifications to the subscriber. A general overview of our approach is shown in Figure 1. A subscription-based forecast query registers at the database system given various parameters, e.g., the *time series* to forecast, a *continuous forecast horizon* and a *threshold* of acceptable forecast deviations. For example, the simple forecast query in Figure 1 requests forecasts of customer energy demand for the next two steps (two forecast values) with a threshold of 10%. The database system itself stores time series data as well as associated forecast models and automatically manages these models [8]. As time proceeds and new time series values arrive models are updated and optionally maintained by the DBMS. This results in better forecast values leading to notifications sent to the subscriber that contain at least the forecast horizon specified by the subscriber (further denoted as *notification length*). For our example query, a notification needs to be sent if new forecasts are available that deviate by more than 10% from old forecasts sent before. The subscriber processes all notifications, where the processing costs of the subscriber often depend on the notification length. These *subscriber costs* can be communicated to the database system to optimize future notifications.

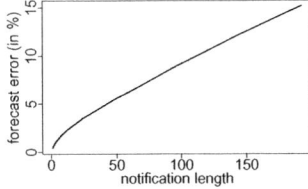 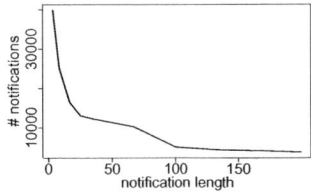

Fig. 2. Influence of Notification Length

One important use case of notification-based forecast queries can be found in the energy domain. Forecasting is crucial for modern energy data managements (EDM) systems to plan renewable energy resources and energy demand of customers. In this use case, the EDM system subscribes at a forecasting component, e.g., the DBMS, to receive forecasts on a regular basis. These forecasts are used to balance demand and supply and are crucial to reduce penalties paid for any kind of imbalances, i.e., remaining differences between demand and supply [2,13,17]. It is important to update the EDM system just with significant new forecast values to ease the computational expensive job of energy balancing.

Therefore, our objective in this paper is the reduction of the processing costs of the subscriber by trading the number of notifications against the notification length. We distinguish two extreme cases (Figure 2). On the one hand, we can choose a short notification length (i.e., a short forecast horizon). With this approach, we achieve a low forecast error, but we need to send notifications more frequently (as we deal with continuous queries). On the other hand, we can choose a large notification length, resulting in much less notifications, but in a high error for forecasts far from current time. Both extreme cases result in a high overhead at the subscriber side. This first one requires repeatedly processing new notifications, the second one needs to process very long notifications as well as reprocess many notifications containing improved forecasts. To solve this problem, we need to increase the notification length just as far as to ensure accurate forecasts that require a small number of notifications. This poses an important but challenging optimization problem of reducing the costs of the subscriber.

Contributions and Outline. Our primary contribution is the introduction of the concept of subscription-based forecast queries, which include a parameter definition, processing model, cost model and optimization objective (Section 2). Then, in Section 3, we propose different computation approaches to minimize the costs of the subscriber according to our optimization objectives defined in Section 2. Our experimental study (Section 4) investigates the performance of the computation approaches, the influence of subscription parameters, the computational costs of our approach as well as the validity of the cost model using real-world datasets. Finally, we survey related work in Section 5 and conclude in Section 6.

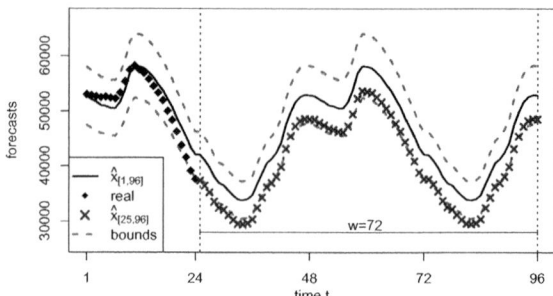

Fig. 3. Development of Forecast Values

2   Foundations of Subscription-Based Forecast Queries

In this section, we set the foundations of subscription-based forecast queries. We start with discussing the parameters of such a query (Section 2.1), followed by introducing our general processing model that leads to two different kind of notifications to the subscriber (Section 2.2). We then explain our cost model, which captures the costs of the subscriber depending on the number of notifications and the notification length (Section 2.3). Finally, we sketch out our optimization objective of reducing the overall subscriber costs (Section 2.4).

2.1   Forecast-Based Subscriptions

We start with defining the parameters of subscription-based forecast queries.

Definition 1 (Forecast-Based Subscriptions). *A forecast-based subscription* $S = (X, h, \alpha, w, g, k_{max})$ *consists of a time series description* X, *a minimum continuous forecast horizon* h, *a threshold* α, *an aggregation window* w, *an aggregation function* g *and a maximum horizon extension* k_{max}.

The time series description $X = x_{[1,t]}$ can be an arbitrary SQL query that specifies the time series to forecast [8]. The continuous forecast horizon h specifies the minimum number of forecast values $\hat{x}_{[t+1,t+h]}$ and implies that at each time t, the subscriber holds at least h forecast values. In addition, each subscription specifies a relative *threshold* α. At time t, we must notify the subscriber with new forecast values if these new forecast values $\hat{x}_{[t+1,t+w]}$ deviate more than α from the old values sent to the subscriber, using a window w and aggregation function g. Examples for aggregation functions are mean (average deviation in window above threshold) or max (maximum in window above threshold), where the decision depends on the intended sensitivity to deviations.

Example 1. *Figure 3 shows a real-dataset example. A forecast model was created up to time* $t = 0$, *using the triple seasonal exponential smoothing model* [19]. *The solid line displays forecast values* $\hat{x}_{[1,96]}$ *created at time* $t = 0$ *with a horizon of* $h = 96$. *At each time step* $t + i$ *with* $i > 0$ *new real data arrives and we*

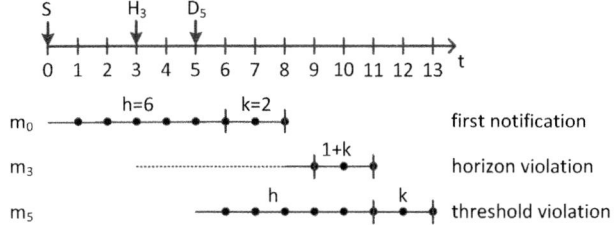

Fig. 4. Processing Forecast-based Subscriptions

can update the forecast model. At time $t = 24$, we can create new forecast values $\hat{x}_{[25,96]}$ that capture the history better and now slightly differ from the original ones (line with crosses). A subscriber wants to be notified whenever new forecast values deviate by more than 10% from the old ones (dashed lines). We see that at time $t = 24$ some forecast values $\hat{x}_{[25,96]}$ are outside this threshold. Using an aggregation window of $w = 72$ and the aggregation function max we would need to send a notification.

Finally, the maximum horizon extensions k_{max} is specific to our processing model and thus explained in the following.

2.2   Processing Model

At subscription registration time, we send at least h forecast values to the subscriber. From that point in time, notifications are caused by one of the following two reasons. First, new forecast values are sent whenever the subscriber has less than h forecast values (*horizon violation* H_t), where we need to send at least the missing values. However, we can additionally send k values—the *horizon extension*—in order to avoid a lot of horizon violations. There, the subscriber specifies the maximum number of additional forecast values k_{max}. Second, we send a notification if the threshold is violated at time t (*threshold violation* D_t). In this case, we consider *all* sent forecast values as invalid and we resend forecast values with a horizon h plus the horizon extension k. A different approach might be to resend only values that violate the threshold. However, this might lead to systematic errors since forecast values are often based on each other.

Example 2. *Figure 4 shows an example subscription. At time $t = 0$, we create a subscription $S = (X, h, \alpha, w, g, k_{max})$ with minimum horizon $h = 6$. The horizon extension is set to $k = 2$. Initially, we send a notification m_0 with $h + k = 6 + 2 = 8$ forecast values. At time $t = 3$, the subscriber has only $h - 1 = 5$ forecast values (horizon violation H_3). Hence, we send a notification m_3 with $1 + k = 3$ forecast values. We do not override sent values as these are still valid (below the subscription threshold) but send missing values (one value plus $k = 2$ values). Then, at time $t = 5$, the subscriber threshold α is violated according to the aggregation window w and function g (threshold violation D_5). We send a notification m_5 with $h + k = 8$ new values, which override all sent values.*

We define the total number of sent values $h + k$ or $k + 1$ as *notification length*. The parameter k has high influence on the number of notifications and individual notification lengths and therefore on the subscriber costs. If we set k quite low, we send many notifications because the horizon is violated. Thus, the subscriber needs to repeatedly process new forecast values leading to high costs. If we set k quite high, the notification length increases and we need to resend a lot of values if the threshold is violated. Thus, the subscriber needs to reprocess many updated forecast values. In the next subsection, we introduce a cost model that allows the quantification of this influence.

2.3   Cost Model

The subscriber cost function can be arbitrary and might be unknown. We therefore use this as a black box function, where we can retrieve the costs for arbitrary notification lengths. Internally, this might be a known analytical function or existing techniques might be used to learn these costs online [18]. However, the costs might be different for $h + k$ new values (threshold violation) or $k + 1$ additional values (horizon violation). Depending on the horizon h and the horizon extension k, we distinguish two cost functions. First, F_C denotes the costs of a complete restart of the subscriber algorithm as necessary for threshold violations, where all forecast values are resent. Second, F_I denotes the costs of an incremental version of the subscriber algorithm that is used for horizon violations, where the subscriber only processes additional values.

These considerations lead to our cost model that calculates the costs between two successive threshold violations D_t and D_{t+i}:

Definition 2 (Threshold Violation Costs). *Assume a horizon extension k. The costs in a threshold violation interval $\Delta D = (D_t, D_{t+i})$ are defined as:*

$$C_{k,\Delta D} = F_C(h + k) + \left\lfloor \frac{\Delta D}{k+1} \right\rfloor \cdot F_I(k+1). \tag{1}$$

As explained before, whenever a threshold violation D_t occurs, we send $h + k$ values leading to complete costs of $F_C(h + k)$. Additionally, a certain number of horizon violations occur until the next threshold violation D_{t+i}, each leading to incremental costs of $F_I(k + 1)$. The number of horizon violations equals the number of times $k + 1$ fits in the threshold violation interval ΔD as we send notifications until the end of the interval. If $k + 1$ is smaller or equal than ΔD no additional incremental costs occur.

Example 3. *Consider again Example 2, the first threshold violation occurs at D_5, so according to our definition $\Delta D = 4$. In this interval, we require once complete costs $F_C(h + k)$ at the start of the interval and once ($\lfloor 4/(2 + 1) \rfloor = 1$) incremental costs $F_I(k)$ until the threshold violation D_5. At D_5 a new threshold violation intervals begins, which again starts with complete costs.*

The total costs $C_{k,\Delta D}$ in one threshold violation interval strongly depend on the subscriber cost functions F_C and F_I.

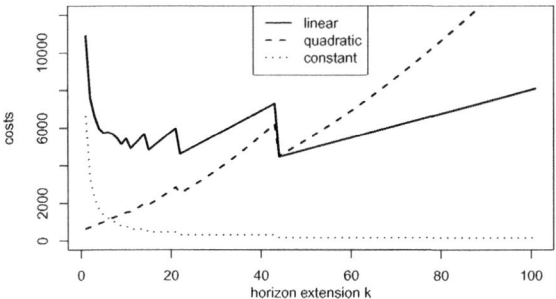

Fig. 5. Example Cost Functions

Example 4. *Figure 5 illustrates the influence of different cost functions. It shows the theoretical costs according to different horizon extensions k for a fixed minimum horizon h = 24 and threshold violation interval ΔD = 44. We used three different cost functions: constant (150), linear (64x + 150) and quadratic (x^2), where the same cost function is applied for F_C and F_I. For a quadratic cost function, long notifications are very expensive, so it is best to send many small notifications (k = 0). If the cost function only contains setup cost (constant), the goal is to reduce the number of notifications. Hence, we would choose the highest possible horizon extension in order to avoid any horizon violation. The linear function shows a possible cost function between these two extremes. The dips in the linear cost function arise when k + 1 is a divisor of ΔD. Thus, all horizon violations fit exactly in the interval and no values are sent unnecessary.*

The threshold violation costs formula makes the simplifying assumption that the threshold violation interval ΔD and the horizon extension k are independent of each other. This might not always be the case as k influences the accuracy of the forecast values and thus also the threshold violation interval. However, preliminary experiments have shown that the impact of k on ΔD is very low.

Based on the defined costs of a *single* threshold violation interval, we are now able to calculate the total costs of a *sequence* of threshold violation intervals.

Definition 3 (Total Subscriber Costs). *Assume a sequence of horizon extensions $\{k_1, \ldots, k_n\}$. Then, the total costs of a sequence of threshold violation intervals $\{\Delta D_1, \ldots, \Delta D_n\}$ is defined as:*

$$C_{total} = \sum_{i=1}^{n} C_{k_i, \Delta D_i}. \tag{2}$$

Thus, the best horizon extension depends on the cost function and frequency of threshold violations, which leads to a hierarchy of optimization problems.

2.4   Optimization Problems

Our general *optimization objective* is to reduce the total subscriber costs with regard to the introduced cost model. Furthermore, our *optimization approach* is to choose the best horizon extension, which can be done for different time granularities (offline-static, offline-dynamic, online). This inherently leads to three different optimization problems. These problems are independent of any computation approach and hence, presented separately. Furthermore, they are complete in the sense that additional problems are conceptual composites of them.

The most coarse-grained problem is to choose a single horizon extension for the whole time series, i.e., the sequence of threshold violation intervals (static).

Optimization Problem 1 (Offline – Static). *Assume a sequence of threshold violation intervals $\{\Delta D_1, \ldots, \Delta D_n\}$ and a maximum horizon extension k_{max}. The objective is to minimize the total subscriber costs by choosing a single horizon extension k:*

$$\phi_1 : \min_{0 \leq k \leq k_{max}} \sum_{i=1}^{n} C_{k, \Delta D_i}. \tag{3}$$

For some datasets, we observe different average threshold violation intervals at different time intervals, where a single horizon extension might fail. For example, often energy demand during week days can be forecasted more accurate than during weekend days, leading to more threshold violations at the weekend. Hence, the second optimization problem is more fine-grained (dynamic) as it aims to find a sequence of horizon extensions for a sequence of time slices.

Optimization Problem 2 (Offline – Time Slice). *Assume a sequence of time slices $\{\Delta T_1, \ldots, \Delta T_m\}$ and a sequence of threshold intervals $\{\Delta D_1, \ldots, \Delta D_n\}$, where $n \geq m$. Each ΔD_i is assigned to exactly one ΔT_j, where l_j is the total number of $\Delta D_i s$ in time slice ΔT_j. The objective is to minimize the total subscriber costs by choosing a sequence of horizon extensions $\{k_1, \ldots, k_m\}$:*

$$\phi_2 : \min_{0 \leq k_j \leq k_{max}} \sum_{j=1}^{m} \sum_{i=1}^{l_j} C_{k_j, \Delta D_i}. \tag{4}$$

Finally, the most-fine-grained optimization problem is an adaptive online formulation.

Optimization Problem 3 (Online). *Assume a history of threshold violation intervals $\{\Delta D_1, \ldots, \Delta D_n\}$, horizon extensions $\{k_1, \ldots, k_n\}$ and related costs. The objective is to minimize the total subscriber costs by choosing the next k_{n+1}:*

$$\phi_3 : \min_{0 \leq k_{n+1} \leq k_{max}} C_{k_{n+1}, \Delta D_{n+1}}. \tag{5}$$

As the cost function is given by the subscriber or monitored, we "only" need to determine ΔD in order to calculate the best horizon extension. However, ΔD depends on many aspects, e.g., time series characteristics, model accuracy, forecast horizon or subscription parameters. In reality, we do not know when the next threshold violation occurs. However, we can analyze past threshold violation intervals and use them to predict future threshold violations.

3   Computation Approaches

Regarding the defined optimization problems, we now discuss related computation approaches. As a foundation, we first present our general subscription maintenance algorithm (Algorithm 1) as a conceptual framework for arbitrary computation approaches. This includes two major procedures:

First, REGISTERSUBSCRIBER is called when creating a new subscription S. We first add the new subscription to the list of subscribers on the requested forecast model (line 2). Such a forecast model is stored and maintained for each time series with at least one subscriber. The type of the model (e.g., exponential smoothing) needs to be chosen by a domain expert or by using an heuristic algorithm [10]. To store the list of subscribers itself, we use a simple array structure. However, more advanced data structure can be utilized if the number of subscriber increases (e.g., [3]). Then, we start an analysis phase (line 3). In an offline context, we evaluate single (*static*) or multiple (*dynamic*) horizon extensions. In an online context, we only determine the initial horizon extension. For all three cases, we implicitly interact over a callback interface—implemented by each subscriber— to retrieve the costs F_C and F_I. Finally, we send the first notification to the subscriber with $h + k$ forecast values (line 4).

Second, PROCESSINSERT is called when a new tuple is added to the time series of the specific model. This requires updating the model to the current state of the time series as well as optional maintenance in form of parameter re-estimation (line 6). For this, we use a simple, but robust, time-based strategy that triggers maintenance after a fixed number of inserts [9]. Then, for each subscriber, we check if the threshold is violated over the aggregation window (lines 8 - 12). If so, we adapt the next horizon extension according to the used strategy (*dynamic*, *online*). Finally, we notify the subscriber with $h + k$ new forecast values. If no threshold violation occurred, we check if the horizon is violated and we notify

---

Algorithm 1. Forecast Subscription Maintenance

Require: *model, subList, currentTime*
 1: **procedure** REGISTERSUBSCRIBER(S)
 2:     $subList \leftarrow$ **add**$(subList, S)$
 3:     analyzeK(S)
 4:     **notify**(**predict**($currentTime$, $S.h + S.k$))
 5: **procedure** PROCESSINSERT$(newtuple)$
 6:     $model \leftarrow$ maintainModel$(newtuple)$
 7:     **for** S **in** $subList$ **do**
 8:         $pmodnew \leftarrow$ **predict**$(S.w)$
 9:         $div \leftarrow$ calculateDiv$(pmodnew, S.forecasts)$
10:         **if** $div > S.threshold$ **then**
11:             updateK(S)
12:             **notify**(**predict**($currentTime$, $S.h + S.k$))
13:         **else if** isHorizonViolated(S) **then**
14:             **notify**(**predict**($currentTime + S.h$, $S.k + 1$))

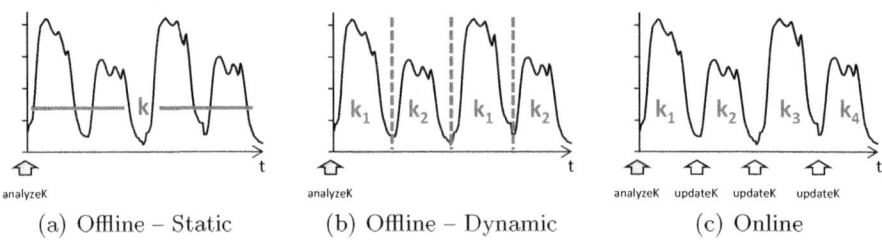

(a) Offline – Static   (b) Offline – Dynamic   (c) Online

Fig. 6. Comparison of Computation Approaches

the subscriber with the missing values plus the horizon extension (lines 13 - 14). Here, the interaction with the subscriber is also done via the callback interface.

The following computation approaches are based on two observations. First, the best horizon extension in the past can be used to determine the best horizon extension for the future. We therefore propose predictive approaches that analyze the history of threshold violation intervals. Second, there is the problem that threshold violation points strongly fluctuate and we would need to predict the trend of prediction errors. We therefore focus on robust and simple approaches rather than highly dynamic analytical approaches. Figure 6 shows an overview of our computation approaches, which are discussed in the following.

Offline – Static. The first computation approach addresses the coarse-grained Optimization Problem 1, where the objective is to choose a single horizon extension. The static approach determines one horizon extension during the analysis phase of our general algorithm (line 3) and uses this during the whole lifetime of a subscription (Figure 6(a)). To determine the horizon extension, we empirically monitor the threshold violation points over the whole history of the time series. For each tuple in the time series history, we execute an adapted version of the procedure PROCESSINSERT (Algorithm 1), where we do not notify the subscriber and just monitor whenever a threshold violation occurs. Given the resulting sequence of threshold violation intervals $\{\Delta D_1, \ldots, \Delta D_n\}$, we calculate the costs for different k's using our cost model (Equation 2). This leads to functions such as shown in Figure 5. Finally, we choose the k with minimum costs.

Offline – Time Slices (Dynamic). The second approach solves Optimization Problem 2 to find a sequence of horizon extensions for a sequence of time slices. This approach is beneficial if (1) the time series shows periodic patterns of threshold violation intervals or (2) if the cost function periodically changes over time. During the analysis phase (line 3), we determine a sequence of horizon extensions for periodic time slices (Figure 6(b)). Whenever a threshold violation occurs during execution (line 11), we set the next horizon extension to the horizon extension of the current time slice. The computation approach is similar to the static approach. We just additionally remember in which time slice the threshold violation occurred and compute a separate horizon extension for each time slice.

To determine the granularity of the time slices, we either use domain knowledge or we empirically evaluate different types of time slices.

Online Approach. This last computation approach solves Optimization Problem 3 of finding the best next horizon extension. An online approach is beneficial if either the time series model evolves leading to different threshold violation points or the cost function evolves leading to changed costs F_C and F_I. The online approach is repeatedly executed during the lifetime of a subscription to determine the next horizon extensions (Figure 6(c)). At registration time, we need to determine an initial k (line 3). This can be either some predefined parameter or we can use our static approach. The main work is done in the function updateK after each threshold violation, where we determine the next horizon extension (line 11) online. Whenever a threshold violation occurs, we monitor the associated ΔD over a predefined window. We then determine the need for recalculating the horizon extension. We analyze two different strategies to trigger recalculation within our experimental evaluation. If recalculation is required, we again compute the costs for different horizon extensions using the monitored sequence of ΔDs and we choose the horizon extension with minimum costs. This requires that the subscriber costs can be retrieved efficiently or we have a (possibly changing) known analytical cost function. Otherwise, we adapt the best k incrementally, i.e., by trying different ks and monitoring the resulting costs.

Computational Costs. The costs of all three approaches depend on (1) the time series length n, to evaluate the history of threshold violations, (2) the number of possible horizon extensions m, to retrieve the subscriber costs, and (3) the number of threshold violations d that occur, as the total costs are calculated by the sum over the cost of one threshold violation interval. For the offline approaches, the number of threshold violations d is calculated over the whole history (n represents the whole time series length); for the online approach only the specified window is used (n equals the window size). Following these considerations, the time complexity of all approaches equals $O(n + m \cdot d)$.

4   Experimental Evaluation

We conducted an experimental study on three real world data sets to evaluate (1) the performance (subscriber costs) of our approaches, (2) the influence of subscription parameters, (3) the computational costs of our approach in relation to the subscriber costs and (4) the validity of our cost model.

4.1   Experimental Setting

Test Environment: We implemented a simulation environment using the statistical computing software environment R. It provides efficient built-in forecast

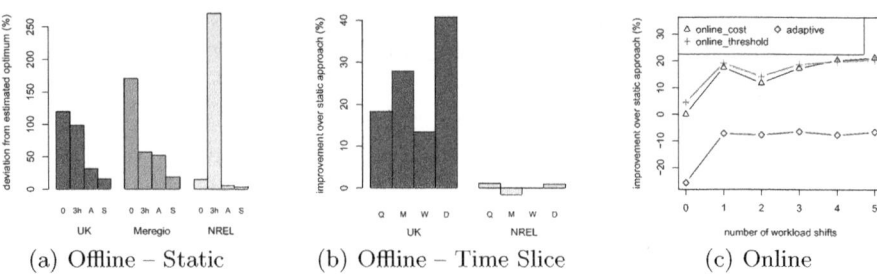

(a) Offline – Static   (b) Offline – Time Slice   (c) Online

Fig. 7. Performance of Computation Approaches

methods and parameter estimators, which we used to build the individual forecast models. All experiments were executed on an IBM Blade (Suse Linux, 64bit) with two processors (each a Dual Core Intel Xeon at 2.80 GHz) and 4 GB RAM.

Dataset Descriptions: We used three real-world energy demand and supply datasets. The first dataset (*UK*) includes energy demand of the United Kingdom and is publicly available [14]. It consists of total energy demand data from April 2001 to December 2009 in a 30 min resolution. The second dataset (*MER*) was provided by EnBW, a MIRABEL project partner. It contains energy demand of 86 customers, from November 2009 to June 2010 in a 1 hour resolution. The third dataset (*NREL*) is energy supply data in the form of the publicly available NREL wind integration datasets [15]. It consists of aggregated wind data from 2004 to 2006 in a 10 min resolution.

Forecast Methods: For all datasets, we used triple exponential smoothing with double or triple seasonality as forecast method [19]. This method is an extension of the robust and widely used exponential smoothing and is tailor-made for short-term energy forecasting. We used three seasonalities (daily, weekly and annual) for the UK and NREL datasets but only two seasonalities (daily and weekly) for the MER dataset. We used the first 6 years for UK, 6 months for MER and 2 years for NREL to train the forecast models. The remaining data was used for forecasting and evaluation of our approaches.

4.2 Evaluation of Computation Approaches

In a first series of experiments, we analyze the performance of our three computation approaches using fixed subscription parameters, i.e., $h = 1$ day, $w = 12$ hours, $\alpha = 0.15$ and $g = $ mean. For the subscriber cost function, we use the linear function from Example 4.

Offline – Static: We start with comparing our static approach to naïve and adaptive approaches [11]. The first naïve approach never sends more forecast values than requested ($k = 0$). The second naïve approach sends as many forecast values as possible ($k = k_{max}$), where $k_{max} = 3h$. The adaptive approach is independent of the time series history but reacts to notification events. It therefore is a representative of an online approach. Obviously, if a notification occurred due to a horizon violation, the horizon extension was too small. Hence, the adaptive approach increases the horizon extension. In contrast, if a threshold violation occurred, the horizon extension was too high and thus, the horizon extension is decreased. We evaluated different strategies to increase/decrease the horizon extension, where a simple strategy performed best that starts with $k = 1$ and doubles or halves the horizon extension depending on the notification event.

Figure 7(a) shows the result of the four approaches for all three datasets. As all datasets exhibit different total costs, we normalized the cost with the estimated best cost if we would exactly know the threshold violation sequence. We first notice that our static approach (S) always outperforms the two naïve approaches (0 and $3h$) and the adaptive approach (A). The reason is that the static approach includes the subscriber cost function into optimization while the other three approaches act independently from the subscriber costs. In addition, we observe that the datasets exhibit different characteristics leading to different performance of the naïve approaches. For the given subscription parameters, the forecast values of the NREL datasets deviate fast from the old values, leading to a small threshold violation interval on average and to a better performance of small horizon extensions. In contrast, the forecast values of the other two datasets are more accurate leading to larger optimal horizon extensions.

Offline – Dynamic (Time Slices): For our second computation approach we only use the UK and NREL dataset as these datasets are long enough to build meaningful time slices. We analyzed four kinds of time slices: quarter (Q), month (M), weekday/weekend (W), and day (D). Thus, we use different horizon extensions for different time slices, e.g., for every quarter of the year. Figure 7(b) shows the results in terms of the improvement over the static approach by subtracting the estimated best costs and normalizing with the costs of the static approach. For our best case, the UK dataset, all four time slice approaches lead to an improvement over the static approach. We observe the highest improvement for daily time slices, which cover the behavior of customers at different days of a week. The NREL dataset does not contain a typical seasonal behavior. It therefore represents a worst-case example and shows no improvement for all four time slice approaches.

Online Approach: We now relax the assumption of static cost functions and change these functions over time. We use the UK dataset and our linear cost function, where we vary the slope of the cost function $[10; 1,000]$. Figure 7(c) shows the improvement of the different approaches over the static approach for different numbers of workload shifts. A workload shift switches from the maximum slope to the minimum slope and vice versa. We analyze two different versions of the online approach. The first one recalculates the horizon extension only if the cost function changes (*online_cost*). The second online approach recalculates the horizon extension after every threshold violation (*online_threshold*). Both of our online approaches clearly outperform the adaptive approach that is even worse than the static approach as it acts independently of the subscriber costs. For

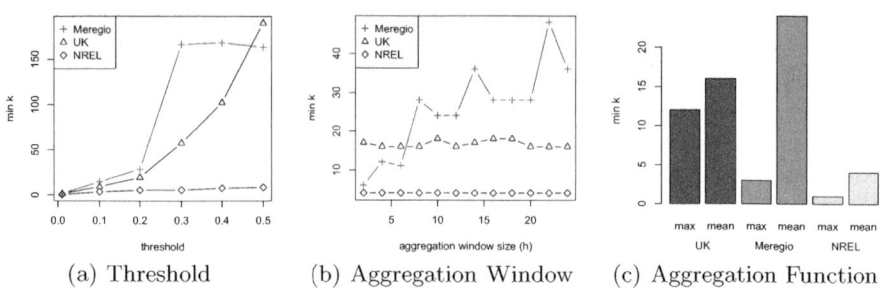

(a) Threshold   (b) Aggregation Window   (c) Aggregation Function

Fig. 8. Influence of Subscription Parameters

no workload shifts, the static and online (cost) approach show exactly the same performance as the online approach is never triggered. In contrast, the online (threshold) approach slightly improves the static approach. For higher number of workload shifts, we yield high improvements over the static approach because the static approach just determines one horizon extension at the beginning. The online (cost) approach performs very similar to the online (threshold) approach for this use case.

4.3   Influence of Subscription Parameters

In a second series of experiments, we investigate the influence of different subscription parameters , i.e., h, α, w, and g. We again use a linear cost function and set the parameters by default to $h = 1$ day, $\alpha = 0.15$, $w = 12$ hours, and $g =$ mean. In the following, we vary one parameter at a time and examine the influence on the best horizon extension k.

Increasing the threshold α leads to longer threshold violation intervals and thus larger horizon extensions. There is a larger increase for the UK and MER datasets than for the NREL dataset. Both data sets constitute more robust forecast values than NREL leading to few threshold violations with high α.

Figure 8(b) shows the influence of the aggregation window w on the horizon extension. We see a strong increase for the MER dataset, while the horizon extensions for the UK and NREL datasets stay almost constant. This is caused by the customer-level granularity of the MER dataset with many fluctuations. Longer aggregation windows weaken this effect and lead to smaller horizon extensions.

Figure 8(c) shows the best k for the aggregation functions mean and max, where it is consistently larger for mean (with $w > 1$). The UK and NREL dataset show only slight differences, while the MER dataset has a larger difference in the best horizon extension because there forecasts exhibit large fluctuations and the use of max triggers threshold violation notifications immediately.

Note that we do not show an experiment for varying horizons as it does not influence the threshold violation interval and thus the horizon extension.

To summarize, the impact of the subscription parameters on k depends strongly on the dataset, which validates our approach of analyzing the time series history to determine the best horizon extension.

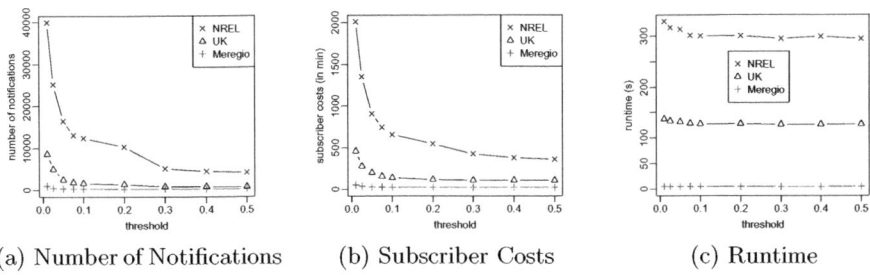

(a) Number of Notifications   (b) Subscriber Costs   (c) Runtime

Fig. 9. Computational Costs

4.4 Computational Costs

In this section, we analyze the computational costs of our approach as well as the overall relationship to the subscriber costs. For this experiment, we use a real cost function from our major use case, the energy domain. This real cost function includes runtime costs of the MIRABEL energy balancing approach [2] and shows a super linear behavior with increasing forecast horizon. We again fix the subscription parameters to $h = 1$ day, $w = 12$ hours and $g =$ mean and vary the subscriber threshold α. To determine the best horizon extension, we use our static approach. Figure 9 shows the trade off between number of sent (or received) notifications, the total resulting subscriber costs and the total time to produce these notifications for all three data sets. Clearly, with increasing threshold α the number of sent notifications decreases as forecasts can have a larger deviation before a notification needs to be sent and thus the notification length increases (Figure 9(a)). In conjunction, the subscribers costs decrease as well as less notifications need to be processed (Figure 9(b)). The runtime of our approach is nearly independent from the number notifications but depends on the data set, i.e., length, granularity and general accuracy (Figure 9(c)). The MER data exhibits the lowest runtime as it consists of hourly data over eight months. In contrast, the NREL data set shows the longest runtime as it is available in a 10 min resolution and hard to forecast. We displayed the overall time to produce all notifications to be comparable with the total subscriber costs. As it can be seen the runtime of our approach is much lower than the total subscriber runtime and thus the subscriber costs form the dominant factor. Note that the average time to produce a single notification equals less than 10 ms for all data sets. This time is measured in a local setting and includes forecast model maintenance, subscription evaluation and sending a notification if required.

4.5 Cost Model Validation

Finally, we validate our cost model, where we use again the real world cost function described in the runtime experiments. To evaluate our cost model we divide the MER data set into two parts of equal size (additionally to the training data

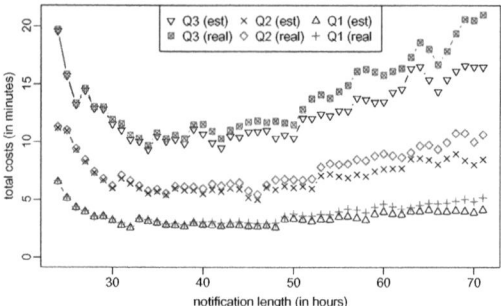

Fig. 10. Cost Model Validation

used for creating the forecast models). On the first part we estimate the total subscribers costs using our cost model defined in Equation 2. On the second part we measure the real subscribers costs for different notification lengths $h + k$. Figure 10 shows the resulting total subscriber costs for three different kind of queries with increasing complexity (Q1 - Q3), i.e., increasing balancing time. Note that the minimum notification length is determined by the forecast horizon, which is $h = 1$ day. For all three queries, our cost model is very accurate for small notification lengths but deviates from the real costs with increasing length. Small notification lengths result in many horizon violations, which are simple to estimate as they are time dependent. With larger notification lengths more threshold violations occurs, where our cost model can never achieve perfect accuracy as we do not know the future. However, most importantly, for all three queries, the minimum of the estimated and real costs are roughly the same leading to the same best horizon extension k. For the UK data set, our cost models performs much better than for the MER data set as this time series, and thus the threshold violation intervals, are easier to predict. In contrast, for the NREL data set our cost model performs slightly worse than for the MER data set as wind data is very fluctuating.

5   Related Work

Existing work has addressed the integration of time series forecasting into DBMS. Approaches to increase speed and accuracy of ad-hoc [7] and recurring queries [8,9] have been proposed. The Fa system [7] also processes continuous forecast queries. However, in contrast to subscription-based forecast queries no notification conditions can be defined and new forecasts are provided in a regular time interval, independently of the actual changes in forecasts.

The concept of notifying users about incoming events generated by data sources has been intensively investigated in the area of publish-subscribe systems [12,3]. Probably mostly related to our approach is the work on value-based subscriptions [3]. There, the subscriber wishes to receive an update if a new value (e.g., price) differs from the old one by more than a specified interval. The

propose different index structures to scale to a large number of subscribers. In contrast, we need to notify the subscriber with multiple values and our focus is the reduction of the costs of the subscriber.

The problem of tracking a value over time is more generalized in the area of function tracking [5,20]. An observer monitors a—possibly multi-valued—function and keeps a tracker informed about the current function value(s) within a predefined distance. Yi and Zhang [20] also suggest to use predictions in order to further reduce communication costs. Our work differs in the sense that we already deal with predictions of arbitrary horizons. Hence, in addition to the number of notifications, we reduce the individual notification lengths in terms of the number of forecast values.

The usage of statistical models to reduce communication between two or more entities is applied in a wide range of areas, e.g., sensor networks [6], bounded approximate caching [16] or mobile objects [4]. For example, within sensor networks energy requirements as well as query processing times are reduced by utilizing statistical models in combination with live data acquisition. In the area of bounded approximate caching, cached copies of data are allowed to become out of date according to specified precision constraints. However, all these approaches significantly differ from our problem statement and solutions. First, we consider only time series data, where future values depend on past ones, requiring different statistical models. Second, we use statistical models to estimate future values of the time series instead of missing real values, which leads to specific notification conditions (horizon- and threshold-based) and notification characteristics (resend all values or only additional ones). Finally, instead of reducing the communication costs between the entities, we reduce the costs of applications that process the forecast values.

6   Conclusion and Future Work

We introduced the concept of subscription-based forecast queries. Their main characteristic is a twofold notification condition: horizon- and threshold-based. This results in two different goals of increasing the notification length to avoid horizon-based notifications and reducing the notification length to avoid resending a lot of values if a threshold-based notification occurs. In this paper, we focused on optimizing these notifications to reduce the costs of the subscriber. We developed different computation approaches for different optimization problems, which all use the time series history to determine a suitable notification length. Our experimental evaluation shows the superiority of our computational approaches over alternatives, a significant reduction of the subscriber costs with low computational overhead as well as the validity of our cost in real world situations. In future work, we plan to extend our initial system and discuss data structures and processing approaches to handle a large number of subscribers.

Acknowledgement. This work has been carried out in the MIRABEL project funded by the EU under the grant agreement number 248195.

References

1. Akdere, M., Cetintemel, U., Riondato, M., Upfal, E., Zdonik, S.: The Case for Predictive Database Systems: Opportunities and Challenges. In: CIDR (2011)
2. Boehm, M., Dannecker, L., Doms, A., Dovgan, E., Filipic, B., Fischer, U., Lehner, W., Pedersen, T.B., Pitarch, Y., Siksnys, L., Tusar, T.: Data management in the mirabel smart grid system. In: EnDM (2012)
3. Chandramouli, B., Phillips, J.M., Yang, J.: Value-based Notification Conditions in Large-Scale Publish/Subscribe Systems. In: VLDB (2007)
4. Chen, S., Ooi, B.C., Zhang, Z.: An Adaptive Updating Protocol for Reducing Moving Object Database Workload. In: VLDB (2010)
5. Cormode, G., Muthukrishnan, S., Yi, K.: Algorithms for Distributed Functional Monitoring. In: SODA (2008)
6. Deshpande, A., Guestrin, C., Madden, S.R., Hellerstein, J.M., Hong, W.: Model-driven Data Acquisition in Sensor Networks. In: VLDB (2004)
7. Duan, S., Babu, S.: Processing Forecasting Queries. In: VLDB (2007)
8. Fischer, U., Rosenthal, F., Lehner, W.: F2db: The flash-forward database system (demo). In: ICDE (2012)
9. Ge, T., Zdonik, S.: A Skip-list Approach for Efficiently Processing Forecasting Queries. In: VLDB (2008)
10. Hyndman, R.J., Khandakar, Y.: Automatic Time Series Forecasting: The forecast Package for R. Journal of Statistical Software (2008)
11. Levis, P., Patel, N., Culler, D., Shenker, S.: Trickle: A Self-Regulating Algorithm for Code Propagation and Maintenance in Wireless Sensor Networks. In: NSDI (2004)
12. Liu, H., Jacobsen, H.-A.: Modeling Uncertainties in Publish/Subscribe Systems. In: ICDE (2004)
13. MeRegio Project (2011), http://www.meregio.de/en/
14. Nationalgrid UK. Demand Dataset (2011),
    http://www.nationalgrid.com/uk/Electricity/Data/Demand+Data/
15. NREL. Wind Integration Datasets (2011),
    http://www.nrel.gov/wind/integrationdatasets/
16. Olston, C., Widom, J.: Offering a Precision-Performance Tradeoff for Aggregation Queries over Replicated Data. In: VLDB (2000)
17. Peeters, E., Belhomme, R., Battle, C., Bouffard, F., Karkkainen, S., Six, D., Hommelberg, M.: Address: Scenarios and Architecture for Active Demand Development in the Smart Grids of the Future. In: CIRED (2009)
18. Shivam, P.: Active and Accelerated Learning of Cost Models for Optimizing Scientific Applications. In: VLDB (2006)
19. Taylor, J.W.: Triple Seasonal Methods for Short-Term Electricity Demand Forecasting. European Journal of Operational Research (2009)
20. Yi, K., Zhang, Q.: Multi-Dimensional Online Tracking. In: SODA (2009)

Minimizing Index Size by Reordering Rows and Columns[*]

Elaheh Pourabbas[1], Arie Shoshani[2], and Kesheng Wu[2]

[1] National Research Council, Roma, Italy
[2] Lawrence Berkeley National Laboratory, Berkeley, CA, USA

Abstract. Sizes of compressed bitmap indexes and compressed data are significantly affected by the order of data records. The optimal orders of rows and columns that minimizes the index sizes is known to be NP-hard to compute. Instead of seeking the precise global optimal ordering, we develop accurate statistical formulas that compute approximate solutions. Since the widely used bitmap indexes are compressed with variants of the run-length encoding (RLE) method, our work concentrates on computing the sizes of bitmap indexes compressed with the basic Run-Length Encoding. The resulting formulas could be used for choosing indexes to build and to use. In this paper, we use the formulas to develop strategies for reordering rows and columns of a data table. We present empirical measurements to show that our formulas are accurate for a wide range of data. Our analysis confirms that the heuristics of sorting columns with low column cardinalities first is indeed effective in reducing the index sizes. We extend the strategy by showing that columns with the same cardinality should be ordered from high skewness to low skewness.

1 Introduction

Bitmap indexes are widely used in database applications [4,16,25,30,34]. They are remarkably efficient for many operations in data warehousing, On-Line Analytical Processing (OLAP), and scientific data management tasks [6,10,22,26,28,29]. A bitmap index uses a set of bit sequences to represent the positions of the values as illustrated in Figure 1. In this small example, there is only a single column $\mathbf{X}$ in the data table, and this column $\mathbf{X}$ has only six distinct values 0, 1, ..., 5. Corresponding to each distinct value, a bit sequence, also known as a bitmap, is used to record which rows have the specific value. This basic bitmap index requires $C \times N$ bits for a column with C distinct values and N rows. In the worst case, where every value is distinct, the value of C is N, this basic bitmap index requires N^2 bits, which is exceedingly large even for modest datasets. To reduce the index sizes, a bitmap index is typically compressed [4,23,30].

---

[*] The authors gratefully acknowledge the suggestion from an anonymous referee that clarifies Lemma 1. This work was supported by the Director, Office of Science, Office of Advanced Scientific Computing Research, of the U.S. Department of Energy under Contract No. DE-AC02-05CH11231.

A. Ailamaki and S. Bowers (Eds.): SSDBM 2012, LNCS 7338, pp. 467–484, 2012.
© Springer-Verlag Berlin Heidelberg 2012

RID	X	bitmaps					
		b_0 $=0$	b_1 $=1$	b_2 $=2$	b_3 $=3$	b_4 $=4$	b_5 $=5$
1	0	1	0	0	0	0	0
2	1	0	1	0	0	0	0
3	1	0	1	0	0	0	0
4	2	0	0	1	0	0	0
5	3	0	0	0	1	0	0
6	4	0	0	0	0	1	0
7	4	0	0	0	0	1	0
8	5	0	0	0	0	0	1

Fig. 1. The logical view of a sample bitmap index with its seven bitmaps shown as the seven columns on the right. In this case, a bit is 1 if the value of **X** in the corresponding row is the value associated with the bitmap.

The most efficient compression techniques for bitmap indexes are based on run-length encoding (RLE) [3, 30], which can be significantly affected by ordering of the rows [21, 28]. The best known strategy for ordering the columns is to place the column with the lowest cardinality first [21]. However, much of the earlier work only analyze the worst case scenario and is only applicable to uniform random data. In this work, we provide an analysis strategy that works for non-uniform distributions and therefore provide more realistic understanding of how to order the columns. We demonstrate that our formulas produce accurate estimates of the sizes. Our analysis also leads to a new ordering strategy: for columns with the same cardinality order the columns with high skew first.

2   Related Work

The size of a basic bitmap index can grow quickly for large datasets. The methods for controlling index sizes mostly fall in one of the following categories, compression [3, 7, 30], binning [19, 33] and bitmap encoding [9, 27, 32]. In this paper, we concentrate on compression. More specifically, how column and row ordering affects the sizes of compressed bitmap indexes. In this section, we provide a brief review of common compression techniques for bitmap indexes, and reordering techniques for minimizing these index sizes.

2.1   Compressing Bitmap Indexes

Any lossless compression technique could be used to compress a bitmap index. However, because these compressed bitmaps need to go through complex computations in order to answer a query, some compression methods are much more effective than others. In a simple case, we could read one bitmap and return it as the answer to a query, for example, the bitmap b_1 contains the answer to query condition "$0 < \mathbf{X} < 2$." However, to resolve the query condition "$\mathbf{X} > 2$," we need to bring bitmaps b_3 through b_5 into memory and then perform two bitwise logical OR operations. In general, we may access many more bitmaps and perform many more operations.

To answer queries efficiently, we need to read the bitmaps quickly and perform the bitwise logical operations efficiently. We can not concentrate on the I/O time and neglect the CPU time. For example, the well-known method LZ77 can compress well and therefore reduce I/O time, however, the total time needed with LZ77 compression is much longer than with specialized methods [17, 31].

Among the specialized bitmap compression methods, the most widely used is the *Byte-aligned Bitmap Code* (BBC) [3], which is implemented in a popular commercial database management system. In tests, it compresses nearly as well as LZ77, but the bitwise logical operations can directly use the compressed bitmaps and therefore require less memory and less time [2, 17]. Another method that works quite well is the *Word-Aligned Hybrid* (WAH) code [30, 31]. It performs bitwise logical operations faster than BBC, but takes up more space on disk. This is because WAH works with 32-bit (or 64-bit) words while BBC works with 8-bit bytes. Working with a larger unit of data reduces the opportunity for compression, but the computations are better aligned with the capability of CPUs. Due to its effectiveness, a number of variations on WAH have also appeared in literature [11, 12, 14].

The key idea behind both BBC and WAH is *Run-Length Encoding* (RLE) that represents a sequence of identical bits with a count. They represent short sequences of mixed 0s and 1s literally, and BBC and the newer variants of WAH also attempt to pack some special patterns of mixed 0s and 1s. In the literature, each sequence of identical values is called a *run*. To enable a concise analysis, we choose to analyze the bitmap index compressed with RLE instead of BBC or WAH. In a straightforward implementation of RLE, one word is used to record each run. Therefore, our analysis focuses on the number of runs.

The commonly used bitmap compression methods such as BBC and WAH are more complex than RLE, and their compressed sizes are more difficult to compute. The existing literature has generally avoided directly estimating the compressed index sizes [18, 20, 21, 28]. In this work, we take a small step away from this general practice and seek to establish an accurate estimation for the sizes of RLE compressed bitmap indexes. We will show that our formulas are accurate and amenable to analysis.

2.2   Data Reordering Techniques

Reordering can improve the compression for data and indexes [1, 21]. Some of the earliest work on this subject was designed to minimize the number of disk accesses needed to answer queries by studying the consecutive retrieval property [13, 15]. Since the bitmap index can be viewed as a bit matrix, minimizing the index sizes is also related to the consecutive ones property [8]. These properties are hard to achieve and approximate solutions are typically used in practice.

A widely used data reordering strategy is to sort the data records in lexicographical order [5, 20, 24]. Many alternative ordering methods exist, one well-known example is the *Gray* code ordering. No matter how sorting is done, a common question is which column to use first. There is a long history of publications on this subject. Here is a brief review of a few of them.

One of the earliest publications on this subject was by Olken and Rotem [24]. In that paper, the authors investigated both deterministic and probabilistic models of data distribution and determined that rearranging data to optimize the number of runs is NP-Complete via reduction to the Traveling Salesman Problem (TSP). Under a probabilistic model, the optimal rearrangement for each attribute has the form of a double pipe organ. The key challenge for implementing their recommendation is that the computational complexity grows quadratically with N, the number of rows in the dataset.

Pinar et al. [28] similarly converted the problem of minimizing the number of runs into a Traveling Salesman Problem. They suggested Gray code ordering as an efficient alternative to the simple lexicographical ordering or TSP heuristics, and presented some experimental measurements to confirm the claim.

Apaydin et al. considered two different types of bitmap indexes under lexicographical ordering and Gray code ordering [5]. They found that Gray code ordering gives slightly better results for the Range encoded bitmap index.

In a series of publications, Lemire and colleagues again proved that minimizing the number of runs is NP hard and affirmed that sorting the columns from the lower cardinality is an effective strategy [18, 20, 21].

All these analyses focus on the bit matrix formed by a bitmap index and consider essentially the worst case scenario, therefore are mostly applicable to uniform random data. In real word applications, the data is hardly ever uniform random numbers. Our work address this nonuniformity by developing an accurate approximation that can be evaluated analytically.

3   Theoretical Analysis

To make the analysis tasks more tractable, we count the number of runs, which is directly proportional to the size of a RLE compressed bitmap index. The key challenge of this approach is that even though the definition of a run only involves values of each individual column, the number of runs for one column depends on the columns sorted before it. To address this challenge, we develop the concept of *leading k-tuple* to capture the dependency among the columns. As we show next this concept can capture the expected number of runs and allows us to evaluate how the key parameters of data affects the expected number of runs and the index sizes.

3.1   Counting k-tuples

Without loss of generality, we concentrate our analysis of a table of integers with N rows and M columns. To avoid the need to construct a bitmap index and count the number of runs, we introduce a quantity that can be directly measured as follows.

X	Y	Z
10	20	30
10	22	33
11	20	31
11	21	30
11	21	32

Z	Y	X
30	20	10
30	21	11
31	20	11
32	21	11
33	22	10

A) Sort **X** first   B) Sort **Z** first

Fig. 2. A small data table sorted in two different ways

Definition 1. *A* chunk *is a sequence of identical values of a column in consecutive rows.*

For the examples shown in Figure 2, in the version that sorted **X** first, the values of **X** form two chunks, one with the value 10 and the other with the value 11. In the version sorted **Z** first, the values of **X** form three chunks, two chunks with the value 10 and one chunk with the value 11. Note that a chunk always includes the maximum number of consecutive identical values. We do not break the three consecutive values into smaller chunks.

The two tables shown in Figure 2 are sorted with different column orders and have different number of chunks. To capture this dependency on column order, we introduce a concept called *leading k-tuple*.

Definition 2. *A* leading k-tuple *is a tuple of k values from the first k columns of a row.*

Depending how the columns are ordered, the leading k-tuples will be different. For example, the first leading 2-tuple in Fig. 2A is (**X**=10, **Y**=20) and the second leading 3-tuple in Fig. 2B is (**Z**=30, **Y**=21, **X**=11). In this paper, when we refer to a *k-tuple*, we only refer to a leading k-tuple, therefore we usually use the shorter term.

Without loss of generality, we refer to the first column in our reordering as column 1 and the kth column as column k. In Fig. 2A, column 1 is **X**, while in Fig. 2B, column 1 is **Z**. Similarly, we refer to the jth smallest value of column k as the value j without regards to its actual content.

An critical observation is captured in the following lemma.

Lemma 1. *The number of chunks for column k is bounded from above by the number of distinct leading k-tuples T_k.*

Typically, T_{k-1} is much smaller than T_k and the number of chunks for column k is very close to T_k. For this reason, we count the number of k-tuples instead of counting the number of chunks. We will discuss the difference between T_k and the number of chunks in Section 3.2.

Assume that the data table was generated through a stochastic process and the probability of a leading k-tuple $(j_1, j_2, \ldots, j_k)$ appearing in the data table is $p_{j_1 j_2 \ldots j_k}$. After generating N such rows, the probability that a particular k-tuple $(j_1, j_2, \ldots, j_k)$ is missing from the data table is $(1 - p_{j_1 j_2 \ldots j_k})^N$. Let C_1

denote the number of possible values for column 1, and C_k denote the number of possible values for column k. The total number of distinct k-tuples is $C_1 C_2 \ldots C_k$. Summing overall all possible leading k-tuples, we arrive at the number of missing k-tuples as $\sum_{j_1 j_2 \ldots j_k} (1 - p_{j_1 j_2 \ldots j_k})^N$, and the number of distinct k-tuples in the data table as

$$T_k = \prod_{i=1}^{k} C_i - \sum_{j_1 j_2 \ldots j_k} (1 - p_{j_1 j_2 \ldots j_k})^N . \tag{1}$$

Note that the above formula works with the probability of k-tuples and is applicable to any data set, even where the columns exhibit correlation. In most cases, the probability of the k-tuples $p_{j_1 j_2 \ldots j_k}$ is harder to obtain than the probability of an individual column. To make use of the probability distribution of the columns, we assume the columns are statistically independent, and $p_{j_1 j_2 \ldots j_k} = p_{j_1} p_{j_2} \cdots p_{j_k}$. The above expression of T_k can be rewritten as:

$$T_k = \prod_{i=1}^{k} C_i - \sum_{j_1 j_2 \ldots j_k} (1 - p_{j_1} p_{j_2} \cdots p_{j_k})^N \tag{2}$$

Our goal is to count the number of runs in the bitmap index for each column of the data table. For the bitmaps shown in Fig. 1, roughly each chunk in the values of $\mathbf{X}$ leads to two runs in some bitmaps. We can generalize this observation as follows.

Lemma 2. *For a column with T_k chunks and C_k distinct values, the number of runs in the bitmap index is $2T_k + C_k - 2$.*

Proof. For each chunk in column k, the corresponding bitmap in the bitmap index will have a sequence of 0s followed by a sequences of 1s. This leads to the term $2T_k$ runs for T_k chunks. For most of the C_k bitmaps in the bitmap index, there is a sequences of 0s at the end of the bitmap. Altogether, we expect $2T_k + C_k$ runs. However, there are two special cases. Corresponding the first chunk in the values of column k, there is no 0 before the corresponding 1s. Corresponding to the last chunk, there is no 0s at the end of the bitmap. Thus, there are two less runs than expected. The total number of runs is $2T_k + C_k - 2$.

3.2   Accidental Chunks

As a sanity check, we next briefly consider the case where all columns are uniformly distributed. This also helps us to introduce the second concept we call the *accidental chunks* which captures the error of using the number of leading k-tuples to approximate the number of chunks for column k.

Assuming the probability of each value is the same, i.e., $p_{j_i} = C_i^{-1}$, we can significantly simplify the above formula as

$$T_k = \left(1 - \left(1 - \prod_{j=1}^{k} C_j^{-1} \right)^N \right) \prod_{j=1}^{k} C_j \tag{3}$$

Table 1. Example of sorting 1 million rows (N=1,000,000) with column cardinality from lowest to highest

C	Max chunks	Exp chunks	Max runs	Exp runs	Actual runs	Error (%)
10	10	10	28	28	28	0
20	200	200	418	418	418	0
40	8000	8000	16038	16038	16038	0
60	480000	420233	960058	840524	841142	0.074
80	38400000	987091	76800078	1974260	1966758	-0.38
100	3840000000	999869	7680000098	1999836	1980078	-1.00

Table 2. Example of sorting 1 million rows (N=1,000,000) with column cardinality from highest to lowest

C	Max chunks	Exp chunks	Max runs	Exp runs	Actual runs	Error (%)
100	100	100	298	298	298	0
80	8000	8000	16078	16078	16078	0
60	480000	420233	960058	840524	840584	0.007
40	19200000	974405	38400038	1948848	1934192	-0.0075
20	384000000	998699	768000018	1997416	1898906	-4.93
10	3840000000	999869	7680000008	1999746	1800250	-9.976

We generated one million rows of uniform random numbers and actually counted the number of chunks and number of runs; the results are shown in Tables 1 and 2. The actual observed chunks include identical values appearing contiguously crossing the k-tuple boundaries. Even though two k-tuples may be different, the values of the last column, column k, could be the same. For example, in Fig. 2B, the three row in the middle all have **X**=1, even though the corresponding 3-tuple are different. This creates what we call *accidental chunks*.

Definition 3. *An accidental chunk in column k is a group of identical values for column k where the corresponding k-tuples are different.*

In general, we count a chunk for column k as an *accidental chunk*, if the values of the kth column are the same, but the leading $(k-1)$-tuples are different. After sorting, the leading $(k-1)$-tuples are ordered and the identical tuples are in consecutive rows. Since the column k is assumed to be statistically independent from the first $(k-1)$ columns, we can compute the number of consecutive identical values as follows.

Starting from an arbitrary row, the probability of the column k being j_k is p_{j_k} and the probability that there is only a single j_k (followed by something else) is $p_{j_k}(1 - p_{j_k})$. The probability that there are two consecutive rows with j_k is $p_{j_k}^2(1 - p_{j_k})$, and the probability for q consecutive j_k is $p_{j_k}^q(1 - p_{j_k})$. The numbers of time j_k appears together is:

$$p_{j_k}(1 - p_{j_k}) + 2p_{j_k}^2(1 - p_{j_k}) + 3p_{j_k}^3(1 - p_{j_k}) + \ldots + (N-1)p_{j_k}^{N-1}(1 - p_{j_k}) + Np_{j_k}^N$$
$$\cong p_{j_k}(1 - p_{j_k}) \sum_{i=1}^{\infty} i p_{j_k}^{i-1} = p_{j_k}(1 - p_{j_k})(1 - p_{j_k})^{-2} = p_{j_k}(1 - p_{j_k})^{-1},$$

where[1] $\sum_{i=1}^{\infty} i p^{i-1} = \sum_{i=1}^{\infty} \frac{\partial p^i}{\partial p} = \frac{\partial (\sum_{i=1}^{\infty} p^i)}{\partial p} = \frac{\partial (1-p)^{-1}}{\partial p} = (1-p)^{-2}$. The average times a value of column k repeats is $\mu_k = \sum_{j_1 j_2 \cdots j_k} \frac{p_{j_k}}{1-p_{j_k}}$.

The set of consecutive values in column k must span beyond the group of identical $(k-1)$-tuples in order to be counted as an accidental chunk. When the majority of the chunks for the first $(k-1)$ columns have only 1 row, the number of chunks for column k is reduced by a factor $1/\mu_k$.

For uniform random data, we can estimate the values of μ_k and check whether they agree with the observations from Tables 1 and 2. These tables show an example of sorting 1 million tuples with column cardinality ordered from the lowest to the highest and from the highest to the lowest, respectively. In Table 1, we see that about 99% of 5-tuples are distinct, more precisely, there are 987091 5-tuples for 1 million rows. This suggests that there might be a noticeable number of accidental chunks for column 6 shown in the last row in Table 1. In this case, the cardinality of column 6 is 100, $p_{j_6} = 1/100$ and $\mu_6 = \sum_{j_6=1}^{100} \frac{1/100}{1-1/100} = \frac{100}{99}$. The number of chunks observed should be T_6/μ_6, which is 1% less than the expected value of T_6. The number of runs in the bitmaps is 1% less the expected value in Table 1, which agree with our analysis. Similarly, in Table 2 about 97% of the 4-tuples are distinct, which suggests that there might be noticeable number of accidental chunks for columns 5 and 6. The cardinality of the last two columns are 20 and 10 respectively, and our formula suggests that the observed number of chunks would be 5% and 10% less than the expected value of T_5, T_6. In Table 2, we again see a good agreement with the predictions[2].

3.3   Asymptotic Case

Assume the value of each $p_{j_1 j_2 \cdots j_k}$ to be very small, say $p_{j_1 j_2 \cdots j_k} \ll 1/N$. In this case, we can approximate the probability that the k-tuple $(j_1, j_2, \ldots, j_k)$ not appearing in our dataset as $(1 - p_{j_1 j_2 \cdots j_k})^N \approx 1 - N p_{j_1 j_2 \cdots j_k}$. This leads to the following approximation for the number of distinct k-tuples.

$$
\begin{aligned}
T_k &\approx \prod_{i=1}^{k} C_i - \sum_{j_1 j_2 \cdots j_k} (1 - N p_{j_1 j_2 \cdots j_k}) \\
&= \prod_{i=1}^{k} C_i - \sum_{j_1 j_2 \cdots j_k} 1 + N \sum_{j_1 j_2 \cdots j_k} p_{j_1 j_2 \cdots j_k} \\
&= N \sum_{j_1 j_2 \cdots j_k} p_{j_1 j_2 \cdots j_k} \quad = \quad N.
\end{aligned}
$$

In other word, every k-tuple will be distinct. If the probability of each individual k-tuple is very small, we intuitively expect all the observed tuples to be distinct. We generalize this observation and state it more formally as follows.

Definition 4. *A tuple $(j_1 j_2 \ldots j_k)$ in a dataset with N tuples is a rare tuple if $p_{j_1 j_2 \cdots j_k} < 1/N$.*

---

[1] http://mathworld.wolfram.com/PowerSum.html

[2] Lemma 1 seems to suggest that the errors in Tables 1 and 2 can only be negative or zero, however there are a few positive numbers. The reason for this is that the number of chunks and runs given are based on the expected number of leading k-tuples, not the number of leading k-tuple observed in the particular test dataset.

In most discussions, we will simply refer to such a tuple as a rare tuple, without referring to the number of rows, N.

Conjecture on rare tuples: *A rare tuple in a dataset will appear exactly once if it does appear in the dataset.*

In a typical case, there is a large number of possible tuples and many of them with very small probabilities while a few tuples with larger probabilities. Therefore, we cannot apply the above estimate to the whole dataset. The implication from the above conjecture is that the rare tuples that do appear in a data set will be different from others. To determine the total number of distinct tuples, we can concentrate on those tuples that appear more frequently, which we call *common tuples*.

Definition 5. *A tuple $(j_1 j_2 \ldots j_k)$ in a dataset with N tuples is a common tuple if $p_{j_1 j_2 \ldots j_k} \geq 1/N$.*

In most discussions, we will simply refer to such a tuple as a common tuple, without referring to the number of rows, N.

3.4   Zipfian Data

In the preceding sections, we have demonstrated that our formulas predict the numbers of runs accurately for uniform data and rare tuples. Next, we consider the more general case involving data with non-uniform distribution and a mixture of rare tuples and common tuples. In order to produce compact formulas, we have chosen to concentrate on data with Zipf distributions. We further assume that each column of the data table is statistically independent from others.

Following the above analysis on rare tuples, we assume that all rare tuples that do appear in a dataset are distinct. A common tuple may appear more than once in a dataset, we say that it has duplicates. More specifically, if a k-tuple appears q times, then it has $(q - 1)$ duplicates. Let D_k denote the number of duplicates, the number of distinct k-tuples in the dataset is $T_k = N - D_k$. This turns the problem of counting the number of distinct values into counting the number of duplicates. To illustrate this process, let us first consider a case of 1-tuple, i.e., one column following the Zipf distribution $p_{j_1} = \alpha_1 j_1^{-z_1}$, where $\alpha_1 = \left(\sum_{j_1=1}^{C_1} j_1^{-z_1} \right)^{-1}$ and z_1 is a constant parameter known the Zipf exponent.

By definition of p_{j_1}, the value j_1 is expected to appear $N p_{j_1}$ times in that dataset. The common values are expected to appear at least once, i.e., $N p_{j_1} = N \alpha_1 j_1^{-z_1} \geq 1$. Since p_{j_1} is a monotonically decreasing function of j_1, common values are smaller than rare ones. Let $\beta_1 \equiv (N \alpha_1)^{1/z_1}$, we see that all j_1 values less than C_1 and $\lfloor \beta_1 \rfloor$ (where $\lfloor . \rfloor$ is the floor operator) are common values. The number of duplicates can be expressed as

$$D_1 = \sum_{j_1=1}^{min(C_1, \lfloor \beta_1 \rfloor)} \left(N \alpha_1 j_1^{-z_1} - 1 \right). \tag{4}$$

For the convenience of later discussions, we define two functions

$$s_1 = \sum_{j_1=1}^{min(C_1,\lfloor \beta_1 \rfloor)} j_1^{-z_1}, \qquad r_1 = \sum_{min(C_1,\lfloor \beta_1 \rfloor)+1}^{C_1} j_1^{-z_1}.$$

By the definition of α_1, we have $\alpha_1 = r_1 + s_1$. Furthermore, the number of common values is $N\alpha_1 s_1$, the number of rare values is $N\alpha_1 r_1$, and the number of distinct values is $T_1 = min(C_1, \lfloor \beta_1 \rfloor) + N\alpha_1 r_1$. Among all possible values of β_1, when $\beta_1 < 1$, there is no common value and $T_1 = N$; when $\beta_1 \geq C_1$, all values are common values and $T_1 = C_1$.

In the more general case where the kth column has cardinality C_k and Zipf exponent z_k, we have $\alpha_k = 1/\sum_{j_k=1}^{C_k} j_k^{-z_k}$, the number of distinct k-tuples and the number of duplicate k-tuples are given by the following expressions,

$$T_k = N - D_k, D_k = \sum_{Np_{j_1 \ldots j_k} > 1} (Np_{j_1 \ldots j_k} - 1), p_{j_1 \ldots j_k} = \prod_{i=1}^{k} p_i = \prod_{i=1}^{k} \alpha_i j_i^{-z_i}. \quad (5)$$

Since the Zipf distribution is a monotonic function, it is straightforward to determine the bounds of the sum in Equation (5) as illustrated in the one-column case above. Given $j_1, \ldots, j_{k-1}$, the upper bound for j_k is given by $\left(N\alpha_k \prod_{i=1}^{k-1} \alpha_i j_i^{-z_i} \right)^{1/z_k}$. In many cases, there are a relatively small number of common tuples, which allows us to evaluate the above expression efficiently. From this expression, we can compute the number of k-tuples and therefore the number of runs in the corresponding bitmap indexes.

Theorem 1. *Let $C_1, C_2, \ldots, C_M$ denote column cardinalities of a data table. Assume all columns have the same skewness as measured by the Zipf exponents. To minimize the total number of runs in the bitmap indexes for all columns with sorting, the lowest cardinality column shall be sorted first.*

Proof. Let's first consider the case of 1-tuple. Given z_1, the summation $\sum_{j_1}^{C_1} j_1^{-z_1}$ increases as C_1 increases. The values of α_1 decreases as C_1 increases (as illustrated in Fig. 3A) which can be expressed as $\partial \alpha_1 / \partial C_1 < 0$. This leads to $\partial \beta_1 / \partial C_1 = \frac{1}{z_1}(N\alpha_1)^{(\frac{1}{z_1}-1)} N \frac{\partial \alpha_1}{\partial C_1} < 0$, which means that β_1 decreases as C_1 increases (as shown in Fig. 3B). In the expression for D_1, the upper bound of the summation is the minimal of C_1 and $\lfloor \beta_1 \rfloor$.

When $C_1 \leq \lfloor \beta_1 \rfloor$, all C_1 values are common values. In these cases, by definition of the Zipfian distribution $\sum_{j_1=1}^{C_1} N\alpha_1 j_1^{-z_1} = N$, the number of duplicates $D_1 = N - C_1$ (see Eq. (4)) and the number of distinct values is C_1.

In cases where $C_1 > \lfloor \beta_1 \rfloor$, the number of distinct values may be less than C_1. Because both α_1 and β_1 decrease with the increase of C_1, each term in the summation for D_1 decreases and the number of terms in the summation may also decrease, all causing D_1 to decrease as C_1 increases, as shown in Fig. 3C. In other words, as C_1 increases, there are fewer duplicates and more distinct values. As shown in Fig. 3D, the value of T_1 increases with C_1.

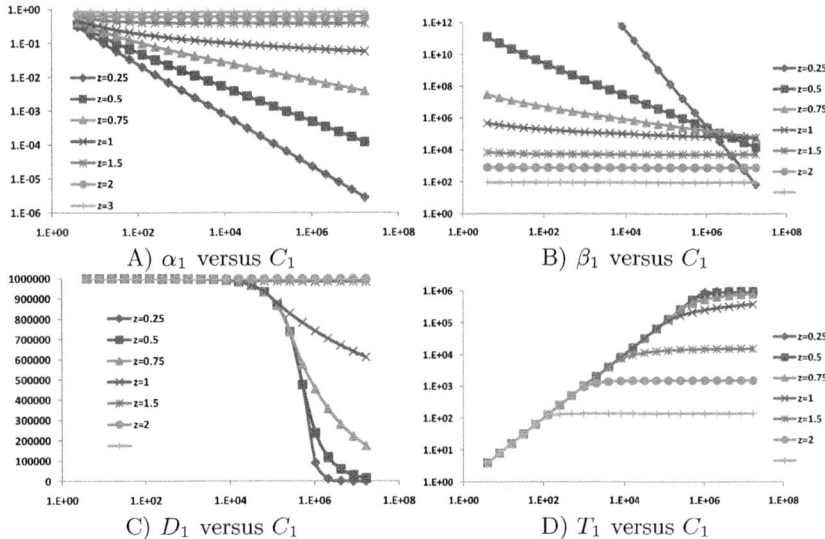

A) α_1 versus C_1 B) β_1 versus C_1

C) D_1 versus C_1 D) T_1 versus C_1

Fig. 3. How the values vary with column cardinality with fixed skewness (assuming 1 million rows)

Now, we consider the case of k-tuple. In this case, the probability for a k-tuple is $p_{j_1 j_2 \ldots j_k} = \alpha_1 \alpha_2 \ldots \alpha_k j_1^{-z} j_2^{-z} \ldots j_k^{-z}$, the common tuples are those with $\alpha_1 \alpha_2 \ldots \alpha_k j_1^{-z} j_2^{-z} \ldots j_k^{-z} \geq 1/N$, or alternatively, $j_1 j_2 \ldots j_k \leq (\alpha_1 \alpha_2 \ldots \alpha_k N)^{1/z}$. Note that the values $j_1 j_2 \ldots j_k$ are positive integers.

Along with the conditions that $1 \leq j_1 \leq C_1, \ldots, 1 \leq j_k \leq C_k$, the number of duplicate tuples can be expressed as follows ($\beta_k \equiv (\alpha_1 \alpha_2 \ldots \alpha_k N)^{1/z}$):

$$
D_k = \sum_{j_1=1}^{min(C_1, \lfloor \beta_1 \rfloor)} \sum_{j_1=1}^{min(C_2, \lfloor \beta_2 j_1^{-1} \rfloor)} \ldots
$$
$$
\sum_{j_1=1}^{min(C_k, \lfloor \beta_k j_1^{-1} j_2^{-1} \ldots j_{k-1}^{-1} \rfloor)} \left(N \alpha_1 \alpha_2 \ldots \alpha_k j_1^{-z} j_2^{-z} \ldots j_k^{-z} - 1 \right). \quad (6)
$$

In the above expression, the order of the column among the k-tuple does not change the number of distinct tuples. Therefore, when all columns are considered together, i.e., $k = M$, it does not matter how the columns are ordered. However, as soon as one column is excluded, say, $k = M - 1$, it does matter which columns are excluded. Assume that we have two columns to choose from, say columns A and B; and the only different between them is their column cardinalities, C_A and C_B. Without loss of generality, assume $C_A > C_B$, consequently, $\alpha_A < \alpha_B$ and $\beta_A < \beta_B$. In the formula D_k, a smaller β_A value indicates that the number of terms in the summation would be no more than that with a larger β_B. For each term, replacing the value of α_k with α_A will produce a smaller value than

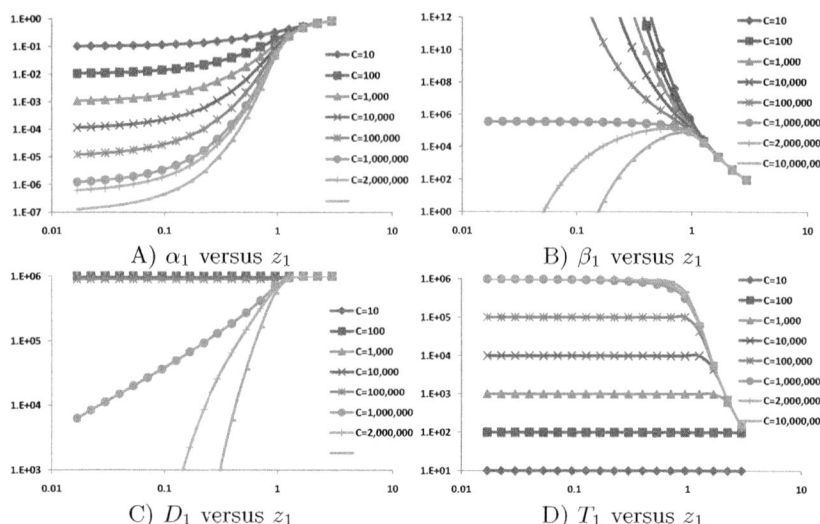

Fig. 4. How the values vary with skewness with fixed column cardinality (assuming 1 million rows)

replacing it with α_B. Therefore, choosing the higher cardinality column decreases D_k, increases the number of distinct tuples and increases the number of runs in the corresponding bitmap index. Thus, sorting the lowest cardinality column first reduces the number of distinct k-tuples and minimizes the number of runs.

Theorem 2. *Let $C_1, C_2, \ldots, C_M$ denote the column cardinalities of a data table. Assume $C_1 = C_2 = \ldots = C_M$. To minimize the total number of runs in the bitmap indexes by sorting, the column with the largest Zipf exponent shall be ordered first.*

Proof. Instead of giving a complete proof here, we will outline the basic strategy. Based on the information shown Fig. 4, particularly Fig. 4B, we need to handle the cases with $C > N$ and $z < 1$ separately from the others. In the normal cases, α_1, β_1, D_1 and T_1 are all monotonic functions and it is clear that sorting columns with larger Zipf exponents first is beneficial.

In the special case with $C > N$ and $z < 1$, where the possible values to use is large and the differences among the probabilities of different values are relatively small, the number of runs for a column is the same as that of a uniformly random column. In which case, which column comes first does not make a difference, and we can follow the general rule derived for the normal cases. Thus, we always sort the column with the largest Zipf exponent first.

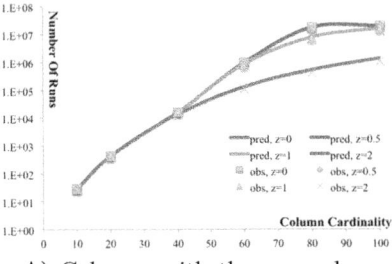

A) Columns with the same skewness ordered from low cardinality to high cardinality

B) Columns with the same skewness ordered from high cardinality to low cardinality

Fig. 5. Predicted numbers of runs (lines) and the observed numbers of runs (as symbols) plotted against the column cardinality (10 million rows). The symbols are very close to their corresponding lines indicating that the predictions agree well with the observations.

4   Experimental Measurements

In the previous section, we took the expected number of distinct k-tuple as an estimate of the number of chunks and therefore the number of runs in a bitmap index. In this section, we report a series of empirical measurements designed to address two issues: (1) how accurate are the formulas for predicting numbers of runs for Zipf data and (2) do the reordering strategies actually reduce index sizes? In these tests, we use a set of synthetic Zipfian data with varying cardinalities and Zipf exponents. The column cardinalities used are 10, 20, 40, 60, 80, and 100; the Zipf exponents used are 0 (uniform random data), 0.5, 1, and 2 (highly-skew data). The test data sets contain 10 million rows, which should be large enough to avoid significant statistical errors.

4.1   Number of Runs

The first set of measurements are the numbers of runs predicted by Eq. (5) and the numbers of runs actually observed on a set of Zipf data. We also collected the expected and the actual numbers of runs with different ordering of the columns. We first organize the synthetic data into four tables where all columns in each table have the same Zipf exponent. In Fig. 5, we display the numbers of runs for each bitmap index (for an individual column) with the data table sorted in two different column orders, the lowest cardinality column first or the highest cardinality column first. In this figure, the discrete symbols denotes the observed values and the lines depict the theoretical predictions developed in the previous section. The number of runs vary from tens to tens of millions. In this large range of values, our predictions are always very close to the actual observations.

Fig. 6 shows similar predicted numbers of runs against observed values for data tables containing columns with the same column cardinality. Again we see

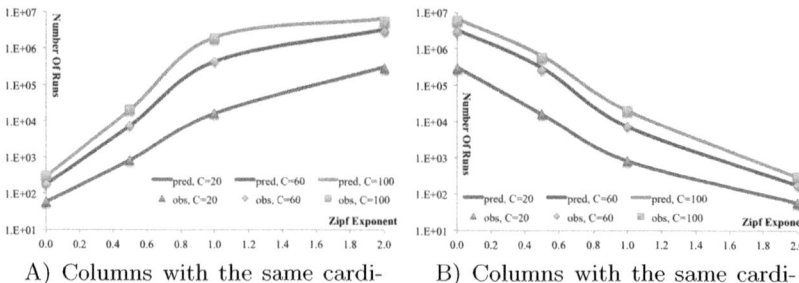

A) Columns with the same cardinalities are ordered from small Zipf exponent to large Zipf exponent

B) Columns with the same cardinalities are ordered from large Zipf exponent to small Zipf exponent

Fig. 6. Predicted numbers of runs (lines) and observed numbers of runs (symbols) plotted against the Zipf exponents (10 million rows). The symbols are very close to their corresponding lines indicating that the predictions agree well with the observations.

Table 3. Total number of runs (in thousands) of columns with the same skew in two different orders (N=10,000,000)

Skew	Total numbers of runs			
z	Low cardinality first (Small per Thm 1)		High cardinality first	
	predicted	observed	predicted	observed
0	38,559	38,382	56,281	53,545
0.5	38,506	35,266	55,904	48,988
1	25,254	22,328	35,629	29,523
2	2,065	1,639	2,557	1,892

that the number of runs vary from tens to millions, and the observed values agree with the predictions very well.

To see exactly how accurate are our predictions, in Table 3 and 4, we listed out the total number of runs for each of the data tables used to generate Fig. 5 and 6. In these two tables, the total numbers of runs are reported in thousands. As we saw in Table 1 and 2 for uniform random data, the predictions are generally slightly larger than the actually observed values. The discrepancy appears to grow as the skewness of the data grows or the column cardinality grows. We believe this discrepancy to be caused by the accidental chunks discussed in Section 3.2. We plan to verify this conjecture in the future.

Since the actual number of runs are different from the expected values, a natural question is whether the predicted advantage of column ordering still observed. In Table 3, we see that the total number of runs of tables sorted with the lowest cardinality column first is always smaller than the same value for the same table sorted with the largest cardinality column first. This is true even for highly-skew data (with $z = 2$), where the predicted total number of

Table 4. Total number of runs (in thousands) of columns with the same column cardinality ordering in two different ways (N=10,000,000)

Cardinality	Total numbers of runs			
C	Low skew first		High skew first (Small per Thm 2)	
	predicted	observed	predicted	observed
10	22	22	22	22
20	326	301	326	301
40	1864	1636	1860	1622
60	3767	3335	3638	3173
80	6016	5274	5472	4776
100	8706	7274	7353	6331

Table 5. The total sizes (KB) of compressed bitmap indexes produced by FastBit under different sorting strategies. Each data table has 10 million rows and four columns with the same column cardinality but different skewness.

C	10	20	40	60	80	100
Low skew first	168	1,540	6,911	11,188	15,146	19,487
High skew first	166	1,393	6,125	11,870	17,913	23,878

runs is nearly 35% higher than the actual observed value ($2065/1639 = 1.26$, $2557/1892 = 1.35$). In this case, sorting the the highest cardinality column first produces about 15% more runs than sorting the lowest cardinality column first ($1892/1639 = 1.15$). The predicted advantage is about 24% ($2557/1639 = 1.24$). Even though it may be worthwhile to revisit the source of error in our predictions, the predicted advantage from Theorems 1 and 2 are clearly present.

From our analysis, we predicted that ordering the columns with the highest skew first is the better than other choices. In Table 4, we show the total numbers of runs from two different sorting strategies, one with the highest skew first and the other with the least skew first. For the columns with relative low column cardinalities, there are enough rows to produce all possible tuples. In this case, the prediction is exactly the same as the observed values. As the column cardinality increases, there are larger differences between the predictions and observations. With each column having 100 distinct values, the difference is almost 20%. However, even in this case, the predicted advantage of sorting the column with the largest Zipf exponent first is still observable in the test.

4.2   FastBit Index Sizes

Even though commonly used bitmap compression methods are based on RLE, they are more complex than RLE and therefore our predictions may have larger errors. To illustrate this point, we show the sizes of a set of bitmap indexes produced by an open-source software called FastBit [29] in Fig. 7 and Table 5. In these tests, our test data is divided into six tables with four columns each to

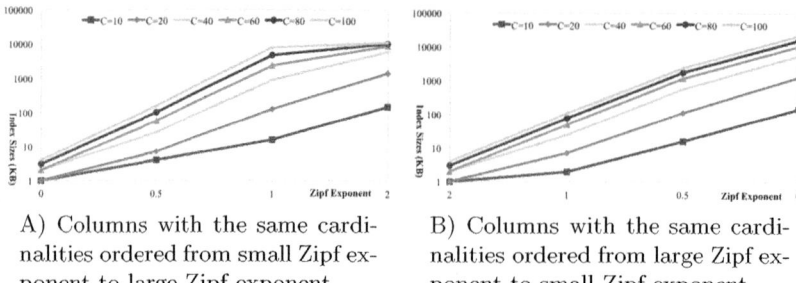

A) Columns with the same cardinalities ordered from small Zipf exponent to large Zipf exponent

B) Columns with the same cardinalities ordered from large Zipf exponent to small Zipf exponent

Fig. 7. Sizes of FastBit indexes ($N = 10,000,000$)

produce the sizes shown in Fig. 6 and Table 4. According to Theorem 2, ordering highly-skewed column first is expected to produce smaller compressed bitmap indexes. In Table 5, we see that the prediction is true for three out of the six data tables, those with $C = 10$, $C = 20$, and $C = 40$.

For data tables with high column cardinalities, the predictions are wrong and the values shown in Fig. 7 offers some clues as why. In general, the last column to be sorted is broken into more chunks, the corresponding bitmap index has more runs and requires more disk space. This index typically requires significantly more space than those of the earlier columns, and therefore dominates the total index size. In Fig. 7B, we see that the index sizes grow steadily from the column being sorted first to the column being sorted last. However, in Fig. 7A, we see the index size for column sorted last did not grow much larger than the previous columns. This is especially noticeable for $C = 100$, $C = 80$ and $C = 60$. More work is needed to understand this trend.

5   Conclusions

In this paper, we developed a set of formulas for the sizes of Run-Length Encoded bitmap indexes. To demonstrate the usefulness of these formulas, we used them to examine how to reorder the rows and columns of a data table. Our analysis extends the reordering heuristics to include non-uniform data. We demonstrated that the formulas are indeed accurate for a wide range of data. We also discussed the limitations of the proposed approach. In particular, because the practical bitmap compression methods are not simple Run-Length Encoding methods, their index sizes deviate from the predictions in noticeable ways. We plan to explore options to capture these deviations in a future study.

References

1. Abadi, D., Madden, S.R., Ferreira, M.C.: Integrating compression and execution in column-oriented database systems. In: SIGMOD. ACM (2006)

2. Amer-Yahia, S., Johnson, T.: Optimizing queries on compressed bitmaps. In: VLDB, pp. 329–338 (2000)
3. Antoshenkov, G.: Byte-aligned bitmap compression. Tech. rep., Oracle Corp. (1994)
4. Antoshenkov, G., Ziauddin, M.: Query processing and optimization in oracle rdb. The VLDB Journal 5, 229–237 (1996)
5. Apaydin, T., Tosun, A.S., Ferhatosmanoglu, H.: Analysis of Basic Data Reordering Techniques. In: Ludäscher, B., Mamoulis, N. (eds.) SSDBM 2008. LNCS, vol. 5069, pp. 517–524. Springer, Heidelberg (2008)
6. Bookstein, A., Klein, S.T.: Using bitmaps for medium sized information retrieval systems. Information Processing & Management 26, 525–533 (1990)
7. Bookstein, A., Klein, S.T., Raita, T.: Simple bayesian model for bitmap compression. Information Retrieval 1(4), 315–328 (2000)
8. Booth, K.S., Lueker, G.S.: Testing for the consecutive ones property, interval graphs, and graph planarity using pq-tree algorithms. Journal of Computer and System Sciences 13(3), 335 – 379 (1976),
http://dx.doi.org/10.1016/S0022-0000(76)80045-1
9. Chan, C.-Y., Ioannidis, Y.E.: Bitmap index design and evaluation. In: SIGMOD, pp. 355–366 (1998)
10. Chaudhuri, S., Dayal, U., Ganti, V.: Database technology for decision support systems. Computer 34(12), 48–55 (2001)
11. Colantonio, A., Pietro, R.D.: Concise: Compressed 'n' composable integer set. Information Processing Letters 110(16), 644–650 (2010),
http://dx.doi.org/10.1016/j.ipl.2010.05.018
12. Deliège, F., Pedersen, T.B.: Position list word aligned hybrid: optimizing space and performance for compressed bitmaps. In: EDBT 2010: Proceedings of the 13th International Conference on Extending Database Technology, pp. 228–239. ACM, New York (2010)
13. Deogun, J.S., Gopalakrishnan, K.: Consecutive retrieval property–revisited. Information Processing Letters 69(1), 15–20 (1999),
http://dx.doi.org/10.1016/S0020-0190(98)00186-0
14. Fusco, F., Stoecklin, M.P., Vlachos, M.: NET-FLi: on-the-fly compression, archiving and indexing of streaming network traffic. Proc. VLDB Endow. 3, 1382–1393 (2010), http://portal.acm.org/citation.cfm?id=1920841.1921011
15. Ghosh, S.P.: File organization: the consecutive retrieval property. Commun. ACM 15, 802–808 (1972), http://doi.acm.org/10.1145/361573.361578
16. Hu, Y., Sundara, S., Chorma, T., Srinivasan, J.: Supporting RFID-based item tracking applications in oracle DBMS using a bitmap datatype. In: VLDB 2005, pp. 1140–1151 (2005)
17. Johnson, T.: Performance of compressed bitmap indices. In: VLDB 1999, pp. 278–289 (1999)
18. Kaser, O., Lemire, D., Aouiche, K.: Histogram-aware sorting for enhanced word-aligned compression in bitmap indexes. In: DOLAP 2008, pp. 1–8. ACM, New York (2008), http://doi.acm.org/10.1145/1458432.1458434
19. Koudas, N.: Space efficient bitmap indexing. In: CIKM, pp. 194–201 (2000)
20. Lemire, D., Kaser, O., Aouiche, K.: Sorting improves word-aligned bitmap indexes. Data & Knowledge Engineering 69(1), 3–28 (2010),
http://dx.doi.org/10.1016/j.datak.2009.08.006
21. Lemire, D., Kaser, O.: Reordering columns for smaller indexes. Information Sciences 181(12), 2550–2570 (2011),
http://dx.doi.org/10.1016/j.ins.2011.02.002

22. Lin, X., Li, Y., Tsang, C.P.: Applying on-line bitmap indexing to reduce counting costs in mining association rules. Information Sciences 120(1-4), 197–208 (1999)
23. MacNicol, R., French, B.: Sybase IQ multiplex-designed for analytics. In: Nascimento, M.A., Tamer Özsu, M., Kossmann, D., Miller, R.J., Blakeley, J.A., Bernhard Schiefer, K. (eds.) Proceedings of 13th International Conference on Very Large Data Bases, VLDB 2004, August 31-September 3, pp. 1227–1230 (2004)
24. Olken, F., Rotem, D.: Rearranging data to maximize the efficiency of compression. In: PODS, pp. 78–90. ACM Press (1985)
25. O'Neil, P.: Model 204 Architecture and Performance. In: Gawlick, D., Reuter, A., Haynie, M. (eds.) HPTS 1987. LNCS, vol. 359, pp. 40–59. Springer, Heidelberg (1989)
26. O'Neil, P.: Informix indexing support for data warehouses. Database Programming and Design 10(2), 38–43 (1997)
27. O'Neil, P., Quass, D.: Improved query performance with variant indices. In: SIGMOD, pp. 38–49. ACM Press (1997)
28. Pinar, A., Tao, T., Ferhatosmanoglu, H.: Compressing bitmap indices by data reorganization. In: ICDE 2005, pp. 310–321 (2005)
29. Wu, K.: FastBit: an efficient indexing technology for accelerating data-intensive science. Journal of Physics: Conference Series 16, 556–560 (2005),
http://dx.doi.org/10.1088/1742-6596/16/1/077
30. Wu, K., Otoo, E., Shoshani, A.: Optimizing bitmap indices with efficient compression. ACM Transactions on Database Systems 31, 1–38 (2006)
31. Wu, K., Otoo, E., Shoshani, A., Nordberg, H.: Notes on design and implementation of compressed bit vectors. Tech. Rep. LBNL/PUB-3161, Lawrence Berkeley National Lab, Berkeley, CA (2001),
http://www-library.lbl.gov/docs/PUB/3161/PDF/PUB-3161.pdf
32. Wu, K., Shoshani, A., Stockinger, K.: Analyses of multi-level and multi-component compressed bitmap indexes. ACM Transactions on Database Systems 35(1), 1–52 (2010), http://doi.acm.org/10.1145/1670243.1670245
33. Wu, K., Stockinger, K., Shoshani, A.: Breaking the Curse of Cardinality on Bitmap Indexes. In: Ludäscher, B., Mamoulis, N. (eds.) SSDBM 2008. LNCS, vol. 5069, pp. 348–365. Springer, Heidelberg (2008); preprint appeared as LBNL Tech Report LBNL-173E
34. Wu, M.C., Buchmann, A.P.: Encoded bitmap indexing for data warehouses. In: ICDE 1998, pp. 220–230. IEEE Computer Society, Washington, DC (1998)

Data Vaults: A Symbiosis between Database Technology and Scientific File Repositories

Milena Ivanova, Martin Kersten, and Stefan Manegold

Centrum Wiskunde & Informatica (CWI), Amsterdam, The Netherlands
{M.Ivanova,Martin.Kersten,Stefan.Manegold}@cwi.nl

Abstract. In this short paper we outline the *data vault*, a database-attached external file repository. It provides a true symbiosis between a DBMS and existing file-based repositories. Data is kept in its original format while scalable processing functionality is provided through the DBMS facilities. In particular, it provides transparent access to all data kept in the repository through an (array-based) query language using the file-type specific scientific libraries.

The design space for data vaults is characterized by requirements coming from various fields. We present a reference architecture for their realization in (commercial) DBMSs and a concrete implementation in MonetDB for remote sensing data geared at content-based image retrieval.

1 Introduction

Unprecedented large data volumes are generated by advanced observatory instruments and demand efficient technology for science harvesting [8]. To date, such data volumes are often organized in (multi-tier) file-based repositories. Navigation and searching for data of interest are performed using metadata encoded in the file names or managed by a workflow system. Processing and analysis are delegated to customized tools, which blend data access, computational analysis and visualization.

Wide adoption of DBMSs in scientific applications is hindered by numerous factors. The most important problems with respect to science file repositories are i) too high up-front cost to port data and application to the DBMS; ii) lack of interfaces to exploit standard science file formats; and iii) limited access to external science libraries for analysis and visualization.

To illustrate the problem we describe a concrete content-based image retrieval application over remote sensing images in the TELEIOS project[1]. The source data are high-resolution TerraSAR-X images in GeoTIFF format [7] accompanied by metadata specification in XML format. The processing pipeline starts with a preparatory phase of tiling an image into smaller chunks, called patches, and applying various feature extraction methods over the patches. The features extracted are stored in a database and used as input for higher level image analysis, such as classification with Support Vector Machines and content-based

---

[1] This work is partly funded by EU project TELEIOS: www.earthobservatory.eu

A. Ailamaki and S. Bowers (Eds.): SSDBM 2012, LNCS 7338, pp. 485–494, 2012.

image retrieval. The DBMS is currently used at a relatively late stage of the pipeline to store the image features, which are of standard data types such as real numbers and strings. Both patching and feature extraction are carried out by customized software tools over image files in file repositories.

Ideally, a user could simply *attach* an external file repository to the DBMS, e.g. using a URI, and let it conduct efficient and flexible query processing over data of interest. However, a state-of-the-art relational database i) requires data to be loaded up-front in the database; ii) cannot naturally understand and support the external format of GeoTIFF images; and iii) provides limited processing capabilities for non-standard data types.

In this paper we outline the *data vault* that provides a true symbiosis between a DBMS and existing (remote) file-based repositories. The data vault keeps the data in its original format and place, while at the same time enables transparent (meta)data access and analysis using a query language. Without pressure to change their file-based archives, scientists can now benefit from extended functionality and flexibility. High level declarative query languages (SQL, SciQL [17]) facilitate experimentation with novel science algorithms. Scientists can combine their familiar external analysis tools with efficient in-database processing for complex operations, for which databases are traditionally good. Transparent, just-in-time load of data reduces the start-up cost associated with adopting a database solution for existing file repositories.

The data vault is developed in the context of MonetDB [11] and its scientific array-query language SciQL. It enables in-database processing of arrays, including raster images. In this work we focus on the data vault aspect and its realization in a real-world test case.

The remainder of this article is organized as follows. We summarize related work in Section 2. An analysis of general data vault requirements is presented in Section 3, followed by a description of our proposal for data vault architecture. Section 5 presents our work in progress on the design and implementation of a remote sensing data vault as a proof of the generic concept. Section 6 concludes the paper.

2   Related Work

Our work on data vaults is related to accessing external data from database systems. The implementation of a remote sensing data vault is a new approach to database support for external remote sensing data.

Access to External Data. The SQL/MED standard (Management of External Data) [15] offers SQL syntax extensions and routines to develop applications that access both SQL and non-SQL(external) data. The standard has limited adoption for applications accessing data from other SQL-server vendors or in CSV files. NoDB [1] proposes advanced query processing over flat data files. Dynamic and selective load and indexing of external data provides low initialization overhead and performance benefits for subsequent queries. A major assumption of the above approaches is, however, that external data are straightforward mapped

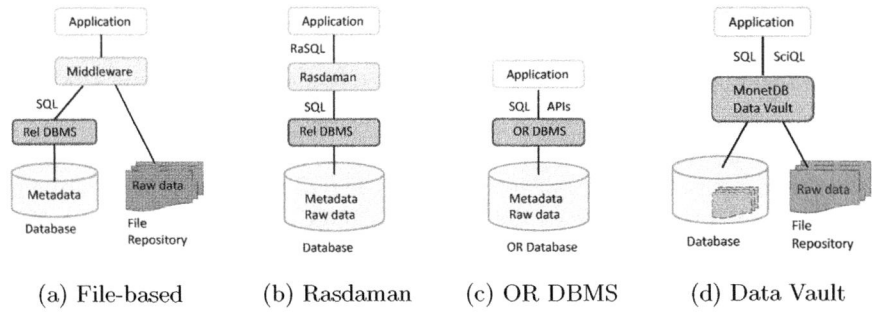

(a) File-based   (b) Rasdaman   (c) OR DBMS   (d) Data Vault

Fig. 1. Software architectures for remote sensing data

to SQL tables. In the domain of scientific applications we take a step further by considering array-based scientific file formats, such as GeoTIFF [7], FITS [6], and mSEED [14] which are typically used in Earth observation, astronomy, and seismology, respectively.

Oracle database file system (DBFS) [10] allows for storage of and access to files with unstructured content in the database. File processing is limited to a rudimentary, file system-like interface to create, read, and write files, and to create and manage directory structures.

Database Support for Remote Sensing Data. Figure 1 illustrates the development of software architectures for remote sensing applications and the role of the database management systems. Figure 1a presents a typical file-based solution where only the metadata are kept in the database (e.g. [16]). Such solutions have some important limitations, such as inflexibility wrt. new requirements, inefficient access, and scalability.

The use of database systems to open the raster data archives for scientific exploration is advocated in [2]. Figures 1b and 1c present architectures where the storage and retrieval of raster data (multi-dimensional arrays) is delegated to the DBMS. RasDaMan [3] array middleware provides applications with database services on multi-dimensional data structures, such as declarative query language, indexing, and optimization. It works on top of a DBMS which provides storage and retrieval for the array tiles and indexes in the form of BLOBs. However, the query processing is not integrated with the database internals.

Raster data management is also offered by some object relational DBMSs in the form of spatial data extenders: Oracle Spatial GeoRaster [12] and PostGIS [13]. They allow storing, indexing, querying, and retrieving raster images and their metadata. However, more complex analysis is delegated to the application or has to be implemented as a UDF. Images have to be loaded explicitly and up-front into the system to enable query processing over them. We propose just-in-time access to data of interest and white-box declarative queries over the sensor data itself. It has been shown that the declarative query paradigm increases productivity and offers flexibility that can be very valuable for ad-hoc science data exploration.

3 Data Vault Requirements

We can summarize the following requirements for the data vault design:

Enhanced Data Model. External file formats use their own specific data models to address the needs of a concrete problem domain. A core concept used in scientific formats is the array, for instance FITS, TIFF, and HDF5 all include arrays. Hence, the database model should be powerful enough to adequately represent the external concepts. We rely on a multi-paradigm data model in which both tables and arrays are first class citizens.

Repository Metadata. The data vault opens up the file repository metadata, encoded in the self-descriptive files, for browsing and sophisticated searching by means of declarative queries. This allows fast identification of data of interest. It is important that access to the metadata is handled separately from the data, so that it does not require loading of the entire data set.

Just-in-Time Load. To deal with the shortcomings of up-front data ingestion, the data vault supports dynamic, just-in-time load driven by the query needs. This is justified by the usage patterns observed over scientific file repositories. It is rarely the case that all raw data i) have good quality and ii) are relevant for the current analysis. Often multiple versions of data, including low quality ones, are kept "just in case", but are never accessed in day-to-day operation. Consequently, there is no big use of putting the burden of low-quality or rarely used external data into the database.

Symbiotic Query Processing. The data vault enables several query processing alternatives. Data requested by a query can be loaded in the database and processed with pure database techniques. The data vault can also use external tools to process files in-situ and capitalize upon the existing support libraries. Symbiotic query processing can combine the benefits of both: use external tools, if efficient ones exist, and carry out operations in the database when the DBMS can perform them better.

A data vault should seamlessly integrate the available science libraries with database query processing. This requires extensions in the query optimization and processing layer of the database software stack. The query optimizer needs to distinguish operations that can be executed by external tools, and, based on a cost model, delegate such operations to be carried out in-situ over the original files.

Cache Management. Once the data vault is set up, the scientists should be able to easily add to and modify files in the repository. To ensure correct functioning, the data vault requires users to refrain from making changes, such as deleting or moving the repository directory. The database management system commits itself to always present an up-to-date repository state to the user.

4   Data Vault Architecture

Next we present the design of the MonetDB [11] data vault architecture, illustrated in Figure 2. It extends MonetDB's software architecture with three components.

The *data vault wrapper* communicates with the external file repository and is built around *data model mapping*. It has components to access data, metadata, and external libraries and tools. The metadata wrapper takes care of accessing the metadata of the external file repository. It populates the *data vault catalog* in the database with a summary of the metadata to facilitate repository browsing, query formulation, and data management.

The data wrapper component accesses external data and creates their internal representation in the *data vault cache*. Similarly, data from the database can be exported to the external format and added to the repository. This facilitates reuse of existing tools to inspect the data products. The functionality wrapper provides access to external libraries. It defines mappings between external functions' input parameters and results to valid database representations.

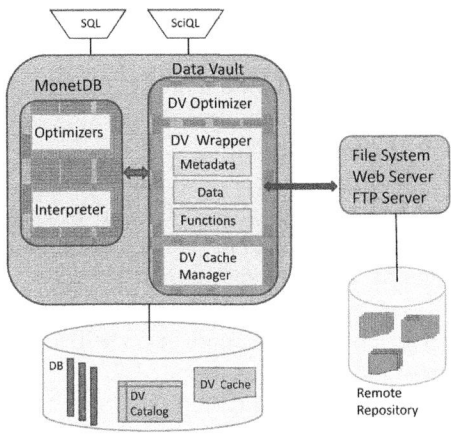

Fig. 2. MonetDB Data Vault Architecture

The data vault cache contains a snippet of the repository data imported into the database. Subsequent analyses over the same external data will have it readily available without the need to repeat the (potentially expensive) importing transformations. This component leverages our work on recycling intermediates [9]. To enable caching of large data sets, the cache is augmented with mechanisms for spilling content onto the disk. The *data vault cache manager* is responsible for keeping the cache content in sync with the repository. The cached data can also be transformed into persistent structures in the database. This transformation can be workload-driven or induced by the user.

The purpose of the *data vault optimizer* is to enable symbiotic query processing. Using the data vault catalog, it detects operations over repository data and makes decisions about their execution locations based on a cost model. It might delegate processing to an external tool and correspondingly inject a call to the functional data wrapper. If in-database processing is expected to be more efficient, the optimizer will inject a call to the cache manager to provide a fresh copy of data under consideration. In turn, the cache manager can call the data wrapper for just-in-time loading of the external data, if they are not already available in the cache, or if the copy is outdated.

Implementation Issues. Our architecture can be used for different data vault installations and applications. The data vault cache manager and query optimizer deal with generic tasks that are common for various data vaults. However, all wrapper components require separate treatment since they are specific for each file format and available software tools. There is an important trade-off between generality and application usability when it comes to the wrapper components implementation.

The most generic way would follow community-agreed specifications, often standards, for a particular data type and its operations. Examples are the object-relational extenders for spatial vector data. At the other end of the spectrum is a tailored implementation for a concrete format and application. An intermediate solution can follow a modular design where the common functionality is available as default modules, but the architecture is easily extensible with modules specific for the application.

To illustrate the problem, consider the design of the data vault catalog for remote sensing data in GeoTIFF format. The application at hand uses lots of metadata encoded in the file names and in auxiliary XML files. Only a small fraction of the standard GeoTIFF tags is used. The generic approach would present all metadata according to the format specifications, such as TIFF tags and geo-keys. Hence, the application will have lots of needless metadata at hand, but may miss the ones encoded in the auxiliary sources. Consequently, the burden lies with the application developer to fill in the missing metadata. Alternatively, an application specific approach would limit itself to just the metadata needed. It would then, however, not be usable for other applications over the same external format. An implementation from scratch, including re-implementation of common functionality, is then called for.

5   Data Vault for Remote Sensing

In this section we describe ongoing work on the implementation of the data vault architecture for the remote sensing (RS) use case.

RS Catalog. The remote sensing data vault catalog stores metadata about the images as required by the application. The catalog is implemented as a set of relational tables in the rs schema. The files table describes the GeoTIFF files in the repository with their location, status (loaded or not), and the timestamp of the last modification. This table is generic for the file-based data vaults.

```
CREATE SCHEMA rs;                      CREATE TABLE rs.catalog (
                                         imageid INT,  fileid INT,  imagewidth INT,
CREATE TABLE rs.files (                  imagelength INT,  resvariant CHAR(4),  mode CHAR(2),
  fileid INT,                            starttime TIMESTAMP,  stoptime TIMESTAMP,
  location CHAR(256),                    sensor VARCHAR(20),  absorbit INT,
  status TINYINT,                        PRIMARY KEY (imageid),
  lastmodified TIMESTAMP );             FOREIGN KEY (fileid) REFERENCES rs.files(fileid) );
```

The catalog table describes image-specific metadata. We chose to explore an application-specific approach. Thus, the catalog table contains image metadata extracted from several sources: the metadata encoded in the GeoTIFF files (image length and width), the auxiliary data in the accompanying XML files (sensor), and properties encoded in the image file names (resolution variant, mode, etc.). The metadata wrapper is a procedure that takes as input parameter the absolute path name of the directory containing the image file repository, e.g., rs.attach('/data/images/geotiff'). It browses the directory, extracts the metadata encoded in different sources, and saves it into the RS catalog.

RS Data. Each image file can be represented in the system as a 2-dimensional array and queried with the proposed SciQL array query language [17]. The set of images available for database processing in the data vault cache is represented as a 3-dimensional array where the 3rd dimension is the associated imageid. The array is pre-defined as a part of the data vault and presents the data vault cache to the users so that they can formulate queries over images of interest in terms of it. However, the array is empty upon attachment of the repository as a data vault and images are not ingested up-front into the system.

```
DECLARE NumCols INT;   SET NumCols = ( SELECT max(imagewidth)  FROM rs.catalog );
DECLARE NumRows INT;   SET NumRows = ( SELECT max(imagelength) FROM rs.catalog );
CREATE ARRAY images (
  id INT DIMENSION,  x  INT DIMENSION [NumCols],  y  INT DIMENSION [NumRows],  v SMALLINT );
```

The data wrapper is responsible for the ingestion of external images when needed. It takes a set of image IDs as determined by the query criteria, locates the corresponding files, and loads them into the data vault cache structure, in this case the 3-dimensional array.

RS Query Processing. To clarify the query processing mechanism over remote sensing images from the data vault we start with a simple SciQL query. It computes image masks by simply filtering pixel values within the range [10,100]. The images are specified through predicates over the remote sensing catalog (e.g. the image resolution variant is spatially enhanced, imaging mode is High-resolution Spotlight, and the start time is in a given time interval).

```
SELECT [id], [x], [y], v   FROM images   WHERE v BETWEEN 10 AND 100   AND id in
(SELECT imageid   FROM rs.catalog
 WHERE resvariant = 'SE__' AND mode = 'HS' AND starttime > TIMESTAMP '2011-12-08 16:30:00');
```

The data vault optimizer recognizes the cache in the form of the images array and injects a call to the cache manager to provide the images as specified by the predicates over the id array attribute. The cache manager checks at run time if the images are available and issues a request to the data wrapper to ingest the missing ones into the images array.

We continue with an example content-based image retrieval application [5]. It takes an example image provided by the user and retrieves all images in the database similar to it according to some similarity measure. The measure used in our application is Fast Compression Distance (FCD) [4]. The main idea is to extract image dictionaries by applying some compression technique and to reason about similarities between two images based on the overlaps between their corresponding dictionaries. The dictionary can be extracted, for instance, with the LZW compression algorithm or by computing N-grams.

An example white-box SciQL function extracting 4-grams by row-wise processing of a 2-dimensional array is depicted below. It is then used to extract the 4-gram dictionaries from patches of a given image *imgid*.

```
CREATE FUNCTION dict4gram ( img ARRAY ( x INT DIMENSION, y INT DIMENSION, v SMALLINT ) )
RETURNS TABLE (dict_elem STRING)
BEGIN
  SELECT DISTINCT CAST(img[x][y  ].v AS STRING) || CAST(img[x][y+1].v AS STRING) ||
                  CAST(img[x][y+2].v AS STRING) || CAST(img[x][y+3].v AS STRING)
  FROM img    GROUP BY img[x][y:y+4];
END;


DECLARE patch_size INT;    SET patch_size = 256;
SELECT id, patch_size, x, y, dict4gram(v)    FROM images WHERE id = imgid
GROUP BY DISTINCT images[id][x:x+patch_size][y:y+patch_size];
```

The dictionaries can be stored in database tables and used to compute the FCD similarity measure as needed by the image retrieval application. The computation is illustrated with the following function that, given an image patch number and maximum distance, calculates the FCD and returns all patches with FCD smaller than the input parameter. We assume that the patch dictionaries computed above are stored in a table `imagedict(patch_id, dict_elem)`.

```
CREATE FUNCTION "FCD_1_n" ( pid INT, dist FLOAT )
RETURNS TABLE ( patch INT, distance FLOAT )
BEGIN
  DECLARE dict_size INT;
  SET dict_size = ( SELECT count(*) FROM imagedicts WHERE patch_id = pid );
  RETURN
    SELECT inter_size.patch AS patch,
      ( ( dict_size - inter_size.cnt ) / CAST( dict_size AS FLOAT ) ) AS distance
    FROM
      ( SELECT d2.patch_id AS patch, count(*) AS cnt
          FROM imagedicts d1 JOIN imagedicts d2 ON d1.dict_elem = d2.dict_elem
          WHERE d1.patch_id = pid
          GROUP BY d2.patch_id
      ) AS inter_size
    WHERE ( dict_size - inter_size.cnt ) / CAST( dict_size AS FLOAT) < dist;
END;
```

Expressing the processing steps of image patching, feature extraction, and similarity computation in SciQL and SQL offers greater flexibility for the user. For instance, it is easy to experiment how a given feature extraction method performs with different patch sizes by simply changing the value of the `patch_size` parameter and re-running the queries. Similarly, the system allows for convenient experimentation with and comparison between different versions of feature extraction methods and similarity measures.

6   Summary and Future Work

The data vault is a symbiosis between database technology and external file-based repositories. It keeps the data in its original format and location, while at the same time transparently opens it up for analysis and exploration through DBMS facilities. Scientists benefit from this functionality, flexibility, and scalability by combining external analysis tools with efficient in-database processing. Transparent, just-in-time load of data of interest reduces the start-up cost associated with adopting a pure database solution for existing file repositories.

A reference architecture to realize data vaults has been presented. A concrete implementation of the data vault using MonetDB has been described. It opens up a large archive of high-quality remote sensing radar images for data mining experiments.

The realization of the data vault provides a vista on different research challenges. Wrapping the external libraries functionality allows us to capitalize upon existing tools for in-situ analysis, but needs careful interface design and cost modeling. Efficient symbiotic query processing requires detection of external data and libraries and extensible cost model and optimizer.

Acknowledgments. We wish to thank our partners in the TELEIOS and COMMIT projects for constructive guidance on the functionality and implementation of the MonetDB Data Vault. In particular, we thank Y. Zhang for her important work on the implementation of SciQL.

References

1. Alagiannis, I., Borovica, R., Branco, M., Idreos, S., Ailamaki, A.: NoDB: Efficient Query Execution on Raw Data Files. In: SIGMOD (2012)
2. Baumann, P.: Large-Scale Earth Science Services: A Case for Databases. In: ER (Workshops), pp. 75–84 (2006)
3. Baumann, P., et al.: The multidimensional database system RasDaMan. SIGMOD Rec. 27(2), 575–577 (1998)
4. Cerra, D., Datcu, M.: Image Retrieval using Compression-based Techniques. In: International ITG Conference on Source and Channel Coding (2010)
5. Dumitru, C.O., Molina, D.E., et al.: TELEIOS WP3: KDD concepts and methods proposal: report and design recommendations, http://www.earthobservatory.eu/deliverables/FP7-257662-TELEIOS-D3.1.pdf
6. FITS. Flexible Image Transport System, http://heasarc.nasa.gov/docs/heasarc/fits.html
7. GeoTIFF, http://trac.osgeo.org/geotiff/
8. Hey, T., Tansley, S., Tolle, K.: The Fourth Paradigm: Data-Intensive Scientific Discovery. Microsoft Research (2009)
9. Ivanova, M., Kersten, M., Nes, N., Gonçalves, R.: An Architecture for Recycling Intermediates in a Column-store. ACM Trans. Database Syst. 35(4), 24 (2010)
10. Kunchithapadam, K., Zhang, W., et al.: Oracle Database Filesystem. In: SIGMOD, pp. 1149–1160 (2011)
11. MonetDB (2012), http://www.monetdb.org/

12. Oracle. Oracle Spatial GeoRaster Developer's Guide, 11g Release 2 (11.2)
13. PostGIS, http://www.postgis.org/
14. SEED. Standard for the exchange of earthquake data (May 2010),
    http://www.iris.edu/manuals/SEEDManual_V2.4.pdf
15. SQL/MED. ISO/IEC 9075-9:2008 Information technology - Database languages -
    SQL - Part 9: Management of External Data (SQL/MED)
16. Stolte, E., von Praun, C., Alonso, G., Gross, T.R.: Scientific data repositories:
    Designing for a moving target. In: SIGMOD Conference, pp. 349–360 (2003)
17. Zhang, Y., Kersten, M., Ivanova, M., Nes, N.: SciQL: Bridging the Gap between
    Science and Relational DBMS. In: IDEAS, pp. 124–133 (2011)

Usage Data in Web Search: Benefits and Limitations

Ricardo Baeza-Yates[1] and Yoelle Maarek[2]

[1] Yahoo! Research, Barcelona, Spain
rbaeza@acm.org
[2] Yahoo! Research, Haifa, Israel
yoelle@ymail.com

Abstract. Web Search, which takes its root in the mature field of information retrieval, evolved tremendously over the last 20 years. The field encountered its first revolution when it started to deal with huge amounts of Web pages. Then, a major step was accomplished when engines started to consider the structure of the Web graph and link analysis became a differentiator in both crawling and ranking. Finally, a more discrete, but not less critical step, was made when search engines started to monitor and mine the numerous (mostly implicit) signals provided by users while interacting with the search engine. We focus here on this third "revolution" of large scale usage data. We detail the different shapes it takes, illustrating its benefits through a review of some winning search features that could not have been possible without it. We also discuss its limitations and how in some cases it even conflicts with some natural users' aspirations such as personalization and privacy. We conclude by discussing how some of these conflicts can be circumvented by using adequate aggregation principles to create "ad hoc" crowds.

Keywords: Web search, usage data, wisdom of crowds, large scale data mining, privacy, personalization, long tail.

1 Introduction

Usage data has been identified as one of the top seven challenges of the Web search technology [2], and is probably one of the major entry barriers for new Web search engines. In fact, usage data has been the key of the last revolution in Web search. In the first generation of search engines, that started in 1994, the main source of ranking signals was the page content. The second revolution came around 1998, with the use of links and their anchor text. The third generation came with the usage of query logs and click-through data to better understand users and implement new functionalities, such as spelling correction, query assistance or query suggestion. This revolution is covered in detail in Section 2.

This last revolution can largely be credited to progress in large scale data mining and to trusting more and more the wisdom of crowds (WoC) principle [22]. The WoC is based on the notion that "the many are smarter than the few"

A. Ailamaki and S. Bowers (Eds.): SSDBM 2012, LNCS 7338, pp. 495–506, 2012.

and that "collective wisdom shapes business, economies, societies and nations". While any individual is obviously a member of the crowd, it is not trivial to derive working solutions from this principle, as demonstrated by all the issues inherent to democracy as a simplistic example. Closer to our topic here, basic ranking techniques work because of this idea and simple algorithms using large amounts of data can beat complex algorithms on small size data. Word-based ranking formulas like BM25 can be seen as a reflection of the wisdom of writers, while a link-based score, like PageRank, could be seen in the early days of the Web as a reflection of the wisdom of webmasters. Similarly, usage data is a reflection of the wisdom of users. Exploiting usage data under its multiple forms, from query logs to clicks, to time spent on a page, brought an unprecedented wealth of implicit information to Web Search. This implicit information can be processed and analyzed at the individual level, for inferring the user's intent and satisfaction, but it is only when it is analyzed at the collective level, that it follows the WoC principle.

Usage data refers to a specific subset of Web data that covers the pattern of usage of Web pages such as "IP addresses, page references and the date and time of accesses" [21]. It is critical to studying all Web applications, yet one application, Web search, is probably the domain for which it has been become the most critical. It is the key source of data in the latest stage of Web search evolution, after on-page data and web graph data [2], and is the focus of our paper here.

We discuss here the benefits of Web search usage data, under the WoC principle, but also its limitations. First by definition, it cannot be applied when not enough data is available, and the poor performance of enterprise search engines as opposed to Web search engines can partially be explained by usage data being either not sufficient or not even exploited in enterprise search settings. Other limitations are due to personalization needs and privacy concerns. In Section 3 we explore the interdependency between these three conflicting factors: size of data, personalization and privacy. Then, we propose a solution to reduce some of these conflicts, by assembling different types of crowds around common tasks or social circles in Section 4. We end with some concluding remarks in Section 5.

2   The Third Web Search Revolution

2.1   Search Usage Data

In the context of Web search, usage data typically refers to any type of information provided by the user while interacting with the search engine. It comes first under its raw form as a set of individual signals, but is typically mined only after multiple signals have been aggregated and linked to the same interaction event. Two major types of such aggregated data relate to user query and click information that are referred to as query logs and click logs.

- **Query logs** include the query string that the user issued, together with the time-stamp of the query, a user identifier[1], possibly the IP of the machine on which the browser runs.
- **Click data** includes the reference to the element the user clicked on the search page together with the timestamp, user identifier, possibly IP, the rank of the link if it is a result, etc. The type and number of such clickable elements vary and keep increasing as the user experience of most major Web search engines get richer and more interactive. Such elements include the actual organic search results (represented by a URL), or sponsored results, a suggested related query, a next page link on the search result page, even a preview button such as offered by some engines, etc. No matter what the element is, a click event is a key signal that usually indicates an expression of interest from the user, and in the best case provides relevance feedback.

One of the most famous query logs is the, now retired, AOL query log, which does indeed include the URL and rank of the search result the user clicked on after issuing a query. Consider the following entry from the AOL log [19]:

AnonID	Query	QueryTime	ItemRank	ClickURL
100218	memphis pd	2006-03-07 09:42:33	1	http://www.memphispolice.org

This row in the log indicates the fact that a given user, identified[2] by id 100218, issued the query ``memphis pd'' and clicked on the first result (ItemRank=1) at 09:42 on March, 7th, which led to the URL http://www.memphispolice.org.

While search logs are rarely being, if at all, released for research anymore, they are extensively used internally in all major commercial search engines. New signals are constantly being considered for the same purpose of getting more diverse feedback. An early example is the signal provided by eye-tracking devices. By aggregating these over many users, researchers generated "heat maps" that allowed them to study, for instance, the impact of changing the length of snippets in search result pages [8]. Another example of signal is the movement of the cursor, which has been tracked by using an instrumented browser or a dedicated toolbar [11]. Signals can also be combined as done by Feild *et al.* [9], who studied searchers' frustration, correlating query log data with sensor information provided by "a mental state camera, a pressure sensitive mouse, and a pressure sensitive chair." Many of these signals cannot be tracked at a large scale as they require additional machinery on the client side and are often reserved to small users' studies that are out of the scope of this paper.

We focus here on usage data that are gathered at a very large scale and typically provided during regular interaction with the search engine. Note however that one can envision that more biometrics signals be provided at a large scale,

---

[1] The same user is usually identified by a browser cookie or a unique id in the case of signed-in users, which are mapped into a unique identifier.

[2] We will discuss in Section 3.3 why replacing a user name by a numeric identifier is far from being sufficient for anonymizing such data.

in the future, if the benefits overcome the cost. A very promising direction in that direction is the study of cursor movement via instrumentation of the search result page, which Huang et al. have demonstrated can be conducted at a large scale [13]. Although these results were limited, if cursor movements analysis proves to be valuable enough, it is not unrealistic to envision that major search engines would absorb the extra cost (which does not seem to be prohibitive as per the same study) in latency and deploy such instrumentation on every search result page.

Even without adding more signals, query logs and click data, can already be credited for many of the key features and improvements in the effectiveness of today's web search engines, as detailed next.

2.2   Benefits

One of the most visible examples of leveraging usage data at a large scale is the query spelling correction feature embodied in "Did you mean" on Google Search page. After years of very sophisticated spell checking research, which modeled typos very carefully [15], Google showed that simply counting similar queries at a small edit distance would, in most cases, surface the most popular spelling as the correct one, a beautiful and simple demonstration of the wisdom of crowds principle. See the now famous "Britney Spears" example in http://www.google.com/jobs/britney.html, where on a query log sample over a three month period, users entered more than 700 variations of Britney Spears' name. The surprising fact was that the correct spelling was more frequent by one order of magnitude than the second most popular. Clearly the "did you mean" feature uses more sophisticated techniques than direct counting, but this simple example illustrates how even trivial "data crunching" over big *real* data brings more value than sophisticated techniques on small datasets.

Query logs have also revolutionized query assistance tools such as related queries and query autocomplete. Unsuccessful searches are typically attributed to either relevant content not being available or the query not adequately expressing the user's needs. Query assistance tools are offered to precisely address the latter and have become more and more popular on Web search sites in recent years. The first type of query assistance tools consists of offering "related queries" after the query is issued. These related queries can be selected then by users who are not satisfied with the current results. A more recent and more revolutionary query assistance tool, launched in 2007 as "Search Assist" on search.yahoo.com and as "Suggest" on google.com, consists of offering query completion to users as they start typing the first few letters of their queries in the search box. Both types of assistance tools heavily rely on usage data, mostly query logs but also click data when the service wants to add a signal of relevance. While, in the past, query assistance tools would analyze the document corpus in order to identify phrases that could serve as alternate queries, the tremendous growth in Web search engine traffic allowed these tools to mostly rely on real user-issued queries. Using query logs as the source corpus significantly improved the quality of suggestions, at least as perceived by the user. There exists a clear vocabulary gap between document and query corpora, as demonstrated in [4], which explains why using the same type of corpus makes sense here. A great deal of research has been conducted on query suggestion and completion, with various approaches that leverage clicks and results in addition to the query logs [3]. In practice, query assistance tools are now deployed on all major Web search engines, and their quality improves as query logs grow.

An additional revolutionary benefit of usage data is to consider clicks as a measure of satisfaction or at least of interest from the user, as if the user, by clicking, had actually voted in favor of the clicked element. A thorough study of the various click models and their relation to search relevance, both for organic and sponsored search, is given in [20]. One major reason of the quick pace of innovation in Web search can be credited to these "pseudo-votes". Instead of testing a new feature, or any sort of change, on a small sample of beta-users in a controlled environment, search engines now use real users on a much larger scale. Some percentage of the real traffic is exposed to the new feature, over a given period of time, under what is called a "1% experiment", "bucket test" or "A/B testing", depending on the terminology. Users who are part of these experiments are, in most cases, not even aware that they are being monitored and therefore act naturally. Various metrics are used to verify whether users react positively or not to the change, thus helping search engines to decide whether to deploy the change to all users. After deployment, user behavior is constantly monitored, not only at pre-launch time, and features for which clicks or other metrics are decreasing might be discontinued or retired as it often happens.

2.3   Today's Entry Barrier?

Usage data at a large scale provides so much information that it is difficult to imagine how a new Web search player could enter the market and hope to be as relevant as its competitors from day one.

The failure of the now retired search engine Cuil, a few years ago, has actually been attributed to the fact that it tried a new model, where instead of "ranking [pages] based on popularity, as Google does, it focuses on the content of each page" [12]. While Cuil's index was reportedly huge, and its engineers and scientists were clearly among the most knowledgeable in the market, it did not have at its disposal any of the implicit signals that can be gathered only after a certain period of regular usage by a critical mass of users. Cuil was discontinued in September 2010, only two years after its launch. Another approach was adopted by Microsoft when it stroke a search alliance with Yahoo! in 2009. As part of the deal, by operating the back end of Yahoo! search, Bing would immediately start tripling search usage data leveraging it into their core ranking algorithms, improving both, Bing and Yahoo! search. From these two examples, we believe that usage data at a large scale now represents the main entry barrier in Web search, and unless a drastically new kind of signal enters the arena, it is going to be more and more difficult for newcomers to compete.

3   Three Conflicting Factors

There are at least three major factors that pull in opposite directions when leveraging usage data:

- **Size of Data:** Large-scale or big usage data is a key pre-requisite to inferring new insights as per the WoC principle. As a consequence, the needs of the crowd dominate long-tail needs, which are expressed by definition by very few individuals. Averaging on the more popular needs can dominate so much, in some cases, that it conflicts with some specific needs that personalization should address.
- **Personalization:** In order for search engines to personalize its services to a specific user, it obviously needs to know more about the persona behind the individual, at the risk of exposing some private aspects. Hence, personalization in many cases might conflict with the user's privacy.
- **Privacy:** One key demand of privacy-protecting activists pertains to not accumulate too much data on a single individual over long periods of time, as they want everyone's past to remain private. However, big data is typically gathered over a given window of time, the longer the window, the bigger the data. Restricting persistence of data to short periods of time clearly reduces the amount of data that can be mined and thus threatens the effectiveness of the WoC principle, in particular for personalization.

In the following, we detail each of these factors, and discuss how to work around these intrinsic conflicts, as depicted in Figure 1.

3.1   Size of Data

The wisdom of crowds principle relies on analyzing large amounts of data. As many signals in the Web follow a power law, it works very well for the head

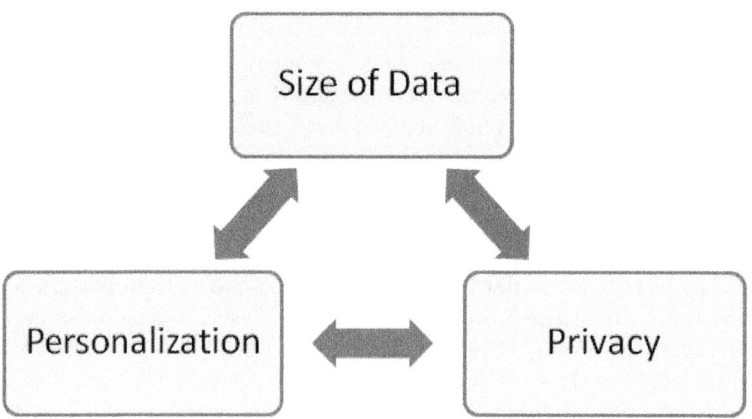

Fig. 1. Three conflicting factors

of the power law distribution without needing so many users. This stops being true however as soon as the long tail is considered. Serving long tail needs is critical to all users, as all users have their shares of head and long tails needs as demonstrated in [10]. Yet, it often happens that not enough data covering the long tail is available. We address this problem in the next section.

We can always try to improve any WoC result by adding more data, if available. Doing so, however, might not always be beneficial. For example, if the added data increases the noise level, results get worse. We could also reach a saturation point without seeing any improvements. Adding data also can challenge the performance of the used algorithms. If the algorithm is linear, doubling the data, without modifying the system architecture, implies doubling the time. This might still be feasible, but for super linear algorithms mostly surely will not. In this case, typical solutions are to parallelize and/or distribute the processing. As all big data solutions already run on distributed platforms, increasing the amount of data requires increasing the number of machines, which is clearly costly. Another solution consists of developing faster algorithms, at the cost of possibly lowering the quality of the solution. This becomes a clear option when the loss in quality is inferior to the improvements obtained with more data. That is, the time performance improvements should be larger than the loss in the solution quality. This opens a new interesting trade-off challenge in algorithm design and analysis.

However, even if more data is added, there will always be cases where the key principle behind WoC of averaging over head needs will alienate users. Take the following example of a user searching for "shwarzeneger" on Google, s/he will automatically be given results for "Arnold Schwarzenegger" without even being asked to click on a "did you mean" suggestion. This will be a good initiative in most cases (*e.g.* head needs), but definitely not good for a user whose family name is weirdly spelled "shwarzeneger". This is just one case for which big data actually conflicts with personalization, a key requirement of Web search as discussed next.

3.2 Personalization

Personalized search has been an open research topic since the early days of search, even in the pre-Web era. The key challenge here is to identify the user-dependent implicit intent behind the query. Given that no engine today can pretend to perfectly map a human being and his/her infinite facets into an exhaustive model, only rough approximations can be conducted, and absolute personalization is simply impossible.

Nevertheless, there are some facets of users that are easy to capture and that do bring obvious value when personalizing. Some of these facets are sometimes considered as belonging to "contextualization" rather than personalization, and include facets such as the geographical location of the user, his or her language of choice, the browser or even the device used, etc. Two clear winning facets among these are language and geographical location, at least at country level. Search results or query suggestions for instance, should clearly be tailored to the language of choice of the user. Similarly location is critical, as it often reflects cultural differences, see the example provided in [14], where Google explains how "liver" would be completed to "liverpool" in UK as opposed to "liver diseases" in the US. Geo location can also be leveraged at a finer level of granularity. Indeed, even if ``pizza in San Francisco'' is significantly more frequent than ``pizza in Little Rock'', a user from Arkansas, would clearly consider the first completion to his query ``pizza'', totally irrelevant. Balancing between popularity and location personalization is however far from being trivial. Consider the same example of ``pizza in Little Rock'' for a resident of El Dorado, Arkansas. That person would probably prefer no pizza recommendation at all to a recommendation for a pizza place located close to 100 miles away. Modeling the precise geo location of every user and deciding whether or not to trigger some query suggestions with a "geo intent" based on the distance between the respective locations is a huge technical challenge that is still being researched.

The other more traditional facets of personalization are semantic facets. Most search engines gather previous queries of signed-in users in order to build such semantic profiles or preferences and tailor search results (sponsored or organic) accordingly. Unlike the previously discussed facets, these personalization facets present privacy risks as we will discuss below. In addition, personalization can sometimes even confuse users, when they see different results whether they are signed in or not. Internet activist, Eli Pariser, blames the "filter bubble" for this and other effects of personalization, in his book [17] and blog[3] of the same name. He actually goes further and claims that personalization imposes some "ideological frames" and influences users in seeing results more and more similar to what they saw in the past. It is clear that abusing personalization as well as the WoC principle has some risks not only in terms of privacy but also in terms of diminishing serendipity. However, most Web search engines are aware of the risk and many use use some sort of "explore/exploit" mechanism to give a chance of exposure to new or less popular results. Such an example of explore/exploit scheme for Web content optimization was introduced by Agrawal et al. in [1], who proposed a multi-armed bandit-based approach to maximize click on a content module, for Web search, online advertising and articles for content publishing. Another alternative to avoid the filter bubble in Web search is result diversification [18].

The final type of facet that we are considering here is what we call the "social facet" that has started to appear in some search results. One such example is the "+1" in Google, which allows user to vote on results they prefer, in a classical relevance feedback approach. More interestingly though, Google search now returns to any signed-in user, content published by members of his/her Google+ circles, such as pictures from Picasa.

While adding more facets is beneficial to personalization, it always poses a threat to privacy. The previously mentioned social facet in particular did and still does worry many users and Google reacted by publishing a revised privacy

---

[3] http://www.filterbubble.com

policy in early 2012, soon after launching its social search results. On the other hand, it is not clear what percentage of queries really benefit from a social facet.

3.3 Privacy

One important concern of users of Web search engines is the fact that their queries expose certain facets of their life, interests, personality, etc. that they might not want to share. This includes sexual preferences, health issues or even some seemingly minor details such as hobbies or taste in movies that they might not be comfortable sharing with all. Queries and clicks on specific pages indeed provide so much information that the entire business of computational advertising rely on these. Query logs and click data reveal so much about users that most search engines stopped releasing logs to the research community due to the AOL incident. In this case, privacy breach in query logs was exposed by two New York Times journalists who managed to identify one specific user in the log, (AnonId 4417749), [5]. They spotted several queries, originating from a same user, referring to the same last name, specific locations such as "landscapers in lilburn, GA" or "homes sold in shadow lake subdivision gwinnett county georgia". Eventually, they could link these queries to an senior woman, who confirmed having issued not only these queries, but also more private ones linked to the same AnonId, such as "60 single men" or "numb fingers". While not all users could necessarily be as easily identified, it revealed what many researchers had realized a while back, namely that simply replacing a user name by a number is not a guarantee of anonymity. Later, it was shown that a combination of few attributes is sufficient to identify most users. More specifically, a triple such as (ZIP code, date of birth, gender) is sufficient to identify 87% of citizens in the US by using publicly available databases [23]. Numerous research efforts have been dedicated to anonymization. A favored one in large repositories or databases is the k-anonymity, introduced by Sweeney [23], which proposes to suppress or generalize attributes until each entry in the repository is identical to at least $k - 1$ other entries.

In practice, Web search engines do aggregate by conducting "data crunching" on such a large scale in order to derive insights under the WoC principle, that k-anonymity is *de facto* respected in most cases. Nevertheless, most users do not differentiate between the exploitation of aggregated and personal usage data, and they need some guarantee that Web search engines will not cross the line and expose their personal usage data. The key challenge for Web Search engines is therefore to keep the trust of their users, while retaining enough data for them to apply large scale data mining methods. In addition, they are accountable to regulators such as the Federal Trade Commission (FTC) in the United States or should comply with the Data Protection Directive legislated in 1995 by the European Union Parliament. Indeed, the FTC has defined several frameworks for protecting consumer privacy, especially online[4]. Recently, the FTC commissioner

---

[4] Protecting Consumer Privacy in an Era of Rapid Change. A Proposed Framework for Business and Policymakers. Preliminary FTC Staff Report, December 2012 (http://www.ftc.gov/os/2010/12/101201privacyreport.pdf).

even threatened to go to congress if privacy policies do not "address the collection of data itself, not just how the data is used" [16].

First, most of the terms of conditions of these systems often guarantee that personal usage data is not shared with third parties. Second, they do employ as much secure communication and storage as possible to promise their users that personal information cannot be stolen away. Finally, they have devised data retention policies to reassure US and EU regulators, the media and, naturally, their users, that they comply with the regulations mentioned before. Indeed, one of the problematic twists of big data is that in many cases a specific user would prefer to forget past facts. A recent New York Times article described how such resurgences from the past can upset users and "how a new breed of Web specialists known as online reputation managers" specialize in erasing one's undesired past from Web search results [6]. Nevertheless, privacy concerns keep rising, especially with the advent of social networks, and in our specific context, social search, where personal content from one's social network can be integrated in search results, as recently done by Google.

4 The Wisdom of "Ad Hoc" Crowds

One solution to many of the previously mentioned issues is to aggregate data in the "right" way. To illustrate the idea, let us consider personalized ranking as an example. The typical personalization approach consists of aggregating data around a specific user using, for instance, query history. This approach suffers from several problems as mentioned before. In order to respect regulations and maintain trust, the user should be aware of and accept the terms of use of the search service, and be signed-in. More problematic, even very active users will not show so many queries that their favorite topics or preferences dominate by orders of magnitude more ephemeral needs. Data will be scarce and the quality of personalization will suffer from it. The approach we advocate for here, is to personalize around the task, around the need, and not around the user so as to increase the amount of relevant data. In some sense, when the WoC principle does not work because the usual straight-forward crowd (inferred by a common query for instance) does not exist, we suggest to form "ad hoc" crowds.

If we identify a clear task behind a query for instance, such as traveling to a remote unknown Pacific island, the idea is to form an "ad hoc" crowd of those users who have tried in the past to travel to the same destination. While the "ad hoc" crowd will be for sure smaller than the usual large crowds encountered in head needs, it is still big enough to generate insights, to learn from others and to guarantee k-anonymity, as long as the task granularity is not too fine. This represents another sub-product of the fact that all users have their own long tail [10]. We believe that "ad hoc" crowds can be generated in a variety of scenarios, as the bulk of user queries belong to a few tasks: users are after all not that different, and many tasks share similarities at some level, for example find a home page, look for information, perform a transaction or download a resource. The main differences among users are not with what they do, but when they do it, how long they do it and how well they do it.

The main challenge behind contextualization and generation of the appropriate "ad hoc" crowd, is the prediction of the user's intent that will characterize the crowd. For that reason, in recent years many researchers have approached this problem by using query log data [7]. The state of the art shows that this can be done with a reasonable size training data, high precision and online performance. We expect to see many types of "ad hoc" crowds arising in the future. Indeed, even Google+ recent social search can be seen as an incarnation of an "ad hoc" crowd, where the crowd is social and is formed by the user's circles rather than by a common need.

5   Final Remarks

In this paper we have shown the importance of usage data in current Web search engines and how the wisdom of crowds was a driving principle in leveraging usage data. However, in spite of its benefits, usage data cannot be fully leveraged due to the conflicting demands of big data, personalization and privacy. We have explored here these conflicting factors and proposed a solution based on applying the wisdom of crowds in a different manner. We propose to build "ad hoc" crowds around common tasks and needs, where a user will be aggregated to different users in various configurations depending on the considered task or need. We believe that more and more "ad hoc crowds" will be considered, in particular in other Web services and inside large Intranets where this idea can impact revenue or/and productivity.

References

1. Agarwal, D., Chen, B.C., Elango, P.: Explore/Exploit Schemes for Web Content Optimization. In: Proceedings of the 2009 Ninth IEEE International Conference on Data Mining, pp. 1–10. IEEE Computer Society, Washington, DC (2009)
2. Baeza-Yates, R., Broder, A., Maarek, Y.: The New Frontier of Web Search Technology: Seven Challenges, ch. 2, pp. 11–23. Springer (2011)
3. Baeza-Yates, R., Maarek, Y.: Web retrieval. In: Baeza-Yates, R., Ribeiro-Neto, B. (eds.) Modern Information Retrieval: The Concepts and Technology behind Search, 2nd edn. Addison-Wesley (2011)
4. Baeza-Yates, R., Saint-Jean, F.: A Three Level Search Engine Index Based in Query Log Distribution. In: Nascimento, M.A., de Moura, E.S., Oliveira, A.L. (eds.) SPIRE 2003. LNCS, vol. 2857, pp. 56–65. Springer, Heidelberg (2003)
5. Barbaro, M., Zeller Jr., T.: A face is exposed for aol searcher no. 4417749. The New York Times, August 9 (2006)
6. Bilton, N.: Erasing the digital past. The New York Times (April 2011), http://www.nytimes.com/2011/04/03/fashion/03reputation.html
7. Brenes, D.J., Gayo-Avello, D., Pérez-González, K.: Survey and evaluation of query intent detection methods. In: Proceedings of the 2009 Workshop on Web Search Click Data, WSCD 2009, pp. 1–7. ACM, New York (2009)
8. Cutrell, E., Guan, Z.: What are you looking for?: an eye-tracking study of information usage in web search. In: Proceedings of the SIGCHI Conference on Human Factors in Computing Systems, CHI 2007, pp. 407–416. ACM, New York (2007)

9. Feild, H.A., Allan, J., Jones, R.: Predicting searcher frustration. In: Proceedings of the 33rd International ACM SIGIR Conference on Research and Development in Information Retrieval, SIGIR 2010, pp. 34–41. ACM, New York (2010)

10. Goel, S., Broder, A., Gabrilovich, E., Pang, B.: Anatomy of the long tail: ordinary people with extraordinary tastes. In: Proceedings of the Third ACM International Conference on Web Search and Data Mining, WSDM 2010, pp. 201–210. ACM, New York (2010)

11. Guo, Q., Agichtein, E.: Exploring mouse movements for inferring query intent. In: Proceedings of the 31st Annual International ACM SIGIR Conference on Research and Development in Information Retrieval, SIGIR 2008, pp. 707–708. ACM, New York (2008)

12. Hamilton, A.: Why cuil is no threat to google. Time.com (Time Magazine Online) (July 2008),
http://www.time.com/time/business/article/0,8599,1827331,00.html

13. Huang, J., White, R.W., Dumais, S.: No clicks, no problem: using cursor movements to understand and improve search. In: Proceedings of the 2011 Annual Conference on Human Factors in Computing Systems, CHI 2011, pp. 1225–1234. ACM, New York (2011)

14. Kadouch, D.: Local flavor for google suggest. The Official Google Blog (March 2009), http://googleblog.blogspot.com/2009/03/
local-flavor-for-google-suggest.html

15. Kukich, K.: Techniques for automatically corecting words in text. ACM Computing Surveys 24(4) (December 1992)

16. Mullin, J.: FTC commissioner: If companies don't protect privacy, we'll go to congress. paidContent.org, the Economics of Digital Content (February 2011)

17. Pariser, E.: The Filter Bubble: What the Internet Is Hiding from You. Penguin Press (2011)

18. Radlinski, F., Dumais, S.: Improving personalized web search using result diversification. In: Proceedings of the 29th Annual International ACM SIGIR Conference on Research and Development in Information Retrieval, SIGIR 2006, pp. 691–692. ACM, New York (2006)

19. Shi, X.: Social network analysis of web search engine query logs. Technical report, School of Information, University of Michigan (2007)

20. Srikant, R., Basu, S., Wang, N., Pregibon, D.: User browsing models: relevance versus examination. In: Proceedings of the 16th ACM SIGKDD International Conference on Knowledge Discovery and Data Mining, KDD 2010, pp. 223–232. ACM, New York (2010)

21. Srivastava, J., Cooley, R., Deshpande, M., Tan, P.-N.: Web usage mining: discovery and applications of usage patterns from web data. SIGKDD Explor. Newsl. 1, 12–23 (2000)

22. Surowiecki, J.: The Wisdom of Crowds: Why the Many Are Smarter Than the Few and How Collective Wisdom Shapes Business, Economies, Societies and Nations. Random House (2004)

23. Sweeney, L.: k-anonymity: a model for protecting privacy. International Journal on Uncertainty, Fuzziness and Knowledge-based Systems 10(5), 557–570 (2001)

Functional Feature Extraction and Chemical Retrieval

Peng Tang, Siu Cheung Hui, and Gao Cong

School of Computer Engineering,
Nanyang Technological University, Nanyang Avenue, Singapore 639798
{ptang1,asschui,gaocong}@ntu.edu.sg

Abstract. Chemical structural formulas are commonly used for presenting the structural and functional information of organic chemicals. Searching for chemical structures with similar chemical properties is highly desirable especially for drug discovery. However, structural search for chemical formulas is a challenging problem as chemical formulas are highly symbolic and spatially structured. In this paper, we propose a new approach for chemical feature extraction and retrieval. In the proposed approach, we extract four types of functional features from Chemical Functional Group (CFG) Graph built from a chemical structural formula, and use them for the first time for chemical retrieval. The extracted chemical functional features are then used for similarity measurement and query retrieval. The performance evaluation shows that the proposed approach achieves promising accuracy and outperforms a state-of-the-art method for chemical retrieval.

Keywords: Functional Feature Extraction, Chemical Structural Retrieval, Chemical Functional Groups.

1 Introduction

Chemical structural formulas are commonly used for presenting the structural and functional information of organic chemicals. In many drug discovery projects, it is often required to search for similar chemical structural formulas of drug-like compounds that are worthy for further synthetic or biological investigation [19]. As such, there is a need to find relevant chemical structures with similar chemical properties for chemical structural queries. Over the past few decades, several chemical structure databases such as ChemSpider [16], PubChem [20], ChEM-BLdb [9] and eMolecules.com [2] have been developed to support structured query retrieval of chemical structures. In addition, a number of chemical structural similarity search methods [1,4,7,8,10,12,17,18] have also been proposed for chemical structural retrieval. The proposed methods are mainly based on the assumption that chemicals which are globally similar in structure to each other are more likely to have similar chemical properties and activities.

However, the current retrieval methods ignore the functional features such as functional groups and interactions between functional groups that are hidden

A. Ailamaki and S. Bowers (Eds.): SSDBM 2012, LNCS 7338, pp. 507–525, 2012.

inside the chemical structure. Take Acetic Acid CH_3COOH as an example, the ChemSpider search engine will return $CH_3COO^-Li^+$ and $(CH_3COO^-)_2Ca^{2+}$ because they have similar structures with CH_3COOH. However, Propionic Acid CH_3CH_2COOH will not be returned. But it has more similar chemical properties with the query CH_3COOH because they have the same functional group $-COOH$.

In a chemical structure, functional features are important to determine the chemical property and activity of the chemicals. Therefore, it is important to consider functional features for chemical structural retrieval. To the best of our knowledge, there is no research work which focuses on extracting functional features and uses the extracted features for similarity measurement.

In this paper, we propose a new approach for chemical feature extraction and retrieval. It constructs Chemical Functional Group (CFG) Graph from the SMILES [21] representation of chemical structure and extracts four types of chemical features (namely the number of Carbons, number of Carbon Chains, Functional Group (FG) feature, and Functional Group Interaction (FGI) feature) for query retrieval. We also propose *ctf-icf* weighting and similarity measure for chemical retrieval. Compared with the existing proposals, the advantage of the proposed approach is that it takes functional features like FG features and FGI features into account in retrieval, and thus the retrieved formulas are more functionally similar to the query. Experimental results show that the proposed approach achieves promising performance, and outperforms a state-of-the-art chemical structural retrieval method.

The rest of the paper is organized as follows. Section 2 reviews the related work on chemical structural retrieval. Section 3 presents our proposed approach. Section 4 describes the method of chemical structural feature extraction and the Chemical Functional Group (CFG) Graph construction for feature extraction. Section 5 describes the query retrieval and ranking process. Section 6 evaluates the performance of the proposed approach in comparison with other methods. Finally, Section 7 concludes the paper.

2   Related Work

ChemSpider [16] is a free online aggregated chemical structure database providing fast text and structural search access to over 26 million structures from hundreds of data sources. PubChem [20] is a free database of chemical structures of small organic molecules and information on its biological activities. ChEM-BLdb [9] is a manually curated chemical database of bio-active molecules with drug-like properties. eMolecules [2] is a search engine for chemical molecules supplied by commercial suppliers. All these publicly available chemical search systems support database query retrieval for chemical structural formulas including exact search, substructure search and similarity search. However, the structured query search method may tend to overfit the chemical structures and fail to recognize chemicals that are more similar in chemical functionalities.

Apart from structured query search, similarity search methods have been under development for decades to find relevant chemical structures. To compute the similarities between chemical structures, different chemical structural representations and similarity measures have been proposed. Currently, there are three major chemical similarity search methods, namely superposition-based similarity methods, histogram-based similarity methods and descriptor-based similarity methods.

The superposition-based similarity methods map one chemical structure onto another. It treats two molecular structures as graphs, and aims to find the correspondence between the atoms in one structure and the atoms in another structure [10,17]. Histogram-based methods [18] transform chemical structures into one or more spectra or histograms, and then calculate the overlapping between the histograms of two chemical structures for similarity measurement. Descriptor-based methods are most popular for chemical structural similarity search. A molecule is represented as a set of *descriptors* or numbers. As such, a molecule can be considered as a point in a multidimensional *descriptor space*. This method is computational efficient. However, in contrast to the superposition-based methods, the equivalence of sub-structures (or parts) between one molecule and another is lost.

In particular, the *fingerprinting* [1,4,7,8,12] method, which is a descriptor-based method, uses a set of user-defined 2D substructures and their frequencies to represent molecules. The substructures are used as the descriptors. In the *fingerprinting* method, only the presence or absence of a descriptor is captured. The substructure descriptors are considered as the fingerprints of the chemical structure. The similarity is defined based on the number of descriptors that the two molecules have in common and normalized by the number of descriptors in each molecule. The *fingerprinting* method is efficient because it is computationally inexpensive to compare two lists of pre-computed descriptors. The Daylight Fingerprint algorithm [1] is one of the most famous fingerprinting methods and therefore it is used for performance evaluation in this paper. It generates the patterns for fingerprints from the molecule itself in the following manner: (1) it generates a pattern for each atom; (2) it generates a pattern representing each atom and its nearest neighbors (plus the bonds that join them); and (3) it generates a pattern representing each group of the atoms and bonds connected by paths up to 2, 3, 4, 5, 6 and 7 bonds long. For example, the molecule Aminoethenol (with the corresponding SMILES representation as OC=CN) would generate the patterns as shown in Figure 1.

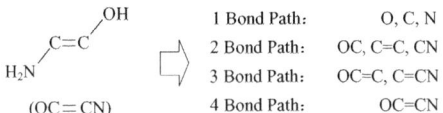

1 Bond Path:	O, C, N
2 Bond Path:	OC, C=C, CN
3 Bond Path:	OC=C, C=CN
4 Bond Path:	OC=CN

Fig. 1. Daylight Fingerprints of Aminoethenol

[7,4,8,12] are variants of the fingerprinting method available for chemical structural search. Some of the newer fingerprints describe the atoms not by their elemental types but by their physiochemical characteristics. This enables the identification of database chemical structures which have similar properties to the query structure but with different sets of atoms in a similarity search.

The current chemical structural similarity search methods focus very much on structural similarities between chemical structures. However, the similarity search should retrieve structures that should have similar chemical properties with the queried structure. Two chemical structures which have similar chemical properties are mainly due to the fact that they have the same functional groups and similar interactions between functional groups. To the best of our knowledge, none of the current similarity search methods have considered the extraction of functional group features from the chemical structures and use these features to search for similar chemical structures in terms of functional properties. Therefore, in this research, we propose a new approach for extracting information on functional groups and their interactions for chemical structural similarity search.

3   Overview of Proposed Approach

Functional groups (FGs) are specific groups of atoms within molecules that determine the characteristics of chemical reactions of those molecules. As the properties and chemical reactions of a specific functional group are quite unique, functional groups can be used for identifying chemical structures. For example, Figure 2(a) shows the chemical structure of Benzyl Acetate which has an Ester functional group and a Phenyl functional group.

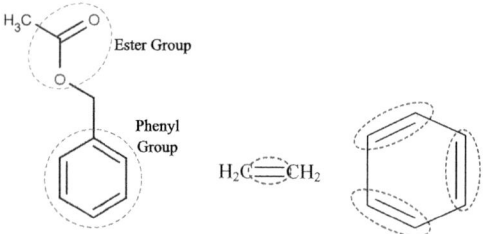

(a) Benzyl Acetate (b) Ethylene (c) Benzene

Fig. 2. Example Chemical Structures

The same functional group will undergo the same or similar chemical reaction(s) regardless of the size of the containing molecule. For example, both Ethanol (CH_3CH_2OH) and n-Propanol ($CH_3CH_2CH_2OH$) have the same Hydroxyl functional group '$-OH$'. Although they have different numbers of Carbon atoms, they can undergo similar reactions such as combustion and deprotonation. They also have similar toxicity and smell, and they are good solvent to other chemicals.

However, the relative activity of a functional group can be changed due to its interaction with a nearby functional group. For example, both Ethylene and Benzene in Figure 2(b) and Figure 2(c) have Alkenyl functional group '$C = C$'. But Benzene has three interconnected Alkenyl functional groups. The interactions among the double bonds of three Alkenyl groups make the double bond more difficult to be broken. Hence, the double bonds in Benzene cannot have the addition reaction as the double bond in Ethylene, unless the reaction takes place under a special condition such as enzyme or UV ray. In addition, the interactions also enable Benzene to have a series of electrophilic substitutions.

The degree of interactions is affected by the distance between two functional groups. The distance between two functional groups is measured by the number of Carbon atoms located between them. The closer the two functional groups is, the stronger the interaction is. If two or more functional groups are connected directly together, they can form a complex functional group due to the strong interaction. For example, connecting the Carbonyl '$-C(=O)-$' and Ether '$-O-$' functional groups will form the Ester functional group. On the other hand, the Phenyl group can be decomposed into three connected double bonds '$C = C$'. Table 1 lists some of the most common basic functional groups.

Table 1. Common Basic Functional Groups

Functional Group	Structural Formula	Functional Group	Structural Formula
Alkenyl	$C = C$	Alkynyl	$C{\equiv}C$
Fluoro	$-F$	Iodo	$-I$
Chloro	$-Cl$	Bromo	$-Br$
Hydroxyl	$-OH$	Ether	$-O-$
Carbonyl	$-C(=O)-$	Aldehyde	$-C(=O)H$
Amine	$-\overset{\vert}{N}-$	Primary Imine	$-C = NH$
Secondary Imine	$-C = N-$	Nitrile	$-C{\equiv}N$
Sulfhydryl	$-SH$	Sulfide	$-S-$
Phosphino	$-\overset{\vert}{P}-$		

In this paper, we propose a new approach for chemical structural feature extraction and retrieval. The proposed approach, which is shown in Figure 3, consists of the following two main processes: Chemical Feature Extraction and Query Retrieval.

4   Chemical Feature Extraction

We extract the following types of chemical features: (1) the number of Carbon atoms; (2) the number of Carbon Chains; (3) Functional Groups; and (4) Functional Group Interactions. It is easier to extract the first two types of chemical

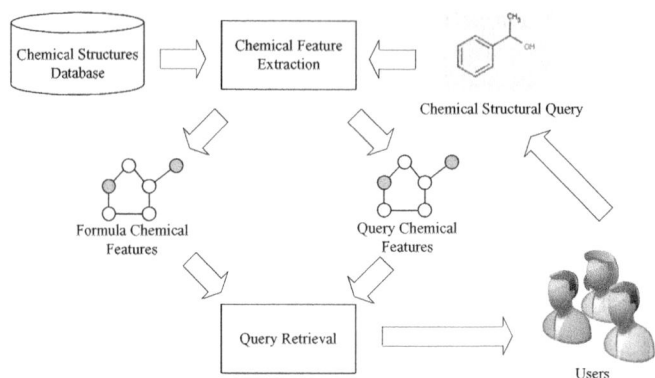

Fig. 3. Proposed Approach

features, and we mainly focus on extracting functional groups and functional group interactions in this section.

We present the structural representation of chemicals in Section 4.1, the proposed approach to extracting functional groups in Section 4.2, the proposed method of building Chemical Functional Group (CFG) Graph that captures chemical structural information stored inside chemical structural formulas in Section 4.3, and the proposed method of extracting chemical features from the CFG Graph of chemical structural formula in Section 4.4.

4.1   Structural Formula Representation

A chemical structure presents the type and spatial arrangement of each atom and bond in the compound. Chemical structures are represented in computers for storage and retrieval. There are two main categories of chemical structure representations. First, chemical structures can be represented as chemical connection table/adjacency matrix/list with additional information on bonds (edges) and atom attributes (nodes). This representation is used by MDL Molfile [6] and CML (Chemical Markup Language) [15]. Second, chemical structures can be represented as a string encoded with structural information. This representation is used by SMILES (simplified Molecular Input Line Entry Specification) [21], InChI (The IUPAC International Chemical Identifier) [14] and InChIKey [11].

Among the five representations, the string representation used by SMILES has the advantages on effective storage and efficient processing. Therefore, in the proposed approach, SMILES is used to represent chemical structural formulas. In SMILES [21], all Hydrogen atoms are omitted. Non-Hydrogen atoms are represented by their atomic symbols such as *C, O, N, P* and *S*. For atomic symbols using two-letters representation, the second letter must be in lower case, e.g. *Cl* and *Br*. Single, double and triple bonds are represented by the symbols −, = and #, respectively. Adjacent atoms are assumed to be connected to each other by a single bond (in this case, the single bond is always omitted).Moreover, branches in chemical structural formulas are specified by enclosing them in parentheses. The implicit connection for a parenthesized expression (a branch) is to the left. Table 2 shows two example SMILES representing structural formulas that involve branches. Furthermore, in SMILES, cyclic or ring structures are represented by breaking one bond in each ring. The broken bond is indicated by a digit number immediately following the atomic symbol at each ring opening. Figure 4 shows the SMILES representation for the ring structure of Cyclohexane.

Table 2. SMILES Representations of Branches

Structural Formula	SMILES
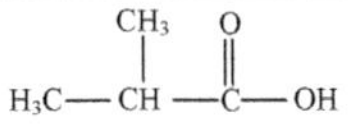	CC(C)C(=O)O
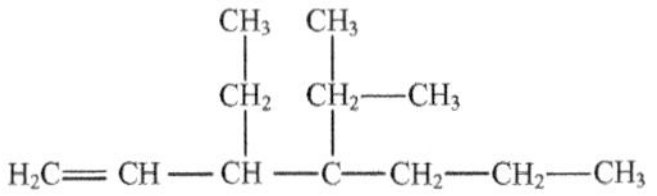	C=CC(CC)C(C(C)C)CCC

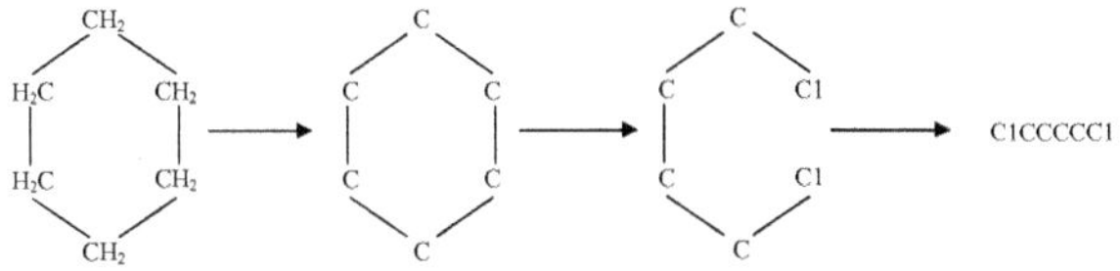

Fig. 4. SMILES Representation of Cyclohexane

4.2   Functional Group Identification

We present the proposed approach to identifying functional groups from the SMILES representations of chemicals. Each functional group corresponds to a string pattern and a SMILES representation. Table 3 gives several examples of commonly used functional groups as well as their SMILES patterns and string patterns.

To identify functional groups for a chemical, we map the SMILES patterns of the functional groups with the SMILES representation of the chemical. Except C (i.e. Carbon) and parentheses, the identified functional groups will be replaced by the corresponding string patterns. For example, $C(=O)$ will be identified as Carbonyl group and '$=O$' will then be replaced by *carbonyl*. The C and parentheses are not replaced as they also contain structural information of the chemicals in the structural formula.

In addition, functional groups are prioritized for the identification process. If a SMILES pattern has more number of symbols (including atomic symbols and operators), the corresponding functional group will have a higher priority. If two SMILES patterns have the same number of symbols, they will have the same priority. Functional groups with higher priority will be identified and replaced first. In other words, for a chemical SMILES we find its longest matching SMILES pattern and replace the matching with the corresponding functional groups, and we find the longest matching pattern in the replaced chemical SMILES; the process is repeated until we cannot find more functional groups.

Table 3. SMILES and String Patterns of Some Example Functional Groups

Functional Group	Structural Formula	SMILES Pattern	String Pattern	Priority
Aldehyde	O‖ —C—H	O=C– or –C=O	*Caldehyde* *aldehydeC*	High
Carbonyl	O‖ —C—	–C(=O)–	*C(carbonyl)*	
Ether	—O—	–O–	ether	↓
Hydroxyl	—OH	O– or –O or –C(O)–	*hydroxyl*	
Alkenyl	⟩C=C⟨	–C=C–	*CalkenylC*	Low

We illustrate the functional group identification process using Benzyl Acetate as an example. The corresponding SMILES representation of Benzyl Acetate is $CC(=O)OCC1=CC=CC=C1$. The identification works in the following three steps.

– In Benzyl Acetate, the Carbonyl group has higher priority because its pattern '=O' contains two atomic symbols. Therefore, the Carbonyl group is identified first and '=O' is replaced by *carbonyl*. The representation becomes $CC(carbonyl)OCC1=CC=CC=C1$.
– Next, the Ether group 'O' is identified and replaced by *ether*. The representation becomes $CC(carbonyl)etherCC1=CC=CC=C1$.
– It follows by the identification of the Alkenyl group 'C=C' with its '=' replaced by *alkenyl*. The string pattern formed will be $CC(carbonyl)etherCC1$ $alkenylCCalkenylCCalkenylC1$.

Finally, we know that Benzyl Acetate contains three types of functional groups: Carbonyl, Ether and Alkenyl.

4.3   Chemical Functional Group (CFG) Graph

To identify the functional group interactions, we propose to construct a CFG Graph from the SMILES representation of the chemical structural formula. We first identify the functional groups and replace them with the corresponding string patterns using the method presented in Section 4.2.

Definition 1 (CFG Graph). *Chemical Functional Group (CFG) Graph* $G = (V, E)$ *is an undirected graph which represents the chemical structure with a set of nodes V and a set of edges E. The set of nodes $V = V_c \cup V_f$, where V_c is a set of carbon nodes $n_c \in V_c$ and V_f is a set of functional group nodes $n_f \in V_f$. Each carbon node n_c represents one Carbon atom in the chemical structural formula. Each functional group node represents one of the basic functional groups. Each edge $e \in E$ indicates one of the following connections:*

- *a connection between two Carbon atoms;*
- *a connection between two functional groups; or*
- *a connection between one functional group and one Carbon atom.*

Figure 5 shows the CFG Graph for Ethylene. In the CFG Graph, there are two carbon nodes and one functional group node (i.e., the double bond).

Fig. 5. CFG Graph for Ethylene

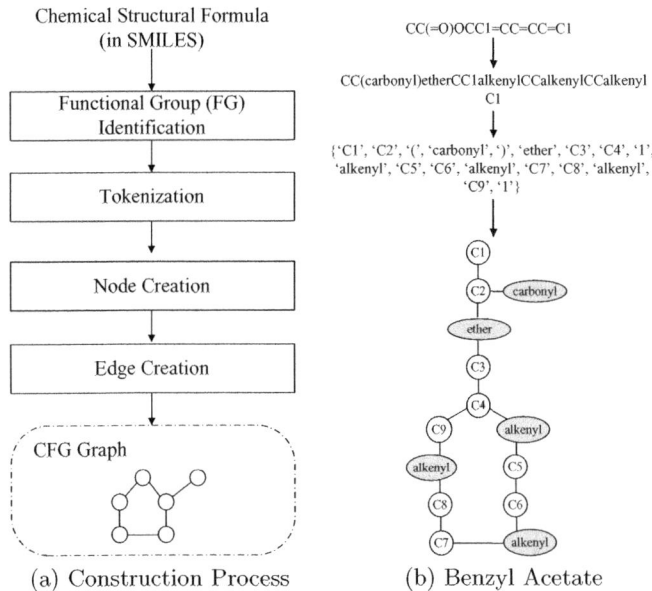

(a) Construction Process   (b) Benzyl Acetate

Fig. 6. CFG Graph Construction

Figure 6(a) shows the proposed construction process for CFG Graph. In the proposed approach, chemical structural formula is first represented using the SMILES representation. To generate the CFG Graph, the formula is processed by the following steps: Functional Group Identification, Tokenization, Node Creation and Edge Creation.

Tokenization. In this step, the generated functional group string pattern will be tokenized into a series of tokens.

Each functional group will become a token; each Carbon atom (capital letter C) in SMILES will be labeled with an integer number starting with 1, and becomes a token; each other letter will also become a token. Note that the order of the extracted tokens will be maintained according to their occurrences in the string. The order indicates the connections of the bonds.

Figure 6(b) shows the CFG Graph construction process for Benzyl Acetate. After Functional Group Identification, the identified string pattern is CC(carbon yl)etherCC1alkenylCCalkenylCCalkenyl C1. It is tokenized into the following list of tokens: ('C1', 'C2', '(', 'carbonyl', ')', 'ether', 'C3', 'C4', '1', 'alkenyl', 'C5', 'C6', 'alkenyl', 'C7', 'C8', 'alkenyl', 'C9', '1').

Node Creation. After tokenization, all the Carbon atoms and functional groups will be identified from the list of tokens. The nodes $V = V_c \cup V_f$ of the CFG Graph will then be created. Take Benzyl Acetate as an example, the carbon node set V_c contains carbon nodes of 'C1','C2',...,'C9', whereas the functional group node set V_f contains functional group nodes of 'carbonyl', 'ether' and three 'alkenyl' groups. This is illustrated in Figure 6(b).

Edge Creation. Graph edges are determined according to the SMILES specification of chemical structures. An edge will be created for each connection identified. It will then be added into the set of edges E of the CFG Graph. The connections between nodes are identified based on the following ways:

(1) If two nodes are created from tokens which are next to each other in the token list, then they are connected. This is derived from the SMILES specification on bonds.
(2) The first node after '(' and the first node after ')' are connected with the first node just before '('. This is derived from the SMILES specification on branches.
(3) If two nodes have the same numeric token following them in the token list, then they are connected. This is derived from the SMILES specification on cyclic structure.

Take Benzyl Acetate shown in Figure 6(b) as an example, $C1$ and $C2$ are connected according to (1). $C2$ and $Carbonyl$, and $C2$ and $Ether$ are connected according to (2). $C4$ and $C9$ are connected according to (3).

Algorithm for Constructing CFG Graph. Algorithm 1 presents the algorithm for the CFG Graph Construction process. The proposed CFG Graph construction approach preserves well the structural information of the functional groups. For example, the ring connection of the Phenyl group structure in Benzyl Acetate is broken when it is represented in SMILES, i.e., the ring is opened between $C4$ and $C9$, according to the cyclic structure specification in SMILES. However, it is restored and preserved during the construction of the CFG Graph. This helps capture correctly the relative distance between functional groups which measures the degree of interactions between functional groups.

Algorithm 1. Functional Group Graph Construction

Input: *CSF*: Chemical structural formula represented in SMILES
Output: $G = (V,E)$: CFG Graph
Process:
  $CSFPattern \leftarrow FGIdentification(CSF)$
  $tokenList \leftarrow tokenization(CSFPattern)$
  for $token \in tokenList$ **do**
    if $token \notin \{ '(', ')' \} \cup ringDigit$ **then**
      $node \leftarrow nodeCreation(token)$
      if $isCarbon(token)$ *is True* **then**
        $V_c \leftarrow V_c \cup node$
      else
        $V_f \leftarrow V_f \cup node$
  $V \leftarrow V_c \cup V_f$
  for $node1, node2 \in V$ **do**
    if $connected(node1, node2)$ *is True* **then**
      $edge \leftarrow edgeCreation(node1, node2)$
      $E \leftarrow E \cup edge$
  return G

Table 4. Carbon Chains

Structural Formula	SMILES	Carbon Chain Number
$H_3C—CH—C—OH$ with CH_3 and O groups	CC(C)C(=O)O	2
$H_2C=CH—CH—C—CH_2—CH_2—CH_3$ with CH_3, CH_2, $CH_2—CH_3$ groups	C=CC(CC)C(C(C)C)CCC	4

4.4   Chemical Feature Extraction

Recall that we extract 4 types of chemical features: (1) the number of Carbon atoms; (2) the number of Carbon Chains; (3) Functional Groups; and (4) Functional Group Interactions.

The number of Carbon atoms (N_C) refers to the number of Carbon atoms in the chemical structural formula. It can be extracted easily by counting the number of Carbon nodes in the CFG Graph. We have presented the approach to extracting functional groups in Section 4.2. Next, we present the extraction of other types of features.

Definition 2 (Carbon Chain). *Carbon Chain is a chain connection of atoms with at least one Carbon atom in it. Carbon Chain can be a backbone or a branch of the chemical structural formula. The longest Carbon Chain is the backbone of the chemical structural formula.*

Table 5. Chemical Features Extracted from Benzyl Acetate

Feature Name	Feature Representation
Number of Carbons	9
Number of Carbon Chains	1
Functional Group	*carbonyl, ether, alkenyl, alkenyl, alkenyl*
Functional Group Interaction	*carbonyl1ether, ether1carbonyl, carbonyl4alkenyl, alkenyl4carbonyl, carbonyl5alkenyl, alkenyl5carbonyl, carbonyl7alkenyl, ...*

Algorithm 2. Chemical Feature Extraction

Input: $G = (V, E)$: CFG Graph
Output:
    N_C - Number of Carbon atoms
    N_{CChain} - Number of Carbon Chains
    FGset - A set of Functional Group features
    FGIset - A set of Functional Group Interaction features
Process:
    $N_C \leftarrow$ *carbonNo(G)*
    $N_{CChain} \leftarrow$ *carbonChainNo(G)*
    for *node* $\in V_f$ **do**
        FG $\leftarrow$ *node.getName(node)*
        FGset $\leftarrow$ *FGset* $\cup$ *FG*
    for *node1, node2* $\in V_f$ **do**
        d $\leftarrow$ *distance(node1, node2)*
        fg1 $\leftarrow$ *node.getName(node1)*
        fg2 $\leftarrow$ *node.getName(node2)*
        FGIset $\leftarrow$ *FGIset* $\cup$ *concat(fg1, d, fg2)*
        FGIset $\leftarrow$ *FGIset* $\cup$ *concat(fg2, d, fg1)*
    return N_C, N_{CChain}, *FGset, FGIset*

Table 4 gives some examples of Carbon Chains. For each chemical structural formula, the number of Carbon Chains (N_{CChain}) can be extracted and counted from the CFG Graph.

For functional group (FG) features, the functional group names of the corresponding functional group nodes will be extracted.

Definition 3 (Functional Group Interaction). *Functional Group Interaction (FGI) indicates the degree of the interaction between any two functional groups. It can be obtained by measuring the distance (i.e. the number of atoms) between two functional group nodes.*

The FGI features are extracted in the form as a combination of the two functional group names and the distance between the two functional group nodes. The distance is calculated by using the A* search algorithm [5].

Algorithm 2 gives the algorithm for Chemical Feature Extraction. Table 5 shows the chemical features that can be extracted from Benzyl Acetate. Although we only identify the basic functional groups in Functional Group Identification, calculating the shortest distance between two functional groups enables us to capture the information on complex functional groups as all complex functional groups can be constructed by the basic functional groups. For example, the Ester functional group can be obtained as a combination feature '*carbonyl1ether*' or '*ether1carbonyl*'.

5   Query Retrieval

When a chemical structural formula query is submitted, the query's CFG Graph will be constructed from its SMILES representation and its chemical features will then be extracted from the CFG Graph. The ranking score for a chemical structural formula will be calculated based on the four types of features extracted. The ranking score is calculated by using the *ctf-icf* weighting which is based on the concept of the *tf-idf* weighting [13]. We proceed to define the terms used in the *ctf-icf* weighting, namely chemical term frequency (*ctf*) and inverse chemical frequency (*icf*), as follows.

Definition 4 (Chemical Term Frequency (*ctf*)). *Chemical term frequency $ctf_{t,c}$ of term t in chemical structure c is defined as the number of times that t occurs in c.*

Definition 5 (Inverse Chemical Frequency (*icf*)). *Chemical frequency cf_t of term t measures the number of chemical structures containing the term t. It is an inverse measure of the informativeness of t because frequent terms are less informative than rare terms. Inverse chemical frequency icf_t is defined by $icf_t = log(1 + N/cf_t)$ where N is the total number of chemical structural formulas.*

Definition 6 (Ctf-icf Weight). *Ctf-icf weight $w_{t,c}$ of a chemical feature term t in chemical structure c is defined as the product of its chemical term frequency ctf and inverse chemical frequency icf:*

$$w_{t,c} = ctf_{t,c} \times log(1 + N/cf_t)$$

Now, each chemical structural formula can be represented by a real-valued vector $\in R^{|V|}$ of ctf-icf weight. $|V|$ is the number of dimensions in the vector space with chemical terms as the axes. Each functional group feature corresponds to a term and so does each functional group interaction feature. Similarly, we treat "Carbons" and "Carbon Chains" as two terms. For "Carbons" feature, its *ctf* is the number of Carbons; for "Carbon Chains" feature, its *ctf* is the number of Carbon Chains.

We can use cosine similarity to compute the relevance of a chemical structural formula to a query using their ctf-icf weights. Cosine similarity is used to measure the similarity between a query and a document in document retrieval.

Two documents are more similar if the relative distribution of terms are more similar. However, it is not always the case for chemical structures. Two chemical structures are similar because they have similar number of chemical functional groups and similar chemical functional group interactions. The absolute term difference is important. Therefore, we use L1 norm distance of two chemical feature vectors to measure the dissimilarity. We proceed to define dissimilarity for the four types of chemical features.

Definition 7 (FG Feature Dissimilarity). *Given a query q and retrieved chemical formula c, FG feature dissimilarity is defined as*

$$Dissim_{FG}(q,c) = \sum_{t_{FG} \in q \cup c} |w_{t_{FG},q} - w_{t_{FG},c}|$$

where t_{FG} is a FG feature term.

Definition 8 (FGI Feature Dissimilarity). *Given a query q and retrieved chemical formula c, FGI feature dissimilarity is defined as*

$$Dissim_{FGI}(q,c) = \sum_{t_{FGI} \in q \cup c} |w_{t_{FGI},q} - w_{t_{FGI},c}|$$

where t_{FGI} is a FGI feature term.

Definition 9 (Carbon Atom Dissimilarity). *If we treat Carbon atom as a feature term for chemical structures, the chemical term frequency ctf for Carbon atom will be N_C. The inverse chemical frequency icf will be $log2$ since every organic chemical structure has at least one Carbon atom (i.e., cf for Carbon atom is 1). Therefore, given a query q and retrieved chemical formula c, Carbon atom dissimilarity is defined as*

$$Dissim_C(q,c) = |N_C(q) - N_C(c)| \times log2$$

Definition 10 (Carbon Chain Dissimilarity). *Similarly, Carbon Chain can be treated as a feature term. Given a query q and retrieved chemical formula c, Carbon Chain dissimilarity is defined as*

$$Dissim_{CChain}(q,c) = |N_{CChain}(q) - N_{CChain}(c)| \times log2$$

Based on the four types of dissimilarity measures, we define chemical structure similarity as follows.

Definition 11 (Chemical Structure Similarity). *The similarity between a chemical structural query q and the retrieved chemical structure c is defined as*

$$Sim(q,c) = \frac{1}{1 + Dissim(q,c)},$$

$$Dissim(q,c) = (1 - \alpha) \times (Dissim_{FG}(q,c) + Dissim_{FGI}(q,c))$$
$$+ \alpha \times (Dissim_C(q,c) + Dissim_{CChain}(q,c)), \quad where \ \alpha \in [0,1].$$

The parameter α is used to discriminate the contributions of different features to the total dissimilarity score, and will be set empirically. The chemical structure similarity will be in the range [0,1]. The higher the chemical structure similarity is, the more similar the two chemical structures are.

6   Performance Evaluation

Objectives The performance study has three main objectives as follows:

- We aim to study the effectiveness of the extracted structural features in chemical retrieval.
- We aim to compare the effectiveness of four types of structural features.
- We aim to evaluate the effectiveness of the proposed retrieval method.

Data and Query. The performance is evaluated using the chemical structure data from the open NCI database [3] which is one of the most complete collections of chemical compounds. A total of 10,646 chemical structures are extracted from the NCI database and converted into the corresponding Canonical SMILES format. In the experiments, 100 chemical structures are used as test queries. Each test chemical structure may contain only single functional group or multiple functional groups. We split the test queries into two sets. One set of the queries is used for tuning the parameter α in the proposed approach, while the other set of test queries is used for performance evaluation. Each set comprises 11 test queries with single functional groups and 39 queries with multiple functional groups. Each set of the test queries covers all the basic functional groups. Thus the set of test queries should be sufficient for evaluating the performance of the proposed approach.

For each method, top 10 formulas will be retrieved for each query. We mix the top 10 results of each method, and then employ a professional to label if the returned result is a *functionally relevant* formula to the query. The performance is measured using Precision@5, Precision@10, Mean Average Precision (MAP) [13] and retrieval time. Precision@n reports the fraction of the top-n chemicals retrieved that are relevant. MAP is the mean of the average precisions for a set of query chemicals; for each query, its average precision is the average of the precision values at the points where each relevant chemical is retrieved. MAP rewards approaches that return relevant chemicals early. Retrieval time is used to evaluate the efficiency of the method. It is measured from the time the query is submitted until the time the result is returned.

Method. We compare with the Daylight Fingerprint method [1]. The Daylight Fingerprint method is one of the most popular descriptor-based methods and a state-of-the-art method for chemical structural similarity search. In addition, we also compare the proposed retrieval approach with the method that uses the same types of features as the proposed approach, but uses cosine similarity to compute the similarity for each type of features.

The experiment is conducted as follows. The functional features of the chemical structures from the data set and the test queries are first extracted using the proposed approach. The performance of the Query Retrieval process is then measured for the proposed approach. The experiment is conducted with Python on a Windows 7 environment running on an Intel Core 2 Duo 2.80GHz CPU with 2.46GB of memory.

Parameter Tuning. We determine the value for the parameter α in the proposed approach using the first set of test queries. The Mean Average Precision

(MAP) is used as the performance measure. During the experiment, the value of α is set from 0 to 1 in intervals of 0.1 and the performance is then measured. Figure 7 shows the performance results based on the different values of α. Based on the performance results, the proposed approach has the best performance of 73.17% when α is set to 0.3. Similarly, we empirically find the best parameter α is 0 when cosine similarity is used. In the subsequent experiment, we set the parameter α as 0.3 and 0 for the proposed method using the proposed similarity measure and cosine similarity respectively.

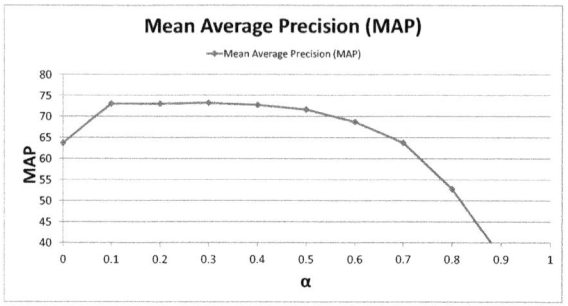

Fig. 7. Performance Results Based on Different Values of α

Table 6. Performance Results for Different Methods

	Precision@5	Precision@10	MAP	Retrieval Time
Proposed Approach	**97.20%**	**89.00%**	**84.53%**	0.5312s
Proposed Approach with Cosine Similarity	73.20%	68.20%	66.71%	0.5357s
Daylight Fingerprint	55.20%	47.80%	46.67%	2.546s

Effectiveness of the Proposed Approach. We evaluate the performance of the proposed approach and other methods using the second set of test queries. Here, we use the Mean Average Precision (MAP) and the precision as the performance measures.

Table 6 gives the performance results for the proposed approach in comparison with other methods. The performance results clearly indicate that the proposed approach outperforms the other methods for the different measures. Specifically, it achieves the performance of 97.20%, 89.00% and 84.53% for Precision@5, Precision@10 and MAP, respectively. This demonstrates that the proposed approach is effective for retrieving chemical structures that have functional similarity with the queries, and the four types of chemical features used in the proposed approach are effective for chemical retrieval. The improvement of the proposed approach over the Daylight Fingerprint method is statistically significant using t test, p-value < 0.01.

Comparing the two versions of the proposed approach: one using the proposed measure to compute similarity for each type of feature, and the other using cosine similarity, we find that the proposed similarity measure is more effective than cosine similarity for chemical retrieval.

Efficiency of the Three Approaches. The two methods of using the four types of chemical features are faster than the one using Daylight Fingerprint features. This is because the Daylight Fingerprint method would generate more features than our proposed feature extraction methods.

Comparing the Four Types of Features. We also evaluate the performance of the proposed approach using different sets of features such as functional groups and functional group interactions. Table 7 gives the performance results of the proposed approach based on the different types of extracted features. The results show that when the proposed approach uses all the four types of extracted chemical features, it achieves the best performance in terms of all the three measures. We observe that MAP is low when using only N_C or N_{CChain}. It is because N_C and N_{CChain} only capture basic structural information of the chemicals. It is difficult to find functionally relevant chemical structures for the query with these structural features alone. However, the performance when using all four types of features is better than the performance when using only the FG and FGI features. The reason is that N_C and N_{CChain} help determine which chemical structure is more relevant to the query when two chemicals have the same FG and FGI features. For example, CH_3COOH, CH_3CH_2COOH and $CH_3CH_2CH_2COOH$ have the same FG and FGI features. But CH_3CH_2COOH is more similar with CH_3COOH since it has less difference on N_C with CH_3COOH. Additionally, we notice that using $FG+FGI+N_C$ performs close to using all four features and it is better than using $FG+FGI+N_{CChain}$. This may be due to the reason that N_C is more discriminative than N_{CChain} for retrieval based on the data set. In summary, the FG and FGI features enable the search of functional relevant chemical structures, and N_C and N_{CChain} help improve the performance.

Table 7. Performance Results based on the Extracted Chemical Features

Feature Components	Precision@5	Precision@10	MAP	Retrieval Time
N_C	50.40%	32.80%	25.75%	0.3529s
N_{CChain}	65.20%	48.00%	41.41%	0.3541s
FG	66.80%	63.80%	64.84%	0.5080s
FGI	72.00%	62.20%	62.98%	0.1750s
$FG + FGI$	83.20%	74.60%	75.10%	0.5290s
$FG + FGI + N_C$	96.80 %	87.20%	83.96%	0.5302s
$FG + FGI + N_{CChain}$	85.70%	77.40%	77.51%	0.5294s
$FG + FGI + N_C + N_{CChain}$	**97.20%**	**89.00%**	**84.53%**	0.5312s

7   Conclusion

This paper proposes a novel and effective approach for chemical feature extraction and retrieval. The proposed approach consists of two major processes: Chemical Feature Extraction and Query Retrieval. In the Chemical Feature Extraction process, it constructs the CFG Graph from the corresponding SMILES representation of a chemical structural formula. The chemical features can then be extracted from the constructed CFG Graph. The Query Retrieval process accepts a user query on a chemical structure, converts the queried structure into the corresponding CFG Graph and extracts chemical features from the CFG Graph. The extracted features are then compared with the extracted features of the stored chemical structures for similarity measurement, ranking and retrieval. In addition, the performance of the proposed approach is evaluated in comparison with other methods. The experimental results have shown that the proposed approach achieves promising performance, and outperforms the state-of-the-art Daylight Fingerprint method. This demonstrates that the extracted structural features are indeed useful for chemical retrieval.

For future work, we are currently investigating additional chemical features that will enhance the performance of the proposed approach. Furthermore, we are also in the process of incorporating the proposed approach into a Web-based Chemical Question-Answering System for supporting chemical structural search.

References

1. Daylight fingerprint,
   http://www.daylight.com/dayhtml/doc/theory/theory.finger.html
2. emolecules.com, http://www.emolecules.com/
3. Nci structure database, http://cactus.nci.nih.gov/download/nci/
4. Brown, R., Martin, Y.: Use of structure-activity data to compare structure-based clustering methods and descriptors for use in compound selection. J. Chem. Inform. Comput. Sci. 36(3), 572–584 (1996)
5. Chow, E.: A graph search heuristic for shortest distance paths. Tech. rep., Lawrence Livermore National Laboratory (2005)
6. Dalby, A., Nourse, J., Hounshell, W., et al.: Description of several chemical structure file formats used by computer programs developed at molecular design limited. J. Chem. Inform. Comput. Sci. 32(3), 244–255 (1992)
7. Ewing, T., Baber, J., Feher, M.: Novel 2d fingerprints for ligand-based virtual screening. J. Chem. Inf. Model. 46(6), 2423–2431 (2006)
8. Fechner, U., Paetz, J., Schneider, G.: Comparison of three holographic fingerprint descriptors and their binary counterparts. QSAR & Combinatorial Science 24(8), 961–967 (2005)
9. Gaulton, A., Bellis, L., Bento, A., et al.: Chembl: a large-scale bioactivity database for drug discovery. Nucl. Acids Res. 40(1), 1100–1107 (2012)
10. Hagadone, T.: Molecular substructure similarity searching: efficient retrieval in two-dimensional structure databases. J. Chem. Inform. Comput. Sci. 32(5), 515–521 (1992)
11. Heller, S., McNaught, A.: The iupac international chemical identifier (inchi). Chemistry International 31(1), 7 (2009)

12. Hert, J., Willett, P., Wilton, D., et al.: Comparison of topological descriptors for similarity-based virtual screening using multiple bioactive reference structures. Org. Biomol. Chem. 2(22), 3256–3266 (2004)
13. Manning, C., Raghavan, P., Schutze, H.: Introduction to information retrieval, vol. 1. Cambridge University Press, Cambridge (2008)
14. McNaught, A.: The iupac international chemical identifier. Chemistry International (2006)
15. Murray-Rust, P., Rzepa, H.: Chemical markup, xml, and the worldwide web. 1. basic principles. J. Chem. Inform. Comput. Sci. 39(6), 928–942 (1999)
16. Pence, H., Williams, A.: Chemspider: an online chemical information resource. J. Chem. Educ. (2010)
17. Rarey, M., Dixon, J.: Feature trees: a new molecular similarity measure based on tree matching. J. Comput. Aided Mol. Des. 12(5), 471–490 (1998)
18. Schuur, J., Selzer, P., Gasteiger, J.: The coding of the three-dimensional structure of molecules by molecular transforms and its application to structure-spectra correlations and studies of biological activity. J. Chem. Inform. Comput. Sci. 36(2), 334–344 (1996)
19. Sheridan, R., Kearsley, S.: Why do we need so many chemical similarity search methods? Drug Discovery Today 7(17), 903–911 (2002)
20. Wang, Y., Xiao, J., Suzek, T., et al.: Pubchem: a public information system for analyzing bioactivities of small molecules. Nucl. Acids Res. 37(2), 623–633 (2009)
21. Weininger, D.: Smiles, a chemical language and information system. 1. introduction to methodology and encoding rules. J. Chem. Inform. Comput. Sci. 28(1), 31–36 (1988)

Scalable Computation of Isochrones
with Network Expiration*

Johann Gamper[1], Michael Böhlen[2], and Markus Innerebner[1]

[1] Free University of Bozen-Bolzano, Italy
[2] University of Zurich, Switzerland

Abstract. An isochrone in a spatial network is the possibly disconnected set of all locations from where a query point is reachable within a given time span and by a given arrival time. In this paper we propose an efficient and scalable evaluation algorithm, termed (MINEX), for the computation of isochrones in multimodal spatial networks with different transportation modes. The space complexity of MINEX is independent of the network size and its runtime is determined by the incremental loading of the relevant network portions. We show that MINEX is optimal in the sense that only those network portions are loaded that eventually will be part of the isochrone. To keep the memory requirements low, we eagerly expire the isochrone and only keep in memory the minimal set of expanded vertices that is necessary to avoid cyclic expansions. The concept of expired vertices reduces MINEX's memory requirements from $\mathcal{O}(|V^{iso}|)$ to $\mathcal{O}(\sqrt{|V^{iso}|})$ for grid and $\mathcal{O}(1)$ for spider networks, respectively. We show that an isochrone does not contain sufficient information to identify expired vertices, and propose an efficient solution that counts for each vertex the outgoing edges that have not yet been traversed. A detailed empirical study confirms the analytical results on synthetic data and shows that for real-world data the memory requirements are very small indeed, which makes the algorithm scalable for large networks and isochrones.

Keywords: spatial network databases, isochrones.

1 Introduction

Reachability analyzes are important in many applications of spatial network databases. For example, in urban planning it is important to assess how well a city is covered by various public services such as hospitals or schools. An effective way to do so is to compute isochrones. An *isochrone* is the possibly disconnected set of all locations from where a query point, q, is reachable within a given time span. When schedule-based networks, such as the public transport system, or time-dependent edge costs are considered, isochrones depend on the arrival time at q. Isochrones can also be used as a primitive operation to answer other spatial network queries, such as range queries, that have to retrieve objects within the area of the isochrone. For instance, by joining an isochrone with an inhabitants database, the percentage of citizens living in the area of the isochrone can be determined without the need to compute the distance to individual objects.

---

* This work is partially funded by the Province and the Municipality of Bozen-Bolzano.

A. Ailamaki and S. Bowers (Eds.): SSDBM 2012, LNCS 7338, pp. 526–543, 2012.

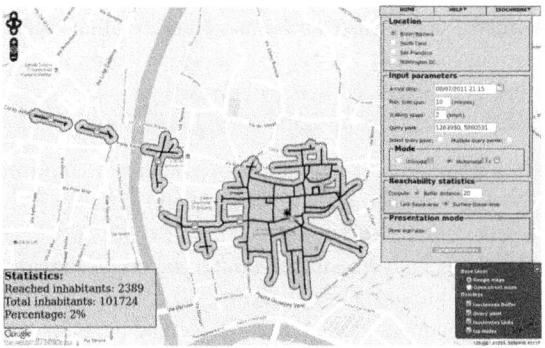

Fig. 1. Screenshot of an Isochrone

Example 1. Figure 1 shows the 10 min isochrone at 09:15 pm for a query point (*) in Bozen-Bolzano. The isochrone consists of the bold street segments that cover all points from where the query point is reachable in less than 10 minutes, starting at 09:05 pm or later and arriving at 09:15 pm or before. A large area around the query point is within 10 minutes walking distance. Smaller areas are around bus stops, from where the query point can be reached by a combination of walking and going by bus. The box in the lower left corner shows the number of inhabitants in the isochrone area.

We focus on isochrones in multimodal spatial networks, which can be classified as continuous or discrete along, respectively, the space and the time dimension. *Continuous space* means that all points on an edge are accessible, whereas in a *discrete space* network only the vertices can be accessed. *Continuous time* networks can be traversed at any point in time, *discrete time* networks follow an associated schedule. For instance, the pedestrian network is continuous in time and space, whereas public transport systems are discrete in both dimensions.

The paper proposes the Multimodal Incremental Network Expansion with vertex eXpiration (MINEX) algorithm. Starting from query point, q, the algorithm expands the network backwards along incoming edges in all directions until all space points that are within d_{max} from q are covered. Since only network portions are loaded that are part of the isochrone, the memory complexity of MINEX is independent of the network size. This yields a solution that scales to GIS platforms where a web server must be able to handle a large number of concurrent queries. The runtime is determined by the incremental loading of the network portions that are part of the isochrone. Our goal in terms of runtime is to be on a par with existing solutions for small to medium-size isochrones.

MINEX eagerly prunes the isochrone and keeps in memory only the expanded vertices that form the expansion frontier and are needed to avoid cyclic network expansions. These vertices must be updated as the expansion proceeds and the frontier moves outwards. Newly encountered vertices are added and vertices that will never be revisited, termed expired vertices, are removed. The removal of expired vertices reduces the memory requirements from $\mathcal{O}(|V^{iso}|)$ to $\mathcal{O}(\sqrt{|V^{iso}|})$ for grid and to $\mathcal{O}(1)$ for spider networks, respectively. Since the isochrone does not contain sufficient information to identify expired vertices, we propose an efficient strategy that counts for each vertex v the outgoing edges that have not yet been traversed. When all these edges have been traversed, v will not be revisited and hence expires.

The technical contributions can be summarized as follows:

– We define *vertex expiration*, which allows to determine a minimal set of vertices that need to be kept in memory to avoid cyclic network expansions. Since an isochrone does not contain sufficient information to identify such vertices, we propose an efficient solution that counts the outgoing edges that are not yet traversed.
– We propose a scalable *disk-based multimodal network expansion algorithm*, MINEX, that is independent of the network size and depends only on the isochrone size. Its runtime is $O(|V^{iso}|)$. The eager expiration of vertices reduces the memory requirements from $O(|V^{iso}|)$ to $O(\sqrt{|V^{iso}|})$ and $O(1)$ for grid and spider networks, respectively.
– We show that MINEX is *optimal* in the sense that only portions of the network are loaded that will become part of the isochrone, and each edge is loaded only once.
– We report the results of an extensive empirical evaluation that shows the scalability of MINEX, confirms the analytical results for the memory requirements on synthetic data, and reveals an even more substantial reduction to a tiny and almost constant fraction of the network size on real-world data.

The rest of the paper is organized as follows. Section 2 discusses related work. In Section 3 we define isochrones in multimodal networks. Section 4 presents the MINEX algorithm. Section 5 reports the results of the empirical evaluation. Section 6 concludes the paper and points to future research directions.

2   Related Work

Different query types have been studied for spatial network databases, e.g., [5,6,19]. Shortest path (SP) queries return the shortest path from a source to a target vertex. Range queries determine all objects that are within a given distance from a query point q. kNN queries determine the k objects that are closest to q. Isochrones are different and return the (possibly disconnected) minimal subgraph that covers all space points in the network from where q is reachable within a given timespan d_{max} and by a given arrival time t (i.e., space points with a SP to q that is smaller than d_{max}). By intersecting the area covered by an isochrone with an objects relation, all objects within an isochrone can be determined without the need to compute the distance to the individual objects. This provides a useful instrument in various applications, such as city planning to analyze the coverage of the city with public services (hospitals, schools, etc.) or market analysis to analyze catchment areas and store locations.

Isochrones have been introduced by Bauer et al. [3]. The algorithm suffers from a high initial loading cost and is limited by the available memory since the entire network is loaded in memory. Gamper et al. [8] provide a formal definition of isochrones in multimodal spatial networks, together with a disk-based multimodal incremental network expansion (MINE) algorithm that is independent of the actual network size but maintains the entire isochrone in memory. This paper proposes a new network expiration mechanism that maintains the minimal set of vertices that is required to avoid cyclic network expansions. The actual memory requirements turn out to be only a tiny fraction of the isochrone size. We conduct extensive experiments and compare our algorithm with other network expansion algorithms. Marciuska and Gamper [18] present two different approaches to determine objects that are located within the area that is covered by an isochrone.

Most network queries, including isochrones, are based on the computation of the shortest path (SP) among vertices and/or objects. Dijkstra's [6] incremental network expansion algorithm is the most basic solution and influenced many of the later works. Its major limitations are the expansion towards all directions and its main memory nature. The A^* algorithm [10] uses a lower bound estimate of the SP (e.g., Euclidean distance) to get a more directed search with less vertex expansions. Other techniques have been proposed to improve the performance of SP and other network queries, including disk-based index structures and pre-processing techniques.

Papadias et al. [19] present two disk-based frameworks for computing different network queries: Incremental Euclidean Restriction (IER) repeatedly uses the Euclidean distance to prune the search space and reduce the number of objects for which the network distance is computed; Incremental Network Expansion (INE) is an adaptation of Dijkstra's SP algorithm. Deng et al. [5] improve over [19] by exploiting the incremental nature of the lower bound to limit the number of distance calculations to only vertices in the final result set. Almeida and Güting [4] present an index structure and an algorithm for kNN queries to allow a one-by-one retrieval of the objects on an edge.

Another strategy takes advantage of partitioning a network and pre-computing all or some of the SPs to save access and computation cost at query time. Examples are the partitioning into Voronoi regions and pre-computing distances within and across regions [16], shortest path quadtrees [21], and the representation of a network as a set of spanning trees with precomputed NN lists [11]. Other works divide a large network into smaller subgraphs that are hierarchically organized together with pre-computed results between boundary vertices [1,13,14].

For networks that are too large for exact solutions in reasonable time, efficient approximation techniques have been proposed, most prominently based on the landmark embedding technique and sketch-based frameworks [9,17,20,22]. For a set of so-called landmark vertices the distance to all other vertices is pre-computed. At query time, the precomputed distances and the triangle inequality allow to estimate the SP.

The work in [7,15] investigates time-varying edge costs, e.g., due to changing traffic conditions. There is far less work on schedule-based transportation networks and networks that support different transportation modalities [12]. Bast [2] describes why various speed up techniques for Dijkstra's SP algorithm are either not applicable or improve the efficiency only slightly in schedule-based networks.

The MINEX algorithm proposed in this paper leverages Dijkstra's incremental network expansion strategy for multiple transportation modes, and it applies eager network expiration to minimize the memory requirements. Most optimization techniques from previous work are not applicable to isochrones in multimodal networks, mainly due to the presence of schedule-based networks (cf. [2]) and the need to explore each individual edge, which makes search space pruning such as in the A^* algorithm more difficult and less effective.

3   Isochrones in Multimodal Networks

In this section we provide a formal definition of isochrones in multimodal spatial networks that support different transport modes.

Definition 1 (Multimodal Network). *A* multimodal network *is a seven-tuple* $N = (G, R, S, \rho, \mu, \lambda, \tau)$. $G = (V, E)$ *is a directed multigraph with a set V of vertices and a multiset E of ordered pairs of vertices, termed edges. R is a set of transport systems. $S = (R, TID, W, \tau_a, \tau_d)$ is a schedule, where TID is a set of trip identifiers, $W \subseteq V$, and $\tau_a : R \times TID \times W \mapsto \mathbb{T}$ and $\tau_d : R \times TID \times W \mapsto \mathbb{T}$ determine arrival and departure time, respectively ($\mathbb{T}$ is the time domain). Function $\mu : R \mapsto \{\text{'csct'}, \text{'csdt'}, \text{'dsct'}, \text{'dsdt'}\}$ assigns to each transport system a transport mode, and the functions $\rho : E \mapsto R$, $\lambda : E \mapsto \mathbb{R}^+$, and $\tau : E \times \mathbb{T} \mapsto \mathbb{R}^+$ assign to each edge transport system, length, and transfer time, respectively.*

A multimodal network permits several transport systems, R, with different modalities in a single network: continuous space and time mode $\mu(.) = \text{'csct'}$, e.g., pedestrian network; discrete space and time mode $\mu(.) = \text{'dsdt'}$, e.g., the public transport system such as trains and buses; discrete space continuous time mode $\mu(.) = \text{'dsct'}$, e.g., moving walkways or stairs; continuous space discrete time mode $\mu(.) = \text{'csdt'}$, e.g., regions or streets that can be passed during specific time slots only. Vertices represent crossroads of the street network and/or stops of the public transport system. Edges represent street segments, transport routes, moving walkways, etc. The schedule stores for each discrete time ('dsdt', 'csdt') transport system in R the arrival and departure time at the stop nodes for the individual trips. For an edge $e = (u, v)$, function $\tau(e, t)$ computes the time-dependent transfer time that is required to traverse e, when starting at u as late as possible yet arriving at v no later than time t. For discrete time edges, the transfer time is the difference between t and the latest possible departure time at u according to the given schedule in order to reach v before or at time t. This includes a waiting time should the arrival at v be before t. For continuous time edges, the transfer time is modeled as a time-dependent function that allows to consider, e.g., different traffic conditions during rush hours.

Example 2. Figure 2 shows a multimodal network with two transport systems, $R = \{\text{'P'}, \text{'B'}\}$, representing the pedestrian network with mode $\mu(\text{'P'}) = \text{'csct'}$ and bus line B with mode $\mu(\text{'B'}) = \text{'dsdt'}$, respectively. Solid lines are street segments of the pedestrian network, e.g., edge $e = (v_1, v_2)$ with $\rho(e) = \text{'P'}$. An undirected edge is a shorthand for a pair of directed edges in opposite directions. Pedestrian edges are annotated with the edge length, which is the same in both directions, e.g., $\lambda((v_1, v_2)) = \lambda((v_2, v_1)) = 300$. We assume a constant walking speed of 2 m/s, yielding a fixed transfer time, $\tau(e, t) = \frac{\lambda(e)}{2\,\text{m/s}}$. Dashed lines represent bus line B. An excerpt of the schedule is shown in Fig. 2(b), e.g., $TID = \{1, 2, \dots\}$, $\tau_a(\text{'B'}, 1, v_6) = \tau_d(\text{'B'}, 1, v_6) = 05{:}33{:}00$. The transfer time of a bus edge $e = (u, v)$ is computed as $\tau(e, t) = t - t'$, where $t' = \max\{\tau_d(\text{'B'}, tid, u) \mid \tau_a(\text{'B'}, tid, v) \leq t\}$ is the latest departure time at u.

A *location* in N is any point on an edge $e = (u, v) \in E$ that is accessible. We represent it as $l = (e, o)$, where $0 \leq o \leq \lambda(e)$ is an offset that determines the relative position of

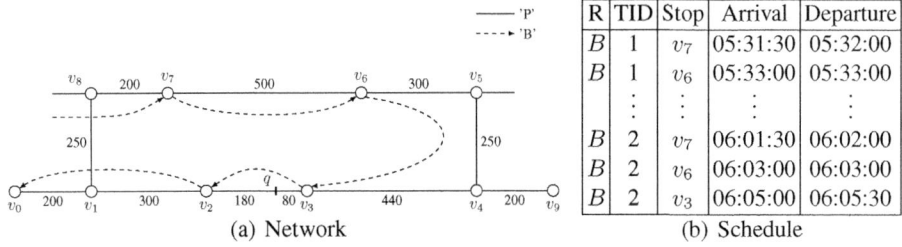

R	TID	Stop	Arrival	Departure
B	1	v_7	05:31:30	05:32:00
B	1	v_6	05:33:00	05:33:00
$\vdots$	$\vdots$	$\vdots$	$\vdots$	$\vdots$
B	2	v_7	06:01:30	06:02:00
B	2	v_6	06:03:00	06:03:00
B	2	v_3	06:05:00	06:05:30

(a) Network   (b) Schedule

Fig. 2. Multimodal Network

l from u on edge e. A location represents vertex u if $o = 0$ and vertex v if $o = \lambda(e)$; any other offset refers to an intermediate point on edge e. In continuous space networks all points on the edges are accessible. Since a pedestrian segment is modeled as a pair of directed edges in opposite direction, any point on it can be represented by two locations, $((u,v),o)$ and $((v,u),\lambda((u,v))-o)$, respectively. For instance, in Fig. 2 the location of q is $l_q = ((v_2, v_3), 180) = ((v_3, v_2), 80)$. In discrete space networks only vertices are accessible, thus $o \in \{0, \lambda(e)\}$ and locations coincide with vertices.

An *edge segment*, (e, o_1, o_2), with $0 \leq o_1 \leq o_2 \leq \lambda(e)$ represents the contiguous set of space points between the two locations (e, o_1) and (e, o_2) on edge e. We generalize the length function for edge segments to $\lambda((e, o_1, o_2)) = o_2 - o_1$.

Definition 2 (Path, Path Cost). *A path from a source location* $l_s = ((v_1, v_2), o_s)$ *to a destination location* $l_d = ((v_k, v_{k+1}), o_d)$ *is defined as a sequence of connected edges and edge segments,* $p(l_s, l_d) = \langle x_1, \ldots, x_k \rangle$, *where* $x_1 = ((v_1, v_2), o_s, \lambda((v_1, v_2)))$, $x_i = (v_i, v_{i+1})$ *for* $1 < i < k$, *and* $x_k = ((v_k, v_{k+1}), 0, o_d))$. *With arrival time* t *at* l_d, *the path cost is*

$$\gamma(\langle x_1, \ldots, x_k \rangle, t) = \begin{cases} \tau(x_k, t) & k = 1, \\ \gamma(\langle x_k \rangle, t) + \gamma(\langle x_1, \ldots, x_{k-1} \rangle, t - \gamma(\langle x_k \rangle, t)) & k > 1. \end{cases}$$

The first and the last element in a path can be edge segments, whereas all other elements are entire edges. Since isochrones depend on the arrival time at the query point, we define the path cost recursively as the cost of traversing the last edge (segment), x_k, considering the arrival time t at the destination l_d, plus the cost of traversing $\langle x_1, \ldots, x_{k-1} \rangle$, where the arrival time at v_k (the target vertex of edge x_{k-1}) is determined as t minus the cost of traversing x_k. The cost of traversing a single edge is the transfer time τ. Edges along a path may belong to different transport systems, which enables the changing of transport system along a path.

Example 3. In Fig. 2, a path from v_7 to q is to take bus B to v_3 and then walk to q, i.e., $p(v_7, l_q) = \langle x_1, x_2, x_3 \rangle$, where $x_1 = (v_7, v_6)$ and $x_2 = (v_6, v_3)$ are complete edges and $x_3 = ((v_3, v_2), 0, 80)$ is an edge segment. With arrival time $t = 06:06:00$, the path cost is $\gamma(p(v_7, l_q), t) = (06:03:00 - 06:02:00) + (06:05:20 - 06:03:00) + 80/2 = 240$ s. To reach q at 06:06:00, the bus must arrive at v_3 no later than 06:05:20. Since the latest bus matching this constraint arrives at 06:05:00, we have a waiting time of 20 s at v_3.

The *network distance*, $d(l_s, l_d, t)$, from a source location l_s to a destination location l_d with arrival time t at l_d is defined as the minimum cost of any path from l_s to l_d with arrival time t at l_d if such a path exists, and ∞ otherwise.

Definition 3 (Isochrone). *Let* $N = (G, R, S, \mu, \rho, \lambda, \tau)$ *with* $G = (V, E)$ *be a multimodal network,* q *be the query point with arrival time* t, *and* $d_{max} > 0$ *be a time span. An isochrone,* $N^{iso} = (V^{iso}, E^{iso})$, *is defined as the minimal and possibly disconnected subgraph of* G *that satisfies the following conditions:*

- $V^{iso} \subseteq V$,
- $\forall l(l = (e, o) \land e \in E \land d(l, q, t) \le d_{max}$
  $$\Leftrightarrow \exists x \in E^{iso}(x = (e, o_1, o_2) \land o_1 \le o \le o_2)).$$

The first condition requires the vertices of the isochrone to be a subset of the network vertices. The second condition constrains an isochrone to cover exactly those locations that have a network distance to q that is smaller or equal than d_{max}. Notice the use of edge segments in E^{iso} to represent edges that are only partially reachable. Whenever an edge e is entirely covered, we use e instead of $(e, 0, \lambda(e))$.

Example 4. In Fig. 3, the subgraph in bold represents the isochrone for $d_{max} = 5$ min and $t = 06{:}06{:}00$. The numbers in parentheses are the network distance to q. Edges close to q are entirely reachable, whereas edges on the isochrone border are only partially reachable. For instance, (v_0, v_1) is only reachable from offset 80 to v_1. Bus edges are not included in the isochrone since intermediate points on bus edges are not accessible. Formally, the isochrone in Fig. 3 is represented as $N^{iso} = (V^{iso}, E^{iso})$ with $V^{iso} = \{v_0, \ldots, v_9\}$ and $E^{iso} = \{((v_0, v_1), 80, 200), ((v_8, v_1), 130, 250), (v_1, v_2), (v_2, v_1), (v_2, v_3), (v_3, v_2), (v_3, v_4), (v_4, v_3), ((v_5, v_4), 170, 250), ((v_9, v_4), 120, 200), ((v_5, v_6), 60, 300), ((v_7, v_6), 260, 500), ((v_6, v_7), 380, 500), ((v_8, v_7), 80, 200)\}$.

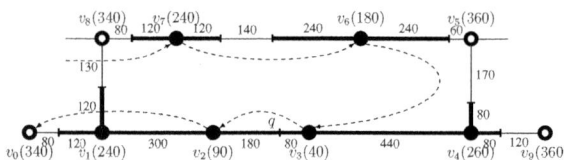

Fig. 3. Isochrone in Multimodal Network

4   Incremental Network Expansion in Multimodal Networks

This section presents the multimodal incremental network expansion algorithm with vertex expiration (MINEX) for computing isochrones in multimodal networks.

4.1   Algorithm MINEX

Consider a multimodal network N, query point q with arrival time t_q, duration d_{max}, and walking speed s. The expansion starts from q and propagates backwards along the incoming edges in all directions. When a vertex v is expanded, all incoming edges $e = (u, v)$ are considered, and the distance of u to q when traversing e is incrementally computed as the distance of v plus the time to traverse e. The expansion terminates when all locations with a network distance to q that is smaller than d_{max} have been reached.

Algorithm 1 shows MINEX which implements this strategy. The multimodal network is stored in a database, and – as the expansion proceeds – the portions of the network that eventually will form the isochrone are incrementally retrieved. The algorithm maintains two sets of vertices: closed vertices (C) that have already been expanded and open vertices (O) that have been encountered but are not yet expanded. For each vertex $v \in O \cup C$, we record the network distance to q, d_v (abbrev. for $d(v, q, t_q)$), and a counter, cnt_v, which keeps track of the number of outgoing edges that have not yet been traversed. C is initialized to the empty set. O is initialized to v with $d_v = 0$ and the number of outgoing edges if q coincides with vertex v. Otherwise, $q = ((u, v), o)$ is an intermediate location, and O is initialized to u and v with the corresponding walking distance to q; the reachable segments of edges (u, v) and (v, u) are output.

Algorithm 1. MINEX($\mathsf{N}, q, t_q, d_{max}$)

```
1   C ← ∅;
2   if q coincides with v then  O ← {(v, 0, cnt_v)};
3   else // q = ((u, v), o) = ((v, u), o')
4   |    O ← {(u, o/s, cnt_u), (v, o'/s, cnt_v)};
5   |    Output ((u, v), max(0, (o/s−d_max)s), o) and ((v, u), max(0, (o'/s−d_max)s), o');
6   while O ≠ ∅ and first element has distance ≤ d_max do
7   |    (v, d_v, cnt_v) ← first element from O;
8   |    O ← O \ {v};
9   |    C ← C ∪ {v};
10  |    foreach e = (u, v) ∈ E do
11  |    |    if u ∉ O ∪ C then O ← O ∪ {(u, ∞, cnt_u)};
12  |    |    d'_u ← τ(e, t_q − d_v) + d_v;
13  |    |    d_u ← min(d_u, d'_u);
14  |    |    cnt_u ← cnt_u − 1;
15  |    |    if u ∈ C ∧ cnt_u = 0 then C ← C \ {u};
16  |    |    if μ(ρ(e)) ∈ {'csct', 'csdt'} then
17  |    |    |    if d'_u ≤ d_max then Output (e, 0, λ(e));
18  |    |    |    else Output (e, o, λ(e)), where d((e, o), q, t_q) = d_max;
19  |    if cnt_v = 0 then C ← C \ {v};
20  return;
```

During the expansion phase, vertex v with the smallest network distance is dequeued from O and added to C. All incoming edges, $e = (u, v)$, are retrieved from the database and considered in turn. If vertex u is visited for the first time, it is added to O with a distance of ∞ and the number of outgoing edges, cnt_u. Then, the distance d'_u of u when traversing e is computed and the distance d_u is updated. If e is a 'csct' or 'csdt' edge, the reachable part of e is added to the result. Edges of type 'dsct' or 'dsdt' produce no direct output, since only the vertices are accessible, and they are added when their incoming 'csct' edges are processed. Finally, cnt_u is decremented by 1; if u is closed and $cnt_u = 0$, u is expired and removed from C (more details on vertex expiration are below). Once all incoming edges of v are processed, the expiration and removal of v

is checked. The algorithm terminates when O is empty or the network distance of the closest vertex in O exceeds d_{max}.

Example 5. Figure 4 illustrates a few steps of MINEX for $d_{max} = 5\,\text{min}$, $t_q = 06{:}06{:}00$, and $s = 2\,\text{m/s}$. Bold lines indicate reachable network portions, solid black nodes are closed, and bold white nodes are open. The numbers in parentheses are the distance and the counter. Figure 4(a) shows the isochrone after the initialization step with $C = \{\}$ and $O = \{(v_2, 90, 3), (v_3, 40, 3)\}$. Vertex v_3 has the smallest distance to q and is expanded next (Fig. 4(b)). The distance of the visited vertices is $d_{v_4} = 40 + 440/2 = 260\,\text{s}$ and $d'_{v_2} = 40 + 260/2 = 140\,\text{s}$, which does not improve the old value $d_{v_2} = 90\,\text{s}$. For the distance of v_6, we determine the required arrival time at v_3 as $t = t_q - d_{v_3} = 06{:}06{:}00 - 40\,\text{s} = 06{:}05{:}20$ and the latest bus departure at v_6 as $06{:}03{:}00$, yielding $d_{v_6} = 40 + (06{:}05{:}20 - 06{:}03{:}00) = 180\,\text{s}$. After updating the counters, the new vertex sets are $C = \{(v_3, 40, 3)\}$ and $O = \{(v_2, 90, 2), (v_6, 180, 2), (v_4, 260, 2)\}$. Next, v_2 is expanded as shown in Fig. 4(c)). Figure 4(d) shows the isochrone after the termination of the algorithm; the gray vertex v_3 is expired.

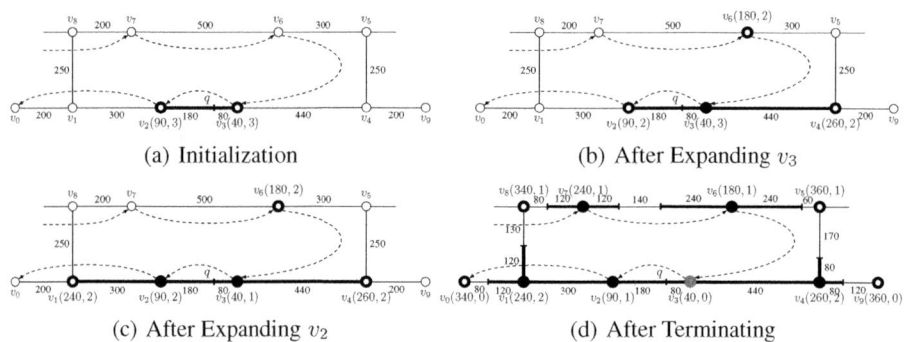

Fig. 4. Stepwise Computation of N^{iso} for $d_{max} = 5\,\text{min}$, $s = 2\,\text{m/s}$, and $t_q = 06{:}06{:}00$

Notice that an algorithm that alternates between (completely) expanding the continuous network and (completely) expanding the discrete network is sub-optimal since many portions of the network would be expanded multiple times. We empirically evaluate such an approach in Sec. 5.

4.2 Expiration of Vertices

Closed vertices are needed to avoid cyclic network expansion. In order to limit the number of closed vertices that need to be kept in memory we introduce expired vertices (Def. 4). Expired vertices are never revisited in future expansion steps, hence they are not needed to prevent cyclic expansions and can be removed (Lemma 1). Isochrones contain insufficient information to handle vertex expiration (Lemma 2). Therefore, MINEX uses a counter-based solution to correctly identify expired vertices and eagerly expire nodes during the expansion (Lemma 3).

To facilitate the discussion we introduce a couple of auxiliary terms. For a vertex v, the term *in-neighbor* refers to a vertex u with an edge (u, v) and the term *out-neighbor* refers to a vertex w with an edge (v, w). Recall that the status of vertices transitions from open (O) when they are encountered first, to closed (C) when they are expanded, and finally to expired (X) when they are expired; the sets O, C, and X are pairwise disjoint.

Definition 4 (Expired Vertex). *A closed vertex, $u \in C$, is expired if all its out-neighbors are either closed or expired, i.e., $\forall v((u, v) \in E \Rightarrow v \in C \cup X)$.*

Example 6. Consider the isochrone in Fig. 4(d). Vertex v_3 is expired since v_2 and v_4 are closed, and v_3 has no other out-neighbors. In contrast, v_2 is not yet expired since the out-neighbor v_0 is not yet closed (and the expansion of v_0 leads back to v_2).

Lemma 1. *An expired vertex u will never be revisited during the computation of the isochrone and can be removed from C without affecting the correctness of MINEX.*

Proof. There is only one way to visit a vertex u during network expansion: u has an out-neighbor v (that is connected via an edge $(u, v) \in E$) and $v \in O$; the expansion of v visits u. Since according to Def. 4 all of u's out-neighbors are closed or expired, and closed and expired vertices are not expanded (line 11 in Alg. 1), u cannot be revisited.

The identification of expired vertices according to Def. 4 has two drawbacks: (1) it requires a database access to determine all out-neighbors since not all of them might already have been loaded, and (2) the set X of expired vertices must be kept in memory.

Lemma 2. *If the isochrone is used to determine the expiration of a closed vertex, $u \in C$, the database must be accessed to retrieve all of u's out-neighbors, and X needs to be stored in memory.*

Proof. According to Def. 4, for a closed vertex u to expire we have to check that all out-neighbors v are closed or expired. The expansion of u loaded all out-neighbors v that have also an inverse edge, $(v, u) \in E$. For out-neighbors v that are not connected by an inverse edge, $(v, u) \notin E$, we have no guarantee that they are loaded. Therefore, we need to access the database to get *all* adjacent vertices. Next, suppose that X is not maintained in memory and there exists an out-neighbor v of u without an inverse edge, i.e., $(v, u) \notin E$. If v is in memory, its status is known. Otherwise, either v already expired and has been removed, or it has not yet been visited. In the former case, u shall expire, but not in the latter case, since the expansion of v (re)visits u. However, with the removal of X we loose the information that these vertices already expired, and we cannot distinguish anymore between not yet visited and expired vertices.

Example 7. The isochrone does not contain sufficient information to determine the expiration of v_2 in Fig. 4(c). While v_1 and v_3 are loaded and their status is known, the out-neighbor v_0 is not yet loaded (and actually violates the condition for v_2 to expire). To ensure that *all* out-neighbors are closed, a database access is needed. Next, consider Fig. 4(d), where v_3 is expired, i.e., $X = \{v_3\}$. To determine the expiration of v_2, we need to ensure that $v_3 \in C \cup X$. If X is removed from memory, the information that v_3 is already expired is lost. Since v_3 will never be revisited, v_2 will never expire.

To correctly identify and remove all expired vertices without the need to access the database and explicitly store X, MINEX maintains for each vertex, u, a counter, cnt_u, that keeps track of the number of outgoing edges of u that have not yet been traversed.

Lemma 3. *Let cnt_u be a counter associated with vertex $u \in V$. The counter is initialized to the number of outgoing edges, $cnt_u = |\{(u, v) \mid (u, v) \in E\}|$, when u is encountered for the first time. Whenever an out-neighbor v of u is expanded, cnt_u is decremented by 1. Vertex u is expired iff $u \in C$ and $cnt_u = 0$.*

Proof. Each vertex v expands at most once (when it is dequeued from O), and the expansion of v traverses all incoming edges (u, v) and decrements the counter cnt_u of vertex u by 1. Thus, each edge in the network is traversed at most once. When $cnt_u = 0$, vertex u must have been visited via all of its outgoing edges. From this we can conclude that all out-neighbors have been expanded and are closed, which satisfies the condition for vertex expiration in Def. 4.

Example 8. In the isochrone in Fig. 4(d), vertex v_3 is expired and can be removed since $cnt_{v_3} = 0$ and $v_3 \in C$. Vertex v_2 expires when v_0 is expanded and counter cnt_{v_2} is decremented to 0. Similar, vertex v_6 expires when v_5 is expanded.

Lemma 4. *Vertices cannot be expired according to an LRU strategy.*

Proof. We show a counter-example in Fig. 5(a), which illustrates a multimodal network expansion that started at q. Although q has been expanded and closed first, it cannot be expired because an edge from vertex v, which will be expanded later, leads back to q (and would lead to cyclic expansions). In contrast, the gray vertices that are expanded and closed after q can be expired safely.

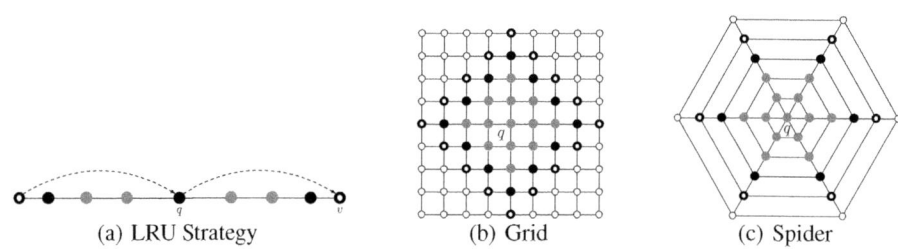

(a) LRU Strategy   (b) Grid   (c) Spider

Fig. 5. Network Expiration

4.3   Properties

Vertex expiration ensures that the memory requirements are reduced to a tiny fraction of the isochrone. Figures 5(b) and 5(c) illustrate the isochrone size and MINEX's memory complexity for grid and spider networks, respectively. Solid black vertices (C) and vertices with a bold border (O) are stored in memory, whereas gray vertices are expired (X) and removed from memory. The following two lemmas provide a bound for the isochrone size and MINEX's memory complexity for these two types of networks. (Only the pedestrian mode is considered, though the results can easily be extended to multimodal networks.)

Lemma 5. *The size of an isochrone, $|V^{iso}|$, is $\mathcal{O}(d_{max}^2)$ for a grid network and $\mathcal{O}(d_{max})$ for a spider network and a central query point q.*

Proof. Consider the grid network in Fig. 5(b). Without loss of generality, we measure the size of an isochrone as the number of its vertices (i.e., open, closed, and expired vertices), and we assume a uniform distance of 1 between connected vertices. The size of an isochrone with distance $d = 1, 2, \ldots$ is given by the recursive formula $|V^{iso}|_d = |V^{iso}|_{d-1} + 4d$ with $|V^{iso}|_0 = 1$; $4d$ is the number of new vertices that are visited when transitioning from distance $d-1$ to d (i.e., the number of vertices at distance d that are visited when all vertices at distance d–1 are expanded). This forms an arithmetic series of second order (1, 5, 13, 25, 41, 61, ...) and can also be written as $|V^{iso}|_d = 1 + \sum_{i=0}^d 4i = 2d^2 + 2d + 1$, which yields $|V^{iso}| = \mathcal{O}(d_{max}^2)$.

Next, consider the spider network in Fig. 5(c). Without loss of generality, we assume a uniform distance of 1 between all adjacent vertices along the same outgoing line from q. It is straightforward to see that the size of the isochrone is $|V^{iso}| = deg(q) \cdot d_{max} + 1 = \mathcal{O}(d_{max})$, where $deg(q)$ is the degree of vertex q.

Lemma 6. *The memory complexity of MINEX is $|O \cup C| = \mathcal{O}(d_{max}) = \mathcal{O}(\sqrt{|V^{iso}|})$ for a grid network and $\mathcal{O}(1)$ for a spider network and a central query point q.*

Proof. Recall that MINEX keeps only the open and closed vertices, $O \cup C$, in memory. Consider the grid network in Fig. 5(b). By referring to the proof of Lemma 5, the cardinality of the open vertices at distance d can be determined as $|O|_d = 4d$ and the cardinality of the closed vertices as $|C|_d = 4(d-1)$. Thus, the memory requirements in terms of d_{max} are $|O \cup C| = \mathcal{O}(d_{max})$.

To determine the memory requirements depending on the size of the isochrone, $|V^{iso}|$, we use the formula for the size of an isochrone from the proof of Lemma 5 and solve the quadratic equation $2d^2 + 2d + 1 - |V^{iso}|_d = 0$, which has the following two solutions: $d_{1,2} = \frac{-2 \pm \sqrt{2^2 - 4 \cdot 2 \cdot (1 - |V^{iso}|_d)}}{2 \cdot 2} = \frac{-1 \pm \sqrt{2|V^{iso}|_d - 1}}{2}$. Since the result must be positive, $d = \frac{-1 + \sqrt{2|V^{iso}|_d - 1}}{2}$ is the only solution. By substituting d in the above formulas for open and closed vertices we get, respectively, $|O|_d = 4d = 4\frac{-1 + \sqrt{2|V^{iso}|_d - 1}}{2}$ and $|C|_d = 4(d - 1) = 4(\frac{-1 + \sqrt{2|V^{iso}|_d - 1}}{2} - 1)$, which proves $|O \cup C| = \mathcal{O}(\sqrt{V^{iso}})$.

Next, we consider the spider network in Fig. 5(c) with the query point q in the centre. It is straightforward to see that the cardinality of the open and closed vertices is $|O \cup C| = 2 \cdot deg(q) = \mathcal{O}(1)$, where $deg(q)$ is the degree of vertex q.

Theorem 1. *Algorithm MINEX is optimal in the sense that all loaded vertices and 'csct'/'csdt' edges are part of the isochrone, and each of these edges is loaded and traversed only once.*

Proof. When a vertex v is expanded, all incoming edges $e = (u, v)$ are loaded and processed (Alg. 1, line 1). If e is a 'csct'/'csdt' edge, the reachable portion of e (including the end vertices u and v) is added to the isochrone (line 1). While u might not be reachable, v is guaranteed to be reachable since $d_v \leq d_{max}$. In contrast, 'dsct'/'dsdt' edges are not added since they are not part of the isochrone; only the end vertices u and

v are accessible, which are added when the incoming '$csct$'/'$csdt$' edges are processed. Therefore, since each vertex is expanded at most once each edge is loaded at most once, and all loaded edges except '$dsct$'/'$dsdt$' edges are part of the isochrone.

5   Empirical Evaluation

5.1   Overview

Setup and Data Sets. In the experiments we measure memory and runtime complexity of MINEX and compare it with the following alternative approaches: (1) Incremental network expansion INE [19], which incrementally loads the network but keeps all loaded vertices in memory. (2) Dijkstra's algorithm [6], which initially loads the entire network in memory. (3) Incremental Euclidean restriction IER [19], which instead of loading the entire network, uses the Euclidean lower bound property to incrementally load smaller network chunks as illustrated in Fig. 6. First, a chunk around q is loaded that contains all vertices

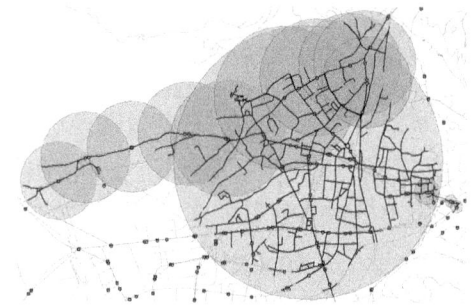

Fig. 6. Network Expansion in IER

that are reachable in walking mode (i.e., within distance $d_{max} \cdot s$). After doing network expansion in memory, for all encountered bus stops a new (smaller) chunk is loaded, etc. Chunks are not re-loaded again if they are completely covered by network portions that are already in memory. (4) PGR, which is based on PostgreSQL's pgRouting and works similar to IER, but uses the network distance instead of the Euclidean distance.

The multimodal networks and the schedules are stored in a PostgreSQL 8.4 database with the spatial extension PostGIS 1.5 and the PG routing extension. All experiments run in a virtual machine (64bits) on a Dual Processor Intel Xeon 2.67 GHz with 3 GB RAM. The algorithms were implemented in Java using the JDBC API to communicate with the database.

We test three real-world networks that are summarized in Table 1. Size is the network size, and $|V|$, $|E|$, $|E_{csct}|$, $|E_{dsdt}|$, and $|S|$ are the number of vertices, edges, pedestrian edges, edges of different means of transport, and schedule entries, respectively. The network size is given in Megabyte, whereas the other columns show the number of tuples in K. We use also synthetic grid and spider networks like the one shown in Fig. 5(b) and 5(c). The walking speed in all experiments is set to 1 m/s. Isochrone size ($|V^{iso}|$) and the memory requirements ($|V^{MM}|$) are measured in terms of number of vertices. The other input parameters vary from dataset to dataset.

Summary of Experiments. The first experiment in Sec. 5.2 measures memory and confirms that thanks to vertex expiration MINEX's memory requirements are only a

Table 1. Real-World Data Sets: Italy (IT), South Tyrol (ST), and San Francisco (SF)

Data	Size	$|V|$	$|E|$	$|E_{csct'}|$	$|E_{dsdt'}|$	$|S|$
IT	2,128	1,372.0	3,633.7	3,633.1	0.6	1.3
ST	137	77.7	197.8	182.4	9.4	179
SF	138	33.6	96.4	90.0	6.4	1,112

small fraction of the isochrone size in all settings, whereas all other algorithms require significantly more. The second experiment in Sec. 5.3 shows that IER loads many edges multiple times, whereas MINEX loads each edge in the isochrone only once. The third experiment in Sec. 5.4 measures the runtime. For isochrones that are smaller than 9% of the network (which is frequently the case, especially for large and skewed regional networks) MINEX outperforms Dijkstra, whereas the latter is better for large isochrones, provided that the entire network fits in memory.

5.2 Memory Consumption

We begin with synthetic networks; only the pedestrian mode is used. The results confirm Lemma 5 and 6 and are shown in Fig. 7, where d_{max} and the isochrone size vary, respectively. For grid networks, MINEX's memory requirements grow linearly in d_{max} and with the square root in $|V^{iso}|$; INE's memory consumption corresponds to the isochrone size, i.e., $|V^{MM}| = |V^{iso}|$, and grows quadratically in d_{max}. In spider networks, the memory complexity is constant for MINEX and linear in d_{max} for INE.

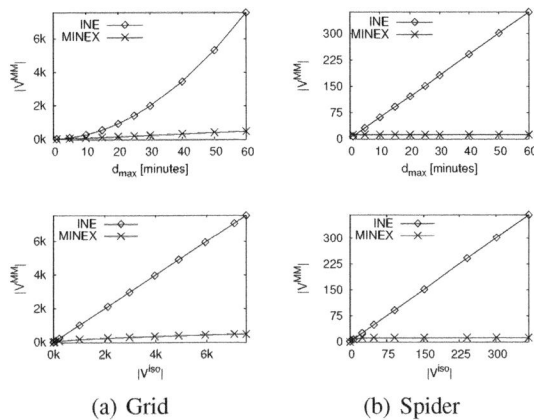

(a) Grid   (b) Spider

Fig. 7. Memory Requirements in Synthetic Networks

Figure 8 shows the memory complexity for the real-world data sets. We compare additionally with Dijkstra, IER, and PGR. As expected, MINEX's memory consumption is only a tiny fraction of the isochrone size, and it further decreases when the isochrone reaches the sparse network boundary. In contrast, the memory of INE, PGR and IER grows quadratically in d_{max} until the isochrone approaches the network border, where the growing slows down. For the IT data set this effect is not visible since d_{max} is too small. The memory of Dijkstra is equal to the size of the entire network, including vertices and edges, which can be inferred from Table 1, e.g., 2.1 GB for IT. To keep small values visible, the memory requirements of Dijkstra are not shown.

Figure 9 shows a 90 min isochrone for the ST network. Such regional or country-wide networks typically have an irregular network structure, varying density of

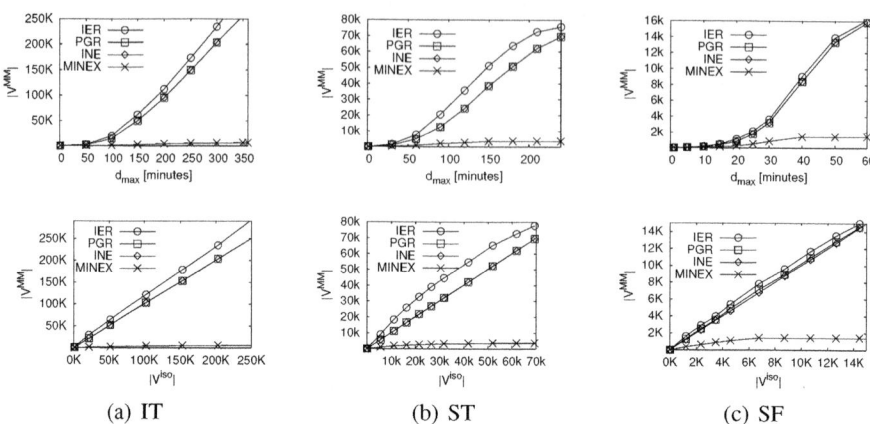

Fig. 8. Memory Requirements in Real-World Data

vertices and edges, and fast wide-area transport systems. In this type of skewed networks, isochrones are characterized by many remote islands, and the size of isochrones is typically only a small fraction of the (comparably very large) network. Thus, Dijkstra is difficult to apply due to its high memory complexity, whereas MINEX requires only a tiny amount of memory (which in practice is even much smaller than the isochrone).

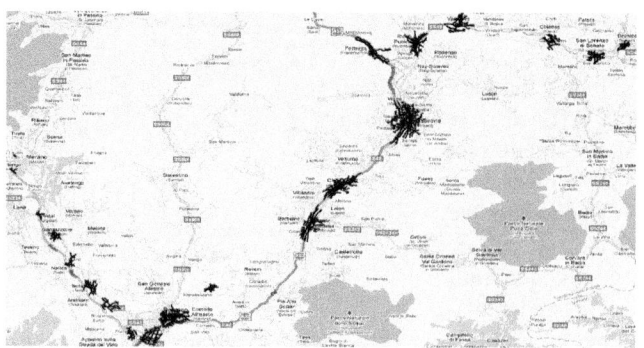

Fig. 9. Isochrone in the Regional ST Network

5.3 Multiple Loading of Tuples

The experiment in Fig. 10 measures the total number of tuples (i.e., network edges) that are loaded from the database, using a logarithmic scale. MINEX and INE load the minimal number of tuples since each edge is loaded only once. They converge towards Dijkstra when the isochrone approaches the network size. In Fig. 10(a) IER and PGR load approximately the same number of tuples as MINEX since there is no overlapping in the areas that are retrieved by the range queries. This is because we have only the high-speed trains with very few stops. In contrast, in Fig. 10(b) and 10(c) the number of overlapping range queries is large, yielding many vertices and edges that are loaded multiple times. The reason here is the high density of public transport stops. Fig. 6 illustrates the overlapping of IER, which can be quite substantial.

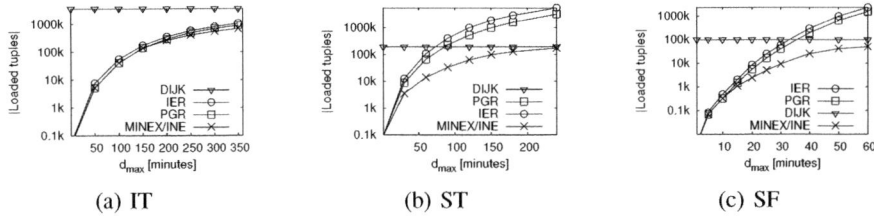

(a) IT   (b) ST   (c) SF

Fig. 10. Number of Loaded Tuples

5.4   Runtime

Figure 11 shows the runtime depending on d_{max} and the isochrone size. For small values of d_{max} and small isochrones, Dijkstra has the worst performance due to the initial loading of the entire network. For large d_{max} and isochrones, Dijkstra (though limited by the available memory) is more efficient since the initial loading of the network using a full table scan is faster than the incremental loading in MINEX and INE, which requires $|V^{iso}|$ point queries. For MINEX we expect a runtime that is comparable to INE since essentially the same expansion strategy is applied (modulo vertex expiration).

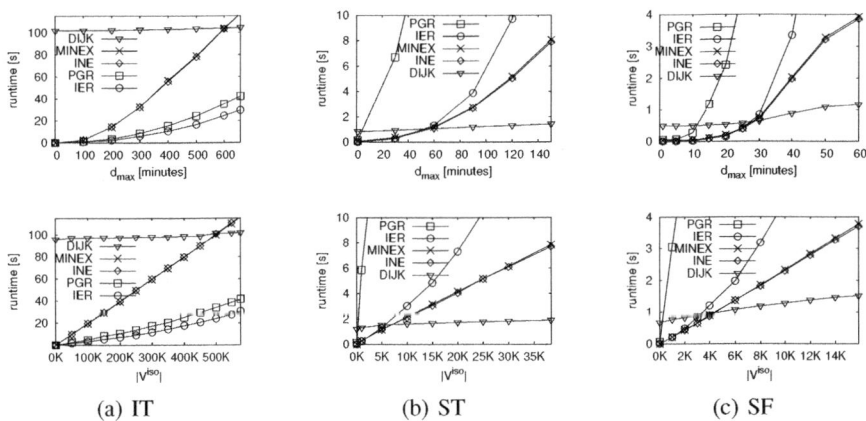

(a) IT   (b) ST   (c) SF

Fig. 11. Runtime in Real World Data

The first experiment in Fig. 11(a) on the IT data set runs on a large skewed network with few and distantly located train stations. IER and PGR outperform Dijkstra since there are almost no overlappings due to the sparse number of train stations. INE and MINEX are slower than IER and PGR because of the larger number of database accesses (one access for each vertex expansion), but they are more efficient than Dijkstra. The break-even point occurs after 600 minutes when the size of the isochrone correspond to 38% of the network. Figure 11(b) shows the runtime of the regional network ST. Dijkstra outperforms INE and MINEX after a d_{max} of 60 minutes when the isochrone size corresponds to 9% of the network. IER and PGR collapse because of the large number of overlapping loaded areas in range queries. Figure 11(c) shows the runtime in an urban network with a duration of one hour. The break-even point is reached at $d_{max} = 30$ min, which corresponds to 11% of the entire network.

In Fig. 12(a) we further analyze the break-even point for the real-world networks and for grid and spider networks with 2k, 6k, and 10k vertices (G2K, S2K, …).

We empirically determined that the break-even point is reached when the size of the isochrone is between 9–38% of the network size. Since the runtime of both Dijkstra and MINEX is dominated by the database accesses and the loading of the (sub)network, the break-even point occurs when the

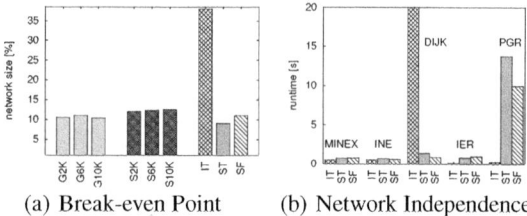

(a) Break-even Point   (b) Network Independence

Fig. 12. Runtime

time for MINEX's incremental loading and the time for Dijkstra's initial loading cross. In terms of d_{max}, the break-even point varies depending on the walking speed and the frequency of the public transportation.

Figure 12(b) confirms that MINEX is independent of the network size. We compute an isochrone of a fixed size $|V^{iso}| = 3.000$ for the different real-world data sets. The runtime of MINEX and INE is almost the same for all data sets. In contrast the runtime of Dijkstra depends directly from the network size. IER and PGR depends on the density of the public transport network.

6   Conclusion and Future Work

In this paper we introduced isochrones for multimodal spatial networks that can be discrete or continuous in, respectively, space and time. We proposed the MINEX algorithm, which is independent of the actual network size and depends only on the isochrone size. MINEX is optimal in the sense that only those network portions are loaded that eventually will be part of the isochrone. The concept of expired vertices reduces MINEX's memory requirements to keep in memory only the minimal set of expanded vertices that is necessary to avoid cyclic expansions. To identify expired vertices, we proposed an efficient solution based on counting the number of outgoing edges that have not yet been traversed. A detailed empirical study confirmed the analytical results and showed that the memory requirements are very small indeed, which makes the algorithm scalable for large networks and isochrones.

Future work points in different directions. First, we will investigate multimodal networks that include additional transport systems such as the use of the car. Second, we will investigate the use of various optimization techniques in MINEX as well as approximation algorithms to improve the runtime performance for large isochrones. Third, we will study the use of isochrones in new application scenarios.

References

1. Balasubramanian, V., Kalashnikov, D.V., Mehrotra, S., Venkatasubramanian, N.: Efficient and scalable multi-geography route planning. In: EDBT, pp. 394–405 (2010)
2. Bast, H.: Car or Public Transport—Two Worlds. In: Albers, S., Alt, H., Näher, S. (eds.) Efficient Algorithms. LNCS, vol. 5760, pp. 355–367. Springer, Heidelberg (2009)
3. Bauer, V., Gamper, J., Loperfido, R., Profanter, S., Putzer, S., Timko, I.: Computing isochrones in multi-modal, schedule-based transport networks. In: ACMGIS, pp. 1–2 (2008)
4. de Almeida, V.T., Güting, R.H.: Using Dijkstra's algorithm to incrementally find the k-nearest neighbors in spatial network databases. In: SAC, pp. 58–62 (2006)
5. Deng, K., Zhou, X., Shen, H.T., Sadiq, S.W., Li, X.: Instance optimal query processing in spatial networks. VLDB J. 18(3), 675–693 (2009)
6. Dijkstra, E.W.: A note on two problems in connexion with graphs. Numerische Mathematik 1(1), 269–271 (1959)
7. Ding, B., Yu, J.X., Qin, L.: Finding time-dependent shortest paths over large graphs. In: EDBT, pp. 205–216 (2008)
8. Gamper, J., Böhlen, M.H., Cometti, W., Innerebner, M.: Defining isochrones in multimodal spatial networks. In: CIKM, pp. 2381–2384 (2011)
9. Gubichev, A., Bedathur, S., Seufert, S., Weikum, G.: Fast and accurate estimation of shortest paths in large graphs. In: CIKM, pp. 499–508. ACM, New York (2010)
10. Hart, P., Nilsson, N., Raphael, B.: A formal basis for the heuristic determination of minimum cost paths. IEEE Trans. on Systems Science and Cybernetics SSC-4(2), 100–107 (1968)
11. Hu, H., Lee, D.-L., Xu, J.: Fast Nearest Neighbor Search on Road Networks. In: Ioannidis, Y., Scholl, M.H., Schmidt, J.W., Matthes, F., Hatzopoulos, M., Böhm, K., Kemper, A., Grust, T., Böhm, C. (eds.) EDBT 2006. LNCS, vol. 3896, pp. 186–203. Springer, Heidelberg (2006)
12. Huang, R.: A schedule-based pathfinding algorithm for transit networks using pattern first search. GeoInformatica 11(2), 269–285 (2007)
13. Jing, N., Huang, Y.-W., Rundensteiner, E.A.: Hierarchical encoded path views for path query processing: An optimal model and its performance evaluation. IEEE Trans. Knowl. Data Eng. 10(3), 409–432 (1998)
14. Jung, S., Pramanik, S.: An efficient path computation model for hierarchically structured topographical road maps. IEEE Trans. Knowl. Data Eng. 14(5), 1029–1046 (2002)
15. Kanoulas, E., Du, Y., Xia, T., Zhang, D.: Finding fastest paths on a road network with speed patterns. In: ICDE (2006)
16. Kolahdouzan, M.R., Shahabi, C.: Voronoi-based k nearest neighbor search for spatial network databases. In: VLDB, pp. 840–851 (2004)
17. Kriegel, H.-P., Kröger, P., Renz, M., Schmidt, T.: Hierarchical Graph Embedding for Efficient Query Processing in Very Large Traffic Networks. In: Ludäscher, B., Mamoulis, N. (eds.) SSDBM 2008. LNCS, vol. 5069, pp. 150–167. Springer, Heidelberg (2008)
18. Marciuska, S., Gamper, J.: Determining Objects within Isochrones in Spatial Network Databases. In: Catania, B., Ivanović, M., Thalheim, B. (eds.) ADBIS 2010. LNCS, vol. 6295, pp. 392–405. Springer, Heidelberg (2010)
19. Papadias, D., Zhang, J., Mamoulis, N., Tao, Y.: Query processing in spatial network databases. In: VLDB, pp. 802–813 (2003)
20. Potamias, M., Bonchi, F., Castillo, C., Gionis, A.: Fast shortest path distance estimation in large networks. In: CIKM, pp. 867–876 (2009)
21. Samet, H., Sankaranarayanan, J., Alborzi, H.: Scalable network distance browsing in spatial databases. In: SIGMOD Conference, pp. 43–54 (2008)
22. Thorup, M., Zwick, U.: Approximate distance oracles. In: STOC, pp. 183–192. ACM, New York (2001)

A Dataflow Graph Transformation Language and Query Rewriting System for RDF Ontologies[*]

Marianne Shaw[1], Landon T. Detwiler[2], James F. Brinkley[2,3], and Dan Suciu[1]

[1] Computer Science & Eng., University of Washington, Seattle WA
[2] Dept. of Biological Structure, University of Washington, Seattle, WA
[3] Dept. of Medical & Biological Informatics, University of Washington, Seattle, WA

Abstract. Users interested in biological and biomedical information sets on the semantic web are frequently not computer scientists. These researchers often find it difficult to use declarative query and view definition languages to manipulate these RDF data sets. We define a language IML consisting of a small number of graph transformations that can be composed in a dataflow style to transform RDF ontologies. The language's operations closely map to the high-level manipulations users undertake when transforming ontologies using a visual editor. To reduce the potentially high cost of evaluating queries over these transformations on demand, we describe a query rewriting engine for evaluating queries on IML views. The rewriter leverages IML's dataflow style and optimizations to eliminate unnecessary transformations in answering a query over an IML view. We evaluate our rewriter's performance on queries over use case view definitions on one or more biomedical ontologies.

1 Introduction

A number of biological and biomedical information sets have been developed for or converted to semantic web formats. These information sets include vocabularies, ontologies, and data sets. They may be available in basic RDF, a data model for the semantic web in which graphs are collections of triple statements, or languages with higher-level semantics, such as OWL. Researchers want to leverage the biomedical information available on the semantic web.

Mechanisms are available for scientists to manipulate RDF ontologies. Researchers can manually modify a copy of the content or develop a custom program to transform the data. Visual editors ([6][5][2]) can be used to modify and augment a copy of the data. Extraction tools such as PROMPT [21] can extract subsets of information; these subsets can be combined and modified by a visual editor. The high-level functionality of these visual tools maps well to researchers' mental model for transforming an ontology. Unfortunately, the data must be locally acquired and the user actions repeated when the data is updated.

Declarative view definition languages such as RVL [18], NetworkedGraphs [24], and vSPARQL [26] allow users to define views that transform an RDF information set. These view definitions can be evaluated on-demand, avoiding stale data

---

[*] This work was funded by NIH grant HL087706.

A. Ailamaki and S. Bowers (Eds.): SSDBM 2012, LNCS 7338, pp. 544–561, 2012.

and expensive user actions. They can also be maintained, evolved, and used to define complex transformations. Unfortunately, non-technical users find it difficult to create declarative view definitions that perform the high-level operations available in visual editors. This problem is exacerbated by the need to reconstruct the unmodified data along with the transformations. Transformation languages enable users to specify only the modifications that need to be applied to data.

In this paper, we make two contributions. First, we present a high-level dataflow view definition language that closely matches users' mental model for transforming ontologies using visual editing tools. The language consists of a small number of graph transformations that can be composed in a dataflow style. Instead of a single declarative query, a user specified sequence of operations defines a transforming view over which queries can be rewritten and evaluated.

Second, we address the query processing challenge for our language. We present a query rewriting system that composes queries with IML view definitions, reducing the high cost of on-demand evaluation of queries on transformed ontologies. The rewriter leverages IML's dataflow style to eliminate transformations that are redundant or unnecessary for answering the query. The rewriter incorporates a set of optimizations to streamline the generated query. It then uses graph-specific statistics to produce an efficient query.

We evaluate our query rewriting engine on queries over a set of use case view definitions over one or more of four biomedical ontologies [26]. We compare the performance of our rewritten queries against the cost of first materializing and then querying a transformed ontology. 60% of the rewritable queries have an execution time at least 60% less than the view materialization time.

The paper is structured as follows. We present a motivating example in Sect. 2. Section 3 presents the InterMediate Language through an example. Section 4 presents the query rewriting engine and its optimizations. After describing our implementation in Sect. 5, we leverage IML definitions of nine use case views and their associated queries to evaluate our query rewriting engine's performance in Sect. 6. We discuss related work in Sect. 7 before concluding in Sect. 8.

2   Motivating Example

A radiologist spends significant time manually inspecting and annotating medical images. He notes normal anatomical objects and anomalous regions and growths; these annotations are used by a patients' doctor to suggest follow-up treatment.

To reduce the time and manual effort consumed by this task, the radiologist wants to develop an application (similar to [1]) that provides a list of medical terms for annotating an image. The radiologist will indicate the region of the body (e.g. gastrointestinal tract) that the image corresponds to, and the application should provide a set of terms that can be used to "tag" visible objects.

The radiologist needs the annotation terms that he associates with objects in the image to be well-defined. This ensures that there will be no misunderstanding by the doctors that read his findings. Therefore, the provided terms should be concepts from established biomedical references. The Foundational Model of Anatomy (FMA) [23], a reference ontology modeling human anatomy, can be used for anatomical terms, and the National Cancer Institute Thesaurus (NCIt) [4], a vocabulary containing information about cancer, can be used to provide terms for anomalous growths.

A visual editor like Protege can be used to identify the objects that might be visible in a medical image. Simply displaying all terms is impractical; the FMA contains 75,000 classes and the NCIt contains 34,000 concepts. This identification process is tedious, time-intensive, and error-prone. For example, identifying visible parts of the gastrointestinal tract in the FMA requires inspection of more than 1300 objects. This manual process must be repeated every time the FMA and NCIt are updated – approximately yearly for the FMA, monthly for the NCIt. Instead, a view definition can be created and evaluated on demand.

3   A Dataflow Language for Transforming RDF Ontologies

We propose the InterMediate Language (IML), a view definition language that both 1) enables non-technical users to define transformations of RDF ontologies, and 2) makes it possible to efficiently answer queries over those transformations.

There are many ways to transform an ontology to create a new ontology. Facts about individual classes, properties, or restrictions can be added, deleted, or modified. Relevant subsets or subhierachies of the ontology can be extracted and combined, or restructured as needed. Unnecessary subsets or subhierarchies can be removed. Multiple ontologies can be merged to create a new one.

IML has been designed to support this range of functionality. The language consists of a small number of high-level graph transformations that correspond to the functionality provided in visual editors. (In Sect. 7, we compare IML to Protege/PROMPT, a popular visual ontology editor.) A sequence of operations produce a transforming view definition where output of one operation flows as input, via a default graph, to the next; this corresponds to making changes, sequentially, to a local copy of an ontology in a visual editor. Transforming view definitions are named and can be referenced by other transforming views.

3.1   IML Transforming Operations

IML contains selection, modification, addition, and utility operations. Table 1 contains the grammar for IML's operations. In their simplest form, IML operations operate on concretely specified resources. For more advanced transformations, IML leverages the syntax of SPARQL, the query language for RDF. IML uses the SPARQL grammar's `WhereClause` for querying graph patterns in an RDF graph and its `ConstructTemplate` for producing new graphs.

We discuss common IML operations in the context of a simplified IML view for the motivating example in Sect. 2. From the FMA, 1) `lines 1-7` extract the partonomy of the gastrointestinal tract, eliminating variants of the "part" property label; 2) `lines 8-11` identify the subclasses of `fma:Cardinal_organ_part` recursively (approximating visibility in images); and 3) `lines 13-15` extract the visible part hierarchy by joining the partonomy and visible organ parts.

Table 1. IML Transforming Operations

extract_edges ConstructTemplate OptClauses
extract_cgraph { *varOrTerm* ImlPropertyList } OptClauses
extract_reachable { *varOrTerm* ImlPropertyList } OptClauses
extract_path { *varOrTerm* ImlPropertyList *varOrTerm* } OptClauses
extract_recursive { ConstructTemplate OptClauses } { ConstructTemplate OptClauses }
add_edge < *varOrTerm verb varOrTerm* > OptClauses
delete_edge < *varOrTerm verb varOrTerm* > OptClauses
delete_node *varOrTerm* OptClauses
delete_property *verb* OptClauses
delete_cgraph { *varOrTerm* ImlPropertyList } OptClauses
replace_edge_property < *varOrTerm verb varOrTerm* > *verb* OptClauses
replace_edge_(subject|object|literal) < *varOrTerm verb varOrTerm* > *varOrTerm* OptClauses
replace_property *verb verb* OptClauses
replace_(node|literal) *varOrTerm varOrTerm* OptClauses
merge_nodes *varOrTermList* CreateNode RetainElimList MergeSourceList OptClauses
split_node *varOrTerm* SplitNodeList OptClauses
union_graph SourceSelectorList
copy_graph SourceSelector
OptClauses := (**graph** SourceSelector)? WhereClause?
SourceSelector := *IRIref*
ImlPropertyList := '[' ImlProperty (',' ImlProperty)* ']'
ImlProperty := (*varOrTerm* |(**outgoing** | **incoming**) (*varOrTerm*)?)
CreateNode := **create** *SkolemFunction*
RetainElimList := (((**retain** | **eliminate**) *varOrTerm* ImlPropertyList)+)?
MergeSourceList := ((**merge_source** *varOrTerm sourceSelector*)+)?
SplitNodeList := CreateNamedNode RetainElimList (CreateNamedNode RetainElimList)+
CreateNamedNode := **create** *IRIref* ',' *SkolemFunction*
VarOrTermList := '[' *varOrTerm* (',' *varOrTerm*)* ']'

```
1) INPUT <http://.../fma>                          # Extract partonomy of GI tract
2) { extract_cgraph { fma:GI_tract [outgoing(fma:regional_part), outgoing(fma:systemic_part),
3)          outgoing(fma:constitutional_part)] } graph <http://.../fma>
4)    replace_property fma:regional_part fma:part
5)    replace_property fma:systemic_part fma:part
6)    replace_property fma:constitutional_part fma:part
7) } OUTPUT <gi_part_hierarchy>

8)  INPUT <http://.../fma>    # Find subclasses of Cardinal_organ_part (visible on images)
9)  { extract_reachable { fma:Cardinal_organ_part [incoming(rdfs:subClassOf)] }
10)       graph <http://.../fma>
11) } OUTPUT <visible>

12) INPUT <gi_part_hierarchy>, <visible> # Select visible parts by joining the part hierarchy
13) { extract_edges { ?a fma:part ?c }   # and the set of visible elements.
14)      where { graph <gi_part_hierarchy> { ?a fma:part ?c } .
15)           graph <visible> { ?x localhost:reaches ?c } . }
16)   add_edge <fma:Appendix fma:part fma:Appendix_tip>
17)   delete_edge <fma:Appendix fma:part fma:Tip_of_appendix>
18) } OUTPUT <visble_hierarchy>
```

Selection. A technique used to create a new ontology is to select subsets of relevant information from an existing one and combine them, via join or union. The ability to extract relevant parts of an ontology is critical because many ontologies contain considerably more information than is needed by a scientist.

IML's `extract_edges` operation enables the selection of specific relevant edges from an ontology. For example, `extract_edges { ?a fma:FMAID ?c }` `where { ?a fma:FMAID ?c }` selects all of the `FMAID`s from the FMA. The operation can also be used to join two RDF graphs. `Lines 13-15` extract the visible parts of the GI tract by joining the part hierarchy with the visible organ parts.

Commonly, users need to select an entire hierarchy from an ontology. Using the `extract_cgraph` operation, a user specifies a starting resource and a set of properties that are recursively followed to extract a connected graph. `Lines 2-3` select the partonomy of the gastrointestinal tract by recursively extracting all of its regional, systemic, and constitutional parts. Similarly, `extract_reachable` produces a list of all nodes that can be reached by recursing over a set of properties. `Line 9` recursively finds all of the subclasses of `fma:Cardinal_organ_part` to identify anatomical elements that may be visible on a medical image.

Modification. When leveraging an existing ontology, scientists often delete, modify or rename some of the content. The `delete_edge` operation can be used to eliminate edges from an ontology; `line 17` deletes the edge specifying that `fma:Appendix` has part `fma:Tip_of_appendix`. All instances of a resource or property in a graph can be eliminated with `delete_node` or `delete_property`.

Users may need to modify triples from an RDF graph. The `replace_node` and `replace_property` operations replace all instances of a resource in a graph. Lines 4-6 replace all extracted part edges in the GI tract partonomy with a uniform `fma:part` edge. For more fine-grained replacements, the `replace_edge_*` operations can change the subject, property, or object of a RDF triple.

Addition. Users can add new facts to an ontology using IML's `add_edge` operation, which adds new triples to an RDF graph. `Line 16` adds a new edge indicating that `fma:Appendix` has part `fma:Appendix_tip`.

4   IML Query Rewriting and Optimization

A typical IML program is a dataflow diagram consisting of a sequence of IML operations. A naive approach to evaluate an IML program is to evaluate each operation sequentially. This is inefficient, as each operation produces an intermediate result that may be comparable in size to the input RDF data.

We have developed a system for rewriting queries over IML view definitions. The rewriter leverages IML's dataflow style to combine operations and eliminate transformations in the view definition that are unnecessary for answering the query, thus reducing query evaluation time.

The query rewriting engine is depicted in Fig. 1. The IML view definition and query are parsed into an abstract syntax tree. Individual operations are converted into a set of Query Pattern Rules (QPRs) and combined to create a rewritten QPR set representing the query. During this process, optimizations eliminate unnecessary or redundant transformations. Performance optimizations are applied before the query is converted to vSPARQL and evaluated.

vSPARQL is an extension to the SPARQL1.0 standard that enables transforming views through the use of (recursive) subqueries, regular-expression styled property path expressions, and dynamic node creation using skolem functions. We describe a vSPARQL view definition (below) for the IML view in Sect. 3.

vSPARQL supports `CONSTRUCT`-style subqueries to generate intermediate results. The subquery on `lines 14-17` creates an RDF graph `<visible>` listing all

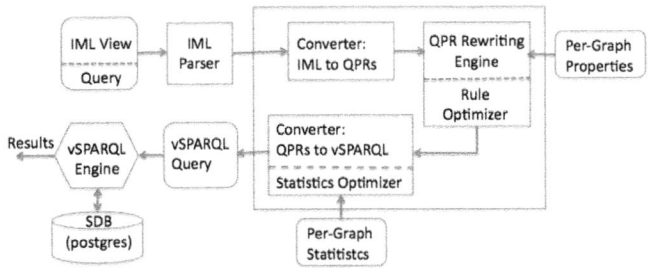

Fig. 1. Overview of IML query rewriting system

of the subclasses of `fma:Cardinal_organ_part` using a recursive property path expression. Recursive subqueries consist of two or more CONSTRUCT queries, a base case and a recursive case; the recursive case is repeatedly evaluated until a fixed point is reached. Lines 2–13 define a recursive query that extracts the partonomy of the gastrointestinal tract into `<gi_part_hierarchy>`. The results of the two subqueries are joined on lines 19–20.

```
1) construct { ?a fma:part ?c . fma:Appendix fma:part fma:Appendix_tip }        # Add_edge
2) from namedv <gi_part_hierarchy> [                          # Extract partonomy of GI tract
3)    construct { fma:Gastrointestinal_tract fma:part ?c }              # Base case
4)    from <http://.../fma>
5)    where { fma:Gastrointestinal_tract ?p ?c  .
6)      filter((?p=fma:regional_part)||(?p=fma:systemic_part)||(?p=fma:constitutional_part)) }
7)    union
8)    construct { ?prev fma:part ?c }                              # Recursive case
9)    from <http://.../fma>
10)   from namedv <gi_part_hierarchy>
11)   where { graph <gi_part_hierarchy> { ?a fma:part ?prev }
12)      ?prev ?p ?c .
13)      filter((?p=fma:regional_part)||(?p=fma:systemic_part)||(?p=fma:constitutional_part)))}]
14) from namedv <visible> [    # Find subclasses of Cardinal_organ_part (visible on images)
15)   construct { fma:Cardinal_organ_part lcl:reaches ?b }
16)   from <http://.../fma>
17)   where { ?b rdfs:subClassOf* fma:Cardinal_organ_part } ]
18) where {                        # Select visible parts by joining the part hierarchy
19)      graph <gi_part_hierarchy> { ?a fma:part ?c }       # and the set of visible elements.
20)      graph <visible> { ?x lcl:reaches ?c }
21)      FILTER((?a != fma:Tip_of_appendix) && (?c != fma:Tip_of_appendix)) # Delete_edge }
```

The working draft for SPARQL1.1 includes property path expressions, embedded subqueries, and skolem functions. Many, but not all, vSPARQL recursive subqueries can be expressed using property path expressions, including the subquery in lines 2–13. SPARQL1.1's property path expressions cannot require multiple constraints on nodes along a path, nor recursively restructure a graph.

4.1   Query Pattern Rule (QPR) Sets

During rewriting each IML operation is converted into a set of Query Pattern Rules (QPRs). Each QPR consists of a graph **pattern**, a list of **constraints**, and a graph **template**. The QPR indicates that if the constrained pattern (which

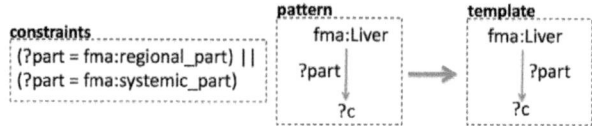

Fig. 2. Example QPR: If the default graph contains a triple matching (fma:Liver ?part ?c), where ?part is either fma:regional_part or fma:systemic_part, add it to our output.

may contain disallowed triples), is found in the graph to which it is applied, the corresponding `template` should be added to the output. A QPR is in Fig. 2.

For each IML operation, a *set* of QPRs is generated. The result of an IML operation is the union of all of the QPRs in the QPR set. Many operations permit the optional specification of a `WhereClause` for defining variables via query pattern bindings. The `WhereClause` is separated into graph pattern and `FILTER` elements; these define a QPR's `pattern` and `constraints`, respectively. `UNION` and `OPTIONAL` statements inside of `WhereClause`s cause multiple QPR sets to be generated, one for each combination of possible `WhereClause` patterns.

4.2 Query Rewriting Process

The rewriting engine starts with the last IML operation in the query and proceeds, operation by operation, towards the top of the query block. At each step, the IML operation is first converted into a QPR set; this QPR set is then combined with the working QPR set to produce a new working QPR set. If an IML operation `iml_op_x` references a subquery block via the `GRAPH` keyword, the rewriting engine recursively rewrites `iml_op_x`'s `pattern` over the named subquery block to produce a QPR set for `iml_op_x`. The rewriter combines the QPR set for `iml_op_x` with the current working set and continues to the next preceding IML operation. After the first IML operation in the query's subquery block is processed (i.e. the top of the IML block is reached), the working QPR set represents the overall rewritten query that must be evaluated.

As each IML operation `iml_op_x` is encountered, the rewriting engine combines `iml_op_x`'s QPR set with the working QPR set. It does this by combining *each* QPR in the working set with *each* of the QPRs in `iml_op_x`'s QPR set. If a QPR pattern has more than one element (i.e. triple pattern), each element is unified with each of iml_op_x's QPRs.

QPRs are combined by unifying the `pattern` of one QPR with the `template` of the other. If no unification is found, then no QPR is produced. If a unification is found, then the unifying values are substituted in to produce a new QPR.

Unification Example: We illustrate this process with an example. Figure 3 depicts two IML operations, `extract_edges` and `add_edge`, and their QPR sets. Each of add_edge's two QPRs must be unified with `extract_edges`' single QPR. Unifying add_edge's `QPR1.pattern` and extract_edges' `QPR1.template` produces a QPR equivalent to `extract_edges`' QPR1. Figure 3(c) presents the unification found for add_edge's `QPR2.pattern` and extract_edges' `QPR1.template`, producing Fig 3(d)'s QPR. The final QPR set contains both of these QPRs.

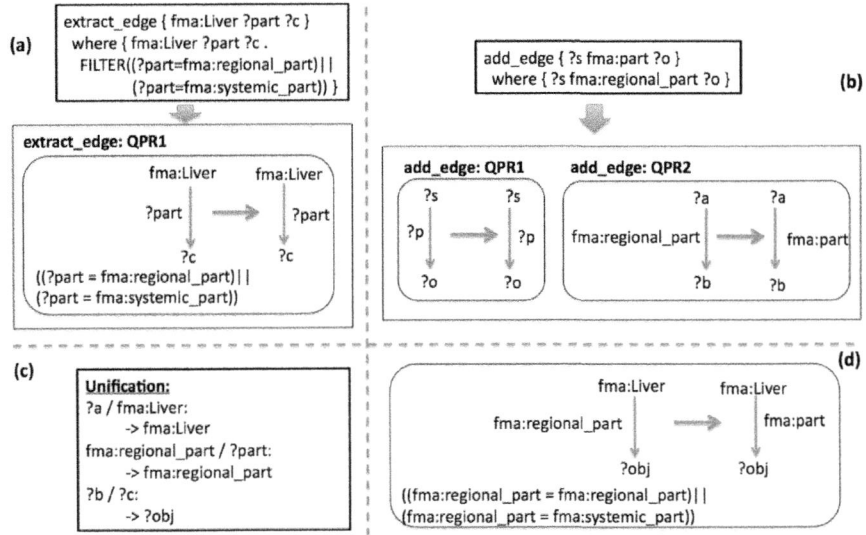

Fig. 3. Unifying (a) extract_edge's QPR1.template and (b) add_edge's QPR2.pattern produces the unification set (c) and QPR (d)

More generally, for two IML operations `iml_op_1` and `iml_op_2`, we combine the two QPRs by finding a possible unification of `pattern2` with `template1`. If a unification is found, we substitute those unifying values into `pattern1` and `constraint1` (producing `pattern1'` and `constraint1'`), and `constraint2` and `template2` (producing `constraint2'` and `template2'`) to generate a QPR (`result_QPR`) that can be added to the new working set.

```
iml_op_1: pattern1 + constraint1 -> template1
iml_op_2: pattern2 + constraint2 -> template2
result_QPR: pattern1' + constraint1' + constraint2' -> template2'
```

4.3   Rewriting Optimizations

Rewriting can produce working sets that have a large number of QPRs. At every step in the rewriting process, *every* QPR in the current working set is combined with *every* QPR in the preceding operation's QPR set. If a QPR's `pattern` has multiple elements, each of these elements must be combined with *every* QPR in the preceding operation's QPR set.

This rule explosion is compounded by the fact that many IML operations generate QPR sets with multiple rules. The `add_edge`, `union_graphs`, and `replace_*` operations all produce a minimum of 2 QPRs, while the `merge_nodes` and `split_node` operations each produce a minimum of 5 QPRs. UNION and OPTIONAL statements, as well as multiple result triples for the `extract_edges` operation, can cause additional QPRs to be added to an operation's QPR set.

Many of the QPRs in the rewritten query's QPR set may be invalid (i.e., will never produce a valid result) or redundant, yet increase the cost of query evaluation. In the next two sections, we describe optimizations to both slow this rule explosion, by eliminating invalid or redundant rules, and make evaluation of the generated rules more efficient. We use "QPR" and "rule" interchangeably.

4.4   Rule-Based Optimizations

Our rewriting engine applies rule-based optimizations to the working QPR set after rewriting over an IML operation; we describe them in this section. Several of these optimizations use RDF characteristics to eliminate invalid rules.

Constraint Simplification. The Constraint Simplification optimization identifies and eliminates those rules whose `constraints` will *always* be false. This optimization simplifies a rule by evaluating expressions in its `constraints`, focusing on equality and inequality expressions.

Bound Template Variables. The Bound Template Variables optimization eliminates rules where the variables in the `template` are not used in the `pattern`; these rules will never produce a valid result. This scenario often occurs when an operation's `WhereClause` contains a `UNION` or `OPTIONAL` statement.

RDF Literal Semantics. RDF requires both subjects and predicates to be URIs, not literals. The RDF Literal Semantics optimization eliminates rules that have a `pattern` or `template` with a literal in the subject or predicate position.

Literal Range Paths. RDF Schema enables ranges to be associated with specific properties in an RDF graph. Thus, if a property p has a literal range (e.g. integer, boolean), any triple with predicate p will have a literal as the object. RDF restricts literals to the object position in a triple. The Literal Range Paths optimization eliminates any QPR whose `pattern` tries to match a path of length 2, when the first triple in the path has a property with a literal range. Literal range properties are provided on a per-ontology basis and only applied to `WhereClauses` evaluated against an unmodified input ontology.

Observed Paths. The Observed Paths optimization relies on the fact that not all pairs of properties will be directly connected via a single resource. If a QPR pattern contains a path that is *never* observed in the underlying RDF ontology, the pattern will never match the data and the rule can be eliminated. This optimization uses a per-ontology list of property pairs to eliminate QPRs; it is applied to patterns that are matched against the unmodified RDF ontology.

Query Containment. The Query Containment optimization determines which, if any, of the rules in a QPR set are redundant and eliminates them. If `QPR1` is contained by `QPR2`, then `QPR1` will contribute a subset of the triples generated by `QPR2`; `QPR1` is redundant and can be removed from the QPR set.

This optimization checks if a QPR is contained by another QPR by determining if there is a homomorphism between the two QPRs. Let `q1` and `q2` represent two QPRs, and let t_{q1} and t_{q2} represent triples that are produced by `q1` and `q2`, respectively. A homomorphism $f : q2 \rightarrow q1$ is a function $f : variables(q2) \rightarrow variables(q1) \cup constants(q1)$ such that: (1) $f(body(q2)) \subseteq body(q1)$ and (2) $f(t_{q1}) = t_{q2}$. The homomorphism theorem states that $q1 \subseteq q2$ iff there exists a homomorphism $f : q2 \rightarrow q1$.

To determine if there is a homomorphism between two QPRs, we map their `patterns` to a SAT problem. If a possible solution is found by a SAT solver, meaning that QPR1 is contained by QPR2, we substitute the values from the homomorphic mapping into both QPRs' `constraints` to determine if QPR1 is indeed contained by QPR2.

The scenario in which one query is contained by another often occurs when an operation's `WhereClause` contains an `OPTIONAL` statement. The query containment optimization is only applied when rewriting over a view with this property.

4.5   Performance Optimizations

When converting the final rewritten QPR set into a vSPARQL query, we use optimizations to reduce the cost of evaluating the rewritten query.

Query-Template Collapse. The Query Template Collapse optimization identifies QPRs with identical `patterns` and `constraints`, and produces a single vSPARQL subquery with multiple triples in the `CONSTRUCT` template.

Query Minimization. Rewriting a query over a view can produce QPRs with redundant `pattern` elements. This often happens when multiple elements in the `WhereClause` need to be rewritten over the same IML view. The Query Minimization optimization detects and eliminates redundant elements in a `pattern`.

This optimization builds upon our Query Containment optimization. For a given QPR q, we remove a single `pattern` element wc to create q'. If q' is contained by the original query q, we can remove wc permanently from q; if not, wc is not redundant and must remain in q. We repeat this process for each element in the QPR's `pattern`.

Statistics-based Query Pattern Reordering. IML's rewriting engine converts QPRs into vSPARQL subqueries. Each QPR's `pattern` and `constraints` are converted into a `WHERE` clause. During conversion, the rewriter has two goals, to minimize the cost of evaluating: 1) individual QPRs, and 2) all QPRs.

The rewriter uses per-ontology statistics, shown in Table 2, to achieve these two goals. These statistics are used to assign an expected triple result set size to individual elements in a QPR `pattern` based upon Table 3. Elements connected via a shared variable are grouped and then ordered with the more selective elements first. For query patterns containing property path expressions, we estimate the triple set size using the estimated fan in (fan out) of the path. The `fanInPath(pE)` function recursively calculates the fan in of a path expression.

```
fanInPath(pE):                    // Uses statistics in Table 2
   Property(prop): FanIn(prop)
   Inverse(Property(prop)): FanOut(prop)
   Alternate(lpath,rpath): fanInPath(lpath) + fanInPath(rpath)
```

Table 2. Per RDF Graph Statistics

Overall Graph Statistics		For Each Property p	
Total Triples	Average Fan In Degree	# Total Triples	Fan Out(p)
# Distinct Subjects	Average Fan Out Degree	# Distinct Subjects	Fan In(p)
# Distinct Predicates	Average PPlus Length	# Distinct Objects	Fan Out(p+)
# Distinct Objects			Fan In(p+)

Table 3. Query pattern vs. estimated triple set size vs. estimated variable cardinality

Query Pattern	Est. Tuple Set Size	Est. Variable Cardinality
?a ?b ?c	TotalTriples	?a = #DistSubjects, ?b =#DistProperties ?c=#DistObjects
?a ?b z	AvgFanIn	?a = AvgFanIn, ?b = AvgFanIn
?a y ?c	TotalTriples(y)	?a = #DistSubjects(y), ?c = #DistObjects(y)
?a y z	FanIn(y)	?a = AvgFanIn
x ?b ?c	AvgFanOut	?b = AvgFanOut, ?c = AvgFanOut
x ?b z	min(AvgFanIn, AvgFanOut)	?b = min(AvgFanIn, AvgFanOut)
x y ?c	FanOut(y)	?c = FanOut(y)
x y z	1	
?a (pE) ?c	TotalTriples-1	?a = ?c = max(#DistSubjects, #DistObjects)
?a (pE) y	fanInPath(pE)	?a = fanInPath(pE)
x (pE) ?c	fanOutPath(pE)	?c = fanOutPath(pE)
x (pE) y	1	

```
Sequence(lpath, rpath): fanInPath(lpath) * fanInPath(rpath)
Modified(path):  if( ZeroOrMore(path) || OneOrMore(path) )
                     fanInModPath(path)

fanInModPath(pE):                        // For p+ and p* paths
   Property(prop): FanInPlus(prop)
   Inverse(Property(prop)): FanOutPlus(prop)
   Alternate(lpath, rpath): fanInPath(lpath+)+fanInPath(rpath+)
   Sequence(lpath, rpath): fanInPath(lpath+)*fanInPath(rpath+)
   Modified(path):  c = fanInPath(path.subpath)
                    if( ZeroOrMore(subPath) || OneOrMore(subPath) )
                       c *= AvgPPlusLength
```

The rewriter repeatedly chooses the most selective WHERE clause element, with preference for those containing already visited variables, and adds it to the query pattern. Individual constraint expressions are ordered after their corresponding variables have been initially bound. As elements are added to the WHERE clause, the rewriter updates the expected cardinality of its variables. The updated cardinalities are used to recalculate triple set sizes and choose the next WHERE clause element to add to the BGP. This is similar to the algorithm in [27]; we do not use specialized join statistics.

Property Path Expression Direction Using estimated variable cardinalities and the fanInPath(pE) routine, the rewriter estimates the number of graph nodes that will be touched by evaluating a property path expression in the forward and reverse directions. If the cost of evaluating a path expression is decreased by reversing its direction, the rewriter reverses the path before adding it to the rewritten WHERE clause.

Anchored Property Path Subqueries The same property path expression may appear multiple times in the same QPR or in many different QPRs. If these expressions have a shared constant as either the subject or object value, we call them anchored property path expressions. The rewriter determines if it is more efficient to evaluate and materialize a repeated anchor property path expression once in a subquery, or to evaluate each property path expression individually.

The rewriter creates a subquery for an anchored path expression in two situations: 1) if an anchored property path expression occurs in more than three different QPRs; and 2) if a single QPR has multiple instances of the same anchored path expression and their evaluation cost is greater than three times the cost of evaluating the anchored path expression in a subquery.

5   Implementation

Our system, developed in Java, produces vSPARQL queries that can be evaluated by the execution engine – an extension to Jena's ARQ – described in [26]. The rewriting engine converts constraints to conjunctive normal form when they are added to a QPR, eliminating the need for refactoring during rewriting. MiniSAT [3] is used for solving the SAT problems generated for our Query Containment optimization. To reduce the impact of repeated property path expressions in multiple QPRs, we have added a LRU path cache to the vSPARQL query engine. The path cache stores the result of evaluating individual property path expressions, keyed on the (source URI, property path expression) pair.

The query rewriter does not rewrite all IML operations; most notably, we do not rewrite `extract_recursive` and certain property path expressions. The operation and its dependencies are converted to nested vSPARQL subqueries.

6   Evaluation

We evaluate our query rewriting system on the use case view definitions described in [26]. These view definitions transform one or more of four RDF biomedical ontologies: NCI Thesaurus [4], Reactome [7], Ontology of Physics for Biology [12], and the Foundational Model of Anatomy [23]. Although IML can express all of the transformations, the presence of `extract_recursive` operations late in two view definitions prevent beneficial query rewriting.

Table 4 presents the RDF triple size statistics for our views and queries. We use vSPARQL queries evaluated over on-demand, in-memory materialized views as our baseline query performance. These queries and views incorporate the improvements identified through query rewriting; thus rewriting performance benefits are the result of eliminating unnecessary transformations.

For this work, all view and query combinations are executed on a Intel Xeon dual quad core 2.66GHz 64-bit machine with 16GB of RAM. The computer runs a 64-bit SMP version of RedHat Enterprise Linux, kernel 2.6.18. PostgreSQL 8.3.5 is used for backend storage of the Jena SDB and is accessed using PostgreSQL's JDBC driver version 8.3-603. We use 64-bit Sun Java 1.6.0.22.

Table 4. Use case view and query size statistics. Entries are the number of RDF triples in the input ontology, view, or query result. Numbers in parenthesis are the number of new triples added to the base ontology by the view. Individual queries are referred to by their view and query number; for example, v2q1 is Craniofacial view query1.

	Mitotic Cell Cycle (v1)	Cranio-facial (v2)	Organ spatial location (v3)	Neuro FMA Ontology (v4)	NCI Thesaurus Simplification (v5)	Bio-simulation model editor (v6)	Blood contained in heart (v7)	Radiologist liver ontology (v8)	Blood fluid properties (v9)
FMA		1.67M	1.67M			1.67M	1.67M	1.67M	1.67M
FMA*				1.7M					
NCIt					3.37M				
Reactome	3.6M								
OPB									1,992
view	37	4,104	175	72,356	3.37M (180)	38	72	413	3,016 (1024)
query1	6	1	2	2	21	13	3	4	64
query2	4	4	3	10	9	1	1	9	16
query3	2	2	6	59	1	5	4	6	10
query4	5	24	2	17	6	3	2	3	41
query5	4	2	2	1		0	6	9	1
query6	5	3	1			1	13	5	
query7		592	1			3		1	
query8		520	1					1	
query9								46	
query10								1	
# subq in view def	1	3	2	17	0	4	3	7	2
time to materialize view (secs)	3.54	23.84	23.22	254.35	392.58	5.01	25.23	110.53	4.81

We evaluate each view and query combination five[1] times and the smallest execution time is reported. Between each run we stop the PostgreSQL server, sync the file system, clear the caches, [2] and restart the PostgreSQL server.

6.1 Rule Explosion and Rule Optimizations

Our optimizations reduce rule explosion during rewriting, decreasing evaluation time of rewritten queries. Table 5 presents the number of rules in the QPR set generated for each view and query combination, with and without optimizations.[3] Rule optimizations are able to curb the size of QPR sets for many rewritten queries. For example, the Bound Template Variable, Query Containment, and Constraint Simplification optimizations offset the impact of OPTIONAL and multiple CONSTRUCT templates in the Organ Spatial Location's queries 5 and 6.

---

[1] Due to its long execution time, the Organ Spatial Location view, optimized with no path cache, is evaluated 3 times.

[2] Caches are cleared by writing "3" to /proc/sys/vm/drop_caches

[3] The reported number of rules using optimizations is the number of vSPARQL subqueries in the generated query. N/A indicates we were unable to rewrite a query.

Table 5. Number of generated rules for each view and query (fewer is better). The first number is rules generated without optimizations. The second parenthesized number is rules generated using all optimizations. N/A indicates we could not rewrite the query.

	Mitotic Cell Cycle (v1)	Cranio-facial (v2)	Organ spatial location (v3)	Neuro FMA Ontology (v4)	NCI Thesaurus Simplification (v5)	Bio-simulation model editor (v6)	Blood contained in heart (v7)	Radiologist liver ontology (v8)	Blood fluid properties (v9)
view	3 (1)	6 (2)	4 (2)	N/A	10 (2)	N/A	4 (2)	192 (72)	5 (2)
query1	4 (4)	16 (4)	20 (4)	N/A	40 (3)	N/A	3 (3)	256 (62)	4 (1)
query2	N/A	2 (2)	16 (4)	N/A	32 (2)	N/A	3 (3)	12 (6)	20 (7)
query3	N/A	2 (2)	4 (2)	N/A	16 (1)	N/A	4 (2)	1296 (324)	N/A
query4	4 (4)	N/A	64 (2)	N/A	16 (1)	N/A	16 (14)	2208 (300)	160 (17)
query5	2 (2)	72 (16)	8192 (3)	N/A		N/A	4 (3)	N/A	4 (1)
query6	1 (1)	6 (4)	4096 (3)			N/A	12 (12)	N/A	
query7		8 (4)	64 (2)			N/A		64 (6)	
query8		8 (4)	64 (2)					60 (18)	
query9								2592 (180)	
query10								2208 (300)	

6.2   Best Rewritten Query Performance

The time needed to materialize each use case's vSPARQL view definition is presented in Table 4. We compare times for rewriting and evaluating IML queries to the on-demand evaluation of queries over these vSPARQL views.

We evaluate each query with several different optimizations. First we rewrite each query using all of the performance optimizations described in Section 4.5 and evaluate for path cache sizes of 0MB and 4MB. Next, for 4MB path caches, we rewrite and evaluate the query without anchored property path subqueries, and we rewrite and evaluate the query without Query Minimization.

Figure 4 compares the best baseline query and rewritten query execution times. The chart displays the percentage difference from the baseline vSPARQL execution time. If a rewritten query takes the same time as the baseline query, it will have value 0 on the chart; a query that takes two times the baseline execution time will have a value of 100.

Most queries are able to benefit substantially from query rewriting. 29 of the 41 queries (71%) achieve at least a 10% improvement over the baseline execution times; 25 of the 41 queries (60%) have execution times that are 60% less than the baseline. These results indicate that rewriting can significantly improve performance for a majority of our queries.

13 queries' evaluation times do not improve by more than 10%; 10 of these queries take longer to evaluate. 8 of these queries are over the Blood Fluid Properties (v9) and Mitotic Cell Cycle (v1) views and have baseline execution times of less than 5 seconds; 3 of these queries have small improvements but 5 cannot overcome the cost of rewriting. 3 of the remaining queries' (v3q2,v8q9,v8q3) patterns do not specify URIs to limit the portions of the view that they should be applied to and must be evaluated against the entire transformed view; rewriting introduces redundancy and increases execution time. Query v7q5 introduces a concrete URI in a FILTER expression; our rewriter does not yet benefit from

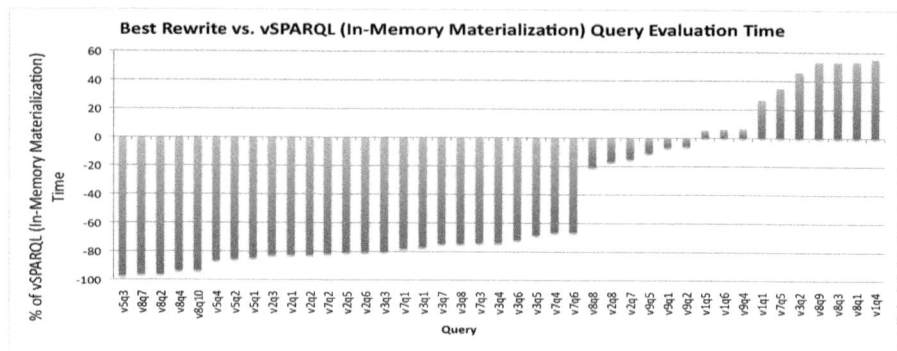

Fig. 4. Best case rewritten query vs. vSPARQL (in-memory materialization) execution time. We plot the rewritten query's execution time as the percentage difference from the baseline execution. The best query (v5q3) was 97% faster than materializing the baseline vSPARQL view in memory. Some queries (on the right) performed worse.

URIs introduced in this manner. Query v8q1 does not benefit from rewriting. The view extracts the subclass hierarchy of `fma:Organ` and replaces two direct subclasses with their (four) children; we then extract the subclass hierarchy for each of these four children, instead of once, thus increasing execution time.

6.3   Impact of rewriting options

We consider the impact of the rewriting options on our results. For space reasons, we only discuss the Organ Spatial Location view's performance, seen in Fig. 5.

Our Anchored Property Path optimization defines subqueries to prevent expensive property path expressions from being repeatedly evaluated. If a query does not introduce a URI not seen in the view, anchored subqueries can prevent repeated evaluation of a property path expression; this is seen with query 2. However, in the case of selective queries like query 5 and 6, rewriting generates rules whose WHERE clauses do not need to be completely evaluated; the first few

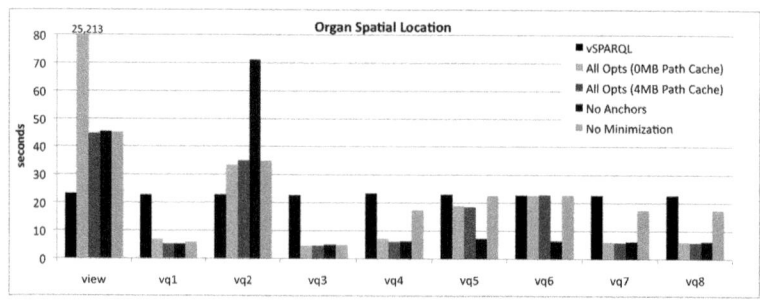

Fig. 5. Effect of rewriting options on the Organ Spatial Location view

elements in the WHERE clause determine that it will never match the data, and expensive path expressions are not evaluated. For these queries, anchored paths, which are always evaluated, increase query execution time.

Query minimization can eliminate redundancy in rewritten queries. It is needed to improve evaluation of queries 4-8. These queries contain multiple WHERE clause elements joined by a shared variable; when the elements are rewritten over a common view, duplicate WHERE clause elements are created. Query minimization eliminates these duplicates and reduces execution time.

The path cache can eliminate or reduce repeated evaluation of the same property path expression. In Fig. 5, the absence of a path cache results in a large increased execution time for the rewritten view.

The rewriter provides significant benefits for queries that introduce URIs that are not in the view definition. Rewriting can improve queries by determining the most efficient direction for evaluating property path expressions. Anchored path subqueries can be used when new URIs are not introduced by the query to prevent expensive property path expressions from being repeatedly evaluated. Query minimization should be used when multiple query pattern elements are joined by a shared variable.

7   Related Work

Scripting and visual pipe transformation languages [8][17][9] allow users to specify a sequence of operations to create mashups and transform RDF. However, users must develop modules that provide the functionality available in visual editors. Evaluating queries over these transformations can be expensive; typically the entire transformed ontology is materialized and the query evaluated on it.

Visual editors are often used for transforming existing ontologies. Table 6 compares IML with the functionality provided by a visual ontology editor commonly used by bioinformatics researchers: Protege and its plug-in PROMPT. Protege is a visual editor for creating and modifying ontologies. It centers development around the subclass (i.e. "is a") hierarchy, with additional properties and values assigned to classes in this hierarchy. PROMPT provides functionality for extracting information from an ontology by traversing specified paths or combinations of paths. It also supports merging and comparing ontologies.

IML applies a sequence of transformations to a data set. These transformations can also be achieved using nested queries. Until recently SPARQL has not included support for subqueries. Schmidt [25] has developed a set of equivalences for operations in the SPARQL algebra that can be used for rewriting and optimizing queries; this work pre-dates subqueries. There has been considerable research on optimizing nested queries for relational databases[11]. Optimization of XQuery's nested FLWOR statements has focused on the introduction of a groupby operator to enable algebraic rewriting [19][22][28], including elimination of redundant navigation[13]. Most related to our work is [16], which minimizes XQuery queries with nested subexpressions whose intermediate results are queried by other subexpressions. The rewrite rules recursively prune

Table 6. Protege/PROMPT vs. IML Functionality

Protege & PROMPT	IML Operations
Extract edges, hierarchies	extract_edges, extract_cgraph
Delete resources, properties, values, (IS_A)hierarchies	delete_node, delete_property, delete_edge, delete_cgraph
Move resources, (IS_A) hierarchies	replace_edge_subject, replace_edge_property, replace_edge_object, replace_edge_literal
Rename resources, properties	replace_node, replace_property
Add resource, property, value	add_edge
Merge resources	merge_nodes
Combine ontologies	union_graphs
	extract_reachable, extract_path, extract_recursive
	split_node
	copy_graph

nested queries, eliminating the production of unnecessary intermediate results, and creates a simplified, equivalent Xquery query.

The XQuery Update Framework (XQUF)[10] extends XQuery with transformation operations. Bohannon[14] presents an automaton-based technique for converting XQUF transform queries, and user queries composed with transform queries, into standard XQuery; the generated query only accesses necessary parts of the XML document. This work only addresses queries containing a single update expression. Fegaras[15] uses XML schemas to translate XQUF expressions to standard XQuery, relying on the underlying XQuery engine for optimization.

Several works have investigated rewriting SPARQL queries for efficient evaluation. Stocker[27] uses statistics on in-memory RDF data to iteratively order query pattern edges based on their minimum estimated selectivity. RDF-3X[20] optimizes execution plans using specialized histograms and frequent join paths in the data for estimating selectivity of joins. Our algorithm for producing efficient SPARQL queries is similar to an approach in[27], but adds in statistics specifically for property-path expressions.

8   Conclusions

We have presented a transforming view definition language IML for manipulating RDF ontologies. The language consists a small set of graph transformations that can be combined in a dataflow style. Our rewriting system for IML leverages the language's dataflow and compositional characteristics to rewrite queries over transforming views. We evaluated our rewriting system by defining transforming views over a set of use cases over RDF biomedical information sets.

References

1. Annoteimage, http://sig.biostr.washington.edu/projects/AnnoteImage/
2. Knoodl, http://knoodl.com/
3. The minisat page, http://minisat.se
4. Ncithesaurus, http://nciterms.nci.nih.gov

5. Neon toolkit, http://neon-toolkit.org/
6. The protege ontology editor and knowledge acquisition system, http://protege.stanford.edu
7. Reactome, http://www.reactome.org
8. Sparqlmotion, http://www.topquadrant.com/products/SPARQLMotion.html
9. Sparqlscript, http://www.w3.org/wiki/SPARQL/Extensions/SPARQLScript
10. Xquery update facility 1.0, http://www.w3.org/TR/xquery-update-10/
11. Chaudhuri, S.: An overview of query optimization in relational systems. In: PODS 1998, pp. 34–43. ACM, New York (1998)
12. Cook, D.L., Mejino, J.L., Neal, M.L., Gennari, J.H.: Bridging biological ontologies and biosimulation: The ontology of physics for biology. In: American Medical Informatics Association Fall Symposium (2008)
13. Deutsch, A., Papakonstantinou, Y., Xu, Y.: The next framework for logical xquery optimization. In: VLDB 2004, vol. 30, pp. 168–179. VLDB Endowment (2004)
14. Fan, W., Cong, G., Bohannon, P.: Querying xml with update syntax. In: SIGMOD 2007, pp. 293–304. ACM, New York (2007)
15. Fegaras, L.: A Schema-Based Translation of XQuery Updates. In: Lee, M.L., Yu, J.X., Bellahsène, Z., Unland, R. (eds.) XSym 2010. LNCS, vol. 6309, pp. 58–72. Springer, Heidelberg (2010)
16. Gueni, B., Abdessalem, T., Cautis, B., Waller, E.: Pruning nested xquery queries. In: Conf. on Information and Knowledge Management, pp. 541–550. ACM, New York (2008)
17. Le-Phuoc, D., Polleres, A., Hauswirth, M., Tummarello, G., Morbidoni, C.: Rapid prototyping of semantic mash-ups through semantic web pipes. In: WWW 2009, pp. 581–590. ACM, New York (2009)
18. Magkanaraki, A., Tannen, V., Christophides, V., Plexousakis, D.: Viewing the semantic web through rvl lenses. Web Semantics 1(4), 359–375 (2004)
19. May, N., Helmer, S., Moerkotte, G.: Strategies for query unnesting in xml databases. ACM Trans. Database Systems 31, 968–1013 (2006)
20. Neumann, T., Weikum, G.: Rdf-3x: a risc-style engine for rdf 1, 647–659 (2008)
21. Noy, N.F., Musen, M.A.: Specifying Ontology Views by Traversal. In: McIlraith, S.A., Plexousakis, D., van Harmelen, F. (eds.) ISWC 2004. LNCS, vol. 3298, pp. 713–725. Springer, Heidelberg (2004)
22. Re, C., Simeon, J., Fernandez, M.: A complete and efficient algebraic compiler for xquery. In: ICDE 2006. IEEE Computer Society (2006)
23. Rosse, C., Mejino Jr., J.L.V.: A reference ontology for biomedical informatics: the foundational model of anatomy. Journal of Biomedical Informatics 36(6) (2003)
24. Schenk, S., Staab, S.: Networked graphs: a declarative mechanism for sparql rules, sparql views and rdf data integration on the web. In: WWW 2008, pp. 585–594. ACM, New York (2008)
25. Schmidt, M., Meier, M., Lausen, G.: Foundations of sparql query optimization. In: ICDT 2010, pp. 4–33. ACM, New York (2010)
26. Shaw, M., Detwiler, L.T., Noy, N., Brinkley, J., Suciu, D.: vsparql: A view definition language for the semantic web. Journal of Biomedical Informatics 44(1) (February 2011)
27. Stocker, M., Seaborne, A., Bernstein, A., Kiefer, C., Reynolds, D.: Sparql basic graph pattern optimization using selectivity estimation. In: WWW 2008, pp. 595–604. ACM, New York (2008)
28. Wang, S., Rundensteiner, E.A., Mani, M.: Optimization of nested xquery expressions with orderby clauses. Data Knowledge Eng. 60 (February 2007)

Sensitive Label Privacy Protection on Social Network Data

Yi Song[1], Panagiotis Karras[2], Qian Xiao[1], and Stéphane Bressan[1]

[1] School of Computing
National University of Singapore
{songyi,xiaoqian,steph}@nus.edu.sg
[2] Rutgers Business School
Rutgers University
karras@business.rutgers.edu

Abstract. This paper is motivated by the recognition of the need for a finer grain and more personalized privacy in data publication of social networks. We propose a privacy protection scheme that not only prevents the disclosure of identity of users but also the disclosure of selected features in users' profiles. An individual user can select which features of her profile she wishes to conceal. The social networks are modeled as graphs in which users are nodes and features are labels. Labels are denoted either as sensitive or as non-sensitive. We treat node labels *both* as background knowledge an adversary may possess, *and* as sensitive information that has to be protected. We present privacy protection algorithms that allow for graph data to be published in a form such that an adversary who possesses information about a node's neighborhood cannot safely infer its identity and its sensitive labels. To this aim, the algorithms transform the original graph into a graph in which nodes are sufficiently indistinguishable. The algorithms are designed to do so while losing as little information and while preserving as much utility as possible. We evaluate empirically the extent to which the algorithms preserve the original graph's structure and properties. We show that our solution is effective, efficient and scalable while offering stronger privacy guarantees than those in previous research.

1 Introduction

The publication of social network data entails a privacy threat for their users. Sensitive information about users of the social networks should be protected. The challenge is to devise methods to publish social network data in a form that affords utility without compromising privacy. Previous research has proposed various privacy models with the corresponding protection mechanisms that prevent both inadvertent private information leakage and attacks by malicious adversaries. These early privacy models are mostly concerned with identity and link disclosure. The social networks are modeled as graphs in which users are nodes and social connections are edges. The threat definitions and protection

A. Ailamaki and S. Bowers (Eds.): SSDBM 2012, LNCS 7338, pp. 562–571, 2012.

mechanisms leverage structural properties of the graph. This paper is motivated by the recognition of the need for a finer grain and more personalized privacy.

Users entrust social networks such as Facebook and LinkedIn with a wealth of personal information such as their age, address, current location or political orientation. We refer to these details and messages as features in the user's profile. We propose a privacy protection scheme that not only prevents the disclosure of identity of users but also the disclosure of selected features in users' profiles. An individual user can select which features of her profile she wishes to conceal.

The social networks are modeled as graphs in which users are nodes and features are labels[1]. Labels are denoted either as sensitive or as non-sensitive. Figure 1 is a labeled graph representing a small subset of such a social network. Each node in the graph represents a user, and the edge between two nodes represents the fact that the two persons are friends. Labels annotated to the nodes show the locations of users. Each letter represents a city name as a label for each node. Some individuals do not mind their residence being known by the others, but some do, for various reasons. In such case, the privacy of their labels should be protected at data release. Therefore the locations are either sensitive (labels are in red italic in Figure 1[2]) or non-sensitive.

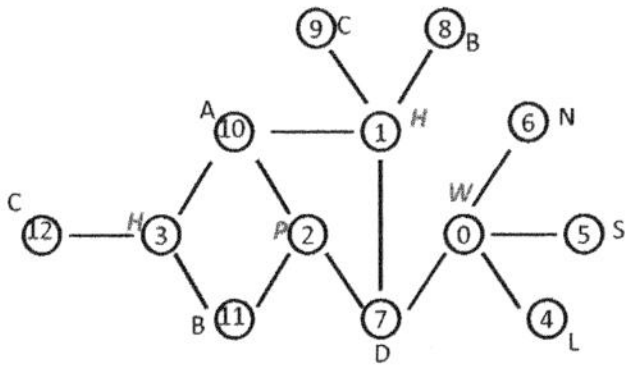

Fig. 1. Example of the labeled graph representing a social network

The privacy issue arises from the disclosure of sensitive labels. One might suggest that such labels should be simply deleted. Still, such a solution would present an incomplete view of the network and may hide interesting statistical information that does not threaten privacy. A more sophisticated approach consists in releasing information about sensitive labels, while ensuring that the identities of users are protected from privacy threats. We consider such threats as *neighborhood attack*, in which an adversary finds out sensitive information based on *prior knowledge* of the number of neighbors of a target node and the labels of these neighbors. In the example, if an adversary knows that a user has three friends and that these friends are in A (Alexandria), B (Berlin) and C (Copenhagen), respectively, then she can infer that the user is in H (Helsinki).

We present privacy protection algorithms that allow for graph data to be published in a form such that an adversary cannot safely infer the identity and

---

[1] Although modeling features in the profile as attribute-value pairs would be closer to the actual social network structure, it is without loss of generality that we consider atomic labels.

[2] W: Warsaw, H: Helsinki, P: Prague, D: Dublin, S:Stockholm, N: Nice, A: Alexandria, B: Berlin, C: Copenhagen, L: Lisbon.

sensitive labels of users. We consider the case in which the adversary possesses both structural knowledge and label information.

The algorithms that we propose transform the original graph into a graph in which any node with a sensitive label is indistinguishable from at least $\ell-1$ other nodes. The probability to infer that any node has a certain sensitive label (we call such nodes *sensitive nodes*) is no larger than $1/\ell$. For this purpose we design ℓ-diversity-like model, where we treat node labels as *both* part of an adversary's background knowledge *and* as sensitive information that has to be protected.

The algorithms are designed to provide privacy protection while losing as little information and while preserving as much utility as possible. In view of the tradeoff between data privacy and utility [16], we evaluate empirically the extent to which the algorithms preserve the original graph's structure and properties such as density, degree distribution and clustering coefficient. We show that our solution is effective, efficient and scalable while offering stronger privacy guarantees than those in previous research, and that our algorithms scale well as data size grows.

The rest of the paper is organized as follows. Section 2 reviews previous works in the area. We define our problem in Section 3 and propose solutions in Section 4. Experiments and result analysis are described in Section 5. We conclude this work in Section 6.

2   Related Work

The first necessary anonymization technique in both the contexts of micro- and network data consists in removing identification. This nave technique has quickly been recognized as failing to protect privacy. For microdata, Sweeney et al. propose k-anonymity [17] to circumvent possible identity disclosure in naively anonymized microdata. ℓ-diversity is proposed in [13] in order to further prevent attribute disclosure.

Similarly for network data, Backstrom et al., in [2], show that naive anonymization is insufficient as the structure of the released graph may reveal the identity of the individuals corresponding to the nodes. Hay et al. [9] emphasize this problem and quantify the risk of re-identification by adversaries with external information that is formalized into structural queries (node refinement queries, subgraph knowledge queries). Recognizing the problem, several works [5,11,18,20,21,22,24,27,8,4,6] propose techniques that can be applied to the naive anonymized graph, further modifying the graph in order to provide certain privacy guarantee. Some works are based on graph models other than simple graph [12,7,10,3].

To our knowledge, Zhou and Pei [25,26] and Yuan et al. [23] were the first to consider modeling social networks as labeled graphs, similarly to what we consider in this paper. To prevent re-identification attacks by adversaries with immediate neighborhood structural knowledge, Zhou and Pei [25] propose a method that groups nodes and anonymizes the neighborhoods of nodes in the same group by generalizing node labels and adding edges. They enforce a *k-anonymity* privacy constraint on the graph, each node of which is guaranteed to have the same immediate neighborhood structure with other $k-1$ nodes. In [26], they improve the privacy guarantee provided by k-*anonymity* with the idea of ℓ-diversity, to protect labels on nodes as well. Yuan et al. [23] try to be more practical by considering users' different privacy concerns. They divide privacy requirements into three levels, and suggest methods to generalize labels and modify structure corresponding to every privacy demand. Nevertheless, neither Zhou and Pei, nor Yuan et al. consider labels as a part of the background knowledge. However, in case adversaries hold label information, the methods of [25,26,23] cannot achieve the same privacy guarantee. Moreover, as with the context of microdata, a graph that satisfies a k-anonymity privacy guarantee may still leak sensitive information regarding its labels [13].

3   Problem Definition

We model a network as $G(V, E, L^s, L, \Gamma)$, where V is a set of nodes, E is s set of edges, L^s is a set of sensitive labels, and L is a set of non-sensitive labels. Γ maps nodes to their labels, $\Gamma : V \rightarrow L^s \cup L$. Then we propose a privacy model, ℓ-*sensitive-label-diversity*; in this model, we treat node labels *both* as part of an adversary's background knowledge, *and* as sensitive information that has to be protected. These concepts are clarified by the following definitions:

Definition 1. *The **neighborhood information** of node v comprises the degree of v and the labels of v's neighbors.*

Definition 2. *(ℓ-**sensitive-label-diversity**) For each node v that associates with a sensitive label, there must be at least $\ell-1$ other nodes with the same neighborhood information, but attached with different sensitive labels.*

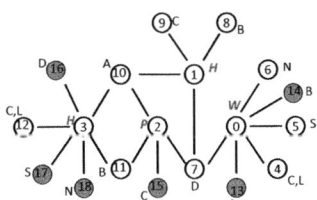

Fig. 2. Privacy-attaining network example

In Example 1, nodes 0, 1, 2, and 3 have sensitive labels. The neighborhood information of node 0, includes its degree, which is 4, and the labels on nodes 4, 5, 6, and 7, which are L, S, N, and D, respectively. For node 2, the neighborhood information includes degree 3 and the labels on nodes 7, 10, and 11, which are D, A, and B. The graph in Figure 2 satisfies 2-*sensitive-label-diversity*; that is because, in this graph, nodes 0 and 3 are indistinguishable, having six neighbors with label A, B, {C,L}, D, S, N separately; likewise, nodes 1 and 2 are indistinguishable, as they both have four neighbors with labels A, B, C, D separately.

4   Algorithm

The main objective of the algorithms that we propose is to make suitable group-ing of nodes, and appropriate modification of neighbors' labels of nodes of each group to satisfy the *l-sensitive-label-diversity* requirement. We want to group nodes with as similar neighborhood information as possible so that we can change as few labels as possible and add as few noisy nodes as possible. We propose an algorithm, Global-similarity-based Indirect Noise Node (GINN), that does not attempt to heuristically prune the similarity computation as the other two algo-rithms, Direct Noisy Node Algorithm (DNN) and Indirect Noisy Node Algorithm (INN) do. Algorithm *DNN* and *INN*, which we devise first, sort nodes by degree and compare neighborhood information of nodes with similar degree. Details about algorithm *DNN* and *INN* please refer to [15].

4.1   Algorithm GINN

The algorithm starts out with group formation, during which all nodes that have not yet been grouped are taken into consideration, in clustering-like fashion. In the first run, two nodes with the maximum similarity of their neighborhood labels are grouped together. Their neighbor labels are modified to be the same immediately so that nodes in one group always have the same neighbor labels. For two nodes, v_1 with neighborhood label set (LS_{v_1}), and v_2 with neighborhood label set (LS_{v_2}), we calculate neighborhood label similarity (NLS) as follows:

$$NLS(v_1, v_2) = \frac{|LS_{v_1} \cap LS_{v_2}|}{|LS_{v_1} \cup LS_{v_2}|} \tag{1}$$

Larger value indicates larger similarity of the two neighborhoods.

Then nodes having the maximum similarity with any node in the group are clustered into the group till the group has ℓ nodes with different sensitive labels. Thereafter, the algorithm proceeds to create the next group. If fewer than ℓ nodes are left after the last group's formation, these remainder nodes are clustered into existing groups according to the similarities between nodes and groups.

After having formed these groups, we need to ensure that each group's mem-bers are indistinguishable in terms of *neighborhood information*. Thus, neigh-borhood labels are modified after every grouping operation, so that labels of nodes can be accordingly updated immediately for the next grouping opera-tion. This modification process ensures that all nodes in a group have the same *neighborhood information*. The objective is achieved by a series of modification operations. To modify graph with as low information loss as possible, we devise three modification operations: *label union, edge insertion* and *noise node addi-tion*. Label union and edge insertion among nearby nodes are preferred to node addition, as they incur less alteration to the overall graph structure.

Edge insertion is to complement for both a missing label and insufficient degree value. A node is linked to an existing nearby (two-hop away) node with that label. Label union adds the missing label values by creating super-values shared among labels of nodes. The labels of two or more nodes coalesce their values to a single super-label value, being the union of their values. This approach maintains data integrity, in the sense that the true label of node is included among the values of its label super-value. After such edge insertion and label union operations, if there are nodes in a group still having different neighborhood information, noise nodes with non-sensitive labels are added into the graph so as to render the nodes in group indistinguishable in terms of their neighbors' labels. We consider the unification of two nodes' neighborhood labels as an example. One node may need a noisy node to be added as its immediate neighbor since it does not have a neighbor with certain label that the other node has; such a label on the other node may not be modifiable, as its is already connected to another sensitive node, which prevents the re-modification on existing modified groups.

Algorithm 1: Global-Similarity-based Indirect Noisy Node Algorithm

---

Input: graph $G(V, E, L, L^s)$, parameter l;
Result: Modified Graph G'

1 **while** $V_{left} > 0$ **do**
2   **if** $|V_{left}| \geq l$ **then**
3   compute pairwise node similarities;
4   group $\mathcal{G} \leftarrow v_1, v_2$ with $Max_{similarity}$;
5   Modify neighbors of $\mathcal{G}$;
6   **while** $|\mathcal{G}| < l$ **do**
7             $dissimilarity(V_{left}, \mathcal{G})$;
8   group $\mathcal{G} \leftarrow v$ with $Max_{similarity}$;
9   Modify neighbors of $\mathcal{G}$ without actually adding noisy nodes ;
10   **else if** $|V_{left}| < l$ **then**
11   **for** *each* $v \in V_{left}$ **do**
12             $similarity(v, \mathcal{G}s)$;
13             $\mathcal{G}_{Max\_similarity} \leftarrow v$;
14   Modify neighbors of $\mathcal{G}_{Max\_similarity}$ without actually adding noisy nodes,
15 Add expected noisy nodes;
16 **Return** $G'(V', E', L')$;

In this algorithm, noise node addition operation that is expected to make the nodes inside each group satisfy ℓ-sensitive-label-diversity are *recorded*, but *not* performed right away. Only after *all* the preliminary grouping operations are performed, the algorithm proceeds to process the *expected node addition* operation at the final step. Then, if two nodes are expected to have the same labels of neighbors and are within two hops (having common neighbors), only one node is added. In other words, we merge some noisy nodes with the same label, thus resulting in fewer noisy nodes.

5   Experimental Evaluation

We evaluate our approaches using both synthetic and real data sets. All of the approaches have been implemented in Python. The experiments are conducted on an Intel core, 2Quad CPU, 2.83GHz machine with 4GB of main memory running Windows 7 Operating System. We use three data sets. The first data set [1] is a network of hyperlinks between weblogs on US politics. The second data set that we use is generated from the Facebook dataset proposed in [14]. The third data set that we use is a family of synthetic graphs with varying number of nodes. The first and second datasets are used for the evaluation of effectiveness (utility and information loss). The third data set is used to measure runtime and scalability (running time). (Please refer to [15] for more information.)

5.1   Data Utility

We compare the data utilities we preserve from the original graphs, in view of measurements on degree distribution, label distribution, degree centrality [19], clustering coefficient, average path length, graph density, and radius. We show the number of the noisy nodes and edges needed for each approach.

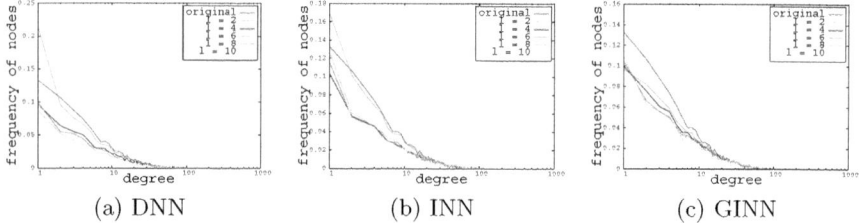

(a) DNN   (b) INN   (c) GINN

Fig. 3. Facebook Graph Degree Distribution

Figure 3 shows the degree distribution of the Facebook graph both before and after modification. Each subfigure in Figure 3 shows degree distributions of graphs modified by one algorithm. We can see that the degree distributions of the modified graphs resemble the original ones well, especially when l is small.

To sum up, these measurements (for other results please refer to [15]) show that the graph structure properties are preserved to a large extent. The strong resemblance of the label distributions in most cases indicates that the label information, another aspect of graph information, is well maintained. They suggest as well that algorithm *GINN* does preserve graph properties better than the other two while these three algorithms achieve the same privacy constraint.

5.2   Information Loss

In view of utility of released data, we aim to keep information loss low. Information loss in this case contains both structure information loss and label information loss. We measure the loss in the following way: for any node $v \in V$,

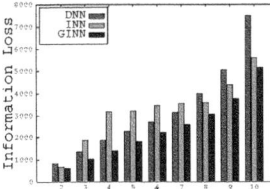

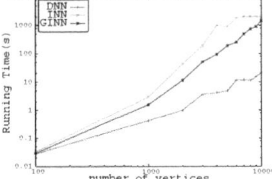

Fig. 4. Information Loss   **Fig. 5.** Running Time

label dissimilarity is defined as: $\mathcal{D}(l_v, l'_v) = 1 - \frac{|l_v \cap l'_v|}{|l_v \cup l'_v|}$, where l_v is the set of v's original labels and l'_v the set of labels in the modified graph. Thus, for the modified graph including n noisy nodes, and m noisy edges, information loss is defined as

$$IL = \omega_1 n + \omega_2 m + (1 - \omega_1 - \omega_2) \sum \mathcal{D}(l_v, l'_v) \qquad (2)$$

where ω_1, ω_2 and $1 - \omega_1 - \omega_2$ are weights for each part of the information loss. Figure 4 shows the measurements of information loss on the synthetic data set using each algorithm. Algorithm *GINN* introduces the least information loss.

5.3 Algorithm Scalability

We measure the running time of the methods for a series of synthetic graphs with varying number of nodes in our third dataset. Figure 5 presents the running time of each algorithm as the number of nodes increases. Algorithm *DNN* is faster than the other two algorithms, showing good scalability at the cost of large noisy nodes added. Algorithm *GINN* can also be adopted for quite large graphs as follows: We separate the nodes to two different categories, with or without sensitive labels. Such smaller granularity reduces the number of nodes the anonymization method needs to process, and thus improves the overall efficiency.

6 Conclusions

In this paper we have investigated the protection of private label information in social network data publication. We consider graphs with rich label information, which are categorized to be either sensitive or non-sensitive. We assume that adversaries possess prior knowledge about a node's degree and the labels of its neighbors, and can use that to infer the sensitive labels of targets. We suggested a model for attaining privacy while publishing the data, in which node labels are *both* part of adversaries' background knowledge *and* sensitive information that has to be protected. We accompany our model with algorithms that transform a network graph before publication, so as to limit adversaries' confidence about sensitive label data. Our experiments on both real and synthetic data sets confirm the effectiveness, efficiency and scalability of our approach in maintaining critical graph properties while providing a comprehensible privacy guarantee.

Acknowledgement. This research was partially funded by the A*Star SERC project "Hippocratic Data Stream Cloud for Secure, Privacy-preserving Data Analytics Services" 102 158 0037, NUS Ref:R-702-000-005-305.

References

1. Adamic, L.A., Glance, N.: The political blogosphere and the 2004 U.S. election: divided they blog. In: LinkKDD (2005)
2. Backstrom, L., Dwork, C., Kleinberg, J.M.: Wherefore art thou R3579X?: anonymized social networks, hidden patterns, and structural steganography. Commun. ACM 54(12) (2011)
3. Bhagat, S., Cormode, G., Krishnamurthy, B., Srivastava, D.: Class-based graph anonymization for social network data. PVLDB 2(1) (2009)
4. Campan, A., Truta, T.M.: Data and Structural k-Anonymity in Social Networks. In: Bonchi, F., Ferrari, E., Jiang, W., Malin, B. (eds.) PinKDD 2008. LNCS, vol. 5456, pp. 33–54. Springer, Heidelberg (2009)
5. Cheng, J., Fu, A.W.-C., Liu, J.: K-isomorphism: privacy-preserving network publication against structural attacks. In: SIGMOD (2010)
6. Cormode, G., Srivastava, D., Yu, T., Zhang, Q.: Anonymizing bipartite graph data using safe groupings. PVLDB 19(1) (2010)
7. Das, S., Egecioglu, Ö., Abbadi, A.E.: Anonymizing weighted social network graphs. In: ICDE (2010)
8. Francesco Bonchi, A.G., Tassa, T.: Identity obfuscation in graphs through the information theoretic lens. In: ICDE (2011)
9. Hay, M., Miklau, G., Jensen, D., Towsley, D., Weis, P.: Resisting structural re-identification in anonymized social networks. PVLDB 1(1) (2008)
10. Li, Y., Shen, H.: Anonymizing graphs against weight-based attacks. In: ICDM Workshops (2010)
11. Liu, K., Terzi, E.: Towards identity anonymization on graphs. In: SIGMOD (2008)
12. Liu, L., Wang, J., Liu, J., Zhang, J.: Privacy preserving in social networks against sensitive edge disclosure. In: SIAM International Conference on Data Mining (2009)
13. Machanavajjhala, A., Gehrke, J., Kifer, D., Venkitasubramaniam, M.: ℓ-diversity: privacy beyond k-anonymity. In: ICDE (2006)
14. MPI, http://socialnetworks.mpi-sws.org/
15. Song, Y., Karras, P., Xiao, Q., Bressan, S.: Sensitive label privacy protection on social network data. Technical report TRD3/12 (2012)
16. Song, Y., Nobari, S., Lu, X., Karras, P., Bressan, S.: On the privacy and utility of anonymized social networks. In: iiWAS, pp. 246–253 (2011)
17. Sweeney, L.: K-anonymity: a model for protecting privacy. International Journal of Uncertainty, Fuzziness and Knowledge-Based Systems 10(5) (2002)
18. Tai, C.-H., Yu, P.S., Yang, D.-N., Chen, M.-S.: Privacy-preserving social network publication against friendship attacks. In: SIGKDD (2011)
19. Tore, O., Filip, A., John, S.: Node centrality in weighted networks: generalizing degree and shortest paths. Social Networks 32(3) (2010)
20. Wu, W., Xiao, Y., Wang, W., He, Z., Wang, Z.: K-symmetry model for identity anonymization in social networks. In: EDBT (2010)

21. Ying, X., Wu, X.: Randomizing social networks: a spectrum perserving approach. In: SDM (2008)
22. Ying, X., Wu, X.: On Link Privacy in Randomizing Social Networks. In: Theeramunkong, T., Kijsirikul, B., Cercone, N., Ho, T.-B. (eds.) PAKDD 2009. LNCS, vol. 5476, pp. 28–39. Springer, Heidelberg (2009)
23. Yuan, M., Chen, L., Yu, P.S.: Personalized privacy protection in social networks. PVLDB 4(2) (2010)
24. Zhang, L., Zhang, W.: Edge anonymity in social network graphs. In: CSE (2009)
25. Zhou, B., Pei, J.: Preserving privacy in social networks against neighborhood attacks. In: ICDE (2008)
26. Zhou, B., Pei, J.: The k-anonymity and ℓ-diversity approaches for privacy preservation in social networks against neighborhood attacks. Knowledge and Information Systems 28(1) (2010)
27. Zou, L., Chen, L., Özsu, M.T.: K-automorphism: a general framework for privacy-preserving network publication. PVLDB 2(1) (2009)

Trading Privacy for Information Loss
in the Blink of an Eye

Alexandra Pilalidou[1,*] and Panos Vassiliadis[2]

[1] FMT Worldwide, Limassol,Cyprus
apilalid@gmail.com
[2] Dept. of Computer Science, Univ. of Ioannina, Hellas
pvassil@cs.uoi.gr

Abstract. The publishing of data with privacy guarantees is a task typically performed by a data curator who is expected to provide guarantees for the data he publishes in quantitative fashion, via a privacy criterion (e.g., k-anonymity, l-diversity). The anonymization of data is typically performed off-line. In this paper, we provide algorithmic tools that facilitate the negotiation for the anonymization scheme of a data set in user time. Our method takes as input a set of user constraints for (i) suppression, (ii) generalization and (iii) a privacy criterion (k-anonymity, l-diversity) and returns (a) either an anonymization scheme that fulfils these constraints or, (b) three approximations to the user request based on the idea of keeping the two of the three values of the user input fixed and finding the closest possible approximation for the third parameter. The proposed algorithm involves precomputing suitable histograms for all the different anonymization schemes that a global recoding method can follow. This allows computing exact answers extremely fast (in the order of few milliseconds).

1 Introduction

The area of privacy-preserving data publishing serves the purpose of allowing a data curator publish data that contain sensitive information for persons in the real world while serving the following two antagonistic goals: (a) allow well-meaning data miners extract useful knowledge from the data, and, (b) prevent attackers from linking the published, anonymized records to the individuals to which they refer. Frequently, the method of choice for the anonymization method involves the *generalization* of the values of the published tuples to values that are more abstract. This creates the possibility to *hide tuples in the crowd of similar tuples*. Typically, the privacy guarantee per tuple is expressed as a *privacy criterion*, e.g., k-anonymity or l-diversity ([1], [2]) that quantitatively assesses each of the groups of similar tuples (k-anonymity for its size, l-diversity also for the variance of sensitive values within the group).

Overall, *the data curator has to negotiate antagonistic goals: (a) the request to avoid exceeding a certain threshold of deleted (suppressed) tuples, (b) maximum*

---

* Work performed while in the Univ. of Ioannina.

A. Ailamaki and S. Bowers (Eds.): SSDBM 2012, LNCS 7338, pp. 572–580, 2012.
© Springer-Verlag Berlin Heidelberg 2012

tolerated generalization heights per attributes that are acceptable by the end users, and, (c) the curator's constraint on the privacy value. Of course, it might not be possible to achieve a consensus on the parameters of the problem. So, the desideratum is that the curator interactively explores different anonymization possibilities. If, for example, the curator sets a suppression threshold too low for the data set to sustain while respecting the rest of the user criteria, then the system should ideally respond very quickly with a negative answer, along with a set of proposals on what possible generalizations, close to the one that he originally submitted, are attainable with the specific data set. As opposed to the state of the art methods that operate off-line, *our method informs the curator (a) in user time (i.e., ideally with no delay that a person can sense), (b) with the best possible solution that respects all the constraints posed by the involved stakeholders, if such a solution exists, or, (c) convenient suggestions that are close to the original desiderata around generalization, suppression and privacy.*

The research community has focused on different directions, complementary but insufficient for the problem. We refer the interested reader to the excellent survey of Fung et al. [3] for further probing on the state of the art. The most related works to our problem are (a) [4] who deal with the problem of finding the local recoding with the least suppressed cells (not tuples) without any hierarchies in mind, and, (b) [5], where the author works in a similar setting and prove that the probability of achieving k-anonymity tends to zero as the number of dimensions rises to infinite; the theoretical analysis is accompanied with a study of generated data sets (one of which rises up to 50 dimensions) that supports the theoretical claim. Still, compared to these works, our method is the first to simultaneously combine a focus to on-line response, generalization hierarchies, different values of k or l, and different privacy criteria.

Our approach is based on a method that involves precomputing statistical information for several possible generalization schemes. A *generalization scheme* is determined by deciding the level of generalization for every quasi-identifier – in other words, a generalization scheme is a vector characterizing every quasi-identifier with its level of generalization. To efficiently compute the amount of suppression for a given pair of (i) value for the privacy criterion, and, (ii) a generalization scheme, we resort to the *precalculation of a histogram* per generalization scheme that allows us to calculate the necessary statistical information. For example, in the case of k-anonymity, we group the data by the quasi identifier attribute set in their generalized form and we count how many groups have size 1, 2, ... etc. So, given a specific value of k, we can compute how many tuples will be suppressed for any generalization scheme. Similarly, in the case of naive l-diversity, for each group we count the number of different sensitive values and the size of the group too.

We organize generalization schemes as nodes in a lattice. A node v is lower than a node u in the lattice if u has at least one level of generalization higher than v for a certain quasi-identifier and the rest of the quasi-identifiers in higher or equal levels. The main algorithm exploits the histograms of the nodes in the lattice and checks whether there exists a node of height less than **h** that satisfies

k (or, l) without suppressing more than $MaxSupp$ tuples. This is performed by first checking the *top-acceptable-node* v_{max} defined with generalization levels $\mathbf{h}=[h_1^c, \ldots, h_n^c]$. If a solution exists, then we exploit a monotonicity property of the lattice and look for possible answers in quasi identifiers with less or equal generalization levels than the ones of the top acceptable node. In the case that no solution exists in the top acceptable node, the algorithm provides the user with 3 complementary suggestions as answers:

- The first suggested alternative satisfies k (or, l) and $\mathbf{h}$ but not $MaxSupp$. In fact, we search the space under the top acceptable node and provide the solution with the minimum number of suppressed tuples. In typical situations, we prove that the answer is already found in the top acceptable node and thus, provide the answer immediately.
- The second alternative is a solution that satisfies k (or, l) and $MaxSupp$ but violates $\mathbf{h}$. This means that we have to explore the space of quasi identifiers that are found in generalization levels higher or equal than the top acceptable node. We exploit the lattice's monotonicity properties to avoid unnecessary checks and utilize a binary search exploration of heights on the lattice.
- The third suggested alternative is a solution that finds the maximum possible k (or, l) for which $\mathbf{h}$ and $MaxSupp$ are respected for the quasi identifiers of the top acceptable node. Similarly to the first case, this answer can be provided immediately at the top acceptable node.

In Fig. 1 we depict the lattice for the Adult data set where we constrain the quasi identifier set to *Age*, *Work class* and *Education*. Each node of the lattice is a generalization scheme; underneath each node you can see the number of tuples that violate the constraint of 3-anonymity. The label of each node shows the height of the generalization for each of the QI attributes. We pose two queries to the lattice, both constraining k to 3 and the height to level 1 for age, level 2 for *Work class* and level 1 for *Education* (coded as 121 for short). Then, the *top-acceptable-node* v_{max} is 121 and the lattice it induces is depicted as the blue diamond over the lattice. We hide nodes 030 and 102 from the figure as they are not members of the sublattice induced by v_{max}. The first query involves a tolerance for 20 suppressed tuples; in this case, v_{max} suppresses less than 20 tuples and we know that its sub-lattice will produce an exact answer. Out of all candidates, node 111 provides the exact answer with the lowest height (and since we have a tie in terms of height with node 120, the one with the least suppression among the two). The second query requires a maximum suppression of 8 tuples; in this case, since v_{max} fails the constraint, we provide two suggestions concerning the relaxation of suppression and privacy directly at v_{max}, and, a third suggestion for the relaxation of the height constraint by exploring the full lattice (and ultimately resulting in node 400).

2 Anonymity Negotiation over a Full Lattice

In this section, we present Algorithm *SimpleAnonymityNegotiation*. The proposed algorithm takes as input a relation R to be generalized, a set of hierarchies

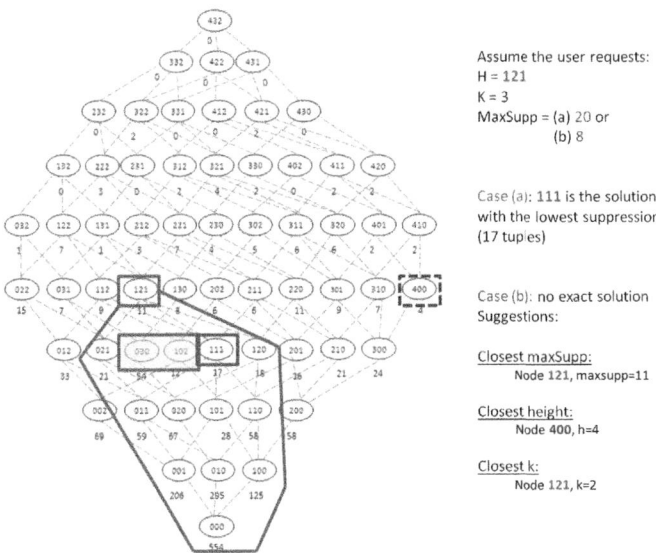

Assume the user requests:
H = 121
K = 3
MaxSupp = (a) 20 or
          (b) 8

Case (a): 111 is the solution
with the lowest suppression
(17 tuples)

Case (b): no exact solution
Suggestions:

Closest maxSupp:
    Node 121, maxsupp=11

Closest height:
    Node 400, h=4

Closest k:
    Node 121, k=2

Fig. 1. Example of lattice and query answers. The lattice is annotated with suppressed tuples for |QI|= 3 and k=3. The QI is Age, Work class, Race.

H for the quasi-identifier attributes, the histogram lattice L for all possible combinations of the generalization levels, and the requirements for the maximum desirable generalization level per quasi-identifier (**h**), the maximum tolerable number of tuples to be suppressed ($maxSupp$) and the least size of a group (k or l), as the privacy constraint. The outputs of the algorithm are (a) either a node of the lattice (i.e., a generalization scheme) that provides the best possible *exact* solution to the user requirements (with best possible being interpreted as the one with the *lowest height*, and, if more than one candidate solutions have this lowest height, the one with the minimum suppression), or, (b) three suggestions for approximate answers to the user request, the first relaxing the number of suppressed tuples, the second relaxing the constraints on the heights per dimension and the third relaxing the minimum acceptable privacy criterion (e.g., k in k-anonymity).

Algorithm *SimpleAnonymityNegotiation* starts by identifying a reference node in the lattice, to which we refer a v_{max}. The node v_{max} is the node that satisfies all the constraints of **h** for the quasi-identifiers, at the topmost level; in other words, v_{max} is the highest possible node that can obtain an exact answer to the user's request. We will also refer to v_{max} as the *top-acceptable* node. Then, two cases can hold: (a) v_{max} is able to provide an exact solution (the *if* part), or (b) it is not, and thus we have to resort to approximate suggestions to the user (the *else* part). The check on whether a node can provide an exact solution is given by function *checkExactSolution* that looks up the histogram of a node v and performs the appropriate check depending on the privacy criterion.

```
Algorithm SimpleAnonymityNegotiation(L,k,h,MaxSupp)
In: Lattice L with the histograms for R,H, constraints for k, h, MaxSupp
Out: an exact solution s[v,k,h,supp] or s1,s2,s3, si=[v_i,k_i,h_i,supp_i]
Var: a 2D vector of candidate solutions Candidates[hmax][]
Begin
Let v_max be the node that corresponds to the constraint h ;
if v_max is visited then exit;
mark v_max as visited;
if (checkExactSolution(v_max,L,k,h,MaxSupp) == true){
        Candidates[height(v_max)] = Candidates[height(v_max)] ∪ {v_max};
        for all v_c in lower(v_max)

        ExactSublatticeSearch(v_c,L,k,h,MaxSupp,Candidates);
//when the recursion is over, the Candidates has the full list of nodes
//that can serve as candidate solutions
        minHeight = minimum height having Candidates[minHeight] != {};
        v_win = v in Candidates[minHeight] with the lowest possible suppression for k;
        return(v_win,k,minHeight,suppressed(v_win,k));
}
else{
        approxSol_1 = ApproximateMaxSupp(L,v_max,k,h,MaxSupp);
        approxSol_2=ApproximateH(L,v_max,height(v_max),height(top),k,h,MaxSupp);
        approxSol_3 = ApproximateK(L,v_max,k,h,MaxSupp);

        return approxSol_1, approxSol_2, approxSol_3;
}
End.
```

```
ExactSublatticeSearch(v,L,k,h,MaxSupp,Candidates){
        if v is visited then exit;
        mark v as visited;
        if (checkExactSolution(v,L,k,h,MaxSupp) == true){
                Candidates[height(v)] = Candidates[height(v)] U {v};
                for all v_c in lower(v)
                        ExactSublatticeSearch(v_c,L,k,h,MaxSupp,Candidates);
        }
}
```

```
checkExactSolution(v,L,k,h,MaxSupp){
        lookup histogram of v in L;
        if suppressed(v,k) <= MaxSupp && height(v) <= h  return true;
        else return false;
}
```

Fig. 2. Algorithm Simple Anonymity Negotiation and accompanying functions

Exact Answer. When an exact answer can be provided by the top-acceptable node v_{max}, then we can be sure that the sublattice induced by v_{max} contains an exact answer; however, we need to discover the one with the minimum possible height and, therefore, we need to descend down the lattice to discover it. The auxiliary variable *Candidates* holds all the nodes that conform to the user request, organized per height. Each time such a node is found, it is added to *Candidates* at the appropriate level (Line 5) and its descendants (returned via the function *lower()*) are recursively explored via the call of function *ExactSublatticeSearch*. When the lattice is appropriately explored we need to find the lowest level with a solution in the lattice, and, among the (several candidate) solutions of this level we must pick the one with the least suppression.

If node v_{max} fails to provide an answer that conforms to the user request, then we are certain that it is impossible to derive such a conforming answer from our lattice and we need to search for approximations. So, we provide the users with suggestions on the possible relaxations that can be made to his criteria.

Suppression and Privacy Relaxations. Function *ApproximateMaxSupp* respects the privacy criterion k and the max tolerable height **h**, and returns the

```
ApproximateMaxSupp(L,v,k,h,MaxSupp){
    find the minimum amount of suppressed tuples, approxSupp, s.t.
checkExactSolution(v,L,k,h,approxSupp) returns true;
        if no such value exists, return {};
        else{
                for all v_c in sublattice(v) (recursively){
                        checkExactSolution(v_c,L,k,h,approxSupp)
                            break when a whole level fails to produce a solution;
                }
                let v_win be the node with the lowest height that satisfies k,h,approxSupp
                (with arbitrary tie resolution)
                return v_win,k,h,approxSupp;
        }
}
```

```
ApproximateH(L,v,h_low,h_high,k,h,MaxSupp){
    while(h_low <= h_high){
        h_current = middle between h_low and h_high;
        flag = checkIfNoSolutionInCurrentHeight(L,h_current,k,MaxSupp);
        if (flag == true){
            low = current + 1;
        }
        else{
            currentMinHeight = current;
            high = current - 1;
        }
    }
    for all v_c in currentMinHeight, find the one v_win, with the minimum suppressed(v_c,k);
        //exception: this fails only if k > |R|, else top of the lattice always answers
        return v_win,k,height(v_win),MaxSupp;
}
```

```
checkIfNoSolutionInCurrentHeight(L,h_current,k,MaxSupp){
        for all v_c in h_current
        if suppressed(v_c,k) <= MaxSupp return false;
        return true;
}
```

Fig. 3. Approximation Functions

best possible relaxation with respect to the number of suppressed tuples. Since **h** is to be respected, we are restricted in the sub-lattice induced by v_{max}. Since v_{max} has failed to provide a conforming answer, no node in the sublattice can provide such an answer, either. So, we assess the number of tuples that have to be suppressed if we retain k fixed and stay at the highest candidate node v_{max}. Observe that any node in the sublattice of v_{max} will result in higher or equal number of suppressed tuples – remember that the lower we go, the smaller the groups are and the higher the suppression. In other words, it will either be v_{max} that will give the answer or one of its descendants in the rare case that the groups of the descendant are mapped one to one to the groups of v_{max}, thus resulting in exactly the same number of suppressed tuples. The relaxation of privacy is identical (omitted for lack of space).

Height Relaxation. Function *ApproximateH* retains the maximum tolerable number of suppressed tuples *MaxSupp* and the privacy criterion of k and tries to determine what is the lowest height h that can provide an answer for these constraints. This time, we operate outside the borders of the sublattice of v_{max} since **h** is not to be respected. The function *ApproximateH* performs a binary search on the height between the height of v_{max} and the upper possible height (the top of the lattice). Every time a level is chosen, we start to check its nodes for possible solutions via the function *checkIfNoSolutionInCurrentHeight*. If the function explores a height fully and fails to find an answer, this is an indication that we should not search lower than this height (remember: failure to find a solution signals for ascending in the lattice). Every time the function finds a node that can answer, then we must search in the lower heights for possibly lower solutions. When the binary search stops, the value *currentMinHeight* signifies the lowest possible height where a solution is found. Then, we explore this height fully to determine the node with the minimum suppression.

3   Experiments

We present our results over the Adult data set [6] over two privacy criteria: k-anonymity and naive l-diversity. We have assessed the performance in terms of time and visited nodes as we vary the value for the privacy criterion, the maximum tolerable generalization height and the maximum tolerable amount of suppressed tuples. We performed 28 user requests: each time we keep the two out of the three parameters fixed in their middle value and vary QI with 3, 4, 5, 6 attributes as well as the parameters under investigation. The values of k range in 3, 10, 50. The values for l are smaller (3,6,9) in order to avoid suppressing the entire data set. The values for **h** are: (a) low heights having levels with heights 1 and 0, (b) middle heights having mostly 2 and few 1 level heights and, (c) middle-low in between (remember that they vary per QI). In all our experiments we have used a Core Duo 2.5GHz server with 3GB of memory and 300GB hard disk, Ubuntu 8.10, and MySQL 5.0.67. The code is written in Java.

Effectiveness and Efficiency. The *increase of the privacy criterion* (Fig. 14, k) has divergent effects. When QI is small, there is an exact answer and the search is directed towards lower heights. Consequently, as k increases the solution is found earlier. On the contrary, for larger QI sizes and relaxations to user request, the increase of k sublinearly increases the search space.

The *increase of the maximum tolerable height* (Fig 14, **h**) has a consistent behavior. When the QI size is small, we can have exact solutions; in this case, when we increase the maximum tolerable height, this increases the search space too. In contrast, when relaxations are sought, the higher the constraint, the faster a solution is found.

The *constraint on the maximum tolerable suppression* (Fig. 14, $MaxSupp$) is similar: the higher the constraint is set, the faster an approximate solution is found (except for low QI sizes where exact answers are possible and the behavior is inverse due to the exploration of $L(v_{max})$).

In all experiments, it is clear that the costs are dominated by the QI size. *Finally, in all experiments, the times ranged between 1 and 8 msec, thus facilitating the online negotiation of privacy with the user, in user time.* The

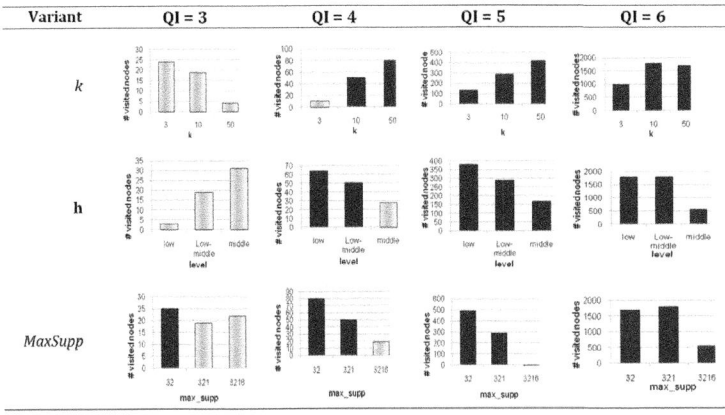

Variant	QI = 3	QI = 4	QI = 5	QI = 6

Fig. 4. Number of visited nodes for different QI, k, h, MaxSupp. *All times range between 1 and 8 msec.* Light coloring is for exact matches and dark coloring is for approximate matches.

experiments with l-diversity demonstrate a similar behavior as the experiments of k-anonymity. See [7] for a detailed report of all our experiments.

The Price of Histograms. The lattice of generalization schemes and most importantly, the histograms with which the lattice is annotated come with a price, both in terms of space and in terms of construction time. The lattice construction is negligible in terms of time; however, this does not hold for its histograms: observe that as the QI size increases, the time for the histogram construction increases exponentially (Fig. 5). The reason for this phenomenon is depicted in Fig. 5 where the number of nodes per QI size is also depicted. Although the time spent to construct the histograms is significant, the amount of memory that is needed to keep the histograms in main memory is quite small (e.g., we need approx. 3 MB for the largest QI size for k-anonymity; the value drops to approx. 1 MB for l-diversity, since the number of discrete values that the histograms take are much lower in this case).

QI size	No. nodes	Avg constr. time for k-anon (min)	Avg constr. time for l-div (min)
QI=3	60	0,141	0,1858
QI=4	180	0,602	0,702
QI=5	900	3,7	4,53
QI=6	3600	19,02	21,12

Fig. 5. Lattice size in no. nodes (left); average construction time (min.) for the full lattice and the respective histograms for k-anonymity (middle) and l-diversity (right) over the Adult data set

4   Summary and Pointers for Further Probing

In this paper, we report on our method that allows a data curator trade information loss (expressed as tuple suppression and increase in generalization levels) for privacy (expressed as the value of a privacy criterion like k-anonymity or naive l-diversity). *The full version of this paper [7] includes material that is omitted here for lack of space.* Specifically:

- We theoretically prove that *our method is guaranteed to provide the best possible answers* for the given user requests.
- We provide an extensive discussion on the validity of the problem. To the best of our knowledge, [7] is the first to perform a systematic study and report on the interdependency of suppression, generalization and privacy in a quantitative fashion.
- We provide extensive experimental results, in full detail for all the different combinations of QI size, k or l. Moreover, all the experiments have also been performed on the IPUMS data set, and the reported results demonstrate a similar behavior with the Adult data set.
- To handle the issue of scale (as the off-line lattice-and-histogram construction is dominated by both the QI size and the data size) we provide a method for the selection of a small subset of characteristic nodes of the lattice to be annotated with histograms, based on a small number of tests that rank QI levels for the grouping power.

Acknowledgments. We would like to thank X. Xiao and Y. Tao for explanations concerning the IPUMS data set.

References

1. Samarati, P.: Protecting respondents' identities in microdata release. IEEE Trans. Knowl. Data Eng. 13(6), 1010–1027 (2001)
2. LeFevre, K., DeWitt, D.J., Ramakrishnan, R.: Incognito: Efficient full-domain k-anonymity. In: Proceedings of the ACM SIGMOD International Conference on Management of Data, Baltimore, Maryland, USA, June 14-16, pp. 49–60 (2005)
3. Fung, B.C.M., Wang, K., Chen, R., Yu, P.S.: Privacy-preserving data publishing: A survey of recent developments. ACM Comput. Surv. 42(4) (2010)
4. Park, H., Shim, K.: Approximate algorithms for k-anonymity. In: Proceedings of the ACM SIGMOD International Conference on Management of Data, Beijing, China, June 12-14, pp. 67–78 (2007)
5. Aggarwal, C.C.: On k-anonymity and the curse of dimensionality. In: Proceedings of the 31st International Conference on Very Large Data Bases (VLDB), Trondheim, Norway, August 30-September 2, pp. 901–909 (2005)
6. U.C. Irvine Repository of Machine Learning Databases: Adult data set (1998), http://www.ics.uci.edu/~mlearn
7. Pilalidou, A.: On-line negotiation for privacy preserving data publishing. MSc Thesis. MT 2010-15, Dept. of Computer Science, Univ. of Ioannina (2010), http://www.cs.uoi.gr/~pvassil/publications/2012_SSDBM/

Extracting Hot Spots from Satellite Data

Hideyuki Kawashima, Chunyong Wang, and Hiroyuki Kitagawa

University of Tsukuba
Tennodai, Tsukuba, Ibaraki, Japan
{kawasima,chunyong.wang,kitagawa}@kde.cs.tsukuba.ac.jp

1   Introduction

With the advances of sensing technologies, massive amount of scientific data have been collected and analyzed in these days. LSST generates 20 TB per night, LHC generates 15 PB per year, and an earth observation satellite sensor data related to GEO-Grid [1] has generated more than 170 TB so far. ASTER sensor data are high resolution image data provided GEO-Grid.

ASTER data provides three kinds of data. One of them is thermal infrared radiometer, which is shortly denoted as TIR. Originally, TIR data are utilized to discover mineral resources and to observe status of atmosphere and the surface of ground and sea. The size of each data is almost 800×800 pixels, and each pixel covers $90\ m \times 90\ m$ of area.

Unlike the above cases, we believe that TIR can be used to detect hot spots all over the world. The meaning of "hot spots" is places that generate thermals which include steel plants or fires. This paper proposes a threshold based method and a statistics based method to discover hot spots from TIR data. We implement our methods with SciDB [2]. All of procedures in our methods are implemented by array manipulation operators which are natively supported by SciDB and UDFs. The result of experiments which detect steel plants shows that statistics based method outperforms threshold based method as for recall.

The rest of this paper is organized as follows. Section 2 describes GEO-Grid project which provides ASTER data. Section 3 describes our proposal which includes statistics and threshold based methods and then it describes a hot spot detection system with SciDB. Section 4 describes evaluation of our proposal methods. Finally, section 5 concludes this paper.

2   GEO-Grid

In this section, we describe GEO-Grid and ASTER sensor data archived in GEO-Grid. Our research group participates in GEO-Grid project. GEO-Grid (Global Earth Observation Grid) is a system for archiving and high-speed processing satellite large quantities of satellite observation data by using grid technique.

GEO-Grid introduces the design concept called VO (Virtual Organization), where necessary data or service is provided depending on a demand from a research community (e.g., disaster prevention, environmental conservation, geological research). Our research group belongs to "BigVO", in which we can get data sensed by MODIS and ASTER.

A. Ailamaki and S. Bowers (Eds.): SSDBM 2012, LNCS 7338, pp. 581–586, 2012.

2.1   MODIS

MODIS is the name of an optical sensor on NASA's earth observation satellite "TERRA/AQUA". MODIS sensor is mounted on both TERRA satellite and AQUA satellite, and the observation cycle is once a day. Spatial resolutions of MODIS are 250m (band 1, 2), 500m (band 3-7) and 1000m (band 8-36). MODIS can observe waveband of 0.4-14 μm with 36 channels. From satellite images from MODIS, cloud, radiated energy, aerosol, ground coverage, land use change, vegetation, earth surface temperature, ocean color, snow cover, temperature, humidity, sea ice, etc., can be observed with the use of 36 channels.

2.2   ASTER

ASTER is one of optical sensors on TERRA satellite, and can sense waveband from visible to thermal infrared. The observation cycle of ASTER is 16 days. ASTER consists of three independent sensors (VNIR, SWIR and TIR).

VNIR is an optical sensor which can sense reflected light of geosphere from visible to near infrared, and intended to do resource survey, national land survey, vegetation, and environment conservation. SWIR is an optical sensor which can sense short wavelength infrared region from 1.6 μm to 2.43 μm with multiple bands, and intended to conduct resource survey, environment conservation such as vegetation, volcanic action with a rock or mineral distinction. TIR is an optical sensor with five bands which can sense thermal infrared radiation on earth surface. Its sensing region is from 8 μm to 12 μm. Primary purposes of TIR are to discover distinct mineral resources or to observe air, geo sphere, or sea surface. In this paper we show that TIR data can be utilized to discover hot spots on the earth, in reality.

3   Detecting Hot Spots

3.1   Hot Spots

In this paper we refer geographical points which have higher temperatures compared with neighboring points to as "hot spots". Hot spots include steel plants, cement plants, volcanos in eruption. This paper focuses on to discover steel plants in Japan from TIR data.

3.2   Computing Radiance Temperature

To discover hot spots such as steel plants from TIR data, we first need to compute the temperature of ground surface from TIR data. It is because digital numbers in TIR data do not express surface temperature. We show equations for the computations in the following. This computation is based on Susaki's method [3]. B_λ means spectral radiance from a black body, T^* means radiance temperature, T means temperature, respectively.

$$B_\lambda = (DN - 1) \times UCC \tag{1}$$

$$T^* = \frac{hc/_{k\lambda}}{\ln\left(1/_{B_\lambda} \times 2hc^2/_{\lambda^5} + 1\right)} \qquad (2)$$

$$T = T^* - 273.15 \qquad (3)$$

TIR data includes five bands. They are from band 10 to band 14. Each pixel of data has five values which correspond to the five bands. Since band 13 has the best quality, we use only the values of band 13. *DN* means digital number which is stored in each pixel of array data. *UCC* means a conversion factor, and it is set to 0,005693. *h* means Planck's constant, and it is set to 6.626×10^{-34}. *c* means light speed and it is set to 2.988×10^8. *k* means Boltzmann's constant and it is set to 1.38×10^{-23}. λ means wave length. Band 13 provides wave length from 10.25 to 10.95, and we use intermediate wavelength 10.60.

3.3 Threshold Based Method

Hot spots should have high temperatures by its definition. Based on this simple intuition, we present Algorithm 1. Inputs include point information (P_i) which has latitude (lat), longitude (lon), and digital number (D_{13i}). Inputs also include threshold which is given by user. Output includes a set of hot spots. Each hot spot should have point information. The algorithm is quite simple. It first translates digital number (DN) to temperature data (line 3—6). Then, if an array has a pixel of which value which exceeds threshold, then the pixel is added to the set of hot spots.

```
1.     INPUT:   InputSet : {{Pos_i,DN_13i},...}, Threshold
2.     OUTPUT:  HotSpots : {Pos_i,...}
3.
4.     for all {Pos_i,DN_13i} ∈ InputSet do
5.         Compute Temp_13i from DN_13i by equations 1-3;
6.         Add Temp_13i to (Pos_i,DN_13i) in InputSet;
7.     end
8.     for all {Pos_i,DN_13i,Temp_13i} ∈ InputSet do
9.         if (Temp_13i > Threshold) then
10.            Add Pos_i to HotSpots;
11.        end
12.    end
13.    return HotSpots;
```

Algorithm 1. Threshold based Method

3.4 Statistics Based Method

Normal distributions are widely seen in natural and social phenomenon. On normal distribution, it is widely known that mean plus or minus 3σ is rare. We focus attention on this feature, and apply it to hot spot detection. We show the statistics based method on Algorithm 2. In the algorithm, "$k\sigma$" means a parameter given by a user. When k is 3, the algorithm detects points with temperature which is more than $\mu + 3\sigma$ as hot spots. It should be noted that this algorithm deletes points of which ground height is zero since it should be in the sea.

```
1.    INPUT:   InputSet: {{Pos_i, DN_13i, Elv_i},…}, kσ
2.    OUTPUT:  HotSpots: {Pos_i,…}
3.
4.    TempSet ← φ;
5.    for all {Pos_i,DN_13i,Elv_i} ∈ InputSet do
6.        Compute Temp_13i from DN_13i by equations 1-3;
7.        Add Temp_13i to {Pos_i,DN_13i,Elv_i}
8.        if (Elv_i ≠ 0 and Temp_13i > 0) then
9.            Add {Pos_i,DN_13i,Elv_i,Temp_13i} to TempSet;
10.       end
11.   end
12.   Compute μ and σ with all of Temp_13i in TempSet;
13.   for all {Pos_i,DN_13i,Elv_i,Temp_13i} ∈ TempSet do
14.       if (Temp_13i > μ+kσ) then
15.           Add Pos_i to HotSpots;
16.       end
17.   end
18.   return HotSpots;
```

Algorithm 2. Statistics based Method

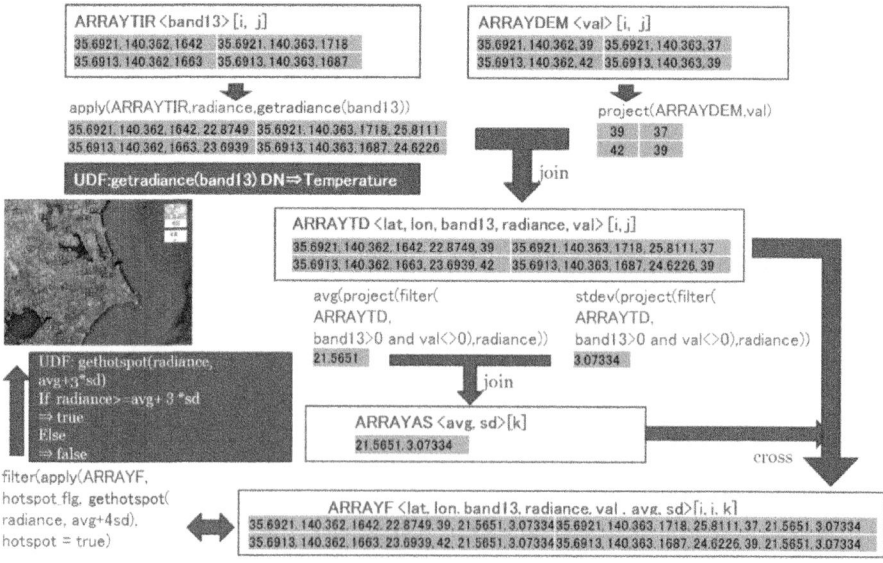

Fig. 1. Proposed Hot Spot Detection System

3.5   Implementation of Hot Spot Detections

We implemented the above two algorithms on SciDB [2] array database system as shown in Figure 1. The reason why we adopted SciDB is its data model. Array data model is appropriate to manipulate satellite image data compared with relational data model.

Red characters mean operations: apply, project, join, avg, stddev, cross, and filter are supported by SciDB. Brown characters mean UDF: getradiance returns radiance by computing TIR data (band13) with equations in Section 2.1, and gethotspot returns the locations of hotspots with Algorithm 1 and 2.

4   Evaluation

We used TIR data for eight famous steel plants in Japan. We show the details in Table 1. We used 7 to 14 TIR data for each plant from 2001 to 2011.

Table 1. Target Steel Plants for Experiment

Steel Plants	# data	Observation duration
Kashima	9	2002—2011
Sumitomo Yahata	9	2000—2011
Sumitomo Metal Engineering (Wakayama)	7	2002—2011
Nippon Steel Corporation (Nagoya)	9	2000—2011
Kobe Steel Group	8	2002—2011
JFE East Japan	14	2000—2011
JFE West Japan	12	2001—2009
Nisshin Steel (Kure)	9	2002—2011

We show the result of experiments in Table 2.Threshold-based method has three patterns while statistics based method has two patterns. As for precision, threshold based method shows better result while statistics based method shows dramatically better result as for recall.

One of the reasons why recall shows low value is considered as seasonal change of temperatures. We observed temperature data around JFE East Japan steel plant from 2004 to 2006. The temperature shows more than 50 degree at July 5th, 2004 while it shows 23 degrees at December 27th , 2006. Statistics based method (algorithm 2) detects points with relatively high temperatures while threshold based method (algorithm 1) detects all of points which have higher temperatures. Therefore threshold based method shows lower recalls, especially TH=60.

Table 2. Result of Hot Spot Detection Experiment

	Threshold-based			Statistics-based	
	TH=60	TH=45	TH=30	3σ	4σ
Precision	1.000	0.835	0.701	0.684	0.832
Recall	0.052	0.195	0.597	0.935	0.870

5   Conclusions and Future Work

This paper described our simple algorithms to detect hot spots with ASTER TIR data and a system for the detection with SciDB. Result of experiments that detect Japanese steel plants showed that statistical based method outperformed threshold based method as for recall.

In future work, we plan to develop more sophisticated detection algorithms and its scalable extension for massive satellite image data, and we also plan to detect outlier events such as natural fire. In addition, we plan to construct an event detection infrastructure from satellite images including digital elevation model (DEM) [4].

Acknowledgement. This work is partially supported by KAKENHI (#20240010, # 22700090). This research used ASTER data β. The data was generated by GEO-Grid managed by The National Institute of Advanced Industrial Science and Technology (AIST). Original ASTER data are managed by Ministry of Economy, Trade and Industry (METI).

References

[1] GeoGRID, http://www.geogrid.org
[2] SciDB, http://www.scidb.org
[3] Susaki, J.: Remote Sensing Image Data Processing by GRASS (in Japanese), http://www.envinfo.uee.kyoto-u.ac.jp/user/susaki/grass/grass3.html
[4] Takagi, T., Kawashima, H., Amagasa, T., Kitagawa, H.: Providing constructed buildings information by ASTER satellite DEM images and web contents. In: Proc. DIEW Workshop in Conjunction with DASFAA (2010)

A Framework for Enabling Query Rewrites when Analyzing Workflow Records*

Dritan Bleco and Yannis Kotidis

Athens University of Economics and Business
76 Patission Street, Athens, Greece
{dritanbleco,kotidis}@aueb.gr

Abstract. Workflow records are naturally depicted using a graph model in which service points are denoted as nodes in the graph and edges portray the flow of processing. In this paper we consider the problem of enabling aggregation queries over large-scale graph datasets consisting of millions of workflow records. We discuss how to decompose complex, ad-hoc aggregations on graph workflow records into smaller, independent computations via proper query rewriting. Our framework allows reuse of precomputed materialized query views during query evaluation and, thus, enables view selection decisions that are of immense value in optimizing heavy analytical workloads.

1 Introduction

A Workflow Management application in a Customer Support Call Center manages massive collections of different trouble (issue) tickets generated daily by an Issue Tracking System (ITS). Such an application tracks all activities from the creation of the trouble ticket till its completion. A trouble ticket is commonly composed by different flows of tasks that may be serviced in parallel or sequentially by distinct agencies within the company's domain called service points.

A natural way to depict the workflow followed by a trouble ticket is to utilize a graph model in which service points are denoted as nodes and edges depict the flow of processing. Figure 1 depicts the graph instance associated with a particular trouble ticket. Numeric labels on the nodes depict processing time (in days) within the service point. The values on the edges depict delays for propagating the associated task(s) from one service point to another (e.g. handoffs latencies or wait times).

In this paper we consider the problem of enabling analytical queries over large-scale graph datasets related to workflow management applications such as those serving ITS. We describe a comprehensive framework for modeling analytical queries that range over the structure of the graph records. For example, queries like "find the average ticket completion time" in an ITS application are naturally captured by our framework.

As will be explained, in our framework we decompose complex, ad-hoc aggregations on graph workflow records into smaller, independent computations via proper query rewriting. In the context of a large-scale data warehouse, our framework allows re-use

---

* This work was partially supported by the Basic Research Funding Program, Athens University of Economics and Business.

A. Ailamaki and S. Bowers (Eds.): SSDBM 2012, LNCS 7338, pp. 587–590, 2012.

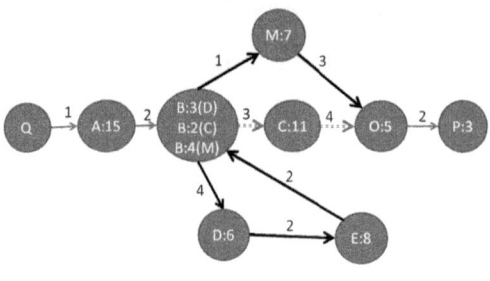

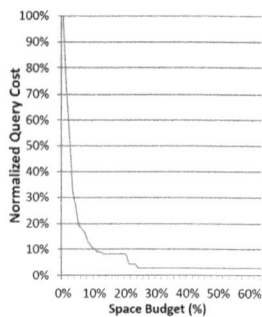

Fig. 1. A sample workflow record

Fig. 2. Benefits of Materialization

of precomputed materialized query views during query evaluation and, thus, enables view selection decisions [1]. Our framework is not an attempt to provide a new algebra for graph analytics. All computations discussed in our work, can be naturally expressed in relational algebra, given a decomposition of the graph records in a relational back-end. Our query rewriting framework allows us to optimize ad-hoc aggregations over workflow records by utilizing precomputed views, independently of the technological platform used for storing and querying such records. In our experimental evaluation, we demonstrate that our techniques can be used to expedite costly queries via query rewriting and available precomputations in the data warehouse.

2   Motivation and Basic Concepts

Figure 1 captures information related to a single trouble ticket workflow that we will refer to as a *record* henceforth. Service point B is a special type of node called *splitting node*, where processing of the ticket is split between two tasks that are processed in parallel following different *flows* in particular $[B, C, O]$ (or $[BCO]$ for brevity) and $[BDEBMO]$. These independent flows are merged (synchronized) at node O, which is a *merging node*. From there, the ticket is moved to node P where it is marked as completed. There can be recordings of different cost metrics, in addition to the time we consider in this example. A decision maker would like to analyze data according to all these attributes over different parts of the graph for thousands of such records. In what follows we describe a formal way to analyze such data.

In a workflow management application a flow is simply a sequence of nodes resulting from the concatenation of adjacent edges. When a flow is uniquely identified by its endpoints, for brevity, we omit the internal nodes. For example flow $[ABC]$ is depicted as $[AC]$. In the record of Figure 1, flow $[BDEBMO]$ contains a cycle, $[BDEB]$ in our case. This cycle shows that this flow, for some reason, goes back to node B for further proccessing and after that to node M. On each node there is a cost related with each input/output edge. For example, to proceed from B to M the flow was processed locally on B for four days.

When we analyze the costs on a flow in a trouble ticket often we want to omit the costs on the two side nodes. Borrowing notation from mathematics, we denote this "open-ended" flow similarly to an interval whose endpoints are excluded. For example (BCO) denotes our intention to look at the processing of the flow excluding cost related to its starting and ending points. Similarly, a flow can be opened in only one of its side nodes, i.e. $[BO)$.

In a large ITS application, a lot of tickets are processed daily, creating a massive collection of such data. Given these primitive data an analyst should be able to answer queries like

- Q_1: What is the total wall-clock time for each trouble ticket from its initialization till its completion?
- Q_2: What is the total waiting time at a certain merging node?
- Q_3: What are the total processing time and total delay time?
- Q_4: What is the wasting time due to a non approved task (a circle during the flow or a flag over a node depicts such a situation)?

These queries will be executed over a large collection of tickets. Primitive statistics computed at a per-ticket basis can then be combined to compute aggregate statistcs like average completion time per type of ticket etc. Query Q_1 in this example requires us to find the *longest flow* between nodes Q and P. During the computation, attention should be paid to merging nodes that synchronize execution of parallel flows. In our running example two flows reach merging node O; the fastest one waits for the other flow. For the sample record of Figure 1 the longest flow is $[QABDEBMOP]$ with a total time of 68 days. In this example, query Q_1 spans over the whole record. Depending on the scenario being analyzed, a user may restrict the analysis over parts of the input records (e.g. between two specific service points). Query Q_2 calculates the *longest and shortest flow* among tasks starting at splitting node B (including the cost(s) on this node) and ending at the merging node O. The waiting time is the difference between the returned values of the two subqueries. For our example this is 40-20=20 days. Query Q_3 sums up the measures on the edges (the delay time) and the values on the nodes (the processing time). In this record, the processing is 64 days and delay time 24 days. Query Q_4 first needs to identify the non approved - circles on the ticket and then sum up the total time of the flow related to these circles. For the depicted record the wasting time is related with the cirlce B, D, E, B and equals to 25. Obviously the cost of the second proccess on B is not included because this value is related with the new flow from B to M.

Our framework allows us to model such queries in an intuitive manner via a decomposition of a record into flows and the use of two basic operators we introduce next.

3   Query Rewriting and Evaluation

In order to allow composition of flows we introduce the *merge-join* operator ($\triangleright$) that concatenates two flows f_1 and f_2 that run in parallel when they have the same starting and ending nodes. No merging node should be present in the two flows except the starting or/and the ending node. For flows that are running sequentially, we use a second operator called *union* operator ($\oplus$) that concatenates two flows f_1 and f_2 when the ending node of f_1 is the same as the starting node of f_2 and one of the two flows is open-ended at the common end-point. Using the two operators the ticket depicted in Figure 1 can be rewritten as $[QP]= [QAB) \oplus \{[BDEBMO] \triangleright [BCO]\} \oplus (OP]$.

To compute different statistics on these measures we can use a *Flow Level Aggregation* function $F_f(r)$ which takes as input a flow f and a record r. The function F is applied on the measures of the flow and returns f along with the computed aggregate. As an example, in the record r of Figure 1, $SUM_{[QABD)}(r)$ returns flow $[QABD)$ and its duration, i.e. 25 days (denoted as $[QABD):25$). In case more than one flows are given, the function is computed over the (existing) individual flows and the result is returned along with each respective flow.

In a subsequent step a *Flow Set Level Aggregation* function can be used in order to aggregate the results of the previous step and to return a unique value for a set of flows. As an example, function $MAX(SUM_{[QP]}(r))$ computes the total wall-clock time for the ticket depicted in record r along with the longest flow. The aggregate function over a union among two or more flows can be written also as a union among the aggregate function results of these flows. The aggregate function over the merge join operator among flows can be written as the aggregate function over the set of flows that were related with the merge join. Thus, for our record r we have

$$MAX(SUM_{[QAB)\oplus\{[BDEBMO]\triangleright[BCO]\}\oplus(OP]}(r)) =$$

$$MAX(SUM_{[QAB)}(r) \oplus_{SUM} SUM_{\{[BDEBMO],[BCO]\}}(r) \oplus_{SUM} SUM_{(OP]}(r))$$

The union operator concatenates flows with common ending and starting nodes and, additionally consolidates their measures. In this example, we need to add their measures, and this is indicated with the use of function SUM underneath the operator. In general, the rewrites for pushing flow level aggregation (for the union and merge-join respectively) on a flow are of the form (F,H are appropriate aggregate functions)

$$F_{f=f_1\oplus f_2}(r) = F_{f_1}(r) \oplus_H F_{f_2}(r) \text{ and } F_{f=f_1\triangleright f_2}(r) = F_{f_1,f_2}(r) = \{F_{f_1}(r), F_{f_2}(r)\}$$

Furthermore the Flow Set Level Aggregation can be pushed inside each Flow Level Aggregation Function and can be omitted in case the latter is executed over a simple flow. So continuing the above example we have $MAX(SUM_{[QP]}(r))=[QAB)$: $18 \oplus_{SUM} [BDEBMO] : 45 \oplus_{SUM} (OP] : 5=[QABDEBMOP] : 68$.

In order to demonstrate the effectiveness of our rewrite techniques, we synthesized a random graph consisting of 10000 nodes and 15000 edges and generated 120 million workflow records by selecting random subgraphs from it. The records were stored in a relational backend using a single table storing their edges and appropriate indexes. We created 100 random queries on these records. Half of the queries used the SUM flow level aggregation function. The rest performed, additionally, the MAX flow set level aggregation. We used the Pick By Size (PBS) algorithm [2] for selecting materialized views. In Figure 2 we depict the reduction in query execution cost (compared to running these queries without rewrites) with respect to the available space budget of the views. The results demonstrate that even a modest materialization of 10%, via the use of our rewrites provides substantial savings (up to 90%).

References

1. Kotidis, Y., Roussopoulos, N.: A Case for Dynamic View Management. ACM Transactions on Database Systems 26(4) (2001)
2. Shukla, A., Deshpande, P., Naughton, J.F.: Materialized View Selection for Multidimensional Datasets. In: VLDB, pp. 488–499 (1998)

Towards Enabling Outlier Detection in Large, High Dimensional Data Warehouses*

Konstantinos Georgoulas and Yannis Kotidis

Athens University of Economics and Business
76 Patission Street, Athens, Greece
{kgeorgou,kotidis}@aueb.gr

Abstract. In this work we present a novel framework that permits us to detect outliers in a data warehouse. We extend the commonly used definition of distance-based outliers in order to cope with the large data domains that are typical in dimensional modeling of OLAP datasets. Our techniques utilize a two-level indexing scheme. The first level is based on Locality Sensitivity Hashing (LSH) and allows us to replace range searching, which is very inefficient in high dimensional spaces, with approximate nearest neighbor computations in an intuitive manner. The second level utilizes the Piece-wise Aggregate Approximation (PAA) technique, which substantially reduces the space required for storing the data representations. As will be explained, our method permits incremental updates on the data representation used, which is essential for managing voluminous datasets common in data warehousing applications.

1 Introduction

Assuring quality of data is a fundamental task in information management. It becomes even more critical in decision making applications, where erroneous data can mislead to disastrous reactions. The data warehouse is the cornerstone of an organization's information infrastructure related to decision support. The information manipulated within a data warehouse can be used by a company or organization to generate greater understanding of their customers, services and processes. Thus, it is desirable to provision for tools and techniques that will detect and address potential data quality problems in the data warehouse.

It is estimated that as high as 75% of the effort spent on building a data warehouse can be attributed to back-end issues, such as readying the data and transporting it into the data warehouse [1]. This is part of the Extract Transform Load (ETL) processes, that extract information pieces from available sources to a staging area, where data is processed before it is eventually loaded in the data warehouse local tables. Processing at the data staging area includes cleansing, transformation, migration, scrubbing, fusion with other data sources etc.

In this paper, we propose a novel framework for identifying outliers in a data warehouse. Outliers are commonly defined as rare or atypical data objects that do not behave

---

* This work was partially supported by the Basic Research Funding Program, Athens University of Economics and Business.

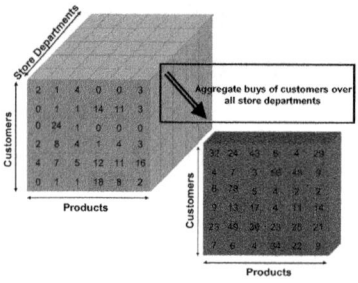

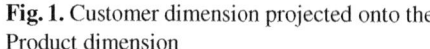

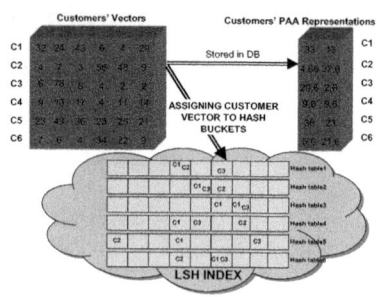

Fig. 1. Customer dimension projected onto the Product dimension

Fig. 2. Framework's Overview

like the rest of the data. Often, erroneous data points appear as outliers when projected on a properly derived feature space. In our work, we exploit the dimensional modeling used in a data warehouse and let the user examine the data under selected dimensions of interest. This way, our definition of what constitutes an outlier has a natural interpretation for the policy makers that interact with this data. Moreover, our techniques are tailored for the massive and periodic schedule of updates that occur with each ETL process. Clearly, techniques that require substantial pre- or post- processing of data are not suitable for handling massive datasets such as those in a data warehouse.

2   A Framework for Detecting Outliers in a Data Warehouse

Given a data warehouse with multiple dimensions $d_1, d_2, \ldots, d_n$, each organized by different hierarchy levels h_k the data warehouse administrator may select a pair $(d_{aggr}, h_{aggr\_level})$ so as to define the requested aggregation and, similarly, a pair $(d_{proj}, h_{proj\_level})$ in order to denote the space that these aggregates should be projected upon. For instance the aggregate dimension can be customer at the hierarchy level of customer-type and the projected dimension product at the product-brand level. These pairs indicate our intention to compare different customer types based on cumulative sales of the brand of products they buy in order to search for outliers. Clearly, a data warehouse administrator may define multiple such pairs of dimensions in order to test the data for outliers. An example presented in Figure 1 where the customer (aggregate) dimension is projected onto the product dimension. This projection leads to a high dimensional vector for each customer that summarizes all his buys over the whole list of products.

An $O(D, M)$ distance based outlier is defined [2] as a data item O in a dataset with fewer than M data items within distance D from O. The definition, in our domain, suggests that range queries need to be executed in the data space defined by the projected dimension in order to compute the number of data items that lay inside a range of D from item O. As has been explained, the projected space can have very high dimensionality (i.e. equal to the number of all products in the data warehouse, which is in the order of thousands), which renders most multidimensional indexing techniques ineffectual, due to the well documented curse of dimensionality [3].

In order to address the need to compare data items on a high-dimensionality space when looking for outliers, we adapt a powerful dimensionality reduction technique called LSH [4]. LSH generates an indexing structure by evaluating multiple hashing functions over each data item (the resulting vector when projecting a customer on the space of products she buys). Using the LSH index, we can estimate the k nearest neighbors of each customer and compute outliers based on the distances of each customer from its k neighbors. We thus propose an adapted approximate evaluation of distance-based outliers that treats a data item O as an outlier if less than M of its k nearest neighbors are within distance M from O. Please notice that this alternative evaluation permits us to utilize the LSH index for a k-NN query (with $k > M$) and restrict the range query on the k results retrieved from the index. Thus, the use of the LSH index permits effective evaluation of outliers, however it introduces an approximation error, because of collisions introduced by the hashing functions. There have been many proposals on how to tune and increase performance of LSH (e.g. [5,6]), however such techniques are orthogonal to the work we present here.

The use of LSH enables computation of outliers by addressing the curse of dimensionality. Still, an effective outlier detection framework needs to address the extremely high space required for storing the resulting data vectors. The size of these vectors is proportional to the size of a data cube slice on the selected pair of dimensions. Moreover, these vectors need to be updated whenever the data warehouse is updated with new data. We address both these issues (space overhead, update cost) using the PAA representation instead of the original vectors. Utilizing PAA, we store vectors of lower dimensionality than the real ones, thus gaining in space. We can also compute the distances between each data item and its nearest neighbours through their PAA representations much faster than using the real data items without losing too much in accuracy as we will show in our experimental evaluation. PAA [7] represents a data item of length n in $\mathbb{R}^N$ space (where $n > N$) achieving a dimensionality reduction ratio $N{:}n$. Given that each data item X is a vector with coordinates $x_1, .., x_n$, its new representation will be a new vector $\bar{X}$ of length N and coordinates the mean values of the N equisized fragments of vector X. So according to PAA a vector $X = x_1, .., x_n$ is represented by

$$\bar{X} = \bar{x}_1, .., \bar{x}_N \text{ where } \bar{x}_i = \frac{N}{n} \sum_{j=\frac{n}{N}(i-1)+1}^{\frac{n}{N}i} x_j.$$

Beside the space savings provided by PAA, its adaptation has another important advantage in our application. Because of its definition, PAA representations are linear projections that permit incremental updates whenever new data arrives at the data warehouse. Let PAA_{old} denote the PAA vector of a customer and PAA_{delta} the PAA representation of the customer's buys in the newly acquired updates. Then, in order to compute the new representation PAA_{new} for this customer we can simply add the two vectors, i.e. $PAA_{new}{=}PAA_{old}{+}PAA_{delta}$. This property is vital for data warehouses, where incremental updates are of paramount importance [8].

3 Experiments and Concluding Remarks

In our experimental evaluation, we used a clustered synthetic dataset that represents orders of 10,000 customers over a list of 1200 products. Each cluster contains customers

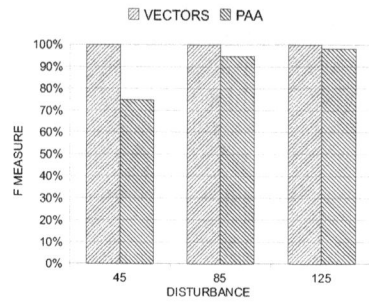

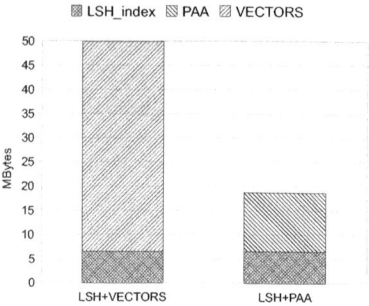

Fig. 3. F-Measure

Fig. 4. Space for the LSH Index, the Vectors or PAAs

with similar behavior. In particular, the customers within each cluster have a randomly (pre)selected set of "hot" products that represent 20% of the whole product list. For a customer in the cluster, 80% of her orders are on that 20% subset of products. Different clusters have different set of hot products. The frequencies of customers' orders follow the normal distribution with different mean values per cluster. In order to evaluate the performance of our method we injected in the dataset outliers in the form of spurious orders. We created three infected datasets. In the first one, the spurious orders add low disturbance (measured by the number of spurious orders) to the original data. while in the second medium and in the third large. In Figure 3 we depict the f-measure ($\frac{2 \times recall \times precision}{recall + precision}$) in detecting the injecting outliers for the tree datasets. We compare two variants. The first uses the LSH index and the data vectors generated by projecting the customers on the product dimension. The second setup, instead of the original vectors, it only stores their PAA representations of one quarter of the original vector length. In the Figure we observe that the accuracy of the PAA method is quite similar to the other method that stores the actual vectors, while the storage of PAA is much smaller, as it shown in Figure 4. Moreover, the size of the LSH index is very small, while its accuracy in computing distance-based outliers is at least 98%.

References

1. Kimball, R.: The Data Warehouse Toolkit. John Wiley & Sons (1996)
2. Subramaniam, S., Palpanas, T., Papadopoulos, D., Kalogeraki, V., Gunopulos, D.: Online Outlier Detection in Sensor Data Using Non-Parametric Models. In: VLDB (2006)
3. Korn, F., Pagel, B.U., Faloutsos, C.: On the 'Dimensionality Curse' and the 'Self-Similarity Blessing'. IEEE Trans. Knowl. Data Eng. 13(1), 96–111 (2001)
4. Andoni, A., Indyk, P.: Near-optimal hashing algorithms for approximate nearest neighbor in high dimensions. In: FOCS, pp. 459–468. IEEE Computer Society (2006)
5. Lv, Q., Josephson, W., Wang, Z., Charikar, M., Li, K.: Multi-Probe LSH: Efficient Indexing for High-Dimensional Similarity Search. In: VLDB, pp. 950–961 (2007)
6. Georgoulas, K., Kotidis, Y.: Distributed Similarity Estimation using Derived Dimensions. VLDB J. 21(1), 25–50 (2012)
7. Keogh, E.J., Chakrabarti, K., Pazzani, M.J., Mehrotra, S.: Dimensionality Reduction for Fast Similarity Search in Large Time Series Databases. Knowl. Inf. Syst. 3(3), 263–286 (2001)
8. Roussopoulos, N., Kotidis, Y., Roussopoulos, M.: Cubetree: Organization of and Bulk Updates on the Data Cube. In: SIGMOD Conference, pp. 89–99 (1997)

Multiplexing Trajectories of Moving Objects

Kostas Patroumpas[1], Kyriakos Toumbas[1], and Timos Sellis[1,2]

[1] School of Electrical and Computer Engineering
National Technical University of Athens, Hellas
[2] Institute for the Management of Information Systems, R.C. "Athena", Hellas
{kpatro,timos}@dbnet.ece.ntua.gr, toumbask@gmail.com

Abstract. Continuously tracking mobility of humans, vehicles or merchandise not only provides streaming, real-time information about their current whereabouts, but can also progressively assemble historical traces, i.e., their evolving trajectories. In this paper, we outline a framework for online detection of groups of moving objects with approximately similar routes over the recent past. Further, we propose an encoding scheme for synthesizing an indicative trajectory that collectively represents movement features pertaining to objects in the same group. Preliminary experimentation with this multiplexing scheme shows encouraging results in terms of both maintenance cost and compression accuracy.

1 Motivation

As smartphones and GPS-enabled devices proliferate and location-based services penetrate into the market, managing the bulk of rapidly accumulating traces of objects' movement becomes all the more crucial for monitoring applications. Apart from effective storage and timely response to user requests, data exploration and trend discovery against collections of evolving trajectories seems very challenging. From detection of flocks [9] or convoys [4] in fleet management, to similarity joins [1] for car-pooling services, or even to identification of frequently followed routes [2,8] for traffic control, the prospects are enormous.

We have begun developing a stream-based framework for *multiplexing trajectories* of objects that approximately travel together over a recent time interval. Our perception is that a symbolic encoding for sequences of trajectory segments can offer a rough, yet succinct abstraction of their concurrent evolution. Taking advantage of inherent properties, such as heading, speed and current position, we can continuously report groups of objects with similar motion traces. Then, we may regularly construct an indicative path per detected group, which actually epitomizes spatiotemporal features shared by its participating objects.

Overall, such a scheme could be beneficial for:

- *Data compression*: collectively represent traces of multiple objects with a single "delegate" that suitably approximates their common recent movement.
- *Data discovery*: find trends or motion patterns from real-time location feeds.
- *Data visualization*: estimate significance of each multiplexed group of trajectories and illustrate its mutability across time (e.g., on maps).

A. Ailamaki and S. Bowers (Eds.): SSDBM 2012, LNCS 7338, pp. 595–600, 2012.

– *Query processing*: utilize multiplexed traces for filtering when it comes to evaluation of diverse queries (range, k-NN, aggregates etc.) over trajectories.

We believe that our ongoing work fuses ideas from trajectory clustering [5] and path simplification [7], but proceeds even further beyond. Operating in a geostreaming context, not only can we identify important motion patterns in online fashion, but we may also provide concise summaries without resorting to sophisticated spatiotemporal indexing. Symbolic representation of routes was first proposed in [1] for filtering against trajectory databases. Yet, our encoding differs substantially, as it attempts to capture evolving spatiotemporal vectors using a versatile alphabet of tunable object headings instead of simply compiling timestamped positions in a discretized space. Finally, this scheme may be utilized in applications that handle motion data (navigation, biodiversity, radar etc.).

The remainder of this paper is organized as follows. In Section 2, we introduce a framework for multiplexing evolving trajectories in real time and explain the basic principles behind our encoding scheme. In Section 3, we report indicative performance results from a preliminary experimental validation of the algorithm. Section 4 concludes the paper with a brief discussion of perspectives and open issues for further investigation.

2   A Multiplexing Framework against Trajectory Streams

In this section, we first present the specifications of the problem and then outline a methodology for multiplexing similar trajectories, which effectively provides almost instant, yet approximate results.

2.1   Problem Formulation

Without loss of generality, trajectory T_o is abstracted as a sequence of pairs $\{\langle p_1, \tau_1 \rangle, \langle p_2, \tau_2 \rangle, \ldots, \langle p_{\text{now}}, \tau_{\text{now}} \rangle\}$ for a given moving object o. Positions $p_k \in \mathbb{R}^d$ in Euclidean space have d-dimensional coordinates measured at discrete, totally ordered timestamps $\tau_k \in \mathbb{T}$, hence $o(\tau_k) \equiv p_k$. Note that $\mathbb{T}$ is regarded as an infinite set of discrete time instants with a total order $\leq$. Then:

Definition 1. *Trajectories of two objects o_i and o_j are considered similar along interval ω up to current time $\tau_{now} \in \mathbb{T}$, iff $L_2(o_i(t), o_j(t)) \leq \epsilon, \forall\, t \in (\tau_{now} - \omega, \tau_{now}]$, where ϵ is a given tolerance parameter and L_2 the Euclidean distance norm.*

Hence, pairs of concurrently recorded locations from each object should not deviate more than ϵ during interval ω. This notion of similarity is confined within the recent past and does not extend over the entire history of movement. However, it can be easily generalized for multiple objects with pairwise similar trajectory segments (Fig. 1a). Given specifications for proximity in space (within distance ϵ) and simultaneity in time (over range ω), our objective is not just to identify such groups of trajectories, but also to incrementally refresh them periodically (every β time units) adhering to the *sliding window* paradigm [6]. More concretely, a framework for online trajectory multiplexing must:

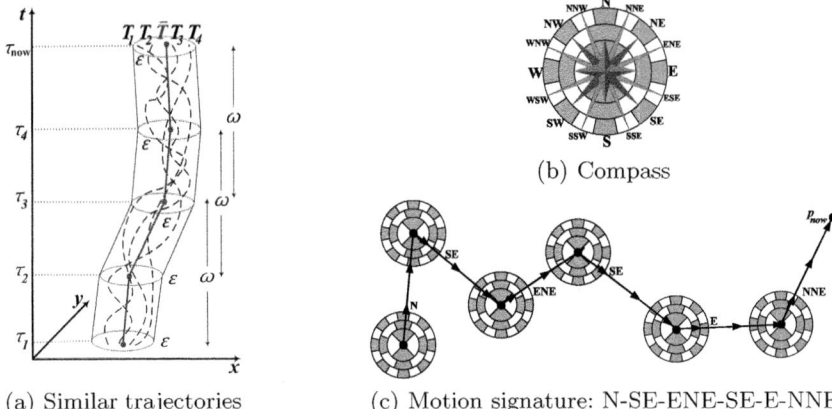

(a) Similar trajectories   (b) Compass

(c) Motion signature: N-SE-ENE-SE-E-NNE

Fig. 1. Orientation-based encoding of streaming trajectories

(i) distinguish objects into groups $\{g_1, g_2, \ldots\}$, each containing synchronized, pairwise similar trajectories during interval ω given a tolerance ϵ.

(ii) create an indicative *"delegate"* trajectory $\bar{T}_k$ for each group g_k with more than n members. For any sample point $\bar{o}(t) \in \bar{T}_k, \forall\, t \in (\tau_{\text{now}} - \omega, \tau_{\text{now}}]$, it holds that $L_2(o_i(t), \bar{o}(t)) \le \epsilon, \forall\, o_i \in g_k$.

(iii) insert, remove or adjust groups regularly (at execution cycles with period β) in order to reflect changes in objects' movement.

2.2   Trajectory Encoding

Checking similarity of trajectory segments according to their timestamped positions soon becomes a bottleneck for escalating numbers of moving objects or wider window ranges. To avoid this, we opt for an approximative representation of traces based on consecutive velocity vectors that end up at the current location of each respective object (Fig. 1c). Every vector is characterized by a symbol that signifies the orientation of movement using the familiar notion of *compass* (Fig. 1b, for movement in $d = 2$ dimensions), which roughly exemplifies an object's course between successive position messages.

Effectively, compass resolution α determines the degree of motion smoothing; when $\alpha = 4$, orientation symbols $\{N, S, E, W\}$ offer just a coarse indication, but finer representations are possible with $\alpha = 16$ symbols (Fig. 1c) or more. Instead of original positions, only the last $\lceil \frac{\omega}{\beta} \rceil$ symbols and speed measures need be maintained per trajectory thanks to the sliding window model, thus offering substantial memory savings. Typically, once the window slides at the next execution cycle, an additional symbol (marking motion during the latest β timestamps) will be appended at the tail of this FIFO sequence, while the oldest one (i.e., at the head) gets discarded.

2.3   Group Detection

Symbolic sequences are more amenable to similarity checks since they act as *motion signatures*. Presently, we identify objects with common signatures through a hash table. Objects with identical symbolic sequences might have almost "parallel" courses, but can actually be very distant from each other. So, the crux of our approach is this:

Proposition 1. *Objects with identical signatures that are currently within ϵ distance from each other, most probably have followed similar paths recently.*

Therefore, identifying groups of at least n objects with a common signature can be performed against their current locations p_{now} through a point clustering technique. We provisionally make use of DBSCAN [3] to detect groups of similar trajectories with proximal current positions. Afterwards, a delegate path $\bar{T}_k$ per discovered cluster with sufficient membership ($\geq n$ objects) can be easily created. Reconstruction of $\bar{T}_k$ starts by calculating centroid $\bar{o}(\tau_{now})$ of its constituent p_{now} locations; apparently, this will be the point at the tail of the sequence. In turn, preceding points are derived after successively averaging respective speeds retained in participating motion signatures. This can be easily accomplished by simply rewinding the symbolic sequence backwards up to its head, i.e., reversely visiting all samples within the sliding window frame.

3   Preliminary Evaluation

To assess the potential of our framework for data reduction and timely detection of trends, we have conducted some preliminary simulations against synthetic trajectories. Next, we present the experimental settings and we discuss some indicative results concerning performance and approximation quality.

Experimental Setup. We generated traces of 10 000 vehicles circulating at diverse speeds along the road network of greater Athens (area $\sim$250 km^2). After calculating shortest paths between randomly chosen network nodes (i.e., origin and destination of objects), we took point samples at 200 concurrent timestamps along each such route. Typically, most trajectories originate from the outskirts of the city, pass through the center and finish up in another suburb.

Table 1. Experiment parameters

Parameter	Values		
Number N of objects	**10 000**		
Window range ω (in timestamps)	10	**20**	50
Window slide β (in timestamps)	2	**4**	10
Tolerance ϵ (in meters)	**100**, 200		
Cluster threshold n	10, **20**, 50, 100		
Compass resolution α	8, **16**, 32		

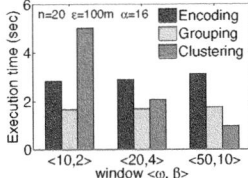 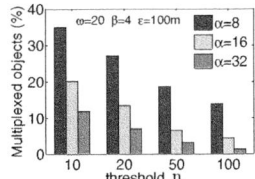 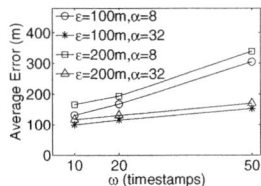

Fig. 2. Per stage cost   **Fig. 3.** Multiplex effect   **Fig. 4.** Result quality

Algorithms have been implemented in C++ and experiments were performed on a conventional laptop machine with an Intel Core 2 Duo CPU at 2.4 GHz and 4GB RAM. All figures show calculated averages of the measured quantities over 200 time units. Table 1 summarizes experimentation parameters and their respective ranges; the default value is shown in bold.

Experimental Results. Admittedly, the proposed algorithm performs a lossy approximation susceptible to errors. The primary causes are crude compass resolutions or speed variations amongst objects placed in the same group. By and large, our empirical validation confirms this intuition, and also expectations for prompt detection of groups.

As Fig. 2 indicates, execution cost per cycle fulfils real-time objectives for varying window specifications, with almost stable overhead for encoding and grouping. Note that the encoding phase is only marginally affected when increasing the window range, although more positions per object must be handled at each execution cycle (every β timestamps). The cost of grouping remains practically stable, as it basically depends on the total count N of monitored objects. In contrast, the clustering overhead fluctuates, but drops sharply for wider windows as less objects tend to share motion signatures for too long. For shorter ω, more objects appear to move together lately and thus create more candidate groups; accordingly, the clustering cost escalates as it requires distance calculations among all members of each group.

Our next experiment attempts to appraise how effective this method is. Figure 3 plots the multiplexing degree, i.e., the fraction of objects assigned into identified groups of sufficient size. Clearly, distinguishing important trends is sensitive to threshold n. In case that membership into a group falls below limit n, its trajectories are not multiplexed at all. Besides, approximation gets more pronounced with a coarser resolution α. But for finer resolutions, less trajectory matchings are identified, as objects tend to retain particular features of their course and cannot easily fit into larger groups.

Still, average error between a delegate path and its contributing trajectories is tolerable (Fig. 4), especially for less smoothed signatures (larger α). When probing longer intervals, this deviation may well exceed the desired ϵ. This phenomenon must be attributed to the relaxed notion of "density-reachability" in DBSCAN [3], which does not dictate that all cluster members be within distance

ϵ from its centroid, but only pairwise. For less detailed trajectory representations (i.e., encodings based on small resolution α), this deviation propagates backwards when probing retained sequences to reconstruct a delegate path, so error may exacerbate ever more. However, the algorithm seems to achieve more reliable approximations in case of finer motion signatures, particularly for $\epsilon = 200$m.

4   Outlook

In this work, we set forth a novel approach for multiplexing trajectory features that get frequently updated from streaming positions of moving objects. We have been developing a methodology for detecting groups of objects that approximately travel together over the recent past. Thanks to an encoding scheme based on velocity vectors, this process can be carried out in almost real time with tolerable error, as our initial empirical study indicates.

We keep working on several aspects of this technique and we soon expect more gains in terms of scalability and robustness. In particular, we intend to take advantage of intermediate symbolic sequences in order to improve clustering with higher representativeness and less recalculations, even in presence of massive positional updates. We also plan to further investigate the grouping phase with more advanced schemes, like those used in string and sequence matching. Last but not least, it would be challenging to study this technique as an optimization problem, trying to strike a balance between similarity tolerance and resolution of the encoding scheme.

References

1. Bakalov, P., Hadjieleftheriou, M., Keogh, E., Tsotras, V.J.: Efficient Trajectory Joins using Symbolic Representations. In: MDM, pp. 86–93 (2005)
2. Chen, Z., Shen, H.T., Zhou, X.: Discovering Popular Routes from Trajectories. In: ICDE, pp. 900–911 (2011)
3. Ester, M., Kriegel, H.-P., Sander, J., Xu, X.: A Density-Based Algorithm for Discovering Clusters in Large Spatial Databases with Noise. In: KDD, pp. 226–231 (1996)
4. Jeungy, H., Yiu, M.L., Zhou, X., Jensen, C.S., Shen, H.T.: Discovery of Convoys in Trajectory Databases. PVLDB 1(1), 1068–1080 (2008)
5. Lee, J., Han, J., Whang, K.: Trajectory Clustering: a Partition-and-Group Framework. In: ACM SIGMOD, pp. 593–604 (2007)
6. Patroumpas, K., Sellis, T.: Maintaining Consistent Results of Continuous Queries under Diverse Window Specifications. Information Systems 36(1), 42–61 (2011)
7. Potamias, M., Patroumpas, K., Sellis, T.: Sampling Trajectory Streams with Spatiotemporal Criteria. In: SSDBM, pp. 275–284 (2006)
8. Sacharidis, D., Patroumpas, K., Terrovitis, M., Kantere, V., Potamias, M., Mouratidis, K., Sellis, T.: Online Discovery of Hot Motion Paths. In: EDBT, pp. 392–403 (2008)
9. Vieira, M., Bakalov, P., Tsotras, V.J.: On-line Discovery of Flock Patterns in Spatiotemporal Data. In: ACM GIS, pp. 286–295 (2009)

On Optimizing Workflows
Using Query Processing Techniques

Georgia Kougka and Anastasios Gounaris

Department of Informatics, Aristotle University of Thessaloniki, Greece
{georkoug,gounaria}@csd.auth.gr

Abstract. Workflow management systems stand to significantly benefit from database techniques, although current workflow systems have not exploited well-established data management solutions to their full potential. In this paper, we focus on optimization issues and we discuss how techniques inspired by database query plan compilation can enhance the quality of workflows in terms of response time.

1 Introduction

Workflow management takes the responsibility for executing a series of interconnected tasks in order to fulfill a business goal or implement semi- or fully-automated scientific processes. Typically, most workflow models emphasize the control flow aspect employing well-established models, such as BPEL. Nevertheless, data flow plays a crucial role in the effective and efficient execution and thus is equally significant. Data-centric workflows take a complementary approach and regard data management as a first class citizen, along with the control of activities within the workflow (e.g., [1,4]). An example of strong advocates of the deeper integration and coupling of databases and workflow management systems has appeared in [7]. Earlier examples of developing data-centric techniques of manipulating workflows include the Grid-oriented prototypes in [5,4] and the work in [3]. Those prototypes allow workflow tasks to be expressed on top of virtual data tables in a declarative manner in order to benefit from database technologies; however, they do not proceed to the application of query optimization techniques with a view to speeding up the workflow execution.

The contribution of this work is as follows. We demonstrate how the performance of data-centric workflows can further benefit from techniques inspired by databases. We target workflows that either process unnecessary data or contain services the relevant order of which is flexible, i.e., some activities within the workflow can be invoked in an arbitrary order while producing the same results; we term these services as being commutative. We discuss query optimization techniques that build on top of algebraic laws and the application of those techniques to such workflows with a view to modifying their structure without affecting their semantics in order to improve performance. We focus on fully automated workflows, i.e., workflows that do not require human intervention; in such workflows, the execution of constituent tasks may be cast as (web)

A. Ailamaki and S. Bowers (Eds.): SSDBM 2012, LNCS 7338, pp. 601–606, 2012.

service calls as very commonly encountered in a wide range of workflow management systems. In summary, our proposal aims to bring in a novel dimension in workflow optimization. Currently, the vast majority of workflow optimization efforts deal with scheduling and resource allocation (e.g., [8]); in addition, database-inspired proposals refer to specific applications only (e.g., [2]).

The remainder of the paper is structured as follows. The core part of our proposal is in Section 2. Section 3 contains case studies along with initial insights into the performance gains, and the conclusions appear in Section 4.

2 Database-Inspired Solutions to Workflow Optimization

Workflow optimizations based on structure modifications have not be analyzed to an adequate extent to date. Query optimization techniques are well suited to fill this gap, at least partially. Query plans consist of operators from the relational algebra (e.g., joins, selects, projects), the commutative and associative properties of which are well understood. The theoretic background of query optimization is based on algebraic laws that specify equivalence between expressions. On top of such laws, several optimization techniques can be built. Our idea is to apply similar laws to workflow structures. In other words, we treat workflows as query plans, and the constituent services as query operators in order to allow the application of query optimization rules. In order to apply the techniques suggested hereby, we assume that the precedence constraints between services are known and the input to each invoked activity is a list of data values. Similarly, the service output is another list of values with potentially different number of elements and element size. Such a model is followed by systems such as [6].

Note that in arbitrary business and scientific scenarios, it is common to employ services for data filtering, duplicate removal, transformation and remote method calls, joining inputs, merging inputs, and so on. All those operations correspond to operators such as selects and projects, duplicate elimination, user-defined functions, joins, unions, respectively. More specifically, in our work, we regard the evaluation of a selection predicate, including the case where it contains user-defined functions, as analogous to any workflow service that process a list of data items and produces another list with potentially different number and size of items. Projection complements selection in the sense that selection may filter rows, whereas projection filters columns in database tables. Duplicate elimination in query plans is analogous to services that remove duplicates from lists of items. Grouping is analogous to services that receive a list of input elements, group those elements into groups and process each group as a new element. The join operator is analogous to services that accept multiple inputs and combine them according to some criteria. Furthermore, operators such as unions and intersections directly apply to workflow services operating on datasets.

The fact that workflow services can be mapped to operators implies that common algebraic laws, such as selection and join reordering, commutativity of selections and duplicate elimination, distribution of selections over joins and unions, and pushing selections and projections to operators upstream are applicable to workflows as well. As such, two traditional optimization techniques that we employ in a workflow context are i) to reorder workflow services so that the services that are more selective are executed as early in the execution plan as possible; and ii) to introduce new filtering services in a workflow, when the data eliminated do not contribute to the result output. Reducing the size of data that we have to process leads to much faster computations. Finally, since the notions of selectivity and cost can be extended for services, we may apply more sophisticated cost-based optimizations, too.

3 Case Studies

We present two representative case studies with real scientific workflows, and we aim to demonstrate the performance benefits when the input data is dirty in the sense that it contains duplicates.

Case Study I: The first case study deals with a simple workflow titled as *"Link protein to OMIM disease"*, which is shown in Fig. 1[1]. Its purpose is to find diseases that are related to certain user-defined keywords. The first two services, *search* and *split OMIM results* correspond to the processes of searching and linking the input to a set of diseases from the OMIM human diseases and genes database[2], and presenting the results as individual lists. When a disease is found, it is extracted and then labeled, with the help of the *extract diseases from OMIM* and *label OMIM disease* services, respectively. The labeling service performs data transformation only, so that

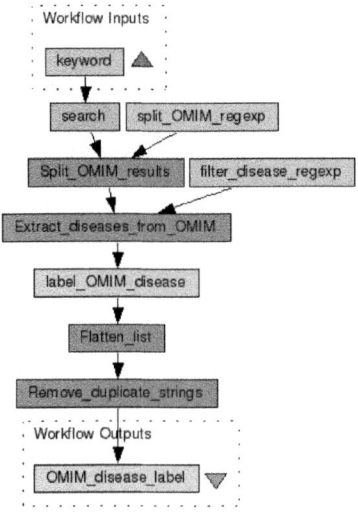

Fig. 1. A workflow that links proteins to diseases

XML tags are inserted. The next two services in the workflow are *flatten list* and *remove duplicate strings*. The former removes one level of nesting, whereas the other performs duplicate elimination. If we want to apply a string elimination service to data that appear more than once, the procedure of flattening list is a necessary task because orders datasets. These services are very important because of position and role they have to this workflow.

Based on the above description, we can easily deduce (even if not stated explicitly in the workflow description) that the flattening and duplicate elimination services are commutative with the labeling one. In addition, duplicate elimination is the service that may lead to the most significant reductions in the data set manipulated by the workflow, i.e., it is the one with the lower selectivity; actually, the labeling service does not filter data at all and has selectivity 1. As

---

[1] Taken from http://www.myexperiment.org/workflows/115.html

[2] http://www.ncbi.nlm.nih.gov/sites/entrez?db=omim

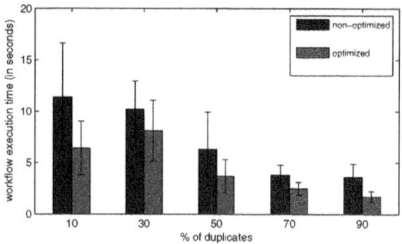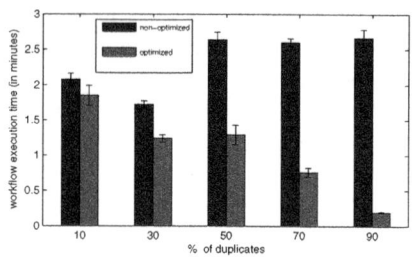

Fig. 2. Performance improvements with 100 Protein/Gene IDs as input for a local workflow (left) and for a workflow accessing a remote service (right)

such, we can improve the execution time if we apply duplicate elimination just before the *extract diseases from OMIM*, in the spirit of optimizations discussed in Sec. 2, or even earlier, i.e., just after the input submission. More specifically, we assume a simplified version of the workflow, where the workflow starts with the *split OMIM results* service and the workflow runs locally only. We consider two flavors, a non-optimized and an optimized one; the non-optimized one performs labeling before duplicate elimination, whereas the optimized one reorders the services and performs duplicate elimination at the very initial stage. Although the modification is simple, there are tangible performance gains. We experimented with an input of 100 OMIM records and a variable proportion of duplicates. Fig. 2 (left) summarizes the results. We used the Taverna 2.3 workbench environment on a Intel Core(TM) 2Duo T7500 machine with 3GB of RAM. The results correspond to the average and the standard deviation of 10 runs after removing the two highest values as outliers to decrease standard deviations. The main observation drawn from the figure is that, if duplicates exist in the input, the decrease in the response time can be higher than 50%.

Case Study II: In the second case study, we experimented with a more intensive subworkflow of a workflow named *"Get Kegg Gene information"*[3]. This sub-workflow gets as input a list of KEGG Genes IDs and according to this list extracts from Kyoto Encyclopedia of Genes and Genomes (KEGG)[4] database information relating to these KEGG genes, such as Gene Description. For example, an input of KEGG Gene ID could be *hsa:400927* which corresponds to the *"hsa:400927 TPTE and PTEN homologous inositol lipid phosphatase pseudogene"* description. The first service of this sub-workflow is *split by regex* and it is necessary in order to reformulate the input data for the workflow services that follow. The key service of this workflow is *get gene description GenomeNet*. This workflow service corresponds to the process of linking KEGG Gene IDs with KEGG Genes database and presenting a brief description for each of the genes. The *get gene description GenomeNet* service is followed by other two services, *merge descriptions* and *remove nulls*, which are responsible for merging the nested content of the previous output and eliminating possible null outputs,

---

[3] Taken from http://www.myexperiment.org/workflows/611.html
[4] http://www.genome.jp/kegg/

respectively. This service, for each KEGG Gene ID input, connects with the KEGG database in order to search and return the corresponding description of each gene, which is quite time-consuming procedure. In case of having an input data set consisting of several duplicate values, we could assume that the required time to execute a large-scale data set is significantly high. This provides room for optimization, since we can modify the workflow process so that *get gene description GenomeNet* service connects and extracts only unique values of KEGG Gene IDS. Therefore, we can use the techniques used for inserting selections in query plans to perform this optimization task. Specifically, we introduce a new workflow service *remove duplicate gene ids* which has the role to eliminate duplicate values, before the *get gene description GenomeNet* service. In this manner, the service will contact the database only for unique values without repeating the same costly requests. Fig. 3 and 4 depict the original and the optimized version of the sub-workflow examined, respectively.

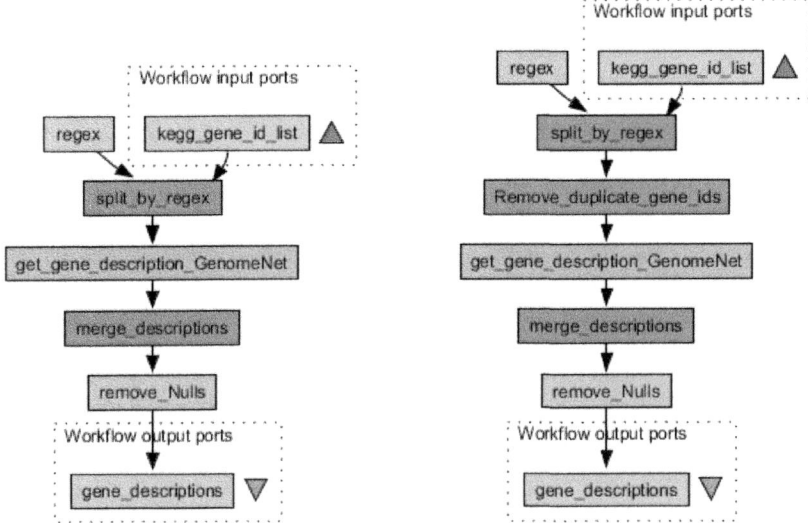

Fig. 3. A sub-workflow of *"Get Kegg Gene information"* that provides description of KEGG Genes

Fig. 4. An optimized sub-workflow that provides description of KEGG Genes

The results shown in Fig. 2 (right) support our intuition that the performance benefits are far more significant and the optimized version runs several times faster. In the second case study, we experimented with a more intensive workflow that involves calls to a remote service, and we inserted duplicate elimination before the service invocation. Specifically, the workflow execution cost is improved gradually while the duplicates of an input data set increases. This means that, for an input data of 100 Gene IDs, which consists of 70% and 90% duplicated values, the optimized version of *"Get Kegg Gene information"* workflow results in 70% and over 90% decrease on response time, respectively.

4 Conclusions

In this paper we discussed the application of query optimization techniques to workflow structure reformations. Our methodology can yield significant performance improvements upon existing data-centric workflows; our preliminary quantitative results in workflows that receive as input data containing duplicates provide promising insights into this aspect.

Our work can be extended in several ways; actually we just scratched the surface of the potential of query optimization techniques for workflows. The most important directions for future work are the development of workflow management systems that fully implement our proposal in line with the points mentioned above, the thorough assessment of performance improvements in a wide range of realistic scenarios, and the investigation of optimization opportunities in more complex workflow patterns. Finally, note that query optimization techniques that seem relevant to workflows are not limited to reordering of commutative services and insertion of early duplicate removal. Complementary aspects include the investigation of equivalent execution plans of different shape and the choice of the physical implementation of logically equivalent workflow services.

References

1. Bhattacharya, K., Hull, R., Su, J.: A data-centric design methodology for business processes. In: Handbook of Research on Business Process Modeling, ch. 23, pp. 503–531 (2009)
2. Dayal, U., Castellanos, M., Simitsis, A., Wilkinson, K.: Data integration flows for business intelligence. In: EDBT, pp. 1–11 (2009)
3. Ioannidis, Y.E., Livny, M., Gupta, S., Ponnekanti, N.: Zoo: A desktop experiment management environment. In: Proceedings of 22th International Conference on Very Large Data Bases, VLDB 1996, Mumbai (Bombay), India, September 3-6, pp. 274–285. Morgan Kaufmann (1996)
4. Liu, D., Franklin, M.: The design of griddb: A data-centric overlay for the scientific grid. In: VLDB, pp. 600–611 (2004)
5. Narayanan, S., Catalyrek, U., Kurc, T., Zhang, X., Saltz, J.: Applying database support for large scale data driven science in distributed environments. In: Proc. of the 4th Workshop on Grid Computing (2003)
6. Oinn, T., Greenwood, M., Addis, M., Alpdemir, N., Ferris, J., Glover, K., Goble, C., Goderis, A., Hull, D., Marvin, D., Li, P., Lord, P., Pocock, M., Senger, M., Stevens, R., Wipat, A., Wroe, C.: Taverna: lessons in creating a workflow environment for the life sciences. Concurrency and Computation: Practice and Experience 18(10), 1067–1100 (2006)
7. Shankar, S., Kini, A., DeWitt, D., Naughton, J.: Integrating databases and workflow systems. SIGMOD Rec. 34, 5–11 (2005)
8. Xiao, Z., Chang, H., Yi, Y.: Optimization of Workflow Resources Allocation with Cost Constraint. In: Shen, W., Luo, J., Lin, Z., Barthès, J.-P.A., Hao, Q. (eds.) CSCWD. LNCS, vol. 4402, pp. 647–656. Springer, Heidelberg (2007)

Optimizing Flows for Real Time Operations Management

Alkis Simitsis, Chetan Gupta, Kevin Wilkinson, and Umeshwar Dayal

HP Labs, Palo Alto, CA, USA
{alkis.simitsis,chetan.gupta,kevin.wilkinson,umeshwar.dayal}@hp.com

Abstract. Modern data analytic flows involve complex data computations that may span multiple execution engines and need to be optimized for a variety of objectives like performance, fault-tolerance, freshness, and so on. In this paper, we present optimization techniques and tradeoffs in terms of a real-world, cyber-physical flow that starts with raw time series sensor data and external event data, and through a series of analytic operations produces automated actions and actionable insights.

1 Modern Analytic Flows

With the ever increasing instrumentation of physical systems, real time operations management has become increasingly relevant. Real time operations management implies developing, deploying, and executing applications that mine and analyze large amounts of data collected from multiple data sources such as sensors and operational logs, in order to help decision makers take informed decisions during operations management.

In our approach, these applications are specified as analytic flow graphs and are executed over streaming and historical data to raise alerts, produce actionable insights, and recommend actions for the operations staff. Such flows comprise complex event processing (CEP) operations, extract-transform-load (ETL) operations, traditional SQL operations, advanced analytic operations, visualization operations, and so on. Given the diversity of data sources and analysis engines, creating correct, complex analytic flows can be a time-consuming and labor-intensive task. Executing such flows efficiently such that real time response requirements are met in a cost effective fashion, requires optimization. Optimizing these flows is a significant problem especially in an environment where the analysis requirements and the data are evolving.

Our QoX optimizer is an in-house developed tool that optimizes analytic flows for a variety of objectives and across multiple engines. Performance used to be the sole objective for flows, but today, we see the need for other objectives as well, for example, cost, fault-tolerance, freshness, energy usage, and so on. Additionally, complex flows often access data from a variety of sources, involving different execution engines (e.g., DBMSs, ETL like Informatica or PDI, and map-reduce engines like Hadoop) and data centers (e.g., enterprise data warehouse, WWW repositories, Amazon EC2).

A. Ailamaki and S. Bowers (Eds.): SSDBM 2012, LNCS 7338, pp. 607–612, 2012.

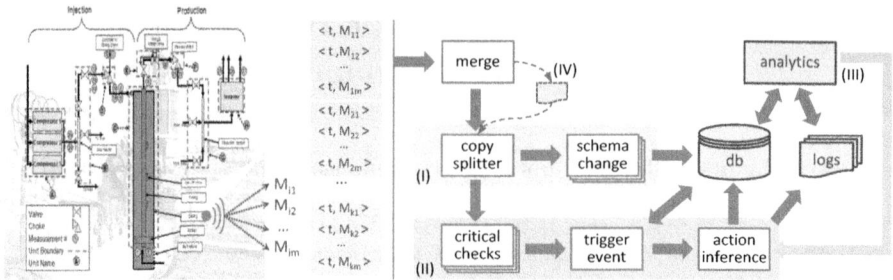

Fig. 1. A graphical representation of a drilling well and an example analytic flow

Next, we describe optimization choices and tradeoffs through a real-world operations management application.

2   A Cyber-Physical Flow

Figure 1 illustrates a cyber-physical flow derived from our experience with oil and gas production [1]. The oil and gas industry collects massive amounts of data from drilling operations via sensors installed in the oil wells, other measurement devices, and textual operational logs that contain observations and actions taken in free-from text. For the purpose of analysis the well may be divided into various components where, from each component, certain physical quantities such as temperature, pressure, and flow rate are monitored with sensors (see the left part of Figure 1).

Assume k sensors and a set of measures $M=\{M_1, M_2, \ldots, M_k\}$, where the measure j for a sensor s_i is denoted as M_{ij}. Then, the data from an individual sensor at time t can be thought of as a tuple $<t,M_{ij_t}>$. (We skip the subscript t for ease of notation where the context is clear.) As shown in Figure 1 (right part), to understand the happenings across a component or across the well requires us to build a tuple that merges sensor readings recorded at the same time (assuming clock synchronization) to get a tuple of the form: $<t, M_{i1}, M_{i2}, \ldots, M_{im}>$ or even $<t, M_{11}, M_{12}, \ldots, M_{1m}, M_{21}, M_{22}, \ldots, M_{2m}, \ldots, M_{km}>$. Next, these tuples are routed through a copy-splitter operation to create separate data flows for historical analysis (Phase I) and real time response (Phase II).

In Phase I, we change the schema before storage using a series of joins and aggregations. This part resembles a typical ETL flow.

In Phase II, the measurements are checked for threshold violations and other anomalies ('critical checks'). In Figures 2 and 3, we describe in more detail two fragments of the flow presented in Figure 1.

Figure 2 illustrates in more detail an example of a critical checks subflow (it checks for threshold violations). Such a flow fragment involves the following operations:

CMA. First, we compute a moving average for P values, where $P \in M$, based on the formula: $(\sum_{i=T-w}^{T} M_{ji})/N$, where the summation is over a window

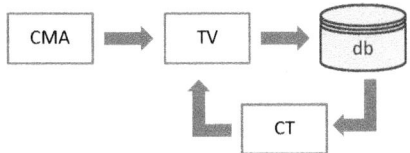

Fig. 2. Critical checks subflow

of size w and is over current time T to $T - w$ and N is the number of observations in the window from time T to $T - w$.

CT. Then, we compute thresholds with a user-defined function that performs a database lookup and dynamically computes appropriate thresholds θ_i based on a moving window measured in terms of hours.

TV. Finally, we perform threshold validation, for validating the measures against threshold values; as for example in: $\forall P_i \in P$ check if $f(P_i) > \theta_i$. The validation result is kept in the database.

A threshold violation generates an event e_i, which is typically in the form $<time, component, event\_type>$, where event types are pre-defined. The event is stored for future reference and leads to an automated action or inference for the operator –possibly leading to a manual action. Such inference and action require data from historical event logs and the subsequent action (either automatic or manual) is written to operator logs. Typically, action inference operations are composite operations like:

Bayesian inference for predicted events (BI). It is implemented as a user-defined function and gives a prediction of future events based on the historical data stored in the database (a database lookup is needed too).

Automated action (AA). Based on the event type (requires a database lookup operation) the flow may automatically decide upon an action and the result of this action is registered in the log files. This process typically runs inside a CEP engine.

Suggested action (SA). This composite operation results in a suggestion for suitable actions need to be taken based on the event observed.

As a more detailed example of an action inference operation, Figure 3 shows example user-defined functions and database operations that realize a *suggested action (SA)* operation. Before suggesting an appropriate action, the flow searches into the database (shown as *db* in Figures 1 and 3) for previous occurrences of an event and also into the log files for corresponding actions to similar events. The flow is realized as follows.

Compute states (CS). First, we need to compute the current system state. A state is often a synopsis (that preserves distance) of the measurements at all time points. In some scenarios, the Bayesian inference step may also require state computation.

Find times (FT). Then, we query the database to find timestamps of previous occurrences of an event, T_{e_i}.

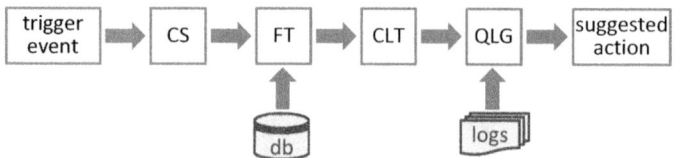

Fig. 3. Suggested action inference subflow

Time of closest state (CLT). Of all the times found, we identify the time $t^* \in T_{e_i}$ where the state was closest to the current state.

Find similar actions (QLG). Finally, we query the log files with t^* to find if there was a similar action; if there is one, we suggest it to the system's operator.

As such applications and resulting complex flows become common, we need strategies to optimize them. Beside the need to optimize for various objectives, such as response time and cost, a big challenge is the presence of different execution engines. We discuss these issues in the next section.

3 Optimization Techniques and Tradeoffs

We model analytic flows as the one in Figure 1 as graphs, where nodes represent operations and data stores and edges represent data flows. In this model, an operation may have multiple implementations for one or more execution engines. For example, filtering, merging, splitting, aggregation are operations supported by many systems. However, some operations may have only a single implementation, which limits the choices of the optimizer.

The designer may first create an analytic flow from scratch or import an existing one from another tool (e.g., an ETL tool or a workflow management system like Taverna [2]) and then optimize it. The latter –i.e., flow optimization– is the focus of our work.

The optimization problem is formulated as a state space search problem. Based on the objectives we want to satisfy (i.e., the objective function), our QoX optimizer uses a set of transitions to create alternative designs, called states. Example transitions for optimizing for multiple objectives include operation reordering, partitioning parallelism, redundancy or adding recovery points, and so on (e.g., see [3]). Example transitions for optimizing flows spanning different engines are function shipping, data shipping, flow restructuring, decomposition, and so on (e.g., see [4]). Hence, using such transitions, the optimizer creates a state space and searches for an optimal (or near optimal) state. Obviously, an exhaustive search of the state space is not realistic for real-world applications, so the optimizer has to use heuristic and greedy algorithms.

The state space search algorithms are driven by a cost model. Each operation comes with a cost formula, which is based on both operational characteristics like the processing load and data characteristics like cardinality and data skew that determine the operation's throughput. For traditional operations, like relational operations, a cost formula is relatively easy to be defined; e.g., a sort operation has a processing cost similar to $n\times log(n)$, for an input of size n. For user-defined functions or operations with black-box semantics, we must inspect the code (if it is available) and/or perform extensive micro-benchmarks, regression analysis, and interpolation to come up with an approximate cost formula. Or, we may estimate their cost using their execution history (e.g., see [5,6]). We have done all three of the above for a fairly rich set of operations currently supported by our optimizer. Since, the QoX optimizer is extensible to new operations and additional implementations of existing operations, we require a cost formula during the registration of an operation implementation. Having at hand the costs of individual operations, we are able to compute the flow cost and compare one state to another.

Example optimization choices. Next, we pinpoint tradeoffs that the optimizer considers when it optimizes the flow of Figure 1 for performance and reliability across multiple engines: a DBMS and an ETL tool for Phase I, a stream processing and a CEP engines for Phase II, and a map-reduce engine for Phase III. (The logs in our example are processed using Hadoop/Pig, but this task can be done with Awk scripts or some other scripting method.)

- There are parallelization opportunities at several points of the flow. For example, when we perform critical tests or in the action interference subflow. In the latter, we could run in parallel the state computation (CS) and the event time determination (FT).

- Analytics computations may be performed either as on-demand querying (i.e., searching for information closer to the sources) or as computations over materialized data (e.g., data stored in the database or in logs).

- The 'schema changes' subflow involves design choices, like deciding where and how to do the schema change operations; e.g., inside or outside the database.

- Another challenge is to choose an engine to execute a flow fragment. For an operation with implementations on multiple engines, the optimizer decides where to execute this operation (function shipping); e.g., perform a join or a user-defined function calculation inside the database or in Hadoop. On the other hand, based on cost, the optimizer may choose to move data from one engine to another and perform the operation there (data shipping).

- Different flow fragments may be optimized for different objectives. For example, Phase II may be optimized for latency and reliability, while Phases I and III may be optimized for cost and/or energy consumption. Similarly, Phase II may need to be delayed if it requires more resources at a peak time (e.g., for responding to an event) and, in such a case, since the incoming data are not persistent, they should be temporarily stored; i.e., enable or disable Phase IV. Deciding when and where in the flow (i.e., how close to the database) to put a storage point is an optimization choice.

- Individual operations may be optimized too. For example, the optimizer may consider varying the degree of merging for the merge operation; i.e., less merging means finer granularity of measures that flow through Phase II. Then, we can optimize for accuracy too; i.e., varying accuracy with level of merging.

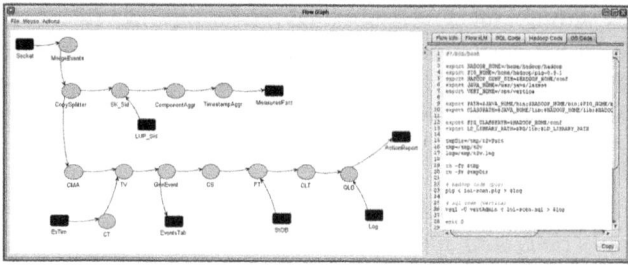 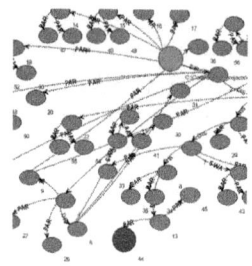

Fig. 4. An example state with part of its execution code and a state space fragment

Figure 4 (left) shows a snapshot of the QoX optimizer and in particular, a state that the optimizer picked for our example flow. The figure also shows part of the code needed for its execution and a fragment of the state space (green node is the initial graph state, red the optimal state, and arcs denote transitions between states). The optimizer continuously monitors the flow and needs a few seconds (~30sec for this flow) to adapt to changes and to come up with an optimal solution.

4 Conclusion

We presented a real-world, cyber-physical flow that performs critical checks and analytics on top of data streaming from sensors placed at different depths of drilling wells. The flow starts from raw time series sensor data and external event data, and continues through a series of operations producing automated actions and actionable insights. Typical challenges include the heterogeneity of data that stream at fast rates, while expensive computations should be performed in real-time. For dealing with such problems, we use an in-house crafted tool that optimizes flows that span different engines for a variety of quality objectives.

Collecting, archiving, analyzing, and visualizing such data requires a comprehensive framework which is the focus of our on-going work.

References

1. Gupta, C., et al.: Better drilling through sensor analytics: A case study in live operational intelligence. In: SensorKDD, pp. 8–15 (2011)
2. Missier, P., Soiland-Reyes, S., Owen, S., Tan, W., Nenadic, A., Dunlop, I., Williams, A., Oinn, T., Goble, C.: Taverna, Reloaded. In: Gertz, M., Ludäscher, B. (eds.) SSDBM 2010. LNCS, vol. 6187, pp. 471–481. Springer, Heidelberg (2010)
3. Simitsis, A., Wilkinson, K., Dayal, U., Castellanos, M.: Optimizing ETL workflows for fault-tolerance. In: ICDE, pp. 385–396 (2010)
4. Simitsis, A., Wilkinson, K., Castellanos, M., Dayal, U.: Optimizing Analytic Data Flows for Multiple Execution Engines. In: SIGMOD Conference (2012)
5. Gupta, C., Mehta, A., Dayal, U.: Pqr: Predicting query execution times for autonomous workload management. In: ICAC, pp. 13–22 (2008)
6. Verma, A., Cherkasova, L., Campbell, R.H.: Aria: automatic resource inference and allocation for mapreduce environments. In: ICAC, pp. 235–244 (2011)

(In?)Extricable Links between Data and Visualization: Preliminary Results from the VISTAS Project

Judith Cushing[1], Evan Hayduk[1], Jerilyn Walley[1], Lee Zeman[1], Kirsten Winters[2], Mike Bailey[2], John Bolte[2], Barbara Bond[2], Denise Lach[2], Christoph Thomas[2], Susan Stafford[3], and Nik Stevenson-Molnar[4]

[1] The Evergreen State College, [2] Oregon State University, [3] University of Minnesota
[4] Conservation Biology Institute (Corvallis OR)
judyc@evergreen.edu

1   Introduction

Our initial survey of visualization tools for environmental science applications identified sophisticated tools such as *The Visualization and Analysis Platform for Ocean, Atmosphere, and Solar Researchers* (VAPOR) [http://www.vapor.ucar.edu], and *Man computer Interactive Data Access System* (McIDAS) and *The Integrated Data Viewer* (IDV) [http://www.unidata.ucar.edu/software]. A second survey of ours (32,279 figures in 1,298 articles published between July and December 2011 in 9 environmental science (ES) journals) suggests a gap between extant visualization tools and what scientists actually use; the vast majority of published ES visualizations are statistical graphs, presenting evidence to colleagues in respective subdisciplines. Based on informal, qualitative interviews with collaborators, and communication with scientists at conferences such as AGU and ESA, we hypothesize that visualizations of natural phenomena that differ significantly from what we found in the journals would positively impact scientists' ability to tune models, intuit testable hypotheses, and communicate results. If using more sophisticated visualizations is potentially so desirable, why don't environmental scientists use the available tools?

We suggest two barriers to using sophisticated scientific visualization: lack of the desired visualizations, addressed elsewhere [1, 2], and difficulty preparing data for visualization, addressed here. As David Maier remarks in his Keynote Address to this conference: Big data [3] "implies a big variety of data sources, e.g., multiple kinds of sensors…on diverse platforms…coming in at different rates over various spatial scales…. Few individuals know the complete range of data holdings, much less their structures and how they may be accessed" [4]. In this paper, we identify two major data issues that scientists face relevant to their use of visualization tools: (1) the complexity of their own data and (2) complex input data descriptors for visualization software and perceived (or actual) difficulty of transforming data prior to using those tools.

We briefly describe our own project (VISTAS), articulate our collaborators' data structures, and report on data requirements for a subset of visualization software identified as useful for ES. We conclude that the complexity of input data formats is so daunting for most scientists and even for visualization researchers and developers,

A. Ailamaki and S. Bowers (Eds.): SSDBM 2012, LNCS 7338, pp. 613–617, 2012.

that visualizations that might advance science, and would certainly advance communication of research results by scientists to colleagues and the public, remain undone. We thus strongly encourage the scientific database community to address the problems scientists face in characterizing and transforming their data.

2   VISTAS Project Overview

Our prior work suggested that visual analytics can help scientists more effectively use large data sets and models to understand and communicate complex phenomena. We hypothesized that cross-scale visualization of natural phenomena would enable scientists to better deal with massive data stores and understand ecological processes. Visualization of ecosystem states, processes, and flows across topographically complex landscapes should enhance scientists' comprehension of relationships among processes and ecosystem services, and the posing of testable hypotheses [5]. Scientific visualization is not new and much excellent work exists, but few tools easily integrate complex topography with visualizing diverse data [6, 7]. Fewer still allow viewers to scale up or down in space and time, critical to ES grand challenges [8].

The recently National Science Foundation funded VISTAS (VISualization of Terrestrial-Aquatic Systems) project aims to develop and test visualizations so scientists better understand and communicate ES. Objectives include: 1) research visualization needs for our science collaborators and develop a proof of concept tool that meets those needs, 2) conduct ES research with the tool, 3) use social science methods to document development processes and to assess the visualizations, answering the questions: which visualizations are most effective, for what purposes, and with which audiences. VISTAS' ES research focuses on one geographical area west of the crest of the Oregon Cascades, on 3D representations of land use, and on process-based models that simulate cycling and transport of water and nutrients, problems similar to other ES grand challenges.

3   VISTAS Architecture and Data Structures

VISTAS' close collaborators include Bob McKane and Allen Brookes who model the ecohydrology of watersheds and basins with **VELMA** [9]; John Bolte, author of a land use model **ENVISION**, a GIS-based decision support tool integrating scenarios, decision rules, ecological models, and evaluation indices [10]; and Christoph Thomas who collects spatially distributed point-measurements of air flow, air temperature and humidity using **SODAR** across the landscape, scales from 10s to 100s of meters [11].

VELMA outputs a tiff file per simulation year, two.csv files with daily and annual results, and files with spatial results, in ESRI Grid ASCII format with a single value for each grid cell. File size depends on the number of variables, currently generating 38.5mb/day (10 gb/yr), but in the future to 5x variables, i.e., 50 gb/yr (64 km2). VISTAS now handles VELMA output of 30m cells, 64 km2, 70,000 cells, 100,000 rows, 30 variables, converting csv files to a 2D contiguous grid, speeding processing 10-20x. Elevation is incorporated into each cell, avoiding the need for alignment.

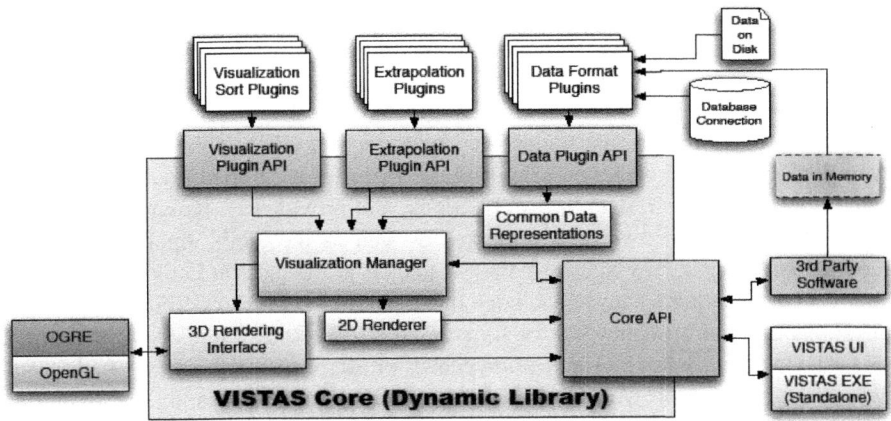

ENVISION stores data as C++ objects, map data as a table, each column a variable, each row mapping to a polygon, and can read NetCDF and rasterize output data. Time steps are yearly. Temporal data stored as delta arrays (map changes) are large, e.g., 10 million elements and designate which polygon changed, start and end values, when change occurred. The ArcGIS-like shape files (points, lines, polygons, a vector data model) are visualized as 2D maps. VISTAS will display ENVISION 3D terrain models, initially 30,000 km2, 180,000 polygons, 10-20 hectares/polygon, a 90m grid.

The **Metek SODAR** sensor [http://www.metek.de] transmits an acoustic pulse at a specified frequency,then listens for a return signal; data are analyzed to determine wind speed and direction and the turbulent character of the atmosphere. Measurements are compiled into a 42x136 matrix per 10-min. increments. Each file contains one 24-hour period (39,168 records). Metek provides simple visualization software, where users can select a custom time period and defined parameters for measured values (spectrum, potential temperature, inversion height, wind direction or speed), but VISTAS will provide capability of overlaying SODAR with VELMA modeled data.

4   Conclusions

This paper concludes with a synopsis of the two previously cited visualization options for environmental science, and barriers scientists face in using such systems – the same barriers we (and other scientific visualization developers) face as we develop and document input data plug-ins for VISTAS.

Tailored for the astro- and geo-sciences, VAPOR provides interactive 3D visualization on UNIX and Windows systems with 3D graphics cards, handling terascale data. VAPOR can directly import WRF-ARW output data with no data conversion, but data must be on the same grid with the same level of nesting. Full access to features is available if users convert data to VAPOR's format (VDC), but VAPOR provides tools to convert common data formats.

McIDAS is an open source tool for visualizing multi- and hyper-spectral satellite data, and handles many data formats, including Unidata's IDV and VisAD; satellite images in AREA, AIRS, HDF, and KLM; and meteorological data in McIDAS-MS, netCDF, or text; and gridded data in NCEP, ECMWF or GRID formats.

With so many data formats supported by these and other sophisticated packages, why don't more environmental scientists use them? Not surprisingly, these sophisticated tools – even if data transformations are provided – require scientists to understand both their own data structures as well as the tool's input data structure requirements. Our survey of information managers at Long Term Ecological Research sites [http://www.lternet.edu] concluded that even for simple ArcGIS or MatLab visualizations, such effort is beyond the time or expertise of many scientists. Alternatives might be to send data to a visualization center, or establish collaborations with visualization specialists. This works for some scientists, with the funds and time, to produce visualizations for a particular purpose, but our collaborators want to use visualizations interactively – to steer their computational models, help intuit new hypotheses, and explain results to collaborators and other stakeholders.

Without software that characterizes data, scientists need to understand arcane formats and many will eschew valuable tools – and developers of tools like VISTAS, VAPOR and McIDAS will continue to spend time and money writing idiosyncratic data transformations. What Howe et al provide for long tail science [12] is a first step towards removing the inextricable links between data syntax and semantics that will enable scientists to use the tools they need. We encourage the SSDBM community to research and develop data descriptor and transformation tools that separate the characterization and transformation of data from the process of creating semantically meaningful analyses and visualizations.

Acknowledgements. The VISTAS project is supported by the National Science Foundation/ BIO/DBI 1062572. Any opinions, findings, and conclusions or recommendations expressed in this material are those of the author(s) and do not necessarily reflect the views of the National Science Foundation. The authors thank the anonymous reviewers whose comments helped sharpen the focus of this preliminary report of VISTAS' findings.

References

1. Cushing, J.B., et al.: What you see is what you get? In: Data Visualization Options for Environmental Scientists. Ecological Informatics Management Conference (2011)
2. Schultz, N., Bailey, M.: Using extruded volumes to visualize time-series datasets. In: Dill, J., et al. (eds.) Expanding the Frontiers of Visual Analytics and Visualization, pp. 127–148. Springer (2012) ISBN 978-1-4471-2803-8
3. Alexander, F., et al.: Big Data. IEEE Computing in Science & Engineering 13, 10–13 (2011)
4. Maier, D.: Navigating Oceans of Data. In: Abstracts of the Conference on Scientific and Statistical Database Management,
   http://cgi.di.uoa.gr/~ssdbm12/keynote1.html

5. Cushing, J.B., et al.: Enabling the Dialogue-Scientist< >Resource-Manager< > Stake-holder: Visual Analytics as Boundary Objects. IEEE Intelligent Systems 24, 75–79 (2009)
6. Smelik, R.M., et al.: Survey of Procedural Methods for Terrain Modeling. In: Egges, A., et al. (eds.) Proc. of the CASA Workshop on 3D Advanced Media in Gaming and Simulation (3AMIGAS), Amsterdam, The Netherlands, pp. 25–24 (2009)
7. Thomas, C.K., et al.: Seasonal Hydrology Explains Interannual and Seasonal Variation in Carbon and Water Exchange in a Semiarid Mature Ponderosa Pine Forest in Central Oregon. Journal of Geophysical Research 114, G04006 (2009)
8. Kratz, T.K., et al.: Ecological Variability in Space and Time: Insights Gained from the US LTER Program. BioScience 53, 57–67 (2003)
9. McKane, R., et al.: Integrated eco-hydrologic modeling framework for assessing effects of interacting stressors on multiple ecosystem services. ESA Annual Meeting (August 2010)
10. Bolte, J.P., et al.: Modeling Biocomplexity - Actors, Landscapes and Alternative Futures. Environmental Modelling & Software 22, 570–579 (2007)
11. Turner, D.P., Ritts, W.D., Wharton, S., Thomas, C.: Assessing FPAR Source and Parameter Optimization Scheme in Application of a Diagnostic Carbon Flux Model. Remote Sensing of Environment 113(7), 1529–1539 (2009)
12. Howe, B., Cole, G., Souroush, E., Koutris, P., Key, A., Khoussainova, N., Battle, L.: Database-as-a-Service for Long-Tail Science. In: Bayard Cushing, J., French, J., Bowers, S. (eds.) SSDBM 2011. LNCS, vol. 6809, pp. 480–489. Springer, Heidelberg (2011)

FireWatch: G.I.S.-Assisted Wireless Sensor Networks for Forest Fires

Panayiotis G. Andreou, George Constantinou, Demetrios Zeinalipour-Yazti, and George Samaras

Department of Computer Science, University of Cyprus, Nicosia, Cyprus
{panic,gconst02,dzeina,cssamara}@cs.ucy.ac.cy

Abstract. Traditional satellite and camera-based systems, are currently the predominant methods for detecting forest fires. Our study has identified that these systems lack immediacy as detected fires must gain some momentum before they are detected. In addition, they suffer from decreased accuracy especially during the night, where visibility is diminished. In this paper, we present FireWatch, a system that aims to overcome the aforementioned limitations by combining a number of technologies including Wireless Sensor Networks, Computer-supported Cooperative Work and Geographic Information Systems in a transparent manner. Compared to satellite and camera-based approaches, FireWatch is able to detect forest fires more accurately and forecast the forest fire danger more promptly. FireWatch is currently scheduled to be deployed at the Cypriot Department of Forests.

1 Introduction

Traditional satellite and camera-based systems [4,5], such as MODerate resolution Imaging Spectroradiometer (MODIS) [5], are currently the predominant methods for detecting forest fires. These systems have the ability to predict global changes in order to assist organizations in making sound decisions concerning the protection of our environment from natural disasters, including fires. Camera-based systems have been popular, especially in Europe [2,3,4], during the past few years. These systems usually consist of terrestrial, tower-based cameras that enable reliable and automated early detection of fires. In most cases, these systems are supplemented by alarm mechanisms that alert the involved users of potential risks thus supporting the final decisions on further actions. Although camera-based systems provide a more fine-grained solution than satellite systems, because of the smaller scanning regions and the continuous human monitoring factor, they lack immediacy as detected fires must gain some momentum before these are detected. In addition, camera-based systems suffer like their satellite-based counterparts from decreased accuracy, especially during the night where visibility is diminished. Both satellite and camera-based systems rely mainly on the human factor to analyze the findings and decide on the course of action.

A. Ailamaki and S. Bowers (Eds.): SSDBM 2012, LNCS 7338, pp. 618–621, 2012.

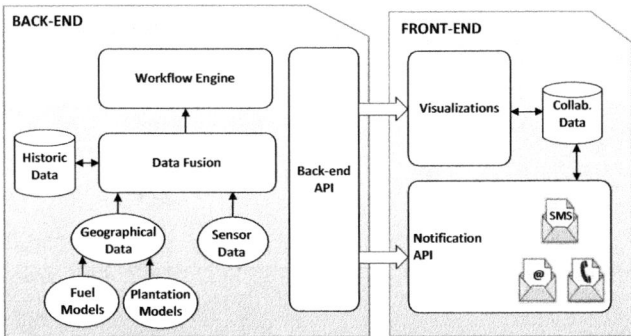

Fig. 1. The FireWatch architecture

In this paper, we present the FireWatch system[1] that aims to overcome the aforementioned limitations by combining a number of technologies including Wireless Sensor Networks (WSNs), Computer-supported Cooperative Work (CSCW) and Geographic Information Systems (GIS). The Wireless Sensor Network incorporates energy efficient algorithms that continuously monitor the environment and provide real-time measurements. The Collaboration Engine employs both snapshot and historic fire detection/prediction algorithms and enables the notification of all involved authorities. The enhanced Geographic Information System, besides managing and presenting all types of geographical data, inserts an additional layer that displays plantation and fuel models thus enabling fire officers to better assess the situation.

Compared to satellite and camera-based approaches, FireWatch is able to detect forest fires more accurately and forecast the forest fire danger more promptly. As a result, Cypriot forest officers, who are currently relying on watch towers and periodic patrolling of the forest, will be provided with more accurate decision-support data, hence they will plan a more refined course of action.

2   System Architecture

An overview of the FireWatch architecture is illustrated in Figure 1. It consists of two tiers, the Back-end, which is responsible for data collection, data analysis and event detection/prediction, and the Front-end, which visualizes the data as well as provides the means to contact appropriate authorities (e.g., voice messages, sms, email) in the case of a fore fire detection. We now provide more detail on each component of the FireWatch architecture.

- The **Sensor Data** component, which stores the real-time data provided by the wireless sensor network deployed in a high risk forest area. It incorporates energy-efficient algorithms [1], which run locally at each sensor node in order to increase the network longevity and minimize maintenance costs.

---

[1] The FireWatch System, http://firewatch.cs.ucy.ac.cy/

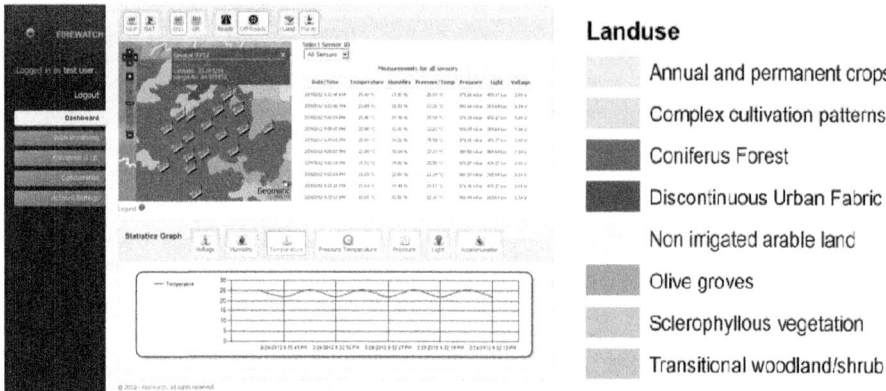

Landuse

Annual and permanent crops

Complex cultivation patterns

Coniferus Forest

Discontinuous Urban Fabric

Non irrigated arable land

Olive groves

Sclerophyllous vegetation

Transitional woodland/shrub

Fig. 2. The FireWatch interface

- The **Geographical Data** component, which stores geographical information as well as road and off-road data. Furthermore, it incorporates fuel and plantation models for estimating fire behavior.
- The **Data Fusion** component, which correlates sensor and geographical data and produces a comprehensive view of the network status. Additionally, this data is stored in a database (Historic Data), in order to by analyzed together with other data utilized by the organization (e.g., meteorological data).
- The **Workflow Engine**, which incorporates information models that analyze the real-time and historic data so as to detect/predict forest fire events.
- The **Visualizations** component displays all data in a visual manner (i.e., map view, tabular format, graphs). Additionally, it provides different interfaces for querying the data of each component.
- The **Notification API** support different notification mechanisms for providing alerts to users (e.g., visual alerts, audio alerts) and involved authorities (e.g., voice messages, SMS, Email).

FireWatch adopts a virtualization paradigm in order to increase flexibility in deployment and provide a dynamic, scalable and robust platform. It incorporates three individual virtual machines that can be maintained independently. The first virtual machine manages the wireless sensor network and includes tools for porting SQL queries to sensor nodes. Additionally, it provides APIs for translating raw measurements to real values. The second virtual machine consists of a GIS server that provides geographical data to the platform. It includes a map visualization component as well as APIs for creating/updating map layers. The third virtual machine comprises the collaboration system. It maintains a dedicated workflow engine where users can create new workflows or update existing ones according to organizational requirements.

3   FireWatch Interface

FireWatch features an innovative web-based interface (see Figure 2) that enables fire officers to overview the status of the network as well as query/retrieve specific information. The FireWatch interface is composed of the following primary visualization components:

- **Dashboard:** Provides an overview of the whole system. Users can observe the latest sensor measurements both in tabular format and charts. Additionally, a map of the high risk forest area and the sensor locations is displayed.
- **WSN Monitoring:** Enables users to execute queries in order to monitor and access real-time and historic sensor measurements (e.g., temperature, humidity, etc.). WSN health and predicted lifetime is also available.
- **Enhanced G.I.S.:** Plots sensor locations on the map as well as provides alerts to users about possible fire detection/prediction events. The map component supports three modes: i) normal view; ii) satellite view; and iii) plantation/fuel models view. Additionally, in the event of a forest fire, it can provide alternative off-road routes to the location of the event.
- **Collaboration:** FireWatch users can search and retrieve appropriate government authorities that can aid in forest fire management. A number of different communication techniques such as alerts, sms, emails are available.

4   Conclusions

In this paper, we have presented FireWatch, a novel fire detection/prediction platform that combines technologies from the areas of Wireless Sensor Networks, Computer-supported Cooperative Work and Geographic Information Systems under a uniform framework. FireWatch virtualization mechanisms increase flexibility in deployment and provide a dynamicity, scalability and robustness. Finally, its multi-modal interface enables users to acquire a holistic view of the network, take informed decisions and coordinate easily with involved authorities.

Acknowledgements. This work is partly supported by the European Union under the project CONET (#224053), the project FireWatch (#0609-BIE/09) sponsored by the Cyprus Research Promotion Foundation, and the third author's startup grant sponsored by the University of Cyprus.

References

1. Andreou, P., Zeinalipour-Yiazti, D., Chrysanthis, P.K., Samaras, G.: Towards a Network-aware Middleware for Wireless Sensor Networks. In: 8th International Workshop on Data Management for Sensor Networks (DMSN 2011), The Westin Hotel, Seattle, WA, USA, August 29 (2011)
2. European Commision Joined Research Center: Forest Fires in Southern Europe. Report No.1 (July 2001)
3. European Forest Fire Information System – EFFIS, http://effis.jrc.it/Home/
4. Hochiki Europe, http://www.hochikieurope.com/
5. NASA MODIS website, http://modis.gsfc.nasa.gov/

AIMS: A Tool for the View-Based Analysis of Streams of Flight Data

Gereon Schüller[1], Roman Saul[1], and Andreas Behrend[2]

[1] Fraunhofer-Gesellschaft, Dept. FKIE-SDF, Neuenahrer Straße 20,
53343 Wachtberg, Germany
{gereon.schueller,roman.saul}@fkie.fraunhofer.de
[2] University of Bonn, Institute of CS III, Römerstraße 164, 53117 Bonn, Germany
behrend@cs.uni-bonn.de

Abstract. The Airspace Monitoring System (AIMS) monitors and analyzes flight data streams with respect to the occurrence of arbitrary, freely definable complex events. In contrast to already existing tools which often focus on a single task like flight delay detection, AIMS represents a general approach to a comprehensive analysis of aircraft movements, serving as an exemplary study for many similar scenarios in data stream management. In order to develop a flexible and extensible monitoring system, SQL views are employed for analyzing flight movements in a declarative way. Their definition can be easily modified, so that new anomalies can simply be defined in form of view hierarchies. The key innovative feature of AIMS is the implementation of a stream processing environment within a traditional DBMS for continuously evaluating anomaly detection queries over rapidly changing sensor data.

1 Introduction

The continuous growth in air space traffic challenges existing monitoring systems for air traffic control. It leads to situations where anomalies or critical events are detected too late or remain undetected at all. Many problems are caused by local deviations from flight plans which may induce global effects to aircraft traffic. One effect is a considerable number of close encounters of planes in airspace occurring every day. Another example is the violation of no-fly zones over cities for noise protection reasons, or over power plants for preventing terrorist attacks.

The Airspace Monitoring System (AIMS) [12] is a prototype of a system for monitoring and analyzing local and global air traffic.[1] It has been developed at the University of Bonn in cooperation with Fraunhofer FKIE and EADS Deutschland GmbH. One aim of this prototype is the detection of anomalies within the movements of single aircrafts (e.g., critical delays, deviations from flight plans or critical maneuvers). Based on that, a global analysis is supported (e.g., critical encounters, zones with high flight density, airport jams) which allows for adjusting air traffic influenced by local phenomena. Currently, the

---

[1] http://idb.informatik.uni-bonn.de/research/aims

A. Ailamaki and S. Bowers (Eds.): SSDBM 2012, LNCS 7338, pp. 622–627, 2012.
© Springer-Verlag Berlin Heidelberg 2012

system is used to monitor the complete German airspace every 4 seconds with up to 2000 flights in peak times. The key innovative feature of AIMS is the use of continuously evaluated SQL queries (stored in the DB as views, too) for declaratively specifying the situations to be detected. To this end, consistent tracks of individual planes have to be derived and complex event occurrences within these tracks have to be found. Our general research aim is the development of efficient DBMS-based methods for real-time gathering and monitoring of streams of track data. In particular we want to answer the following questions with respect to the airspace data: Is it possible to detect aircrafts entering a critical situation like collision course, leaving the flight path or entering bad weather zones using SQL views? How can flights and their behaviour be classified using SQL views defined over track data? How can tracking of aircrafts be improved by using (derived) context information? With respect to data stream management, the following research questions are addressed by AIMS: Which type of continuous query can be efficiently evaluated using a conventional relational DBMS? Up to which frequency/volume is it feasible to use a relational DBMS for evaluating continuous queries over a stream of track data? To which extent can an SQL-based analysis be used for airspace surveillance?

Using SQL queries as executable specifications has the advantage of being able to easily extend the system by additional criteria without having to re-program large amounts of code. In order to continuously re-evaluate the respective queries, data stream management systems (DSMS) like STREAM[2] or Aurora [3] could be used. However, DSMS do not provide all capabilities that are needed for a reliable in-depth analysis, like recovery control, multiuser access and processing of historical as well as static context data. In the practical demonstration, we want to show how commercial data base systems in combination with intelligent rule-rewriting can be used to processes a geospatial data stream. The proposed approach provides insights for the implementation and performance of related applications where geospatial or sensor data streams have to be processed.

2   System Architecture

The AIMS system consists of four main components: (1) A feeder component which takes a geospatial data stream as an input and periodically pushes its data into the database. This track feeder also continuously activates the re-evaluation of the anomaly detection views. (2) A graphical user interface programmed in Java using the NetBeans Platform library. It shows the positions of aircrafts on an OpenStreetMap. This map can be configured in order to show the result of selected queries, only. Additionally, query results are displayed in tabular form. (3) An Oracle server which stores the stream data and performs the continuous evaluation of the user-defined anomaly detection views. (4) An in-memory database that works as a cache and stores the results of selected queries on the client side. This database system forms the basis of a time-shift and video recorder functionality. A graphical representation of the architecture of the system is shown in Fig. 1.

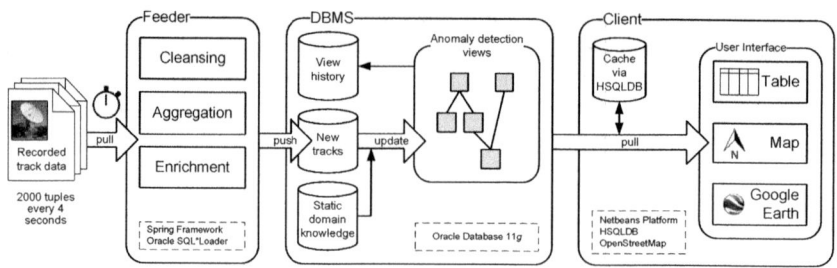

Fig. 1. Architecture of the AIMS system

3   View-Based Flight Analysis

Although there are various commercial implementations of flight tracking services (e.g., AirNav [1], FlightView [7] or FlightStats [6]), they are often limited to a set of predefined tasks like delay detection or identification of basic flight states such as departing, approaching or cruising. In order to develop a flexible and extensible monitoring system, SQL views can be employed for analyzing our track data. The advantage is that the underlying definition can be easily recovered and modified while new anomalies can be simply defined in form of view hierarchies. As a first example for an interesting event in airspace we consider *landing flights*. To this end, the plane must have a negative vertical speed (i.e., it is descending), it has to be below a certain flight level (like 3000 ft), and it must be in the vicinity of an airport, e.g. closer than 20 mi. These criteria can be expressed in SQL as follows:

```
CREATE VIEW vwLanding AS
SELECT * FROM Tracks t, airports a
WHERE t.vertSpeed<0 AND t.flightLevel<3000 AND dist(a.pos, t.pos)<20000;
```

where the function `dist` is a user defined function for calculating the Euclidic distance between two positions on the globe. The base table `airports` stores data about position and names of airports.

Critical encounters represent another interesting anomaly where two planes come closer than the prescribed distance of security. In AIMS a critical encounter is given if two planes are closer than 300 ft:

```
CREATE VIEW vwEncounter AS
SELECT * FROM Tracks t1, Tracks t2
WHERE dist(t1.pos, t2.pos)<300 AND t1.ID<>t2.ID;
```

Since the employed UDF-expression cannot be indexed, this implementation leads to a quadratic run-time. A better performance can be achieved by preselecting all those flights having approximate longitude and latitude values, as the values used can be indexed.

4 Incremental Stream Analysis

It is widely believed that DBMS are not well-suited for dynamically processing continuous queries. We believe, however, that even conventional SQL queries can be efficiently employed for analyzing a wide spectrum of data streams using incremental recomputation strategies.

For the incremental recomputation of our anomaly detection views, specialized view sets are used that reflect the difference between old and new data. The difference is represented by so-called *delta views*, which are denoted by X_i for induced insertions into X and by X_d for induced deletions. For example, the following view vwLandings_i has been derived from the original definition of the view vwLandings and provides all new tracks identified as landing planes:

```
CREATE VIEW vwLanding_i AS
SELECT * FROM Tracks_i t, airports a
WHERE t.vertSpeed<0 AND t.flightLevel<3000 AND dist(a.pos, t.pos)<20000;
```

The original landing view is materialized and update statements such as

```
INSERT INTO vwLanding SELECT * FROM vwLanding_i
```

are used for the incremental maintenance. This kind of incremental evaluation is well-known in literature and applied for efficient integrity checking and materialized view maintenance [9]. As soon as view hierarchies are considered, however, the optimization effect of using specialized delta views may be limited if no generalized selection pushing strategy is considered. As an example consider the following algebra expression for defining the view $P(x)$ based on the relations $Q(x)$, $R(x)$, $S(x, y)$ and $T(x, z)$:

$$P \leftarrow (Q \cup R) - \pi_x(S \bowtie T)$$

The following rule would yield the induced insertions P_i of P resulting from insertions Q_i into Q:

$$P\_i \leftarrow (Q\_i - R) - \pi_r(S \bowtie T)$$

Despite of the focus on changes with respect to Q, no optimization effect is achieved with respect to the evaluation of the right-hand argument of the set difference operator. A possible reordering of operations by using classical rules of algebraic optimization cannot provide a better focus on the changes of Q either. However, another way is to use the small number of tuples in Q_i already for determining all matching join partners by introducing two semi-joins:

$$P\_i \leftarrow (Q\_i - R) - \pi_x((Q\_i \ltimes S) \bowtie (Q\_i \ltimes T))$$

Under the assumption that S and T are quite large in comparison to the size of Q_i and that there is a low selectivity of the tuples in Q_i, the argument sizes of the join and difference operator are considerably reduced. Thus, the resulting incremental expression provides a much better focus on the changes to Q.

This improved delta expression results from applying the Magic Sets rewriting technique [4] which uses auxiliary relations (these are the 'Magic Sets') in order to store and to dynamically apply generated constants from delta views [8,10,11]. In AIMS, we used an extension of this technique, called Magical Updates, i. e., the application of Magic Sets to the transformed rule set for incremental evaluation. As an example, consider the following SQL statement that defines our view for detecting deviations from the flight plan:

```
CREATE RECURSIVE VIEW DELAY AS
(SELECT RealFlight.dist-(FlightPlan.dist+OldDelay.dist)
 FROM RealFlight rf, FlightPlan fp, Delay as OldDelay
 WHERE OldDelay.time   = fp.time - 1  AND fp.time = rf.Time
  AND  rf.Time = max (SELECT Time FROM RealFlight)
  AND  rf.Code = fp.Code AND fp.Code = OldDelay.Code);
```

Applying our transformation would yield the following rule for computing induced insertions from `RealFlight`:

```
CREATE RECURSIVE_i VIEW DELAY AS
(SELECT RealFlight.dist-(FlightPlan.dist+OldDelay.dist)
 FROM  RealFlight_i rf_i, FlightPlan fp, Delay as OldDelay ...
```

Despite of the focus on changes with respect to `RealFlight`, no optimization effect is achieved with respect to the evaluation of the join between RealFlight and FlightPlan and RealFlight and old Delay, respectively. Magic Updates, however, allows to use the small number of tuples in `RealFlight_i` already for determining all matching join partners by introducing the Magic Set `m_Delay_b` applied in two semi-joins:

```
CREATE VIEW m_Delay_i AS            CREATE VIEW m_FlightPlan_i AS
SELECT Delay_i.*                    SELECT Flightplan_i.*
FROM (m_RealFlight_b JOIN Delay_i)  FROM (m_RealFlight_b JOIN FlightPlan_i)
```

Under the assumption that `Delay` and `Flight` plan are quite large as compared to `Realflight_i` and that there is a low selectivity of the tuples in `Realflight_i`, the argument sizes of the joins are considerably reduced. Thus, the resulting incremental expression provides a much better focus on the changes to `Realflight`.

5   First Results

AIMS could successfully identify critical situations like close encounters or deviations from flight plans. It was interesting to notice the high number of large deviations (sometimes more than 50 mi). In addition, the high number of critical approaches was very surprising, as there were far more close encounters than expected. We could also show violations of no-fly zones and determine zones with a critically high number of aircraft movements. Currently we are working on the determination of abnormal landing approaches, and even these detection views can be efficiently evaluated.

Another result is that our incremental evaluation of SQL views provides indeed a suitable approach for analyzing this real world stream scenario. AIMS is capable of monitoring the entire German airspace by processing $\approx$1400 tuples (12 attributes) in real-time (every 3 - 4 seconds). For performance measurements, the system has been tested with recorded data which were periodically fed into the system with increasing update frequency. Our continuous queries such as critical encounters, landings and region violations, etc. can be processed in parallel in less than 0.3 seconds. The test system was a standard desktop PC with an Intel(R) Xeon(R) W3530 (2.8 GHz) processor, 12 GB RAM, a 256GB SSD. The software was run under Windows 7, Oracle 11g (ver. 11.2.0.10.0) and Java 64 bit, version 1.6.0_22.

6   Demonstration

In the demonstration, we first present the monitoring of the German airspace in real time. To this end, predefined views can be selected by the user and the respective monitoring results are visualized on a zoomable map. We show how the track feeder frequency can be adjusted and the respective performance of the selected anomaly detection views. We highlight how easily users can define new anomaly detection views in SQL without any 'traditional' programming. To this end, the audience may freely choose new criteria which are to be continuously monitored. In addition, users are allowed to search for close encounters and may switch back in time in order to see the development of this kind of anomalies.

References

1. AirNav Systems (2012), http://www.airnavsystems.com/
2. Arasu, A., et al.: STREAM: The Stanford Stream Data Manager. In: SIGMOD 2003, p. 665 (2003)
3. Abadi, D., et al.: Aurora: A Data Stream Management System. In: SIGMOD 2003, p. 666 (2003)
4. Bancilhon, F., Maier, D., Sagiv, Y., Ullman, J.D.: Magic sets and other strange ways to implement logic programs. In: PODS 1986, pp. 1–15 (1986)
5. Behrend, A., Manthey, R.: Update Propagation in Deductive Databases Using Soft Stratification. In: Benczúr, A.A., Demetrovics, J., Gottlob, G. (eds.) ADBIS 2004. LNCS, vol. 3255, pp. 22–36. Springer, Heidelberg (2004)
6. FLIGHTSTATS (2012), http://www.flightstats.com/
7. FlightView (2012), http://www.flightview.com/
8. Grant, J., Minker, J.: The impact of logic programming on databases. CACM 35(3), 66–81 (1992)
9. Gupta, A., Mumick, I.S.: Materialized Views: Techniques, Implementations, and Applications. MIT Press (1999)
10. Mumick, I.S., Finkelstein, S.J., Pirahesh, H., Ramakrishnan, R.: Magic is relevant. In: SIGMOD 1990, pp. 247–258 (1990)
11. Mumick, I.S., Pirahesh, H.: Implementation of Magic-Sets in a Relational Database System. SIGMOD Record 23(2), 103–114 (1994)
12. Schüller, G., Behrend, A., Manthey, R.: AIMS: an SQL-based system for airspace monitoring. In: ACM SIGSPATIAL IWGS 2010, pp. 31–38 (2010)

TARCLOUD: A Cloud-Based Platform to Support miRNA Target Prediction

Thanasis Vergoulis[1,2], Michail Alexakis[2], Theodore Dalamagas[2],
Manolis Maragkakis[4], Artemis G. Hatzigeorgiou[3], and Timos Sellis[1,2]

[1] NTUA, Athens, Greece
[2] IMIS, "Athena" R.C., Athens, Greece
[3] DIANA-Lab, B.S.R.C. "Alexander Fleming", Athens, Greece
[4] University of Pennsylvania, Philadelphia, USA

Abstract. Micro RNAs (miRNAs) are small RNA molecules that target protein coding genes and inhibit protein production. Since experimental identification of miRNA targets poses difficulties, computational miRNA target prediction is one of the key means in deciphering the role of microRNAs in development and disease. However, these computational methods are CPU-intensive. For example, the predictions for a single miRNA molecule on the whole human genome according to a popular target prediction method require about 30 minutes. Such performance is a hindrance to the biologists' requirement for near-real time target prediction. In this paper, we present TARCLOUD, a Cloud-based target prediction solution built on Microsoft's Azure platform. TARCLOUD is a highly-scalable solution based on distributed programming models that provides near-real time predictions to its users through an easy and intuitive interface. The work has been selected as one of the pilot use cases for the VENUS-C FP7 Research Infrastructures Program.

Keywords: Cloud Computing, miRNA target prediction.

1 Introduction

For many years, biologists used to consider that only the regions of genome which translate into proteins are important for life. This has dramatically changed after the discovery, in the late 1990s, of regions in the "non-translated" genome playing a key role in several life functions. Among those functions, one of the most important is the silencing of genes by small RNA molecules, called micro RNAs (*miRNAs*). In brief, each miRNA *targets* particular genes, destroying their transcripts and, consequently, prohibiting the production of the encoded protein.

Knowing the miRNAs that target a particular gene helps to understand the causes of human diseases and searching for treatments. However, the laboratory experiments that determine targets are costly and time-consuming. Therefore, computational methods to predict miRNA targets have been proposed. The first *miRNA target prediction methods* were developed back in 2003. During the last

A. Ailamaki and S. Bowers (Eds.): SSDBM 2012, LNCS 7338, pp. 628–633, 2012.

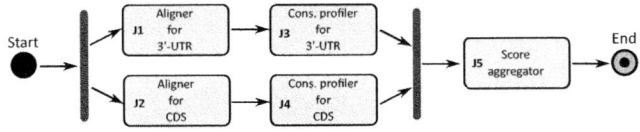

Fig. 1. The workflow of `TARCLOUD` jobs

decade, more than a dozen of such methods were proposed, making the field of miRNA target prediction one of the most active in bioinformatics. An excellent survey of the field can be found in [1].

The `microT` [6] method is one of the most cited and applied miRNA target prediction techniques. Its Web interface[1] provides access to the targets of the most known miRNAs[2] for two species ("Homo sapiens" and "Mus musculus"). Furthermore, it allows users to upload miRNA sequences and apply the `microT` method on the desired genome.

Unarguably, the most requested feature by `microT` users is near-real time target prediction. However, `microT` is extremely computationally intensive, and in its current single server-based implementation requires about 30 minutes to execute. In order to achieve near-real time performance, we have designed and developed `TARCLOUD`, a Cloud-based target prediction method based on `microT`. `TARCLOUD` utilizes Microsoft's Azure distributed architecture to improve efficiency, and provides an easy and intuitive interface. Moreover, it has been selected as one of the pilot use cases for *VENUS-C*[3], which is a European Commission's 7th Framework Programme project with the goal to provide Cloud computing infrastructures for scientific applications in Europe.

2   The TARCLOUD Solution

First, Section 2.1 discusses the implementation platform of `TARCLOUD`. Then, Section 2.2 details its architecture, and Section 2.3 presents its user interface.

2.1   Framework

`TARCLOUD` is implemented using the *Microsoft Azure*[4] Cloud platform. More specifically, this platform consists of various services commoditized through three product brands. These are (a) *Windows Azure*, an operating system providing scalable compute and storage facilities, (b) *SQL Azure*, a Cloud-based version of SQL Server, and (c) *Windows Azure AppFabric*, a collection of services supporting Cloud applications. The platform provides an API built on REST, HTTP and XML that allows a developer to interact with the Azure services.

---

[1] http://diana.cslab.ece.ntua.gr/microT/
[2] These are stored in the current version of MirBase (http://www.mirbase.org/)
[3] http://www.venus-c.eu
[4] http://www.windowsazure.com

2.2 Architecture

Before discussing the TARCLOUD components, we first present an overview of its processes, which are based on the microT method [6].

Biologists have established that nucleotides close to the 5'-end of the miRNA are crucial for recognizing a target gene sequence and binding to it [2,3,4,5]. We refer to the miRNA subsequence defined by these nucleotides as the *seed* of the miRNA. For each miRNA, TARCLOUD locates, in the genome, approximate occurrences of its seed. We call these occurrences *"miRNA recognition elements" (MREs)*. Note, that as miRNA targets have been verified only in the 3'-untranslated regions (3'-*UTR*s) or in the coding regions (*CDS*s) of gene transcripts[5], TARCLOUD seeks MREs only in 3'-UTR and CDS regions of the genes.

After identifying MREs, TARCLOUD computes a score for each of them. This score is computed considering many parameters, such as the alignment quality, the binding energy, the conservation of the area in a set of species, etc. Then, TARCLOUD aggregates these scores for the MREs of each gene (taking into consideration the molecular folding) and produces a *"target prediction" score* for each gene. This score indicates how possible is for each gene to be a target of a particular miRNA molecule.

Based on the previous, a TARCLOUD task consists of five distinct jobs. Figure 1 presents the workflow showing the dependencies among these jobs. Each rectangle represents one independent TARCLOUD job. The arrows depict the data flow between the jobs. The grey vertical bars represent a fork or a join. A fork ignites concurrent activities, while a join merges the output of concurrent activities. These jobs, labelled J1 through J5, are performed by the following TARCLOUD components.

Aligner (J1 and J2). The *aligner* component is responsible for (a) locating the MREs, (b) computing the alignment score of the seed, and (c) computing the binding energy of the miRNA to the gene (by using RNAhybrid [7]). For more details about how the alignment is done see also [6]. The aligner executes the job J1 for the 3'-UTR regions of all genes of the selected genome, and J2 for the CDS regions. Note that these two jobs are executed concurrently.

Conservation Profiler (J3 and J4). The *conservation profiler* component is responsible for checking how particular MREs are preserved in the genomes of several species. Currently, TARCLOUD uses up to 27 species to assess the MRE conservation profile, taking into account both conserved and non-conserved MREs for the estimation of the final score. Similar to the aligner, the conservation profiler executes the job J3 for the 3'-UTR regions of all genes of the selected genome, and J4 for the CDS regions. These jobs are also executed concurrently.

Score Aggregator (J5). The *score aggregator* component computes, for each gene, the aggregated score of all its MREs. The aggregation is a weighted sum

---

[5] CDS is the region of the gene transcript which encodes the protein, while the 3'-UTR is the region preceded by CDS and which contain several "regulatory sequences".

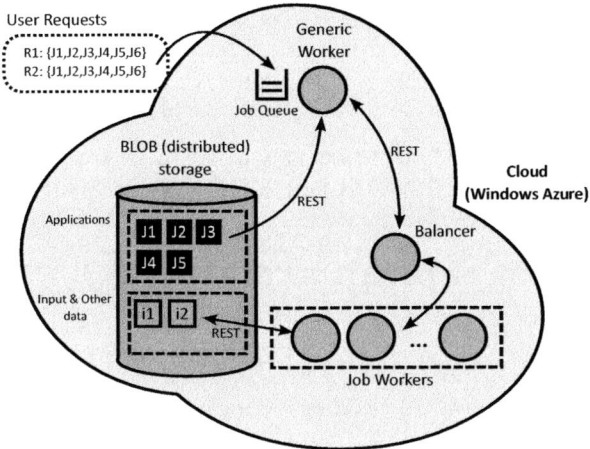

Fig. 2. The TARCLOUD architecture

that also considers the molecular folding (i.e., the 3-D structure of the molecules). The score aggregator executes the job J5.

The TARCLOUD architecture, illustrated in Figure 2, involves Virtual Machines (VMs), called *Job Workers*, fully capable to perform any of the previous jobs. Each Job Worker listens for HTTP (REST) requests and executes the TARCLOUD jobs described by these requests. Any input and output data are stored mostly as BLOBs in the distributed Cloud storage of Microsoft Azure. We refer to this storage space as *the BLOB storage*. The number of Job Worker instances deployed for a particular configuration of TARCLOUD, is a design parameter crucial for the efficiency of the system. In order to distribute the HTTP requests to the Job Worker instances, we also deploy a VM, called *Balancer*. Note that it is not essential that the Balancer is located in a separate VM. One of the Job Workers can also host the Balancer.

For each of the TARCLOUD jobs, a separate Azure application[6] is implemented and stored in the BLOB storage. These applications are just thin clients who create HTTP (REST) requests. The requests tell the Balancer to occupy some of the Job Workers to execute a particular TARCLOUD job. Another VM, called *Generic Worker (GW)*, has the responsibility to load the previous applications from the BLOB storage, and execute them in the order determined by the workflow of Figure 1, after a user request. User requests are accumulated into a *job queue*. To achieve synchronization, GW just checks if the input for each job is ready. In brief, for each job, GW knows the URIs of its input files (in the BLOB storage) and, before the job execution, performs polling to check if the input files are complete. The execution starts only when the input files are ready. For instance, for the workflow of Figure 1, GW will not execute J3, until the output files of J1 (which are input for J3) are ready.

---

[6] Azure applications are 32-bit Windows executables.

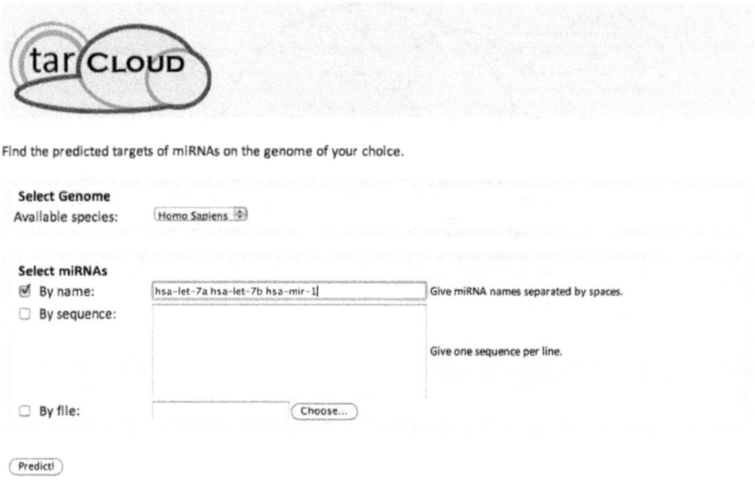

Fig. 3. A snapshot of the TARCLOUD search interface

2.3   User Interface

We also implemented a Web interface, to give to TARCLOUD users an easy and intuitive way to start miRNA target prediction tasks. Figure 3 illustrates a snapshot of this interface. The user just selects the desired species from a drop-down list. Then, she selects one or more miRNA sequences. She has 3 options: (a) select already known miRNA sequences by inserting their names in the "Select miRNA by name" field (we store all miRNAs in the latest version of miRBase), (b) select unknown miRNAs by determining their sequences in the "Select miRNA by sequence" field, or (c) select a set of unknown miRNAs by uploading a file containing a name and a sequence for each of them. When the user clicks on the "Predict!" button, a request for the previously determined job is sent to the Generic Worker.

When the output is ready and stored in the BLOB storage, the TARCLOUD Web interface renders a webpage containing the output. The user has two options: (a) browse the predicted targets by using our build-in results renderer (see Figure 4), or (b) download the output text files to his own computer.

3   Demonstration Scenarios

We demonstrate the functionality of TARCLOUD by executing queries provided by us and by members of the audience, and by explaining the benefits of our architecture. Next, we discuss two demonstration scenarios.

Scenario 1. In this use case, the user can test several configurations of TARCLOUD and see in practice how do they affect the efficiency of our system. In particular,

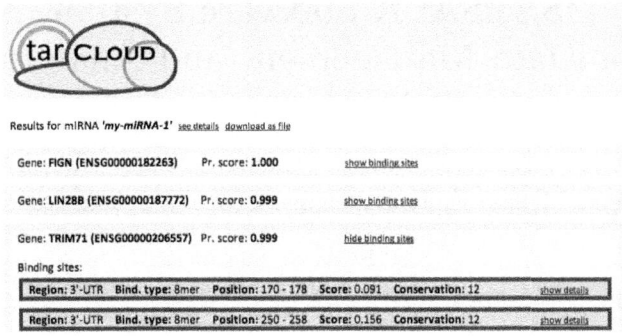

Fig. 4. A snapshot of the TARCLOUD results renderer for miRNA "my-miRNA-1"

for the same species and miRNA sequences, the user selects to execute TARCLOUD by deploying, each time, a different number of Job Worker instances.

Scenario 2. In this case, the user can see the efficiency of TARCLOUD in comparison to this of the original single server based implementation of microT. The user selects the desired species and miRNA sequences in both systems, and, then, she requests the predictions. This scenario demonstrates the superiority of TARCLOUD, as it often responds within a few seconds, being the only option to produce near-real time predictions.

References

1. Alexiou, P., Maragkakis, M., Papadopoulos, G.L., Reczko, M., Hatzigeorgiou, A.G.: Lost in translation: an assessment and perspective for computational microrna target identification. Bioinformatics 25(23), 3049–3055 (2009)
2. Doench, J.G., Sharp, P.A.: Specificity of microrna target selection in translational repression. Genes Dev. 18(5), 504–511 (2004)
3. Kiriakidou, M., Nelson, P.T., Kouranov, A., Fitziev, P., Bouyioukos, C., Mourelatos, Z., Hatzigeorgiou, A.G.: A combined computational-experimental approach predicts human microrna targets. Genes Dev. 18, 1165–1178 (2004)
4. Krek, A., Grün, D., Poy, M.N., Wolf, R., Rosenberg, L., Epstein, E.J., MacMenamin, P., da Piedade, I., Gunsalus, K.C., Stoffel, M., Rajewsky, N.: Combinatorial microrna target predictions. Nature Genetics 37, 495–500 (2005)
5. Lewis, B.P., Burge, C.B., Bartel, D.P.: Conserved seed pairing, often flanked by adenosines, indicates that thousands of human genes are microrna targets. Cell 120, 15–20 (2005)
6. Maragkakis, M., Reczko, M., Simossis, V.A., Alexiou, P., Papadopoulos, G.L., Dalamagas, T., Giannopoulos, G., Goumas, G., Koukis, K., Kourtis, K., Vergoulis, T., Koziris, N., Sellis, T., Tsanakas, P., Hatzigeorgiou, A.G.: Diana-microt web server: elucidating microrna functions through target prediction. Nucleic Acids Research 37(suppl. 2), W273–W276 (2009)
7. Rehmsmeier, M., Steffen, P., Hochsmann, M., Giegerich, R.: Fast and effective prediction of microrna/target duplexes. RNA 10, 1507–1517 (2004)

SALSA: A Software System for Data Management and Analytics in Field Spectrometry⋆

Baljeet Malhotra[1], John A. Gamon[2], and Stéphane Bressan[3]

[1] SAP Research, Singapore
baljeet.malhotra@sap.com
[2] University of Alberta, Canada
gamon@ualberta.ca
[3] National University of Singapore
steph@nus.edu.sg

Abstract. Field spectrometry is emerging as an important tool in the study of the dynamics of the biosphere and atmosphere. Large amounts of data are now collected from spectrometers mounted on towers, robotic trams and other platforms. These data are crucial for verifying not only the optical data captured by satellites and airborne systems but also to validate the flux measurements that track ecosystem-atmosphere gas exchanges, the "breathing of the planet" critical to regulating our atmosphere and climate. There is a need for readily available systems for the management, processing and analysis of field spectrometry data. In this paper we present SALSA, a software system for the management, processing and analysis of field spectrometry data that also provides a platform for linking optical data to flux measurements. SALSA is demonstrated using real data collected from multiple research sites.

1 Introduction

Generic data management and geographical information systems are natural candidates for the management, processing and analysis of data in earth and atmospheric sciences. Yet, because of the specificity of workflows and data, existing systems do not seem to meet the researchers' needs. The bottom-up construction of systems for particular tasks helps understand the nature of the original requirements and shall eventually lead to the extension and tuning of generic platforms to the specific needs of these applications. We consider here the case of *field spectrometry* that is used for measuring the optical properties of ecosystems. For instance, spectral reflectance (a specific form of optical measurement that expresses a surface's reflectance as a function of wavelength) is commonly used to compute the Normalized Difference Vegetation Index (NDVI) [7] to capture the greenness of the vegetation being monitored. NDVI is then used to quantify the exchanges of gasses (fluxes) between the biosphere and atmosphere,

---

⋆ This research was partially funded by NSERC Canada, SAP Research, Singapore and the A*Star SERC project "Hippocratic Data Stream Cloud for Secure, Privacy-preserving Data Analytics Services" 102 158 0037, NUS Ref:R-702-000-005-305.

A. Ailamaki and S. Bowers (Eds.): SSDBM 2012, LNCS 7338, pp. 634–639, 2012.

i.e., the breathing of vegetation [8]. To that end large amounts of optical data are collected from multi-angular towers, robotic trams and other systems equipped with optical sensors. These data are then used to calibrate or validate similar data collected from aircraft or satellite, which are then used for regional or global estimates of carbon flux that relate to the breathing of the planet.

While NASA's MODIS (Moderate Resolution Imaging Spectroradiometer) project [6] allows the collection management and sharing of satellite data, less effort has been put in resolving the data management, processing and analysis issues for field spectrometry. Some of these issues indeed are specific and compel the design and implementation of specific solutions. There are two main reasons for this. First, field spectrometry is performed locally at various spatial and temporal resolutions, and by groups of researchers that operate in isolation from each other resulting in various systems of field spectrometry. Second, due to the many brands and configurations in use, field spectrometry leads to unique workflows that do not subscribe to a common set of standards, and hence produce data in various structures and formats.

In this paper we discuss the data management issues arising for such field spectrometry based systems and present SALSA, a software system for the processing and analysis of field spectrometry data. SALSA provides a platform for the linking of optical data with flux measurements that is crucial to our understanding of the breathing of the planet. We demonstrate the effectiveness and efficiency of the system with large volumes of raw data collected from the robotic trams used for field spectrometry.

2   Field Spectrometry

Spectrometry refers to the measurement of light reflectance in the visible, near and short-wave infrared wavelengths (approx. 400–2500 nm). In ecological applications, it helps understand and quantify vegetation composition, phenology (seasonality), changing surface albedo and biosphere-atmosphere fluxes. Field spectrometry is done from the ground and can be compared with remote optical sensing from scientific satellites or airborne systems that operate above ground. The MODIS project, a part of NASA's Earth Observing System Data and Information System (EOSDIS), collects, store and share satellite spectroradiometer images and data. EOSDIS is arguably the largest and most sophisticated scientific database for earth observation [5].

While very consistent, satellite and airborne systems can be confounded by changing sky conditions (e.g., clouds) in optical wavelengths. By sampling under the atmosphere and correcting for changing sky conditions field spectrometry avoids these problems, providing a valuable means of "ground truthing" satellite and aircraft measurements. A recent trend (past 5-10 years) has been to automate the instruments from stationary (e.g. multi-angular towers [4]) and mobile (robotic trams [2]) platforms that operate on the ground by Photosynthetically Active Radiation (PAR) sensing. A primary reason for this development and automation is that it allows effective and efficient measurement and calculation of various vegetation indices, which can then be used to relate to the gas exchange within individual sampling regions ("footprint") of the eddy covariance towers (flux towers) [1]. Automated field spectrometry is essential for the validation of fluxes modelled from satellite [8], and can assist in scaling up from local flux measurements to regional and global measurements. Efficient data management is a key to successfully automate the field spectrometry workflows and processes, which is the precise problem that we address in this paper. Before presenting our software system, for data management in field spectrometry, next we briefly describe a robotic tram system for field spectrometry.

2.1 A Robotic Tram System

A *tram system* consists of a cart with instruments riding on a rail [2], and can be automated (robotic) to simplify field sampling. A typical robotic tram is equipped with two spectral detectors (channels A and B), one to measure spectral reflectance or simply radiance (light reflected from the vegetation on ground) and one to sample irradiance (light from the sky). The two spectrometers are mounted on the tram, one each *looking* downward and upward to measure radiance and irradiance, respectively. The tram moves (back and forth) on a fixed path (of approx. 100 m long) while taking optical samples with the two spectrometers operating simultaneously. This process is repeated several times throughout a day for several weeks or even months. Irradiance measurements are needed to *correct* radiance measurements according to the sky conditions that may change during the operation of the tram. Furthermore, to obtain reflectance, radiance measurements (from vegetation) need to be corrected by radiance measurements taken from a *white panel.* Essentially, a tram has two sampling modes one each for taking measurements from (i) target vegetation and (ii) white panel. Given $\rho_{r-X}^{(\omega)}$ and $\rho_{i-X}^{(\omega)}$ radiance and irradiance measurements, respectively, corresponding to wavelength, ω, where X represents one of the two sampling modes, i.e., *target* or *panel*, the sampling procedure can be summarized in the following equation.

$$\rho^{(\omega)} = \frac{\rho_{r-target}^{(\omega)}}{\rho_{i-target}^{(\omega)}} \times \frac{\rho_{i-panel}^{(\omega)}}{\rho_{r-panel}^{(\omega)}} \tag{1}$$

In a typical situation, system generates raw data in a tabular format that generally contain three columns and multiple rows. The top portion of the file contains the metadata of the captured optical data. After the metadata description, the left column in the file contains the wavelengths (typically between 300 to 1130 nm) of the spectrum being scanned. The next two columns in the file contain the radiance and irradiance measurements corresponding to the wavelengths mentioned in the left column. Typically, one such file is generated per scan during the movement of the robotic tram. A typical field study spanning several weeks could easily contain hundreds or even thousands of scans performed by the tram system producing that many numbers of files containing raw optical data. One primary reason for measuring the spectral reflectance (corrected and cross-calibrated) is to calculate *vegetation indices,* which are then used in models to estimate carbon fluxes [8]. A commonly used index is NDVI [2]. Given optical wavelengths, ω_1 and ω_2, NDVI is computed as follows.

$$NDVI^{(\omega_1,\omega_2)} = \frac{\rho^{(\omega_1)} - \rho^{(\omega_2)}}{\rho^{(\omega_1)} + \rho^{(\omega_2)}} \tag{2}$$

3 A Software System

One of the main functionalities of the software system that we intend to build lies at automatically computing the corrected reflectance as mentioned in Eq. 1 and various indices such as the one shown in Eq. 2. It is a non-trivial task not only because large volumes of raw data are scattered in multiple files, but automation essentially requires logical separation of data into multiple clusters representing radiance and irradiance measurements taken from either target vegetation or white panels. To that end a tram system poses two main challenges.

First: The column for radiance (or irradiance) measurements in raw data files is not fixed. For instance, in one file the radiance measurements could lie in the middle column and in another file they could be in the right column. The format of raw data files actually depends on users' choice of a particular detector (out of two) for recording radiance (or irradiance) measurements. Since users may (knowingly or unknowingly) choose different settings of a tram, the raw data may be produced in various formats.

Second: Each raw data file may either belong to the target vegetation or white panel. Recall that to obtain reflectance the radiance measurements from the target vegetation need to be cross-calibrated against the radiance measurements from the white panel as shown in Eq. 1. Though researchers take notes (e.g., which particular scan and the corresponding raw data file belongs to which type of target, i.e., vegetation or white panel) during the field measurements, these notes are susceptible to errors or misplacement.

To address the above issues, we designed and developed a software system, SmArt Light Spectrum Analyzer or SALSA for short, that *smartly* detects irradiance and radiance measurements and also senses whether they correspond to white panels or target vegetation. Apart from the problems discussed above data cleaning, interpolation and noise detection are some of the other issues SALSA addresses. Next, we present some technical details of SALSA.

3.1 Functionalities and Technicalities

SALSA's main functionalities are grouped into three main categories: (i) raw data processing, (ii) index generation, and (iii) visualization/reporting. Users, first upload raw data files using "Extract Raw Data" option from the main menu of SALSA. As raw data are extracted from multiple files several sequential steps are performed, e.g., radiance and irradiance data are separated, noisy samples are detected and *removed* from further processing based on users' confirmation. Users are also given choice for data interpolation to discrete (e.g. 1-nm) intervals (a necessary step for calculating vegetation indices or comparing different instruments with different calibrations).

An important step toward automatically computing the corrected reflectance is to separate the irradiance and radiance data. This is achieved by exploiting the fact that the irradiance measurements are usually greater than the radiance measurements, particularly in a specific wavelength range. To that end we used a binary classification procedure that is used to detect the channels used for irradiance and radiance measurements. The input to the procedure is a set of samples, S. Every sample, $s \in S$, consists of two sets of measurements, i.e., $\rho_A{}^{(\Omega)}$ and $\rho_B{}^{(\Omega)}$ recorded simultaneously on the channels A and B, respectively. The procedure selects only a subset of the recorded measurements corresponding to a particular set of wavelengths, Ω. For a given sample, if all the measurements recorded (for the specified wavelengths Ω) on channel A are greater than the measurements recorded on channel B, then the counter ccA is incremented. This process is repeated for all the samples. Finally, channel A is classfied as irradiance channel if majority of samples recorded by it has greater spectrum values than the channel B; otherwise channel B is classified as irradiance channel. In our experiments, we found that the above simple procedure classified the channels correctly in most of the cases. In these experiments Ω typically consisted of wavelenghts in the range of 330~500 nm.

Through the white panel correction menu of SALSA users can complete the cross-calibration and obtain reflectance. Users first choose a set of white panel samples they wish to use for cross-calibration. To that end user can *manually* select white panel samples, e.g., by visualization in SALSA or by referring to their field notes. Or users can also invoke SALSA's feature for automatically detecting white panel samples. SALSA uses a clustering algorithm to distinguish vegetation samples from white panel samples based on their spectral *shape*, which are usually different. To be precise, after irrandiance and radiance measurements are seperated (as described previously), the raw reflectances are computed using the formula as follows: $\frac{\rho_{r-X}^{(\omega)}}{\rho_{i-X}^{(\omega)}}$. Note that the raw reflectance can be computed without knowing whether measurements are actually from traget or white panel samples, i.e., without knowing X in the above formula. A sample is classified as white panel sample if it satisfies the following inequality.

$$\sum_{\omega=p}^{q} \frac{\rho_{r-X}^{(\omega)}}{\rho_{i-X}^{(\omega)}} > \sum_{\omega=r}^{s} \frac{\rho_{r-X}^{(\omega)}}{\rho_{i-X}^{(\omega)}}; \tag{3}$$

otherwise the sample is classified as a target sample. The typical values that we used in our experiments for p, q, r, and s were 325, 345, 875, and 895 nm, respectively. By exploiting the spectral shape of the samples within the above wavelengts range, SALSA is able to distinguish the white panel samples from the target samples in most of the cases. After detecting white panel samples, users can choose the target samples that need to be corrected against the detected (and chosen) white panel samples. All the choices that are made by the users are tracked and stored using the metadata management module of SALSA. In this way, processing steps can be traced, providing greater transparency.

SALSA has a data analytics module, which computes important indices including the one mentioned in Eq. 2. One useful feature of this module is its ability to link other forms of optical data, such as the ones produced by satellites, with field spectrometry data. Oftentimes, researchers *simulate* satellite remote sensing data using data from on or above ground sensors for crop and resource management [3]. In SALSA, simulation of various satellites' sensors is provided through field spectrometry data. Currently, SALSA simulates MODIS, LANDSAT-7, LANDSAT-MSS, QUICKBIRD, ASTER and SPOT-5 satellites.

Another useful feature of SALSA is its visualization and reporting module, which allows users to visualize and generate various reports based on raw and processed data. Using this menu users can plot raw radiance and irradiance measurements as well as processed optical data, i.e., corrected reflectance and various indices. Through the reporting menu users can generate reports in various formats. This feature is particularly useful when researchers need to exchange processed data for further analysis and data integration purposes.

4   Conclusions

In this paper we presented an overview of SALSA, a software system that we designed, developed and used for efficient processing and analysis of field spectrometry data. SALSA software and some sample datasets are available at: http://www.ualberta.ca/~salsa/. SALSA offers a platform for linking optical data from various sources such as MODIS and some other satellites as well as with flux measurements for the better understanding of the breathing of the planet. In future we would like to include new modules of SALSA that will facilitate the linking of flux measurements with the optical data in various forms and structures in a cloud computing environment.

References

1. Baldocchi, D., et al.: Fluxnet: A new tool to study the temporal and spatial variability of ecosystem- scale carbon dioxide, water vapor, and energy flux densities. Bulletin of the American Meteorological Society 82(1), 2415–2434 (2001)
2. Gamon, J.A., et al.: A mobile tram system for systematic sampling of ecosystem optical properties. Remote Sensing of Environment 103(3), 246–254 (2006)
3. Harma, P., et al.: Detection of water quality using simulated satellite data and semi-empirical algorithms in Finland. The Science of The Total Environment 268(1–3), 107–121 (2001)
4. Hilker, T., et al.: Tracking plant physiological properties from multi- angular tower-based remote sensing. Oecologia 165(4), 865–876 (2011)
5. Marshall, E.: Fitting plant earth into a user-friendly database. Science 261(13), 846–848 (1993)
6. MODIS, http://modis.gsfc.nasa.gov/
7. Rouse, J.W., et al.: Monitoring vegetation systems in the great plains with ERTS. In: 3rd ERTS Symposium, NASA, vol. 1, pp. 309–317 (1973)
8. Running, S., et al.: A continuous satellite-derived measure of global terrestrial primary production. Bioscience 54, 547–560 (2004)

Incremental DNA Sequence Analysis in the Cloud

Romeo Kienzler[1], Rémy Bruggmann[2], Anand Ranganathan[3], and Nesime Tatbul[1]

[1] Department of Computer Science, ETH Zurich, Switzerland
romeok@student.ethz.ch, tatbul@inf.ethz.ch
[2] Bioinformatics, Department of Biology, University of Bern, Switzerland
remy.bruggmann@biology.unibe.ch
[3] IBM T.J. Watson Research Center, NY, USA
arangana@us.ibm.com

Abstract. In this paper, we propose to demonstrate a "stream-as-you-go" approach that minimizes the data transfer time of data- and compute-intensive scientific applications deployed in the cloud, by making them incrementally processable. We describe a system that implements this approach based on the IBM InfoSphere Streams computing platform deployed over Amazon EC2. The functionality, performance, and usability of the system will be demonstrated through two DNA sequence analysis applications.

1 Introduction

In many areas of science, huge amounts of data is being generated at rates that outrun the ability of researchers to store, transmit, and analyze it. For example, in DNA sequence analysis, complex workflows need to be efficiently executed over digital DNA fragments that are now being generated much faster and cheaper owing to the recently invented Next Generation Sequencing (NGS) methods [17]. For such data- and compute-intensive scientific applications, researchers are increasingly turning to cloud computing as a scalable and cost-effective solution. In this case, raw input data that is generated by special scientific devices (e.g., NGS machines) outside the cloud must first be shipped into the cloud. However, due to limited bandwidth between the client and the cloud, transferring large data sets into the cloud can introduce significant latencies and may even become a bottleneck that hinders the scalability advantage of the cloud.

In our recent work, we have proposed an incremental data access and processing approach for data- and compute-intensive cloud applications that can hide data transfer latencies while maintaining linear scalability [7], [8]. In our approach, data is accessed in a "stream-as-you-go" fashion instead of in whole batches, making a stream-based data management architecture a suitable base for implementation. In this demonstration, we propose to show the functionality, performance, and usability of our approach in action through two practical applications of DNA sequence analysis:

1. **Read alignment:** This application involves a very basic and common process in DNA sequence analysis workflows: aligning digital DNA fragments, called *reads*, against a reference genome. In the demo, we will show how a well-known read aligner package (SHRiMP [13]) as part of a more complex workflow can be transparently replaced with our stream-as-you-go version to incrementally run in the

A. Ailamaki and S. Bowers (Eds.): SSDBM 2012, LNCS 7338, pp. 640–645, 2012.

cloud, both producing early results as well as significantly reducing the total processing time of the whole workflow.

2. **SNP detection:** This application additionally involves detecting SNPs (Single Nucleotide Polymorphisms [3]) as reads are being aligned against a reference genome. Different from the first demo, we will show how a complete workflow can be replaced with our stream-as-you-go version and can be pushed into the cloud through an easy-to-use client interface. In this demo, we will use Bowtie [10] (instead of SHRiMP) as the read alignment package and SOAPsnp [12] as the SNP detection package, making our approach directly comparable to the state of the art performance-wise (i.e., the MapReduce-based approach of Crossbow [9]). Thus, we will also report on our performance improvements.

In the rest of this paper, we describe in more detail, our stream-as-you-go approach and the applications that will be used to demonstrate its key features.

2   The Stream-as-you-go Approach

Our key idea to address the data upload latency of scientific applications deployed in the cloud is to enable useful data processing as soon as the first piece of the data set hits the cloud rather than waiting until the arrival of the whole data set. This way, the data transfer latency can be hidden by overlapping it with data processing time. To realize this idea, we propose to use a stream-based data management platform. Our main motivation to do so is to exploit the incremental and in-memory data processing model of Stream Processing Engines (SPEs) (in our specific implementation, the IBM InfoSphere Streams engine [2]). More specifically, we bring (parts of) existing scientific workflows (algorithms/software) into the cloud in a way that they can work with their input data in an incremental fashion. One generic way of realizing this is to use command line tools provided by most of these software. They commonly read and write to standard Unix pipes, which we can exploit by building custom streaming operators that wrap the relevant Unix processes. Then the SPE essentially acts as the middleware to handle all system-level requirements such as inter-process communication, data partitioning and dissemination, operator distribution, and dynamic scaling.

Figure 1 illustrates our general approach. As seen, the main goal is to provide partitions of the source data to the analysis processes running in parallel on different slave cloud nodes in a streaming fashion. This way, data transfer time can be hidden and early results can be generated. Furthermore, incremental processing of streaming data also allows in-memory processing, eliminating the latency of disk access.

3   The DNA Sequence Analysis Use Case

Determining the order of the nucleotide bases in DNA molecules and analyzing the resulting sequences have become very essential in biological research and applications. With the invention of the NGS methods in 2004 [17], higher amounts of genetic data can be read in much less time and at lower cost [7], which has led to the generation of very large datasets to be efficiently analyzed. The output of NGS machines are random

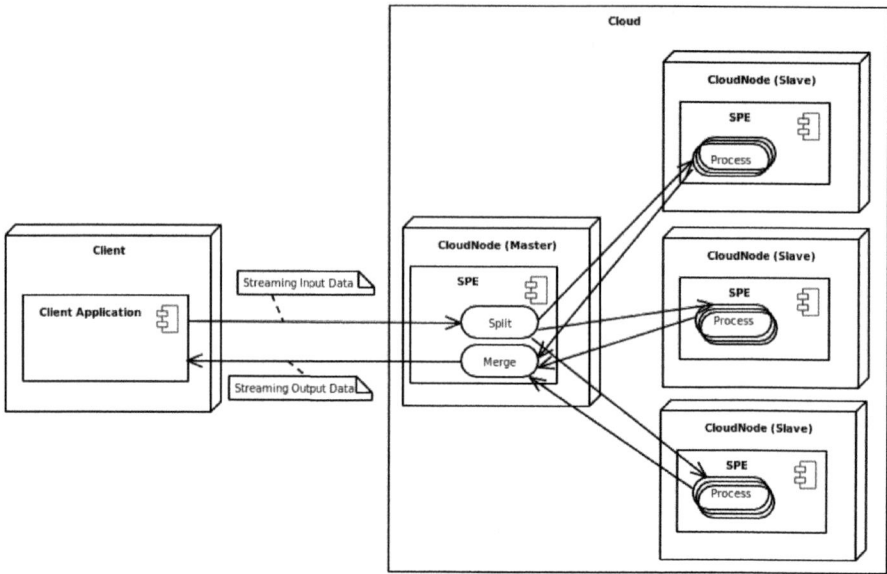

Fig. 1. The stream-as-you-go approach

DNA fragments (reads) of short length. Therefore, they must first be aligned into a complete sequence by mapping them back to a reference genome [11]. The alignment can also highlight the differences against the reference. Such a difference is called a polymorphism. The polymorphism of a single DNA letter is called *Single Nucleotide Polymorphism (SNP)*. SNPs are important to identify, since they are recognized as the main cause of human genetic variability [5]. As such, read alignment and SNP detection are two common, computationally-intensive applications in this domain.

Researchers have recently started using cloud infrastructures for various DNA sequence analysis applications. The current state of the art in massively parallel analysis of large genomic data sets is mainly based on using MapReduce [6] or other similar frameworks [15]. Prominent examples include CloudBurst [14] and CloudAligner [16] for read alignment, and Crossbow [9] for the complete SNP detection process. Despite providing basic scalability, all these solutions suffer from the data transfer latency, since the cloud frameworks that they are based on are primarily designed for batch processing of data stored in a distributed file system in the cloud. In the following, we describe how read alignment and SNP detection can be modeled and implemented using our stream-as-you-go approach, which overcomes this bottleneck.

3.1   Read Alignment a la Stream-as-you-go

Figure 2 shows our stream-as-you-go implementation of the read alignment application. The complete workflow consists of an input data format conversion process, the SHRiMP read aligner process, and an output data format conversion process. Only the SHRiMP part of the workflow is replaced with a stream-as-you-go version deployed

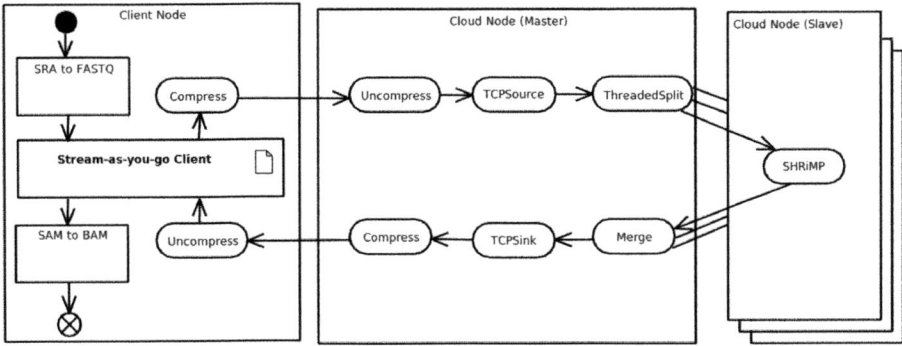

Fig. 2. Stream-as-you-go implementation of read alignment

in the cloud, while the original data conversion processes continue to run at the client node. The client application sends compressed read data to the cloud. After being uncompressed, data gets submitted to a Streams application which first splits it across the available cluster nodes. Split reads are then aligned in parallel using the SHRiMP read aligner package. The output from each SHRiMP instance which are incrementally generated on each processing node in parallel are finally merged, compressed, and sent back to client, where they are uncompressed before the final format conversion. Further details about this implementation can be found in our earlier publication [7].

3.2   SNP Detection a la Stream-as-you-go

Figure 3 shows our stream-as-you-go implementation of the SNP detection application. The client application sends compressed read data to the cloud. After being uncompressed, data gets submitted to a Streams application which first splits it across the available cluster nodes. Split reads are then aligned in parallel using the Bowtie read aligner package. The output from each Bowtie instance is further partitioned by genome position to be then sorted using a distributed in-memory insertion sort algorithm. After some data conversion steps, the sorted data is fed into the SOAPsnp SNP detection package for SNP calling. Results which are incrementally generated on each processing node in parallel are finally merged and sent back to the client over a TCP connection. Further details about this implementation can be found in our earlier publication [8], where we also show almost an order of magnitude reduction in total processing time compared to the state of the art (MapReduce-based Crossbow [9]).

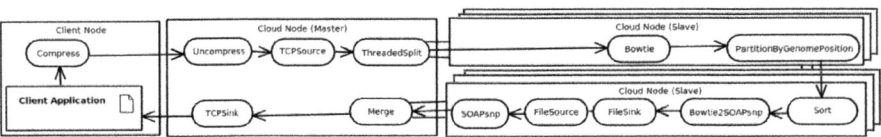

Fig. 3. Stream-as-you-go implementation of SNP detection

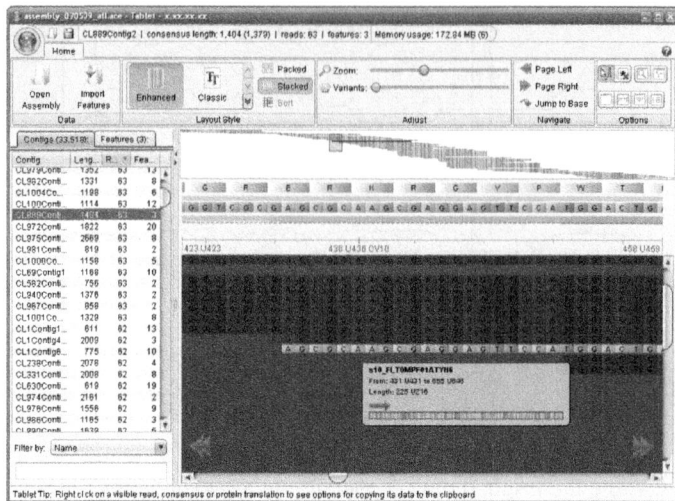

Fig. 4. Tablet-based visualization of aligned reads in comparison to the reference genome

4   Demonstration Details

In order to demonstrate the key features of our stream-as-you-go approach, we will use the two application scenarios whose implementations are described in the previous section.

In the read alignment demo, a partial workflow (i.e., the SHRiMP part) will be converted into an incremental version and will be transparently pushed to the cloud. We will contrast the speed of a local run of the whole workflow against its cloud-enabled counterpart. We will show that, besides a significant speedup, nothing else changes. The process as well as the results stay the same.

For illustration purposes, we will sniff the traffic between the client and the cloud. On the outgoing link, we will see the raw read data, whereas on the incoming link, already processed results will be seen. Data sets will be visualized and explained using the Tablet assembly viewer software [4]. For example, Figure 4 displays the results of an experiment, in which 30000 reads of Streptococcus suis (an important pathogen of pigs) have been aligned against its reference genome. This read data set is taken from the CloudBurst project [14].

In the SNP detection demo, a complete workflow will be converted into a Streams-based incremental version to be deployed in the cloud. We will show an easy-to-use graphical user interface, which allows researchers to run a complete SNP calling process on cloud resources without worrying about the details (Figure 5). The interface allows to select the source and the target data files (for reads and the reference genome) as well as the predefined data analysis process to be used. They can also configure their cloud cluster by selecting the number of nodes to be used based on a corresponding time and price estimation. Once the analysis completes, we will display the detected SNPs. We are planning to use the "E. Coli Small Example" dataset provided at the Crossbow website [1]. The read file in this dataset is taken from an E. Coli experiment and contains 8922730 reads with a total size of 1.4 GB. The process aligns these reads

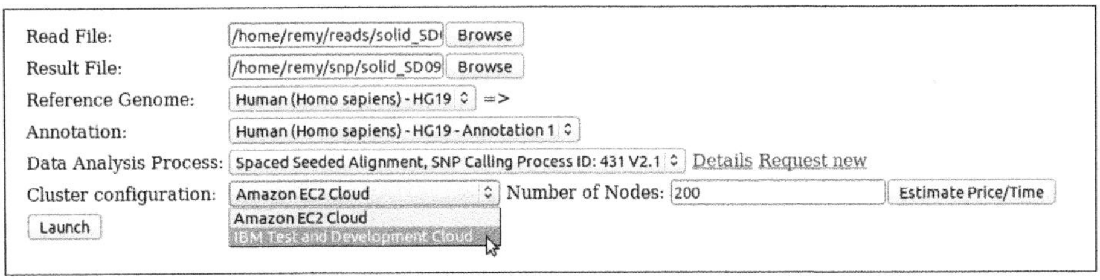

Fig. 5. Easy-to-use cloud deployment interface

against the E. Coli reference genome (NC_008253.1) containing 5594158 base pairs with a total size of 5.4 MB.

Acknowledgements. This work has been supported in part by an IBM faculty award.

References

1. Crossbow, http://bowtie-bio.sourceforge.net/crossbow/
2. IBM InfoSphere Streams,
   http://www.ibm.com/software/data/infosphere/streams/
3. SNP,
   http://en.wikipedia.org/wiki/Single-nucleotide_polymorphism
4. Tablet Assembly Viewer, http://bioinf.scri.ac.uk/tablet
5. Collins, F.S., Guyer, M., Chakravarti, A.: Variations on a Theme: Cataloging Human DNA Sequence Variation. Science 278(5343) (1997)
6. Dean, J., Ghemawat, S.: MapReduce: Simplified Data Processing on Large Clusters. In: OSDI Conference (2004)
7. Kienzler, R., Bruggmann, R., Ranganathan, A., Tatbul, N.: Large-Scale DNA Sequence Analysis in the Cloud: A Stream-Based Approach. In: Alexander, M., D'Ambra, P., Belloum, A., Bosilca, G., Cannataro, M., Danelutto, M., Di Martino, B., Gerndt, M., Jeannot, E., Namyst, R., Roman, J., Scott, S.L., Traff, J.L., Vallée, G., Weidendorfer, J. (eds.) Euro-Par 2011, Part II. LNCS, vol. 7156, pp. 467–476. Springer, Heidelberg (2012)
8. Kienzler, R., Bruggmann, R., Ranganathan, A., Tatbul, N.: Stream As You Go: The Case for Incremental Data Access and Processing in the Cloud. In: ICDE DMC Workshop (2012)
9. Langmead, B., Schatz, M.C., Lin, J., Pop, M., Salzberg, S.L.: Searching for SNPs with Cloud Computing. Genome Biology 10(11) (2009)
10. Langmead, B., Trapnell, C., Pop, M., Salzberg, S.: Ultrafast and Memory-efficient Alignment of Short DNA Sequences to the Human Genome. Genome Biology 10(3) (2009)
11. Li, H., Homer, N.: A Survey of Sequence Alignment Algorithms for Next Generation Sequencing. Briefings in Bioinformatics 11(5) (2010)
12. Li, R., Li, Y., Fang, X., Yang, H., Wang, J., Kristiansen, K., Wang, J.: SNP Detection for Massively Parallel Whole-Genome Resequencing. Genome Research 19(6) (2009)
13. Rumble, S.M., Lacroute, P., Dalca, A.V., Fiume, M., Sidow, A., Brudno, M.: SHRiMP: Accurate Mapping of Short Color-space Reads. PLoS Computational Biology 5(5) (2009)
14. Schatz, M.C.: CloudBurst: Highly Sensitive Read Mapping with MapReduce. Bioinformatics 25(11) (2009)
15. Taylor, R.: An Overview of the Hadoop/MapReduce/HBase Framework and its Current Applications in Bioinformatics. BMC Bioinformatics 11(suppl. 12) (2010)
16. Tung, N., Weisong, S., Douglas, R.: CloudAligner: A Fast and Full-featured Map Reduce-based Tool for Sequence Mapping. BMC Research Notes 4 (2011)
17. Voelkerding, K.V., Dames, S.A., Durtschi, J.D.: Next Generation Sequencing: From Basic Research to Diagnostics. Clinical Chemistry 55(4) (2009)

AITION: A Scalable Platform for Interactive Data Mining

Harry Dimitropoulos, Herald Kllapi, Omiros Metaxas, Nikolas Oikonomidis,
Eva Sitaridi, Manolis M. Tsangaris, and Yannis Ioannidis

MaDgIK Lab, Dept. of Informatics & Telecommunications,
University of Athens, Ilissia GR15784, Greece

Abstract. AITION is a scalable, user-friendly, and interactive data
mining (DM) platform, designed for analyzing large heterogeneous
datasets. Implementing state-of-the-art machine learning algorithms, it
successfully utilizes generative Probabilistic Graphical Models (PGMs)
providing an integrated framework targeting feature selection, Knowledge Discovery (KD), and decision support. At the same time, it offers advanced capabilities for multi-scale data distribution representation, analysis & simulation, as well as, for identification and modelling
of variable associations.

AITION is built on top of Athena Distributed Processing (ADP) engine, a next generation data-flow language engine, capable of supporting
large-scale KD on a variety of distributed platforms, such as, ad-hoc
clusters, grids, or clouds. On the front end, it offers an interactive visual
interface that allows users to explore the results of the KD process. The
end result is that users not only understand the process that led to a
statistical conclusion, but also the impact of that conclusion on their
hypotheses.

In the proposed demonstration, we will show AITION in action at
various stages of the knowledge discovery process, showcasing its key features regarding interactivity and scalability against a variety of
problems.

1 AITION Description

PGMs are a popular and well-studied framework for compact representation
of a joint probability distribution over a large number of interdependent variables, as well as, for efficient reasoning about such a distribution [5,6]. AITION
(Fig. 1) is one of the latest and most advanced systems in this area. Developed
as part of an EC project [1], AITION implements state-of-the-art algorithms
& techniques (exact or approximate) for Bayesian Network (BN) Structure &
Parameter Learning, Markov Blanket induction, and real-time inference. Furthermore, ontologies and *a-priori* knowledge can be incorporated with the BN,
defining topological constraints, in order to automate causal discovery & feature
selection and provide semantic modelling under uncertainty. This way, AITION
presents a rich 'natural' framework for imposing structure and prior knowledge,
providing the domain expert with the ability to seed the learning algorithm with
knowledge about the problem at hand.

A. Ailamaki and S. Bowers (Eds.): SSDBM 2012, LNCS 7338, pp. 646–651, 2012.
© Springer-Verlag Berlin Heidelberg 2012

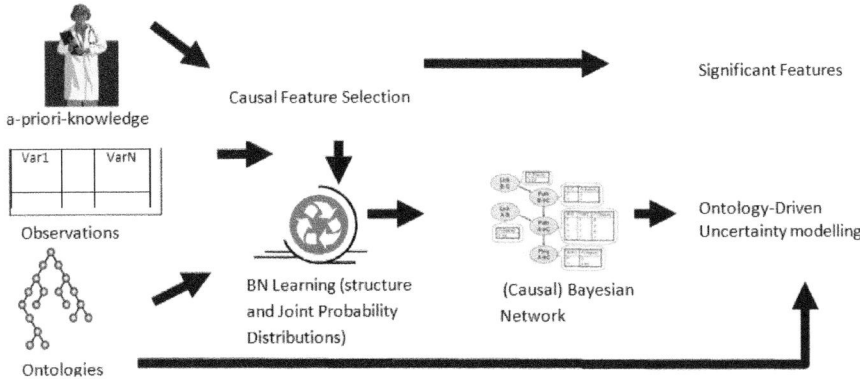

Fig. 1. Illustration of the AITION Framework

2   KDD Workflow

Knowledge Discovery in Databases (KDD) is the non-trivial process of identifying valid, novel, potentially useful, and ultimately, understandable patterns in data [2]. In other words, KDD is the process of discovering useful information and patterns in data, converting raw data into useful information. The KDD workflow consists of a series of transformation steps starting from data pre-processing to model building, reasoning and knowledge extraction. AITION supports the whole KDD workflow, as shown in Figure 2.

The goal of the pre-processing is to validate, curate and transform (e.g. discretize) raw data to facilitate the application of a Data Mining (DM) algorithm. The model building step is related to the construction of a BN based on data & prior knowledge and consists of two subtasks: (a) Structure learning (or qualitative analysis), where the goal is to build a directed acyclic graph (DAG) encoding the assertions of conditional independence between variables & (b) Parameter learning (quantitative analysis), assessing conditional probability distributions. Finally, the goal of inference in a BN is to answer queries about unobserved variables, given values of some observed variables targeting either the most probable configuration - Maximum a Posteriori (MAP) - or estimating Posterior Marginal Densities.

3   System Architecture

The AITION system consists of several components as seen in Fig. 3, including the *User Interface (UI)* and the *backend*. The heart of the backend is the *ADP Engine* [7] (providing distributed query processing) and a *Relational Database* (for storing original data and knowledge models).

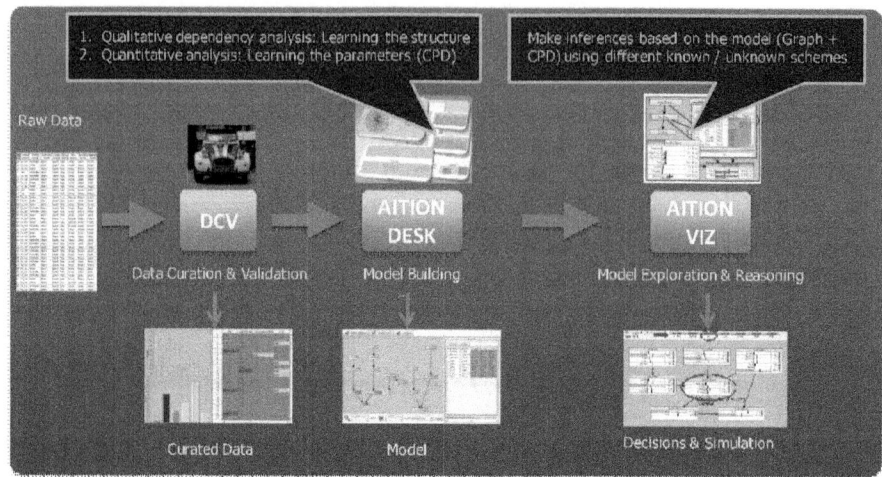

Fig. 2. The KDD workflow

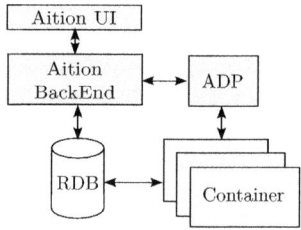

Fig. 3. AITION System Architecture

A collection of DM algorithms, most of them from WEKA [8], have been adopted and run as ADP operators, giving us the opportunity to express them as ADP queries. The optimizer facilitates "optimal" execution using all available resources, or by meeting certain cost-performance objectives. AITION applications need no modification to run over grids, ad-hoc clusters or cloud platforms.

The AITION UI Engine is a *thick client* connecting to the backend and managing all user interaction. It enables the user to execute the DM workflow. It also provides visualisation and analysis of the Bayesian knowledge models, utilizing the GraphViz toolkit of AT&T Research [3].

4   Demonstration Overview

In the demonstration, we will show AITION in a typical data mining session, as a sequence of steps: examining data samples, building a knowledge model from them, testing its validity, and finally, visualizing & exploring the end result performing a set of interesting inference scenarios.

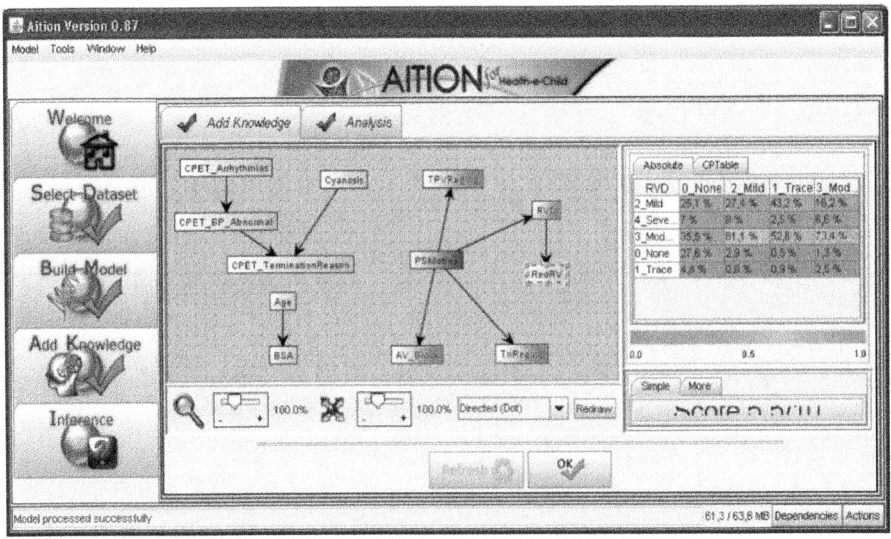

Fig. 4. A typical screen of the model building process

In more detail, we will start with an introductory example illustrating the basic notions of a KDD flow based on Bayesian Networks, as well as, familiarize the audience with the AITION workspace. Afterwards, we will show a real-world case from the medical domain, focusing on AITION's knowledge extraction and reasoning capabilities. Based on those two examples, we will cover some of the key aspects of the system, including:

Model Building: To learn the structure of the graph, AITION first performs a *qualitative dependency analysis* of the data; a repetitive process, in order to generate and evaluate several models in parallel using different training parameters. The user can then inspect the resulting graph (where nodes correspond to data features and the links/edges connecting the nodes indicate that there are probability relationships between them) and modify it (e.g., by adding or removing edges between nodes), before the next stage of *quantitative dependency analysis*, where AITION learns the parameters of the model (the conditional probability distribution). A typical screen of the model building process is show in Figure 4.

Reasoning and Visualization: An interactive workspace enables the user to perform *reasoning* using *inference* in graphs. *A-posteriori* probabilities can be computed for a specific node given some evidence: e.g., in a medical application, we can perform diagnostic, predictive, and inter-causal inference. The inference capabilities of AITION are highly interactive, including the ability to *perturbate* the values of a selected node and visually see the degree by which the other nodes in the graph are affected. A typical screen from this analysis is shown in Figure 5, where we set a value to a specific node (*RVD*), and marginal distributions for all related nodes are estimated. Finally, given a pre-computed model, the

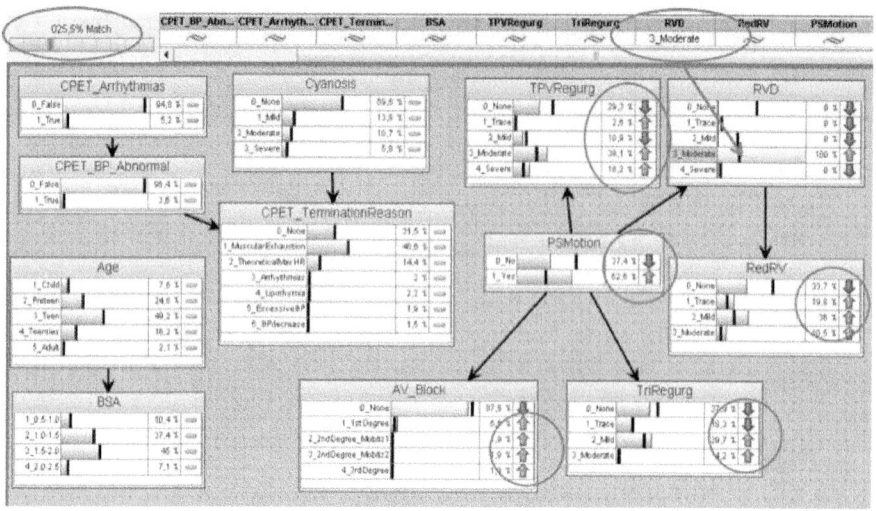

Fig. 5. A snapshot of a Knowledge Model for a medical problem during Reasoning. By setting the value of node *RVD* to Moderate, the marginal distributions for all related nodes are estimated.

user can load another set of instances (a test dataset) to perform classification, decision support, or predict missing values.

5   Conclusion and Future Work

We have demonstrated AITION applied on different domains. Solving these problems required some of its key features, including the parallel processing aspect in order to compute an appropriate PGM, and its visualization in order to make both the model & the DM process better understood by a non-technical audience. We plan to adopt more advanced algorithms for model learning & inference, while also enhancing the analytical capabilities of the tool, including the automatic generation of reports.

We also plan to further extend AITION incorporating advanced Statistical Relational Learning (SRL) and Graph Mining techniques. This way, we will create a comprehensive reasoning and simulation framework able to provide multi-scale and multi-entity predictive models. SRL [4] is an emerging area of research at the intersection of machine learning, graph mining, relational data mining, and inductive logic programming, aiming at combining statistical learning and probabilistic reasoning within logical or relational (frame-based) representations. Implementing this framework, we will be able to represent complex situations involving a variety of entities/objects, as well as, relations between them; something not possible using the simpler propositional or feature vector based representations.

References

1. http://www.health-e-child.org (2010)
2. Fayyad, U.M., Piatetsky-Shapiro, G., Smyth, P.: From data mining to knowledge discovery: An overview. In: Advances in Knowledge Discovery and Data Mining, pp. 1–34 (1996)
3. Gansner, E.R., North, S.C.: An open graph visualization system and its applications to software engineering. Softw., Pract. Exper. 30(11), 1203–1233 (2000)
4. Getoor, L., Taskar, B.: Introduction to Statistical Relational Learning. MIT Press (2007)
5. Koller, D., Friedman, N.: Probabilistic Graphical Models. MIT Press (2009)
6. Pearl, J.: Probabilistic Reasoning in Intelligent Systems, 2nd revised edn. Morgan Kaufmann, San Mateo (1988)
7. Tsangaris, M.M., Kakaletris, G., Kllapi, H., Papanikos, G., Pentaris, F., Polydoras, P., Sitaridi, E., Stoumpos, V., Ioannidis, Y.E.: Dataflow processing and optimization on grid and cloud infrastructures. IEEE Data Eng. Bull. 32(1), 67–74 (2009)
8. Witten, I.H., Frank, E.: Data mining: practical machine learning tools and techniques, 2nd edn. Elsevier, Morgan Kaufman, Amsterdam (2005)

Author Index

GPSR Compliance

The European Union's (EU) General Product Safety Regulation (GPSR) is a set of rules that requires consumer products to be safe and our obligations to ensure this.

If you have any concerns about our products, you can contact us on ProductSafety@springernature.com

In case Publisher is established outside the EU, the EU authorized representative is:

Springer Nature Customer Service Center GmbH
Europaplatz 3
69115 Heidelberg, Germany

Batch number: 09490866

Printed by Printforce, the Netherlands